Process Modeling in Pyrometallurgical Engineering

Process Modeling in Pyrometallurgical Engineering

Editors

Henrik Saxén
Marco A. Ramírez-Argáez
Alberto N. Conejo
Abhishek Dutta

MDPI • Basel • Beijing • Wuhan • Barcelona • Belgrade • Manchester • Tokyo • Cluj • Tianjin

Editors
Henrik Saxén
Abo Akademi University
Finland

Marco A. Ramírez-Argáez
Universidad Nacional
Autónoma de México (UNAM)
México

Alberto N. Conejo
University of Science and
Technology Beijing (USTB)
China

Abhishek Dutta
Izmir Institute of Technology
Turkey

Editorial Office
MDPI
St. Alban-Anlage 66
4052 Basel, Switzerland

This is a reprint of articles from the Special Issue published online in the open access journal *Processes* (ISSN 2227-9717) (available at: https://www.mdpi.com/journal/processes/special_issues/pyrometallurgical_engineering).

For citation purposes, cite each article independently as indicated on the article page online and as indicated below:

LastName, A.A.; LastName, B.B.; LastName, C.C. Article Title. *Journal Name* **Year**, *Volume Number*, Page Range.

ISBN 978-3-0365-0654-8 (Hbk)
ISBN 978-3-0365-0655-5 (PDF)

Contents

About the Editors

Henrik Saxén is Professor in Heat Engineering and has been working at the Process and Systems Engineering Laboratory of Faculty of Science and Engineering, Åbo Akademi University, Finland, since September 1997. Prior to this, he held different positions at the (former) Department of Chemical Engineering of the same university. He completed his M.Sc. in Chemical Engineering on Process Control in 1983 and his doctoral work on mathematical modeling of blast furnaces in 1988 at the same department. Prof. Saxén's primary research is on modeling, simulation, interpretation, and optimization of industrial processes and energy systems, with particular focus on sustainable and efficient iron- and steelmaking technology. He has collaborated extensively with universities and companies, and has participated in many national and international R&D projects on iron- and steelmaking, e.g., within the framework of EU Research Funds for Coal and Steel. He has published around 170 international peer-reviewed journal papers, and has supervised more than 20 doctorate and 90 master's students. He has or is currently serving on the editorial boards of numerous journals, including *Metals*, *ISIJ International*, and *Heliyon*, and serves as reviewer for more than 60 scientific journals. He has also edited several Special Issues of international journals on modeling in iron- and steelmaking. Prof. Saxén's teaching activities are mainly in the field of process engineering, including courses on heat and mass transfer, thermodynamics, refrigeration, energy technology, process modeling and optimization, as well as AI and neural networks. He was Dean of the Faculty of Chemical Engineering of Åbo Akademi University in 1998–2000 and Vice Rector (Research) of the same university in 2006–2009. He has served as Visiting Professor of Shanghai University, China, since 2017.

Marco A. Ramírez-Argáez is a Metallurgical Chemical Engineer and graduate of UNAM, where he received his master's degree in Metallurgical Engineering from the Saltillo Unit CINVESTAV, and doctorate in Materials Science and Engineering from MIT in the USA. Since receiving his Ph.D. degree, his main line of research has been focused on metallurgical process engineering, specialized in mathematical and physical modeling. He has worked in the foundry of nodular and gray irons as well as aluminum, TIG electric arc welding, in steel processes such as the electric arc furnace and steel ladle, and in the desalination of crude oil developing fundamental mathematical models based on first principles as the transport of mass, energy, and momentum are solved with numerical methods such as control volume and finite element. His academic production consists of 56 articles published in international peer-reviewed journals indexed in the JCR, 49 bachelor students and 27 master's students, and three completed doctorates. He has directed 13 academic projects as well as 7 industrial projects. He has applied for 6 patents, 3 in Mexico, 1 in the USA, 1 in Germany, and 1 in Brazil. He has served as a referee in 34 international journals in the field of materials science and metallurgy. He is currently a Full-Time Professor at the Faculty of Chemistry of the UNAM, as well as Level 2 of the National Autonomous System of Mexico. He has received several international and national recognitions, which includes the following highlights: Geoffrey Belton Award given by the United States Steelworkers Association for the best doctoral thesis in 2001, the RUNJA (National University Recognition for Young Academics) award in the teaching area of Exact Sciences in 2009, the Gabino Barreda Medal for the best average of the IQM career at UNAM and the medal for the best student in Mexico given by the Diario de México, first place in patents from UNAM in 2018 from the Program for the Promotion for Patenting and Innovation, as well as second place in the national technology award of the National Chamber of the Steel Industry (CANACERO) in 2011.

Alberto N. Conejo (Cuitzeo Mexico, 25 September1959) has served as Professor at the University of Science and Technology Beijing (USTB) at the School of Metallurgical and Ecological Engineering since September 2018. Prior to this, he was Professor of Ironmaking and Steelmaking in Mexico for 30 years. He also worked for several metallurgical plants from 1982 until 1988 in several positions. His PhD in Metallurgy was completed at Colorado School of Mines in Golden CO, USA. His Ph.D. thesis was related to gas–solid reactions and the production of iron carbide from iron oxides. During more than 15 years, he had a strong collaboration with the steel industry in Mexico, in particular, with ArcelorMittal Lazaro Cardenas. He has provided more than 1200 hours of technical training on ironmaking and steelmaking to union workers, supervisors, and process engineers. Has served as Visiting Professor at Tokoku University; ArcelorMittal laboratories in Aviles, Spain; University of Science and Technology Beijing (USTB), China; and Indian Institute of Technology Kanpur, India. He has received several national and international awards, including Charles W. Briggs award 2002, Iron and steel society, USA; Best paper award on EAF steelmaking; Michoacán State award—Technology award, 2005, for work on EAF slag foaming; National award (2nd place)—Technology and Science award granted by the Mexican Steel Producers Association (Canacero), 2010/2011 for work on EAF modeling. He has been a member of the National System of Researchers (SNI), at Level II (top 18% nationwide) in 2018–2021 and was Level II member during 2014–2017. He has served as an editorial board member of *Metallurgical Research and Technology* (previously *Revue de Metallurgie*) since January 2017 and *International Journal of Minerals, Metallurgy and Materials* (Beijing, China) since July 2019. He has published more than 60 technical papers in journals, with h-index = 15 (Scopus). He has supervised or co-supervised 18 master's and 2 Ph.D. students.

Abhishek Dutta is a Chemical Engineer and graduate of the University of Madras and Jadavpur University, both in India. He obtained his Ph.D. in Chemical Engineering from Ghent University, Belgium, in mathematical modeling using population balances for FCC catalysts. He then continued postdoctoral research in Bioengineering at the same University, again using the concept of population balances for cell heterogeneity. Since 2013, he has served as a Full-Term Visiting Assistant Professor at the Faculty of Engineering Technology at KU Leuven, Campus Groep T. He became involved in academic–industrial projects as a part of capacity development in translational research from February 2020. His academic production consists of around 50 articles in international peer-reviewed journals indexed in the JCR, 20 master's students, and three ongoing doctorates under his co-supervision. He is currently Associate Professor, full-time, at Izmir Institute of Technology in Turkey. His research includes development of energy-efficient processes focusing on the improvement of existing and development of new chemical (and biochemical) process technologies, resource-efficient high-temperature processes involving complex hydrodynamic and reactive phenomena, and cost-efficient experimentally driven mathematical models valorizing waste resource recovery. He has won one European research grant, been co-awarded one Belgian VLIR-UOS grant, and has obtained numerous grants from industries. He is a Life Member of the Indian Institute of Chemical Engineers.

Editorial

Special Issue on "Process Modeling in Pyrometallurgical Engineering"

Henrik Saxén [1,*], Marco A. Ramírez-Argáez [2], Alberto N. Conejo [3] and Abhishek Dutta [4]

1   Process and Systems Engineering Laboratory, Faculty of Science and Engineering, Åbo Akademi University, Biskopsgatan 8, FI-20500 Åbo, Finland
2   School of Chemistry, National Autonomous University of Mexico (UNAM), Mexico City C.P. 04510, Mexico; marco.ramirez@unam.mx
3   School of Metallurgical and Ecological Engineering, University of Science and Technology, 30 Xueyuan Road, Haidian District, Beijing 100083, China; aconejonava@hotmail.com or aconejo@ustb.edu.cn
4   Department of Chemical Engineering, Izmir Institute of Technology, Gülbahçe Campus, Urla, Izmir 35430, Turkey; abhishekdutta@iyte.edu.tr
*   Correspondence: hsaxen@abo.fi

Citation: Saxén, H.; Ramírez-Argáez, M.A.; Conejo, A.N.; Dutta, A. Special Issue on "Process Modeling in Pyrometallurgical Engineering". *Processes* **2021**, *9*, 252. https://doi.org/10.3390/pr9020252

Received: 26 January 2021
Accepted: 26 January 2021
Published: 29 January 2021

Publisher's Note: MDPI stays neutral with regard to jurisdictional claims in published maps and institutional affiliations.

This Special Issue on "Process Modeling in Pyrometallurgical Engineering" consists of 39 articles, including two review papers, and covers a wide range of topics related to process development and analysis based on modeling in ironmaking, steelmaking, flash smelting, casting, rolling operations, etc. The approaches include small-scale experiments and experimental design, first-principles modeling, detailed modeling based on CFD or DEM, and statistical and machine-learning-based methods. In the following paragraphs the issue is briefly scanned, presenting the papers in the order roughly following the route from raw materials processing to rolling and heat treatment.

1. Coal

Lignite is a low-grade coal with the lowest carbon content. Liu et al. [1] studied the structural changes of Huolinhe lignite when it was modified by microwave and ultrasound treatment and reported the chemical composition of the products.

2. Sintering

Sintering is an important agglomeration process that is a prerequisite for downstream processing. Liu et al. [2] studied the combustion kinetics during sintering at a partial replacement of coke powder by anthracite and found improvements in the combustion rate. Two combustion models, the volumetric and random pore models, were compared and the latter was demonstrated to yield better agreement with the measurements.

Wang et al. [3] studied a partial replacement of coke with biomass in sintering. The results indicated a higher combustion rate using charcoal due to a higher surface area, in spite of a lower fixed carbon content. It may therefore be viable to partially replace coke by charcoal in the sintering process.

Cheng et al. [4] investigated the removal of arsenic from iron ore through roasting in air and nitrogen. They reported higher removal ratios using a reducing atmosphere.

3. Blast Furnace

The ironmaking blast furnace (BF) is an extremely complicated industrial unit process, and this is a reason why many computational approaches have been made to study its behavior or to predict its performance under new operation conditions.

An analysis of the process based on a state-of-the-art 3D static model is presented by Kou et al. [5], addressing the problem of estimating the extent of the solution loss, or Boudouard, reaction. The model is applied to shed light on how operation factors, such as blast oxygen enrichment and scrap charging, affect the solution-loss in the furnace. Another global study of the BF is provided by Ouyang et al. [6], with the goal of detecting abnormal

1

events in the process. This entirely data-driven approach is based on a multidimensional Gated Recurrent Unit (GRU) neural network that, using its feedback nodes, can consider temporal evolutions of signals.

The charging of burden is treated in three papers. Wei et al. [7] studied the fundamental features of BF burden materials, including the angles of repose and bed porosity, by small-scale experiments in combination with computational analysis based on the Discrete Element Method (DEM). Mio et al. [8] used a unique 1:3 pilot model of the BF top to study the radial distribution of the rings of burden materials and used the results for validation of a DEM-based model of the system. Li et al. [9] outlined a model of the burden distribution in the BF by considering the falling trajectories, layer formation, and descent. Radar measurements of the burden profile provided supporting evidence.

The flow of molten iron and slag in the coke bed of the lower BF is very complex and traditional modeling techniques cannot describe the flow patterns. Natsui et al. [10] presented an interesting attempt based on smoothed particle hydrodynamics (SPH) simulation, demonstrating the role of wettability on the arising flow rivulets of iron and slag, shedding light on the complex liquid flow patterns. The role of the dead man, and how it influences the conditions in the lower part of the BF, is reviewed by Shao et al. [11], presenting approaches to model the dead-man state and the way in which it affects hearth performance, e.g., the flow paths of hot metal, lining wear, and liquid levels.

The combustion region, or raceway, formed in front of the tuyeres is studied in two papers. Peng et al. [12] developed a three-dimensional transient Eulerian multiphase flow model of the raceway to study how the raceway size, pressure distribution, and flow field are affected by the blast parameters. The use of natural gas as auxiliary reductant is studied by Okosun et al. [13] using a CFD model of the furnace from the raceway to the top. Special attention is focused on means to counteract the cooling of the raceway associated with high injection rates of natural gas.

The last stage of ironmaking in the BF is the tapping operation. The complex flow patterns in the main trough were studied by Ge at al. [14], using transient 3D CFD based on the volume of fluid (VOF) model. As for the outflows of iron and slag, Roche et al. [15] developed a strategy for compressing the information about the outflow rates from a large BF by principal component analysis.

4. Iron Smelting

Alternative ironmaking technologies have also been developed and some are analyzed in this Special Issue. Xie et al. [16] carried out water modeling experiments to study the mixing time in a smelter as a function of nozzle position, nozzle diameter, nozzle immersion depth, gas flow rate, and liquid properties. Applying dimensional analysis, the authors derive expressions for the mixing time and compare the results with practical findings.

Many studies of the burden distribution in the blast furnace have been undertaken, but much less work has been reported for the COREX process, which has a more complex charging system. Li et al. [17] reported results from mathematical modeling of burden distribution in the COREX melter–gasifier. Based on experimental results from a pilot rig, the model was found to accurately predict the DRI/coal ratio as a function of the radial position.

Sun et al. [18] described a mathematical model predicting the raceway geometry in the melter–gasifier as a function of time and gas velocity. By dynamic simulations, the authors concluded that the final shape is reached in a short time (<1 s). Increasing the velocity of the gas increases the depth of the raceway. For a normal blowing speed of 250 m/s and a tuyere angle of 4°, a raceway depth of 950 mm was predicted.

5. Copper Smelting

Smelters used for other metals, e.g., copper and nickel, are also complex processes, where mathematical modeling can provide valuable information used for enhancing the process or for improved control. Jylhä et al. [19] developed a CFD–DEM model to study the settling of copper particles in a flash smelter settler, applying a population balance model to describe the settling and coalescence of the droplets. The modeling confirmed that small particles (< 100 mm) remain in the slag, suggesting that an operation with a thinner slag layer would increase the yield of the process.

Wang et al. [20] found a higher elimination rate of arsenic in copper smelters by controlling the oxygen/sulfur potential, reporting a decrease in As from 0.07% to 0.02%.

Navarra et al. [21] discussed the application of sensors and process control systems for process automation of copper smelters, and stressed the potential of data-driven models and discrete-event simulation for smelter optimization.

6. BOF

The basic oxygen furnace that converts hot metal into liquid steel is also characterized by harsh conditions that justify model-based analysis. Jiang et al. [22] introduced a novel hybrid model integrating multiple linear regression (MLR) and Gaussian process regression (GPR) to predict the oxygen consumption for optimization of the energy requirement of the BOF. The model was validated with actual data collected from a steel factory in China.

Rahnama et al. [23] reported correlations between the operating parameters and rate of decarburization (dc/dt) in a pilot plant. A positive correlation was found between the decarburization rate and the oxygen flow as well as the temperature and CO_2 content in the waste gas, while a negative correlation was found with the lance height. A neural network was trained to predict the decarburization in a full-scale plant, yielding satisfactory performance.

Dering et al. [24] described a first principles-based dynamic model of the BOF. The model considers an energy balance, slag formation, as well as decarburization rate. The authors estimated a set of parameters to adapt the model to data reported in the literature and from a reference BOF, and the model was demonstrated to capture the dynamics of the process.

7. EAF

The high temperatures and complex melting phenomena in the electrical arc furnace are reasons why many model-based studies of the process have been undertaken. Carlsson et al. [25] determined the effect of scrap shape and density on the energy consumed to melt the scrap in the EAF by using a statistical model and process optimization algorithms validated through plant trials. The results provide significant evidence that a well-chosen scrap categorization is important to predict the electric energy demand.

A simulator of an EAF based on a dynamic model was developed by Hay et al. [26], to be used as an automatic control tool for assessment of multiple scenarios in the operation. The model can also be used for training furnace operators.

Chen et al. [27] developed a 3D mathematical model of the interaction of the supersonic coherent jet with the steel bath. The model predicts the decarburization kinetics, including the distribution of the in-bath components, flow patterns, and bath temperatures, and can be used to optimize the refining process.

8. Ladle Furnace

Ladle treatment is an important step for adjusting the final composition and temperature of the metal before it is cast. Two papers deal with the simultaneous optimization of mixing and slag open-eye area in ladle furnaces. Jardón-Pérez et al. [28] validated a CFD model against PIV measurements and applied the model to analyze the mixing time and open-eye area in gas-stirred ladles using two plugs with equal (50%/50%) or differentiated (75%/25%) flows. Yang et al. [29] applied a physical model to measure the mixing times and interface slag entrainment under different conditions, including the injection modes, gas flow rates, and top slag thicknesses. The authors suggested an optimum argon flow rate between 36 m^3/h and 42 m^3/h with two plugs.

Zhao et al. [30] reported fundamental research on a water–oil–air physical model to study the dynamics occurring when bubbles pass through the liquid–liquid interface for different oil viscosities at various gas flow rates. They found that bubble movement is greatly influenced by the viscosity of the oil and that the water-oil interface stability was enhanced with increased viscosity of the oil phase.

Lei et al. [31] computed, based on the Ion-Molecule Coexistence Theory, the titanium distribution ratio in ladle furnace slags (CaO–SiO$_2$–Al$_2$O$_3$– MgO–FeO–MnO–TiO$_2$) at 1853 K for tire cord steel production, and found a good agreement of the model with the measurements. The structural unit CaO was found to play a pivotal role in the slags.

Finally, Conejo [32] presented an extensive and exhaustive review of physical and mathematical models of mass transfer in gas-stirred ladles, stressing the effects of the process variables on the mass transfer coefficient. The review noted that currently there is a lack of means to simultaneously keeping both liquid phases (steel and slag) well mixed in the ladles.

9. Casting and Solidification

The solidification of a metal is a complex and gradual process that is difficult to control. Niaz et al. [33] reported numerical predictions of the Horizontal Single Belt Casting (HSBC) process, which avoids multiple hot-rolling steps by directly producing a thin sheet. Results from an experimental rig were compared with findings from a CFD model, and the latter was applied to study non-uniformity and other undesired conditions, as well as means to address these, e.g., by appropriate design of the metal feeding system.

Precipitation behavior of inclusions was studied by both Wang et al. [34] and Li et al. [35], the former for titanium nitride (TiN) during solidification and the latter for chromium spinels in stainless steel slags. A systematic study was made to clarify the mechanism of TiN precipitation to guide the development of ultra-high strength grade steels. The stability of chromium in stainless steel slag was found to have a positive correlation with spinel particle size and a negative correlation with the calcium content of the spinel. However, both groups of authors stress that further experimental work and theoretical analysis are needed to understand the precipitation behavior of the inclusions to improve the quality of the finished steel.

10. Rolling

Most steel products are manufactured by hot rolling, because it is one of the most efficient plastic-forming processes. On the basis of comparative studies on the temperature distribution during hot plate rolling and rod rolling, Hwang [36] concluded that the temperature distribution and variation of a rod during shape rolling are different from those of a plate during flat rolling. The higher variation in effective stress of the rod along the circumferential direction induces a higher temperature difference of the rod. The same author [37] also investigated the effect of roll design on the strain distribution of the flat surface, lateral spreading, and the strain inhomogeneity of a flat-rolled wire, proposing a new strategy for fabricating high-quality flat-rolled wires through a cambered roll of a small radius.

Hu et al. [38] developed an optimization model for hot rolling based on the time-of-use (TOU) electricity pricing using a genetic algorithm. Jumps between adjacent slabs in width, hardness, and thickness were avoided by including penalties in the objective function. The method was verified on batch results from the hot rolling of 240 slabs of different sizes and was demonstrated to reduce the cost of power required for rolling.

Chen et al. [39] found that the hot working ability of a nickel-based GH4698 alloy markedly decreased under lower temperatures and higher strain rates in isothermal compressions: this alloy is extremely sensitive to thermal processing parameters and cracking may easily occur. An Arrhenius model was used to estimate the flow stresses and profiles for processing under various thermal conditions.

Funding: There is no funding support.

Conflicts of Interest: The authors declare no conflict of interest.

References

1. Liu, J.; Qiu, S.; He, Z.; Yu, Y. Experiments and 3D molecular model construction of lignite under different modification treatment. *Processes* **2020**, *8*, 399.
2. Liu, J.; Yuan, Y.; Zhang, J.; He, Z.; Yu, Y. Combustion kinetics characteristics of solid fuel in the sintering process. *Processes* **2020**, *8*, 475. [CrossRef]
3. Wang, Z.; Ohno, K.; Nonaka, S.; Maeda, T.; Kunitomo, K. Temperature distribution estimation in a Dwight-Lloyd Sinter machine based on the combustion rate of charcoal quasi-particles. *Processes* **2020**, *8*, 406. [CrossRef]
4. Cheng, R.; Zhang, H.; Ni, H. Arsenic removal from arsenopyrite-bearing iron ore and arsenic recovery from dust ash by roasting method. *Processes* **2019**, *7*, 754.
5. Kou, M.; Zhou, H.; Wang, L.P.; Hong, Z.; Yao, S.; Xu, H.; Wu, S. Numerical simulation of effects of different operational parameters on the carbon solution loss ratio of coke inside blast furnace. *Processes* **2019**, *7*, 528. [CrossRef]
6. Ouyang, H.; Zeng, J.; Li, Y.; Luo, S. Fault detection and identification of blast furnace ironmaking process using the gated recurrent unit network. *Processes* **2020**, *8*, 391. [CrossRef]
7. Wei, H.; Li, M.; Li, Y.; Ge, Y.; Saxén, H.; Yu, Y. Discrete Element Method (DEM) and experimental studies of the angle of repose and porosity distribution of pellet pile. *Processes* **2019**, *7*, 561. [CrossRef]
8. Mio, H.; Narita, Y.; Nakano, K.; Nomura, S. Validation of the burden distribution of the 1/3-Scale of a blast furnace simulated by the Discrete Element Method. *Processes* **2020**, *8*, 6. [CrossRef]
9. Li, M.; Wei, H.; Ge, Y.; Xiao, G.; Yu, Y. A mathematical model combined with radar data for bell-less charging of a blast furnace. *Processes* **2020**, *8*, 239.
10. Natsui, S.; Tonya, K.; Nogami, H.; Kikuchi, T.; Suzuki, R.O.; Ohno, K.; Sukenaga, S.; Kon, T.; Ishihara, S.; Ueda, S. Numerical study of binary trickle flow of liquid iron and molten slag in coke bed by smoothed particle hydrodynamics. *Processes* **2020**, *8*, 221. [CrossRef]
11. Shao, L.; Xiao, Q.; Zhang, C.; Zou, Z.; Saxén, H. Dead-man behavior in the blast furnace hearth—a brief review. *Processes* **2020**, *8*, 1335. [CrossRef]
12. Peng, X.; Wang, J.; Zuo, H.; Xue, Q. Evolution and physical characteristics of a raceway based on a transient Eulerian multiphase flow model. *Processes* **2020**, *8*, 1315. [CrossRef]
13. Okosun, T.; Nielson, S.; D'Alessio, J.; Ray, S.; Street, S.; Zhou, C. On the impacts of pre-heated natural gas injection in blast furnaces. *Processes* **2020**, *8*, 771. [CrossRef]
14. Ge, Y.; Li, M.; Wei, H.; Liang, D.; Wang, X.; Yu, Y. Numerical analysis on velocity and temperature of the fluid in a blast furnace main trough. *Processes* **2020**, *8*, 249. [CrossRef]
15. Roche, M.; Helle, M.; Saxén, H. Principal component analysis of blast furnace drainage patterns. *Processes* **2019**, *7*, 519.
16. Xie, J.; Wang, B.; Zhang, J. Parametric dimensional analysis on a C-H2 smelting reduction furnace with double-row side nozzles. *Processes* **2020**, *8*, 129. [CrossRef]
17. Li, H.; Zou, Z.; Luo, Z.; Shao, L.; Liu, W. Model study on burden distribution in COREX melter gasifier. *Processes* **2019**, *7*, 892.
18. Sun, Y.; Chen, R.; Zhang, Z.; Wu, G.; Zhang, H.; Li, L.; Liu, Y.; Li, X.; Huang, Y. Numerical simulation of the raceway zone in melter gasifier of COREX process. *Processes* **2019**, *7*, 867. [CrossRef]
19. Jylhä, J.-P.; Khan, N.A.; Jokilaakso, A. Computational approaches for studying slag–matte interactions in the flash smelting furnace (FSF) settler. *Processes* **2020**, *8*, 485. [CrossRef]
20. Wang, Q.; Wang, Q.; Tian, Q.; Guo, X. Simulation study and industrial application of enhanced arsenic removal by regulating the proportion of concentrates in the SKS copper smelting process. *Processes* **2020**, *8*, 385. [CrossRef]
21. Navarra, A.; Wilson, R.; Parra, R.; Toro, N.; Ross, A.; Nave, J.-C.; Mackey, P.J. Quantitative methods to support data acquisition modernization within copper smelters. *Processes* **2020**, *8*, 1478. [CrossRef]
22. Jiang, S.L.; Shen, X.; Zheng, Z. Gaussian process-based hybrid model for predicting oxygen consumption in the converter steelmaking process. *Processes* **2019**, *7*, 352. [CrossRef]
23. Rahnama, A.; Li, Z.; Sridhar, S. Machine learning-based prediction of a BOS reactor performance from operating parameters. *Processes* **2020**, *8*, 371. [CrossRef]
24. Dering, D.; Swartz, C.; Dogan, N. Dynamic modeling and simulation of basic oxygen furnace (BOF) operation. *Processes* **2020**, *8*, 483. [CrossRef]
25. Carlsson, L.S.; Samuelsson, P.B.; Jönsson, P.G. Modeling the effect of scrap on the electrical energy consumption of an electric arc furnace. *Processes* **2020**, *8*, 1044. [CrossRef]
26. Hay, T.; Echterhof, T.; Visuri, V.-V. Development of an electric arc furnace simulator based on a comprehensive dynamic process model. *Processes* **2019**, *7*, 852. [CrossRef]
27. Chen, Y.; Silaen, A.K.; Zhou, C.Q. 3D integrated modeling of supersonic coherent jet penetration and decarburization in EAF refining process. *Processes* **2020**, *8*, 700. [CrossRef]
28. Jardón-Pérez, L.E.; González-Rivera, C.; Ramírez-Argáez, M.A.; Dutta, A. Numerical modeling of equal and differentiated gas injection in ladles: Effect on mixing time and slag eye. *Processes* **2020**, *8*, 917. [CrossRef]

29. Yang, F.; Jin, Y.; Zhu, C.; Dong, X.; Lin, P.; Cheng, C.; Li, Y.; Sun, L.; Pan, J.; Cai, Q. Physical Simulation of molten steel homogenization and slag entrapment in argon blown ladle. *Processes* **2019**, *7*, 479. [CrossRef]
30. Zhao, H.; Wang, J.; Zhang, W.; Xie, M.; Liu, F.; Cao, X. Bubble motion and interfacial phenomena during bubbles crossing liquid-liquid interfaces. *Processes* **2019**, *7*, 719. [CrossRef]
31. Lei, J.; Zhao, D.; Feng, W.; Xue, Z. Titanium distribution ratio model of ladle furnace slags for tire cord steel production based on the ion-molecule coexistence theory at 1853 K. *Processes* **2019**, *7*, 788. [CrossRef]
32. Conejo, A.N. Physical and mathematical modelling of mass transfer in ladles due to bottom gas stirring: A review. *Processes* **2020**, *8*, 750. [CrossRef]
33. Niaz, U.; Isac, M.M.; Guthrie, R.I.L. Numerical modeling of transport phenomena in the horizontal single belt casting (HSBC) process for the production of AA6111 aluminum alloy strip. *Processes* **2020**, *8*, 529. [CrossRef]
34. Wang, L.; Xue, Z.-L.; Chen, Y.-L.; Bi, X.-G. Understanding TiN precipitation behavior during solidification of SWRH 92A tire cord steel by selected thermodynamic models. *Processes* **2020**, *8*, 10. [CrossRef]
35. Li, J.; Mou, Q.; Zeng, Q.; Yu, Y. Experimental study on precipitation behavior of spinels in stainless steel-making slag under heating treatment. *Processes* **2019**, *7*, 487. [CrossRef]
36. Hwang, J.-K. Thermal behavior of a rod during hot shape rolling and its comparison with a plate during flat rolling. *Processes* **2020**, *8*, 327. [CrossRef]
37. Hwang, J.-K. Effect of cambered and oval-grooved roll on the strain distribution during the flat rolling process of a wire. *Processes* **2020**, *8*, 876. [CrossRef]
38. Hu, Z.; He, D.; Song, W.; Feng, K. Model and algorithm for planning hot-rolled batch processing under time-of-use electricity pricing. *Processes* **2020**, *8*, 42. [CrossRef]
39. Chen, R.; Xiao, H.; Wang, M.; Li, J. Flow behavior and hot processing map of GH4698 for isothermal compression process. *Processes* **2019**, *7*, 491. [CrossRef]

Article

Experiments and 3D Molecular Model Construction of Lignite under Different Modification Treatment

Jihui Liu [1], Shuang Qiu [1], Zhijun He [1,*] and Yaowei Yu [2,*]

[1]   School of Materials and Metallurgy, University of Science and Technology Liaoning, Anshan 114051, China; gtyj66@126.com (J.L.); daiwha_baosteel@163.com (S.Q.)
[2]   State Key Laboratory of Advanced Special Steel, Shanghai Key Laboratory of Advanced Ferrometallurgy, School of Materials Science and Engineering, Shanghai University, Shanghai 201900, China
*   Correspondence: hezhijun@ustl.edu.cn (Z.H.); Yaowei.yu@Hotmail.com (Y.Y.)

Received: 25 February 2020; Accepted: 23 March 2020; Published: 29 March 2020

Abstract: In this paper, Huolinhe lignite was selected as the lignite experimental sample, using microwave modification and ultrasonic modification separately as improvement methods. The three-dimensional molecular models of HLH before and after modification were established base on the parameters obtained by ^{13}C NMR, X-ray photoelectron spectroscopy (XPS), Raman spectroscopy (Raman), and Fourier transform infrared (FTIR). After the microwave treatment, the methylene carbon in the HLH coal sample structure mostly exists in the form of long straight chains, and after microwave and ultrasonic treatment, the -OH content of oxygen atoms in the coal sample increases, and form the CO- and the COO-. The proportion is decreasing. The models were adjusted and tested by the covalent bond concentration method and carbon chemical shift spectra calculation using Chemdraw software. A new method is proposed to study the structure and physicochemical properties of lignite modification from the molecular point of view through this study.

Keywords: Lignite; microwave and ultrasound modification; structural characterization; 3D molecular model; structural simulation

---

1. Introduction

In recent years, the shortage of high-rank coal resources has gradually become a prominent problem in industrial development [1,2]. Lignite is widely used in energy fields, such as pyrolysis, combustion, gasification and liquefaction [3,4]. In order to make more effective use of lignite resources, many scholars carried out much research on modification treatment processes for lignite characteristics. Arash Tahmasebi [5] discovered that the content of some functional groups in pulverized coal particles decreased significantly after microwave irradiation, but the content of aromatic carbon and aromatic ring in lignite was not affected by microwave pyrolysis. Sun Qiang [6] selected coal samples were treated with water and heat treatment and found that the rate of re-absorption decreased with the increase of temperature, and the lignite quality could improve most in high temperature and low humidity. Ge Lichao [7] found the rank of lignite increased after microwave modification and the combustion reaction process moved to high temperature zone by Thermogravimetry (TG) analysis.

The existing research focuses more on the optimization of modification processes and proposes new modification processes. The mechanism of these processes were difficult to study by experimental methods due to innumerable coupling reaction pathways during the utilization of lignite [8–10]. Therefore, Huolinhe lignite (HLH) was selected as the experimental sample, using microwave modification (MM) and ultrasonic modification (UM) as improvement methods separately. The two-dimensional molecular models of HLH before and after modification were established based on the parameters obtained by a series of detection methods, and three-dimensional model is constructed based on molecular mechanics and molecular dynamics. A new method is proposed to study the structure and physicochemical properties of lignite modification from the molecular point of view through this study.

2. Experiment

2.1. Modification of Lignite

HLH was low degree of coalification was selected as the experimental sample. HLH sample was ground to 109–180 μm and dried in a vacuum drying chamber at 40 °C for 24 h. Take HLH sample in a crucible, added 100 mL distilled water and blended fully. The crucible contained HLH sample was placed in the ultrasonic oscillator to water bath oscillation for 60 s. The crucible was placed in the drying oven for drying treatment with 85 °C for 4 h. The crucible containing the HLH sample was placed in the a quartz reaction tube of the microwave reactor, the modification parameters set as 200 W and 60 s, and the microwave activated. Industrial analysis and elemental analysis of lignite samples before and after treatment were carried out and the results are shown in Table 1.

Table 1. Proximate and ultimate analyses of HLH lignite.

Sample (wt. %), ad	Proximate Analysis				Ultimate Analysis				
	M_{ad}	A_{ad}	FC_{ad}	V_{daf}	C	H	O	N	S
HLH	16.32	21.97	24.20	37.51	78.38	6.24	13.35	1.51	0.52
MM	15.22	22.32	26.03	36.43	80.67	5.81	11.29	1.66	0.57
UM	15.46	22.64	25.06	36.84	81.77	5.47	10.31	1.85	0.60

Note: ad: air-dry basis; daf: dry-and-ash-free basis. M: moisture; A: ash; FC is fixed carbon; V: volatile matter content.

2.2. FTIR and Structural Parameters Analysis

2.2.1. FTIR Results Analysis

Infrared spectroscopy is closely related to the chemical structure of the substance. It can be confirmed the aromatic structure, oxygen-containing structure and fat structure of coal by FTIR detection [11]. Infrared spectra of all samples are shown in Figure 1, with curves smoothed and baselines corrected.

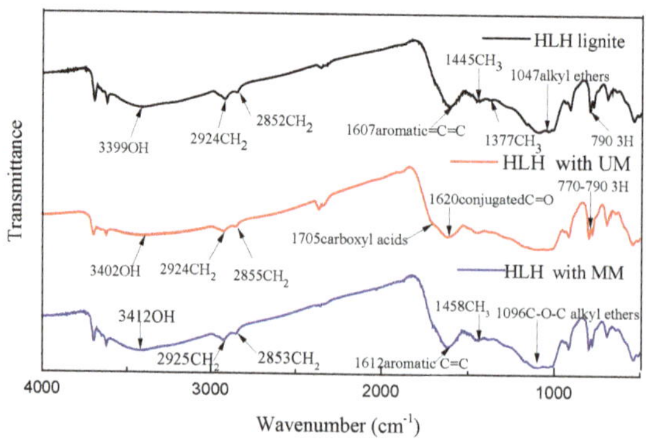

Figure 1. FTIR spectra of lignite before and after modification.

It can be observed that selected coal samples contain similar functional groups, hence the absorption of sample to infrared spectrum occurs at same wavenumber positions [11]. These obtained spectra are comprehensive curves of many independent peaks, which have to be deconvoluted to achieve. All spectra were divided into 4 regions, namely 700–900 cm^{-1} (aromatic structures), 1000–1800 cm^{-1} (oxygen-containing structures), 2800–3000 cm^{-1} (aliphatic structures) and 3000–3600 cm^{-1} (hydroxyl structure). Consequently, the area values of specific peak can be derived. The corresponding relationships between peak position and functional group are shown in Tables 2–5 [12,13].

Table 2. Aromatic structure of HLH before and after modification.

Assignment	Relative Area		
	HLH	MM	UM
4H	11.051	13.573	12.106
3H	86.498	86.427	87.894
2H	2.459	—	—

Table 3. Oxygen-containing functional group of HLH before and after modification.

Assignment	Relative Area		
	HLH	MM	UM
Alkyl ethers	16.406	14.231	13.321
C-O phenols, ethers	48.679	40.349	51.372
C-O in aryl ethers	9.023	11.452	8.213
Symmertric CH_3-Ar, R	1.612	1.823	—
Asymmertric CH_3-, CH_2-	2.462	7.435	9.468
Aromatic C=C	10.954	9.097	6.039
Conjugated C=O	9.633	10.979	10.534
Carboxyl acids	1.231	4.634	1.053

Table 4. Fatty structure of HLH before and after modification.

Assignment	Relative Area		
	HLH	MM	UM
Sym. R_2CH_2	13.118	20.622	17.705
R_3CH	47.68	28.051	36.046
Asym. R_2CH_2	24.814	30.553	27.132
Asym. RCH_3	14.388	20.774	19.118

Table 5. Hydroxyl structure of HLH before and after modification.

Assignment	Relative Area		
	HLH	MM	UM
OH-N	4.419	3.118	8.304
Ring hydroxyl	45.653	41.322	31.922
Phenol OH	35.271	41.202	42.944
OH-π	14.662	14.358	16.829

It can be found that there are 3 substitution modes of hydrogen atoms on benzene ring in HLH structure. The proportion of triple substituted aromatics (3H) is the largest among all the samples. The proportion of triple substituted aromatics of HLH raw coal is 86.498%, the MM sample is 86.427% in, the UM sample is 87.893%. The tetrasubstituted hydrocarbons (2H) of HLH lignite has also changed significantly. Tetrasubstituted hydrocarbons (2H) are not found in the infrared spectra of HLH lignite after both modification. In the process of modification, the substitution reaction of aromatic hydrocarbons may cause structural changes, which is due to the instability of other atoms and aliphatic side chains in benzene rings.

It is also found that the form and proportion of oxygen elements in HLH sample changed a lot after modification, however the carbonyl (C=O), alkyl ether (C-O-C), and phenol hydroxyl (-OH) groups are still the main existing forms of oxygen-containing functional group of HLH.

2.2.2. FTIR Structural Parameters Analysis

FTIR structural parameters could be obtained by the area of peak with peak fitting [14,15].
(1) Ratio of hydrogen to carbon H/C.

$$\frac{H}{C} = \frac{H_{ad}}{\frac{C_{ad}}{12}} \tag{1}$$

(2) The aromatic carbon ratio $f_{ar\text{-}F}$: On the premise of ignoring carbonyl carbon, assuming that coal only contains aromatic carbon and aliphatic carbon, the formula for calculating aromatic carbon rate is as follows:

$$\frac{H_{al}}{H} = \frac{A(3000-2800)\text{cm}^{-1}}{A(3000-2800)\text{cm}^{-1} + A(900-700)\text{cm}^{-1}} \tag{2}$$

$$f_{ar-F} = 1 - \frac{C_{al}}{C} = 1 - \frac{\frac{H_{al}}{H} \times \frac{H}{C}}{\frac{H_{al}}{C_{al}}} \tag{3}$$

where aromatic hydrogen ratio H_{ar}/H: C_{al}/C is the ratio of aliphatic carbons to the total number of carbons, H/C represents the ratio of hydrogen to carbon atoms, H_{al}/C_{al} is 1.8 for all coal samples, and represents the atomic ratio between hydrogen and carbon in aliphatic groups.
(3) Fat carbon ratio $f_{al\text{-}F}$:

$$f_{al-F} = 100 - f_{ar-F} \tag{4}$$

(4) Lipid chain length and branching degree of coal I_1: According to the ratio between CH_2 and CH_3, namely the area ratio of $A(CH_2)/A(CH_3)$, the aliphatic group length and the branched chain degree were calculated to determine the aliphatic structural parameters. The intensity ratio of CH_2/CH_3 was determined by Equation (5):

$$I_1 = \frac{CH_2}{CH_3} = \frac{A_{2852\text{cm}^{-1}} + A_{2924\text{cm}^{-1}}}{A_{2957\text{cm}^{-1}}} \tag{5}$$

(5) Alkane branching degree:

$$\delta_F > \frac{R_3CH}{A_{(3000-2800\text{cm}^{-1})}} \tag{6}$$

Compared with the alkane branching degree of the 3 coal samples from Table 6, the δ_F of HLH without modification is 47.68% which indicate there were much tertiary carbon and quaternary carbon in HLH raw coal. There were many branching structures in HLH raw coal. After microwave modification, the main structure of HLH is methylene carbon with long straight chain.

Table 6. FTIR structural parameters of HLH before and after modification.

Parameter	HLH	MM	UM
H/C	0.97	0.89	0.83
H_{al}/H	0.56	0.44	0.45
$f_{ar\text{-}F}$	61.38	78.49	79.84
$f_{al\text{-}F}$	38.62	21.52	16.16
I_1	2.64	2.85	2.35
δ_F	47.68%	25.05%	36.05%

2.3. ^{13}C NMR Results Analysis

Figure 2 shows ^{13}C NMR spectra of HLH lignite. It is including 2 main peaks. The lipid-carbon peak area with chemical shift of $0-90 \times 10^{-6}$ and the aromatic-carbon peak area with chemical shift of $90-165 \times 10^{-6}$. The sample also contains a small amount of carbonyl carbon, with a chemical shift of $165-220 \times 10^{-6}$ in the peak area [16,17]. The carbon spectra obtained before and after modification were fitted by peak-splitting method, and 9 carbon skeleton structural parameters were obtained. The results are shown in Table 7.

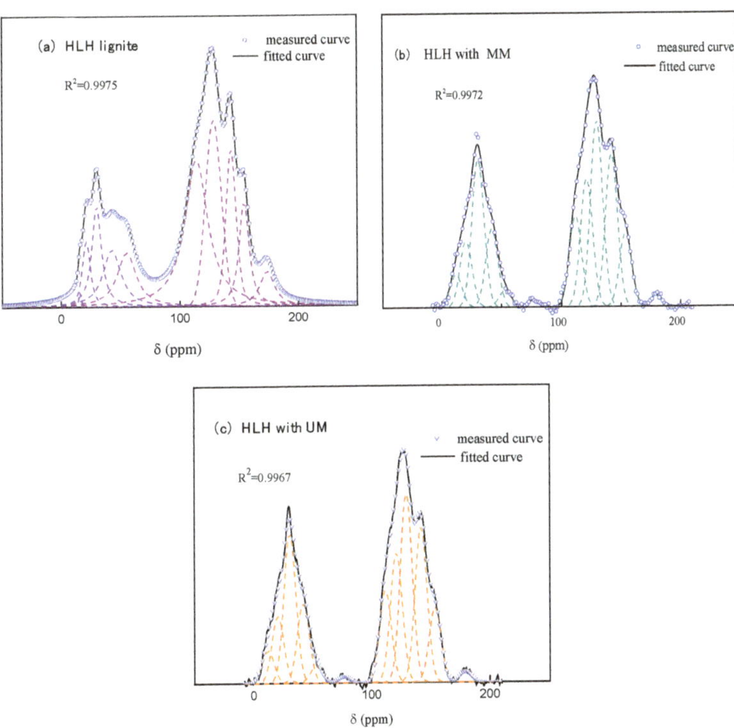

Figure 2. Peak fit of ^{13}C NMR of HLH before and after modification.

Table 7. Structure attribution and relative content of chemical shifts in ^{13}C NMR spectra of HLH before and after modification.

Chemical Shift	Structural Fragments	Symbols	Carbon Distribution	Sample		
				HLH	MM	UM
14–16	Aliphatic CH$_3$	f_{al}^M				
16–22	Aromatic CH$_3$	f_{al}^A	Aliphatic carbon	30.48	32.246	32.21
22–50	Methylene	f_{al}^H				
50–90	Oxy- aliphatic carbon	f_{al}^O				
100–129	Aromatic protonated	f_a^H				
129–137	Aromatic bridgehead	f_a^B	Aromatic carbon	64.983	63.719	64.703
137–148	Aromatic branched	f_a^S				
148–165	Oxy- aromatic carbon	f_a^O				
>165	Carbonyl carbon	f_a^C	Carbonyl carbon	4.319	3.135	3.087

$X_{BP} = f_a^B / (f_a^H + f_a^O + f_a^S + f_a^B)$, the ratio of aromatic bridge carbon to peripheral carbon of HLH before and after modification is an important parameter to construct macromolecular structure model of lignite, which represent condensation degree of polycyclic aromatic hydrocarbons as well as reflecting the size of aromatic cluster. According the parameters shown above, the X_{BP} of 3 samples could be calculated, and the value of HLH is 0.26, 0.29 for sample with microwave modification, and 0.28 for sample with ultrasonic modification.

2.4. XPS Results Analysis

2.4.1. Oxygen Element Analysis

XPS is often used to characterize the existence of oxygen, nitrogen and other heteroatoms in coal, which provides an important basis for the construction of macromolecular structure model of HLH [18]. XPS tests of HLH raw coal, microwave modified HLH and ultrasonic modified HLH were carried out, and the XPS maps of 3 samples were processed by peak fitting. The peak-splitting diagram is shown in Figure 3, and the form and content of nitrogen elements are shown in Table 8.

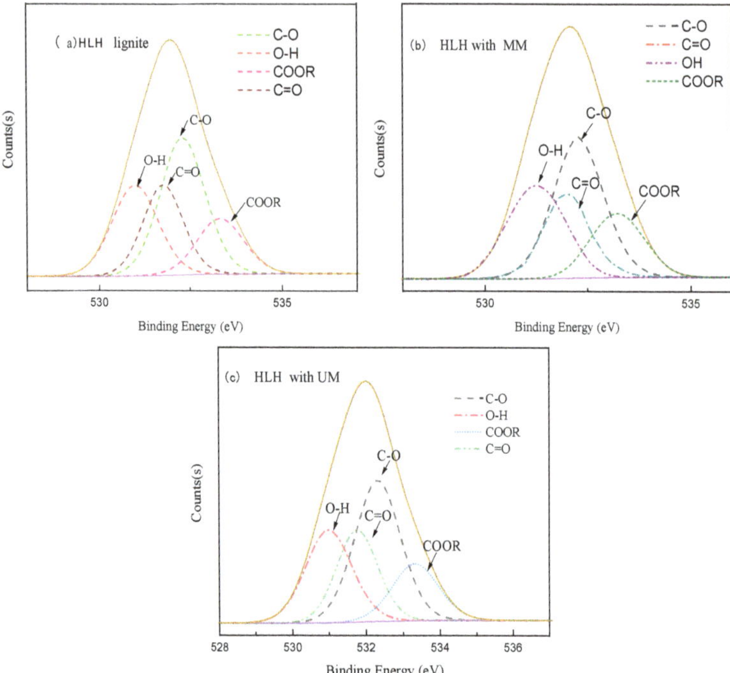

Figure 3. XPS peak fitting of oxygen atoms of HLH before and after modification.

The main forms of oxygen in HLH coal samples are C=O, C-O, -OH and COO-, and it can be found in Table 8 that oxygen exists in most of the four forms of C-O in structure. The content of C-O in HLH samples after microwave modification decreases 2%, but the content of COO-structure form decreases from 14.37% to 35.25% of raw coal. 12.37%. The reason for this change may be that microwave treatment destroys the oxygen structure of HLH raw coal. The content of oxygen-OH and C=O in HLH coal samples treated by ultrasonic wave also increased obviously, while the content of CO-form in opposite direction was still decreasing. Therefore, the microwave and ultrasonic treatment of HLH has different effects on the existing forms of oxygen elements. The stability of COO-and C-O structure forms is relatively poor and easy to be destroyed in the process of modification. On the contrary, the stability of C=O structure makes it not easy to be destroyed in the process of ultrasonic and microwave treatment, which increases the proportion of C=O in the structure.

Table 8. XPS detection and analysis of oxygen composition forms of HLH before and after modification.

Elemental Peak	Functionality	Binding Energy (eV)	Molar Content (%)
HLH	C-O	532.87	37.25
	C=O	531.27	24.71
	-OH	529.89	23.67
	COO-	533.28	14.37
MM	C-O	532.08	35.25
	C=O	531.87	26.47
	-OH	530.08	25.91
	COO-	533.28	12.37
UM	C-O	532.08	36.56
	C=O	531.87	26.21
	-OH	530.08	25.35
	COO-	533.28	11.88

2.4.2. Nitrogen Element Analysis

Nitrogen in coal mainly comes from coal-forming plants, and most of it exists in the form of organic matter, which mainly includes pyridine nitrogen (N-6), pyrrole nitrogen (N-5), nitrogen oxides (N-X) and proton nitrogen (N-Q) [19,20]. In order to characterize the forms of nitrogen elements in HLH before and after modification, the XPS spectra of HLH lignite were fitted by peak-splitting method. The peak-splitting diagram is shown in Figure 4, and the forms and contents of nitrogen elements are shown in Table 9.

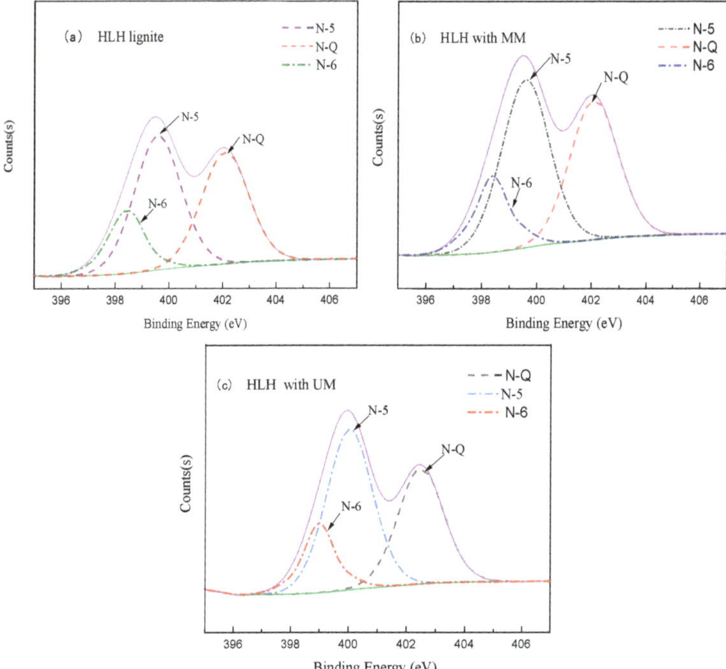

Figure 4. XPS peak fitting of nitrogen atoms of HLH before and after modification.

Table 9. XPS detection and analysis of nitrogen composition forms of HLH before and after modification.

Elemental Peak	Functionaliy	Binding Energy (eV)	Molar Content (%)
HLH	N-Q	402.21	32.78
	N-5	396.75	42.81
	N-6	399.81	24.41
MM	N-Q	402.21	31.28
	N-5	396.75	43.02
	N-6	399.81	25.70
UM	N-Q	402.21	30.78
	N-5	396.75	43.31
	N-6	399.81	25.91

By comparing the XPS data of N element in HLH before and after modification, it is obvious that the percentages of N-5, N-6 and N-Q in HLH have changed significantly. The percentage of N-Q in pulverized coal decreased to 31.28% and 30.78% respectively after modification, and the corresponding proportion of N-5 and N-6 increased to varying degrees. The total amount of N-5 and N-6 of modified HLH is nearly 70%. It can be seen that microwave and ultrasonic modification methods mainly play a role in N-Q, while N-5 and N-6 form of nitrogen bond are relatively stable, and the above modification methods have little effect on it. Therefore, in order to make the model representative, N-5 and N-6 nitrogen bonds are often used in the construction of macromolecule HLH lignite structure model.

2.5. Raman Results Analysis

There are two relatively broad D and G peaks in the Raman spectra of HLH, and two vibration peaks have abundant information, the G peak in lignite does not really represent its crystal structure. It mainly reflects the strength of the stretching vibration bond of aromatic rings [21,22]. Raman spectra of coal samples are exhibited in Figure 5.

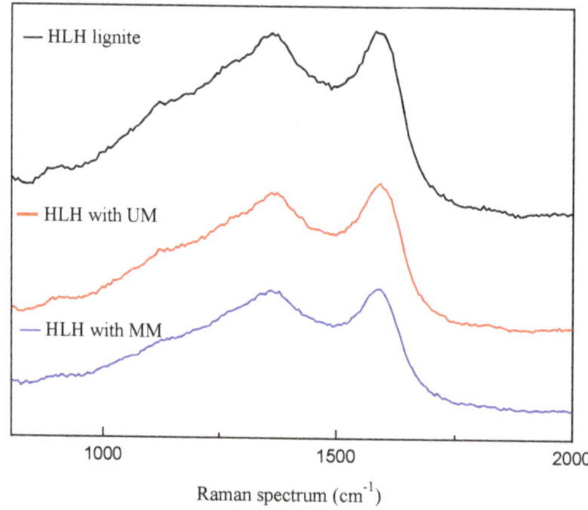

Figure 5. Raman spectra of HLH coal samples before and after modification.

There are many overlapping peaks between D and G peaks in Raman spectra of coal. In order to obtain more accurate functional group information of HLH lignite before and after modification, Position and area values of D and G peaks of coal samples obtained by deconvolution process of Raman spectra shown in Figure 6. Raman fitting parameters of HLH coal samples before and after modification are shown in Table 10.

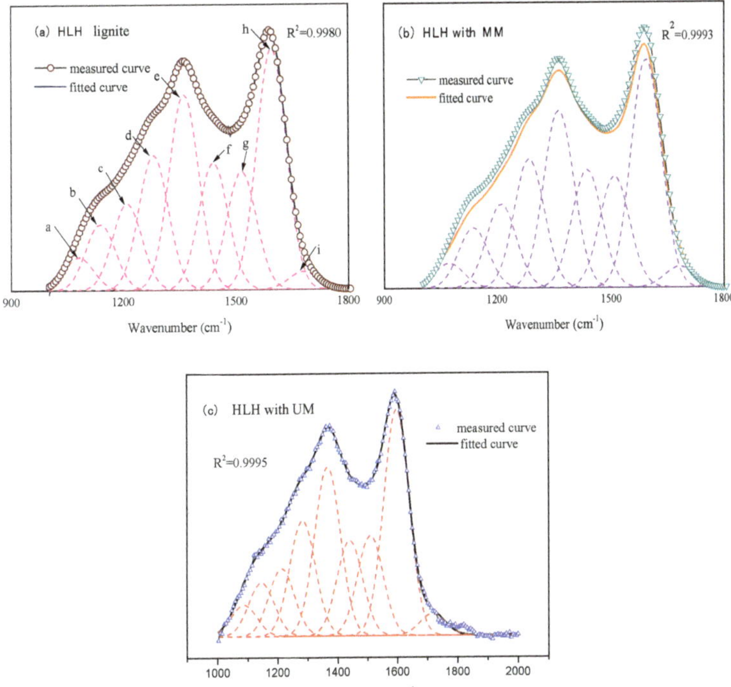

Figure 6. Raman peak separation fitting charts of HLH coal samples before and after reformation.

Table 10. Raman spectrum structural parameters of different coal samples.

Sample	Peak Position		Peak Area		Peak Height		$P_{G\text{-}D}$	A_D/A_G	I_D/I_G
	P_D	P_G	A_D	A_G	I_D	I_G			
HLH	1361.1	1594.3	47,030	57,990	4782.3	5926.3	233.2	0.81	0.81
MM	1362.1	1591.8	35,770	45,858	3105.5	3997.8	229.7	0.78	0.78
UM	1365.3	1595.6	41,480	55,500	3867.6	5173.1	230.3	0.75	0.75

From the A_D and A_G values of coal samples before and after modification, it can be found that the D peak area of coal samples after microwave and ultrasonic treatment decreases, especially the D peak area of coal samples after microwave treatment decreases significantly, which also shows that microwave treatment makes HLH structure more complete. Compared with G peak area A_G, the A_G value of HLH structure after microwave modification is significantly smaller than that of HLH. It can be seen that the total number of aromatic rings in macromolecular structure of HLH treated by microwave is the smallest, and the enrichment degree of aromatic carbon is the lowest, followed by the macromolecular structure of HLH coal treated by ultrasound, while the content of aromatic rings in the macromolecular structure of HLH coal is the highest and the enrichment degree of aromatic carbon is the largest.

I_D/I_G is usually used to evaluate the disordering degree of carbon materials, and it decreases with the increase of graphitization degree. After microwave treatment, the I_D/I_G value of HLH was reduced from 0.81 to 0.78. Similarly, after ultrasonic treatment, the I_D/I_G value of HLH was also reduced to 0.75, and the I_D/I_G value was reduced. This indicated that the ordering degree of aromatic ring layers in the structure increased and the content of fat chain and side chain decreased after microwave and ultrasonic treatment.

3. HLH Molecular Model Before and After Modification

3.1. Determination of the Type and Number of Aromatic Ring

The average molecular formula of HLH is $C_{167}N_3O_{27}H_{149}$ by elemental analysis. The average molecular formula of microwave modified HLH is $C_{148}H_{129}N_3O_{20}$, and that of ultrasonic modified HLH is $C_{155}H_{131}O_{23}N_{23}$. Combined with XPS and elemental analysis, it is found that S content in HLH lignite is extremely small. S element was added to the macromolecular model, but it was found that the percentage of S atom in the analysis of experimental elements was about 1.5%, which was obviously inconsistent with the actual results. Therefore, a small amount of S content was neglected in the construction of the macromolecular model of HLH.

X_{BP} is calculated by using the twelve structural parameters, which is calculated by ^{13}C NMR. The ratio of aromatic bridge carbon to periphery carbon of HLH raw coal is 0.26, that of microwave modified HLH is 0.29, and that of ultrasonic modified HLH is 0.28. The X_{BP} of naphthalene ring with two rings is 0.25, and that of anthracene ring with three rings is 0.40. Therefore, the aromatic framework of HLH raw coal structure model is mainly composed of benzene ring and naphthalene ring. After HLH with microwave modification model and HLH with ultrasonic modification aromatic framework is mainly composed of naphthalene ring, with anthracene ring and benzene ring supplemented. In order to make the X_{BP} of HLH raw coal and HLH model after microwave and ultrasonic modification close to 0.26, 0.29 and 0.28, the combination number of benzene, naphthalene and anthracene rings in its structural model needs to be adjusted continuously. The type and number of aromatic rings in the three structural models are determined. The results are shown in Table 11.

Table 11. Type and quantities of aromatic rings of HLH before and after modification.

Type of Aromatic Unit Structure	Number		
	HLH	**MM**	**UM**
(naphthalene)	6	4	5
(benzene)	5	3	3
(quinoline)	2	2	2
(pyrrole)	1	1	1
(anthracene)	—	1	1

Comparing the previous ^{13}C NMR and FTIR spectra, it is found that the proportion of oxygen bonded by carbon-oxygen double bond is much smaller than that bonded by carbon-oxygen single bond, which is consistent with the XPS test results. This shows that the main forms of oxygen in macromolecular structure before and after HLH modification are mostly in the form of ether bond and hydroxyl bond of carbon-oxygen single bond, and the other oxygen-containing structures are in the second place. The final form of oxygen in the structure is determined by constantly adjusting the structure of oxygen.

The main forms of nitrogen in HLH macromolecular structure are pyridine nitrogen and pyrrole nitrogen. According to XPS analysis, the number of nitrogen atoms is always 3 of HLH before and after modification, and the main forms of nitrogen elements were pyridine nitrogen and pyrrole nitrogen the content ratio was 2:1. Therefore, two pyridine rings and one pyrrole ring were added to the HLH structure model before and after modification.

3.2. Model Construction

Wiser coal chemical model is used as the basis, which is widely accepted and considered to be comprehensive and reasonable. Combining with the above calculation results, the existing forms and proportions of each part of the macromolecular structure model before and after modification of HLH are summarized and analyzed. Finally, the two-dimensional molecular model of HLH before and after modification is preliminarily established according to the chemical structure characteristics obtained from the experiments. As shown in Figures 7–9.

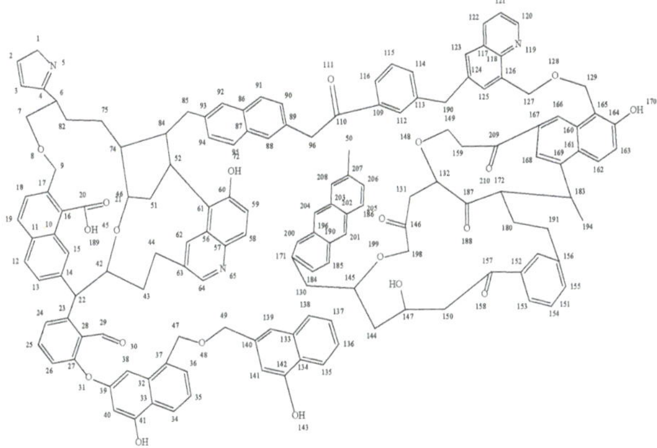

Figure 7. Final two-dimensional model macromolecular structure diagram of HLH.

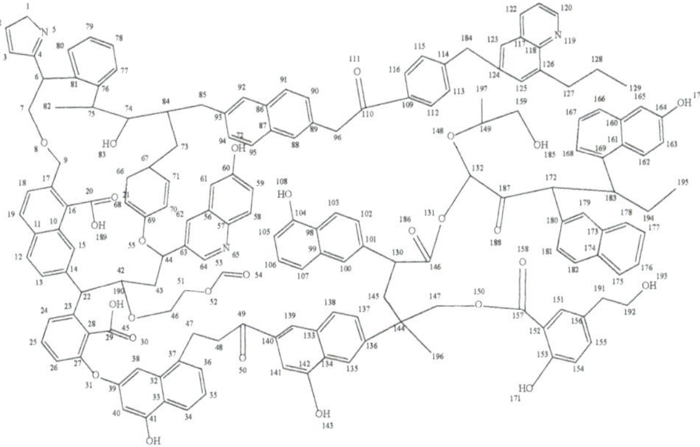

Figure 8. Final two-dimensional model macromolecular structure of HLH after microwave modification.

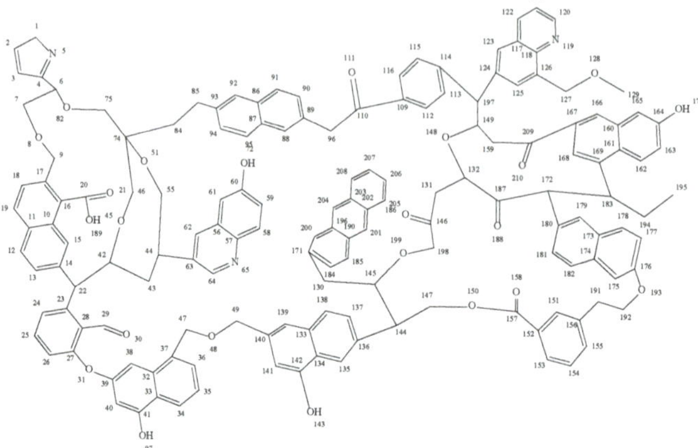

Figure 9. Final two-dimensional model macromolecular structure of HLH after Ultrasonic modification.

3.3. Verification of Model

The ^{13}C NMR chemical shifts of the three models were calculated and compared with the experimental ^{13}C NMR chemical shifts. Because of the complexity and diversity of coal's macromolecule structure, it is necessary to constantly adjust the types and quantities of its structural units in the construction process, so as to make its ^{13}C NMR simulation spectra better consistent with the experimental spectra. The comparison of ^{13}C NMR simulation spectra and experimental spectra of the three models is shown in Figure 10.

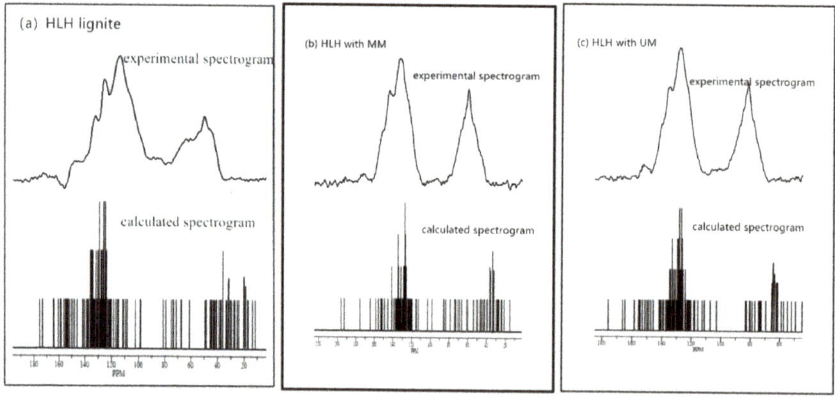

Figure 10. Comparison of ^{13}C NMR computational spectra and experimental spectra of HLH final two-dimensional model before and after modification: (**a**) HLH, (**b**) HLH after microwave modification, (**c**) HLH after Ultrasonic modification.

It can be seen that the ^{13}C NMR simulation spectra of the three models are in good agreement with the experimental spectra. There are some errors between the two spectra due to some uncontrollable factors in the experimental process, but it is considered acceptable and will not affect the characterization of average macromolecular structure of HLH coal samples.

In order to adjust the structure of the model to approximate the experimental data, most researchers adopted the simulation of ^{13}C NMR spectrum, but which cannot avoid the choice of isomers [23,24]. At the same time, the model obtained by this method is only a conceptual structure, which cannot reflect the properties of chemical reactions. By using the main covalent bond concentration instead of the ^{13}C NMR simulation spectrum, the molecular model can be better corrected. According to the 11 structural parameters (f_a, f_a^H, f_a^O, f_a^B, f_a^S, f_a^C, f_{al}, f_{al}^A, f_{al}^M, f_{al}^H, f_{al}^O) obtained from ^{13}C NMB data, nine covalent bond concentrations of HLH lignite (Car-Car, Car-Cal, Cal-Cal, Car-H, Cal-H, Car-O, Cal-O, Cal=O, and O-H.) can be obtained. This covalent concentration method reflects the essence of ^{13}C NMR simulation spectroscopy. By comparing the simulated concentration of the main covalent bond with the experimental results, the preliminary two-dimensional molecular model is modified. The concentrations of nine types covalent bonds in coal can be determined by Equations (7)–(15).

$$con_{C_a-C_a} = \frac{1}{2}\left[\frac{C\%}{12}\left(3f_a - f_a^H - f_a^S - f_a^O\right)\right] \tag{7}$$

$$con_{C_a-C_{al}} = \frac{C\%}{12}f_a^S \tag{8}$$

$$\begin{aligned}con_{C_{al}-C_{al}} &= \frac{1}{2}\left[\frac{C\%}{12}\left(4f_{al} + 2f_a^C - f_a^C - f_{al}^O\right) - n_{C_{al}-H}\right]\\ &= -\frac{H\%}{2} + \frac{O\%}{16} + \frac{1}{2}\left[\frac{C\%}{12}\left(4f_{al} + f_a^H - f_a^S - f_a^O - 2f_{al}^O - f_a^C\right)\right]\end{aligned} \tag{9}$$

$$con_{C_a-H} = \frac{C\%}{12}f_a^H \tag{10}$$

$$\begin{aligned}con_{C_{al}-H} &= H\% - con_{O-H} - con_{C_a-H}\\ &= H\% - 2\frac{O\%}{16} + \frac{C\%}{12}\left(f_a^O + f_{al}^O + 2f_a^O + 3f_a^C - f_{ar}^H\right)\end{aligned} \tag{11}$$

$$con_{C_{ar}-O} = \frac{C\%}{12}f_{ar}^O \tag{12}$$

$$con_{C_{al}-O} = \frac{C\%}{12}\left(f_{al}^O + f_a^C\right) \tag{13}$$

$$con_{C_{al}=o} = \frac{C\%}{12}\left(f_a^C\right) \tag{14}$$

$$con_{O-H} = \frac{20\%}{16} - \frac{C\%}{12}\left(f_a^O + f_{al}^O + 3f_a^C\right) \tag{15}$$

Based on the formula, the information of covalent bond concentration of the three final planar model structures is calculated and summarized in Table 12, in order to more intuitively compare and analyze the difference between the covalent bond concentration obtained from the experiment and that calculated from the model. Then, the molecular models of the three structures obtained from HLH raw coal, microwave, and ultrasonic upgrading of HLH coal, respectively, were adjusted.

Table 12. Covalent bond concentrations of 3 structural models of HLH before and after modification.

Sample	con_{Ca-Ca}	con_{Ca-Cal}	$con_{Cal-Cal}$	con_{Ca-H}	con_{Cal-H}	con_{Ca-O}	con_{Cal-O}	$con_{Cal=O}$	con_{O-H}
HLH	46.701	6.943	18.037	16.410	35.768	5.303	5.389	2.828	7.208
HLH Model	46.233	6.822	17.354	17.231	35.154	5.036	5.673.	2.32	7.053
MM	47.669	10.053	19.488	14.499	36.133	4.851	3.621	2.113	5.252
MM Model	47.276	10.997	17.376	15.547	34.453	4.784	3.435	2.349	5.127
UM	47.095	9.934	20.374	15.093	38.309	4.793	2.133	2.005	6.407
UM Model	46.967	9.322	19.763	15.216	37.867	4.452	2.243	2.105	6.395

According to the covalent bond concentration obtained from the experiment and the concentration calculated by the model, it was found that the concentration of the nine covalent bonds is not different from that calculated by the model, but the difference between the experimental concentration of $con_{Cal-Cal}$ and con_{Car-H} and the calculated concentration of the model is relatively large. The reason for this may be that there is a certain difference between the proportion of elements in the model and that in the experimental measurement, which makes the calculation of the nine covalent bonds concentration more intensive.

3.4. Adjustment of Model

After the final model of molecular structure is obtained, the molecular formulas and element contents of the three models of HLH under different modification conditions are calculated. By comparing the results of Tables 13 and 14, it can be seen that the element contents of the three models are very close to the experimental values, and there is no significant difference, which verifies the reliability of the model.

Table 13. Element content measured by experiments of HLH before and after modification.

Sample	Ultimate Analysis (wt. %), ad			
	C	H	O	N
HLH	78.56	6.36	13.48	1.61
MM	80.87	5.99	11.39	1.76
UM	81.97	5.67	10.41	1.95

Table 14. Molecular formula and element content of the model of HLH before and after modification.

Sample	Molecular Formula	Ultimate Analysis (wt. %)			
		C	H	O	N
HLH	$C_{167}H_{151}N_3O_{27}$	76.21	5.78	16.41	1.6
MM	$C_{148}H_{129}N_3O_{20}$	78.52	5.73	14.10	1.85
UM	$C_{155}H_{131}N_3O_{23}$	77.45	5.49	15.31	1.75

3.5. Construction of 3- Dimensional Model

The three-dimensional model is constructed for the 2- dimensional structure of HLH before and after modification, whose three-dimensional geometric optimization configuration is calculated by MM and MD of Forcite module in Materials Studio 8.0 software. The structure model of HLH before and after optimization is shown in Figures 11–13, and the energy change in the process of optimization is shown in Table 15.

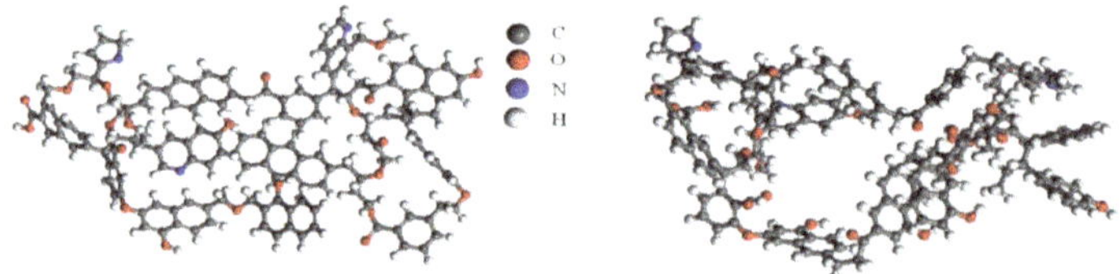

(**a**) Initial 3-dimensional structure model   (**b**) Optimized by MM and MD

Figure 11. 3-dimensional structure of HLH lignite model before and after geometric optimization.

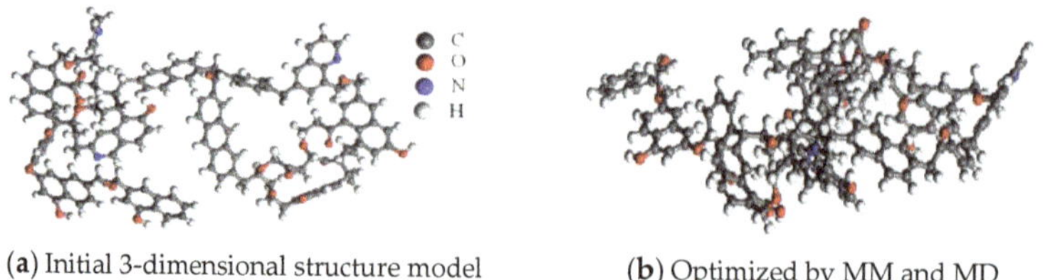

(**a**) Initial 3-dimensional structure model   (**b**) Optimized by MM and MD

Figure 12. 3-dimensional structure of HLH lignite model modified by microwave before and after geometric optimization.

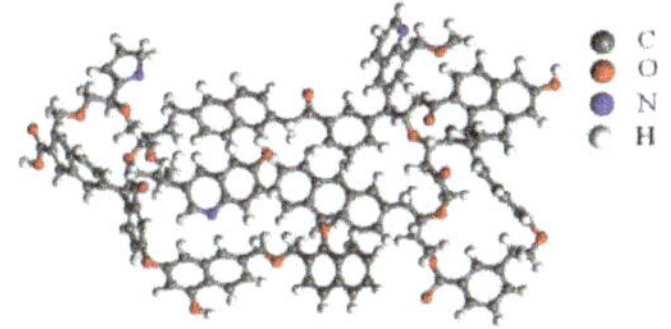

(**a**) Initial 3-dimensional structure model

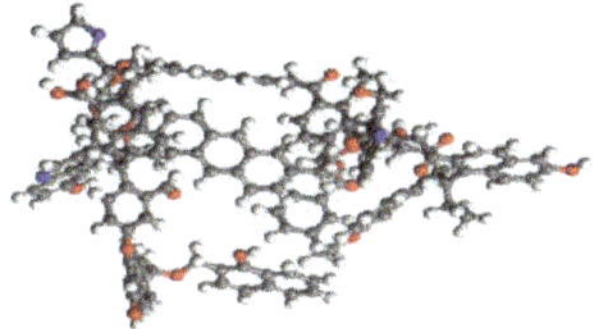

(**b**) Optimized by MM and MD

Figure 13. 3-dimensional structure of HLH lignite model modified by ultrasound before and after geometric optimization.

Table 15. Energy comparison before and after HLH structural model optimization.

Sample		Total Energy (Kcal·mol^{-1})	Valence Energy (Kcal·mol^{-1})				Non-Bond Energy (Kcal·mol^{-1})		
			E_B	E_A	E_T	E_I	E_H	E_{van}	E_E
HLH	Initial	6030.56	2602.16	65.21	90.59	2.05	0	3270.55	0
	Final	810.65	204.89	197.06	119.08	18.42	0	342.05	−70.85
MM	Initial	6310.80	2225.16	153.04	89.24	1.68	0	3841.68	0
	Final	850.20	191.04	217.58	139.81	24.99	0	344.30	−67.52
UM	Initial	7608.20.	2402.78	142.34	122.83	1.56	0	4938.69	0
	Final	846.37	187.50	189.76	134.07	17.57	0	348.01	−30.54

It can be seen from this that the total energy of the minimum energy structures of the three structures decreases sharply. Compared with other terms, the value of Van der Waals energy (E_{Van}) in the non-bonding energy is the largest, which constitutes the most important part of the potential energy. Therefore, the decrease of the inter-molecular Van der Waals energy (E_{Van}) is also the main factor that makes the HLH macromolecular structure model stable in space.

4. Conclusions

(1) The molecular formula of the three structures was determined by elemental analysis. The average molecular formula of HLH raw coal configuration was $C_{167}N_3O_{27}H_{149}$, and the aromatic part consists of five benzene rings, six naphthalene rings, two pyrrole rings and one pyridine ring. The average molecular formula of Mm coal configuration was $C_{148}H_{129}N_3O_{20}$, the aromatic structure includes three benzene rings, four anthracene rings, four naphthalene rings, one anthracene ring, one pyridine ring and two pyrrole rings. And the average molecular formula of Um coal configuration was $C_{155}H_{131}O_{23}N_{23}$, the aromatic structure includes three benzene rings, five anthracene rings, four naphthalene rings, one anthracene ring, one pyridine ring, and two pyrrole rings.

(2) It was found that after microwave and ultrasonic treatment, the orderliness of aromatic ring layer arrangement increased. The content of fat chain and side chain decreased, and the existence form of oxygen atoms also changed, in which the proportion of C-O and COO- form shows a decreasing trend. It was found that the total energy of the three structures decreased significantly after optimization. The chemical bonds between the atoms are obviously bent and distorted, and the space configuration is more stereoscopic. Although the model constructed in this paper is not the most comprehensive and optimized configuration before and after HLH modification, some problems such as isomers are considered in the process of construction, which has a certain reference value for better understanding and application of lignite upgrading.

Author Contributions: J.L. provided methodology and original draft writing. Z.H. and Y.Y. conceived and designed the study. S.Q. performed the experiments. Y.Y. provided the samples. S.Q. wrote the paper. Z.H. reviewed and edited the manuscript. All authors have read and agreed to the published version of the manuscript.

Funding: This work was financially supported by National Natural Science Foundation of China (51874171) and and University of Science and Technology Liaoning Talent Project Grants (No. 601011507-05).

Conflicts of Interest: The authors declared that they have no conflict of interest to this work. We declare that we do not have any commercial or associative interest that represents a conflict of interest in connection with the work submitted.

References

1. He, Q.Q.; Wan, K.; Hoadley, A.; Yeasmin, H.; Miao, Z.Y. TG–GC–MS study of volatile products from Shengli lignite pyrolysis. *Fuel* **2015**, *156*, 121–128. [CrossRef]
2. Liu, P.; Zhang, D.X.; Wang, L.L.; Zhou, Y.; Pan, T.Y.; Lu, X.L. The structure and pyrolysis product distribution of lignite from different sedimentary environment. *Appl. Energy* **2016**, *163*, 254–262. [CrossRef]
3. Zhao, H.; Guo, F.; Yang, J. Adsorption characteristic of Indonesia lignite and dewater experiment. *J. China Coal Soc.* **2008**, *33*, 799–802.
4. Hua, Z.; Qin, Z. Solubilization rules of small molecules with different states under ultrasonic extraction. *J. China Univ. Min. Technol.* **2012**, *41*, 91–94.
5. Tahmasebi, A.; Yu, J.; Li, X.; Meesri, C. Experimental study on microwave drying of Chinese and Indonesian low-rank coals. *Fuel Process. Technol.* **2011**, *92*, 1821–1829. [CrossRef]
6. Sun, Q.; Zhang, Y.; Li, Q. Physical and chemical characteristics and gasification reactivity of lignite fast pyrolysis char. *J. Zhejiang Univ. (Eng. Sci.)* **2016**, *50*, 2045–2051.
7. Ge, L. Basic Research on Typical Low-Quality Coal Upgrading and Poly-Generation System Based on the Cascade Utilization of Coal. Ph.D. Thesis, Zhejiang University, Hangzhou, China, 2014; pp. 1–10.
8. Castro-Marcano, F.; Russuo, M.F., Jr.; van Duin, A.C.T.; Mathews, J.P. Pyrolysis of a large-scale molecular model for Illinois no. 6 coal using the ReaxFF reactive force field. *J. Anal. Appl. Pyrolysis* **2014**, *109*, 79–89. [CrossRef]
9. Cui, W.; Li, M.; Gong, G.; Lu, K.; Sun, S.; Dong, F. Guided trilateral filter and its application to ultrasound image despeckling. *Biomed. Signal Process. Control* **2020**, *55*, 101625. [CrossRef]
10. Mokhtar, N.M.; Omar, R.; Idris, A. Microwave pyrolysis for conversion of materials to energy: A brief review. *Energy Sources Part A Recovery Util. Environ. Eff.* **2012**, *34*, 2104–2122. [CrossRef]
11. Han, F.; Meng, A.H.; Li, Q.H.; Zhang, Y. Thermal decomposition and evolved gas analysis (TG-MS) of lignite coals from Southwest China. *J. Energy Inst.* **2016**, *89*, 94–100. [CrossRef]
12. Li, X.; Zeng, F.G.; Wang, W.; Dong, K.; Cheng, L.Y. FTIR characterization of structural evolution in low-middle rank coals. *J. China Coal Soc.* **2015**, *40*, 2900–2908.
13. Yang, F.; Hou, Y.; Wu, W.; Niu, M.; Ren, S.; Wang, Q. A new insight into the structure of Huolinhe lignite based on the yields of benzene carboxylic acids. *Fuel* **2017**, *189*, 408–418. [CrossRef]
14. Xu, F.; Pan, S.; Liu, C.; Zhao, D.; Liu, H.; Wang, Q.; Liu, Y. Construction and evaluation of chemical structure model of Huolinhe lignite using molecular modeling. *RSC Adv.* **2017**, *7*, 41512–41519. [CrossRef]
15. Liang, H.Z.; Wang, C.G.; Zeng, F.G.; Li, M.F.; Xiang, J.H. Effect of demineralization on lignite structure from Yinmin coalfield by FT-IR investigation. *J. Fuel Chem. Technol.* **2014**, *42*, 129–137.
16. Guo, D.Y.; Ye, J.W.; Wang, Q.B.; Guo, X.J. FT-IR and 13C-NMR characterizations for deformed coal in Pingdingshan mining. *J. China Coal Soc.* **2016**, *41*, 3040–3046.
17. Wang, Y.G.; Zhou, J.L.; Chen, Y.J.; Hu, X.; Zhang, S.; Lin, X. Contents of O-containing functional groups in coals by 13C-NMR analysis. *J. Fuel Chem. Technol.* **2013**, *41*, 1422–1426.
18. Wang, J.; He, Y.; Li, H.; Yu, J.; Xie, W.; Wei, H. The molecular structure of Inner Mongolia lignite utilizing XRD, solid state 13 C NMR, HRTEM and XPS techniques. *Fuel* **2017**, *203*, 764–773. [CrossRef]
19. Wang, Q.; Liu, Q.; Wang, Z.C.; Liu, H.P.; Bai, J.R.; Ye, J.B. Characterization of organic nitrogen and sulfur in the oil shale kerogens. *Fuel Process. Technol.* **2017**, *160*, 170–177. [CrossRef]
20. Kelemen, S.R.; Afeworki, M.; Gorbaty, M.L.; Sansone, M.; Kwiatek, P.J.; Walters, C.C.; Freund, H.; Siskin, M. Direct characterization of Kerogen by X-ray and solid-state ^{13}C nuclear magnetic resonance methods. *Energy Fuels* **2007**, *21*, 1548–1561. [CrossRef]
21. Zerda, T.W.; John, A.; Chmura, K. Raman studies of coals. *Fuel* **1981**, *60*, 375–378. [CrossRef]
22. Li, M.F.; Zeng, F.G.; Qi, F.H.; Sun, B.L. Raman spectral characteristics of different coal grades and their relationship with XRD structural parameters. *Spectrosc. Spectr. Anal.* **2009**, *29*, 2446–2449.

23. Takanohashi, T.; Kawashima, H. Construction of a model structure for upper Freeport coal using ^{13}C NMR chemical shift calculations. *Energy Fuels* **2002**, *16*, 379–387. [CrossRef]
24. Xiang, J.H.; Zeng, F.G.; Liang, H.Z.; Sun, B.L.; Zhang, L.; Li, M.F.; Jia, J.B. Model construction of the macromolecular structure of Yanzhou coal and its molecular simulation. *J. Fuel Chem. Technol.* **2011**, *39*, 481–488. [CrossRef]

Article

Combustion Kinetics Characteristics of Solid Fuel in the Sintering Process

Jihui Liu [1], Yaqiang Yuan [1], Junhong Zhang [1], Zhijun He [1,*] and Yaowei Yu [2,*

[1] School of Materials and Metallurgy, University of Science and Technology, Liaoning 114051, China; gtyj66@126.com (J.L.); as2013h@163.com (Y.Y.); gtyj0412@163.com (J.Z.)
[2] State Key Laboratory of Advanced Special Steel, School of Materials Science and Engineering, Shanghai University, Shanghai 200444, China
* Correspondence: hezhijun@ustl.edu.cn (Z.H.); yaoweiyu@shu.edu.cn (Y.Y.)

Received: 23 February 2020; Accepted: 11 April 2020; Published: 17 April 2020

Abstract: In order to systematically elucidate the combustion performance of fuel during sintering, this paper explores the influence of three factors, namely coal substitution for coke, quasi-particle structure and the coupling effect with reduction and oxidation of iron oxide, on fuel combustion characteristics, and carries out the kinetic calculation of monomer blended fuel (MBF) and quasi-granular fuel (QPF). The results show that replacing coke powder with anthracite can accelerate the whole combustion process. MBF and QPF are more consistent with the combustion law of the double-parallel random pore model. Although the quasi-particle structure increases the apparent activation energy of fuel combustion, it can also produce a heat storage effect on fuel particles, improve their combustion performance, and reduce the adverse effect of diffusion on the reaction process. In the early stage of reaction, the coupling between combustion of volatiles and reduction of iron oxide is obvious. The oxidation of iron oxide will occur again when the combustion reaction of fuel is weakened.

Keywords: quasi-particle structure; monomer blended fuel; quasi-particle fuel; apparent activation energy; coupling effect

1. Introduction

At present, the supply of energy is highly reliant on fossil fuel [1]. The rising demand for energy around the globe is leading to many economic and environmental problems [2]. Ironmaking in China is still dependent on the blast furnace [3]. The direct reducibility and intensity of high-basicity sinter, which is the main raw material for blast furnace ironmaking, markedly influence the production efficiency of ironmaking [4,5]. The combustion characteristic of fuel in sintering mixture materials plays a decisive role in sinter quality [6,7]. Therefore, it is urgent to explore the combustion characteristics of solid fuel in the sintering process in order to provide a theoretical basis for improving the combustion efficiency, improving the quality of sinter and reducing the consumption of solid fuel [8]. Coke is made from natural bituminous coal heated between 950 °C and 1050 °C in an airless environment. It is the main fuel in the sintering site [9]. Previously, the research on sintering fuel combustion was mostly based on coke breeze [10,11]. However, with the development of the steel industry and the continuous improvement of the grog ratio in the blast furnace, the supply of coke powder for sintering has been short. Therefore, many sintering sites use anthracite wholly or partially to replace coke powder as fuel for sintering [12,13].

In the traditional sense, the experimental samples for studying the combustion characteristics of fuels consist mainly of monomer fuel. However, during the sintering process, the solid fuel in the material layer is usually distributed in a dispersed manner. Hence, the combustion law of sintering fuel should generally be different from the monomer fuel and the fuel layer [14,15]. In recent years, researchers have begun to use a quasi-particle structure to describe the existence of fuel inside the sintering mixture [16,17]. In order to exclude the interference brought by other reactions, Al_2O_3 pure powder reagent is often used in many studies on quasi-particles sinter to replace other materials used in production for experimental exploration [18].

Combustion of sintering fuel at low temperature is generally considered as a chemical reaction control process. However, the introduction of a quasi-particle structure will greatly improve the diffusion resistance of internal fuel combustion. Therefore, the extent to which diffusion controls combustion is greatly increased [19,20]. This paper innovatively introduced the double parallel reaction volume model (DVM) and double parallel random pore model (DRPM) to conduct a comparative analysis of the combustion characteristics of monomer blended fuel and quasi-particle fuel and calculated the related kinetic parameters, which are widely used in the kinetic calculation of co-combustion of multiple fuels [21], single-fuel gasification [22] and co-gasification of multiple fuels [23,24]. They can not only obtain the suitable dynamic models for describing the combustion process of two kinds of fuel, but also characterize the effect of quasi-particle structure on the combustion characteristics of sintering fuel. There are many physical and chemical reactions in the sintering process. Due to the influence of sintering temperature and atmosphere, iron oxides undergo different degrees of reduction and oxidation reactions. The occurrence of these reactions is coupled with fuel combustion, which greatly affects the quality of sinter [25]. Therefore, this study systematically explores the influence of factors such as the substitution of anthracite for coke powder, quasi-particle structure and the coupling effect of reduction and oxidation of iron oxide on the combustion characteristics of sintering fuel in order to provide a certain guiding significance for improving fuel efficiency and reducing sintering production cost.

2. Materials and Methods

2.1. Materials

The fuel used in the experiment was 25% anthracite blended with 75% coke powder by weight and the particle size was less than 0.105 mm. The specific industrial analysis, elemental analysis and calorific value of the raw materials are shown in Table 1. Because there are many types of reactions of the sintering process of iron ore and the coupling between the reactions is strong, in order to investigate the influence of the quasi-particle structure on the combustion kinetic parameters of the sintering fuel, Fe_2O_3 and Al_2O_3 pure powder reagents with a particle size of less than 0.147 mm were used instead of the iron ore and flux applied in an industrial setting. In addition to single anthracite and coke powder, there are also three-blended fuels, namely monomer blended fuel (MBF), quasi-particle fuel (QPF) and sintered mixture (SDM). The schematic diagrams of them are shown in Figure 1, and the ratios of raw materials are shown in Table 2.

Table 1. Proximate and ultimate analysis of the fuel (in dry basis).

Samples	Proximate Analysis/%			Ultimate Analysis/%					$Q_{gr}/(MJ\ g^{-1})$
	FC	V	A	C	H	O	N	S	
Anthracite	76.51	6.94	16.58	75.86	1.69	2.25	0.81	0.36	29.31
Coke	86.68	0.23	11.80	87.17	0.77	0.64	0.91	0.51	35.23

FC, V and A represent fixed carbon, volatile and ash, respectively; subscript gr means gross calorific value.

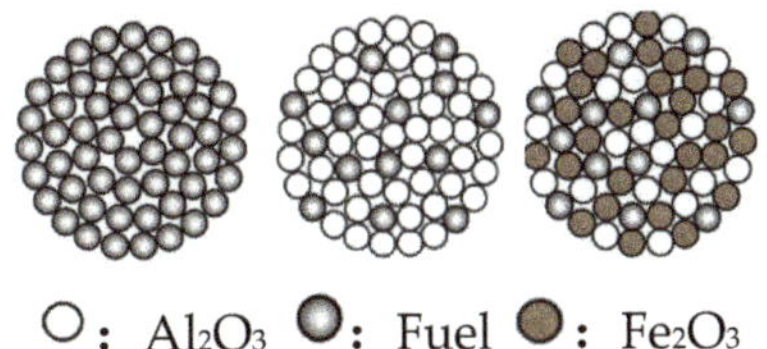

Figure 1. Schematic diagram of monomer blended fuel (MBF), quasi-granular fuel (QPF) and sintered mixture (SDM).

Table 2. Raw materials and proportions used in the experiment, wt%.

Samples	Fe_2O_3	Al_2O_3	Anthracite	Coke
MBF	-	-	25	75
QPF	-	60	10	30
SDM	76	20	1	3

2.2. Thermogravimetric Experiment

Combustion experiments were conducted using the HCT-4 thermal analyzer. The usage of anthracite, coke powder and MBF was 1.2 ± 0.1 mg for each group, and 30 ± 0.1 mg for QPF and SDM respectively. Each sample was loaded into a crucible for thermal analysis with a height of 4 mm and diameter of 5 mm. The samples were heated from room temperature (25 °C) to 1000 °C at heating rates of 5.0, 10.0, 15.0, and 20.0 °C/min, respectively. The rate of heating is expressed as β. At the same time, the air was injected at a flow rate of 100 mL/min during the heating to provide an oxidizing atmosphere for the heating. To ensure the accuracy of all experimental results, each experiment was repeated at least three times under the same conditions.

The conversion of the sample (α) was calculated with the mass loss data collected during the combustion

$$\alpha = \frac{m_0 - m_t}{m_0 - m_\infty} \tag{1}$$

where m_0 is the original mass of the sample; m_t is the mass at time t; m_∞ is the final mass of the sample after the reaction.

In the thermogravimetric combustion experiment, the parameters of the sample can be determined by using the thermal analysis curve (TG-DTG), including ignition temperature (T_i), combustion temperature (T_j), peak temperature (T_p), combustion reaction time (t), flammability index (C) and combustion characteristic index (S). The determination method of characteristic parameters is shown in Figure 2.

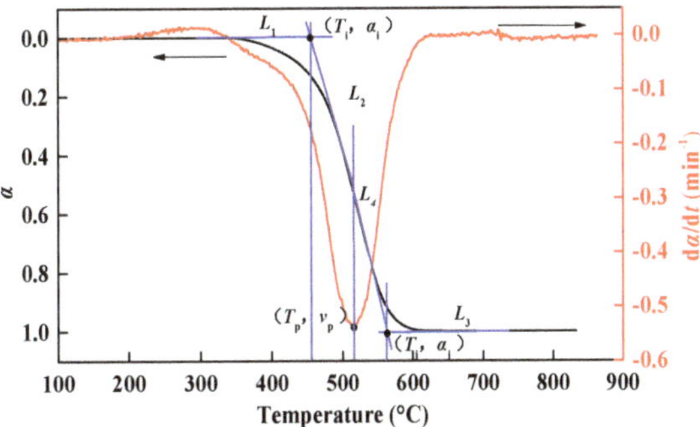

Figure 2. Schematic diagram of characteristic parameter determination method of the thermogravimetric curve.

The flammability index reflects the ability of the sample to react at the beginning of combustion. This index can measure the ignition stability of the sample during combustion,

$$C = v_{\mathrm{p}}/T_{\mathrm{i}}^2 \tag{2}$$

where v_{p} is the maximum combustion reaction rate in min^{-1}.

The combustion characteristic index reflects a combined characteristic of the ignition and combustion of the sample. If the value of S is larger, the combustion performance of the sample is better,

$$S = v_{\mathrm{p}} \times v_{\mathrm{m}}/(T_{\mathrm{i}}^2 \times T_{\mathrm{j}}) \tag{3}$$

where v_{m} is the average burning rate of the sample from T_{i} to T_{j} in min^{-1}.

2.3. Thermal Analysis Kinetic

The combustion process of the sintering fuel can be regarded as a gas–solid heterogeneous reaction. The total combustion reaction consists of two independent chemical reactions:

$$\text{Anthracite} + (x/2 + y + z/2)O_2 \rightarrow xCO + yCO_2 + zH_2O \tag{4}$$

$$\text{Coke} + (x/2 + y + z/2)O_2 \rightarrow xCO + yCO_2 + zH_2O \tag{5}$$

In order to further clarify the combustion reaction mechanism of the pure mixed fuel and the quasi-particle fuel, we introduced two kinetic models to study the combustion behavior of the sample:

$$\frac{d\alpha}{dt} = \sum_{i=1}^{2} c_i k_i f(\alpha_i) \tag{6}$$

where t is the reaction time, c_i is the proportion of a reaction to the total response, k_i is the combustion reaction rate constant, and $f(\alpha_i)$ is a function of the differential reaction mechanism.

The relationship between apparent reaction rate and temperature can be derived from the Arrhenius equation,

$$k = Ae^{-E/RT} \tag{7}$$

where E is reaction activation energy, A is the pre-exponential factor, R is the universal gas constant, and T is the temperature.

Currently, volumetric models (VM) and random pore models (RPM) are widely used to describe various coal char combustion reactions and are used to calculate kinetic parameters,

$$\frac{d\alpha_{\mathrm{VM}}}{dt} = A_{\mathrm{VM}}e^{-E_{\mathrm{VM}}/RT}(1 - \alpha_{\mathrm{VM}}) \tag{8}$$

$$\frac{d\alpha_{\mathrm{RPM}}}{dt} = A_{\mathrm{RPM}}e^{-E_{\mathrm{RPM}}/RT}(1 - \alpha_{\mathrm{RPM}})\sqrt{1 - \psi \ln(1 - \alpha_{\mathrm{RPM}})} \tag{9}$$

where ψ is the parameter of particle structure,

$$\psi = \frac{4\pi L_0(1 - \varepsilon_0)}{S_0^2} \tag{10}$$

where S_0 is the pore surface area, L_0 is the pore length, and ε_0 is the porosity of particles.

Since the experimental materials use two types of fuels with different combustion performances, it is necessary to optimize the VM and RPM. The expressions of DVM and DRPM are obtained by combining Equations (6)–(9),

$$\frac{d\alpha_{\text{DVM}}}{dt} = \sum_{i=1}^{2} c_i A_i e^{-E_i/RT} (1 - \alpha_i)$$

(11)

$$\frac{d\alpha_{\text{DRPM}}}{dt} = \sum_{i=1}^{2} c_i A_i e^{-E_i/RT} (1 - \alpha_i) \sqrt{1 - \psi_i \ln(1 - \alpha_i)}$$

(12)

In the non-isothermal analysis experiment, in order to determine the kinetic parameters and improve the calculation accuracy, three or more types of heating rates are usually selected. Thus, this experiment adopts four different heating rates to calculate kinetic parameters. Under the constant heating rate of the experiment, the reaction temperature can be obtained from the initial temperature and reaction time,

$$T = T_0 + \beta t$$

(13)

where β is the heating rate, and T_0 is the starting temperature of 25 °C. After $t = (T - T_0)/\beta$ is substituted into Equations (11) and (12), the formulas can be integrated to give

$$\alpha_{\text{DVM}} = \sum_{i=1}^{2} c_i \left(1 - \exp\left(-\frac{A_i RT^2}{\beta E_i} \cdot \exp\left(\frac{-E_i}{RT}\right)\right)\right)$$

(14)

$$\alpha_{\text{DRPM}} = \sum_{i=1}^{2} c_i \left(1 - \exp\left(-\exp\left(\frac{-E_i}{RT}\right) \cdot \frac{A_i RT^2}{\beta E_i} \cdot \left(1 + \exp\left(\frac{-E_i}{RT}\right) \cdot \frac{\psi_i A_i RT^2}{4\beta E_i}\right)\right)\right)$$

(15)

The combustion kinetic parameters were calculated by the above two kinetic models at different heating rates. The experimental data of the reaction rate ($d\alpha/dt$) and conversion rate (α) were fitted in 1stop software using a nonlinear least-squares method. Then, the parameters A, E and ψ that are obtained are substituted into Equations (14) and (15) to obtain the relationship between the sample conversion rate (α) and the temperature (T) during combustion. At the same time, due to the possible deviation between the actual value and the calculated value of the model, the root mean square error ($RMSE$) is introduced to evaluate the error between the fitted data and the actual value of the DVM and DRPM models,

$$RMSE(\alpha) = \frac{\sqrt{\sum_{i=1}^{N} \left(\alpha_{\text{exp}}^i - \alpha_{\text{cal}}^i\right)^2}}{N} \times 100\%$$

(16)

$$RMSE\left(\frac{d\alpha}{dt}\right) = \frac{\sqrt{\sum_{i=1}^{N} \left(\frac{d\alpha}{dt}_{\text{exp}}^i - \frac{d\alpha}{dt}_{\text{cal}}^i\right)^2}}{N} \times 100\%$$

(17)

where α_{exp}^i and α_{cal}^i are the experimental and calculated values of the conversion rate at points $i = 1, 2, 3, \ldots$; $\frac{d\alpha}{dt}_{\text{exp}}^i$ and $\frac{d\alpha}{dt}_{\text{cal}}^i$ are the experimental and calculated values of reaction rate at some points, and N is the number of data points.

3. Results and Discussion

3.1. FTIR Analysis

In order to compare and analyze the combustion characteristics of anthracite and coke powder used in sintering site, this experiment adopts Fourier transform infrared spectroscopy to detect the functional group structure of the two fuels and compare the differences between the structures, thus providing a theoretical basis for studying their combustion characteristics and laws. In infrared spectrum detection, the absorption peak of each functional group has a specific spectral position.

The corresponding functional group can be found according to the peak of the characteristic peak on the spectrum. According to relevant studies [26,27], the infrared spectral curves of coal samples can be divided into hydroxyl (3700~3000 cm^{-1}), aliphatic hydrocarbon (3000~2800 cm^{-1}), oxygen-containing functional group (1800~1000 cm^{-1}) and aromatic hydrocarbon (900~700 cm^{-1}) according to the spectral wave number, the structure and properties of functional groups. Therefore, in this paper, the infrared spectral curves of anthracite and coke powder were divided into nine points A~I in order to better distinguish the differences in the structure of their functional groups. It can be seen from the Figure 3 that there is a big gap between the two spectral curves.

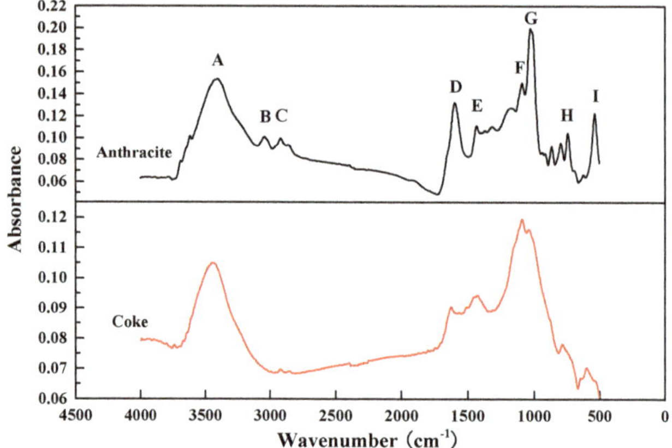

Figure 3. Fourier infrared spectra of anthracite and coke.

It can be seen from Table 3 that there are significant differences between coke powder and anthracite at B and H, which represent the –CH group on aromatic ring and aliphatic hydrocarbon in coal respectively. As the content of –CH will increase with the increase of volatile substances, compared with coke powder, anthracite still contains a small amount of volatile substances, so the content of –CH group in anthracite is significantly higher than that of coke powder. At the same time, the comparison of the absorbance of the two at point A in the infrared spectrum shows that the content of hydroxyl in the pyrolysis process of coking coal is greatly reduced, that is, the number of active groups and the activity of coal are reduced, so the reactivity of coke powder is significantly lower than that of anthracite. This is also the reason why the mixed combustion process of anthracite and coke powder for sintering will present a multi-stage weightless reaction.

Table 3. Infrared spectrum absorption peak classification of anthracite and coke powder.

Peak Position	Wavenumber/cm^{-1}		Functional Group
	Anthracite	Coke	
A	3414.4	3439.1	–OH
B	3042.2	-	–CH (Aromatic hydrocarbon)
C	2923.3	2914.7	–CH$_3$, –CH$_2$–
D	1604.1	1626.3	–C=C–
E	1439.1	1444.1	–CH$_2$–
F	1089.9	1094.2	C–O (Phenol, alcohol, ether, ester)
G	1023.2	1037.7	–Si–O–
	870.1	-	Carbonate minerals
H	793.1	786.2	Substituted benzene class
	746.9	-	–CH$_2$–
I	542.3	540.6	–S–S–

3.2. Thermogravimetric Characteristics

3.2.1. Combustion Characteristics of Anthracite and Coke

In order to better characterize the combustion process of anthracite and coke powder for sintering, this study conducted a TG-DTG analysis on the two fuels by means of thermogravimetric experiments at the same heating rate. It can be seen from Figure 4 that the combustion interval of anthracite and coke powder is concentrated in a certain region, and there is only one section of weightlessness. Because the volatile content of anthracite and coke powder is low, most of the reaction weight loss can be considered as the result of fixed carbon combustion. However, the volatile content of anthracite is 6.94%, which is significantly higher than that of coke powder. Therefore, the combustion interval of anthracite is obviously smaller than that of coke powder.

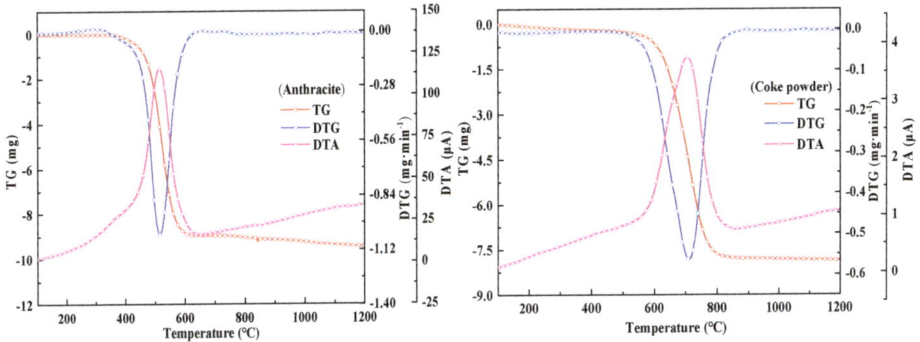

Figure 4. Thermogravimetric curves of anthracite and coke.

Combined with Equations (2) and (3), the combustion characteristic parameters of anthracite and coke powder for sintering can be obtained. As shown in Table 4, the average combustion rate, ignition stability C value and combustion characteristic index S value of anthracite were higher than that of coke powder, while the combustion time was lower than that of coke powder. Therefore, anthracite has a better combustion performance than coke powder. This is because the volatile content of coke powder and the strength of hydroxyl absorption peak are low, which directly leads to the ignition temperature of anthracite being lower than that of coke powder, resulting in significant differences in the combustion performance of the two. Therefore, the coal coke mixed fuel used in sintering raw materials will have multi-stage weightless reactions during the combustion process.

Table 4. Combustion characteristics of anthracite and coke powder.

Samples	T_i/(°C)	T_p/(°C)	v_p/(min^{-1})	T_j/(°C)	v_m/(min^{-1})	$C \times$ 10^{-6}/(min^{-1}·°C^{-2})	$S \times$ 10^{-9}/(min^{-2}·°C^{-3})	t/(min)
Anthracite	353.7	516.6	1.0529	619.3	0.0974	8.4562	1.3237	10.27
Coke	443.5	708.7	0.5672	846.4	0.0726	2.8837	0.2473	13.77

3.2.2. Combustion Characteristics of MBF and QPF

Figure 5 shows the conversion (α) and the reaction rate ($d\alpha/dt$) of MBF and QPF at different heating rates. The improving trend of the reaction rates gradually slows down as the heating rate increases and all the reaction curves have a common feature that the weight loss reaction of the two samples is divided into two parts, which is because that the different combustion characteristics of anthracite and coke divide the whole combustion process into two reaction zones. The first reaction zone is mainly the combustion reaction of anthracite, which occurs in the temperature range of about 400–600 °C. The second reaction zone is the combustion of coke breeze, and the temperature range of this reaction is about 550–950 °C. Because of the difference in the amount of anthracite and coke powder added, the rate of change peak of the latter is significantly higher than that of the former.

As shown in Table 5, when the heating rate (β) is increased from 5 °C/min to 20 °C/min, T_i, T_j, $T_{p\text{-}1}$ and $T_{p\text{-}2}$ all increase, and the conversion rate and reaction rate curves are shifted to the high-temperature region, which indicates that the amount of heat transferred from the surrounding environment to the inside of the sample per unit time increases. The increase of the heat will greatly improve the combustion rate of the fuel but also shortens the reaction time (t) of samples at the same temperature, so the phenomenon of the curve's movement to the high-temperature area will occur. By comparing the $T_{p\text{-}2}$ and T_j values of MBF in Table 5 at 5 °C/min with the T_p and T_j values in Table 4, it can be seen that the maximum reaction rate and the corresponding temperature at the end reaction of the coke powder in the mixed fuel are significantly lower than the corresponding temperature when the coke powder is burned alone. This indicates that the combustion of anthracite before coke powder provides heat for the latter's oxidation, thus speeding up the reaction process of coke powder. Therefore, when the sintering site adopts anthracite to partially replace coke powder, the migration speed of combustion zone will be accelerated and the sintering efficiency will be improved.

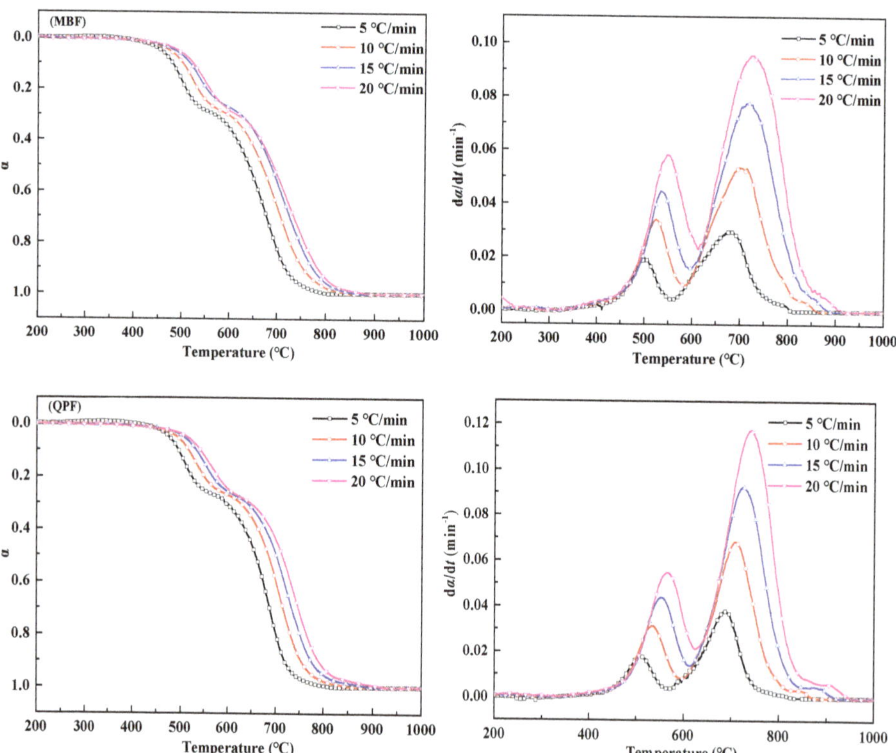

Figure 5. Fractional conversion and reaction rate-conversion curves of MBF and QPF at different heating rates.

Table 5. Combustion characteristic parameters of MBF and QPF at different heating rates.

Sample	β/(°C min^{-1})	T_i (°C)	$T_{p\text{-}1}$ (°C)	$v_{p\text{-}1}$ (min^{-1})	$T_{p\text{-}2}$ (°C)	$v_{p\text{-}2}$ (min^{-1})	T_j (°C)	v_m (min^{-1})	$C \times 10^{-7}$ (min^{-1}·°C^{-2})	$S \times 10^{-12}$ (min^{-2}·°C^{-3})	t (min)
MBF	5	399.2	501.2	0.020	676.1	0.029	802.8	0.012	1.63	2.52	60.32
	10	423.1	523.2	0.034	698.6	0.054	856.1	0.023	2.62	7.07	33.29
	15	419.4	536.3	0.045	717.3	0.078	876.9	0.033	3.79	14.18	22.71
	20	436.0	543.3	0.057	717.3	0.094	895.1	0.044	4.32	20.05	17.59
QPF	5	414..3	509.0	0.018	681.9	0.038	826.1	0.012	1.86	2.73	63.42
	10	432.6	531.5	0.031	707.4	0.068	838.6	0.025	3.06	8.99	30.71
	15	447.8	549.9	0.044	719.9	0.092	901.1	0.033	3.89	14.29	23.41
	20	456.8	561.4	0.055	732.2	0.116	917.5	0.043	5.29	24.81	17.81

It can be seen that the QPF flammability index, combustion characteristic index and combustion reaction time are significantly higher than MBF, while the average burning rate of both is about the same. This indicates that the quasi-particle structure extends the reaction time of fuel combustion, but it does not inhibit the burning rate and in fact, improves the combustion performance. It is probable that the inert particles around the fuel reduce the loss of heat generated by the combustion reaction. At the same time, as the combustion reaction progresses, the particle size of the fuel gradually shrinks, so that a "regenerator" is formed around the wrapped fuel. The specific schematic diagram is shown in Figure 6. This is also the reason for the heat storage in the combustion zone during sintering.

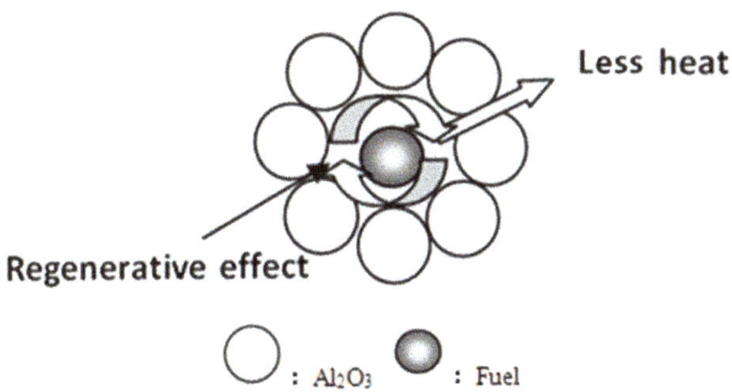

Figure 6. Thermal storage of quasi-particle sintering.

3.3. Combustion Kinetic Parameters

The relationship between the conversion rates and reaction rates of the two kinds of fuel at different heating rates is shown in Figure 7. The whole combustion reaction process is divided into two stages. The earlier stage is mainly anthracite combustion, while the latter stage is mainly composed of coke-powder combustion. The trend of the curves describing the two combustion stages is roughly the same. The rate in the initial stage of the reaction increases rapidly with an increase in the conversion rate and then decreases rapidly after reaching a certain peak value. The combustion reaction of anthracite and coke breeze follow the law of non-uniform reaction. Therefore, the combustion reaction of the fuel in the low-temperature condition is in the "chemical-controlled zone", where the combustion rate is greatly affected by the temperature, and the reaction rate increases with an increase of temperature; As the reaction proceeds, the ash content on the fuel surface gradually increases and adheres to the particle surface, which limits the diffusion of gas to the solid boundary layer and the desorption of gaseous reaction products from the solid surface to a certain extent, then greatly slows down the rate of late combustion reactions. After the peak, the reaction enters the "diffusion-controlled zone". Because of the dependence of reaction rate on the diffusion rate of the gas, the burning rate of the fuel decreases as the conversion rate increases.

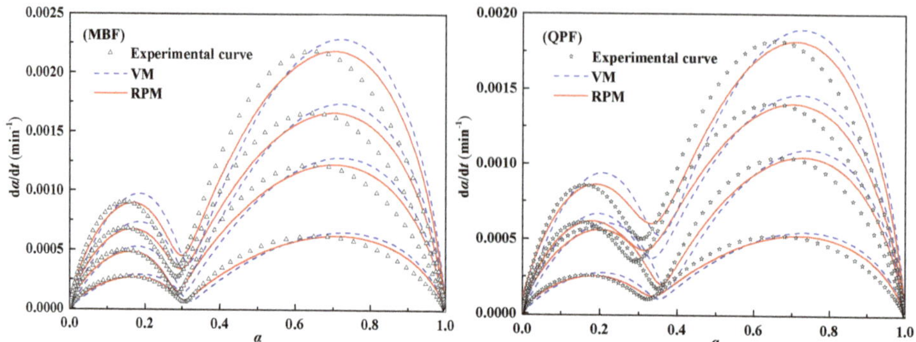

Figure 7. Combustion rates of MBF, QPF and fitting curves of double parallel reaction volume model (DVM) and double parallel random pore model (DRPM).

Table 6 lists the kinetic parameters and correlation coefficient R^2 of MBF and QPF calculated according to Equations (14) and (15). It can be seen from the data that the correlation coefficients of the two samples calculated by the DVM model are all ≤ 0.9993, and the correlation coefficients calculated by DRPM are all ≥ 0.9994. Whether it is MBF or QPF, the correlation coefficient of the data calculated by the latter is slightly higher than by the former. Table 7 lists the *RMSEs* of conversion and combustion rates calculated according to Equations (16) and (17). The results show that the *RMSEs* of conversion and combustion rate calculated by DRPM model at different heating rates are less than 0.8 and 0.01, respectively, which are far less than the *RMSEs* of the conversion rate and combustion rate calculated by the DVM model. Therefore, the combustion behavior of the MBF and QPF follows the combustion law of the double parallel reaction random pore model.

To further verify the reliability of the data calculated by the DRPM model, the corresponding kinetic parameters in Table 6 were brought into Equation (15), and the resulting conversion curves were compared with the experimental curves. As shown in Figure 8, α_1 and α_2 increase with temperature at different heating rates, representing the conversion curves of anthracite and coke powder. The remaining curves are experimental and calculated results of MBF and QPF. It can be seen from the figure that they have a high degree of fit. The experimental results are equivalent to the sum of the anthracite combustion conversion rate and the coke powder combustion conversion rate.

It can be seen from Table 6 that the activation energy of the MBF calculated by DRPM is slightly lower than that of QPF. In order to better reflect the height of barriers, Equation (18) was introduced to calculate the apparent activation energy of samples,

$$E_\alpha = c_1 E_1 + c_2 E_2 \tag{18}$$

where E_α is the apparent activation energy of samples in kJ·mol^{-1}, c_1 is the proportion of the anthracite combustion reaction to the total response, E_1 is the activation energy of the anthracite combustion reaction in kJ·mol^{-1}, c_2 is the proportion of the coke combustion reaction to the total response, and E_2 is the activation energy of the coke combustion reaction in kJ·mol^{-1}.

The calculation results are shown in Figure 9.

Table 6. Kinetic parameters of samples calculated from different models.

Simple	Model	$\beta/(°C·min^{-1})$	c_1	$E_1/(kJ·mol^{-1})$	A_1/min^{-1}	Ψ	c_2	$E_2/(kJ·mol^{-1})$	A_2/min^{-1}	Ψ_2	R^2
MBF	DVM	5	0.3279	123.5	2.76×10^5	-	0.6721	187.3	3.20×10^7	-	0.9992
		10	0.3528	137.2	2.54×10^6	-	0.6472	207.6	3.81×10^8	-	0.9993
		15	0.3023	114.1	8.32×10^4	-	0.6977	175.4	6.15×10^6	-	0.9991
		20	0.3188	117.3	1.38×10^5	-	0.6812	177.6	9.14×10^6	-	0.9991
	DRPM	5	0.3279	68.6	0.028	2.74×10^5	0.6721	112.6	0.434	8.27×10^5	0.9994
		10	0.3528	73.1	0.105	5.97×10^5	0.6472	125.8	2.318	7.51×10^5	0.9995
		15	0.3023	114.3	8.32×10^4	8.17×10^{-14}	0.6977	107.4	0.302	3.69×10^5	0.9994
		20	0.3188	117.5	13.76×10^4	0	0.6812	109.8	0.433	8.34×10^5	0.9995
QPF	DVM	5	0.3078	146.6	9.60×10^6	-	0.6922	221.3	2.29×10^9	-	0.9991
		10	0.2891	149.8	1.41×10^7	-	0.7109	224.2	2.80×10^9	-	0.9989
		15	0.2908	140.1	3.40×10^6	-	0.7092	209.9	4.14×10^8	-	0.9991
		20	0.2926	143.4	4.64×10^6	-	0.7074	212.5	5.52×10^8	-	0.9992
	DRPM	5	0.3078	67.3	0.058	1.44×10^5	0.6922	130.5	4.026	9.39×10^5	0.9994
		10	0.2891	81.7	0.292	8.31×10^5	0.7109	136.4	6.858	9.82×10^5	0.9995
		15	0.2908	92.6	0.285	2.03×10^5	0.7092	125.9	2.781	9.82×10^5	0.9994
		20	0.2926	73.1	0.163	5.47×10^5	0.7074	98.4	0.809	2.97×10^5	0.9994

Table 7. Deviation between the experimental and calculated curves.

Sample	$\beta/(°C·min^{-1})$	RMSE (α)/(%)		RMSE (dα/dt)/(%)	
		DVM	DRPM	DVM	DRPM
MBF	5	1.12	0.72	3.3×10^{-3}	2.29×10^{-3}
	10	1.04	0.66	6.72×10^{-3}	4.62×10^{-3}
	15	1.12	0.74	9.37×10^{-3}	6.57×10^{-3}
	20	1.19	0.77	1.25×10^{-2}	8.68×10^{-3}
QPF	5	1.07	0.68	3.66×10^{-3}	2.52×10^{-3}
	10	1.17	0.68	7.65×10^{-3}	4.94×10^{-3}
	15	1.14	0.74	1.02×10^{-2}	7.09×10^{-3}
	20	1.13	0.74	1.33×10^{-2}	9.24×10^{-3}

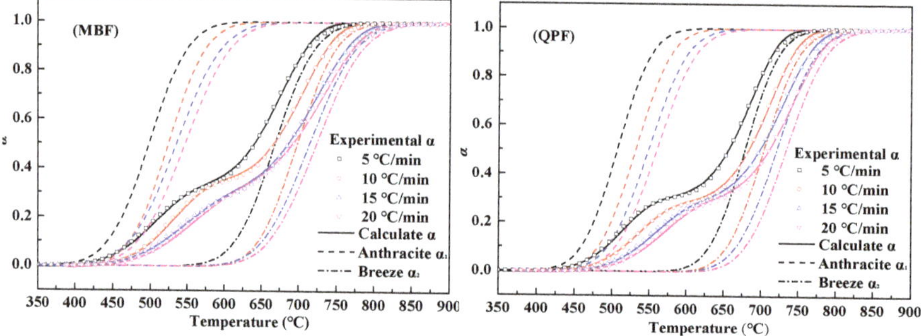

Figure 8. Comparison the correlation between experimental data and calculation results of MBF and QPF at different heating rates.

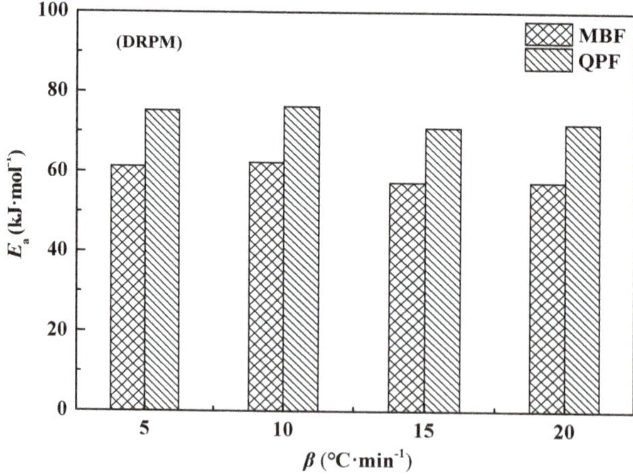

Figure 9. Comparison the apparent activation energy of MBF and QPF calculated by DRPM.

At different heating rates, the apparent activation energy of QPF is significantly higher than that of MBF, and the difference between them is maintained between 13.62 and 14.53 kJ·mol^{-1}, which does not change with the increase of the heating rate. This indicates that the difference in apparent activation energy between QPF and MPF is the reaction energy barrier provided by the diffusion resistance during the combustion reaction. The inert particles in the quasi-particle structure hinder the contact of the active part on the fuel particles with the air, thereby improving the reaction energy barrier of internal fuel. Compared with the conventional simple fuel particle structure, an increased number of particles in the quasi-particle structure increase the tortuosity of the internal pores, which greatly reduces the diffusion coefficient of the gas in the heterogeneous reaction and then increases the diffusion resistance of anthracite and coke powder combustion and the activation degree of the diffusion control reaction. However, by comparing the combustion performance parameters of QPF and MBF, it can be seen that the quasi-granular structure can produce a certain heat storage effect on the fuel during the combustion reaction process, which not only improves the combustion performance of the fuel but also increases its combustion rate, thus reducing the adverse effect of diffusion on fuel combustion.

3.4. Kinetic Analysis of Quasi-Particle Fuel Combustion

In order to clarify the coupling effect between the redox reaction of iron oxide and the combustion of fuel in the sintering mixture, we carried out SDM in the thermal analysis experiments at different heating rates, as shown in Figure 10. The increase of heating rate accelerates the reaction speed of each reaction in the sintering process and the weight variation of SDM increases with the increase of heating rate. Since the reaction is complicated in the sintering process, a reaction overlap is highly likely to occur. In order to better distinguish the redox reaction and explore the reasons for the change of sample weight, we carried out experiments using a DTA analysis under different heating rates and found that the second derivative of the DTA curve obtained the hidden information of the peak structure in the overlapping region. The results of the analysis are shown in Figure 11.

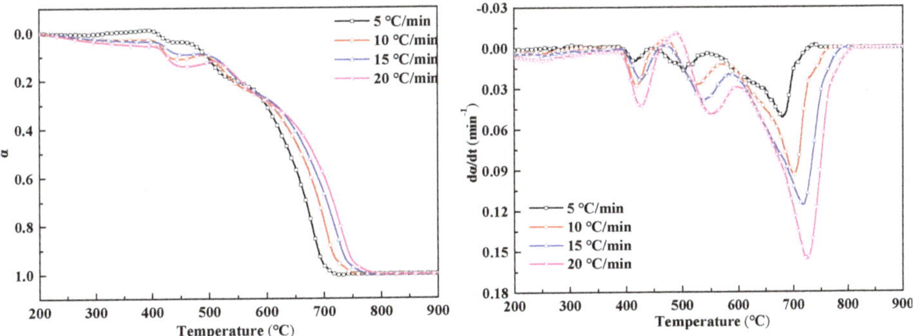

Figure 10. Fractional conversion and reaction rate-conversion curves of the sintered mixture at different heating rates.

As can be seen from the figure, SDM exhibits a small change in exothermic → endotherm → exothermic below 500 °C. The volatile first undergoes combustion and exothermic reaction, which provides a reducing atmosphere for Fe_2O_3 in the sample and a small amount of CO is sufficient to completely reduce Fe_2O_3 to Fe_3O_4. Due to the continuous renewal of the reaction gas, the reducing atmosphere is replaced by an oxidizing atmosphere. The Fe_3O_4 produced by the reduction is oxidized and the experimental curve shows a slight exothermic weight gain,

$$Volatile + (x/2 + y + z/2)O_2 \rightarrow xCO + yCO_2 + zH_2O \tag{19}$$

$$3Fe_2O_3 + CO = 2Fe_3O_4 + CO_2 \tag{20}$$

$$2Fe_3O_4 + 1/2O_2 = 3Fe_2O_3 \tag{21}$$

The reactions can be regarded as a series of reactions at this stage. Among them, CO and Fe_3O_4 are both the product of the former reaction and the reactant of the latter reaction. This indicates that the increase of the latter reaction rate will promote the previous reaction, and the slowing of the previous reaction rate will inhibit the latter reaction, thus forming a significant coupling relationship.

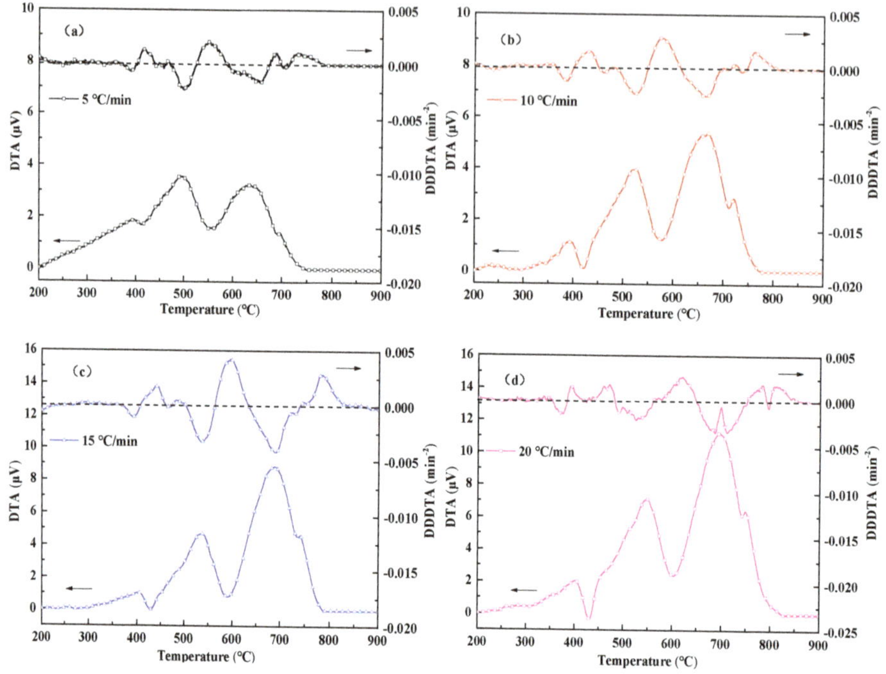

Figure 11. Differential thermal analysis curves of the sintered mixture at different heating rates. (a) 5 °C/min, (b) 10 °C/min, (c) 15 °C/min, (d) 20 °C/min.

Above 500 °C, anthracite and coke breeze burn, and the exothermic is intense, which may mask the reduction reaction of Fe_2O_3. However, the DTA curve shows a significant reduction of the endothermic peak in the 550–630 °C range. This is because this stage is the combustion interval between anthracite and coke breeze and the oxidative exothermic is replaced by the reduced endotherm. However, due to the large weight loss caused by the combustion and reduction reactions, a weight gain phenomenon is not obvious at this stage. The coupling phenomenon of each reaction in the sintering mixture is very obvious and is complicated.

4. Conclusions

There are many –CH groups and hydroxyl groups in anthracite, which makes the combustion performance of anthracite significantly better than that of coke powder. When the coke powder is partially replaced by anthracite, the whole combustion process will be accelerated. DVM and DRPM models were used to calculate the dynamics of MBF and QPF at different heating rates. Through comparative analysis of the relevant kinetic parameters and root mean square error, it was found that the DRPM model was more suitable for describing the combustion process of MBF and QPF. Although quasi-particles increase the apparent activation energy of fuel combustion, they also produce a heat storage effect on fuel particles, improve their combustion performance, and reduce the adverse effect of diffusion effect on the combustion reaction process. According to the differential thermal analysis of SDM samples, the coupling between volatiles combustion and redox reaction of iron oxides is obvious in the early combustion period and the oxidation of iron oxides will occur again when the combustion reaction of fuel is weakened.

Author Contributions: In this paper, the experimental work and thesis writing were mainly undertaken by J.L. and Y.Y. (Yaqiang Yuan), while J.Z., Z.H. and Y.Y. (Yaowei Yu) were mainly responsible for the review, guidance and revision of the manuscript. All authors have read and agreed to the published version of the manuscript.

Funding: This work was financially supported by National Natural Science Foundation of China (51874171) and University of Science and Technology Liaoning Talent Project Grants (No. 601011507-05).

Conflicts of Interest: The authors declare no conflict of interest.

References

1. Usman, M.; Farooq, M.; Naqvi, M.; Saleem, M.W.; Hussain, J.; Naqvi, S.R.; Jahangir, S.; Jazim Usama, H.M.; Idrees, S.; Anukam, A. Use of Gasoline, LPG and LPG-HHO Blend in SI Engine: A Comparative Performance for Emission. *Processes* **2020**, *8*, 74. [CrossRef]
2. Heidari, M.; Salaudeen, S.; Norouzi, O.; Acharya, B.; Dutta, A. Numerical Comparison of a Combined Hydrothermal Carbonization and Anaerobic Digestion System with Direct Combustion of Biomass for Power Production. *Processes* **2020**, *8*, 43. [CrossRef]
3. Liu, Y.L.; Wang, J.S.; Zhang, H.J. Reduction behavior of ferrous burden under simulated oxygen blast furnace conditions. *Ironmak. Steelmak.* **2015**, *5*, 358–365. [CrossRef]
4. Loo, C.E.; Leung, W. Factors Influencing the Bonding Phase Structure of Iron Ore Sinters. *ISIJ Int.* **2003**, *9*, 1393–1402. [CrossRef]
5. Zhang, G.L.; Wu, S.L.; Chen, S.G.; Zhu, J.; Fan, J.X.; Su, B. Optimization of Dolomite Usage in Iron Ore Sintering Process. *ISIJ Int.* **2013**, *9*, 1515–1522. [CrossRef]
6. Ellis, B.G.; Loo, C.E.; Witchard, D. Effect of ore properties on sinter bed permeability and strength. *Ironmak. Steelmak.* **2007**, *2*, 99–108. [CrossRef]
7. Kawachi, S.; Kasama, S. Effect of Micro-particles in Iron Ore on the Granule Growth and Strength. *ISIJ Int.* **2011**, *7*, 1057–1064. [CrossRef]
8. Xu, H.L.; Pan, G.Y.; Shao, Y.J.; Ye, L.D.; Fan, X.G. Analysis of energy consumption evaluation indicator for iron and steel production. *Energy Metall. Ind.* **2017**, *2*, 3–7.
9. Zhou, M.S.; Han, S.F.; Wang, L.; Jiang, X.; Xu, L.B.; Zhai, L.W.; Liu, J.; Zhang, H.; Qin, X.L.; Shen, F.M. Effect of Size Distribution of Coke Breeze on Sintering Performance. *Steel Res. Int.* **2015**, *11*, 1242–1251. [CrossRef]
10. Umadevi, T.; Deodhar, A.V.; Kumar, S.; Prasad, C.S.; Ranjan, M. Influence of coke breeze particle size on quality of sinter. *Ironmak. Steelmak.* **2008**, *8*, 567–574. [CrossRef]
11. Yuki, A.; Kiyonori, Y.; Yutake, S. Effect of Coke Breeze Addition Timing on Sintering Operation. *ISIJ Int.* **2013**, *9*, 1523–1528.
12. Ri, D.W.; Chung, B.J.; Choi, E.S. Effects of anthracite replacing coke breeze on iron ore sintering. *Rev. Metall.* **2008**, *5*, 248–254. [CrossRef]
13. Yang, W.; Choi, S.; Choi, E.S.; Ri, D.W.; Kim, S. Combustion characteristics in an iron ore sintering bed—Evaluation of fuel substitution. *Combust. Flame* **2006**, *3*, 447–463. [CrossRef]
14. Zhao, J.P.; Loo, C.E.; Dukino, R.D. Modelling fuel combustion in iron ore sintering. *Combust. Flame* **2015**, *4*, 1019–1034. [CrossRef]
15. Zhao, J.P.; Loo, C.E.; Zhou, H.; Yuan, J.L.; Li, X.B.; Zhu, Y.Y.; Yang, G.H. Modelling and analysis of the combustion behavior of granulated fuel particles in iron ore sintering. *Combust. Flame* **2018**, *3*, 257–274. [CrossRef]
16. Oyama, N.; Higuchi, T.; Machida, S.; Sato, H.; Takeda, K. Effect of High-phosphorous Iron Ore Distribution in Quasi-particle on Melt Fluidity and Sinter Bed Permeability during Sintering. *ISIJ Int.* **2009**, *5*, 650–658. [CrossRef]
17. Ma, P.N.; Cheng, M.; Zhou, M.X.; Li, Y.W.; Zhou, H. Combustion characteristics of different types of quasi-particles during iron ore sintering. *J. Eng. Sci.* **2019**, *3*, 316–324.
18. Ogi, H.; Maeda, T.; Ohno, K.; Kunitomo, K. Effect of Coke Breeze Distribution on Coke Combustion Rate of the Quasi-particle. *ISIJ Int.* **2015**, *12*, 2550–2555. [CrossRef]
19. Ohno, K.; Noda, K.; Nishioka, K.; Maeda, T.; Shimizu, M. Combustion Rate of Coke in Quasi-particle at Iron Ore Sintering Process. *ISIJ Int.* **2013**, *9*, 1588–1593. [CrossRef]
20. Ohno, K.; Noda, K.; Nishioka, K.; Maeda, T.; Shimizu, M. Effect of Coke Combustion Rate Equation on Numerical Simulation of Temperature Distribution in Iron Ore Sintering Process. *ISIJ Int.* **2013**, *9*, 1642–1647. [CrossRef]

21. Wang, G.W.; Zhang, J.L.; Shao, J.G.; Liu, Z.J.; Zhang, G.H.; Xu, T.; Guo, J.; Wang, H.Y.; Xu, R.S.; Lin, H. Thermal behavior and kinetic analysis of co-combustion of waste biomass/low rank coal blends. *Energy Convers. Manag.* **2016**, *124*, 414–426. [CrossRef]

22. Wang, G.W.; Zhang, J.L.; Shao, J.G.; Liu, Z.J.; Wang, H.Y.; Li, X.Y.; Zhang, P.C.; Geng, W.W.; Zhang, G.H. Experimental and modeling studies on CO_2 gasification of biomass chars. *Energy* **2016**, *114*, 143–154. [CrossRef]

23. Wang, G.W.; Zhang, J.L.; Huang, X.; Liang, X.H.; Ning, X.J.; Li, R.P. Co-gasification of petroleum coke-biomass blended char with steam at temperatures of 1173–1373 K. *Appl. Therm. Eng.* **2018**, *137*, 678–688. [CrossRef]

24. Wang, G.W.; Zhang, J.L.; Shao, J.G.; Zhang, P.C. Experiments and Kinetics Modeling for Gasification of Biomass Char and Coal Char under CO_2 and Steam Condition. *Miner. Met. Mater. Soc.* **2016**, *145*, 375–382.

25. Tobu, Y.; Nakano, M.; Nakagawa, T. Effect of Granule Structure on the Combustion Behavior of Coke Breeze for Iron Ore Sintering. *ISIJ Int.* **2013**, *9*, 1594–1598. [CrossRef]

26. Zhang, J.; Zheng, N.; Wang, J. Two-stage hydrogasification of different rank coals with a focus on relationships between yields of products and coal properties or structures. *Appl. Energy* **2016**, *173*, 438–447. [CrossRef]

27. Li, D.; Li, W.; Li, B. A New Hydrogen Bond in Coal. *Energy Fuels* **2003**, *17*, 791–793. [CrossRef]

 *processes*

Article

Temperature Distribution Estimation in a Dwight–Lloyd Sinter Machine Based on the Combustion Rate of Charcoal Quasi-Particles

Ziming Wang [1,*], **Ko-ichiro Ohno** [2,*], **Shunsuke Nonaka** [3], **Takayuki Maeda** [2] **and Kazuya Kunitomo** [2]

[1] Department of Materials Process Engineering, Kyushu University, 744 Motooka, Nishi-ku, Fukuoka 819-0395, Japan
[2] Department of Materials Science and Engineering, Faculty of Engineering, Kyushu University, 744 Motooka, Nishi-ku, Fukuoka 819-0395, Japan; maeda@zaiko.kyushu-u.ac.jp (T.M.); kunitomo@zaiko.kyushu-u.ac.jp (K.K.)
[3] Department of Materials Process Engineering, Graduate School of Engineering, Kyushu University, Now at JFE Steel, 1 Mizushima-kawasaki-dori, Kurashiki, Okayama Prefecture 712-8511, Japan; sh-nonaka@jfe-steel.co.jp
* Correspondence: ou.shimei.515@s.kyushu-u.ac.jp (Z.W.); ohno.ko-ichiro.084@m.kyushu-u.ac.jp (K.-i.O.)

Received: 27 February 2020; Accepted: 25 March 2020; Published: 31 March 2020

Abstract: The coke combustion rate in an iron ore sintering process is one of the most important determining factors of quality and productivity. Biomass carbon material is considered to be a coke substitute with a lower CO_2 emission in the sintering process. The purpose of this study was to investigate the combustion rate of a biomass carbon material and to use a sintering simulation model to calculate its temperature profile. The samples were prepared using alumina powder and woody biomass powder. To simplify the experimental conditions, alumina powder, which cannot be reduced, was prepared as a substitute of iron ore. Combustion experiments were carried out in the open at 1073 K~1523 K. The results show that the combustion rates of the biomass carbon material were higher than that of coke. The results were analyzed using an unreacted core model with one reaction interface. The kinetic analysis found that the k_c of charcoal was higher than that of coke. It is believed that the larger surface area of charcoal may affect its combustion rate. The analysis of the sintering simulation results shows that the high temperature range of charcoal was smaller than that of coke because of charcoal's low fixed carbon content and density.

Keywords: coke combustion rate; charcoal combustion rate; iron ore sintering process; temperature distribution; biomass; quasi-particle

---

1. Introduction

CO_2 emission from Japan's industrial sector is much higher than that from other sectors. In 2017, it accounted for approximately 37.2% of the total emission. In the industrial sector, the iron and steel-making industry accounts for approximately 39.4% of energy consumption. The steel industry emits approximately 13% of the CO_2 in Japan [1]. The demands of global environmental conservation require a greenhouse gas reduction. Currently, approximately 80 million tons of pig iron are produced by blast furnace annually in Japan. Coal and coke are used as reducing materials and heat sources, respectively, and a large amount of CO_2 is emitted in the iron-making process. Therefore, the development of innovative technologies is required to reduce the CO_2 emission.

Trees absorb carbon dioxide during photosynthesis. When wood from forests is burned as fuel, carbon dioxide is generated. If the forest is renewed after tree cutting, the carbon dioxide will be absorbed by the trees again during the growth process. Thus, the use of wood for energy is carbon neutral. Therefore, using wood instead of fossil fuels makes it possible to reduce carbon dioxide emissions and contribute to the prevention of global warming [2].

It is increasingly difficult to prepare a sintering iron ore due to the high price of raw materials, environmental regulations and inferior quality raw materials. Limited work [3–6] has been conducted to investigate the application of biomass in the sintering process to replace coke breeze, with this work mainly focused on its environmental impacts and low substitution rates. Therefore, it is necessary to improve the sintering iron ore method. In the sintering process, the coke combustion rate is one of the most important determining factors of quality and productivity.

However, little research has been conducted on the combustion behavior of biomass carbon material quasi-particles during the sintering process. The purpose of this study was to investigate the combustion rate of a biomass carbon material and use a sintering simulation model to calculate its temperature profile.

2. Experimental Sample and Procedure

To simulate the test particles, the samples were prepared using alumina powder and woody biomass powder. The woody biomass powder used in this study was commercial mangrove charcoal, which is normally used for barbecues. To simplify the experimental conditions, alumina powder was prepared as a substitute for iron ore. Alumina eliminated the effects of melt formation, reduction and re-oxidation of iron ore during coke combustion. Charcoal powder with a particle diameter of −125 μm and 125~250 μm was used in this experiment. The particle size of iron ore was to simulate the adhere powder layer, but not the coke particle. Coke powder with the same particle diameter was used to compare the results. Alumina powder with a particle diameter of −250 μm was prepared to match the particle size of the iron ore. The analysis of the results of prepared carbon material are listed in Table 1. Compared with coke, charcoal has a lower ash ratio and higher volatile matter content and therefore charcoal has a lower fixed carbon. The surface area of charcoal is higher than that of coke. This was also observed by an SEM, as shown in Figure 1. Figure 2 shows the overall view of raw materials and samples. After the alumina and coke powders were mixed, 0.5 mass % flour was mixed as a binder. The flour was just used to enable keeping the tablet shape until it was inserted to the platinum basket. It was thought that the effect of the flour could be negligible, because flour evaporates at a lower temperature—600 K—than the experimental temperature in this study. Then, the mixture was pressed into 10 mm diameter tablets by stainless dies. The height of the tablet was 10 mm with a void ratio of 35%. The void ratio was decided from the information on the tablet volume and the true density of the sample mixture. Each true density of the sample materials was measured by pycnometer. The weight ratio of coke in each sample was fixed at 20 mass %. To ensure that the volume ratio of the samples was the same as the hematite and alumina samples, 22.1 mass % coke with 77.9 mass % alumina and 20.8 mass % charcoal with 79.2 mass % alumina was also prepared.

Table 1. Properties of carbon materials.

	Ash (mass %)	V.M. (mass %)	Fix.C. (mass %)	Specific Surface Area (m²/g)
Charcoal (−125 μm)	1.84	29.2	69.0	61.0
Charcoal (125~250 μm)				28.5
Coke (−125 μm)	10.1	1.71	88.2	2.59
Coke (125~250 μm)				0.92

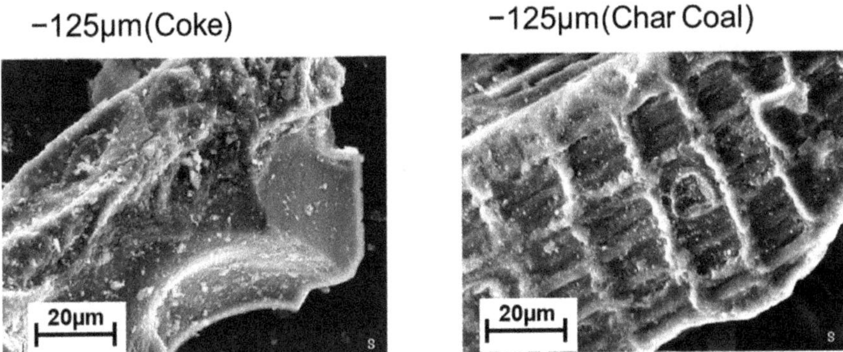

Figure 1. Scanning electron micrograph of coke and charcoal (−125 μm).

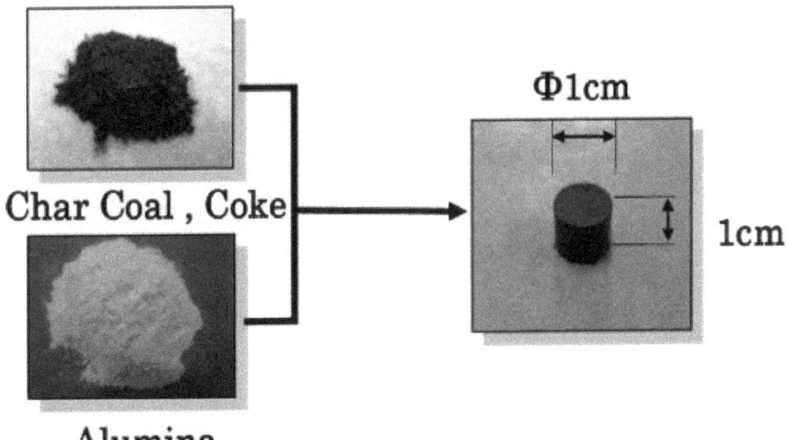

Figure 2. Overall view of the raw materials and the sample.

The measurement of the sample weight loss during coke combustion was done by the thermobalance shown in Figure 3. The sample was placed in a platinum basket. A vertical electric resistance furnace was used to do isothermal heating. The isothermal zone was heated up to 1073 K, 1223 K, 1373 K and 1523 K. Before the combustion experiment, heat treatment of the samples was carried out at each of the given temperatures in a N_2 atmosphere for 30 mins to remove water, Volatile matter (V.M.) and the binder from the samples. Then, air was passed through the reaction tube. The air flow rate was 4 NL/min. When a weight change in the sample was not observed, the experiment was terminated. It was hypothesized that coke ash did not influence the weight loss of the sample because the amount of coke in every sample stayed the same.

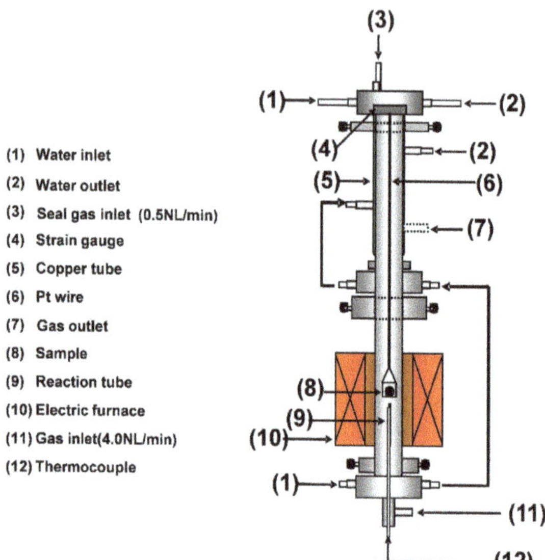

Figure 3. Schematic of the device used in the experiment.

3. Results

The reaction ratio in the study was defined as the removal ratio of fixed carbon from the sample. A carbon combustion reaction can be described using the following chemical reaction, if CO gas formation is ignored:

$$C(s) + O_2(g) = CO_2(g) \tag{1}$$

In the combustion experiment, the sample weight loss was attributed to the decrease in the amount of fixed carbon. Therefore, the reaction ratio (F) at a reaction time can be described by Equation (2):

$$F = \frac{\Delta w_t}{W} \tag{2}$$

Fractional reaction curves at 1073 K are shown in Figure 4. The figure shows that combustion rates of charcoal were quicker than those of coke. This tendency was also observed at 1223 K, 1373 K and 1523 K.

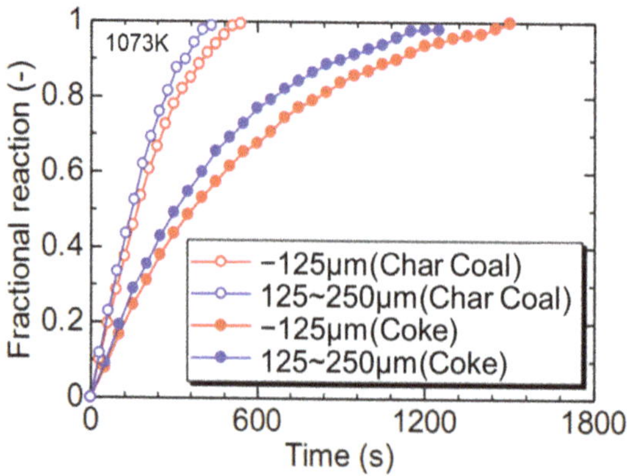

Figure 4. Fractional reaction curves of coke and charcoal combustion at 1073 K.

To verify the reaction mechanism of the combustion reaction, samples with a reaction ratio of 50% were also prepared under 1073 K and 1523 K and cross-sectional and microscopic observations were made. Figures 5 and 6 show the cross-sectional observation and the microstructure at the reaction interfaces of each sample. It was clear that the combustion reaction was a topochemical reaction.

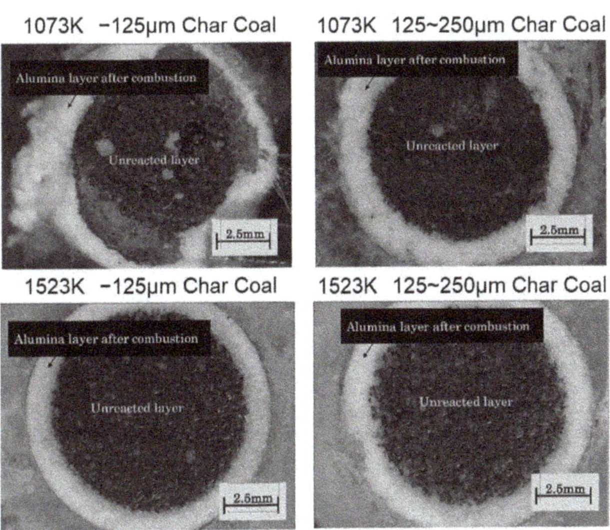

Figure 5. Cross-sectional view of the samples.

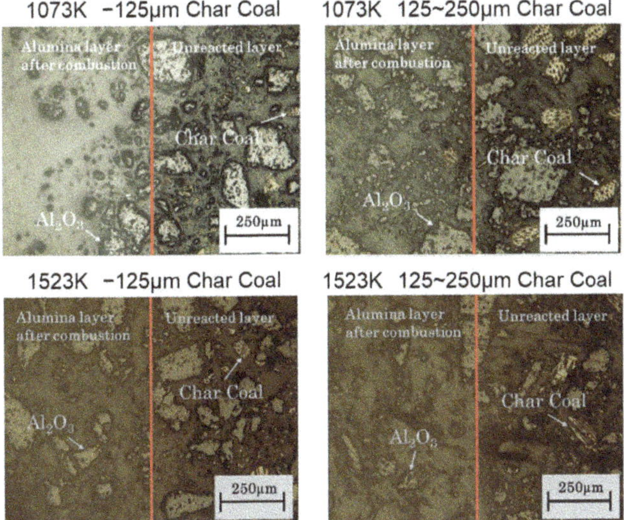

Figure 6. Microstructure of the samples at the reaction interface.

4. Kinetic Analysis

4.1. Unreacted Core Model for Coke

From the fractional reaction curves obtained from the combustion experiment, the combustion reaction rate constant was determined using the unreacted core model [7]. The combustion reaction has five processes [8].

1. O_2 transport from the gas phase to the particle surface through the gas film:

$$- \dot{n}_{g \cdot O_2} = 4\pi r_0^2 k_f \left(C_{O_2} - C_{O_2 \cdot s} \right)$$
(3)

2. O_2 transport from the particle surface to the reaction interface through the alumina powder layer after coke combustion:

$$- \dot{n}_{d \cdot O_2} = (D_{O_2})_{eff} \frac{4\pi r_0 r_i}{r_0 - r_i} \left(C_{O_2 \cdot s} - C_{O_2 \cdot i} \right)$$
(4)

3. The combustion reaction at the reaction interface:

$$- \dot{R} = 4\pi r_i^2 k_c \left(C_{O_2 \cdot i} - \frac{C_{CO_2 \cdot i}}{K} \right)$$
(5)

4. CO_2 transport from the reaction interface to the particle surface through the alumina powder layer after coke combustion:

$$\dot{n}_{d \cdot CO_2} = (D_{CO_2})_{eff} \frac{4\pi r_0 r_i}{r_0 - r_i} \left(C_{CO_2 \cdot i} - C_{CO_2 \cdot s} \right)$$
(6)

5. CO_2 transport from the particle surface to the gas phase through the gas film:

$$\dot{n}_{g \cdot CO_2} = 4\pi r_0^2 k_f \left(C_{CO_2 \cdot s} - C_{CO_2} \right)$$
(7)

The overall rate equation can be described by the quasi-steady state analysis method below:

$$- \dot{n} = \frac{4\pi r_0^2 \left(\frac{K}{1+K} \right) \left(C_{O_2} - C_{CO_2} \right)}{\frac{1}{k_f} + \frac{1}{D_e} \cdot \frac{r_0 (r_0 - r_i)}{r_i} + \frac{1}{k_c} \cdot \frac{K}{1+K} \left(\frac{r_0}{r_i} \right)^2}$$
(8)

$$\frac{1}{D_e} = \frac{K}{1+K} \left(\frac{1}{(D_{O_2})_{eff}} + \frac{1}{K(D_{CO_2})_{eff}} \right)$$
(9)

Equation (8) can be expressed by the following equation, assuming that the combustion reaction of coke was an irreversible reaction and the equilibrium constant K infinite:

$$- \dot{n} = \frac{4\pi r_0^2 C_{O_2}}{\frac{1}{k_f} + \frac{1}{D_e} \cdot \frac{r_0 (r_0 - r_i)}{r_i} + \frac{1}{k_c} \cdot \left(\frac{r_0}{r_i} \right)^2}$$
(10)

$\dot{n}$ can be replaced by the following equation:

$$- \dot{n} = - \frac{d}{dt} \left(\frac{4}{3} \pi r_i^3 \rho_{Cm} \right) = -4\pi r_i^2 \rho_{Cm} \cdot \frac{dr_i}{dt}$$
(11)

The reaction ratio F is expressed by Equation (12):

$$F = 1 - \left(\frac{r_i}{r_0} \right)^3$$
(12)

When Equations (10)–(12) are combined and integrated under boundary conditions; $r = r_0$ at $t = 0$ and $r = r_i$ at $t = t$, Equation (13) is obtained:

$$t = \frac{\rho_{Cm} r_0}{C_{O_2}} \cdot \left[\frac{F}{3k_f} + \frac{d_r}{16 D_e} \left\{ 3 - 3(1 - F)^{\frac{2}{3}} + 2F \right\} + \frac{1}{k_C} \left\{ 1 - (1 - F)^{\frac{1}{3}} \right\} \right]$$
(13)

The gas film mass transfer coefficient, k_f, can be calculated from Ranz–Marshall's Equation [9]. The value of the effective diffusion coefficient in the alumina layer, D_e, and the interfacial reaction rate coefficient of coke, k_C, was obtained by parameter-fitting using the nonlinear least-squares method to the fractional reaction curves.

D_e, and k_C can be expressed by substituting the coefficients in Arrhenius' equation as shown:

$$k_C = A_{(k_C)} exp\left(-\frac{E_{a(k_C)}}{RT}\right) \tag{14}$$

$$D_e = A_{(D_e)} exp\left(-\frac{E_{a(D_e)}}{RT}\right) \tag{15}$$

Equations (13) and (14) can be transformed into the following equations:

$$lnk_C = -\frac{E_{a(k_C)}}{R} \cdot \frac{1}{T} + lnA_{(k_C)} \tag{16}$$

$$lnD_e = -\frac{E_{a(D_e)}}{R} \cdot \frac{1}{T} + lnA_{(D_e)} \tag{17}$$

Figure 7 shows the Arrhenius plot of k_c. The values of k_c are at the same level in all samples. The temperature dependence of k_c is expressed as

Coke	$(-125\ \mu m)$	$k_c = 6.02 \times 10^{-2}\ exp(-9.32 \times 10^3/RT)$	(m/s)
	$(125\text{~}250\ \mu m)$	$k_c = 4.51 \times 10^{-2}\ exp(-5.12 \times 10^3/RT)$	(m/s)

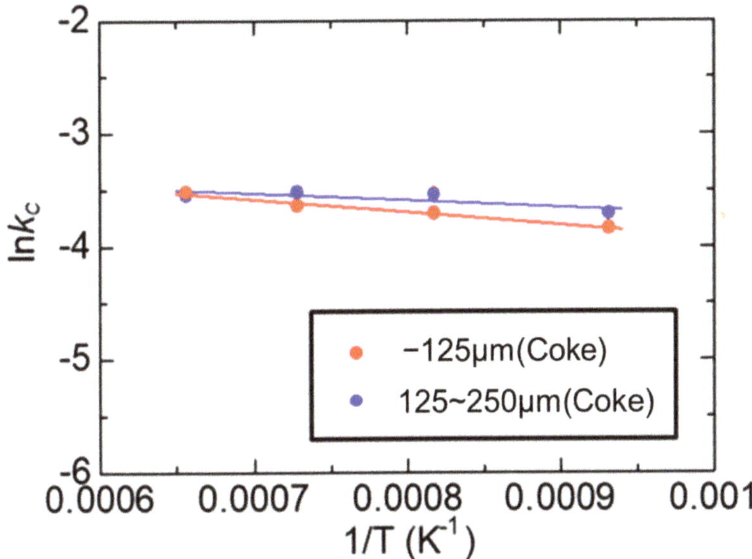

Figure 7. Temperature dependence of the reaction rate constants k_c.

Figure 8 shows the Arrhenius plot of D_e. The temperature dependence of D_e can be expressed as

Coke	$(-125\ \mu m)$	$D_e = 1.79 \times 10^{-3}\ exp(-36.4 \times 10^3/RT)$	(m/s)
	$(125\text{~}250\ \mu m)$	$D_e = 3.06 \times 10^{-3}\ exp(-36.6 \times 10^3/RT)$	(m/s)

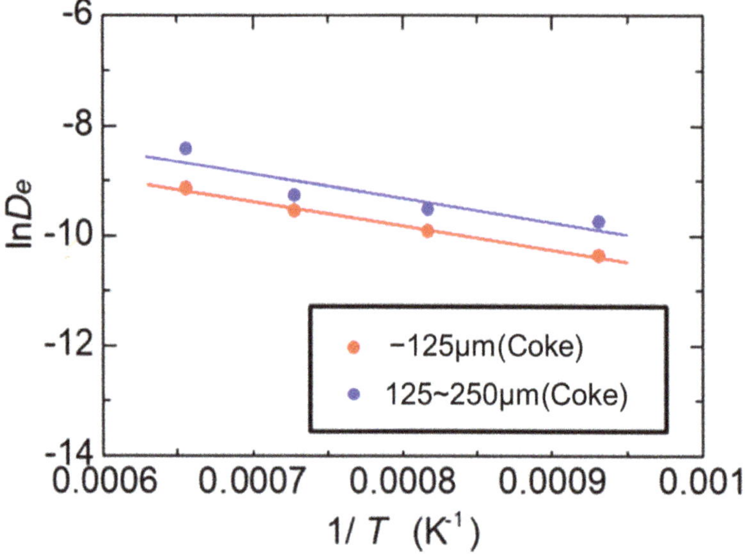

Figure 8. Temperature dependence of the effective diffusivities D_e.

4.2. Chemical Reaction Control Step for Charcoal

Because the surface area of charcoal is larger than that of coke, when a combustion reaction takes place, the reaction area of charcoal will also be larger. Moreover, due to the low ash ratio in the charcoal, the O_2 transportation rate in the alumina layer will be large. Therefore, the reaction is based on the chemical reaction control step.

For the charcoal samples, the chemical reaction control step can be used in the analysis method shown below:

The combustion reaction at the reaction interface can be expressed by Equation (5).

The combustion rate can be expressed as

$$-\dot{n} = \frac{4\pi r_0^2 \left(\frac{K}{1+K}\right)\left(C_{O_2} - \frac{C_{CO_2}}{K}\right)}{\frac{1}{k_c} \cdot \frac{K}{1+K}\left(\frac{r_0}{r_i}\right)^2} \tag{18}$$

Under boundary conditions, $r = r_0$ at $t = 0$ and $r = r_i$ at $t = t$, and this gives Equation (19):

$$\left[1 - (1-F)^{\frac{1}{3}}\right] = \frac{C_{O_2}}{\rho c_m r_0} k_c t \tag{19}$$

Based on the reaction curves obtained by the experiments, k_c was determined using the unreacted core model [3].

k_C can be substituted in Arrhenius' equation as shown by Equation (14) which also can be transformed as Equation (16).

Figure 9 shows the Arrhenius plot of k_c.

The temperature dependence of k_c can be expressed as

Coke	$(-125\ \mu m)$	$kc = 8.24 \times 10^{-3} \exp\left(-10.6 \times 10^3 / RT\right)$	(m/s)
	$(125{\sim}250\ \mu m)$	$kc = 8.54 \times 10^{-3} \exp\left(-10.3 \times 10^3 / RT\right)$	(m/s)

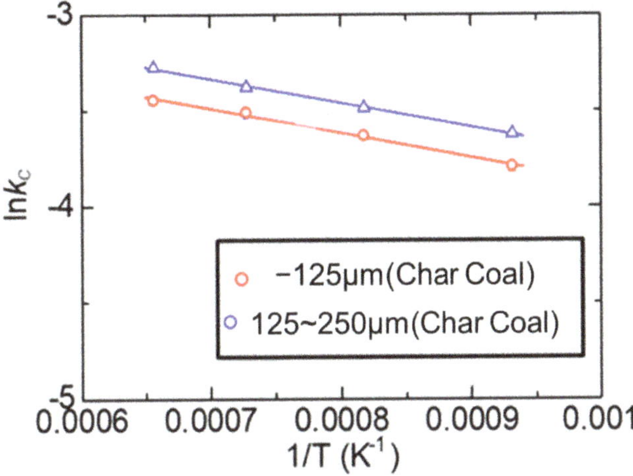

Figure 9. Temperature dependence of the reaction rate constants k_c.

5. Sintering Simulation Model

5.1. Simulation Method

The simulation condition was based on the study results using Ohno's model [10]. The model has S'-type, C-type and P-type quasi-particles as shown in Figure 10. The S' type was calculated using Hottel's equation [11–13], while the C and P types were calculated based the results obtained in this study.

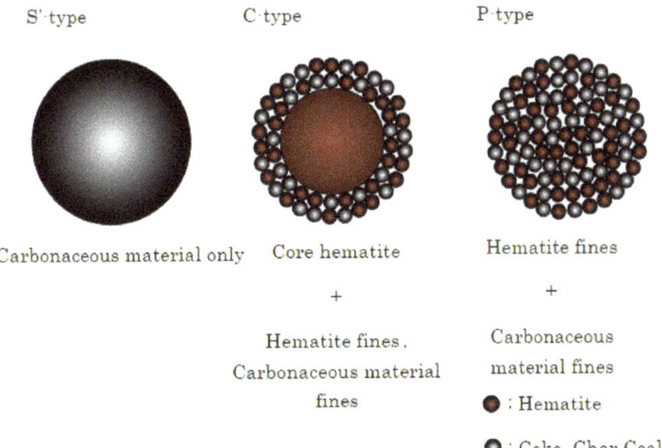

Figure 10. Classification of the quasi-particles.

Our mathematical model is based on the Dwight–Lloyd sinter machine and the calculation range is from the ignition point on the pallet to the discharge of the sinter ore.

The numerical analysis was based on the control volume method shown in Figure 11. The control volume method is obtained by dividing the analysis target region into equal minute portions. Various basic equations, representing phenomena, such as the continuous and energy conservation equations, govern the inside of the analysis target area and are relational equations to be established in each control volume. Assuming that these control volumes are in a sufficiently small area, there would not be a large error even if the changes in various quantities inside the control volumes were linear or approximated to be constant values. In other words, changes in various quantities in a certain control volume can be represented using values at a representative point in the control volume adjacent to the representative point. In our mathematical model, the sintering material layer was divided into minute control volumes in one dimension, the basic governing equations were discretized, and each difference approximation equation was solved using an explicit method.

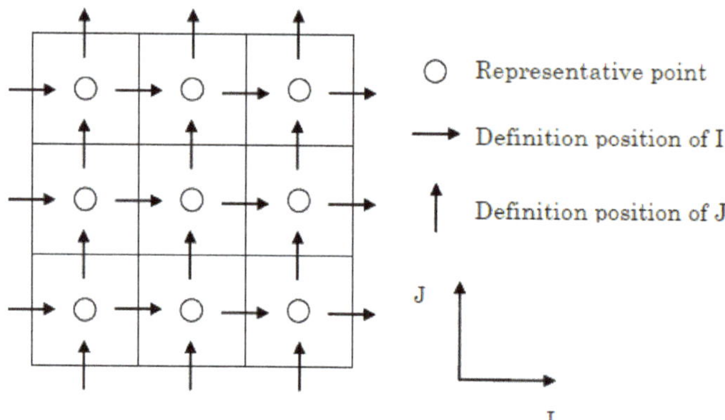

Figure 11. Pattern diagram of the control volume.

In the model used, the temperature distribution was estimated considering the combustion of carbonaceous materials, the decomposition reaction of $CaCO_3$, the evaporation and condensation of water, and the formation and solidification of calcium ferrite melt according to Ohno's model [10].

D_e depends on coke distribution.

The combustion reaction rate of the quasi-particles is expressed by Equations (20) and (21):

$$r^*_{Quasi-particle} = 4\pi r^2_{Quasi-particle} k' C_{O_2} \tag{20}$$

$$k' = 1 / \left(\frac{1}{k_f} + \frac{r_0(r_0 - r_i)}{D_e r_i} + \frac{r_0^2}{k_c r_i^2} \right) \tag{21}$$

In this equation, D_e has a value of 10^8 because the resistance of the diffusion can be ignored. The material balance is calculated using Equation (22):

$$\frac{\rho_{i\cdot x}|_{t+\Delta t} - \rho_{i\cdot x}|_t}{\Delta t} = -\frac{(\rho_{Z+\Delta Z}u_i)|_{Z+\Delta Z} - (\rho_Z u_{i-1})|_Z}{\Delta Z} + r^*_{i,x} \tag{22}$$

The material balance equations of N_2, O_2, CO_2 and H_2 respectively, can be expressed as follows:

$$\frac{\partial(\rho_{N_2}u)}{\partial Z} = -\frac{\partial\rho_{N_2}}{\partial t} \tag{23}$$

$$\frac{\partial(\rho_{O_2}u)}{\partial Z} = -\frac{\partial\rho_{O_2}}{\partial t} - r^*_{Coke} \tag{24}$$

$$\frac{\partial(\rho_{CO_2}u)}{\partial Z} = -\frac{\partial\rho_{CO_2}}{\partial t} + r^*_{Coke} + r^*_{CaCO_3} \tag{25}$$

$$\frac{\partial(\rho_{H_2O}u)}{\partial Z} = -\frac{\partial\rho_{H_2O}}{\partial t} + r^*_{H_2O} \tag{26}$$

Because convection did not occur in the solid phase, the thermal budget of the phase considering heat transfer and reaction heat can be represented as

$$\rho_s C_{P\cdot s}\frac{\partial T_s}{\partial t} - \frac{6(1-\varepsilon_a)}{\dot{d}}h\left(T_g - T_s\right) + H_{Coke}\left(r^*_{Coke}n_{Coke} + r^*_{Quasi-paritcle}n_{Quasi-particle}\right)$$
$$+H_{CaCO_3}r^*_{CaCO_3}n_{CaCO_3} + H_{H_2O\cdot V}r^*_{H_2O\cdot V} + H_{CF\cdot G}r^*_{CF\cdot G} + H_{CF\cdot S}r^*_{CF\cdot S} = k\frac{\partial^2 T_s}{\partial Z^2} \tag{27}$$

The thermal budget of the gas phase factoring the heat transfer and combustion reaction heat can be expressed as

$$\rho_g C_{P\cdot g}\frac{\partial T_g}{\partial t} - \frac{6(1-\varepsilon_a)}{\dot{d}}h\left(T_s - T_g\right) + C_{P\cdot g}\frac{\partial \rho_g u T_g}{\partial Z} = k\frac{\partial^2 T_g}{\partial Z^2} \tag{28}$$

The particles were charged in the control volume. Therefore, the pressure loss of the fluid also needed to be considered. The pressure loss of a laminar-turbulent transition area can be represented by Ergun's equation as shown:

$$\frac{\Delta P}{\Delta Z} = \frac{150(1-\varepsilon_a)^2}{(\varphi d)^2 \varepsilon_a{}^3} \cdot \frac{\mu_g}{\rho_g}U + 1.75\frac{1-\varepsilon_a}{\varphi d \varepsilon_a{}^3}U^2 \tag{29}$$

5.2. Calculation Conditions

Table 2 lists the common calculation conditions for the sintering process. The composition of raw materials was set to simplify the calculation condition. The influence of MgO was not considered in this study. The particle size of hematite was set to 2.5 mm and 0.25 mm. It was assumed that 2.5 mm and 0.25 mm were the sizes of the nuclear particle and the adhering fine ores, respectively, in the quasi-particle. The 5.1 mass % charcoal calculation was compared with the 4 mass % coke calculation when the fixed carbon content is the same which means that the combustion heat of coke and charcoal during this process is the same. However, in this study, the effect of V.M. was not discussed. As a thought, the gas generated when V.M. is heated may improve the permeability and affect the temperature profile in the same way as a gas fuel injection, mentioned by Oyama [14]. Further research is needed to clarify this factor.

Table 2. Common calculation conditions.

· Sinter Bed	
Bed depth	450 mm
Porosity of sinter bed	35%
· Composition of Raw materials	
Hematite	85.0 mass %
Lime (CaO)	10.0 mass %
Moisture	5.0 mass %
Coke, Charcoal	4.0 mass % (additionally)
Charcoal	5.1 mass % (additionally)
· Diameter of Raw Materials	
Hematite	2.5 mm:0.25 mm ≈ 88.6:11.4
Lime (CaO)	2.0 mm
· Others	
Initial temperature	298 K
Ignition temperature	1573 K
Ignition time	90 s
Gas flow rate (outlet)	0.6 m/s
Calculation cell	5 mm
Time step	0.001 s
Courant number	0.2

Table 3 lists the state of the coke quasi-particles in the sinter bed for calculation using the date of the sinter pot test based on Hida's study [15].

Table 3. Existing state of the coke quasi-particles in the sinter bed.

Existing State of Coke (%)			Total	Amount of Coke and Charcoal in Sinter Bed (%)
S′-Type	C-Type	P-Type		
40.0	30.0	30.0	100	4.0, 5.1

5.3. Calculation Results

Figures 12 and 13 show the temperature profiles of the cases of coke and charcoal at a depth of 20 cm and 40 cm, respectively. The basic case was 4 mass% coke mixing condition. It was compared with cases of 4 mass % and 5.1 mass % charcoal mixing conditions. The influence of particle size is not apparent. Compared with the results for coke, charcoal has a lower temperature profile and a shorter holding time at a high temperature. This is thought to be due to the low fixed carbon content and the density of charcoal which leads to the presence of unreacted charcoal. As a result, the temperature of charcoal in all the sinter cakes was lower than that of coke.

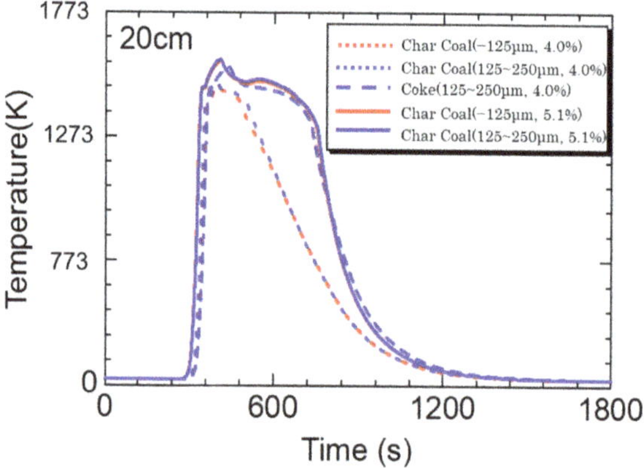

Figure 12. Calculation results of the temperature profile (20cm).

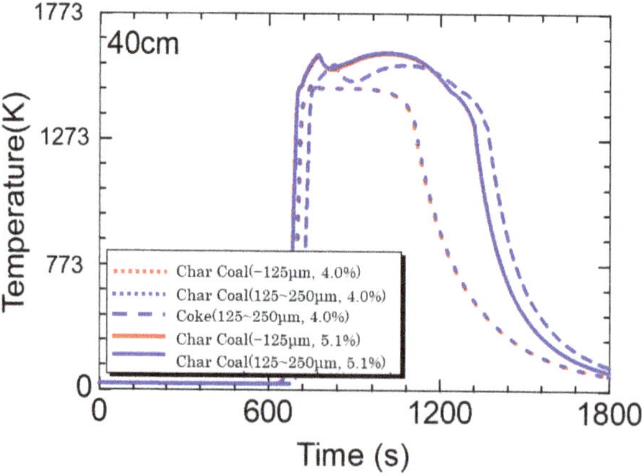

Figure 13. Calculation results of the temperature profile (40cm).

However, when the amount of charcoal was 5.1 mass %, the holding times at a high temperature of both kinds of carbon material were at the same level. The highest temperature of charcoal was higher than that of coke. The rate of the temperature increase of charcoal was also faster than that of coke. This is because the combustion rate of charcoal is higher than that of coke, i.e., if the fixed carbon content of charcoal is equivalent to that of coke, the temperature profile of the same level can be obtained. Therefore, the sintering simulation results of this study show that there are probabilities that charcoal can replace coke in the sintering process.

6. Conclusions

The study aimed to understand the combustion rate of a biomass carbon material using a sintering model to calculate its temperature profile. The following conclusions were made:

- Compared with coke, the reaction curves of charcoal combustion show that the combustion reaction of charcoal is faster.
- The interfacial chemical reaction rate coefficient of charcoal for the experimental data was calculated as follows:

Coke	$(-125\ \mu m)$	$kc = 8.24 \times 10^{-3} \exp\left(-10.6 \times 10^3 / RT\right)$	(m/s)
	$(125\text{\textasciitilde}250\ \mu m)$	$kc = 8.54 \times 10^{-3} \exp\left(-10.3 \times 10^3 / RT\right)$	(m/s)

- Calculations using the rate equation obtained in the sintering simulation model found that the high temperature range of charcoal is smaller than that of coke due to charcoal's low fixed carbon content and density.
- If the fixed carbon content of charcoal is the same as that of coke, which means that the combustion heat of carbon materials is the same, a temperature profile of the same level can be obtained.
- The sintering simulation results suggest that there are probabilities that biomass carbon materials can replace coke in the sintering process.

Author Contributions: Conceptualization, K.-i.O., and K.K.; Methodology, T.M.; Validation, K.-i.O., T.M., and Z.W.; Formal Analysis, S.N., and T.M.; Investigation, S.N.; Data Curation, Z.W., S.N., and T.M.; Writing-Original Draft Preparation, Z.W., and S.N.; Writing-Review and Editing, K.-i.O., and T.M.; Visualization, Z.W., and S.N.; Supervision, K.K.; Project Administration, K.-i.O. All authors have read and agreed to the published version of the manuscript.

Funding: This research received no external funding.

Conflicts of Interest: The authors declare no conflict of interest.

Nomenclature

$A_{(kc,De)}$	Frequency factor (m/s)
$C_{(O2,\ CO2)}$	O_2 or CO_2 concentration in the gas phase (mol/m^3)
$C_{(O2,\ CO2)\cdot i}$	O_2 or CO_2 concentration at the reaction interface (mol/m^3)
$C_{(O2,\ CO2)\cdot s}$	O_2 or CO_2 concentration at the particle surface (mol/m^3)
C_p	Specific heat (J/kg/K)
d	Particle size (m)
D_e	Effective diffusion coefficient in the Al_2O_3 powder layer (m^2/s)
$(D_{O2,\ CO2})_{eff}$	Effective diffusion coefficient of O_2 or CO_2 in the Al_2O_3 powder layer (m^2/s)
$E_{a(kc,De)}$	Activation energy (J/mol)
F	Reaction ratio (–)
H	Reaction heat of each reaction (J/mol)
h	Convection heat transfer coefficient (J/m^2/s/K)
K	Equilibrium constant (–)

k_C	Interfacial chemical reaction rate coefficient (m/s)
k_f	Mass transfer coefficient in the gas film (m/s)
k	Heat conductivity (J/m/s/K)
k'	Overall reaction rate (m/s)
$n_{(Coke,\ Quasi\text{-}particle,\ Lime,\ Ore)}$	The amount of coke, quasi-particle, lime and ore among unit volume (-)
ΔP	Pressure loss (atm)
r_0	Initial radius (m)
r_i	Radius of the non-reaction nucleus (m)
$r_{i,x}{}^*$	Generation rate of the component x in the number i cell (kg/s/m^3)
$r_{Quasi\text{-}particle}$	Distance from the left of the particle to the reaction interface of the quasi-particle (m)
r^*	Reaction ratio of the component in the sample (–)
$r^*_{Quasi\text{-}particle}$	Reaction rate per one particle of the quasi-particle (mol/s)
T_g	Temperature of gas in the control volume (K)
T_s	Temperature of solid in the control volume (K)
Δw_t	Sample weight change (kg)
U	Superficial velocity (m/s)
u	Gas flow rate (m/s)
W	Weight change of the sample during the experiment (kg)
ΔZ	Length of the control volume (m)
P_x	Density of component x (kg/m^3)
ρ_{Cm}	Carbon concentration in the sample (mol/m^3)
$\rho_{N2,\ O2,CO2,\ H2O}$	Density of gas in the sample(kg/m^3)
ε_a	Porosity (-)
Φ	(Surface area of a ball which has the same volume with the particle)/(Surface area of the particle) (-)
μ_g	Viscosity of gas (Pa·s)

References

1. *Japan's National Greenhouse Gas Emissions in Fiscal Year 2017 (Final Figures)*; Ministry of the Environment: Tokyo, Japan, 2019.
2. Takeshima, S. Biomass, Carbon-neutral and Renewable Energy. K. *J. JIME* **2012**, *47*, 133. [CrossRef]
3. Lovel, R.R.; Vining, K.R.; Dell'Amico, M. The influence of fuel reactivity on iron ore sintering. *ISIJ Int.* **2009**, *49*, 195–202. [CrossRef]
4. Zandi, M.; Martinez-Pacheco, M.; Fray, T. Biomass for iron ore sintering. *Miner. Eng.* **2010**, *23*, 1139–1145. [CrossRef]
5. Ooi, T.; Aries, E.; Ewan, B.; Thompson, D.; Anderson, D.R.; Fisher, R.; Fray, T.; Tognarelli, D. The study of sunflower seed husks as a fuel in the iron ore sintering process. *Miner. Eng.* **2008**, *21*, 167–177. [CrossRef]
6. Gan, M.; Fan, X.; Chen, X.; Ji, Z.; Lv, W.; Wang, Y.; Yu, Z.; Jiang, T. Reduction of pollutant emission in iron ore sintering process by applying biomass fuels. *ISIJ Int.* **2012**, *52*, 1574–1578. [CrossRef]
7. Yagi, T.; Ono, Y. A method of analysis for reduction of iron oxide in mixed-control kinetics. *Trans. ISIJ* **1968**, *8*, 377.
8. Ohno, K.; Noda, K.; Nisioka, K.; Maeda, T.; Shimizu, M. Combustion rate of coke in quasi-particle at iron ore sintering process. *ISIJ Int.* **2013**, *53*, 1588–1593. [CrossRef]
9. Ranz, W.E.; Marshall, W.R. Evaporation from drops. *Chem. Eng. Prog.* **1952**, *48*, 173–180.
10. Ohno, K.; Noda, K.; Nisioka, K.; Maeda, T.; Shimizu, M. Effect of coke combustion rate equation on numerical simulation of temperature distribution in iron ore sintering process. *ISIJ Int.* **2013**, *53*, 1642–1647. [CrossRef]
11. Tu, C.M.; Davis, H.; Hottel, H.C. Combustion rate of carbon-combustion of spheres in flowing gas streams. *Ind. Eng. Chem.* **1934**, *26*, 749–757. [CrossRef]
12. Davis, H.; Hottel, H.C. Combustion rate of carbon-combustion at a surface overlaid with stagnant gas. *Ind. Eng. Chem.* **1934**, *26*, 889–892. [CrossRef]
13. Parker, A.S.; Hottel, H.C. Combustion rate of carbon study of gas-film structure by microsampling. *Ind. Eng. Chem.* **1936**, *28*, 1334–1341. [CrossRef]

14. Oyama, N.; Iwami, Y.; Yamamoto, T.; Machida, S.; Higuchi, T.; Sato, H.; Sato, M.; Takeda, K.; Watanabe, Y.; Shimizu, M.; et al. Development of secondary-fuel injection technology for energy reduction in the iron ore sintering process. *ISIJ Int.* **2011**, *51*, 913–921. [CrossRef]
15. Hida, Y.; Sasaki, M.; Enokido, T.; Umezu, Y.; Iida, T.; Uno, S. Effect of the existing state of coke breeze in quasi-particles of raw mix on coke combustion in the sintering process. *Tetsu-to-Hagané* **1982**, *68*, 400–409. [CrossRef]

Article

Arsenic Removal from Arsenopyrite-Bearing Iron Ore and Arsenic Recovery from Dust Ash by Roasting Method

Rijin Cheng [1,2,3], **Hua Zhang** [1,3,*] **and Hongwei Ni** [1,*]

1   The State Key Laboratory of Refractories and Metallurgy, Wuhan University of Science and Technology, Wuhan 430081, China
2   Technical Center, HBIS Group Hansteel Company, Handan 056001, China; chengrijin@wust.edu.cn
3   Key Laboratory for Ferrous Metallurgy and Resources Utilization of Ministry of Education, Wuhan University of Science and Technology, Wuhan 430081, China
*   Correspondence: huazhang@wust.edu.cn (H.Z.); nihongwei@wust.edu.cn (H.N.); Tel.: +86-27-6886-2811 (H.Z. & H.N.)

Received: 12 September 2019; Accepted: 13 October 2019; Published: 16 October 2019

Abstract: In most cases, arsenic is an unfavorable element in metallurgical processes. The mechanism of arsenic removal was investigated through roasting experiments performed on arsenopyrite-bearing iron ore. Thermodynamic calculation of arsenic recovery was carried out by FactSage 7.0 software (Thermfact/CRCT, Montreal, Canada; GTT-Technologies, Ahern, Germany). Moreover, the arsenic residues in dust ash were recovered by roasting dust ash in a reducing atmosphere. Furthermore, the corresponding chemical properties of the roasted ore and dust ash were determined by X-ray diffraction, inductively coupled plasma atomic emission spectrometry, and scanning electron microscopy, coupled with energy-dispersive X-ray spectroscopy. The experimental results revealed that the arsenic in arsenopyrite-bearing iron ore can be removed in the form of $As_2O_3(g)$ in an air or nitrogen atmosphere by a roasting method. The efficiency of arsenic removal through roasting in air was found to be less than that in nitrogen atmosphere. The method of roasting in a reducing atmosphere is feasible for arsenic recovery from dust ash. When the carbon mass ratio in dust ash is 1.83%, the arsenic removal products is almost volatilized and recovered in the form of $As_2O_3(g)$.

Keywords: arsenopyrite; arsenic removal; mechanism; roasting; arsenate; dust ash; arsenic recovery

---

1. Introduction

Arsenic content in the earth's crust is up to 5 mg·kg^{-1}, and more than 300 arsenic species occur in nature. Arsenic is mainly associated with minerals such as pyrite, arsenopyrite, or enargite [1,2]. In most cases, arsenic is an unfavorable element in metallurgical processes. For example, arsenic reduces the quality of raw materials, affects the extraction of metal, interferes with the purity of the product, and poses serious environmental hazards [1]. Arsenic has an adverse effect on steel; for instance, the surface hot shortness increases, and the reduction of area and impact toughness decrease with the increase of arsenic content in steel [3–6]. Under the hot rolling or welding conditions, the arsenic in the steel leads to the increase in the content of arsenic at grain boundaries and the expansion of welding cracks [4,7–9]. Moreover, as the oxidability of arsenic is less than that of iron, it is difficult to remove arsenic by oxidation in the ironmaking or steelmaking process. It is theoretically possible to remove arsenic from molten iron by using excessive Al and Ca–Fe alloys or rare earth elements, but it also needs deep deoxidation and desulfurization before arsenic removal can be achieved, so the cost of arsenic removal is too high to be feasible in realistic production [10]. However, the price of arsenic-bearing iron ore is cheaper than that of high-grade iron ore. Moreover, the total amount of high-grade iron ore is decreasing greatly in the earth, and arsenic-bearing ore needs to be comprehensively utilized through ore blending. Some metallurgical enterprises at home and abroad, such as Peru, Chile, Philippines, France, Mexico, and China, have adopted some arsenic-bearing ore in the metallurgical industry [11]. When these arsenic-bearing ores are used, they are faced with the problem of arsenic removal from ores and the problem of arsenic-bearing dust treatment from the point of view of environmental protection.

Arsenic can be removed from arsenic-bearing ore by a roasting or sintering method due to the volatile nature of arsenic and its compounds [12]; thus, some scholars have explored the appropriate arsenic removal conditions by performing roasting or sintering experiments. Yin et al. and Lu et al. studied arsenic removal from copper–silver ore using a roasting method [13,14], and the impact of different parameters (e.g., temperature, atmosphere, and roasting time) on the arsenic removal ratio was also evaluated. Lu et al. investigated arsenic removal from arsenic-bearing iron ore during the sintering process [15,16]. The effects of temperature, bed depth, gas pressure, and coal ratio on arsenic removal during the sintering process were studied, and the reasonable technical parameters were obtained. In addition to the arsenic removal tests by roasting and sintering methods, a number of scholars also carried out thermodynamic calculations on arsenic removal, and discussed the arsenic-bearing products and suitable conditions for arsenic removal. Chakraborti and Lynch analyzed the As–S–O vapor system [17] and further identified the importance of bed depth in the rapid release of arsenic from arsenical materials under an oxidizing atmosphere. Contreras et al. evaluated the impact of various factors, such as trace element concentration, flue gas composition, temperature, and pressure, on the equilibrium composition based on the arsenic interactions in the co-combustion processes [18]. Nakazawa et al. and Zhang et al. studied the thermodynamics of arsenic removal from arsenic-containing copper ore during roasting [19] and arsenic-bearing iron ores during sintering [20–23], respectively, and obtained the equilibrium components containing arsenic and the arsenic removal rate.

The arsenic removed by roasting is mixed in the dust ash. For arsenic existing in dust, chemical adsorption of gaseous arsenic by CaO or $CaCO_3$ can effectively control arsenic content in flue gas and prevent arsenic pollution [24–26], but arsenate is easy to remain in dust, which is not conducive to the use of dust ash as raw material for sintering or roasting ore. Therefore, recovery of As_2O_3 is another feasible method.

Although arsenic removal experiments and thermodynamic calculations have been carried out by many scholars [12–23], and the volatilization behavior of arsenic in the process of roasting has also been explored [13,22], the mechanism of arsenic removal by roasting has never been reported. Exploration of the mechanism of arsenic removal by a roasting process and investigation of the residual form of arsenic in the roasted ore are important for controlling the arsenic removal efficiency and the arsenic component. Moreover, it is important to recover arsenic from dust ash to prevent further mobilization of arsenic and arsenic contamination. Therefore, these problems were attempted in this research.

2. Materials and Methods

2.1. Experiment on Arsenic Removal from Roasting Iron Ore

The composition of mixed ore used in the test is listed in Table 1.

Table 1. Composition and proportion of roasting ore.

Ore Name	Component/wt %							Proportion/%
	Fe_2O_3	SiO_2	CaO	Al_2O_3	MgO	FeAsS	S	
Iron ore	71.36	10.64	0.40	7.39	0.31	0.05	0.06	90
Arsenopyrite	/	/	/	/	/	92.71	/	10
Mixed ore	64.22	9.58	0.36	6.65	0.28	9.37	0.05	100

Note: "/" Represents "the chemical composition is not determined".

The mixed ore can be obtained by mixing 90% iron ore with 10% arsenopyrite. Mixed ore was crushed by an F77-1 sealed sample grinder for 1 min, then screened by a 75 mesh sieve. The ore powder larger than 75 mesh was crushed again until all powders were less than 75 mesh, and then the crushed powder was mixed. Mixed ore was blended with water to make iron ore balls with a diameter of 10 ± 2 mm. Furthermore, the iron ore balls were heated in an oven at 110 °C for 3 h, until they were completely dry. Then, the balls were taken out and reserved for further experiments.

The roasting test was carried out in a horizontal resistance furnace with 60 mm i.d. quartz tube, and the constant temperature zone of the resistance furnace was controlled using a thermocouple. The experiments were carried out in an air atmosphere and nitrogen atmosphere (1 L·min^{-1}, STP), respectively. The roasting temperature was 700, 800, 900, and 1000 °C, respectively, and the roasting time was 60 min. The porcelain boat loaded with the ore ball was put into the constant temperature zone of the tube furnace. When the roasting time was over, the power was turned off, and the sample was cooled down to room temperature and analyzed by various techniques. The schematic illustration of the roasting method is shown in Figure 1.

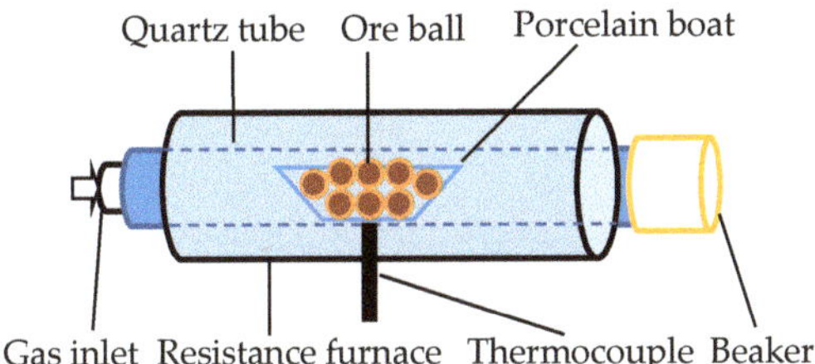

Figure 1. Schematic illustration of roasting test.

2.2. Experiment on Recovery of Arsenic by Roasting Dust Ash

The composition of dust ash of roasting iron ore from an iron plant is shown in Table 2. After adding 10% As_2O_3 chemically pure powder to the dust ash, the XRD pattern of the mixed dust ash powder is shown in Figure 2.

Table 2. Composition of dust ash from roasted iron ore.

Fe	CaO	MgO	SiO$_2$	Al$_2$O$_3$	TiO$_2$	S	K$_2$O	Na$_2$O	Cl$^-$
42.34	7.08	0.82	4.80	3.04	0.72	0.65	1.16	0.15	1.06

The arsenic recovery experiment by roasting was carried out in two different atmospheres with 10% As_2O_3 powder in the dust ash. The experiment was carried out in a horizontal resistance furnace with a 60 mm i.d. quartz tube. The dust ash was first placed in the constant temperature zone of the quartz tube. When the roasting experiment was carried out under an anaerobic atmosphere, the quartz tube was evacuated and washed with high-purity nitrogen 3 times before heating up the furnace. The test process was protected by 100 mL/min of high-purity nitrogen. The dust ash was heated to 600 °C and then cooled to room temperature after an hour of constant temperature. The experimental steps of roasting in a reducing atmosphere with 2% graphite powder as raw material are the same for the anaerobic atmosphere. When the roasting time was over, the sample was cooled down to room temperature and analyzed by the following various tests.

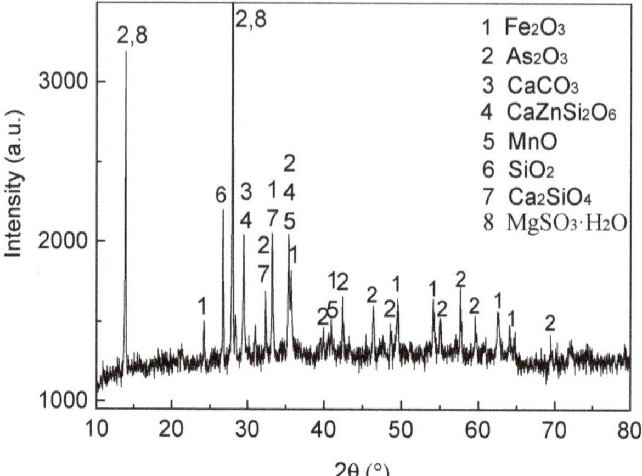

Figure 2. XRD spectra of the mixed dust ash powder.

2.3. Sample Analysis and Testing

The phase composition of iron ore was evaluated by PANalytical XPert PRO MPD XRD (Panaco, Almelo, Netherlands)with Cu target, K radiation, and 40 kV operating voltage. The chemical component of roasted ore was determined by IRIS Advantage Radial inductively coupled plasma atomic emission spectrometry (ICP-AES, Thermo Elemental, Massachusetts, America), and the physicochemical properties of the roasted ore were analyzed by FEI Nova NanoSEM400 (FEI, Hillsboro, America) scanning electron microscopy (SEM), coupled with energy-dispersive X-ray spectroscopy (EDS, FEI, Hillsboro, America).

3. Results

The white volatile that condensed in the cold beaker at the end of the quartz tube during the roasting process is shown in Figure 3a, b, and its XRD pattern is presented in Figure 3c. The XRD pattern and ICP-AES analysis indicated that the white volatiles corresponded to As_2O_3 powder with 89.12% purity, which confirms that the arsenic removal in an oxygen atmosphere is mainly carried out via Equation (1). The arsenic content in the ore after roasting at different temperatures is listed in Table 3.

$$2FeAsS + 5O_2(g) = Fe_2O_3 + As_2O_3(g) + 2SO_2(g) \tag{1}$$

Figure 3. (**a**) Arsenic removal from mixed ore by roasting in resistance furnace with quartz tube under air atmosphere and 1000 °C; (**b**) Collected As_2O_3 powers condensed in cold beaker; (**c**) XRD spectra of collected As_2O_3 powers condensed in cold beaker.

Table 3. Arsenic removal rate of ore subjected to roasting.

Temperature/°C	Arsenic Removal Rate in Air Atmosphere/wt %			Arsenic Removal Rate in Nitrogen Atmosphere/wt %		
	Before Roasting	After Roasting	Arsenic Removal Rate	Before Roasting	After Roasting	Arsenic Removal Rate
700	4.31	3.78	12.30	4.31	1.000	76.80
800	4.31	3.86	10.44	4.31	0.186	95.68
900	4.31	0.88	79.58	4.31	0.051	98.82
1000	4.31	0.57	86.77	4.31	0.027	99.37

Table 3 summarizes how arsenic can be removed from arsenopyrite-bearing iron ore by roasting in an air atmosphere or nitrogen atmosphere. The arsenic removal rate increases with the increase of temperature from 700 to 1000 °C. The arsenic removal rate by roasting method in an air atmosphere is less than that in the nitrogen atmosphere. The arsenic removal rate in the air atmosphere is poor at 700–800 °C, and the arsenic removal rate is about 12%, while the rate is 76.8–95.68% in the nitrogen atmosphere.

4. Discussion

4.1. Thermodynamic Calculation of Mixed Ore Subjected to Roasting

In order to explain the reason for the increase in arsenic removal rate with increasing temperature during roasting, the thermodynamic calculations of the roasting mixed ore in the air atmosphere were carried out by FactSage 7.0 (version 7.0, Thermfact/CRCT, Montreal, Canada, GTT-Technologies, Ahern, Germany) thermodynamic software. The effect of partial pressure of oxygen on the residual arsenic rate at different temperatures was calculated as shown in Figure 4. The results show that arsenate is the residual product in air roasting of arsenic-bearing ores at 700–1000 °C. Figure 4 shows that excessive partial pressure of oxygen is not beneficial to arsenic removal. Moreover, with the decrease of roasting temperature from 1000 to 700 °C, arsenic removal requires lower partial pressure of oxygen, which is the reason why the arsenic removal rate at 700 °C is lower than that at 1000 °C.

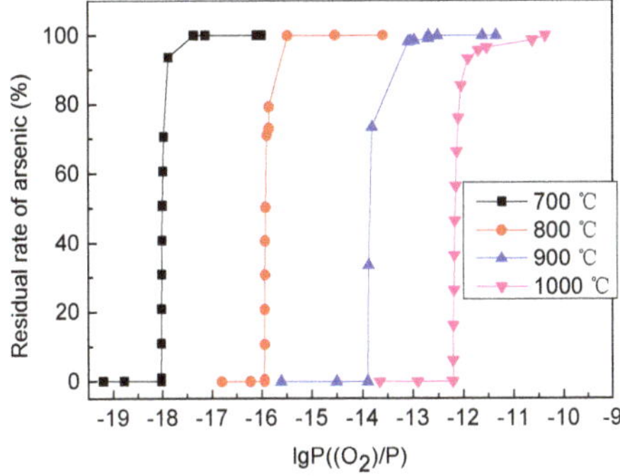

Figure 4. Effect of partial pressure of oxygen on residual rate at different temperatures.

The arsenic removal rate by roasting method in the air atmosphere is poor, which is probably attributed to the reaction between As_2O_3 and oxygen to generate As_2O_5, and then the As_2O_5 reacts with other oxides (Fe_2O_3, Al_2O_3, CaO) via Equations (2) to (4) to generate arsenate. Thus, the roasting product contains a variety of arsenic residues. The arsenic removal rate at 900–1000 °C increases to 79.58–86.77% in air; the formation rate of As_2O_3 is accelerated at high temperature, and a large amount of gas escapes rapidly.

$$As_2O_3(g) + O_2(g) + Fe_2O_3 = 2FeAsO_4 \tag{2}$$

$$As_2O_3(g) + O_2(g) + Al_2O_3 = 2AlAsO_4 \tag{3}$$

$$As_2O_3(g) + O_2(g) + 3CaO = Ca_3(AsO_4)_2 \tag{4}$$

4.2. X-Ray Diffraction Analysis of the Roasted Ore and Dust in Different Atmospheres

The XRD spectra of the roasted ore in air and nitrogen atmospheres are shown in Figure 5a,b, respectively. In the raw material ore, arsenic exists in the form of FeAsS (Figure 5a, bottom). Figure 5a shows the disappearance of peaks of FeAsS due to its decomposition when the ore was roasted at 700–1000 °C, and a small amount of peaks of AlAsO$_4$ at 800 °C and As$_2$O$_3$ at 1000 °C appear for the roasted ore, indicating that FeAsS underwent decomposition via Equation (1) and As$_2$O$_3$(g) underwent reaction via Equation (3). Figure 5b demonstrates that the arsenic removal by roasting in the nitrogen atmosphere is more thorough, and the peaks of arsenates and As$_2$O$_3$(s) are not found in XRD spectra of roasted ore. The reaction of arsenic removal is mainly carried out via Equation (1). The investigation of the mechanism on arsenic removal by roasting method proves that arsenic is mainly removed in the form of gaseous As$_2$O$_3$(g) in the oxidation or nitrogen atmosphere, while the residual arsenic is mostly arsenate.

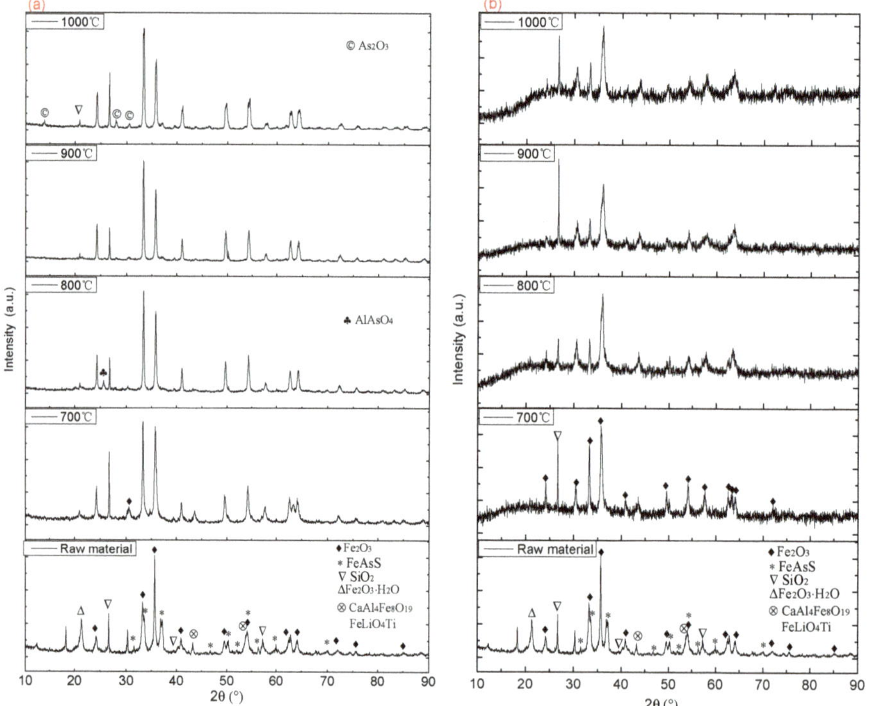

Figure 5. XRD spectra of the roasted ore and dust in: (**a**) Air and (**b**) nitrogen atmosphere.

4.3. Mechanism Research on Arsenic Removal by Roasting Method and Scanning Electron Microscopy and Energy-Dispersive X-Ray Spectroscopy Analysis

Arsenic removal efficiency in the air atmosphere was found to be poor. Therefore, the arsenic residual form in the roasted sample under an air atmosphere was further studied by SEM coupled with EDS. Figure 6 shows the SEM images of the roasted ore under an air atmosphere at different roasting temperatures. Chemical composition of the raw ore and roasted ore is shown in Table 4. Figure 6a demonstrates that arsenic and sulfur occur simultaneously in the raw material, and arsenic is present in the form of FeAsS. Figure 6b shows that arsenic is found in the samples roasted at 700 °C; however, sulfur is not found, which indicates the decomposition of FeAsS, where arsenic is present in $FeAsO_4$ and $Ca_3(AsO_4)_2$. Figures 5 and 6c show that the arsenic in the roasted ore is present as $AlAsO_4$ and $FeAsO_4$ at 800 °C, respectively. The results exhibited in Figure 6d,e are similar to those shown in Figure 6c. The abovementioned SEM and EDS results further confirm that arsenic is removed in the form of $As_2O_3(g)$ by roasting in the air atmosphere, and the residual arsenic reacts with oxide in the ore to generate arsenates.

Table 4. Chemical composition of the raw ore and roasted ore.

Figure No.	Point No.	Atomic Ratio of Elements/at %											
		Mg	Al	Ca	Cr	Si	S	Mn	Fe	Ni	As	Nb	O
6a (raw ore)	A	-	-	-	-	5.40	-	-	33.53	-	-	-	61.08
	B	5.97	-	-	-	10.86	-	-	22.20	-	-	-	60.98
	C	-	9.18	-	-	6.73	-	-	22.75	-	-	-	61.35
	D	-	-	-	-	7.50	0.26	0.43	28.61	-	1.64	-	61.57
	E	-	-	-	-	-	-	-	40	-	-	-	60
	F	-	-	-	-	-	5.91	-	32.42	-	0.48	-	61.18
	G	-	-	-	-	7.23	2.14	-	22.82	-	5.08	-	62.73
6b (700 °C)	A	-	-	-	-	7.30	-	-	24.31	-	0.082	4.36	63.21
	B	-	-	-	-	11.16	-	1.15	24.35	0.80	0.70	-	61.84
	C	-	-	-	-	3.12	-	-	35.84	-	0.42	-	60.62
	D	-	-	0.46	-	10.45	-	-	25.06	0.42	1.70	-	61.92
	E	-	-	0.64	-	8.38	-	-	28.03	-	0.81	-	61.65
	F	-	-	-	0.36	7.73	-	-	29.08	-	1.28	-	61.55
	G	-	-	-	-	2.10	-	-	37.48	-	-	-	60.42
6c (800 °C)	A	-	-	-	0.36	10.00	-	0.38	27.33	-	-	-	61.92
	B	-	-	-	0.34	8.42	-	-	29.56	-	-	-	61.68
	C	-	-	-	-	22.59	-	-	12.55	-	0.34	-	64.52
	D	-	-	-	-	11.86	-	-	25.76	-	-	-	62.37
	E	-	-	-	-	8.02	-	3.49	27.58	-	-	-	60.91
	F	2.59	-	-	-	19.04	-	-	15.08	-	-	-	63.29
	G	-	-	-	-	24.37	-	0.40	9.85	-	0.58	-	64.79
	H	-	-	-	-	12.95	-	-	19.00	-	0.25	3.72	64.08
6d (900 °C)	A	-	-	-	-	9.79	-	-	28.25	-	-	-	61.96
	B	-	-	-	-	7.12	-	-	23.55	-	2.00	4.22	63.11
	C	-	-	-	-	9.72	-	-	25.08	-	3.25	-	61.94
	D	-	7.01	-	0.39	3.87	-	-	27.95	-	-	-	60.77
	E	-	-	-	0.46	6.49	-	-	31.75	-	-	-	60.30
	F	-	-	-	0.34	5.44	-	-	32.65	-	0.49	-	61.09
6e (1000 °C)	A	-	-	-	0.96	7.34	-	-	30.23	-	-	-	61.47
	B	-	1.16	-	-	1.75	-	-	36.74	-	-	-	60.35
	C	-	2.93	-	-	2.50	-	-	26.14	-	-	5.66	62.76
	D	-	1.46	-	-	-	-	-	38.54	-	-	-	60.00
	E	-	-	-	-	2.45	-	-	35.38	-	1.68	-	60.49
	F	-	0.89	-	0.49	16.73	-	-	18.54	-	-	-	63.35
	G	-	-	-	1.02	8.67	-	-	26.13	-	2.44	-	61.73

Note: "-" Represents "below the detection limit".

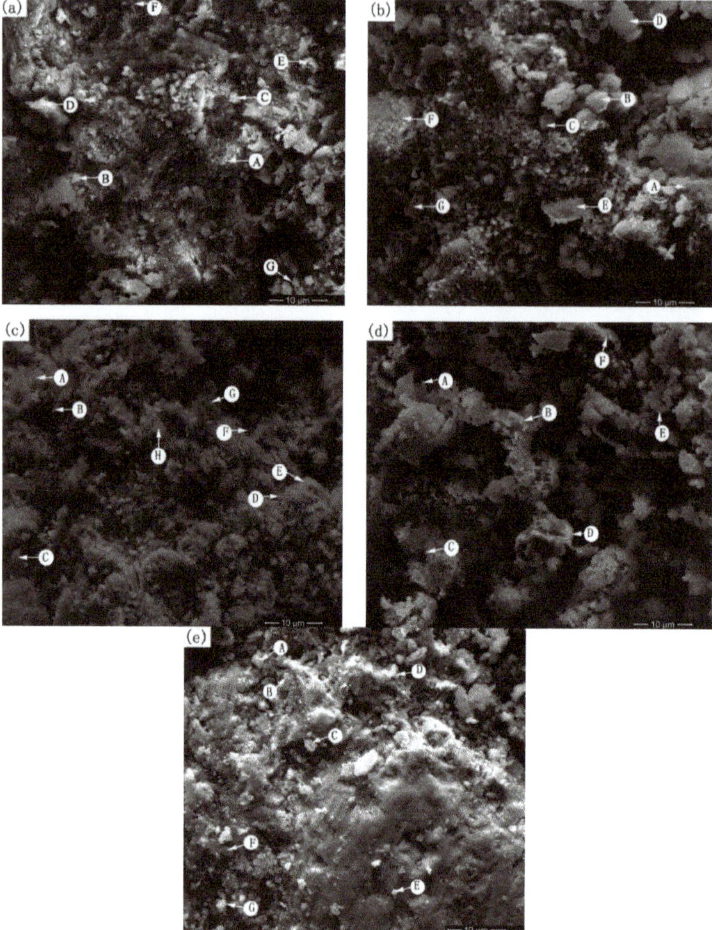

Figure 6. SEM images of the roasted ore in air atmosphere and at different roasting temperatures: (**a**) Raw material; (**b**) 700 °C; (**c**) 800 °C; (**d**) 900 °C; (**e**) 1000 °C.

4.4. Route for the Recovery of Arsenic from Arsenic-Bearing Dust Ash

The arsenic recovery experiment by roasting in the atmosphere of air, anaerobic and reducing, was carried out with the dust ash containing 10% As_2O_3 powder [20]. The experiment confirms that arsenic recovery from dust ash by roasting in the atmosphere of air or an anaerobic atmosphere is difficult, and that arsenate easily remains in the dust ash. Figure 7 shows the effect of temperature on arsenic-containing products in dust ash by reduction roasting. Thermodynamic calculation of arsenic recovery by roasting under a reduction environment shows that arsenic is not easy to be removed from dust and that arsenate is easy to be formed when the roasting temperature is below 390 °C. The arsenic is easily volatilized and recovered when the dust is roasted above 390–890 °C, but it is easy to produce arsenate when the dust ash is roasted above 890 °C, which affects the recovery of arsenic.

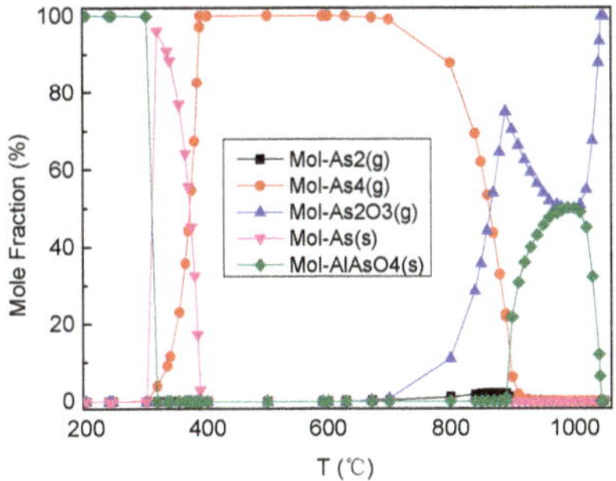

Figure 7. Effect of temperature on arsenic-containing products in dust ash by reduction roasting.

4.5. Effect of Carbon Mass Ratio on Arsenic Removal Products of Dust Ash by Roasting

According to the dust ash in Table 2, the thermodynamic calculation of the effect of carbon powder on arsenic removal products was done using FactSage 7.0. Figure 8 shows the results of the effect of carbon ratio on arsenic removal products of dust ash by roasting using thermodynamic calculations. When the arsenic-bearing dust ash was roasted with a carbon mass ratio increasing from 0 to 1.83%, the percentage of residual solid $AlAsO_4(s)$ in dust ash gradually decreased from 100% to 0, and the percentage of gaseous $As_2O_3(g)$ gradually increased to 100%. When the arsenic-bearing dust ash was roasted with a carbon mass ratio was below 1.63%, the arsenic removal products were the majority of $AlAsO_4(s)$ and a small amount of $As_2O_3(g)$. When the carbon mass ratio was 1.83%, the arsenic removal product was almost volatilized in the form of $As_2O_3(g)$. Subsequently, with the increase of carbon mass ratio, the percentage of volatile $As_2O_3(g)$ gradually decreased, while the percentages of $As_2(g)$ and $As_4(g)$ gradually increased. When the carbon mass ratio increased to 5%, arsenic was almost removed by volatilization of $As_2(g)$ and $As_4(g)$.

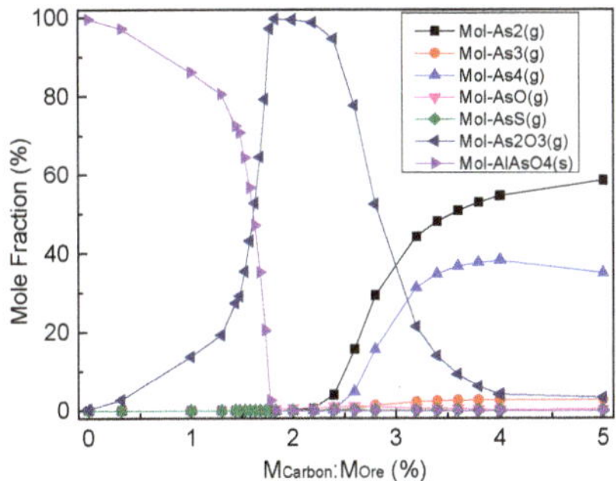

Figure 8. Effect of carbon ratio on arsenic removal products of dust ash by roasting.

The arsenic recovery experiment by roasting in the reducing atmosphere was carried out with dust ash containing 2% carbon powder. Figure 9a shows the recovered products condensed on the edge wall of the quartz tube. Figure 9b exhibits the XRD spectrum of the recovered products. The recovered products are almost $As_2O_3(g)$. Figure 9c is the XRD spectrum of the roasted dust ash, and the peak of arsenic is almost invisible in the XRD spectrum. By comparing dust ash before roasting in Figure 2, it can be seen that almost all arsenic in the dust ash has been volatilized and recovered in the low-temperature section at the end of the quartz tube. Thermodynamic calculation of Figure 8 shows that arsenic volatilized in the form of gaseous $As_2O_3(g)$ when dust ash was roasted with 2% carbon. The results demonstrate that the thermodynamic calculation results are in good agreement with the experimental results.

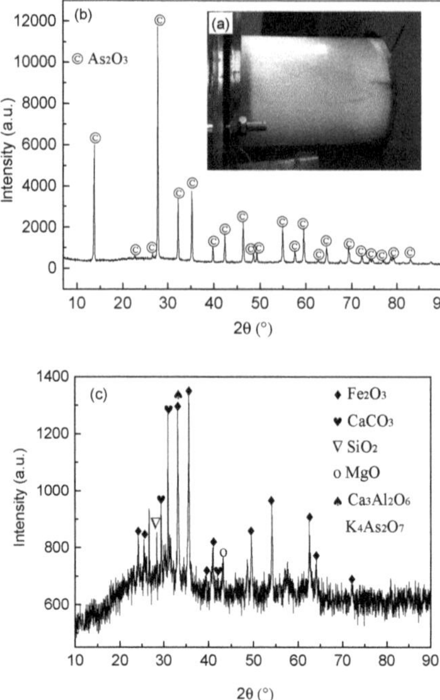

Figure 9. (**a**) Volatile compounds photograph, (**b**) XRD spectrum of volatile compounds collected during roasting dust ash, and (**c**) the dust ash after roasting.

5. Conclusions

(1) Arsenic in arsenopyite-bearing iron ore can be removed by roasting method in an air or nitrogen atmosphere;

(2) The mechanism of arsenic removal by roasting method indicates that the efficiency of arsenic removal by roasting in air is less than that in nitrogen atmosphere. The poor arsenic removal efficiency at low temperature and in an air atmosphere is due to the formation of arsenates by the reaction of $As_2O_3(g)$ with other oxides in the strong oxidizing atmosphere. Lower partial pressure of oxygen is required to ensure an effective arsenic removal rate when arsenic-bearing ore is roasted at lower temperatures. Arsenic is removed in the form of $As_2O_3(g)$ by the roasting method, and residual arsenic reacts with oxides in the ore to generate arsenates.

(3) The arsenic recovery from dust ash by roasting in the atmosphere of an air or anaerobic atmosphere is difficult, and arsenic easily reacts with oxides to form arsenate and remains in the dust ash. The method of roasting in a reducing atmosphere is feasible for arsenic recovery from dust ash. When the arsenic-bearing dust ash is roasted with a carbon mass ratio below 1.63%, the arsenic removal products are the majority of $AlAsO_4(s)$ and a small amount of $As_2O_3(g)$. When the carbon mass ratio is 1.83%, the arsenic removal product is almost volatilized and recovered in the form of $As_2O_3(g)$.

Author Contributions: R.C., H.Z., and H.N. conceived this contribution; R.C., and H.N. wrote this paper; R.C. and H.Z. conducted experiments and analyzed the data. R.C. and H.Z. received funding acquisition.

Funding: This research was funded by the Open Youth Fund of The State Key Laboratory of Refractories and Metallurgy, Wuhan University of Science and Technology (Grant No. 2018QN03), National Key R and D Program of China (2018YFC1900602) and the Research Project of Hubei Provincial Department of Education (D20171104).

Conflicts of Interest: The authors declare no conflict of interest.

References

1. Riveros, P.A.; Dutrizac, J.E.; Spencer, P. Arsenic disposal practices in the metallurgical industry. *Can. Metall. Q.* **2001**, *40*, 395–420. [CrossRef]
2. Díaz, J.A.; Serrano, J.; Leiva, E. Bioleaching of Arsenic-Bearing Copper Ores. *Minerals* **2018**, *8*, 215. [CrossRef]
3. Xin, W.B.; Song, B.; Yang, Z.B.; Yang, Y.H.; Li, L.F. Effect of Arsenic and Copper+Arsenic on high temperature oxidation and hot shortness behavior of C–Mn steel. *ISIJ Int.* **2016**, *56*, 1232–1240. [CrossRef]
4. Xin, W.B.; Song, B.; Huang, C.G.; Song, M.M.; Song, G.Y. Effect of Arsenic content and quenching temperature on solidification microstructure and Arsenic distribution in iron-arsenic alloys. *Int. J. Miner. Metall. Mater.* **2015**, *22*, 704–713. [CrossRef]
5. Yin, L.; Seetharaman, S. Effects of residual elements Arsenic, Antimony, and Tin on surface hot shortness. *Metall. Mater. Trans. B* **2011**, *42*, 1031–1043. [CrossRef]
6. Huang, C.G.; Song, B.; Xin, W.B.; Jia, S.J.; Yang, Y.H. Influence of rare earth La on hot ductility of low carbon steel containing As and Sn. *Heat Treat. Met.* **2015**, *40*, 1–6.
7. Zhu, Y.Z.; Li, B.L.; Liu, P. Effect of annealing and hot rolling on grain boundary segregation of Arsenic in an Mn-steel microalloyed by Ti, Cr and Nb. *J. Iron Steel Res. Int.* **2013**, *20*, 67–72. [CrossRef]
8. Xin, W.B.; Song, B.; Song, M.M.; Song, G.Y. Effect of cerium on characteristic of inclusions and grain boundary segregation of arsenic in iron melts. *Steel Res. Int.* **2015**, *86*, 1430–1438. [CrossRef]
9. Geng, M.S.; Xiang, L.; Wang, X.H.; Zhang, J.M.; Xiao, J.G. Effect of residual elements on continuous cast slab and surface quality of hot rolled plate. *J. Iron Steel Res.* **2009**, *21*, 19–21. (In Chinese)
10. Wang, J.J.; Luo, L.G.; Kong, H.; Zhou, L. The Arsenic removal from molten steel. *High Temp. Mater. Proc.* **2011**, *30*, 171–173. [CrossRef]
11. Valenzuela, A. Arsenic Management in the Metallurgical Industry. Master's Thesis, University of Laval, Quebec, QC, Canada, 2000.
12. Mihajlovic, I.; Strbac, N.; Nikolic, D.; Živkovic, Z. Potential metallurgical treatment of Copper concentrates with high Arsenic contents. *J. S. Afr. Inst. Min. Metall.* **2011**, *111*, 409–416.
13. Yin, Z.L.; Lu, W.H.; Xiao, H. Arsenic removal from copper-silver ore by roasting in vacuum. *Vacuum* **2014**, *101*, 350–353. [CrossRef]
14. Lu, W.H.; Yin, Z.L. Study on thermal decomposition and Arsenic removal of a Silver bearing Copper ore. *Int. J. Miner. Process.* **2016**, *153*, 1–7. [CrossRef]
15. Lv, Q.; Zhang, S.H.; Hu, X. Study on removal Arsenic from iron ore with Arsenic in sintering process. *Adv. Mater. Res.* **2011**, *284*, 238–241.
16. Cheng, R.J.; Ni, H.W.; Zhang, H.; He, H.Y.; Yang, H.H.; Xiong, S. Experimental study on arsenic removal from low arsenic-bearing iron ore with sintering process. *Sinter. Pelletizing* **2016**, *41*, 13–16. (In Chinese)
17. Chakraborti, N.; Lynch, D.C. Thermodynamic analysis of the As–S–O vapor system. *Can. Metall. Q.* **1985**, *24*, 39–45. [CrossRef]
18. Contreras, M.L.; Arostegui, J.M.; Armesto, L. Arsenic interactions during co-combustion processes based on thermodynamic equilibrium calculations. *Fuel* **2009**, *88*, 539–546. [CrossRef]
19. Nakazawa, S.; Yazawa, A.; Jorgensen, F.R.A. Simulation of the removal of Arsenic during the roasting of Copper concentrate. *Metall. Mater. Trans. B* **1999**, *30*, 393–401. [CrossRef]
20. Zhang, S.H.; Lü, Q.; Hu, X. Thermodynamics of arsenic removal from arsenic-bearing iron ores. *Chin. J. Nonferrous Met.* **2011**, *21*, 1705–1712. (In Chinese)
21. Cheng, R.J.; Ni, H.W.; Zhang, H.; Jia, S.K.; Xiong, S. Thermodynamics of arsenic removal from arsenic-bearing iron ores with sintering process and dust ash by roasting. *Iron Steel.* **2017**, *52*, 26–33. (In Chinese)
22. Jiang, T.; Huang, Y.F.; Zhang, Y.B.; Han, G.H.; Li, G.H.; Guo, Y.F. Behavior of arsenic in arsenic-bearing iron concentrate pellets by preoxidizing–weak reduction roasting process. *J. Cent. South Univ. Sci. Technol.* **2010**, *41*, 1–7. (In Chinese)
23. Cheng, R.J.; Ni, H.W.; Zhang, H.; Zhang, X.K.; Bai, S.C. Mechanism research on arsenic removal from arsenopyrite ore during sintering process. *Int. J. Miner. Metall. Mater.* **2017**, *24*, 353–359. [CrossRef]
24. Li, Y.Z.; Tong, H.L.; Zhuo, Y.Q.; Li, Y.; Xu, X.C. Simultaneous removal of SO_2 and trace As_2O_3 from flue gas: Mechanism, kinetics study, and effect of main gases on arsenic capture. *Environ. Sci. Technol.* **2007**, *41*, 2894–2900. [CrossRef] [PubMed]

25. Díaz-Somoano, M.; López-Antón, M.A.; Huggins, F.E.; Martínez-Tarazona, M.R. The stability of arsenic and selenium compounds that were retained in limestone in a coal gasification atmosphere. *J. Hazard. Mater.* **2010**, *173*, 450–454. [CrossRef] [PubMed]
26. Nadia, M.V.; Roberto, B.G.; José, A.R.L.; Miguel, A.B.; Alan, D.C.; Elías, R.F.; Mario, V. Arsenic mobility controlled by solid calcium arsenates-A case study in Mexico showcasing a potentially widespread environmental problem. *Environ. Pollut.* **2013**, *176*, 114–122.

 *processes* 

Article

Numerical Simulation of Effects of Different Operational Parameters on the Carbon Solution Loss Ratio of Coke inside Blast Furnace

Mingyin Kou [1,*], **Heng Zhou** [1], **Li Pang Wang** [2,*], **Zhibin Hong** [1], **Shun Yao** [1], **Haifa Xu** [3] and **Shengli Wu** [1]

1   School of Metallurgical and Ecological Engineering, University of Science and Technology Beijing, Beijing 100083, China
2   Institute of Environmental Engineering and Management, College of Engineering, National Taipei University of Technology, Taipei 10608, Taiwan
3   Baosteel Research Center, No. 889 Fujin Road, Baoshan District, Shanghai 201900, China
*   Correspondence: mingyinkou@gmail.com (M.K.); kuniwang@ntut.edu.tw (L.P.W.)

Received: 12 July 2019; Accepted: 5 August 2019; Published: 9 August 2019

Abstract: Carbon solution loss reaction of coke gasification is one of the most important reasons for coke deterioration and degradation in a blast furnace. It also affects the permeability of gas and fluids, as well as stable working conditions. In this paper, a three dimensional model is established based on the operational parameters of blast furnace B in Bayi Steel. The model is then used to calculate the effects of oxygen enrichment, coke oven gas injection, and steel scrap charging on the carbon solution loss ratio of coke in the blast furnace. Results show that the carbon solution loss ratio of coke gasification for blast furnace B is almost 20% since the results of a model are probably only indicative. The oxygen enrichment and the addition of steel scrap can reduce the carbon solution loss ratio with little effect on the working condition. However, coke oven gas injection increases the carbon solution loss ratio. Therefore, coke oven gas should not be injected into the blast furnace unless the quality of the coke is improved.

Keywords: blast furnace; coke; carbon solution loss; numerical simulation

1. Introduction

CO_2 emission is a global problem that affects many countries. Ironmaking processes contribute significantly to CO_2 emission, and this is the case especially with blast furnace process, which produces more than 90% of the world's pig iron. The blast furnace is a chemical reactor involving counter-current flows of gas and solid [1–3]. Iron-bearing materials and coke are charged in turn at the top of the furnace. Hot air, enriched oxygen, and pulverized coal are blown into the furnace through the tuyeres. Iron-bearing materials are reduced by the reducing gas containing CO and H_2, which are generated from the combustion of coke and coal in the cavity around the exit of a tuyere called the raceway [2–4]. The hot reducing gas flows upward, heats up the iron-bearing materials, and escapes from the top in the form of CO_2, CO, H_2, H_2O and N_2. Coke has several functions in a blast furnace: a fuel providing the heat for chemical reactions and for the melting of iron and slag; a reducing agent in itself and also providing gases for iron oxide reduction; and a permeable skeleton providing a passage for liquids and gases [5]. Therefore, it is very important for the efficiency of the blast furnace process and the quality of the hot metal. The carbon solution loss reaction of coke is a main means of coke gasification for the coke consumption in the upper part of the blast furnace. The strength and the size of the coke will deteriorate with the carbon solution loss reaction when the coke moves towards the lower zones of the blast furnace. This greatly affects the permeability of the bed and the efficiency of the process [1,5]. Therefore, the carbon solution loss of the coke should be restricted.

By conducting experiments under real blast furnace gas-temperature conditions, Babich et al. [6] found that the reaction rate of the coke close to the wall of the blast furnace was higher than that in the center of the furnace. By studying the reactivity of cokes charged in an experimental blast furnace, Lundgren et al. [7] found that the carbon solution loss reaction in the blast furnace was limited by the diffusion rate. Hilding et al. [5] found that the reactivity of coke increased when coke moved from the thermal reserve zone to the cohesive zone of the experimental blast furnace. Having investigated the transformation of mineral matters of cokes in the blast furnace, Gornostayev et al. [8] reported that this reduced coke reactivity by covering the pore wall. Loison et al. [9] stated that the carbon solution loss reaction occurred under a mixed regime in the thermal reserve zone of the blast furnace. When the temperature was raised, the chemical reaction rate increased exponentially and the diffusion rate was limited to the amount of CO_2 gas. Sato et al. [10] stated that the coke pore structure modified the available carbon surface area for the carbon solution loss. Alkalies have catalytic effects on the solution loss reaction, and they can decrease the threshold temperature to approximately 760 °C [5,11,12]. Fe, CaO and MgO in coke ash are also shown to have a catalytic effect on solution loss [13,14]. However, the effects of different operational parameters on the carbon solution loss ratio are seldom studied. Physical experiments and mathematical modelling are common ways of investigating carbon solution loss. Several continuum models have been applied to improve practical operations of the ironmaking process especially in a blast furnace [15–19].

Therefore, in the present work, a three dimensional model of carbon solution loss of coke is established based on the actual production data of blast furnace B in Bayi Steel. The model is then used to analyze the effects of oxygen enrichment, coke oven gas injection and steel scrap charging on the carbon solution loss ratio of coke in the blast furnace. The research provides the theoretical reference and technical direction for Bayi Steel to control the carbon loss solution of coke.

2. Model Establishment

The geometric model of a blast furnace was established based on blast furnace B in Bayi Steel of Baowu Group in China, and it was a one-sixth (60 degree section) sector of the blast furnace. The total height of the model was 27.43 m, as shown in Figure 1, where, for the sake of simplification, the part below the taphole in the blast furnace hearth is not included.

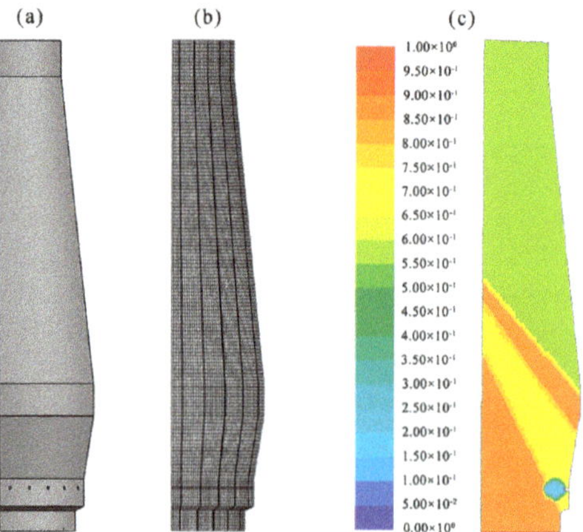

Figure 1. Geometric and mesh models and volume fraction distribution of solid phase of blast furnace B in Bayi Steel of Baowu Group (**a**) geometric model; (**b**) mesh model; (**c**) volume fraction distribution of solid phase.

Both gas and solid phases were considered. The gas phases were the same as those in practice, including CO, CO_2, H_2, H_2O and N_2. The density of gas phase was calculated by the ideal gas law. The solid phase was simplified to include Fe_2O_3, Fe_3O_4, FeO, Fe, and C. Other components were not considered for the sake of simplicity. The density of solid phase was 4190 kg·m^{-3}, the same as the apparent density of the original solid material. Both gas phase and solid phase were considered as continuous phases using the Eulerian method. The mass, energy, and species transfer can be described by Equation (1) under steady state [20,21].

$$\nabla(\alpha_p \cdot \rho_p \cdot \phi \cdot \vec{V}_p) = \nabla(\alpha_p \cdot \Gamma \cdot \nabla(\phi)) + S_\phi \tag{1}$$

where ρ is the phase. Γ and S are the effective diffusivity and the source respectively, varying with respect to the different variables ϕ as listed in Table 1 [21–25].

Table 1. Parameters in Equation (1).

Items.	ϕ	Γ	S_ϕ
continuity	1	0	$M_O \cdot \sum\limits_{n=1}^{N} R_n$ $-M_O \cdot \sum\limits_{n=1}^{N} R_n$
momentum	$\vec{v}_g$ $\vec{v}_s$	0	$\nabla \cdot \overline{\overline{\tau}}_g + \varepsilon_g(-\nabla P + \rho_g \cdot \vec{g}) + \vec{F}_{gs}$ $\nabla \cdot \overline{\overline{\tau}}_s + \varepsilon_s(-\nabla P + \rho_s \cdot \vec{g})$
energy	H_g	$K_g/C_{P,g}$	$E_{gs} + M_O \cdot \sum\limits_{n=1}^{N} (R_n \cdot \Delta H_n^T)$
	H_s	$K_s/C_{P,s}$	$-E_{gs} + M_O \cdot \sum\limits_{n=1}^{N} (R_n \cdot \Delta H_n^T)$

R_n refers to different reactions.

$$\overline{\overline{\tau}}_p = \varepsilon_p \cdot \mu_p \cdot [\nabla \cdot \vec{v}_p + (\nabla \cdot \vec{v}_p)^T] - \frac{2}{3} \cdot \varepsilon_p \cdot \mu_p \cdot (\nabla \cdot \vec{v}_p) \cdot \overline{\overline{I}}$$

$$\vec{F}_{gs} = -[150 \cdot \frac{(1-\varepsilon_g)^2 \cdot \mu_g}{\varepsilon_g^3 \cdot d_s^2} + 1.75 \cdot \frac{\rho_g \cdot \varepsilon_s \cdot |\vec{v}_s - \vec{v}_g|}{d_s}] \cdot (\vec{v}_s - \vec{v}_g)$$

$$E_{gs} = -\frac{6 \cdot k_g \cdot \varepsilon_g \cdot \varepsilon_s}{d_s^2} \cdot (2.0 + 0.6 \cdot Re_s^{1/2} \cdot Pr_g^{1/3}) \cdot (T_g - T_s)$$

The chemical reactions between gas phase and solid phase are as follows, including the indirect reduction of iron oxide (Reactions 1–6), carbon solution loss reaction (Reactions 7), water gas reaction (Reaction 8), water gas shift reaction (Reaction 10), combustion reaction of C and H_2 (Reaction 9 and 11), and so on.

$$3Fe_2O_3 + CO \rightarrow 2Fe_3O_4 + CO_2 \qquad \text{Reaction 1}$$
$$Fe_3O_4 + CO \rightarrow 3FeO + CO_2 \qquad \text{Reaction 2}$$
$$FeO + CO \rightarrow Fe + CO_2 \qquad \text{Reaction 3}$$
$$3Fe_2O_3 + H_2 \rightarrow 2Fe_3O_4 + H_2O \qquad \text{Reaction 4}$$
$$Fe_3O_4 + H_2 \rightarrow 3FeO + H_2O \qquad \text{Reaction 5}$$
$$FeO + H_2 \rightarrow Fe + H_2O \qquad \text{Reaction 6}$$
$$C + CO_2 \longleftrightarrow 2CO \qquad \text{Reaction 7}$$
$$C + H_2O \longleftrightarrow CO + H_2 \qquad \text{Reaction 8}$$
$$C + O_2 \rightarrow CO_2 \qquad \text{Reaction 9}$$
$$CO + H_2O \longleftrightarrow CO_2 + H_2 \qquad \text{Reaction 10}$$
$$H_2 + 0.5O_2 \rightarrow H_2O \qquad \text{Reaction 11}$$

The three-interface unreacted core model was used to calculate the chemical reaction rates, and the physical chemistry data were taken from Perry et al.'s book [26]. The reaction rate constants of the indirect reduction of iron ore by CO or H_2, carbon solution loss reaction, water gas reaction, combustion of carbon and water gas shift reaction were taken from other work [27–30]. The combustion rate of H_2 with O_2 was taken from Kuwabara et al.'s work [31]. The effective diffusion coefficients were taken from other work [27–30]. The viscosity and thermal conductivity of gas were obtained from the literature [18,32–34]. The chemical reaction rates are as follows.

$$R_1 = \frac{A\rho_g}{W} \sum_{m=1}^{3} a_{1,m} \left(K_m \frac{w_{CO,g}}{M_{CO}} - \frac{w_{CO_2,g}}{M_{CO_2}} \right) \tag{2}$$

$$R_2 = \frac{A\rho_g}{W} \sum_{m=1}^{3} a_{2,m} \left(K_m \frac{w_{CO,g}}{M_{CO}} - \frac{w_{CO_2,g}}{M_{CO_2}} \right) \tag{3}$$

$$R_3 = \frac{A\rho_g}{W} \sum_{m=1}^{3} a_{3,m} \left(K_m \frac{w_{CO,g}}{M_{CO}} - \frac{w_{CO_2,g}}{M_{CO_2}} \right) \tag{4}$$

$$R_4 = \frac{A\rho_g}{W} \sum_{m=4}^{6} a_{4,m} \left(K_m \frac{w_{H_2,g}}{M_{H_2}} - \frac{w_{H_2O,g}}{M_{H_2O}} \right) \tag{5}$$

$$R_5 = \left(\frac{6a_s \cdot \varepsilon_s}{\varphi_s \cdot d_s} \right) \frac{\rho_g}{W} \sum_{m=4}^{6} a_{5,m} \left(K_m \frac{w_{H_2,g}}{M_{H_2}} - \frac{w_{H_2O,g}}{M_{H_2O}} \right) \tag{6}$$

$$R_6 = \frac{A\rho_g}{W} \sum_{m=4}^{6} a_{6,m} \left(K_m \frac{w_{H_2,g}}{M_{H_2}} - \frac{w_{H_2O,g}}{M_{H_2O}} \right) \tag{7}$$

$$R_7 = \frac{k_{7,1} P_{CO_2} w_C \rho_s \varepsilon_s}{1 + k_{7,2} P_{CO} + k_{7,3} P_{CO_2}} \tag{8}$$

$$R_8 = \frac{k_{8,4} P_{H_2O} w_C \rho_s \varepsilon_s}{1 + k_{8,2} P_{CO} + k_{8,3} P_{CO_2} + k_{8,5} P_{H_2O}} \tag{9}$$

$$R_9 = \frac{1}{1 + 2500 \exp\left(-\frac{12{,}400}{1.987 T_g}\right)} \left[\frac{\varepsilon_g \rho_g w_{O_2}}{M_{O_2}} \right] \left[\frac{d_s \varphi_s}{A D_{O_2,N_2}^{T_{ave}} Sh} + \frac{1}{k_9} \right]^{-1} \tag{10}$$

$$R_{10} = \varepsilon_s (k_{10,1} + K_{10} k_{10,2}) \left(P_{CO} P_{H_2O} - P_{CO_2} P_{H_2} / K_{10} \right) P_{H_2O} \tag{11}$$

$$R_{11} = \frac{\rho_g w_i}{M_i} \left(\frac{1}{A k_{f11}} + \frac{1}{k_{11,1}} \right)^{-1} \tag{12}$$

In the numerical solution of the arising differential equations, a structured grid was applied. The average mesh size is 40 mm. The sensitivity study of mesh size was carried out with average size of 100 mm, 80 mm, 60 mm, 40 mm, 30 mm, and 20 mm. The difference of top gas temperature between 60 mm and 40 mm is 4.9% while that between 40 mm and 30 mm is within 0.5%. This suggests that the mesh size of 40 mm is reasonable and confirms the mesh independence. The model contains 261,131 nodes and 248,496 hexahedral cells.

The assumptions for this model were as follows: (1) The powder phase was not considered; (2) Other chemical reactions such as flux decomposition and reduction of non-ferrous compounds were ignored; (3) The melting of solids was ignored.

The main operational parameters of the blast furnace in 2018 are listed in Table 2, where HM is the abbreviation of hot metal.

Table 2. Main operational parameters of the blast furnace.

Parameters	Value	Unit
Coke rate	455	kg·ton HM^{-1}
PCI rate	101	kg·ton HM^{-1}
Production	4456	ton HM·d^{-1}
Blast flowrate	4198	Nm^3·min^{-1}
Blast temperature	1122	°C
Oxygen enrichment ratio	0	%
Blast pressure	339	kPa
Blast humidity	5.00	g·m^{-3}
Burden feed amount	7508	ton·d^{-1}

The total Fe content in the mixed iron-bearing materials was 56.20%, and the total C contents in the coke and coal were 87.38% and 62.46%. Based on the material and energy balances, Fe in the mixed iron-bearing materials was converted to Fe_2O_3 and only C in the coke was considered. The coal injection and the oxygen of the blast at the tuyere were converted to CO. Therefore, the mass fraction of Fe_2O_3 and C were 77.05% and 22.95% at the top of blast furnace, respectively. The volume fraction of the solid phase was fixed, as shown in Figure 1c. The burden velocity distribution was firstly calculated with a simple model without taking into account heat and energy transfer. The results of burden velocity were then loaded to the present complex model taking into consideration all the transfers and reactions. The top gas pressure in the blast furnace was 194 kPa. The gas compositions at the tuyere were N_2-74.92 vol%, O_2-15.34 vol%, CO-9.15 vol%, H_2O-0.59 vol%. The conservation equations were solved numerically by the finite volume method with commercial software ANSYS FLUENT (release 17.0) [35]. The Eulerian multiphase module was used in this model. The first order upwind scheme was used for discretization of density, momentum, volume fraction, energy, gas and solid species, and so on. The Phase Coupled SIMPLE method was applied [21,30]. The simulation was considered to have converged when the residuals for each variable were less than 10^{-5}. The solution flow chart is shown in Figure 2.

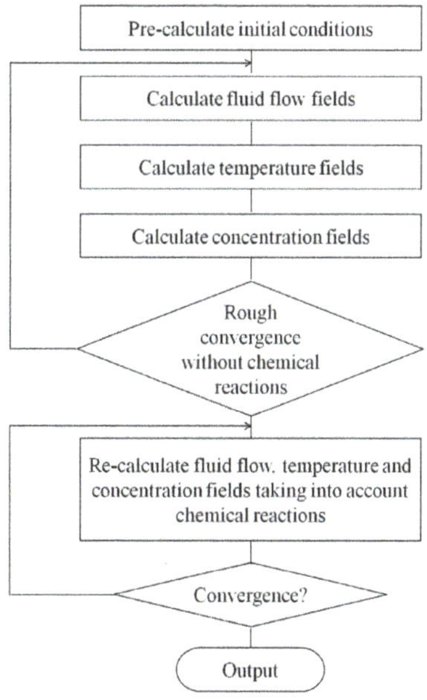

Figure 2. Solution flow chart of the simulation.

3. Results and Discussion

3.1. Base Model

The gas compositions at the top of the blast furnace calculated from this model were verified against some measured results of practical production as shown in Table 3, where H_2O is not listed since it was not measured in the practice. The maximum relative error between the measured and calculated results is 6.3%. Therefore, the present model is considered to be applicable to carry out the further simulation.

Table 3. Comparison between practical values and simulated values.

Parameters	Practical Value	Simulated Value	Relative Error
top gas composition in mole fraction			
CO	22.8%	24.0%	5.3%
CO_2	18.6%	18.9%	1.6%
H_2	2.45%	2.38%	2.9%
top gas temperature	229.6 °C	244.0 °C	6.3%
top gas pressure	194.7 kPa	200.6 kPa	3.0%

Figure 3 shows the mass fraction distribution of carbon in the blast furnace under base operational condition.

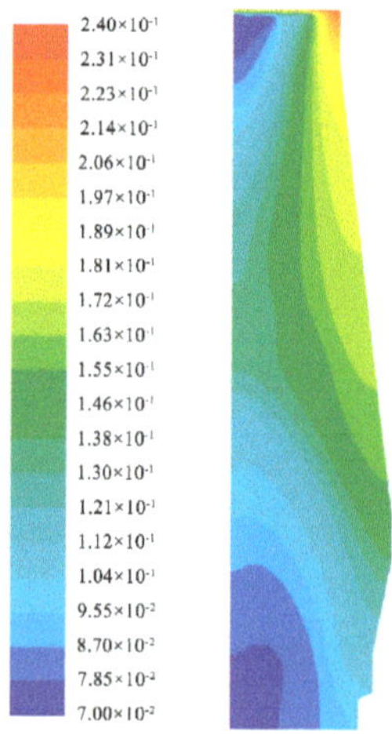

Figure 3. Coke mass fraction distribution in the blast furnace under base operational condition.

It can be seen from Figure 3 that the carbon mass fraction in the blast furnace firstly increases, then decreases after reaching a maximum in the middle. The reason for the increase is that the reduction reactions take place in the middle-upper part of blast furnace, and this leads to the decrease of the ferrite oxide mass fraction. The reason for the decrease of the carbon mass fraction is that carbon solution loss takes place in the lower part of the blast furnace.

The absolute consumption amount of carbon at a certain height of the blast furnace can be calculated based on the material balance. The carbon solution loss ratio of coke can be calculated by

$$y_C = \frac{m_{C-C}}{m_{C-T}} \times 100\%$$

(13)

where m_{C-C} and m_{C-T} are the amounts of carbon consumed by the carbon solution loss reaction and the total carbon input from the blast furnace top, kg.

The carbon solution loss reaction taking place in the middle-upper part of blast furnace further reduces the quality of the coke and results in the poor performance of the blast furnace. Therefore the mean temperature of the cohesive zone is adopted as the boundary of the carbon solution between the middle-upper part and the lower part of the blast furnace. The average temperature of the boundary is 1300 °C. The y_C at this boundary is 18.61%.

3.2. Oxygen Enrichment Ratio

The oxygen enrichment mode in the present work is that the flowrate of the hot blast is constant. Oxygen is added to the blast and forms a mixture of air and oxygen. The amount of coal changes with the oxygen enrichment ratio. Based on the practical data, the coke rate decreases by 5 kg·t^{-1} and produced hot metal increases by 3.3% when the oxygen enrichment ratio increases by 1%. As a result, the amounts of gas, coal, coke, and production can be determined, and the compositions of injected gas and top gas are then calculated based on the balances of Fe, C, H and O. Table 4 presents the typical parameters when the oxygen enrichment ratio varies from 1% to 5% at a step of 1%, where the production, coke rate and coal rate at 1–5% are calculated based on the above balance calculations. The temperature profiles at different oxygen enrichment ratios are shown in Figure 4. The carbon solution loss ratio is also calculated after simulation, as shown in Figure 5.

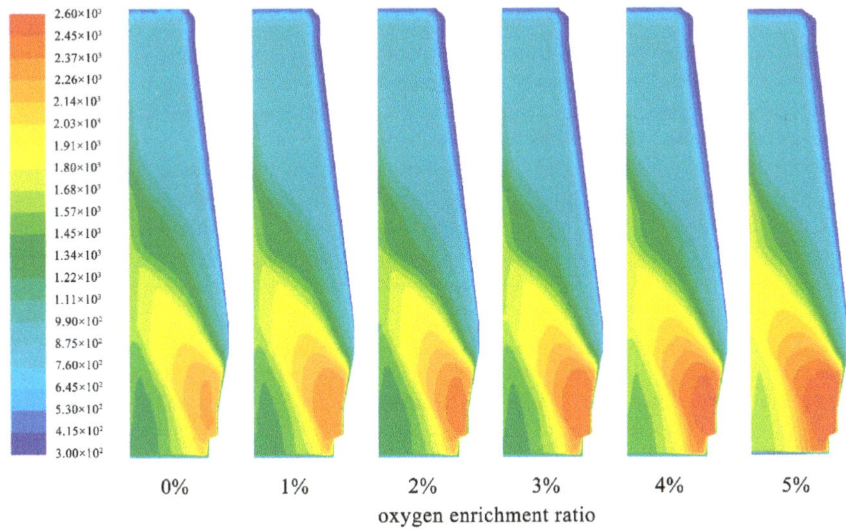

Figure 4. Temperature profiles at different oxygen enrichment ratios.

It can be seen from Figure 4 that the temperature increases in the lower zone of the blast furnace with the increase of oxygen enrichment ratio while decreases a little in the upper zone of the blast furnace. At 5% oxygen enrichment ratio, the facet average temperature at 4 m height (the average temperature of the horizontal plane at 4 m height) increases by 85 °C from the base model. This indicates that the cohesive zone narrows with the increase of oxygen enrichment ratio.

Table 4. Typical parameters when the oxygen enrichment ratio changes.

Oxygen Enrichment Ratio (%)	0 (Base)	1	2	3	4	5
Production (ton HM·day^{-1})	4431	4577	4723	4869	5015	5161
Coke rate (kg·ton HM^{-1})	462	457	452	447	442	437
Coal rate (kg·ton HM^{-1})	97	105	113	121	129	137

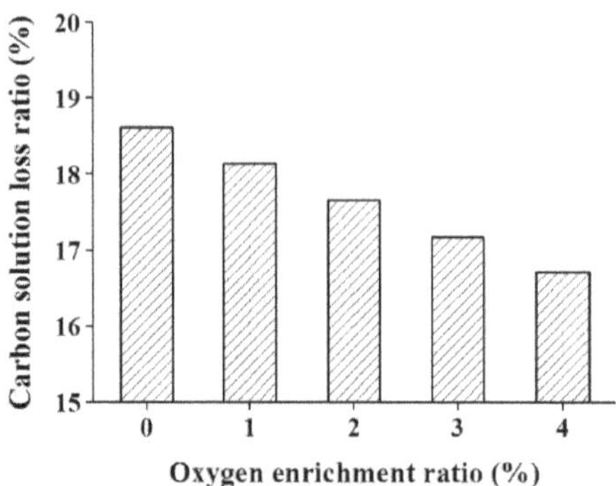

Figure 5. Carbon solution loss ratio at different oxygen enrichment ratio.

It can be seen from Figure 5 that the carbon solution loss ratio decreases gradually as the oxygen enrichment ratio increases. For each 1% increase of the oxygen enrichment ratio, the carbon solution loss ratio decreases by about 0.47%. The reason may be as follows. Firstly, productivity increases while the coke feed rate is kept constant after the oxygen enrichment, which reduces the proportion of coke in the solid. This results in less contact between the gas and the coke. Secondly, the increase of coal injected after the oxygen enrichment leads to an increase of CO content in the lower part of the blast furnace, which limits the carbon solution loss reaction. Thirdly, the burden descending velocity increases after the oxygen enrichment, which causes the residence time of burden to decrease. This makes the carbon loss solution less sufficient.

The carbon consumption decreases and blast furnace productivity increases with the increase of oxygen enrichment. Therefore a higher oxygen enrichment ratio is better for preserving carbon saving and increasing production. However, oxygen enrichment has a great impact on the temperature distribution of the blast furnace, and especially increases the theoretical combustion temperature around the raceway. In the present study, when the oxygen enrichment rate increases by 5%, the theoretical combustion temperature increases by 200 °C, which is still within the normal range of the blast furnace of Bayi Steel. The gas temperature in the top decreases by 44 °C, but is still higher than the 150 °C, which is acceptable for normal production. The actual average oxygen enrichment rate in Bayi Steel is about 3%, so it can be increased to 5% based on the above analysis.

3.3. Coke Oven Gas Injection

In order to decrease the CO_2 emissions of the blast furnace, coke oven gas was mixed with hot blast and injected from tuyeres. Table 5 shows the composition of the coke oven gas in Bayi Steel. The total amount of coke oven gas and blast gas was kept constant when injecting the coke oven gas. The proportion of coke oven gas in the gas mixture was from 3% to 15% at a step of 3%.

Table 5. Composition of coke oven gas.

Element	CH_4	H_2	CO_2	CO	O_2	N_2
Volume fraction(vol%)	26.0	58.0	2.0	8.0	0.5	5.5

The CH_4 and O_2 in coke oven gas were converted to CO and H_2 for simplicity. The replacement ratio of coke oven gas to coke was 0.4 kg·Nm^{-3}. Therefore, for each 3% increase of coke oven gas, the oxygen enrichment rate increased by 0.55%. The production increased by 246 ton HM and the coke ratio decreases by 16kg·ton HM^{-1}. The temperature profiles at different coke oven gas injection ratios are shown in Figure 6. The carbon solution loss ratio under different coke oven gas injection ratios was also calculated after simulation, as shown in Figure 7.

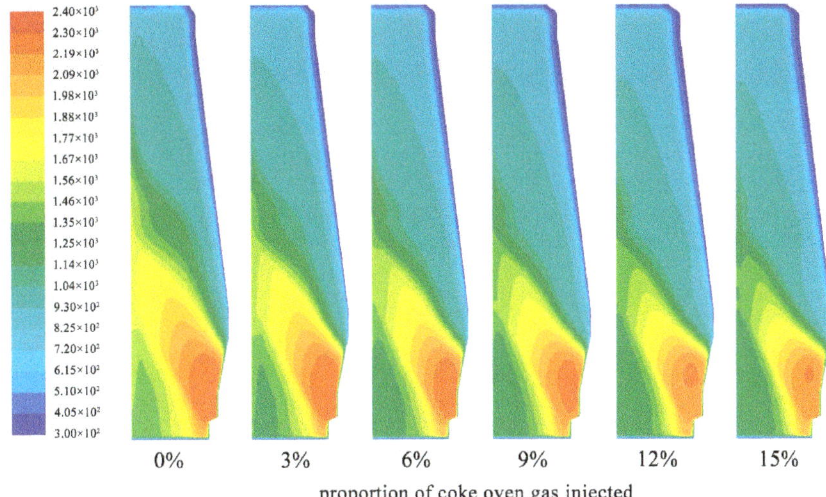

Figure 6. Temperature profiles at different proportions of coke oven gas injected.

It can be seen from Figure 6 that the temperature decreases with the increase of the proportion of coke oven gas injected, and the temperature in the middle zone increase relatively more. When the proportion of coke oven gas injected is 15%, the facet average temperature at 14 m height decreases by 210 °C than the base model. This suggests that the cohesive zone widens with the increase of proportion of coke oven gas injected.

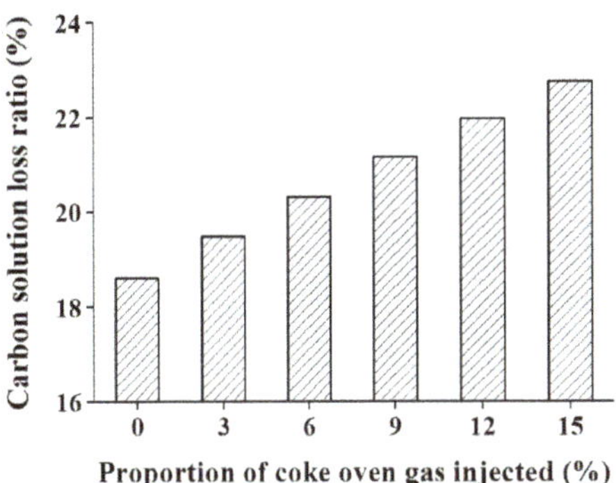

Figure 7. Carbon solution loss ratio at different amount of coke oven gas injected.

It can be seen from Figure 7 that the carbon solution loss ratio of the coke increases gradually as the coke oven gas injection increases. For each 3% increase in the coke oven gas injection ratio, the carbon solution loss ratio of coke increases by an average of 0.83%. The carbon solution loss ratio of the coke increases by 4.15% when the coke oven gas injection is 15%.

The reason may be that the replacement of hot blast with coke oven gas leads to an increase of H_2 in the gas mixture due to the high H_2 content of the coke oven gas. This results in a higher H_2O content in the blast furnace due to the reduction reaction of the H_2 and iron ore. The higher H_2O content may be part in a water gas reaction, which increases the reaction of the carbon solution loss. The temperature of the strong reaction between the coke and the H_2O is from 800 to 1300 °C. The average H_2O content in the gas in the blast furnace increases from 2.45% to 8.67% when the coke oven gas ratio increases from 0% to 15%. Moreover, the high temperature zone shrinks after the injection of the coke oven gas, which leads to the expansion of the zone for the reaction of the carbon solution loss.

Therefore, on the one hand, coke oven gas can reduce the coke rate and CO_2 emission of the blast furnace as a hydrogen-rich gas. On the other hand, the injection of coke oven gas causes the temperature in the lower part of the blast furnace to decrease and the carbon solution loss to increase, which impairs the operation of blast furnace. Therefore, coke oven gas should not be injected into the blast furnace in Bayi Steel in its present condition. Coke oven gas should only be injected only if the carbon solution loss is low enough, and it should be done a higher gas temperature and with better coke quality of a high M40 and a high CSR (coke strength after reaction).

3.4. Steel Scrap Charging

Since the steel market has been strong in China recently, steel scraps are charged so as to increase production. The metallic iron (MFe) of steel scrap is 52.29% and the FeO is 9.07%. The steel scrap is added when the total amount of iron-bearing materials is kept constant. The proportion of steel scrap is from 0% to 10% at a step of 2%, as shown in Table 6, where the absolute amount is also illustrated. For each 2% increase of steel scrap, the oxygen enrichment ratio increases 0.06%, the coke rate decreases 4 kg·ton HM^{-1}, and production increases 70 ton HM while the coal rate and the amount of hot blast remain unchanged. The temperature profiles at different proportions of steel scrap added are shown in Figure 8. The carbon solution loss ratio is calculated after simulation, as shown in Figure 9.

Table 6. Amount of steel scrap added.

Proportion of Steel Scrap Added (%)	2	4	6	8	10
Absolute value(ton·day^{-1})	149.7	299.3	448.9	598.6	748.3

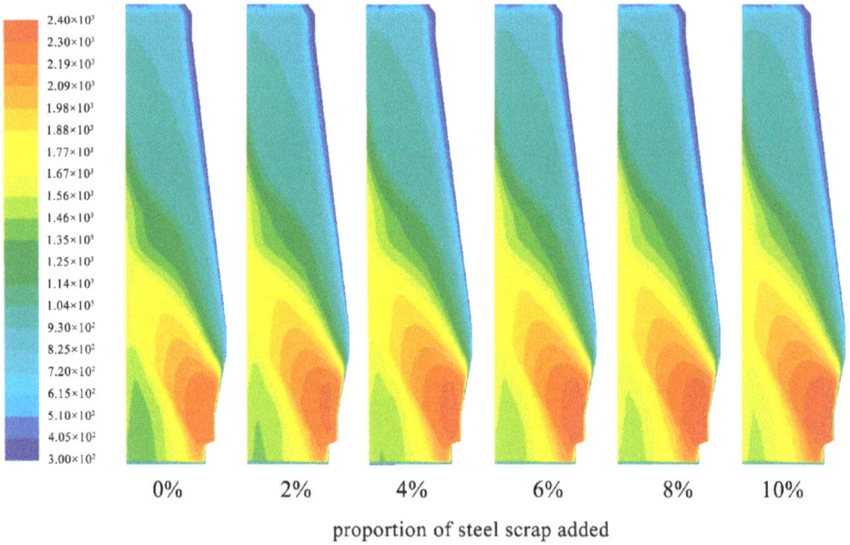

Figure 8. Temperature profiles at different proportions of steel scrap added.

It can be seen from Figure 8 that the temperature increases a little in the lower zone of the blast furnace with the increase of the proportion of steel scrap added, but the degree is very small. When the proportion of steel scrap added is 10%, the facet average temperature at 4 m height increases by 22 °C than the base model. Hardly any difference of temperature in the upper zone of the blast furnace can be found. This indicates that the cohesive zone only narrows a little with the increase of the proportion of steel scrap added.

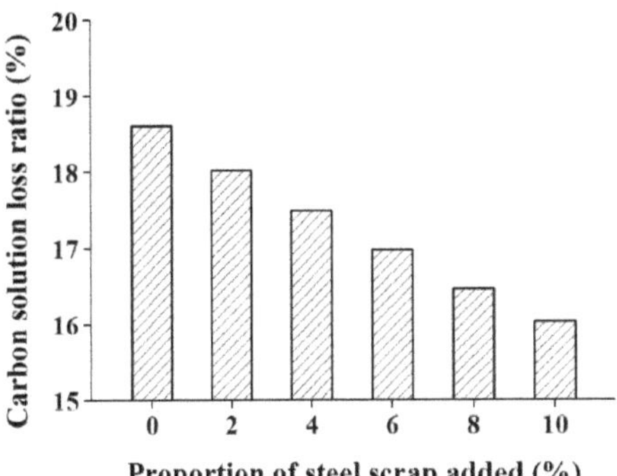

Figure 9. Carbon solution loss ratio at different amounts of steel scrap added.

The carbon solution loss ratio decreases gradually as the proportion of steel scraps added increases; when 10% steel scrap is added, it decreases by 2.6%. The reason may be that the amount of iron ore that needs to be reduced decreases when steel scraps are added in the top of blast furnace since steel scrap provides iron content that does not need reduction. However, this leads the utilization ratio of the gas to decrease. The ratio decreases by 1.32% when the proportion of steel scraps increases to 10%. As a result, the CO content in the gas increases while the CO_2 content decreases. This also leads to a decrease of the carbon solution loss ratio.

It should be noticed that the addition of steel scrap reduces the gas utilization ratio, and also the temperature in the upper part of the blast furnace. But the reduction in the gas utilization ratio and the temperature of the upper part of the blast furnace is small and within acceptable limits. Therefore, the addition of steel scrap is beneficial for the operation of blast furnace in the aspect of carbon solution loss. Steel scrap can be added to the blast furnace in appropriate amounts.

4. Conclusions

A three dimensional model was established to analyze the carbon solution loss of coke based on the practical operational parameters of blast furnace B in Bayi Steel. The model was then used to investigate the effects of the oxygen enrichment ratio, the injection of coke oven gas, and the addition of steel scraps on the carbon solution loss ratio of the coke. The conclusions are as follows.

(1) The carbon solution loss ratio of the blast furnace studied is 18.61% when the coke reaches 1300 °C.

(2) The carbon solution loss ratio decreases with the increase of the oxygen enrichment ratio or when a certain proportion of steel scraps is added, while it increases with an increase in the amount of coke oven gas injected.

(3) The oxygen enrichment ratio and the proportion of steel scraps added can be increased to 5% and 10%, respectively in order to reduce the carbon solution loss without affecting the operation of the blast furnace.

(4) The injection of coke oven gas is not recommended given the current condition of blast furnace B of Bayi Steel.

Author Contributions: Conceptualization, M.K. and S.W.; methodology, H.Z. and Z.H.; Software, M.K and H.Z.; validation, S.Y. and H.X.; formal analysis, L.P.W. and H.X.; investigation, M.K. and H.Z.; resources, H.X. and S.W.; data curation, Z.H. and S.Y.; writing, M.K and Z.H.; supervision, M.K and L.P.W.; funding acquisition, M.K., L.P.W. and H.X.

Funding: The authors would like to acknowledge the financial support by National Key R&D Program of China (grant number 2017YFB0603800, 2017YFB0603803), the National Natural Science Foundation of China (grant number 51804027) and USTB-NTUT Joint Research Program (grant number TW201909, NTUT-USTB-108-06).

Acknowledgments: The authors would like to appreciate much for the anonymous reviewers and editors for the improvement of this work, and Prof. Mark Buck for correcting language.

Conflicts of Interest: The authors declare no conflict of interest.

Nomenclature

A	specific surface area/m^{-1}
D	binary diffusivity for specie i and j/m·s^{-2}
$C_{P,p}$	specific heat capacity of phase p/J·kg^{-1}·K^{-1}
d_s	solid particle diameter/m
E_{gs}	volumetric heat flux/J·m^{-3}
$\vec{F}_{gs}$	gas-solid drag force/N
$\vec{g}$	gravitational acceleration/m·s^{-2}
H_n	specific enthalpy of reaction n/J·kg^{-1}
H_p	specific enthalpy of phase p/J·kg^{-1}
$\overline{\overline{I}}$	identity tensor/-
k	kinetic constant for reaction/-
k_f	film mass transfer resistance/m·s^{-1}
k_p	thermal conductivity of p phase/W·m^{-1}·K^{-1}
K_p	equilibrium constant of reaction n/-
M_i	molecular weight of specie i/kg·kmol^{-1}
$m_{C\text{-}C}$	amount of carbon consumed by carbon solution loss reaction/kg
$m_{C\text{-}T}$	total carbon input from the blast furnace top/kg
P	pressure/Pa
Pr	Prandtl number/-
Re	Reynolds number/-
R_n	rate of reduction reaction n/kmol·m^{-3}·s^{-1}
Sh	Sherwood number/-
S_ϕ	source term for variable ϕ in Equation (1)
T_p	temperature of phase p/K
$\vec{v}_p$	physical velocity of phase p/m·s^{-1}
y_c	carbon solution loss ratio of coke/%
Greek Symbols	
ε_p	volume fraction of phase p/-
ρ_p	density of phase p/kg·m^{-3}
ϕ	general dependent variable in Equation (1)
Γ_ϕ	diffusion coefficient for variable ϕ in Equation (1)
$\overline{\overline{\mu}}_p$	viscosity of phase p/kg·m^{-1}·s^{-1}
$\overline{\overline{\tau}}_p$	stress tensor of phase p/Pa
ω	mass fraction/-
Subscripts	
g	gas
s	solid
p	phase

References

1. Wang, X. *Ferrous Metallurgy (Ironmaking Part)*, 3rd ed.; Metallurgical Industry Press: Beijing, China, 2013.
2. Nogami, H.; Chu, M.; Yagi, J.-I. Multi-dimensional transient mathematical simulator of blast furnace process based on multi-fluid and kinetic theories. *Comput. Chem. Eng.* **2005**, *29*, 2438–2448. [CrossRef]

3.   Yeh, C.-P.; Du, S.-W.; Tsai, C.-H.; Yang, R.-J. Numerical analysis of flow and combustion behavior in tuyere and raceway of blast furnace fueled with pulverized coal and recycled top gas. *Energy* **2012**, *42*, 233–240. [CrossRef]

4.   Zhou, P.; Li, H.-L.; Shi, P.-Y.; Zhou, C.Q. Simulation of the transfer process in the blast furnace shaft with layered burden. *Appl. Therm. Eng.* **2016**, *95*, 296–302. [CrossRef]

5.   Hilding, T.; Gupta, S.; Sahajwalla, V.; Björkman, B.; Wikström, J.-O. Degradation Behaviour of a High CSR Coke in an Experimental Blast Furnace: Effect of Carbon Structure and Alkali Reactions. *ISIJ Int.* **2005**, *45*, 1041–1050. [CrossRef]

6.   Babich, A.; Senk, D.; Gudenau, H.W. Effect of coke reactivity and nut coke on blast furnace operation. *Ironmak. Steelmak.* **2009**, *36*, 222–229. [CrossRef]

7.   Lundgren, M.; Ökvist, L.S.; Björkman, B. Coke Reactivity under Blast Furnace Conditions and in the CSR/CRI Test. *Steel Res. Int.* **2009**, *80*, 396–401. [CrossRef]

8.   Gornostayev, S.S.; Härkki, J.J. Mechanism of Physical Transformations of Mineral Matter in the Blast Furnace Coke with Reference to Its Reactivity and Strength. *Energy Fuels* **2006**, *20*, 2632–2635. [CrossRef]

9.   Loison, R.; Foch, P.; Boyer, A. *Coke: Quality and Production*, 4th ed.; Great Britain at the University Press: Cambridge, UK, 2014.

10.   Sato, H.; Patrick, J.W.; Walker, A. Effect of coal properties and porous structure on tensile strength of metallurgical coke. *Fuel* **1998**, *77*, 1203–1208. [CrossRef]

11.   Mu, J. *Alkali Metals in Blast Furnace Metallurgy*; Metallurgical Industry Press: Hong Kong, China, 1992.

12.   Cai, H. Investigation on Degradation Function of Alkalis on Coke of Blast Furnace. Master's Thesis, University of Science and Technology Beijing, Beijing, China, 2012.

13.   Feng, B.; Bhatia, S.K.; Barry, J.C. Structural ordering of coal char during heat treatment and its impact on reactivity. *Carbon* **2002**, *40*, 481–496. [CrossRef]

14.   Department of Materials Science. *McMaster Cokemaking Course*, 2nd ed.; Department of Materials Science and Engineering, McMaster University: Hamilton, ON, Canada, 1999.

15.   Hou, Q.; Zhou, Z.; Yu, A. Gas-solid flow and heat transfer in fluidized beds with tubes: Effects of material properties and tube array settings. *Powder Technol.* **2016**, *296*, 59–71. [CrossRef]

16.   Sun, J.; Wu, S.; Kou, M.; Shen, W.; Du, K. Influence of Operation Parameters on Dome Temperature of COREX Melter Gasifier. *ISIJ Int.* **2014**, *54*, 43–48. [CrossRef]

17.   Wu, S.; Du, K.; Xu, J.; Shen, W.; Kou, M.; Zhang, Z. Numerical Analysis on Effect of Areal Gas Distribution Pipe on Characteristics Inside COREX Shaft Furnace. *JOM* **2014**, *66*, 1265–1276. [CrossRef]

18.   Xu, J.; Wu, S.; Kou, M.; Du, K. Numerical Analysis of the Characteristics Inside Pre-reduction Shaft Furnace and Its Operation Parameters Optimization by Using a Three-Dimensional Full Scale Mathematical Model. *ISIJ Int.* **2013**, *53*, 576–582. [CrossRef]

19.   Yagi, J.-I. Mathematical Modeling of the Flow of Four Fluids in a Packed Bed. *ISIJ Int.* **1993**, *33*, 619–639. [CrossRef]

20.   Anderson, J.D. *Computational Fluid Dynamic*; McGraw-Hill: New York, NY, USA, 1995.

21.   Ergun, S. Fluid flow through packed columns. *Chem. Eng. Prog.* **1952**, *48*, 89–94.

22.   Ranz, W.E.; Marshall, W.R. Evaporation from drops. *Chem. Eng. Prog.* **1952**, *48*, 141–146.

23.   Chen, J.; Akiyama, T.; Nogami, H.; Yagi, J.-I.; Takahashi, H. Modeling of Solid Flow in Moving Beds. *ISIJ Int.* **1993**, *33*, 664–671. [CrossRef]

24.   Austin, P.R.; Nogami, H.; Yagi, J.-I. A Mathematical Model of Four Phase Motion and Heat Transfer in the Blast Furnace. *ISIJ Int.* **1997**, *37*, 458–467. [CrossRef]

25.   Kou, M.; Wu, S.; Du, K.; Shen, W.; Ma, X.; Chen, M.; Zhao, B. The effect of operational parameters on the characteristics of gas–solid flow inside the COREX shaft furnace. *JOM* **2015**, *67*, 459–466. [CrossRef]

26.   Perry, R.H.; Green, D.W.; Maloney, J.O. *Perry's Chemical Engineers' Handbook*; McGraw-Hill: New York, NY, USA, 1997.

27.   Hara, Y.; Tsuchiya, M.; Kondo, S.-I. Intraparticle Temperature of Iron-Oxide Pellet during the Reduction. *Tetsu-to-Hagane* **1974**, *60*, 1261–1270. [CrossRef]

28.   Negri, E.D.; Alfano, O.M.; Chiovetta, M.G. Moving-bed reactor model for the direct reduction of hematite. Parametric study. *Ind. Eng. Chem. Res.* **1995**, *34*, 4266–4276. [CrossRef]

29.   Takahashi, R.; Takahashi, Y.; Yagi, J.-I.; Omori, Y. Operation and simulation of pressurized shaft furnace for direct reduction. *Trans. Iron Steel Inst. Jpn.* **1986**, *26*, 765–774. [CrossRef]

30. Chu, M.; Yagi, J.; Shen, F. *Modelling on Blast Furnace Process and Innovative Ironmaking Technologies*; Northeastern University Press: Shenyang, China, 2006.

31. Kuwabara, M.; Hsieh, Y.-S.; Muchi, I. A Kinetic Model of Coke Combustion in the Tuyere Zone of Blast Furnace. *Tetsu-to-Hagane* **1980**, *66*, 1918–1927. [CrossRef]

32. Sutherland, W. LII. The viscosity of gases and molecular force. *Lond. Edinb. Dublin Philos. Mag. J. Sci.* **1893**, *36*, 507–531. [CrossRef]

33. Eckert ER, G.; Drake, R.M. *Analysis of Heat and Mass Transfer*; McGraw-Hill: Tokyo, Japan, 1972.

34. Wu, S.; Xu, J.; Yang, S.; Zhou, Q.; Zhang, L. Basic Characteristics of the Shaft Furnace of COREX® Smelting Reduction Process Based on Iron Oxides Reduction Simulation. *ISIJ Int.* **2010**, *50*, 1032–1039. [CrossRef]

35. ANSYS, Inc. *ANSYS FLUENT Tutorial Guide (Release 17.0)*; ANSYS, Inc.: Cannonsburg, PA, USA, 2016.

Article

Fault Detection and Identification of Blast Furnace Ironmaking Process Using the Gated Recurrent Unit Network

Hang Ouyang [1], Jiusun Zeng [1,*] and Yifan Li [1] and Shihua Luo [2]

[1]   College of Metrology and Measurement Engineering, China Jiliang University, Hangzhou 310018, China; sutouyangh@sina.com (H.O.); liyifan@cjlu.edu.cn (Y.L.)

[2]   School of Statistics, Jiangxi University of Finance and Economics, Nanchang 330013, China; luoshihua@aliyun.com

*   Correspondence: jszeng@cjlu.edu.cn; Tel.: +86-139-6800-6265

Received: 22 February 2020; Accepted: 23 March 2020; Published: 27 March 2020

Abstract: It is of critical importance to keep a steady operation in the blast furnace to facilitate the production of high quality hot metal. In order to monitor the state of blast furnace, this article proposes a fault detection and identification method based on the multidimensional Gated Recurrent Unit (GRU) network, which is a kind of recurrent neural network and is highly effective in handling process dynamics. Comparing to conventional recurrent neural networks, GRU has a simpler structure and involves fewer parameters. In fault detection, a moving window approach is applied and a GRU model is constructed for each process variable to generate a series of residuals, which is further monitored using the support vector data description (SVDD) method. Once a fault is detected, fault identification is performed using the contribution analysis. Application to a real blast furnace fault shows that the proposed method is effective.

Keywords: gated recurrent unit; support vector data description; time sequence prediction; fault detection and identification

---

1. Introduction

Maintaining the blast furnace system at a stable status is critical to ensure efficient production of high-quality blast furnace hot metal [1]. Therefore, condition monitoring of the blast furnace ironmaking process becomes a significant issue. During the operation of blast furnace ironmaking process, different kinds of faults may happen, such as hanging, low stockline and abnormal gas flow. If the faults cannot be detected and identified in time and accurately, it may lead to loss in production rate or even a significant accident.

The problem of fault detection and diagnosis for blast furnace ironmaking process is a long lasting and well research topic. Traditional methods like expert knowledge and fuzzy logic have been well developed in different kinds of expert systems [2]. However, constructing and maintaining an up-to-date knowledge base is difficult. Alternatively, classification-based algorithms like support vector machine have been applied to diagnose faults in blast furnaces [3]. Liu et al. [4] proposed a novel strategy based on cost-conscious least squares support vector machine (LS-SVM) to achieve rapid diagnosis of blast furnace faults. An et al. [5] proposed a support vector machine for multiple classification to diagnose blast furnace faults. The main assumption of classification-based methods is that sufficient faulty samples can be collected, which is often not true in a real blast furnace. More recently, multivariate statistical methods became popular in the monitoring of blast furnaces. For example, Vanhatalo applied the principal component analysis (PCA) to monitor the status of an experimental blast furnace [6]. A two-stage PCA is considered to deal with multi-modal distribution in blast furnace data [7]. Shang et al. [8] developed a recursive transformed component statistical analysis (RTCSA)-based algorithms to monitor incipiently happened faults in the iron-making process. In addition, other kinds of PCA-based approaches have been introduced to monitor process faults, such as robust PCA [9] and convex hull-based PCA [10].

In order to deal with process dynamics, Zeng et al. applied a state space model to extract residuals from the process data and used the support vector data description (SVDD) to detect blast furnace faults [11]. Also, Vanhatalo and Kulahci [12] considered the impact of autocorrelation to statistical methods like PCA. Dynamic principal component analysis (DPCA) [13,14] and dynamic linear discriminant analysis (DLDA) [15] are also used to handle dynamic processes. From the above analysis, it can be seen that how to handle process dynamics has become an important task in fault detection and diagnosis of blast furnace.

In this paper, a new process monitoring method based on the GRU network [16] is considered to detect and identify process faults in blast furnace. The GRU network is a new type of recurrent neural network (RNN). Comparing to conventional RNN methods like long-short term memory network, it has comparable capability to handle process dynamics, however with a simpler structure and fewer parameters. In fault detection, a GRU neural network is used to make prediction for each process variable, so that the process dynamics can be filtered and a series of residuals can be generated. The generated residuals are then monitored using the support vector data description (SVDD) method [11]. Faulty variables are then identified by inspecting the deviation of the residuals from normal operation condition (NOC). The benefits of the proposed method can be summarized as: (i) the introduction of GRU network can fully capture the dynamic characteristics of the blast furnace data; (ii) faulty variables can be identified by investigating the residual of each variable, which greatly simplifies subsequent fault diagnosis task.

2. Methodologies

This section describes the methodologies applied in fault detection and identification of blast furnace system. Section 2.1 briefly introduces the GRU network, which is an extension of the LSTM network. Section 2.2 describes the SVDD classifer.

2.1. GRU Neural Network

GRU is a type of recurrent neural network. The main difference between RNN and feed-forward artificial neural network is in their structure. In a feed-forward artificial neural network, signals travel from the inputs to outputs and the flow of information is in the forward direction only. Since there is no backward/feedback flow, the name of "feed-forward" is justified. In contrast, an RNN allows feedback from output to input and hence it is called "recurrent". In addition, the output of the previous time step/state in RNN will be used as the input of the next time step, which is different from feed-forward neural network that considers fixed length input and fixed length output only. With this kind of recurrent structure, RNN can be used to learn the characteristics of the time series and make predictions. A widely used RNN is the LSTM network, which is very suitable to capture long-term dependencies and also able to avoid the vanishing gradient problem. As an improvement of LSTM, GRU network inherits its advantages, whilst having an optimized structure and fewer parameters, resulting in lower computation load and better generalization ability.

2.1.1. The Structure of LSTM Cell

LSTM [17] was originally proposed in 1997, in order to solve the vanishing gradient problem faced by RNN [18]. The main difference between LSTM and standard RNN network is the handling of long-term dependencies. In the RNN network, each cycle involves only the last state and the current input. Because each prediction only involves the state at the last moment, the RNN can only establish a dependency relationship between states in a short time. In contrast, LSTM can establish dependencies between states at arbitrary long intervals, so they are called "Long-Short Term

Memory network". In addition, the LSTM has a cell state update process similar to the conveyor belt structure. The old cell state will remain on the conveyor belt until it needs to be forgotten by structure called "gate". Through this conveyor belt structure, LSTM can take long-term memory from the conveyor belt at any time for learning the characteristics of time series and make predictions. An LSTM unit consists of a cell, an input gate, a forget gate and an output gate. The cell is used to record state values at different time intervals and the three gates are used to control the flow of the information. The introduction of three gates enables LSTM to keep, utilize, or discard a state when necessary.

Let x_t denote a data sample at the tth time instance, C_{t-1} denote the cell value and h_{t-1} the hidden state of each cell at the $t-1$th time instance. The information of previous time is stored in C_{t-1} and h_{t-1}. The input gate regulates to what extent a new value x_t is transferred into the cell, the forget gate controls to what extent C_{t-1} remains in the cell and the output gate regulates to what extent C_{t-1} is used to calculate the output activation. The structure of a standard LSTM cell is shown in Figure 1.

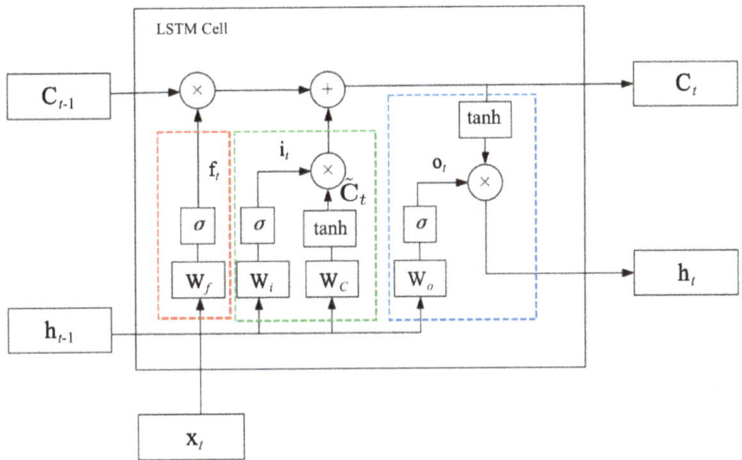

Figure 1. Structure diagram of LSTM cell.

In Figure 1, the green box, blue box and red box correspond to the input gate, the output gate and the forget gate respectively. The mathematic formulation of the forget gate is described as:

$$f_t = \sigma\left(W_f \cdot [h_{t-1}, x_t] + b_f\right) \tag{1}$$

where W_f is the weight matrix of the forget gate; σ is the sigmoid activation function; b_f is the bias vector for the forget gate; $[h_{t-1}, x_t]$ is a vector that merge the previous cell state vector h_{t-1} and the input vector x_t at the current moment. The input gate is used to decide what information will be saved in the cell value. On the other hand, the input gate can be described mathematically as follows.

$$i_t = \sigma\left(W_i \cdot [h_{t-1}, x_t] + b_i\right) \tag{2}$$

$$\tilde{C}_t = \tanh\left(W_c \cdot [h_{t-1}, x_t] + b_i\right) \tag{3}$$

Here, W_i is the weight matrix in the input gate, b_i is the bias vector. The input gate adds new information generated by the current input to the cell value, and creates new memories: i_t and $\tilde{C}_t$. The current state C_t is updated based on the previous cell value C_{t-1}, the new memories i_t and $\tilde{C}_t$ as follows.

$$C_t = f_t \cdot C_{t-1} + i_t \cdot \tilde{C}_t \tag{4}$$

Finally, the hidden state h_t is updated in the output gate as:

$$\mathbf{o}_t = \sigma\left(\mathbf{W}_o \cdot [\mathbf{h}_{t-1}, \mathbf{x}_t] + \mathbf{b}_o\right) \tag{5}$$

$$\mathbf{h}_t = \mathbf{o}_t \cdot \tanh\left(\mathbf{C}_t\right) \tag{6}$$

where $\mathbf{W}_o$ is the weight matrix in the output gate. In this way, the cell value $\mathbf{C}_t$ and hidden state $\mathbf{h}_t$ can be updated whenever a new sample $\mathbf{x}_t$ is available.

2.1.2. The Structure of GRU Cell

The GRU is a refined version of LSTM with a simpler structure [19]. The main difference between GRU and LSTM is in the process of forgetting and updating cell values. In the LSTM network, update of cell values are controlled by two gates, the forget gate and the input gate. Since two gate structures are required, the structure of LSTM is relatively complex. Compared to LSTM, GRU controls both the forgetting coefficient and the update coefficient for the output with one single update gate, so it involves fewer matrix multiplication calculations. Through this simplification, the GRU can retain the functions of the LSTM and reduce network training time. More specifically, it consists of an update gate and a reset gate, which reduces the number of parameters to only one fourth of the LSTM. The reset gate determines how much previous memory is retained and the update gate determines how much new information needs to be combined with the previous memory. The structure of GRU cell is shown in Figure 2.

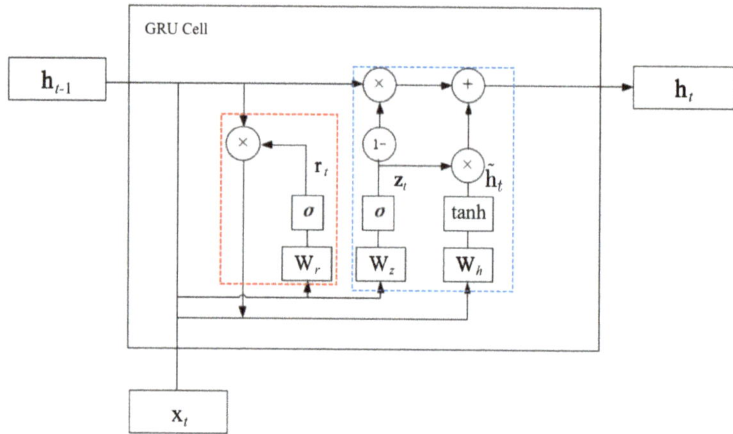

Figure 2. Structure diagram of Gated Recurrent Unit (GRU) cell.

In contrast to LSTM, GRU has only 2 gate functions. The update gate is shown in the blue box in Figure 2 and the reset gate is shown in the red box. The forward transfer formulations of GRU can be calculated as follows.

$$\mathbf{r}_t = \sigma\left(\mathbf{W}_r \cdot [\mathbf{h}_{t-1}, \mathbf{x}_t]\right) \tag{7}$$

$$\mathbf{z}_t = \sigma\left(\mathbf{W}_z \cdot [\mathbf{h}_{t-1}, \mathbf{x}_t]\right) \tag{8}$$

$$\tilde{\mathbf{h}}_t = \tanh\left(\mathbf{W}_{\tilde{h}} \cdot [\mathbf{r}_t \cdot \mathbf{h}_{t-1}, \mathbf{x}_t]\right) \tag{9}$$

$$\mathbf{h}_t = (1 - \mathbf{z}_t) \cdot \mathbf{h}_{t-1} + \mathbf{z}_t \cdot \tilde{\mathbf{h}}_t \tag{10}$$

where $\mathbf{r}_t$ is the reset gate determining how much information in the previous state cell should be forgotten; $\mathbf{z}_t$ is the update gate determining how much information should be brought to the next cell; $\tilde{\mathbf{h}}_t$ is the intermediate state; $\mathbf{h}_t$ is hidden state. For the update gate, a greater value of $\mathbf{z}_t$ means that more new information is brought to the next cell. For the reset gate, a greater value means that more information from the former cell may be ignored [16].

2.2. Support Vector Data Description

SVDD is a kernel method, which maps the data samples into the high-dimensional feature space through a non-linear mapping. In the high-dimensional feature space, a compact hypersphere with the minimum radium while covering the maximum number of data samples is obtained by solution of an optimization problem. The SVDD is generally used in anomaly detection. If a new sample is mapped inside the hypersphere, it is regarded as a normal sample, otherwise it is faulty.

Given a data set $\left\{ \mathbf{x}_i \in \mathcal{R}^d, i = 1, \cdots, N \right\}$ and assume $\mathbf{a} \in \mathcal{R}^d$ the center of the hypersphere, R is the radius of the hypersphere, the following objective function can be obtained for SVDD.

$$\begin{cases} F\left(R, \mathbf{a}, \xi_i\right) = R^2 + C \sum_{i=1}^{N} \xi_i \\ \|\mathbf{x}_i - \mathbf{a}\|^2 \leq R^2 + \xi_i \end{cases} \tag{11}$$

Here, ξ_i is the relaxation factor, and C is the penalty parameter. In Equation (11), ξ_i satisfies $\xi_i \geq 0, \forall i$. The above optimization problem can be transformed as follows using the Lagrangian multipliers.

$$L\left(R, \mathbf{a}, \alpha, \gamma_i, \xi_i\right) = R^2 + C \sum_{i=1}^{N} \xi_i - \sum_{i=1}^{N} \gamma_i \xi_i - \sum_{i=1}^{N} \alpha_i \left[R^2 + \xi_i - \left(\|\mathbf{x}_i\|^2 - 2\mathbf{a}\mathbf{x}_i + \|\mathbf{a}\|^2 \right) \right] \tag{12}$$

where γ_i and α_i is the Lagrange multiplier and they satisfy $\alpha_i \geq 0, \gamma_i \geq 0$. Differentiate Equation (12) with respect to $R, \mathbf{a}$ and ξ_i and make it equal to 0, the following holds:

$$\begin{cases} \frac{\partial L}{\partial R} = 0 \\ \frac{\partial L}{\partial \mathbf{a}} = 0 \\ \frac{\partial L}{\partial \xi_i} = 0 \end{cases} \implies \begin{cases} \sum_{i=1}^{N} \alpha_i = 1 \\ \mathbf{a} = \sum_{i=1}^{N} \alpha_i \mathbf{x}_i \\ C - \alpha_i - \gamma_i = 0 \end{cases} \tag{13}$$

Combining Equation (13) to Equation (12) one can obtain:

$$L = \sum_{i=1}^{N} \alpha_i \left(\mathbf{x}_i \cdot \mathbf{x}_i \right) - \sum_{i=1}^{N} \sum_{j=1}^{N} \alpha_i \alpha_j \left(\mathbf{x}_i \cdot \mathbf{x}_j \right) \tag{14}$$

where α_i is the support vector and $0 \leq \alpha_i \leq C$. Generally, kernel function K is used to calculate whether the distance between the new sample $\mathbf{y} \in \mathcal{R}^d$ and the center of the hypersphere is less than the radius R^2:

$$D^2\left(\mathbf{y}\right) = K\left(\mathbf{y} \cdot \mathbf{y}\right) - 2 \sum_{i=1}^{N} \alpha_i K\left(\mathbf{y} \cdot \mathbf{x}_i\right) + \sum_{i,j=1}^{N} \alpha_i \alpha_j K\left(\mathbf{x}_i \cdot \mathbf{x}_j\right) \leq R^2 \tag{15}$$

The kernel term $K\left(\mathbf{x}_i \cdot \mathbf{x}_j\right)$ is commonly used to replace the inner product $\left(\mathbf{x}_i \cdot \mathbf{x}_j\right)$, which is the Gaussian kernel here:

$$K\left(\mathbf{x}_i \cdot \mathbf{x}_j\right) = \exp\left(-\frac{\|\mathbf{x}_i - \mathbf{x}_j\|^2}{\sigma^2} \right) \tag{16}$$

3. Fault Detection and Identification Strategy

In order to detect and identify a process fault, it is essential to characterize the normal operating condition (NOC). Hence, a training dataset collected under normal operational condition is used to construct the GRU neural network. The GRU neural network generates model residuals, which is further used to construct monitoring statistics using SVDD. As described earlier, the GRU model is capable of extracting the spatial and temporal signatures in the data that are important for characterizing complex ironmaking process. The general framework for fault detection and identification based on GRU-SVDD is described in detail in the following subsections.

3.1. Fault Detection

In order to detect a process fault, it is required to train a model based on the NOC data. In the ironmaking process, this involves training a GRU with multiple time series to model temporal dynamics and correlations between process variables.

The GRU model is trained on historical normal data. Specifically, the GRU model uses the past information captured by its cell value and current observation to predict the next observation. Assume a training set $\left\{ \mathbf{x}_i \in \mathcal{R}^d, i = 1, \cdots, N \right\}$ is collected under NOC, a moving window approach can be applied, with the window length being $n, n \ll N$. Take the first window as an example, the structure of a two-layer GRU is shown in Figure 3.

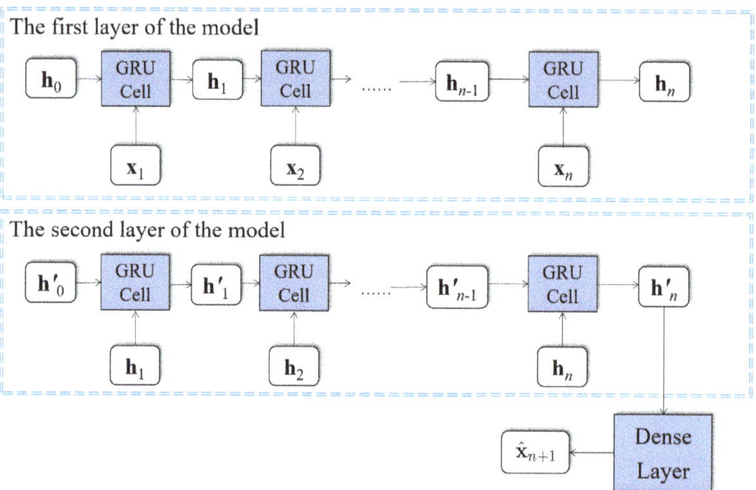

Figure 3. Structure of two layers GRU.

Here, $\mathbf{h}_i \in \mathcal{R}^{d_h}$ denotes the hidden state of the first layer at the ith time, $\mathbf{h}'_i \in \mathcal{R}^{d'_h}$ denotes the hidden state of the second layer and $\hat{\mathbf{x}}_{n+1} \in \mathcal{R}^d$ is the predicted value. The hidden state $\mathbf{h}_i$ of the first layer becomes the input to the second layer of GRU model. The final output is then obtained using the dense layer as follows.

$$\hat{\mathbf{x}}_{n+1} = \mathbf{W}' \mathbf{h}'_n + \mathbf{b}' \tag{17}$$

Here $\hat{\mathbf{x}}_{n+1}$ is the prediction of $\mathbf{x}$ at the $n+1$th time instance, $\mathbf{W}'$ is the weight matrix of the dense layer, $\mathbf{b}'$ is the bias term. A GRU model with more layers can also be used, for the sake of simplicity, however two-layer GRU is considered here.

The model parameters can be trained based on the $N - n + 1$ windows. Once the model parameters are estimated, estimation of model output $\hat{\mathbf{x}}_{n+1}$ can be predicted from the past n samples. A series of residuals can be obtained as $\mathbf{e}_i = |\hat{\mathbf{x}}_i - \mathbf{x}_i|, i = n + 1, \cdots, N$. The residual series obtained from GRU under NOC is then fed into the SVDD to estimate the parameters, namely the center $\mathbf{a}$ and the radius R of the hypersphere. Whenever a new sample is available, the residuals obtained from the GRU can be fed into the SVDD to calculate the the squared distance D^2 according to Equation (15). If D^2 is greater than R^2, it is faulty, otherwise it is normal.

3.2. Fault Identification

Once a fault is detected, the next goal is to identify which variables are the most affected and contribute most to the monitoring statistics. Assume a fault was detected between time t_1 and t_2, let $\mathbf{e}_i = (e_i^1, e_i^2, \cdots, e_i^d)$ denote the residual vector for the l process variables, $i = t_1, \cdots, t_2$. The normalized residuals E_i^l can be used to evaluate the impact of the fault on each variable as:

$$E_i^l = \frac{e_i^l - \mu^l}{\sigma^l} \tag{18}$$

where μ^l is the mean value and σ^l is the standard deviation of the GRU residuals of the NOC training data.

For a clearer exhibition, the deviation E_i^l of each variable is accumulated to get the total contribution rate $CR^l = \sum_{i=t}^{N} E_i^l$. With the contribution rate obtained, operators can know which variables are most sensitive to the process fault. Also, operators can use the contribution plots to identify which kind of fault has occurred.

For completeness, the overall flowchart is summarized in Figure 4, including both the offline training stage (left) and the online monitoring stage (right).

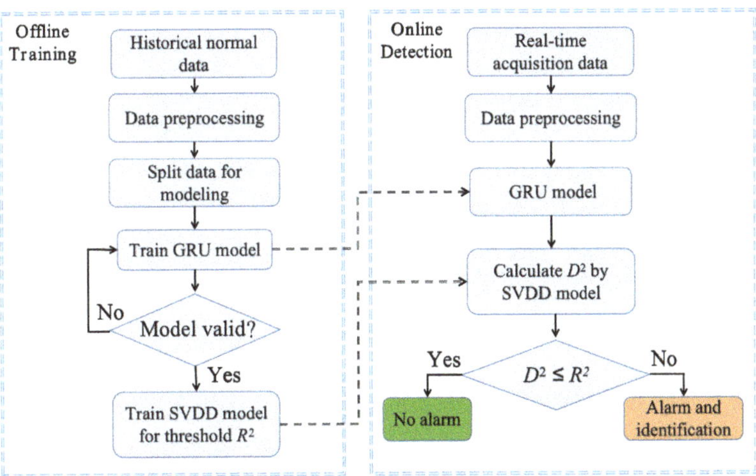

Figure 4. Flowchart of the GRU-support vector data description (SVDD)-based fault detection and identification methodology.

The offline monitoring stage can be summarized as follows:

1. Obtain historical NOC data;
2. Remove extreme values and normalize the training data to have a zero mean and unit variance.
3. Set initial parameters of GRU model and train the model;
4. If the GRU model is valid, the GRU residuals will be fed into the SVDD model, and the threshold R^2 of D^2 statistic is obtained.

The online detection stage can be summarized as follows:

1. Collect online samples;
2. Normalize the online samples;
3. Use the GRU model trained in the offline process to make prediction and get the residuals;
4. Calculate the D^2 statistic using SVDD;

5. Determine whether to alarm by comparing the D^2 statistic and the threshold R^2. If D^2 is greater than R^2, the process is faulty, otherwise it is normal.

6. If the process is faulty, isolate and identify which variables are most severely affected.

4. Application Studies

This section presents the application results of the proposed GRU-based fault detection and identification method to the datasets collected from a blast furnace (with the inner volume of 2500 m^3) in China. Two case studies are studied, with Section 4.1 introduces the application of GRU-based fault detection and identification method to a hanging fault and Section 4.2 presents the application results to a fault involving fluctuation in molten iron temperature.

4.1. Case 1: Hanging Fault

The hanging fault happens in the upper part of the blast furnace, the fault caused a severe drop in the quantity of blast u_1 and pressure of blast u_3, which subsequently resulted in an abnormal change in the composition of flue gas u_5, u_6, u_7. For the purpose of model training, 2000 samples are collected under the normal operation condition and a faulty dataset containing 400 samples is considered. The sampling interval of the data is 20 min. A total of 7 process variables are considered and listed in Table 1. For comparison, the LSTM-SVDD and PCA-SVDD [20] methods are considered.

Table 1. The input variables for Case 1.

No.	Variable
u_1	quantity of blast
u_2	temperature of blast
u_3	pressure of blast
u_4	the quantity of oxygen blasted
u_5	CO concentration in top gas
u_6	CO_2 concentration in top gas
u_7	H_2 concentration in top gas

4.1.1. Residual Generation Using the GRU Network

In order to reduce the impact of extreme values in the process data, the Hampel filter [21] is used to process the training set before feeding into the GRU network. During the training of GRU network, the mean square error loss function and 'Adam' optimizer are used [22]. The length of moving window n is set as 99 by trial and error. The number of hidden states in the first layer d_h and the second layer d_h' are determined in a similar way. Figure 5 shows the modelling errors of the GRU network for u_1 under different combinations of d_h and d_h'.

Considering both the modelling error and structure complexity, the fourth combination in Figure 5 is used so that $d_h = 32$ and $d_h' = 200$. Also, to prevent overfitting, a dropout process is used in the training process by randomly discarding a part of units. Here, the dropout rate of $p_d = 0.2$ is selected. The GRU network uses the past values to predict current values. The predicted values obtained by this model not only contains the past information, but also affected by other related variables. Therefore, when a fault happens, the predictions will deviate from the actual values. The modeling results of the GRU model are shown in Figure 6.

Figure 6 shows that there are some clear changes occurring in several variables (e.g., CO concentration, CO_2 concentration and H_2 concentration). the obtained residuals are then fed into the SVDD model to perform fault detection.

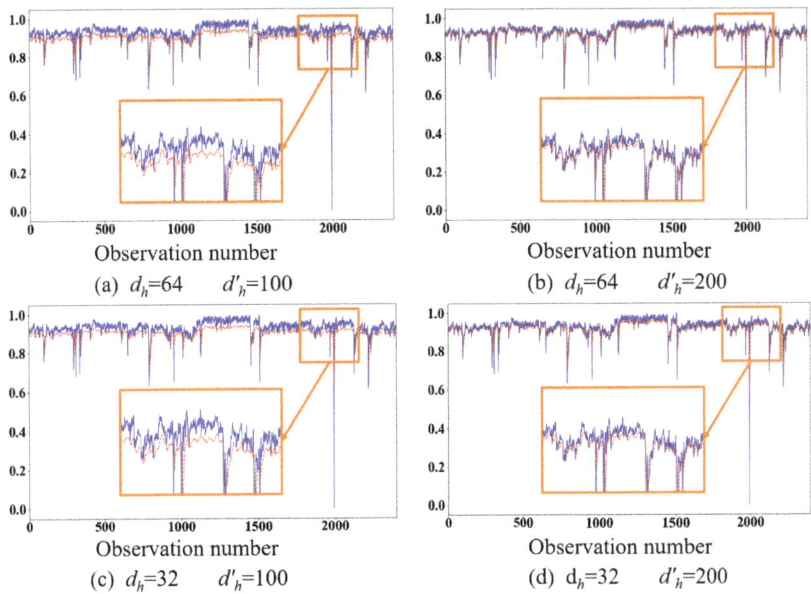

Figure 5. Prediction results for u_1 using models with different parameter settings (the red line corresponding to predictions, the blue line corresponding to true values). (**a**) Prediction results for u_1 with $d_h = 64$ and $d'_h = 100$; (**b**) Prediction results for u_1 with $d_h = 64$ and $d'_h = 200$; (**c**) Prediction results for u_1 with $d_h = 32$ and $d'_h = 100$; (**d**) Prediction results for u_1 with $d_h = 32$ and $d'_h = 100$.

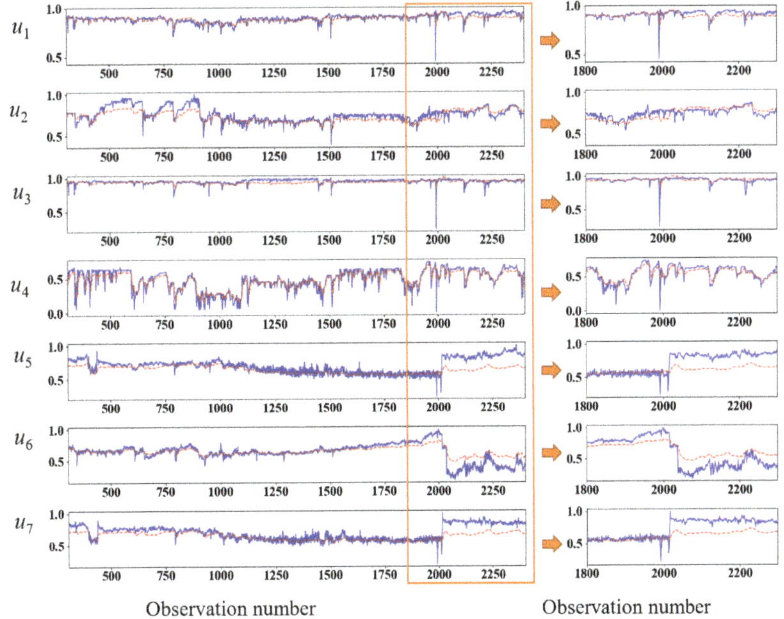

Figure 6. The prediction results of the GRU model in Case 1 (the red lines represent prediction and the blue lines represent actual values).

4.1.2. Fault Detection and Identification

From the previous subsection, the GRU model can be used to generate residuals. As is shown in the Figure 6, there is an obvious change after the 2000th sample. In order to detect this change, SVDD is used here. The parameters of SVDD are set as $\sigma = 10$, $C = 0.01$. With 99% of confidence limit, the monitoring results using SVDD are shown in Figure 7a.

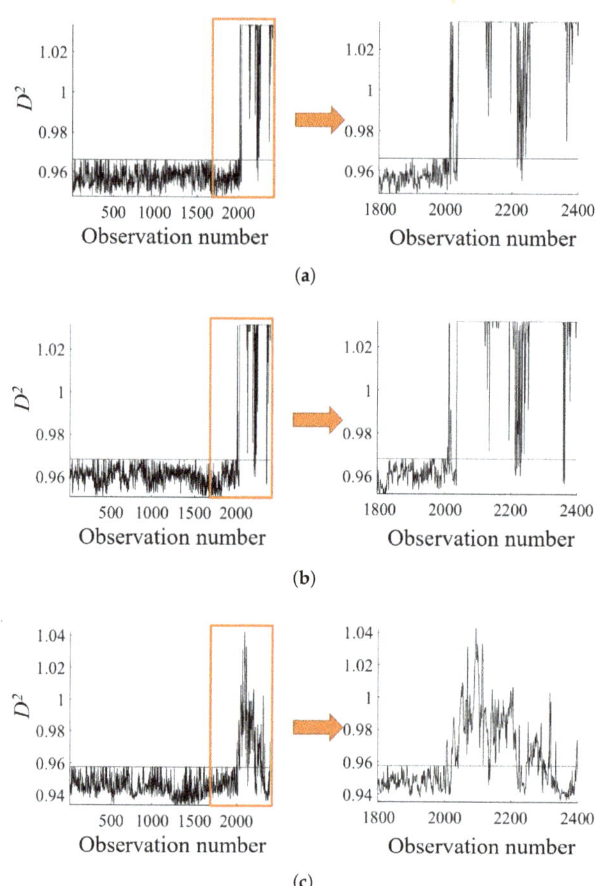

Figure 7. Monitoring results for the hanging fault. (**a**) Monitoring results of GRU-SVDD, (**b**) Monitoring results of LSTM-SVDD, (**c**) Monitoring results of principal component analysis (PCA)-SVDD.

From Figure 7a it can be seen that significant violation of the confidence limit can be observed, indicating that there is a fault happening in the blast furnace system. This is in accordance with the fact that the last 400 samples correspond to a hanging fault. For comparison, the monitoring results using LSTM-SVDD and PCA-SVDD are shown in Figure 7b,c. The LSTM network has the same structure and parameters as GRU-SVDD. For PCA-SVDD, PCA is first performed on the training data and SVDD is used to detect the residual subspace. The number of principal components retained for PCA is 3. Comparing Figure 7a–c, it can be seen that GRU-SVDD and LSTM-SVDD has higher sensitivity than PCA-SVDD. The detection rates of the three methods are shown in Table 2.

Table 2. Comparison of detection rate for different methods.

Methods	Detection Rate
$D^2_{GRU-SVDD}$	93.52%
$D^2_{LSTM-SVDD}$	92.27%
$D^2_{PCA-SVDD}$	72.82%

Table 2 confirms the finding that GRU-SVDD and LSTM-SVDD have better detection rates. Considering the simpler structure of GRU, obviously GRU-SVDD is a better method. After the hanging fault is detected, fault identification is then performed based on the GRU residuals. Figure 8 shows the sample by sample GRU residuals, with deeper color indicating greater residuals.

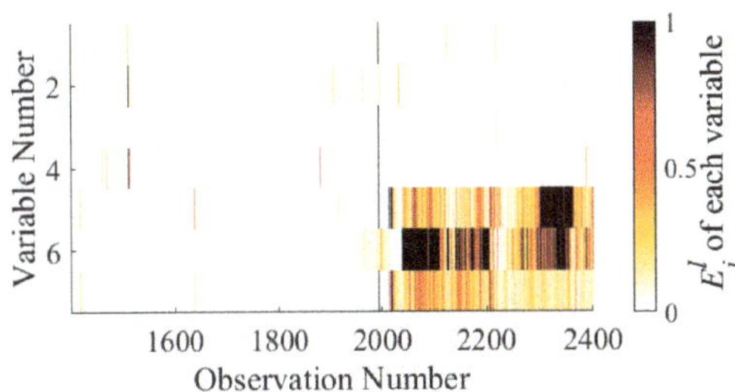

Figure 8. The sample by sample normalized GRU residuals in Case 1.

For a clearer inspection, Figure 9 presents the accumulated normalized GRU residuals. Figure 9 shows that the hanging fault has significant impact on the concentration of flue gas, with the most significant change happening in the CO_2, CO, H_2 concentration. This will lead experienced operators to inspect the gas flow and see whether there is any kind of hanging fault happening in the system.

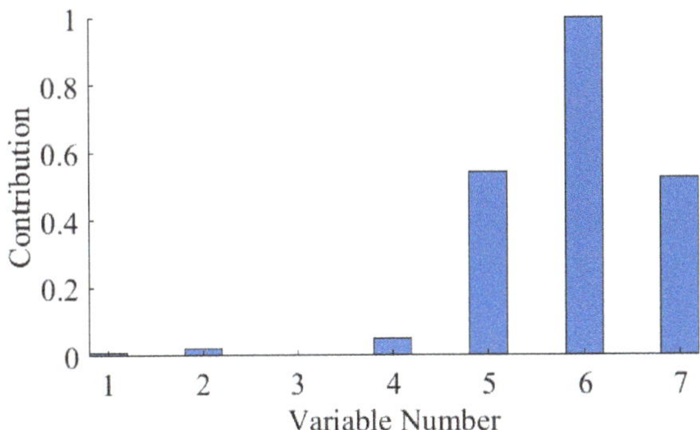

Figure 9. The fault contribution rate for each variable in Case 1.

4.2. Case 2: Abnormal Molten Iron Temperature

In this subsection, a faulty condition from the same blast furnace is considered. The fault involves an abnormal fluctuation of the molten iron temperature, which caused the operators to adjust the quantity of blast u_1 as well as the temperature of blast u_2, resulting in change in a series of variables. In the later stage, the fault was corrected, however the temperature of blast was kept at a relatively low level for the sake of safety. Similar to the hanging fault, 2000 samples were collected under the normal operating conditions for model training, and a faulty dataset containing 1000 samples is considered. The fault involves an abnormal molten iron temperature, which caused reduction in blast quantity, blast temperature and fluctuation in a series of variables related to the gas flow. This time, 10 process variables are considered and listed in Table 3.

Table 3. The input variables for Case 2.

No.	Variable
u_1	quantity of blast
u_2	temperature of blast
u_3	pressure of blast
u_4	quantity of oxygen blasted
u_5	temperature of cold blast
u_6	top pressure
u_7	CO concentration in top gas
u_8	CO_2 concentration in top gas
u_9	H_2 concentration in top gas
u_{10}	pressure of cold blast

Comparing to Table 1, it can be seen that three additional variables, the temperature of cold blast (u_5), the top pressure (u_6) and the pressure of cold blast (u_{10}) are also included. it should be noted some of them are redundant variables (u_5 and u_{10}) that are highly related with other variables. The purpose for introducing these variables is to show the capability of the proposed method in dealing with variable redundancy.

Similar to Section 4.1, the proposed GRU method is applied, with the same parameter values. And the prediction results for the 10 variables are shown in Figure 10. For a clearer exhibition, only the 1000 faulty samples are presented. It can be seen that for the first 200 samples, the prediction accuracy is acceptable. After that, an obvious fluctuation can be observed and the prediction accuracy deteriorated. After the 2450th sample, the prediction accuracy for all variables except u_2 return normal.

After the predictions are obtained, SVDD is applied and the monitoring results are shown in Figure 10b. It can be seen that the fault was successfully detected since significant number of violations can be observed after the 2250th sample. This again indicates the good capability for the proposed method in fault detection. It should be noted that violations can still be observed even after the fault was corrected after 2450th sample. This can be explained, as to avoid further fault, the operators decided to reduce the temperature of blast(u_2), which caused the violations. This can be confirmed by the subsequent fault identification results in Figure 11.

In Figure 11, the first plot involves the fault identification results from samples from 2250 to 2450, while the second plot involves the identification results for samples from 2450 till 3000. As can be seen from the first plot of Figure 11, it can be clearly seen all variables except u_4 have significant contribution to the fault, indicating a significant anomaly arises. This is expected, as to correct the fault, both the quantity of blast and the temperature of blast are reduced, resulting in changes in other variables. From the second plot, it can be seen that after the fault was corrected, the contribution of other variables reduced significantly while that of u_2 remains. This is in accordance with our previous analysis that the operators reduced the temperature of blast to avoid further fault. The application results of the second faulty case also confirmed the performance of the proposed method.

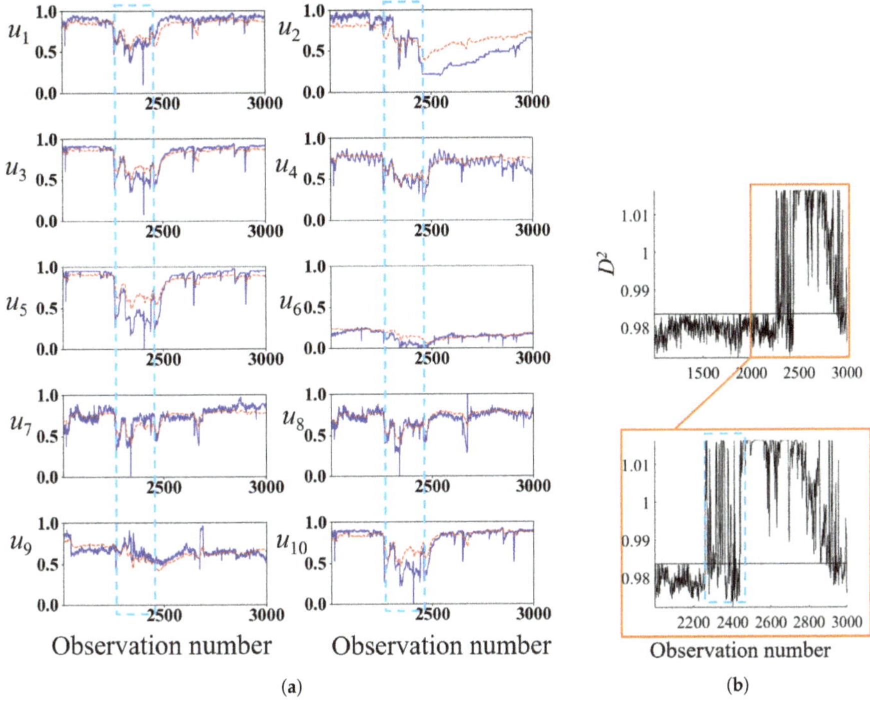

Figure 10. Prediction results using GRU and monitoring results using SVDD for Case 2. (**a**) Prediction results using GRU, (**b**) Monitoring results using SVDD.

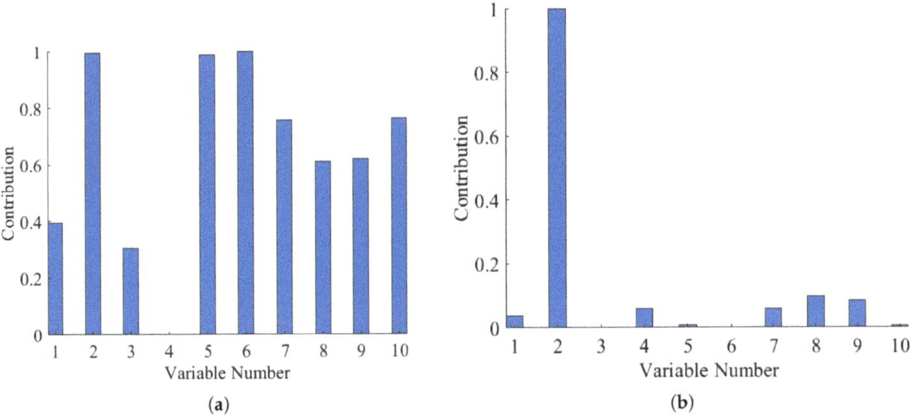

Figure 11. The fault contribution rate for each variable in Case 2. (**a**) The Contribution of different variables in observations 2250 to 2450, (**b**) The Contribution of different variables in observations 2450 to 3000.

5. Conclusions

This paper introduces a fault detection and identification method for blast furnace ironmaking process based on the GRU network and SVDD. The GRU model is capable of handling multi-dimensional inputs to make predictions for future inputs. The residuals between the actual inputs and predictions are then monitored using SVDD. A fault identification method is further developed by inspecting the accumulated normalized residuals. The proposed method is tested on a hanging fault observed in a real blast furnace in China. Application results show that the proposed GRU-SVDD model can successfully detect the hanging fault. Compared with the PCA-SVDD model, GRU-SVDD has a higher detection rate. The method proposed in this article is very suitable for monitoring systems with strong dynamics and non-Gaussianity.

Author Contributions: Conceptualization, J.Z. and Y.L.; methodology, H.O. and J.Z.; validation, H.O. and Y.L.; formal analysis, H.O.; investigation: H.O.; resources, J.Z. and S.L.; data curation, J.Z.; writing—original draft preparation, H.O.; writing—review and editing, J.Z.; visualization, H.O.; supervision, J.Z. and S.L.; project administration, Y.L.; funding acquisition, J.Z. All authors have read and agreed to the published version of the manuscript.

Funding: The authors would like to thank financial support from National Natural Science Foundation of China (Grant Nos. 61673358 and 61973145).

Conflicts of Interest: The authors declare no conflicts of interest.

References

1. Amano, S.; Takarabe, T.; Nakamori, T. Expert system for blast furnace operation at Kimitsu works. *ISIJ Int.* **1990**, *30*, 105–110. [CrossRef]
2. Liao, S. Expert system methodologies and applications–a decade review from 1995 to 2004. *Exp. Syst. Appl.* **2005**, *28*, 93–103. [CrossRef]
3. Tian, H.; Wang, A. A Novel Fault Diagnosis System for Blast Furnace Based on Support Vector Machine Ensemble. *ISIJ Int.* **2010**, *50*, 738–742. [CrossRef]
4. Liu, L.; Wang, A.; Sha, M. Multi-class classification methods of cost-conscious LS-SVM for fault diagnosis of blast furnace. *Ind. J. Iron Steel Res. Int.* **2011**, *18*, 17–23. [CrossRef]
5. An, R.; Yang, C.; Zhou, Z.; Wang, L. Comparison of Different Optimization methods with support vecto machine for blast furnace multi-fault classification. *IFAC-PapersOnLine* **2015**, *48*, 1204–1209.
6. Vanhatalo, E. Multivariate process monitoring of an experimental blast furnace. *Qual. Reliab. Eng. Int.* **2010**, *26*, 495–508. [CrossRef]
7. Zhang, T.; Ye, H.; Wang, W. Fault diagnosis for blast furnace ironmaking process based on two-stage principal component analysis. *ISIJ Int.* **2014**, *54*, 2334–2341. [CrossRef]
8. Shang, J.; Chen, M.; Zhang, H. Increment-based recursive transformed component statistical analysis for monitoring blast furnace iron-making processes: An index-switching scheme. *Control Eng. Pract.* **2018**, *77*, 190–200. [CrossRef]
9. Pan, Y.; Yang, C.; An, R. Robust principal component pursuit for fault detection in a blast furnace process. *Ind. Eng. Chem. Res.* **2017**, *57*, 283–291. [CrossRef]
10. Zhou, B.; Ye, H.; Zhang, H. Process monitoring of iron-making process in a blast furnace with PCA-based methods. *Control Eng. Pract.* **2016**, *47*, 1–14. [CrossRef]
11. Cai, J.; Zeng, J.; Luo, S. A state space model for monitoring of the dynamic blast furnace system. *ISIJ Int.* **2012**, *52*, 2194–2199. [CrossRef]
12. Vanhatalo, E.; Kulahci, M. Impact of autocorrelation on principal components and their use in statistical process control. *Qual. Reliab. Eng. Int.* **2015**, *32*, 1483–1500. [CrossRef]
13. Dong, Y.; Qin, S.J. A novel dynamic pca algorithm for dynamic data modeling and process monitoring. *J. Process Control.* **2018**, *67*, 1–11. [CrossRef]
14. Chiang, L.; Braatz, R.; Russell, E.L. *Fault Detection and Diagnosis in Industrial Systems*; Springer Science & Business Media; Berlin, Germany, 2002; Volume 44, 197–198.
15. Qin, S. Survey on data-driven industrial process monitoring and diagnosis. *Annu. Rev. Control* **2012**, *36*, 220–234. [CrossRef]
16. Wang, J.; Yan, J.; Li, C. Deep heterogeneous GRU model for predictive analytics in smart manufacturing: Application to tool wear prediction. *Comput. Ind.* **2019**, *111*, 1–14. [CrossRef]
17. Hochreiter, S.; Schmidhuber, J. Long Short-Term Memory. *Neural Comput.* **1997**, *9*, 1735–1780. [CrossRef]
18. Funahashi, K.; Nakamura, Y. Approximation of dynamical systems by continuous time recurrent neural networks. *Neural Netw.* **1993**, *6*, 801–806. [CrossRef]

19. Kim, P.; Lee, D.; Lee, S. Discriminative context learning with gated recurrent unit for group activity recognition. *Pattern Recognit.* **2018**, *76*, 149–161. [CrossRef]
20. Li, G.; Hu, Y.; Chen, H. An improved fault detection method for incipient centrifugal chiller faults using the PCA-R-SVDD algorithm. *Comput. Sci.* **2014**, *116*, 104–113. [CrossRef]
21. Pearson, R.; Neuvo, Y.; Astola, J.; Gabbouj, M. Generalized Hampel filters. *EURASIP J. Adv. Signal Process.* **2016**, *87*. [CrossRef]
22. Chang, Z.; Zhang, Y.; Chen, W. Electricity price prediction based on hybrid model of adam optimized LSTM neural network and wavelet transform. *Energy* **2019**, *187*. [CrossRef]

Sample Availability: Samples of the compounds are available from the authors.

Article

Discrete Element Method (DEM) and Experimental Studies of the Angle of Repose and Porosity Distribution of Pellet Pile

Han Wei [1], Meng Li [1], Ying Li [1], Yao Ge [1], Henrik Saxén [2] and Yaowei Yu [1,*]

[1]   State Key Laboratory of Advanced Special Steel, Shanghai Key Laboratory of Advanced Ferrometallurgy, School of Materials Science and Engineering, Shanghai University, Shanghai 200444, China
[2]   Thermal and Flow Engineering Laboratory, Faculty of Science and Engineering, Åbo Akademi University, Biskopsgatan 8, FI-20500 Åbo, Finland
*   Correspondence: yaoweiyu@shu.edu.cn

Received: 12 July 2019; Accepted: 20 August 2019; Published: 23 August 2019

Abstract: The lumpy zone in a blast furnace is composed of piles formed naturally during burden charging. The properties of this zone have significant effects on the blast furnace operation, including heat and mass transfer, chemical reactions and gas flow. The properties of the layers mainly include the angle of repose and porosity distribution. This paper introduces two methods, the Discharging Method and the Lifting Method, to study the influence of the packing method on the angle of repose of the pile. The relationships of the angle of repose and porosity with physical parameters are also investigated. The porosity distribution in the bottom of a pile shows a decreasing trend from the region below the apex to the center. The coordination number of the particles is employed to explain this change. The maximum of the frequency distribution of it was found to show a negative correlation to the static friction coefficient, but becomes insensitive to the parameter as the static friction coefficient increases above 0.6.

Keywords: pellet pile; Discrete Element Method; porosity distribution; angle of repose; coordination number

---

1. Introduction

The lumpy zone of the blast furnace (BF) is composed of layers of piled burden formed naturally during charging. Two significant variables characterize the properties of the layers: The angle of repose and porosity distribution, which reflect the external shape and internal structure, respectively. The former reflects the stability and surface profile of the piles. The latter is a direct reflection of the permeability of the burden, which is closely connected to the gas flow resistance and heat exchange efficiency between the burden and gas in the blast furnace. Therefore, an improved understanding of the formation mechanism and internal state of a pile is important when measures are to be taken to improve the efficiency of the conditions in the upper part of the blast furnace. The coordination number (CN) is an important parameter reflecting the internal structure of the granular pile, which is closely related to porosity. However, it is difficult to gain a deep understanding of the flow and packing of granular materials by experimental methods, due to the complex behavior of granular materials in bulk systems [1,2]. Therefore, numerical simulation has become an interesting and viable option, and, in particular, the discrete element method (DEM). This method can provide estimates of the position, velocity and stress information of each particle in a granular system.

The angle of repose is a fundamental property of a pile, which usually reflects the liquidity potential of it. By simulation, it has been found that the angle of repose is related to DEM parameters, such as the rolling and static friction coefficient [3]. Elperin et al. [4] and Coetzee et al. [5] revealed that the angle of repose is positively correlated with the friction coefficient, but when the friction coefficient increases to a certain value, the angle of repose does not any longer increase or instead even decreases, due to the collapse of the pile. Alizadeh et al. [6] also found that the angle of repose is strongly affected by the particle shape. Furthermore, the particle size [7,8] and packing method [9] also influence the angle of repose.

Porosity is closely related to the permeability of the packed bed. Thus, by adjusting the porosity in the BF can help control the gas distribution, and thus, the heat transfer and reactions in the lumpy zone. It has been observed that the gas permeability of the stock column will deteriorate rapidly when the porosity is reduced to or below 0.3. It is easier to measure the porosity of particles in a container either in an experiment or in simulation than in the real process. Zou and Yu [10] presented experimental research on porosity by using a cylindrical container and found that the initial porosity of a pile is strongly dependent on both particle shape and packing method. Another simulation work by the same authors [11] concluded that particle size also influenced the porosity. However, there are only a few publications on the porosity distribution in a three-dimensional particle pile, which is of interest for the distribution of gas flow in industrial applications, such as in reactors, moving and fluidized beds.

Iron oxide particles are used as the main raw material in the blast furnace. Still, many papers on the simulation of gas-solid two-phase flow in the blast furnace have not considered the radial distribution of porosity, even though the porosity distribution is known to be non-uniform. Therefore, it is an important aspect to consider. The present work studied the effects of DEM parameters and packing method on the properties of pellet piles, with the aim to provide insights that can be used in modeling and further research on the porosity distribution of burden in the blast furnace.

Packing density, as the opposite of porosity, has also been studied by many investigators. Most studies focus on packing density of spherical [12–14] and non-spherical particles [15–20] in a container by dense or loose packing [15]. However, as it is hard to measure the porosity distribution of a conical pellet heap, there are still few publications in this field, due to the anisotropic properties of granular materials, the complexity of pile structure and the opaque mechanism by which the packing evolves in three dimensions [1]. DEM has become a viable choice for studies on the properties of granular piles, and therefore this modeling method has become popular in blast furnace investigations [21–24].

The present work introduces several novel aspects, including the treatment of the porosity distribution in the pile and the fact that industrial-scale pellets are studied, as well as the 1:10 scale-charging system used. The paper studies the angle of repose and porosity distribution of a pile of iron oxide pellet by experimental and numerical methods. Section 2 introduces the experimental work, including the methods and apparatus. The simulation theories and conditions are presented in Section 3. In Section 4, the angle of repose and porosity distribution of a pellet pile are studied to validate the DEM physical parameters determined by the discharging method. The effects of packing method on the angle of repose are also discussed. Finally, the conclusions of the work are proposed.

2. Experimental Work

The experimental study is based on iron oxide pellets. As the shape of the pellet is close to that of a sphere, spherical particles were used in the simulations to be presented. In fact, particle shape affects the porosity of the pile, but since we focus of pellets, particle shape was not considered. The reader is referred to Reference [25] for more information about this matter.

Pellets used in the experiments come from a steel plant in China. About 10,000 pellets were selected after applying sieves with aperture size in the range 13–15 mm. The experimental apparatus is illustrated in Figure 1a, which is a 1:10 scale charging system of a BF. A stable pellet pile was formed on a table by the discharging method. In order to study the profiles of the arising pile, a camera with the lens level along with the desktop was used to take photographs from four different directions of the pile. The angle of repose was obtained by analyzing the profile of the heap using photograph-processing technology.

The same method was used to form a pile for measuring the porosity distribution. Seven 25 mL (plexiglass) breakers placed beforehand to be buried in the pellet pile were slowly removed afterwards to be able to measure the porosity (P), expressed b y

$$P = \frac{V_1}{V} \times 100\%,\tag{1}$$

where V_1 and V denote the void volume (mL) of beakers full of particles and the volume of the empty beaker, respectively. In order to reduce the errors, the experiment was repeated nine times, and the average result was reported.

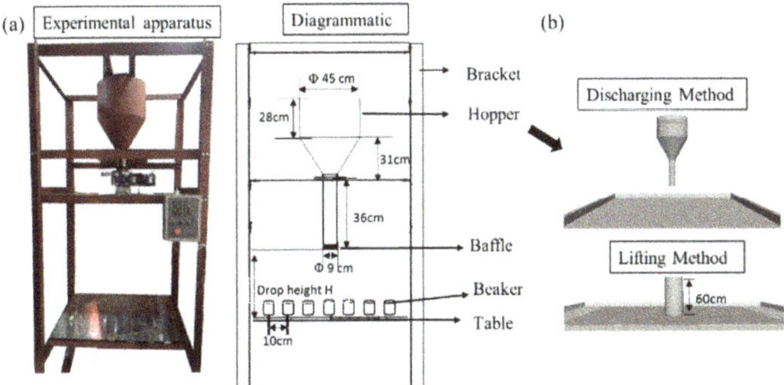

Figure 1. (**a**) Experimental apparatus and schematic of the measurement of pellet pile and (**b**) the geometric models of different packing method in simulation.

3. Simulation Method and Conditions

3.1. Discrete Element Method (DEM)

The simulation part of this research is based on DEM, which was firstly proposed by Cundall and Strack [26]. This method considers two types of motion of a particle, translation and rotation, which are governed by Newton's second law of motion. The elastic contact force expression used in this work is the non-linear Hertz-Mindlin no-slip model [27], which is illustrated in Figure 2. The basic expressions are given in Equations (2) and (3). The former is the translational equation, which is composed of gravitational force, $m_i g$, contact force and viscous contact damping force, where K_n, K_t, γ_n and γ_t express the normal elastic constant, tangential elastic constant, normal damping constant and tangential damping constant, respectively. A particle with the mass of m_i contacts with K particles, and the contact force between them depends on the deformation between particles, δ_n. In the equation, u_i, v_n and v_t represent the translational velocity, and the component of relatively velocity for the normal and tangential directions.

$$m_i \frac{du_i}{dt} = \sum_{j=1}^{K} \left(K_n \delta n_{ij} - \gamma_n v_{nij} \right) + \left(K_t \delta n_{ij} - \gamma_t v_{tij} \right) + m_i g \tag{2}$$

Equation (3) represents the rotational movement of particles, where M_r^k and M_r^d are two torques, which are caused by a tangential force and rolling friction. A Coulomb-type friction law is used to express the friction between the two particles. I_i and ω_i denote the moment of inertia and rotational velocity, respectively.

$$I_i \frac{d\omega_i}{dt} = \sum_{j=1}^{K} \left(M_r^k + M_r^d \right) \tag{3}$$

Table 1 presents the formulas used to calculate the forces and torques between the particles. In this work, we use the open-source software LIGGGHTS 3.5.0 [28] implementation of DEM.

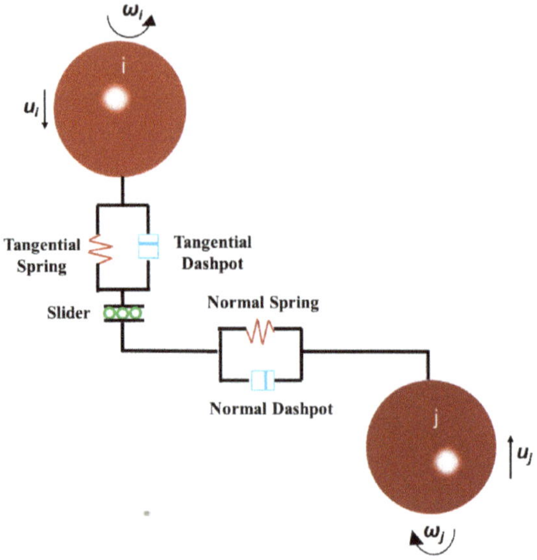

Figure 2. Depiction of forces acting particle *i* in contact with particle *j*.

Table 1. Detailed description of the parameter expressions in discrete element method (DEM).

Parameters	Equations
K_n, K_t	$K_n = \dfrac{4}{3}\gamma^* \sqrt{R^*\delta_n}, K_t = 8G^* \sqrt{R^*\delta_n}$
γ_n, γ_t	$\gamma_n = -2\sqrt{\dfrac{5}{6}}\beta \sqrt{S_n m^*} \gg 0,\ \gamma_t = -2\sqrt{\dfrac{5}{6}}\beta \sqrt{S_t m^*} \gg 0$
S_n, S_t	$S_n = 2\gamma^* \sqrt{R^*\delta_n},\ S_t = 8G^* \sqrt{R^*\delta_n}$
β	$\beta = \dfrac{\ln(e)}{\sqrt{\ln^2(e) + \pi^2}}$
$\dfrac{1}{\gamma^*}$	$\dfrac{1}{\gamma^*} = \dfrac{\left(1-v_1^2\right)}{Y_1} + \dfrac{\left(1-v_2^2\right)}{Y_2}$
$\dfrac{1}{G^*}$	$\dfrac{1}{G^*} = \dfrac{2(2-v_1)(1+v_1)}{Y_1} + \dfrac{2(2-v_2)(1+v_2)}{Y_2}$
$\dfrac{1}{R^*},\ \dfrac{1}{m^*}$	$\dfrac{1}{R^*} = \dfrac{1}{R_1} + \dfrac{1}{R_2},\ \dfrac{1}{m^*} = \dfrac{1}{m_1} + \dfrac{1}{m_2}$
ΔM_r^k	$\Delta M_r^k = -k_r\Delta\theta_r,\ k_r = k_t\cdot R^{*2}$
$M_{r,t+\Delta t}^d$	$\left|M_{r,t+\Delta t}^k\right| \ll M_r^m$ $M_{r,t+\Delta t}^k = M_{r,t}^k + \Delta M_r^k,\ M_r^m = \mu_r R^* F_n$ $M_{r,t+\Delta t}^d = -C_r\dot{\theta}_r\left(\left|M_{r,t+\Delta t}^k\right| < M_r^m\right)$ $M_{r,t+\Delta t}^d = -C_r\dot{\theta}_r\left(\left|M_{r,t+\Delta t}^k\right| = M_r^m\right)$ $C_r = \eta_r C_r^{crit},\ C_r^{crit} = 2\sqrt{I_r k_r}$ $I_r = \left(\dfrac{1}{I_i + m_i r_i^2} + \dfrac{1}{I_j + m_j r_j^2}\right)^{-1}$

In the expressions, G, Y, e and v represent the Shear modulus, Young's modulus, coefficient of restitution and Poisson's ratio, respectively, while μ_r and μ_s represent the rolling and static friction coefficient, respectively.

3.2. Simulation Conditions

The apparatus of the Discharging Method in the simulation consists of a hopper, a baffle and a table, just like the components used in the experiments. In the Lifting Method, the apparatus is only a column barrel. The geometric models are both shown in Figure 1b. DEM parameters of pellet were chosen according to the results of previous work by the authors [29] and are presented in Table 2. In the simulations, we studied the effects of the DEM parameters on the angle of repose and porosity distribution. Furthermore, the effects of different drop heights of the Discharging Method, as well as the effects of lift speed and barrel size in the Lifting Method, on the angle of repose of the pile were also investigated.

Table 2. Physical and contact parameters used in DEM simulation, including pellet particle and walls.

Parameters	Values
Particle number	100,000
Particle density	4837 kg/m^3
Time step	10^{-5} s
Young's modulus	2.5×10^{11} Pa (pellet), 2×10^{11} Pa (steel plane), 7.2×10^{10} (plexiglass)
Poisson ratio (p-p; p-w; p-g)	0.25, 0.3, 0.2
Coefficient of restitution (p-p; p-w; p-g)	0.4, 0.35, 0.2
Coefficient of friction (p-w; p-g)	0.4, 0.25
Rolling friction coefficient (p-w; p-g)	0.4, 0.15
Size of pellet	8 mm, 14 mm, 20 mm

In the table, p-p, p-w and p-g represent the coefficients for pellet-pellet, pellet-wall and pellet-plexiglass (breaker) interaction. Some parameter values were from the literature [17,18].

For the simulation of porosity, we used seven boxes (5 cm × 10 cm × 5 cm) placed along the diameter of the bottom of the pile to measure the bottom porosity distribution (BPD). In determining whether a particle belongs to the box, its central coordinates were used. The porosity of the bed in each box can be calculated by

$$P = \left(1 - \frac{nV_p}{V}\right) \times 100\%, \tag{4}$$

where V is the volume of the box and V_p is the volume of a single pellet, and n is the number of particles in the box.

4. Results and Discussion

4.1. Simulation and Experimental Study of Angle of Repose

4.1.1. Angle of Repose by the Discharging Method

We first studied the influence of DEM parameters on the angle of repose of the pellet pile. As the effect of a physical parameter is studied, the other parameters were kept unchanged at the values reported in Table 2. From work reported in the literature [2], it is known that the angle of repose is sensitive mainly to the rolling and static friction coefficients between the particles. Vertical cross-sections of the pellet pile with different rolling and static friction coefficients are shown in Figure 3. It was observed that when the rolling and static friction coefficients increase from low (0.01) to high (0.99) values, the shape of the pile changed a lot, especially for the latter parameter. The results of contour extractions of the heap are shown in Figure 4. It is obvious that the height of the pile increases and then tends to be stable.

Figure 5 shows the angle of repose with different rolling and static friction coefficients, with error bars indicating the deviation of the angle of repose in different directions of the pile. The angle of repose shows a positive correlation with the friction coefficients. When the static and rolling friction coefficients change from 0.01 to 0.99, the angle of repose of the pellet pile changes about 8° and 20°, respectively, which indicates that static friction coefficient has a stronger impact on the angle of repose.

In general, high static friction is always accompanied by high rolling friction, and the latter depends on the physical properties on the particle surface. However, there is a decreasing trend when $\mu_r > 0.6$ or $\mu_s > 0.8$. A reason may be that the heap has reached the maximum stable angle at this point, and a further increase in the friction coefficient will cause the heap to collapse.

In addition to the effects of the DEM parameters on the angle of repose, the external conditions, such as the drop height cannot be ignored. Figure 6a shows the angle of repose for different drop heights for pellets with different static and rolling friction coefficients. It reveals that when the drop height increases, the angle of repose decreases and this trend will weaken when the friction coefficients increase. It was found that with an increase in the drop height, the bottom size of the heap decreased. Therefore, we define the normalized effective diameter (NED) to express the size, which is the diameter of the bottom circle of the heap where most particles gather, neglecting the particles scattered around the heap because there is an obvious boundary of the high-density particle area and the scattered particle area. Figure 6b shows that for particles with a large coefficient of static friction, the NED of the pile is small. In addition, the NED decreases sharply initially and then levels out when the drop height increases. The reason is that particles will have large kinetic energy when dropping from a high location, and when the particles collide with the packed bed, they more easily bounce and scatter around the heap.

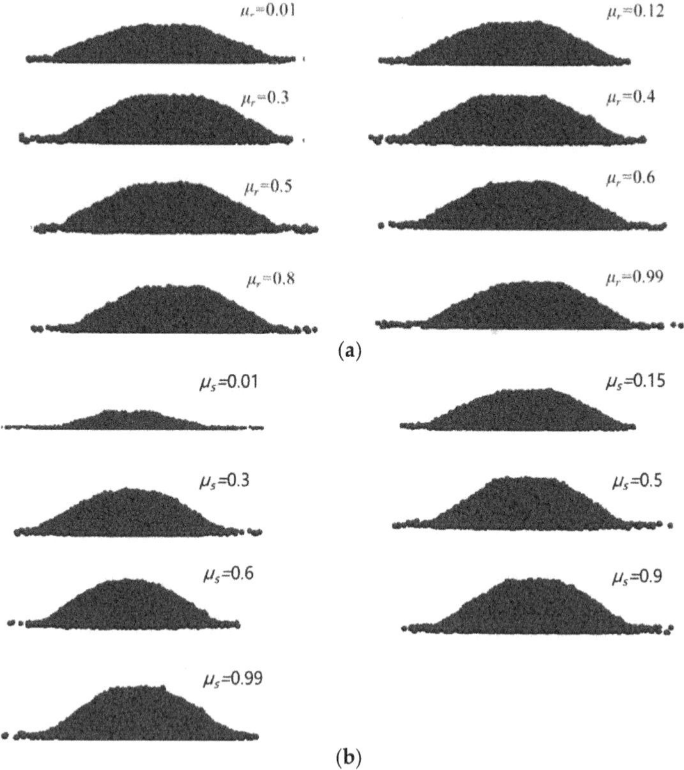

Figure 3. Vertical cross-sections of the pellet pile simulated under different (**a**) rolling and (**b**) static friction (**b**) coefficients.

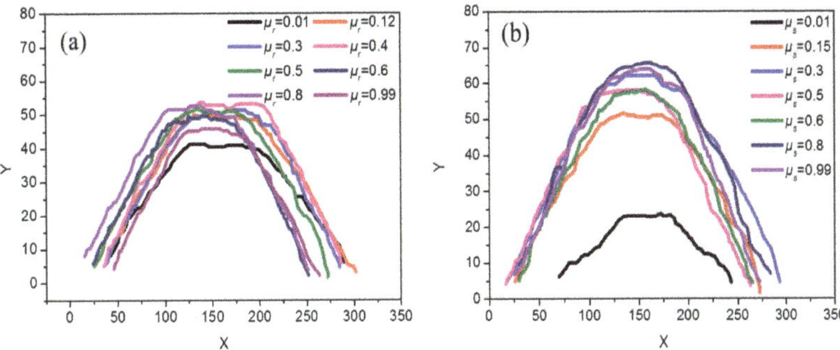

Figure 4. Extracted contours of the pellet pile for different (**a**) rolling and (**b**) static friction coefficients.

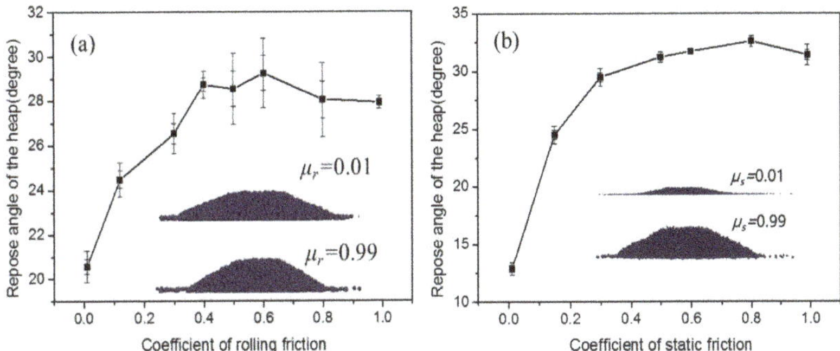

Figure 5. Relationship of the angle of repose of pellet piles and (**a**) rolling friction coefficient with $\mu_s = 0.15$, and (**b**) static friction coefficient with $\mu_r = 0.12$. Error bars indicate the deviation of the angle of repose in different directions of the pile.

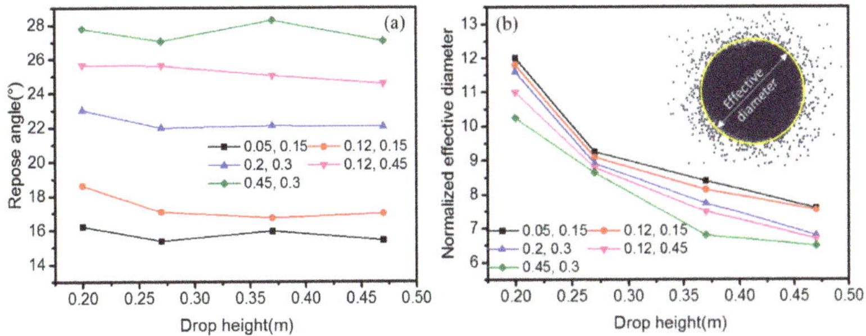

Figure 6. (**a**) Angle of repose and (**b**) and normalized effective diameter of the heap (top view) with different coefficient of static and rolling friction for different drop heights in the Discharging Method. (The numbers in the two figures represent rolling and static friction coefficients, respectively) The inserted subfigure in (**b**) is a top view of the simulated pile.

4.1.2. The angle of Repose by the Lifting Method

In the simulation of the Lifting Method, we designed four different cases (Table 3) to study the influence of the barrel size and lift speed on the angle of repose. Case 1 and Case 2 have the same coefficient of friction, but different barrel size. All the cases were considered with four different lifting velocities (0.005 m/s, 0.01 m/s, 0.02 m/s and 0.03 m/s). Figure 7 shows the angle of repose with different lifting velocities. The angle of repose tends to decrease when the lifting velocity increases, and this trend is weakened as the friction coefficient increases because a small lifting velocity makes it easier to keep the particles in their original positions. Through comparing Cases 1 and 2, it can be seen that the angle of repose will increase if the barrel size gets smaller. The dotted pink line and solid pink line in Figure 7 represent the angle of repose with different packing methods, but the same DEM parameters, which reveals that the angle of repose formed by the Lifting Method is larger than that determined by the Discharging Method.

Table 3. Different cases studied by the Lifting Method.

Cases	Rolling Friction Coefficient	Static Friction Coefficient	Barrel Size (Diameter: m)
Case1	0.05	0.15	0.1
Case2			
Case3	0.12	0.15	0.15
Case4	0.12	0.45	

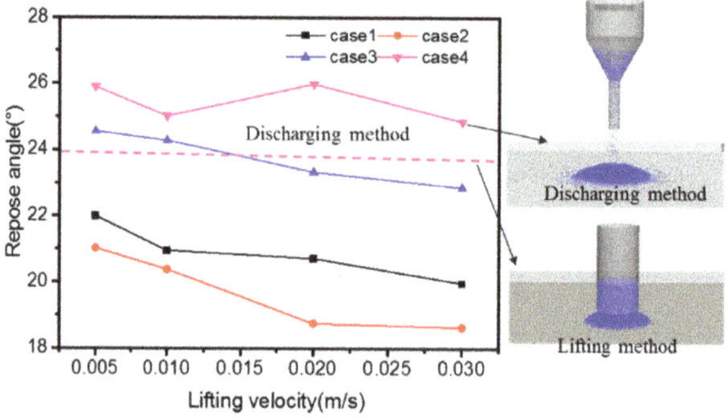

Figure 7. The angle of repose for different lifting velocities. Pink dotted and solid lines represent the angle of repose with different packing methods, but the same DEM parameters.

4.1.3. Simulated vs. Experimental Angles of Repose

The average of the experimentally determined angle of repose of the pellet pile is about 25°. Comparing it with the simulated results in Figure 5, we found that when $\mu_r = 0.12$ and $\mu_s = 0.15$, the experimental and simulated results agree well. Thus, these two values can be used to study the BPD and average porosity of pellet piles. The profiles of the experimental and simulated heaps seen in Figure 8 as black and red lines illustrate the agreement. The figure also shows that the Lifting Method gives a higher heap and a larger angle of repose than the Discharging Method.

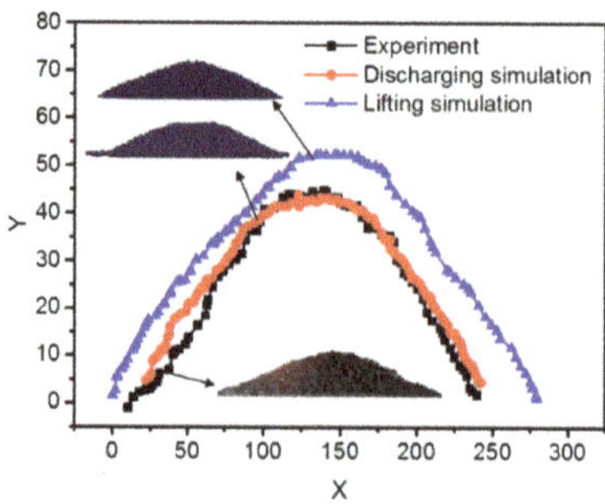

Figure 8. Profile of heap in simulation and experiment.

4.2. Porosity Distribution of Pellet Pile

4.2.1. Simulated vs. Experimental BPD

The BPD of the pile is used to specifically and quantitatively describe the pile porosity. Figure 9 shows the simulated and experimental BPD of the pellet pile measured by the containers. The simulated results include porosity distribution in containers and in-unit boxes. The coordinate 0.00 on the abscissa represents the center position of the pile, and ±0.15 represent the edges. The trends of the curves are seen to be the same in the experiments and simulations. The porosity distribution shows a V-shaped appearance, where the central value is lower than that at the edge. In fact, the experimental and simulated values measured by the containers are all larger than the real ones (without containers) because of the so-called wall effect. The difference of the simulated results (equal value of BPD) with containers and without containers can be used to quantify this effect. We found that the wall effect would result in an error of 3.3 percent points compared with direct measurement without a container. In Figure 9, the experimental porosities are a little larger than the simulated ones because a drainage method was used to measure the porosity in the experiments, and the operation loss of water leads to a larger porosity. In order to further study the distribution of the porosity, the coordination number (CN) was employed, where CN expresses the number of particles in contact with one particle. Figure 10, where different colors represent different values, indicates that CN increases from the edge to the center of the pile, because of the compact packing of particles in the center. These results coincide with the findings on the porosity distribution. The central position of the heap is formed by the vertical falling particles, which have greater kinetic energy, causing a more compact structure of the pile. After the formation of the initial heap, the pile with the continuous falling particles will collapse, and the edges form.

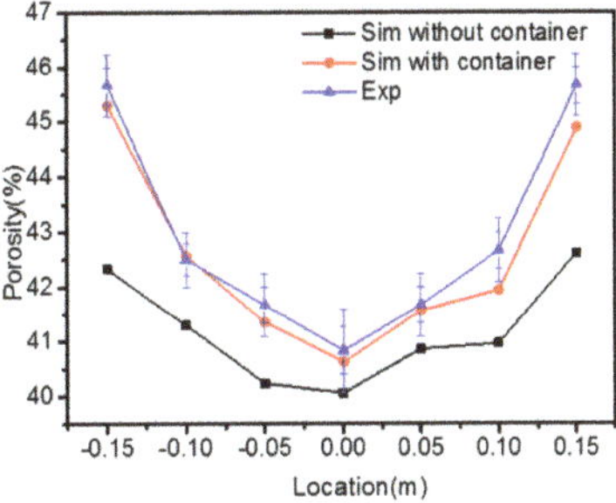

Figure 9. Comparisons of simulated (with and without containers) and experimental bottom porosity distribution (BPD) of the pellet pile.

Figure 10. Coordination number of a vertical cross-section of the simulated pellet pile.

4.2.2. Effects of Rolling and Static Friction on BPD

Figure 11 shows the BPD of the heap when the rolling and static friction coefficients change, where the inserted graphs show the average value of BPD. When the static and rolling friction coefficient change from 0.01 to 0.99, the average porosities change by 7 and 3 percent points, respectively, which means that the static friction coefficient has a greater impact on porosity. This conclusion is also supported by findings reported in the literature [3,6,30–33].

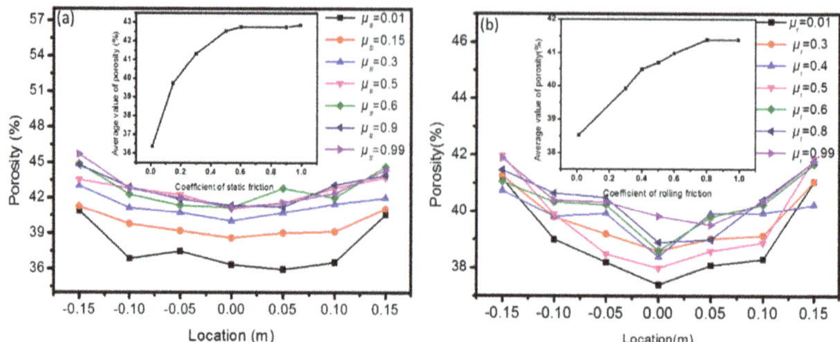

Figure 11. BPD of the pellet pile and the inserted graph is the average porosity (the average value of seven points on a curve) with different (**a**) static and (**b**) rolling friction coefficients.

Figure 12 depicts the frequency distribution of CN for different static and rolling friction coefficients, showing a maximum value of the frequency at CN ≈ 4. Thus, most particles are in contact with four neighboring particles. The static friction coefficient affects the frequency distribution, but mainly for $\mu_s < 0.6$, and the rolling friction seems to have no effect. As seen in Figure 13, when the static friction coefficient increases, the CN of the heap decreases, but the decrease is small for $\mu_s > 0.6$. The reason may be that if the CN is small, the porosity is large. The average CN of the heap is only affected by the rolling friction when the coefficient is very small.

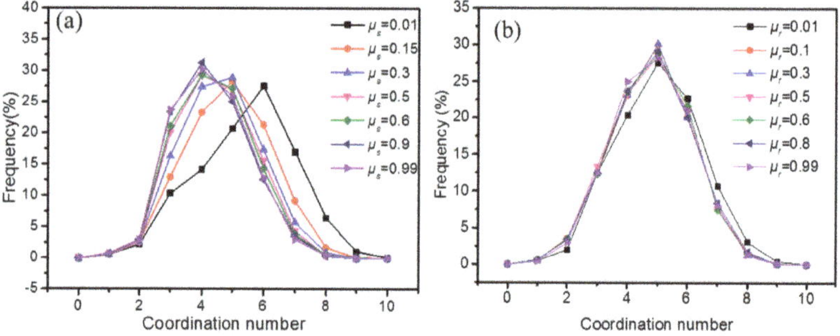

Figure 12. Frequency distribution of coordination number with different (**a**) static and (**b**) rolling friction coefficients.

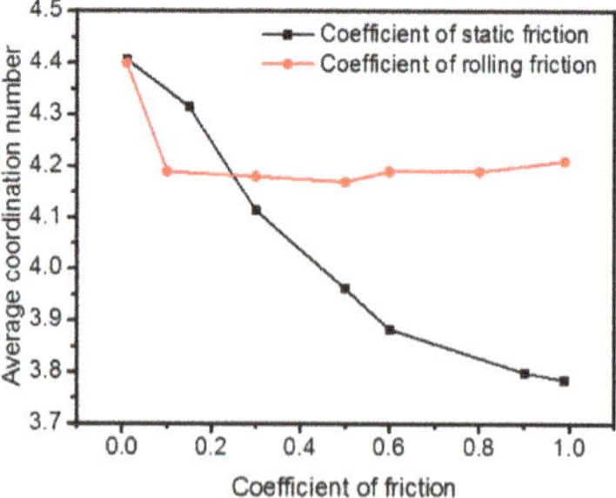

Figure 13. The average coordination number of the whole heap for different static ($\mu_r = 0.12$) and rolling friction ($\mu_s = 0.15$) coefficients.

5. Conclusions

The angle of repose, coordination number (CN) and bottom porosity distribution (BPD) of pellet piles were studied by DEM simulation and experimental methods. A charging system mimicking that of a blast furnace, but in 1:10 scale was designed to simulate the pile formation of iron oxide pellets. The effects of DEM parameters and packing method on the angle of repose were also studied, including the drop height in the Discharging Method and properties (lifting velocity, barrel size) of the Lifting Method. Some of the results are highlighted in the following.

The angle of repose shows a positive correlation with static and rolling friction coefficients. The angle of repose formed by the Lifting Method is bigger than that obtained by the Discharging Method. When the drop height increases, the angle of repose decreases, but this trend will weaken when the static friction coefficient becomes large. In the Lifting Method, the angle of repose tends to decrease with an increase in the lifting velocity or in the barrel size, but the trend is less clear for pellets with large friction coefficients. The size of the bottom circle of the heap is significantly reduced with an increase in the friction coefficient. Appropriate values of the rolling and static friction coefficients for the pellets were found to be 0.12 and 0.15, respectively.

The porosity distribution in the bottom of the heap (BDP) along the heap diagonal shows a V-type behavior, where the value in the center is smaller than those at the edges. The BPD shows an increasing trend with the increase of the friction coefficient. CN is an important parameter reflecting the internal structure of the pile, and expectedly, it shows a negative correlation with porosity. The maximum of the frequency distribution of CN, which occurs at CN ≈ 4, exhibits a negative correlation with the static friction coefficient and eventually remains unchanged when the coefficient grows larger than 0.6. CN is not significantly affected by the rolling friction coefficient.

Author Contributions: Conceptualization, H.W. and Y.Y.; methodology, H.W. and M.L.; Software, H.W. and M.L.; validation, H.W. and Y.L.; formal analysis, Y.G.; investigation, H.W. and Y.Y.; resources, H.W. and M.L.; data curation, H.W. and Y.G.; writing, H.W; supervision, Y.Y. and H.S.; funding acquisition, Y.Y.; Writing-Review and Editing, H.S. and H.W.

Funding: This research was funded by The Program for Professor of Special Appointment (Eastern Scholar) at Shanghai Institutions of Higher Learning grant number TP2015039.

Acknowledgments: The authors are grateful for the financial support from The Program for Professor of Special Appointment (Eastern Scholar) at Shanghai Institutions of Higher Learning (No. TP2015039). The Discrete Element Method was conducted using LIGGGHTS 3.5.0 open source.

Conflicts of Interest: The authors declare no conflict of interest.

References

1. Kou, B.; Cao, Y.; Li, J.; Xia, C.; Li, Z.; Dong, H.; Zhang, A.; Zhang, J.; Kob, W.; Wang, Y. Granular materials flow like complex fluids. *Nature* **2017**, *551*, 360–363. [CrossRef] [PubMed]
2. Zhou, Z.Y.; Zou, R.P.; Pinson, D.; Yu, A.B.; Zhou, Z. Angle of repose and stress distribution of sandpiles formed with ellipsoidal particles. *Granul. Matter* **2014**, *16*, 695–709. [CrossRef]
3. Van Burkalow, A. Angle of Repose and Angle of Sliding Friction: AN Experimental Study. *Geol. Soc. Am. Bull.* **1945**, *56*, 669. [CrossRef]
4. Elperin, T.; Golshtein, E. Comparison of different models for tangential forces using the particle dynamics method. *Phys. A Stat. Mech. Appl.* **1997**, *242*, 332–340. [CrossRef]
5. Coetzee, C. Calibration of the discrete element method and the effect of particle shape. *Powder Technol.* **2016**, *297*, 50–70. [CrossRef]
6. Alizadeh, M.; Hassanpour, A.; Pasha, M.; Ghadiri, M.; Bayly, A. The effect of particle shape on predicted segregation in binary powder mixtures. *Powder Technol.* **2017**, *319*, 313–322. [CrossRef]
7. Dury, A.; Ristow, G.H.; Moss, J.L.; Nakagawa, M. Boundary Effects on the Angle of Repose in Rotating Cylinders. *Phys. Rev. E Stat. Phys. Plasmas Fluids Relat. Interdiscip. Top.* **1997**, *57*, 4491–4497. [CrossRef]
8. Carstensen, J.; Chan, P.-C. Relation between particle size and repose angles of powders. *Powder Technol.* **1976**, *15*, 129–131. [CrossRef]

9. Nan, W.; Wang, Y.; Ge, Y.; Wang, J. Effect of shape parameters of fiber on the packing structure. *Powder Technol.* **2014**, *261*, 210–218. [CrossRef]

10. Zou, R.; Yu, A. Evaluation of the packing characteristics of mono-sized non-spherical particles. *Powder Technol.* **1996**, *88*, 71–79. [CrossRef]

11. Gan, J.; Yu, A.; Zhou, Z. DEM simulation on the packing of fine ellipsoids. *Chem. Eng. Sci.* **2016**, *156*, 64–76. [CrossRef]

12. Bernal, J.D.; Mason, J. Packing of Spheres: Co-ordination of Randomly Packed Spheres. *Nature* **1960**, *188*, 910–911. [CrossRef]

13. Scott, G.D.; Kilgour, D.M. The density of random close packing of spheres. *J. Phys. D Appl. Phys.* **1969**, *2*, 863–866. [CrossRef]

14. Mueller, G.E. Numerically packing spheres in cylinders. *Powder Technol.* **2005**, *159*, 105–110. [CrossRef]

15. Zhao, J.; Li, S.; Jin, W.; Zhou, X. Shape effects on the random-packing density of tetrahedral particles. *Phys. Rev. E* **2012**, *86*, 03131–031306. [CrossRef] [PubMed]

16. Tangri, H.; Guo, Y.; Curtis, J.S. Packing of cylindrical particles: DEM simulations and experimental measurements. *Powder Technol.* **2017**, *317*, 72–82. [CrossRef]

17. Lu, P.; Li, S.; Zhao, J.; Meng, L. A computational investigation on random packings of sphere-spherocylinder mixtures. *Sci. China Ser. G Phys. Mech. Astron.* **2010**, *53*, 2284–2292. [CrossRef]

18. Wouterse, A.; Williams, S.R.; Philipse, A.P. Effect of particle shape on the density and microstructure of random packings. *J. Phys. Condens. Matter* **2007**, *19*, 406215. [CrossRef]

19. Kyrylyuk, A.V.; Philipse, A.P. Effect of particle shape on the random packing density of amorphous solids. *Phys. Status Solidi A* **2011**, *208*, 2299–2302. [CrossRef]

20. Abreu, C.R.; Tavares, F.W.; Castier, M. Influence of particle shape on the packing and on the segregation of spherocylinders via Monte Carlo simulations. *Powder Technol.* **2003**, *134*, 167–180. [CrossRef]

21. Natsui, S.; Ueda, S.; Nogami, H.; Kano, J.; Inoue, R.; Ariyama, T. Analysis on Non-Uniform Gas Flow in Blast Furnace Based on DEM-CFD Combined Model. *Steel Res. Int.* **2011**, *82*, 964–971. [CrossRef]

22. Wei, H.; Zan, L.; Zhang, H.; Saxen, H.; Yu, Y. Two Methods of DEM Application on Charging System of Ironmaking Blast Furnace. In Proceedings of the 2018 37th Chinese Control Conference (CCC), Wuhan, China, 25–27 July 2018; pp. 3412–3415.

23. Yang, W.; Zhou, Z.; Pinson, D.; Yu, A. A New Approach for Studying Softening and Melting Behavior of Particles in a Blast Furnace Cohesive Zone. *Metall. Mater. Trans. B* **2015**, *46*, 977–992. [CrossRef]

24. Geleta, D.D.; Lee, J. Effects of Particle Diameter and Coke Layer Thickness on Solid Flow and Stress Distribution in BF by 3D Discrete Element Method. *Met. Mater. Trans. A* **2018**, *49*, 3594–3602. [CrossRef]

25. Wei, H.; Tang, X.; Ge, Y.; Li, M.; Saxén, H.; Yu, Y. Numerical and experimental studies of the effect of iron ore particle shape on repose angle and porosity of a heap. *Powder Technol.* **2019**, *353*, 526–534. [CrossRef]

26. Cundall, P.A.; Strack, O.D.L. Discussion: A discrete numerical model for granular assemblies. *Geotechnique* **2008**, *29*, 331–336. [CrossRef]

27. Tsuji, Y.; Tanaka, T.; Ishida, T. Lagrangian numerical simulation of plug flow of cohesionless particles in a horizontal pipe. *Powder Technol.* **1992**, *71*, 239–250. [CrossRef]

28. Goniva, C.; Kloss, C.; Hager, A.; Pirker, S. *An Open Source CFD-DEM Perspective*; JKU: Linz, Austria, 2010.

29. Yu, Y.; Saxén, H. Experimental and DEM study of segregation of ternary size particles in a blast furnace top bunker model. *Chem. Eng. Sci.* **2010**, *65*, 5237–5250. [CrossRef]

30. Li, C.; Honeyands, T.; O'Dea, D.; Moreno-Atanasio, R. The angle of repose and size segregation of iron ore granules: DEM analysis and experimental investigation. *Powder Technol.* **2017**, *320*, 257–272. [CrossRef]

31. Zhou, Y.C.; Xu, B.H.; Yu, A.B.; Zulli, P. Numerical investigation of the angle of repose of monosized spheres. *Phys. Rev. E* **2001**, *64*, 021301. [CrossRef] [PubMed]

32. Zhou, Y.; Xu, B.; Yu, A.; Zulli, P. An experimental and numerical study of the angle of repose of coarse spheres. *Powder Technol.* **2002**, *125*, 45–54. [CrossRef]

33. Zhou, Y.; Wright, B.; Yang, R.; Xu, B.; Yu, A. Rolling friction in the dynamic simulation of sandpile formation. *Phys. A Stat. Mech. Appl.* **1999**, *269*, 536–553. [CrossRef]

Article

Validation of the Burden Distribution of the 1/3-Scale of a Blast Furnace Simulated by the Discrete Element Method

Hiroshi Mio [1,*], Yoichi Narita [2], Kaoru Nakano [1] and Seiji Nomura [1]

[1] Ironmaking Research Laboratory, Process Technology Laboratories, Nippon Steel Corporation, 20-1 Shintomi, Futtsu, Chiba 293-8511, Japan; nakano.s6x.kaoru@jp.nipponsteel.com (K.N.); nomura.e9c.seiji@jp.nipponsteel.com (S.N.)

[2] Ironmaking Div., Nagoya Works, Nippon Steel Corporation, 5-3 Tokaimachi, Tokai, Aichi 476-8686 Japan; narita.58d.yoichi@jp.nipponsteel.com

[*] Correspondence: mio.h2s.hiroshi@jp.nipponsteel.com; Tel.: +81-70-3914-4715

Received: 26 November 2019; Accepted: 17 December 2019; Published: 18 December 2019

Abstract: The objective of this paper was to develop a prediction tool for the burden distribution in a charging process of a bell-less-type blast furnace using the discrete element method (DEM). The particle behavior on the rotating chute and on the burden surface was modeled, and the burden distribution was analyzed. Furthermore, the measurements of the burden distribution in a 1/3-scale experimental blast furnace were performed to validate the simulated results. Particle size segregation occurred during conveying to the experimental blast furnace. The smaller particles were initially discharged followed by the larger ones later. This result was used as an input in the simulation. The burden profile simulated using DEM was similar to the experimental one. The terrace was found at the burden surface subsequent to ore-charging, and its simulated position simulated agreed with that of the experimental result. The surface angle of the ore layer was mostly similar. The simulated ore to coke mass ratio (O/C) distribution in the radial direction and the mean particle diameter distribution correlated with the experimental results very well. It can be concluded that this method of particle simulation of the bell-less charging process is highly reliable in the prediction of the burden distribution in a blast furnace.

Keywords: DEM; blast furnace; burden distribution; particle flow; validation

---

1. Introduction

A blast furnace is a reactor with approximately 5000 m^3 of volume to produce pig iron from ore particles. Iron ore (sinter, lump, and pellet) and coke particles are alternately stacked at the topmost layer of the blast furnace, and the hot gas is blown from tuyeres at the bottom of the furnace. Iron ore particles are reduced during the descent, and numerous physical changes and chemical reactions occur between each phase over this period. Thus, it is an extremely complicated system, resulting in the possibility of unfavorable phenomena or serious problems occurring. To avoid these problems, controlling and stabilizing the gas flow in the furnace is of utmost importance because the gas plays a key role for the reduction and the heat source. Thus, keeping the gas flow in the steady state leads to an efficient and low RAR (reducing agent ratio) operations. Therefore, controlling a void fraction in the stacked layer, i.e., the burden distribution at the top of the blast furnace, is the most effective operation for stabilizing the gas flow. Much research has been experimentally conducted and some mathematical models have been proposed to estimate and control the burden distribution [1–5]. These models can give useful information in the daily operation. However, many kinds of particles are usually mixed in ore charging to help the reduction; therefore, it is necessary to analyze the individual solid particles' behavior in the blast furnace for an in-depth analysis of several phenomena that were previously mentioned. The discrete element method (DEM) [6] is one of the most reliable simulation methods for analysis of the solid particle behavior, and an approach using the computational simulation based on DEM is extremely useful to grasp the phenomena found in the charging process of the blast furnace. Some studies on the modeling of solid flow in a blast furnace have been previously reported, for example, the raceway [7,8], solid flow in the blast furnace [9–11], gas–solid flow [12], hopper flow [13,14], and particle trajectory from the rotating chute [15], and authors have developed burden distribution simulators using DEM [16–20]. Validations of the particle trajectory discharged from the rotating chute were studied, and they showed good correlations [19,20]. This simulator still remains a key issue to validate the simulated burden distribution with experimental results in detail.

In this paper, charging tests were performed using a 1/3-scale experimental burden distribution simulator of the blast furnace to validate the simulated results, and the particle size segregation and ore to coke mass ratio (O/C) were investigated. Furthermore, the particle flow during charging into the experimental blast furnace was modeled using DEM, and the results were compared with the experimental ones.

2. Experimental

A 1/3-scale experimental burden distribution simulator, which is shown in Figure 1, was used herein. It is a bell-less-type blast furnace and it has approximately a 3.7-m throat diameter and approximately a 10-m height. Furthermore, it has a surge hopper, parallel hoppers, and a discharging funnel. A horseshoe-shaped rotating chute with a length of 1.7 m is installed at the top of the furnace. A blower and a discharging unit are installed at the bottom of the furnace to assess the effects of the gas flow and the burden descending. Figure 2 shows the detailed schematic illustration of the experimental apparatus.

Figure 1. Picture of the 1/3-scale experimental burden distribution simulator.

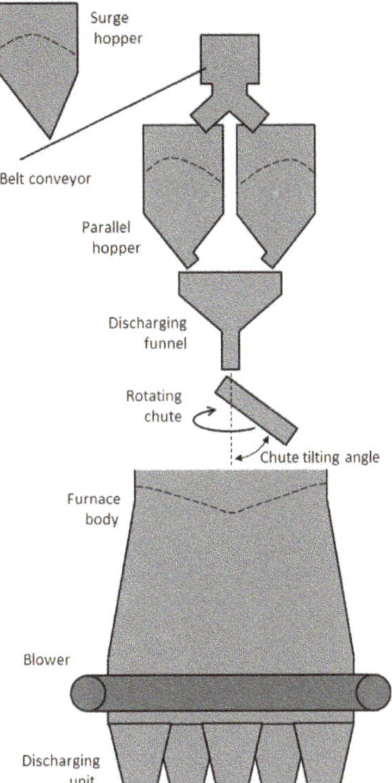

Figure 2. Schematic illustration of the 1/3-scele experimental burden distribution simulator.

Sinter particles were conveyed to the top of the experimental blast furnace via a surge hopper, one of the parallel hoppers, and a discharging funnel, as shown in Figure 2. The sinter, which were sieved in the range 5–20 mm (d_{50} = 11.4 mm), were used in this charging test. The particle size distribution is shown in Figure 3. In total, 5500 kg of sinter was charged into the experimental blast furnace during 15 rotations at 13.4 rpm. Table 1 shows a charging pattern of coke and ore dumps. The gas was not blown in this experiment to simplify the phenomena.

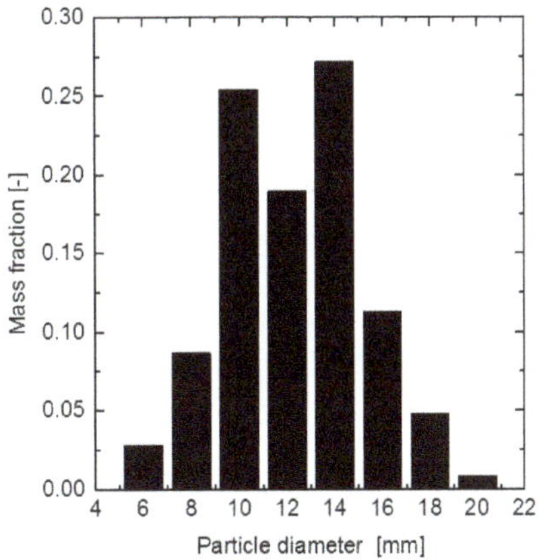

Figure 3. Distribution of the particle diameter of the sinter.

Table 1. Charging conditions for the rotating chute.

Chute Angle [°]	52	50.5	49	47.5	46	44.5	43	41	39	37	35	33	Mass [kg]
Coke		2	2	1	1	1	1	1	1				1250
Ore	2	2	1	1	1	1	1	1	1	1	1	2	5500

Subsequent to charging, a burden surface profile was measured with a laser distance meter. Moreover, the burden was dug up using a vacuum cleaner to obtain the ore to coke mass ratio (O/C) and the particle size distribution. The area that was dug up was rectangular (200 mm × 262 mm) from the furnace wall to its center for seven divisions, and the digging-up depth was every 50 mm, as shown in Figure 4. This was carefully performed to avoid breaking the particle-packing structure. The burden was removed by a large vacuum cleaner, except for the sampling area, before digging up. That is to say, the sampling area became the highest in the burden. The digging up was carried out from the highest position, thus collapse did not occur. Subsequently, the particles were sieved and weighed. The sinter particles, which were discharged from the funnel, were also sampled to check the particle size segregation during conveying to the top of the furnace. This gave the time evolution of the particle size distribution that was used as input of the DEM calculation.

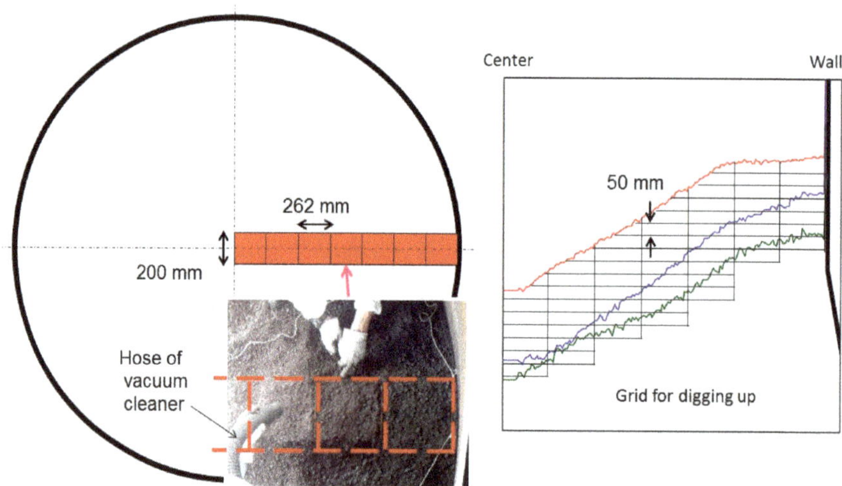

Figure 4. Schematic illustration of the digging-up operation.

3. Simulation

3.1. Discrete Element Method

DEM is one of the most popular and reliable simulation methods for the numerical analysis of particle behavior. This simulation method comprises an idea for determining the kinematic force to each finite-sized particle. The key calculation of DEM comprises three steps; i.e., (1) contact detection, (2) calculation of forces, and (3) updating the trajectories, and these processes are looped until $t = t_{max}$. The contact between two particles is given using Voigt model, which consists of a spring dashpot and a slider for the friction in the tangential component. The contact forces, $\mathbf{F}_n$ and $\mathbf{F}_t$, are calculated using:

$$\mathbf{F}_{n,ij} = \left(K_n \Delta u_{n,ij} + \eta_n \frac{\Delta u_{n,ij}}{\Delta t} \right) \mathbf{n}_{ij}, \tag{1}$$

$$\mathbf{F}_{t,ij} = \min\left\{\mu\left|\mathbf{F}_{n.ij}\right|\mathbf{t}_{ij}, \left[K_t\left(\Delta u_{t,ij} + \Delta\phi_{ij}\right) + \eta_t\left(\frac{\Delta u_{t,ij} + \Delta\phi_{ij}}{\Delta t}\right)\right]\mathbf{t}_{ij}\right\}, \tag{2}$$

where K and η are the spring and damping coefficients, Δu and $\Delta\varphi$ are the relative translational displacement of the gravitational center between two particles and the relative displacement at the contact point caused by the particle rotation, μ is the frictional coefficient, and $\mathbf{n}_{ij}$ and $\mathbf{t}_{ij}$ denote the unit vector from i-th particle to the j-th one in the normal and the tangential components. The subscript "n" and "t" denote the normal and the tangential components. The translational and rotational motions of each particle are updated using:

$$\dot{\mathbf{v}} = \frac{\sum \mathbf{F}}{m}, \tag{3}$$

$$\dot{\omega} = \frac{\sum \mathbf{M}}{I}, \tag{4}$$

where $\mathbf{v}$ is the vector of a particle velocity, $\mathbf{F}$ is the contact force acting on a particle, m and g are the mass of a particle and the gravitational acceleration, ω is the vector of the angular velocity, and $\mathbf{M}$ and I are the moment caused by the tangential force and the moment of inertia.

The shape of granular material in DEM is usually assumed to be spherical to simplify the contact detection and the calculation of the contact force, although the shape of the sinter particle is not spherical. The best solution for considering the particle shape in DEM is to model the exact particle shape using polyhedra. However, this calculation is extremely laborious, and it is unsuitable for the simulation of particle flow in the ironmaking process where the number of particles could be in the billions. Thus, the effect of the particle shape on the motion was considered by setting a proper rolling friction for the particle, and it is given by:

$$\mathbf{M}_{r,i} = -\frac{3}{8}\alpha_i b\left|\mathbf{F}_n\right|\frac{\omega_i}{|\omega_i|}, \tag{5}$$

where b is the radius of the contact area and α_i is the coefficient of the rolling friction. Every particle has different α_i, because the shapes of the sinter are totally different from each other. Its distribution is related with the rollability of the particle [16], and it is shown in Figure 5. The method of having different rolling friction provides a significant agreement with the experimental results [16]; therefore, this method was applied here.

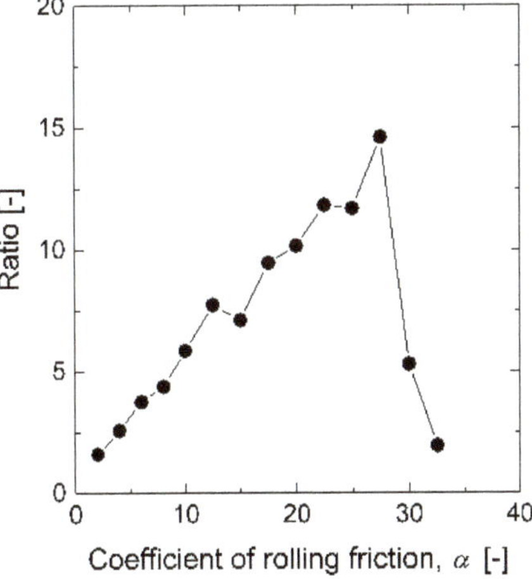

Figure 5. Distribution of the coefficient of the rolling friction [16].

3.2. Simulation Conditions

The particle behavior during charging into the experimental blast furnace was simulated using DEM, and the burden distribution was compared with the experimental results to validate the simulation results. The geometry of the rotating chute and the throat of the furnace were identical to those of the experimental results, and 5500 kg of sinter particles, with a particle density of 3300 kg/m^3, were charged into the furnace. The size of sinter particle was 6 to 20 mm, and its particle size distribution corresponded to Figure 3. The total number of sinter particles was 2,545,086, and the detailed particle condition is tabulated in Table 2. Young's modulus and Poisson's ratio were assumed to be 3.5 GPa and 0.25, respectively. Only the ore-charging process was executed in the simulation, i.e., the coke layer prior to ore charging was arranged at the top of the furnace corresponding to the surface profile subsequent to coke charging in the experimental test. The particle diameter of coke is 7.5 to 30 mm; its density, Young's modulus, and Poisson's ratio are 1050 kg/m^3, 0.54 GPa, and 0.22, respectively; and the number of coke particles is 1,049,686. Table 3 shows the detailed conditions for the coke particles. The input sinter particles were generated at the outlet of the discharging funnel, i.e., the particle behavior from the parallel hopper and the discharging funnel was not simulated to reduce the computing time. The time changes in the mass ratio for each particle of the input sinter corresponded to the sampled results in the experimental work, which was described above, to consider the particle size segregation during conveying to the furnace. The input particles have 5.5 m/s of vertical velocity, which was also measured using high speed video camera [20]. Every particle has a different rolling friction coefficient, which was obtained by generating a random number at the beginning of the simulation. The distribution of the rolling friction coefficient is shown in Figure 5. The charging pattern of the chute tilting angle was the same as that of the experimental one, as shown in Table 1, and the rotational speed was 13.4 rpm. The discrete time was 1.5 μs and the total number of calculation steps was 50 million. The calculation was parallelized using OpenMP.

Table 2. Condition for sinter particle in DEM.

Particle Diameter [mm]	Number of Particle [-]	Mass Fraction [-]
6	414,382	0.028
8	539,490	0.087
10	809,088	0.254
12	349,403	0.190
14	315,434	0.272
16	87,722	0.113
18	26,239	0.048
20	3328	0.008

Table 3. Condition for coke particle in DEM.

Particle Diameter [mm]	Number of Particle [-]	Mass Fraction [-]
7.5	695,453	0.151
12.5	216,471	0.218
17.5	90,753	0.251
22.5	34,048	0.200
30	12,961	0.180

4. Results and Discussions

Figure 6 shows the relation between the normalized discharging time and the mass fraction of each size range of the sinter particles that were discharged from the funnel. It is confirmed that the smaller particles were discharged during the initial period, whereas the larger ones were discharged later. The particle size segregation occurred during charging and discharging at the storages (the surge hopper and the parallel hoppers); therefore, the larger particles tend to be discharged later. This phenomenon affects the radial particle size distribution of the burden. This result of the time change of the particle size during discharging was used as the input particle condition of DEM.

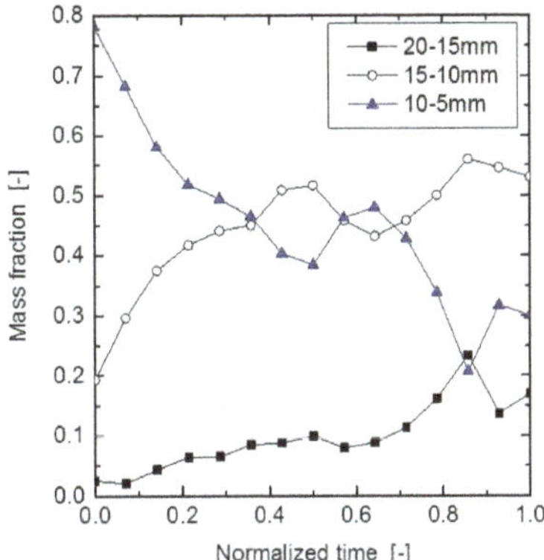

Figure 6. Relation between the mass fraction of each size range of the sinter particles and the normalized discharging time.

Figure 7 shows pictures during ore charging in the experimental test. The black particles in the furnace are coke, and the dark brown ones, which are charged from the rotating chute, are sinter. The sinter particles are stacked near the wall at the beginning of charging, and subsequently, the particles flow toward the center. The sinter covers the coke layer after the 11 rotations. Figure 8 shows the surface profile of the burden after charging. The surface angle of the ore layer is approximately 32.1°. A terrace is found around 1185 mm from the center, and the angle of the terrace is 13.7°. The surface angle of the coke layer is also found to be 36.7°. This value is larger than that of the sinter layer because of the particle shape and the size distribution.

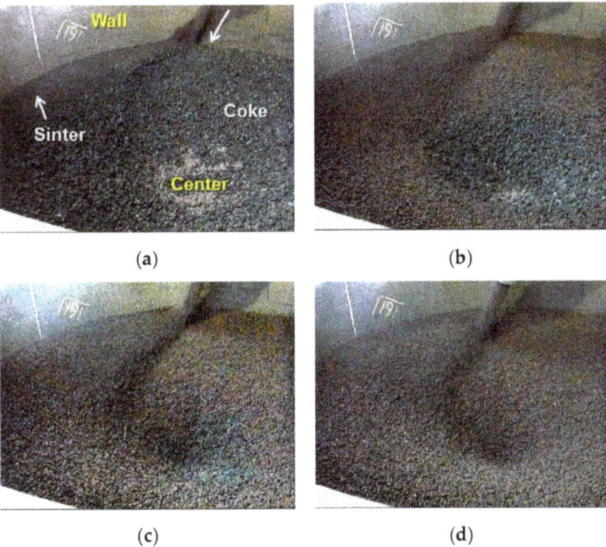

Figure 7. Pictures during ore-charging in the experimental work. (**a**) 1st rotation; (**b**) 4th rotation; (**c**) 7th rotation; (**d**) 11th rotation.

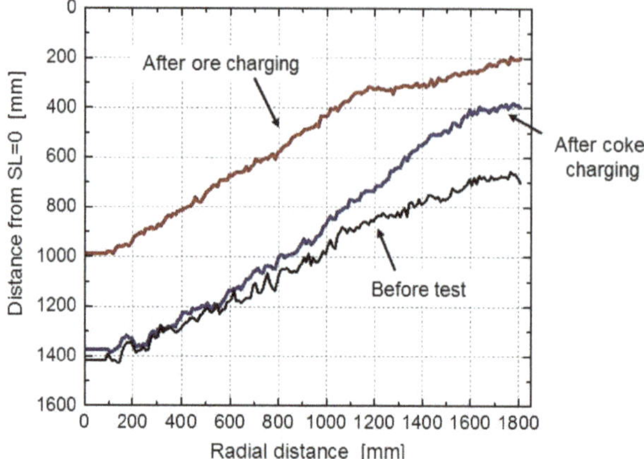

Figure 8. Surface profile of the burden after the charging test.

Figure 9 shows the contour map for the sinter volume fraction in the burden, which was obtained by digging up the burden. The surface profiles subsequent to charging are also drawn in the contour. Most coke particles are situated near the wall, and the thickness of the coke layer around the center is extremely thin. Therefore, it is suggested that a collapse of the coke layer during ore-charging was not significant in this charging condition.

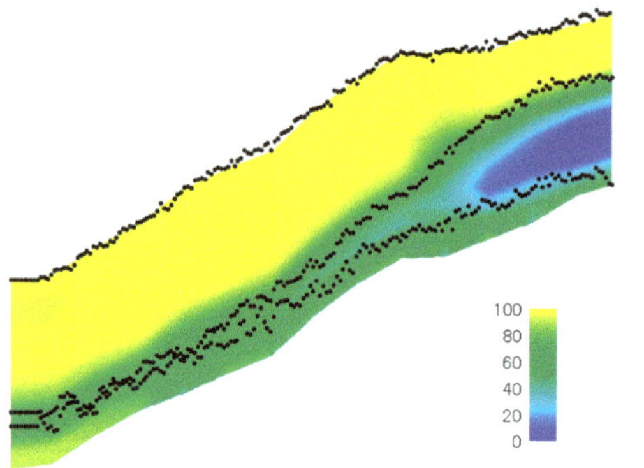

Figure 9. Contour map for the sinter volume fraction in the burden.

Figure 10 shows snapshots of the charging behavior simulated by DEM. The brown particles denote sinter and the blue ones are coke. The sinter particles are charged around the wall toward the center, and they reach the center in approximately 11 chute rotations. Their behavior is found to be very similar to that of the experimental ones, which are shown in Figure 7. Figure 11 shows a cross-section of the burden simulated using DEM. The surface angle is 33.6°, and the position of the terrace is approximately 1250 mm from the center. The terrace angle is approximately 14.2°. Although the profile and the thickness near the center are slightly different, the simulated burden layer is quite similar to the experimental one, especially the shape of the terrace.

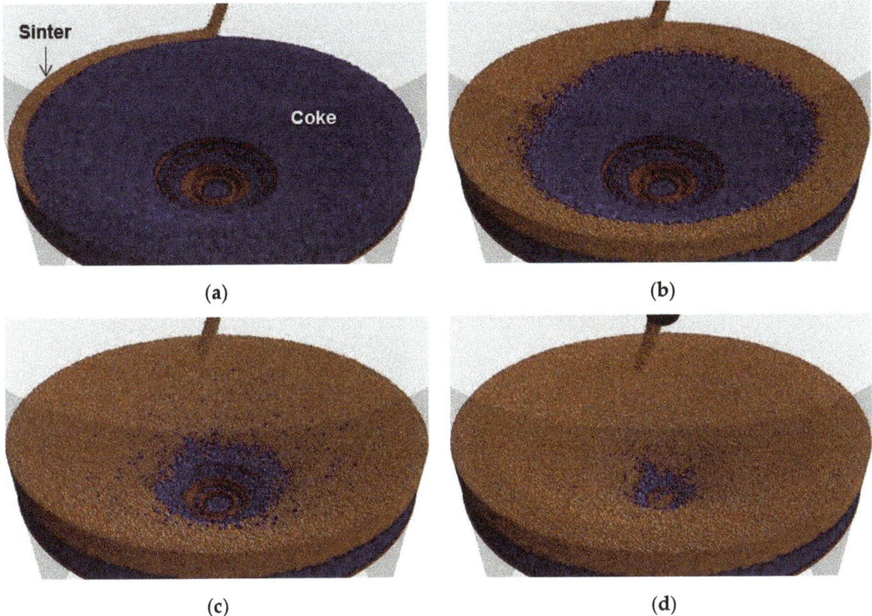

Figure 10. Snapshots of ore charging simulated by DEM. (**a**) 1st rotation; (**b**) 4th rotation; (**c**) 7th rotation; (**d**) 11th rotation.

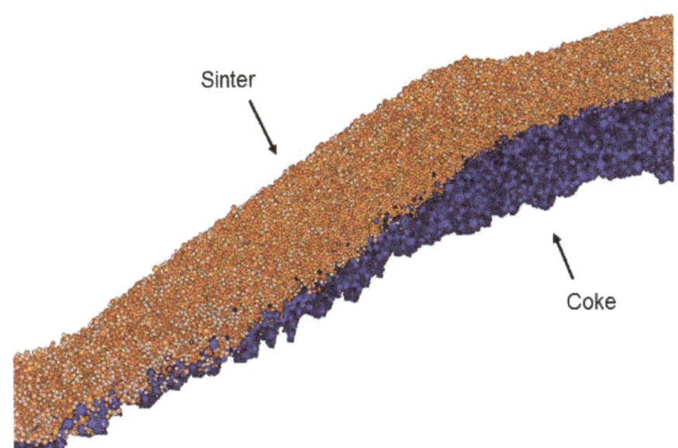

Figure 11. Cross-section of the burden simulated by DEM.

Figure 12 shows the relation between the ore to coke mass ratio (O/C) and the radial distance. The experimental results were obtained by digging up the burden. The value of O/C around the center is enormous because the thickness of the coke layer is thin, and it decreases with the increase in the radial distance because the coke layer becomes thicker. Trends of the O/C for both the experiment and the simulation are excellently correlated. Figure 13 shows the relation between the normalized mean particle diameter of the sinter and the radial distance. The particle diameter at each position was normalized using the mean value of all particles because the mean particle diameter in the experimental work became smaller than that of the initial condition. Some sinter particles had fragmented during the continuous charging into hoppers and discharging. The particle diameter at the terrace is smaller, and it increases while approaching the center due to the particle size segregation during the flow toward the center. A good agreement between the experimental results and the simulated ones was obtained. Therefore, the burden distribution simulated in this work was validated, and it has a high potential to predict the particle behavior during charging. In the simulation, the burden layer formation can be clarified. Figure 14 shows the cross-section of the ore layer, which is color-coded by the chute tilting angle. It is found that the thickness of each layer is thin, and it reaches the center. This is good information for considering the mixing particles in the ore layer. The vertical layer structure for each ring is clarified. Therefore, the discussion of the vertical position of mixed particles is possible. For example, if some particles should be mixed in the purple layer, they should be charged in the blast furnace during the rotation of 39°. It should be noted that this simulation did not consider the effect of the mixing particle and gas flow, but they will be investigated in the future work.

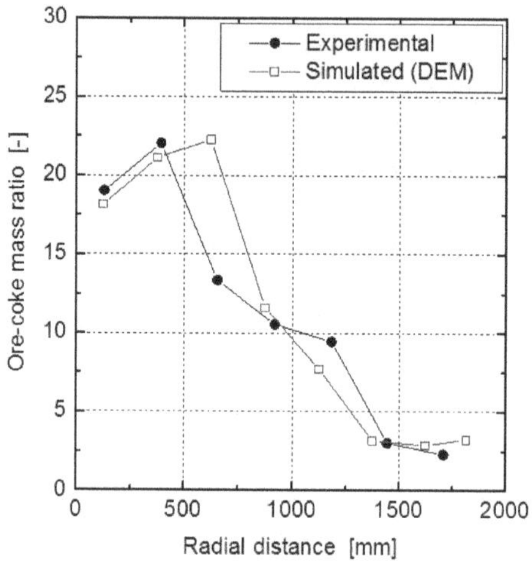

Figure 12. Relation between the ore-coke mass ratio (O/C) and the radial distance.

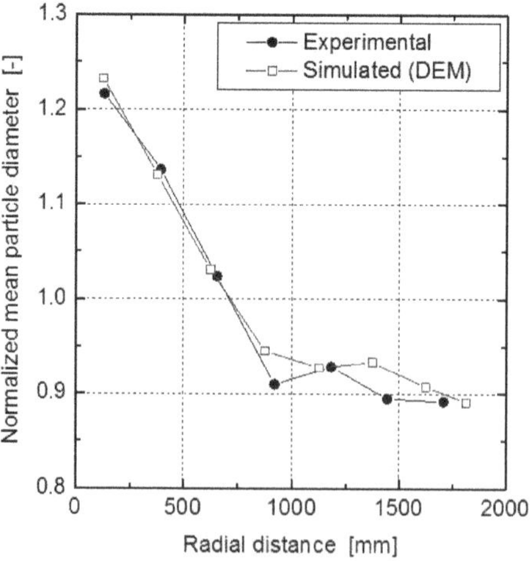

Figure 13. Relation between the normalized mean particle diameter of sinter and the radial distance.

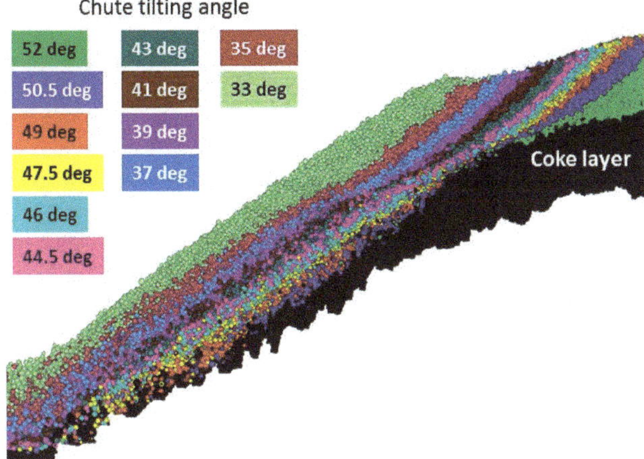

Figure 14. Cross section of the ore layer, which is color-coded by the chute tilting angle.

5. Conclusions

A particle simulation model of the charging process of the bell-less-type blast furnace was developed using DEM. The ore to coke mass ratio (O/C) and the mean particle diameter of the radial direction were compared with the experimental results, which were obtained in the 1/3-scale experimental burden distribution simulator. The following is a summary of this study:

(1) The particle size segregation occurred during conveying to the experimental blast furnace. The smaller particles were initially discharged, whereas the larger ones were discharged later.

(2) The burden profile, which was simulated using DEM, was similar to the experimental one. A terrace was found at the burden surface subsequent to ore charging, and its simulated position agreed with that of the experimental result. The surface angle was mostly similar between them.

(3) The simulated O/C distribution in the radial direction and the mean particle diameter distribution showed excellent correlation with the experimental results.

(4) It can be concluded that this method of particle simulation of the bell-less charging process is highly reliable in the prediction of the burden distribution in the blast furnace.

Author Contributions: Data curation, H.M.; Formal analysis, H.M. and Y.N.; Investigation, H.M.; Supervision, K.N. and S.N.; Validation, H.M.; Writing—original draft, H.M. All authors have read and agreed to the published version of the manuscript.

Funding: This research receive no external funding.

Conflicts of Interest: The authors declare no conflicts of interest.

References

1. Okuno, Y.; Matsuzaki, S.; Kunitomo, K.; Isoyama, M.; Kusano, Y. Development of a mathematical model to estimate burden distribution in bell-less type charging for blast furnace. *Tetsu-to-Hagané* **1987**, *73*, 91–98. [CrossRef]

2. Sawada, T.; Uetani, T.; Taniyoshi, S.; Miyagawa, M.; Sugawara, H.; Yamazaki, M. Blast furnace operation and burden distribution control with bell-less top of the 3 parallel bunker type. *Tetsu-to-Hagané* **1992**, *78*, 1337–1344. [CrossRef]

3. Hattori, M.; Iino, B.; Shimomura, A.; Tsukiji, H.; Ariyama, T. Development of burden distribution simulation model for the bell-less top in a large blast furnace and its application. *Tetsu-to-Hagané* **1992**, *78*, 1345–1352. [CrossRef]

4. Mitra, T.; Saxén, H. Model for fast evaluation of charging programs in the blast furnace. *Metall. Mater. Trans. B* **2014**, *45*, 2382–2394. [CrossRef]

5. Fu, D.; Chen, Y.; Zhou, C.Q. Mathematical modeling of blast furnace burden distribution with non-uniform descending speed. *Appl. Math. Model.* **2015**, *39*, 7554–7567. [CrossRef]

6. Cundall, P.A.; Strack, O.D.L. Discrete numerical model for granular assemblies. *Geotechnique* **1979**, *29*, 47–65. [CrossRef]

7. Nogami, H.; Yamaoka, H.; Takatani, K. Raceway design for the innovative blast furnace. *ISIJ Int.* **2004**, *44*, 2150–2158. [CrossRef]

8. Yuu, S.; Umekage, T.; Miyahara, T. Prediction of stable and unstable flows in blast furnace raceway using numerical simulation methods for gas and particles. *ISIJ Int.* **2005**, *45*, 1406–1415. [CrossRef]

9. Nouchi, T.; Sato, T.; Sato, M.; Takeda, K.; Ariyama, T. Stress field and solid flow analysis of coke packed bed in blast furnace based on DEM. *ISIJ Int.* **2005**, *45*, 1426–1431. [CrossRef]

10. Zhou, Z.; Zhu, H.; Yu, A.; Wright, B.; Pinson, D.; Zulli, P. Discrete particle simulation of solid flow in a model blast furnace. *ISIJ Int.* **2005**, *45*, 1828–1837. [CrossRef]

11. Mio, H.; Yamamoto, K.; Shimosaka, A.; Shirakawa, Y.; Hidaka, J. Modeling of solid particle flow in blast furnace considering actual operation by large-scale Discrete Element Method. *ISIJ Int.* **2007**, *47*, 1745–1752. [CrossRef]

12. Natsui, S.; Nogami, H.; Ueda, S.; Kano, J.; Inoue, R.; Ariyama, T. Simultaneous three-dimensional analysis of gas–solid flow in blast furnace by combining Discrete Element Method and Computational Fluid Dynamics. *ISIJ Int.* **2011**, *51*, 51–58. [CrossRef]

13. Wu, S.; Kou, M.; Xu, J.; Guo, X.; Du, K.; Shen, W.; Sun, J. DEM simulation of particle size segregation behavior during charging into and discharging from a Paul-Wurth type hopper. *Chem. Eng. Sci.* **2013**, *99*, 314–323. [CrossRef]

14. Yu, Y.; Saxén, H. Segregation behavior of particles in a top hopper of a blast furnace. *Powder Technol.* **2014**, *262*, 233–241. [CrossRef]

15. Liu, S.; Zhou, Z.; Dong, K.; Yu, A.; Pinson, D.; Tsalapatis, J. Numerical investigation of burden distribution in a blast furnace. *Steel Res. Int.* **2015**, *86*, 651–661. [CrossRef]

16. Mio, H.; Komatsuki, S.; Akashi, M.; Shimosaka, A.; Shirakawa, Y.; Hidaka, J.; Kadowaki, M.; Matsuzaki, S.; Kunitomo, K. Validation of particle size segregation of sintered ore during flowing through laboratory-scale chute by Discrete Element Method. *ISIJ Int.* **2008**, *48*, 1696–1703. [CrossRef]

17. Mio, H.; Komatsuki, S.; Akashi, M.; Shimosaka, A.; Shirakawa, Y.; Hidaka, J.; Kadowaki, M.; Matsuzaki, S.; Kunitomo, K. Effect of chute angle on charging behavior of sintered ore particles at bell-less type charging system of blast furnace by Discrete Element Method. *ISIJ Int.* **2009**, *49*, 479–486. [CrossRef]

18. Mio, H.; Kadowaki, M.; Matsuzaki, S.; Kunitomo, K. Development of particle flow simulator in charging process of blast furnace by discrete element method. *Miner. Eng.* **2012**, *33*, 27–33. [CrossRef]

19. Mio, H.; Nakauchi, T.; Kawaguchi, Y.; Enaka, T.; Narita, Y.; Inayoshi, A.; Matsuzaki, S.; Orimoto, T.; Nomura, S. High-speed video recording of particle trajectory via rotating chute of Nagoya No.3 blast furnace and its comparison with simulated behavior using DEM. *ISIJ Int.* **2017**, *57*, 272–278. [CrossRef]

20. Mio, H.; Narita, Y.; Matsuzaki, S.; Nishioka, K.; Nomura, S. Measurement of particle charging trajectory via rotating chute of 1/3-scale blast furnace and its comparing with numerical analysis using Discrete Element Method. *Powder Technol.* **2019**, *344*, 797–803. [CrossRef]

Article

A Mathematical Model Combined with Radar Data for Bell-Less Charging of a Blast Furnace

Meng Li [1], Han Wei [1], Yao Ge [1], Guocai Xiao [2] and Yaowei Yu [1,*]

[1] State Key Laboratory of Advanced Special Steel, Shanghai Key Laboratory of Advanced Ferrometallurgy, School of Materials Science and Engineering, Shanghai University, Shanghai 200444, China; L_limeng@outlook.com (M.L.); weihan@shu.edu.cn (H.W.); ge_geyao@163.com (Y.G.)

[2] Jinheng Information Technology Company, Jiangsu Province, Nanjing 210035, China; xiaoguocai@yeah.net

* Correspondence: yaowei.yu@hotmail.com

Received: 8 December 2019; Accepted: 17 February 2020; Published: 20 February 2020

Abstract: Charging directly affects the burden distribution of a blast furnace, which determines the gas distribution in the shaft of the furnace. Adjusting the charging can improve the distribution of the gas flow, increase the gas utilization efficiency of the furnace, reduce energy consumption, and prolong the life of the blast furnace. In this paper, a mathematical model of blast furnace charging was developed and applied on a steel plant in China, which includes the display of the burden profile, burden layers, descent speed of the layers, and ore/coke ratio. Furthermore, the mathematical model is developed to combine the radar data of the burden profile. The above model is currently used in Nanjing Steel as a reference for operators to adjust the charging. The model is being tested with a radar system on the blast furnace.

Keywords: blast furnace; charging system; mathematical model; radar data; burden distribution

1. Introduction

The raw material used in the production of a blast furnace is called burden. It is mainly composed of coke, sinter, and pellet. The prepared burden is loaded into a hopper. After a series of transportation steps on the top of the furnace, it falls onto a rotating chute, when the exit of the hopper opens. Then, the material enters the blast furnace and forms burden layers in the throat. At the bottom, hot air is blown into the furnace to burn the coke, producing carbon monoxide and hydrogen that act as reductants. The rising reductants react chemically with iron oxide in burden and the iron oxide becomes hot metal after the reduction and melting. In the reduction, impurities in iron ore combine with the added flux to form molten slag. Hot metal and slag are discharged from tapholes at the bottom of the furnace. After treatment, molten slag is used as a raw material for cement and hot metal is transported to basic oxygen furnace by torpedo.

Some studies have shown that the charging affects the chemical reaction between the layers and reducing gas in the shaft of the furnace. The layers influence the shape of the cohesive zone and energy utilization efficiency of the furnace [1]. Therefore, the charging system is of great significance for the operation of the blast furnace. Due to the high temperature, high pressure, and dusty environment of the blast furnace, the internal state of the furnace cannot be measured. Even though furnace top temperature detection equipment, cross temperature measurement, and furnace top infrared cameras are already applied in the furnace, they still cannot accurately provide information about the burden layers in the furnace. Simulation technology is a powerful method combining computer, mathematics, physical engineering, and chemical engineering. It is helpful to observe the phenomena that are difficult to be directly measured in practice. Meanwhile, it saves costs, time, and materials in experiments [2].

Many studies have used mathematical models to study the charging system of the blast furnace. Yoshimasa et al. [3] developed a simulation model for the burden distribution of blast furnace charging and studied the trajectory of the raw material, burden descent, and mixing layer, providing important information for the subsequent model development. Pohang Iron and Steel Company proposed a radial distribution function of the burden and applied it in an actual blast furnace to study the distribution characteristics of the burden and to improve the distribution of the gas flow [4]. Krishman et al. [5] developed a mathematical model for the optimization of bell-less charging, and the calculation results were consistent with the actual data. Saxén and Hinnelä [6] developed a bell-less burden distribution model on the basis radar measurement, and the dependence between the layer thickness and charging variables was modeled by neural networks [7]. Nag [8] proposed a mathematical model of the bell-less top to calculate the trajectory of the burden in charging. Park et al. [9] analyzed the blast furnace charging system by developing a burden descent model and a gas flow model, and compared the results with those from a 1/12-scaled model experiment. Samik et al. [10] proposed a general target methodology to estimate the stock profile in the blast furnace, where the burden distribution is based on experiments in different scaled models of a blast furnace with various materials. Shi et al. [11] proposed a new model of stockline profile formation in which equations were developed for the inner and the outer repose angles by considering the influence of the burden's vertical and horizontal flow.

The above mathematical models were developed based on some assumptions and different operating conditions. Therefore, it is quite difficult to apply them to get accurate results of the burden layer for other furnaces. If a charging model can be combined with a reliable burden surface detection method, the reliability of the calculated burden profile information can be increased. For example, rotating radar detection technology can more accurately measure the height of each point of the burden surface even under severe conditions, such as complete darkness and high-dust atmosphere, than a mechanical stock rod [12]. Therefore, in a black-box environment, vibration, and strong airflow, burden distribution online measurement should be stable and accurate. Considering the limits of the radar method, radar data also includes noise and can only reflect the surface shape of the object. Through the combination of radar, data processing, and the charging mathematical model, the error and noise of radar data can be reduced a lot. Furthermore, the shape of the burden surface and the structure of the burden layers can be better estimated [13].

Some scholars have studied the application of radar in blast furnaces, and some achievements have been made. Liu et al. [14] used radar data containing information of the burden surface situation and cross thermometer data reflecting the trend change of the burden surface with time as the training data for a fuzzy neural network to classify and predict the burden surface. Gao et al. [15] suggested that visualization and simulation is a new technology to monitor the charging system and to help the operation of blast furnaces. Based on the real multi-radar data, Zhu et al. [16] estimated the burden profile by a cubic-curve equation at the end of a multi-loop charging. Furthermore, the burden profile before the next multi-loop charging was calculated by considering the impact of the burden descent. Li et al. [17] used fuzzy c-means clustering to classify a large amount of burden surface radar data and proposed a multiple-model set of the burden surface. The real-time burden surface data were matched with the model to produce the expected burden surface. A reconstruction algorithm based on phased array radar data was proposed by Zhang [13] to extract the data of the blast furnace charging line and it was shown to have high efficiency and high accuracy. Tian et al. [18] developed a radar detection-based model for the prediction of the burden surface shape to develop a charging strategy and the results showed that the proposed model had the advantages of higher prediction accuracy for both local details and global shape than mechanical stock rods. In another publication [19], they proposed an innovative data-driven model for predicting the distribution of the burden descent speed. This model has the ability to better characterize the variability in the radial distribution of the burden descent speed than a pure mathematical model based on Newton's second law. Miao et al. [20] proposed a new calculation method of shape fusion of the material line based on a stacking method, which can directly calculate the shape of the material surface and improve the measurement accuracy by 4.8%, compared to the first principle mathematical model. Li et al. [21] recently presented a similar model to improve the measurement accuracy of the burden profile and to use this in modeling of the burden distribution.

From the above literature and statements, there are no publications presenting a combination of radar data and a mathematical model to test, modify, and improve the model. Therefore, this paper will concentrate on this issue.

2. Mathematical Model and Radar Data Treatment

2.1. Mathematical Model Structure

Locations of raw material trajectories, shape of the burden profile, and ratio of ore to coke on the top of the blast furnace are gained by calculating the charging process. This is important for predicting the reducing gas distribution and chemical reactions between layers and the gas in the shaft of the blast furnace. Therefore, combining the mathematical model of the charging with the experience of the production and radar data, the operation of the blast furnace can be optimized to become more stable and efficient.

From the raw material in the hopper to the formation of burden layers in the throat of the furnace, the charging process is decomposed into the trajectory of the burden flow, burden profile, burden layer structure, and burden distribution evaluation. The four steps are calculated by four models: Burden flow trajectory model, burden profile model, burden distribution model, and burden evaluation model, and are shown in Figure 1.

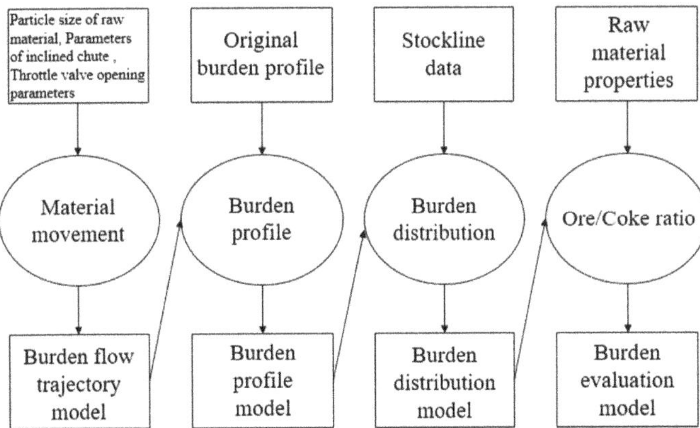

Figure 1. Mathematical model structure.

In Figure 1, after the calculation of the material movement by Newton's second law, the velocity of the burden at the hopper exit, velocity of the burden into the chute, and velocity of the burden leaving the chute tip are obtained successively. Then, based on the inclination angle and rotation speed of the chute, the trajectories of the raw material after leaving the chute tip are calculated. From the trajectories and burden profile model, the coordinates of a new burden profile can be calculated based on the original one. Then, the descent speed of the burden is used to modify the new burden profile and the modified one is saved in the database. In the next calculation, the modified one is considered as an original profile and the procedure is repeated. Finally, the ratio of ore to coke in the last two profiles is obtained and is used as a criterion of evaluation.

2.1.1. Burden Flow Trajectory Model

This part studies the movement of raw material particles and velocities of the material from the hopper to the trajectory of the burden flow. This part is divided into four sections and is shown in Figure 2. It includes the velocity of the burden at the hopper exit, velocity of the burden into the chute, velocity of the burden leaving the chute tip, and trajectory of the burden after leaving the chute tip.

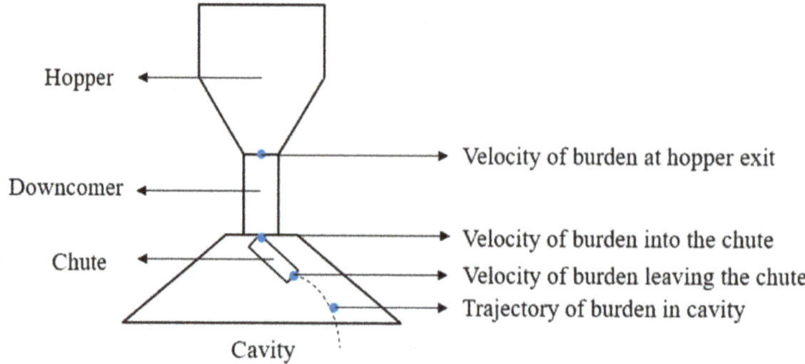

Figure 2. Trajectory of the burden flow passing through the bell-less top.

The raw material flows out from the exit of the hopper in a funnel form, and its velocity (V_0) can be described by the hydraulic formula [22]:

$$V_0 = Q/\pi(2S/C - d_i/2)^2,\tag{1}$$

where S, C, d_i, and Q express the projection area of the throttle valve (m²), circumference of the throttle (m), average particle size of the burden (m), and flow rate of the burden out the throttle valve (t/s), respectively.

According to the literature [23], the relationship between Q and A (throttle valve opening) is as follows:

Ore:

$$Q = 1 \times 10^{-5} A^3 - 7 \times 10^{-4} A^2 + 0.0366A - 0.5;\tag{2}$$

Coke:

$$Q = 3 \times 10^{-5} A^3 - 2.7 \times 10^{-3} A^2 + 0.103A - 1.29.\tag{3}$$

Before reaching the chute, raw material particles free fall with an initial velocity V_0 and collide with the wall of the downcomer (see Figure 2), causing a loss of energy (velocity), which can be calculated by the velocity attenuation factor k. Therefore, the velocity of the particles entering the chute is calculated by:

$$V_1 = \sqrt{k\cos\alpha(V_0^2 + 2g(h + b/\sin\alpha))},\tag{4}$$

where α, h, b, g, and A define the inclination angle of the chute in the vertical direction (°) (see Figure 3), height of the downcomer (m), distance from the chute suspension point to the bottom of the chute (m), acceleration due to gravity (9.81m/s²), and throttle valve opening of the hopper's exit (°), respectively.

The particles fall into the chute at the velocity (V_1) and are mainly subjected to gravitational force (F_1), supportive force (F_2), frictional force (F_3), and centrifugal force (F_4) as shown in Figure 3. The applied forces on particles along the chute can be expressed by:

$$F_1 = mg\tag{5}$$

$$F_2 = mg\sin\alpha - \omega^2 lm\sin\alpha\cos\alpha\tag{6}$$

$$F_3 = \mu F_2\tag{7}$$

$$F_4 = \omega^2 ml\sin\alpha\tag{8}$$

$$\sum F = F_1 + F_2 + F_3 + F_4,\tag{9}$$

where ω, m, l, and μ express the rotation speed of the chute ($r \cdot S^{-1}$), mass of the burden (kg), length of the chute (m), and coefficient of dynamic friction ($-$), respectively.

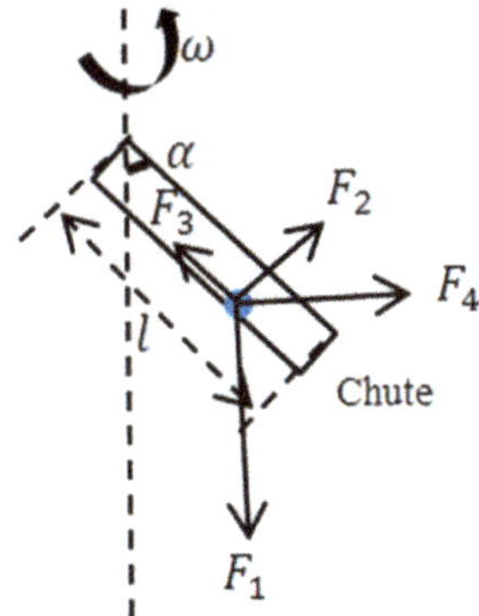

Figure 3. Schematic diagrams of the applied forces on the particle flow along the chute.

According to Newton's second law, the velocity V_2 of particles leaving the chute end can be calculated and it is decomposed into the horizontal velocity V_h, vertical velocity V_v, and tangential velocity V_t as follows:

$$V_2 = \sqrt{\omega^2 \sin\alpha(\sin\alpha + \mu\cos\alpha)l^2 + 2g(\cos\alpha - \mu\sin\alpha)l + (V_1\cos\alpha)^2}, \tag{10}$$

$$V_h = V_2 \sin\alpha \tag{11}$$

$$V_v = V_2 \cos\alpha \tag{12}$$

$$V_t = r\omega \tag{13}$$

After leaving the chute end, the burden moves with the velocity of V_2 in the throat until it falls onto the burden surface. In the movement, burden particles are subjected to gravitational force, buoyancy force, and the drag force of gas. The influence of the latter two forces on the movement of the burden is very small and is ignored [22]. Therefore, the movement of the burden is treated as a slant throw movement with gravity, as is shown in Figure 4.

The slant throw movement of particles can be decomposed into two directions: The radius of the throat (S_r) and the tangential direction of the radius (S_t). Therefore, the distance of particles from the center line of the furnace (S) to the falling point of particles with the burden profile can be calculated by:

$$S_r = r + (V_2 \sin\alpha)t \tag{14}$$

$$S_t = \omega r t \tag{15}$$

$$S = \sqrt{S_r^2 + S_t^2} \tag{16}$$

where r and t are the radial distance from the chute tip to the center line of the blast furnace (m) and the movement time of particles between leaving the chute tip and reaching the burden profile (S).

When the burden moves below the zero value of the stock line, the vertical distance between material particles and the zero value of the stock line can be expressed by:

$$H = h_2 - (h_0 - h_1), \tag{17}$$

where h_0, h_1, h_2, and H define the distance from the chute suspension point to the zero stock line (m), distance from the chute suspension point to the end of the chute (m), vertical distance of the material after leaving the chute tip (m), and distance between the material and zero value of the stock line (m), respectively.

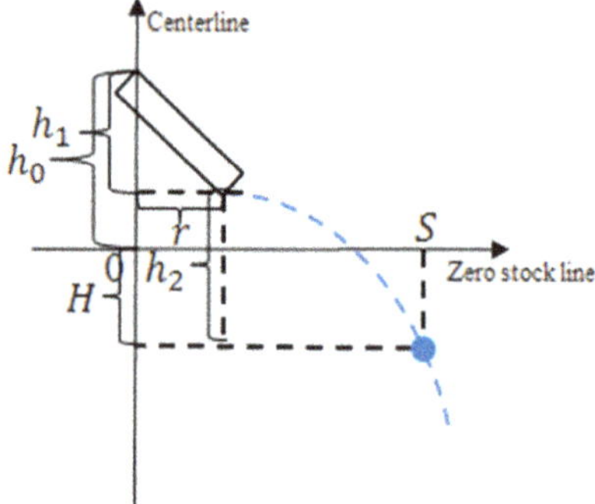

Figure 4. The trajectory of the material in the cavity, where h_0 is the distance from the chute suspension point to the zero value stock of the line (m), h_1 is the distance from the chute suspension point to the end of the chute (m), h_2 is the vertical distance of the material after leaving the chute tip (m), and H is distance between the material and zero value of the stock line (m).

2.1.2. Burden Profile Model

After the raw material moves from the chute tip, it is uniformly dumped to the burden surface of the furnace and forms a new one. For multi-ring charging programs, the material forms a new shape of the burden surface with several concentric piles. There is a certain width of the burden mass flow, after the material leaves the chute tip. Therefore, we describe the mass flow by two trajectories: The main trajectory of the material flow (mass trajectory) and the lower material flow, as shown in Figure 5.

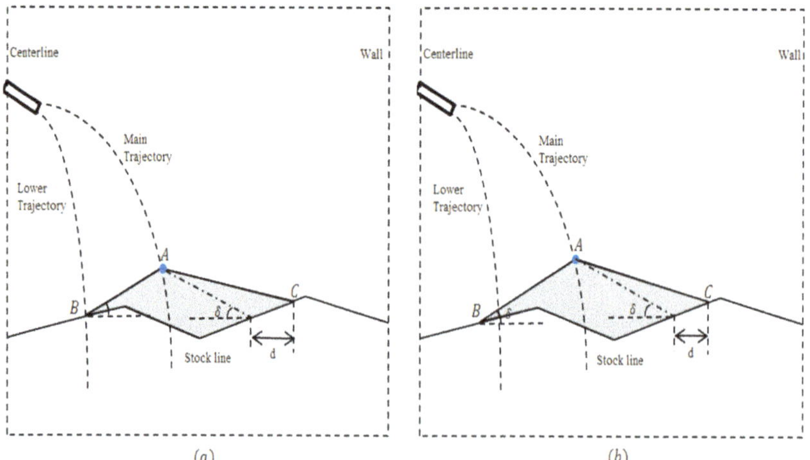

Figure 5. Schematic diagram of a new surface formation on an original one, where δ is the repose angle of the burden. (**a**) the location of the inner foot B is the intersection of the lower trajectory with the previous burden profile; (**b**) the material keeps the repose angle slipping on the inner surface of the profile.

Along the radial direction of the furnace, the surface of a new burden profile can be described by three points as shown in Figure 5: Inner foot B of the pile, outer foot C of the pile, and apex A of the pile.

When the material settles down on the inner surface of a new pile, the angle of the new pile is less than the repose angle of the raw material, and the location of the inner foot B is the intersection of the lower trajectory with the previous burden profile as shown in Figure 5a. When the material slips on the inner surface of the pile, the inner surface reaches the repose angle of the material and will keep the repose angle, as shown in Figure 5b. Therefore, more material moves towards the center of the furnace and the inner foot B is decided by the volume conservation of the raw material. For the apex A calculation, it is given by the intersection of the main trajectory with the previous burden profile. The right-side material of the main trajectory falls to the outer side of the apex A and forms the outer surface of the pile. Because the particles have a velocity component in the radial direction of the furnace, they roll along the outer surface of the pile for a while. The rolling distance of the material is constant for different materials and is 0.7 m for coke and 0.5 m for ore [23].

The new burden profile is described by the A, B, and C points along the radial direction as shown in Figure 6. It is divided into several regions for integration to obtain the volume. Therefore, the new burden volume is calculated as follows and is equal to the volume of the material in each batch:

$$V = 2\pi \int \left(f_{new}(r) - f_{ori}(r) \right) r dr = 2\pi \int_{n_i}^{n_{i+1}} \left(f_m(r) - f_n(r) \right) r dr, \tag{18}$$

where r, $f_{new}(r)$, $f_{ori}(r)$, $f_m(r)$, and $f_n(r)$ are the distance from the center line (m), new burden profile function, original burden profile function, mth burden profile function, and nth burden profile function, respectively.

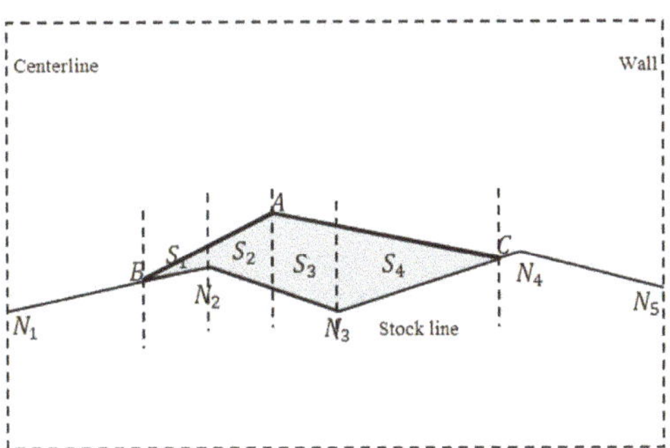

Figure 6. Burden volume partition integral.

2.1.3. Burden Distribution Model

Tapping of hot metal and molten slag from the taphole in the hearth as well as combustion and gasification of coke give the space for the burden to descend in the furnace. The shape of the layer distribution changes during the descent. Therefore, it is necessary to modify the burden profile according to the burden descent. In practice, two to four mechanical stock rods are used to measure the descent of the burden at fixed locations.

According to the measurement, Nishida et al. [24] proposed a velocity distribution of the burden descent along the radial direction of the furnace ($V_{(r)}$), which is measured directly by a profilometer and is a linear assumption based on the measured data. From the velocity distribution, the rate of the burden descent is slower in the center and faster near the wall of blast furnace:

$$V_{(r)} = a + br \tag{19}$$

$$a = \frac{R_1 V_{ch} - \frac{2}{3} R_0 V_{rod}}{R_1 - \frac{2}{3} R_0} \tag{20}$$

$$b = \frac{V_{rod} - V_{ch}}{R_1 - \frac{2}{3} R_0} \tag{21}$$

where $V_{(r)}$ is the velocity component of the burden descent at the radial position r (mm/s), R_0 is the throat radius (m), R_1 is the distance of the profilometer from the center line (m), V_{ch} is the average velocity of the burden descent (mm/s), and V_{rod} is the descent velocity at the profilometer (mm/s).

The descent velocity of the burden is only vertical in the furnace throat (V_r), and divides into vertical (V_h) and horizontal (V_r) components in the shaft region as follows:

$$V_h = V_{(r)} \sin\beta, \quad V_{r=} V_{(r)} \cos\beta, \tag{22}$$

where β defines the shaft angle of the furnace (°).

When different kinds of raw materials are charged into the throat of the furnace, the structure of alternating layers (ore layer and coke layer) forms. The burden still keeps its layer structure in the descent. Therefore, the layer structure of the burden can be used to study the distribution of the ratio of ore to coke and the permeability of the burden in the shaft.

2.1.4. Burden Evaluation Model

In order to gain a reasonable material distribution and help the operation of the blast furnace, it is necessary to evaluate the distribution of the layers. The particle average size and strength of coke are much larger and higher than those of sinter. Coke remains in a solid state to 1500 °C while sinter softens and melts below the cohesive zone in the furnace. In the blast furnace, coke has better permeability than sinter. Therefore, the mass ratio of ore to coke is used to evaluate the burden of the coke state and is calculated as follows:

$$K_{O/C} = \frac{\Delta M_O}{\Delta M_C} = \frac{[f_O(r)_n - f_C(r)_n]\rho_O}{[f_C(r)_n - f_O(r)_{n-1}]\rho_C}, \tag{23}$$

where ΔM_O is the mass of the ore layer (kg), ΔM_C is the mass of the coke layer (kg), $f_O(r)_n$ is the function of ore in the nth layer, and $f_C(r)_n$ is the function of coke in the nth layer. ρ_O and ρ_C are the ore bulk density (kg/m^3) and coke bulk density (kg/m^3), respectively.

2.2. Radar Detection Measurement

Rotating radar detection measurement is an integration system, including a mechanical radar device, signal transmission device, and signal processing system. Two rotating radars are installed on the top of the blast furnace (see Figure 7). One of them only measures half of the burden surface profile from the centerline of the furnace to the periphery. They are symmetrically distributed in the top of the furnace and can detect the full material surface together.

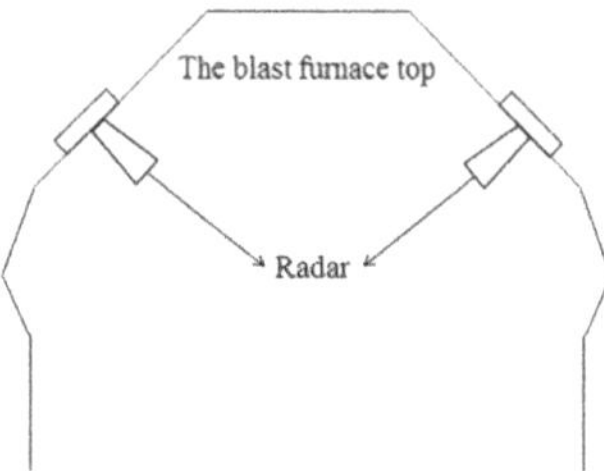

Figure 7. Schematic diagram of the radar installation in the top of the blast furnace.

After a new burden layer has been formed, the two radars are used to get the radar of the full surface. A combination of the mathematical model and radar data is divided into radar data processing and mathematical model calculation and is shown in Figure 8. After radar data processing, the burden profile function is obtained. The burden descent velocity is gained by two methods: The calculation descent function from the mathematical model and the fitting function of radar data. After the descent function, the burden profiles will present the layer structure. Then, we can evaluate the burden distribution through calculation of the ratio of ore to coke.

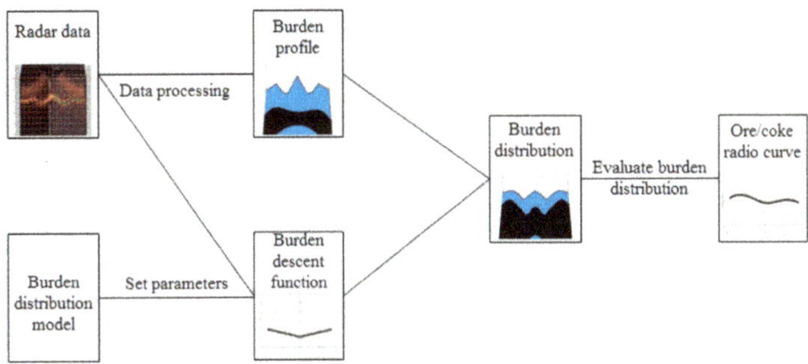

Figure 8. Procedure of the combination of the mathematical model and radar data in a software implementation.

2.2.1. Radar Data Collection

When a rotating radar works, it rotates around the axis of itself to detect the radial data of the burden profile. The radial data includes the height and the radius of the burden surface. Dust, chute shield, and airflow in the throat all interfere with the radar data, producing noise and making the measurement values deviate from the real ones.

2.2.2. Processing of Radar Data

Radar data were collected from a blast furnace of Nanjing Steel. In order to ensure the accuracy and authenticity of the data, the K nearest neighbor algorithm was firstly employed to remove the noise. Then, the Delaunay triangulation algorithm was used to realize the visualization of the 3D surface. The burden profile takes the average values of multiple coordinates extracted from the data. After the above processing, 40 group points were selected as the coordinates of the burden profile. The intersection of the furnace center line and the zero line of the stock line was defined as the origin of the coordinates. These 40 group points were converted into two-dimensional coordinates of the burden profiles.

The 40 groups of radar data were treated and stored in a database, which included the radar data time, burden profile coordinates (x, y), and material type as shown in Figure 9.

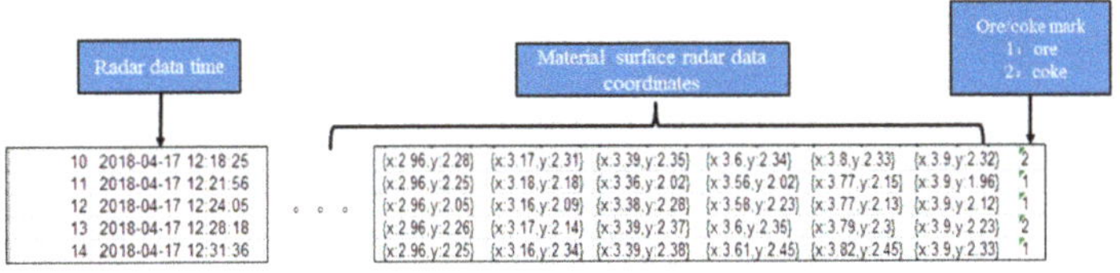

Figure 9. Radar data structure in the database.

From these 40 group of radar data, burden profiles were extracted and illustrated, as depicted in Figure 10. The radar data are independent points and discontinuous in the radial direction (red points in Figure 10). Therefore, a polyfit regression method was used to find the most consistent curve for these radar data. This not only retains the characteristics of the radar data but also makes the burden profile look continuous and smooth.

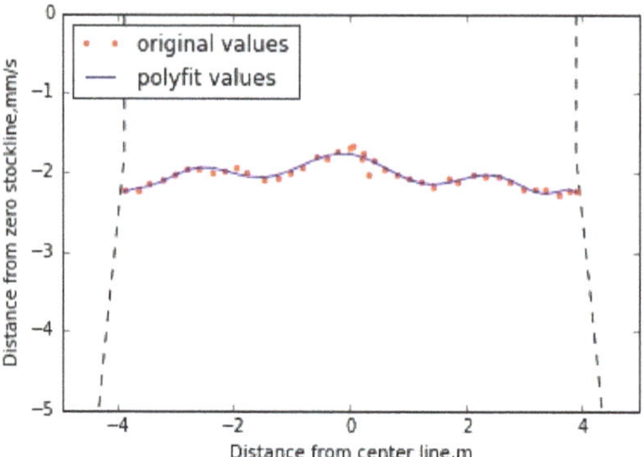

Figure 10. Radar data calculation of the burden profile function.

When the burden profile function is known, it is modified by the burden descent function. There are two methods to obtain the descent function: The one from the mathematical model (Equations (19)–(21)) and the other from the radar data fitting method.

Considering the difference between these two measurements, the descent velocity function can be calculated by:

$$V_d = \frac{f(r)_n - f(r)_{n-1}}{t_n - t_{n-1}},$$

(24)

where $f(r)_n$ and $f(r)_{n-1}$ are the nth burden distribution and the (n − 1)th one after the descent, respectively. t_n and t_{n-1} define the charging times of the nth and (n − 1)th burden distribution, respectively.

A polyfit regression method was used to obtain the descent functions show in Figure 11. The distribution of the descent function is symmetrical at the centerline of the furnace (x = 0). The red dots in Figure 11 and blue curve express the radar data and the fitted curve, respectively.

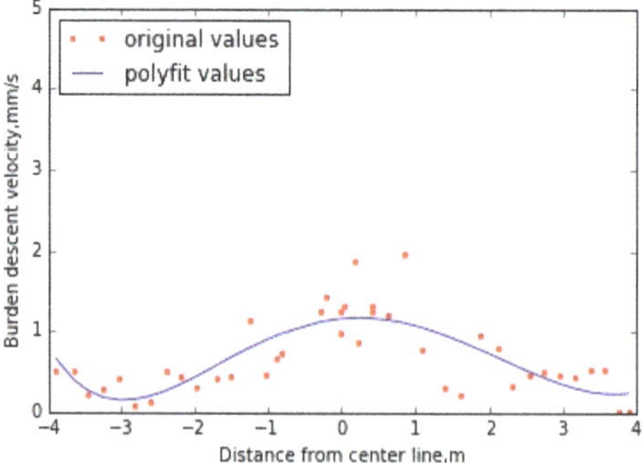

Figure 11. Method of the burden descent velocity from radar data.

2.2.3. Burden Distribution Calculation

From the above, the function of the burden profile and the function of the descent velocity were obtained. Then, the type of the charging material is identified by comparing the last batch with the previous one. If the type is different, radar data are used to calculate the burden profile function. Otherwise, it is used to calculate the descent velocity and to modify the last material layer. After many layer calculations, the eburden forms a structure layer by layer as shown in Figure 12. Then, the coke to ore ratio curve of the burden can be calculated for the last two batches.

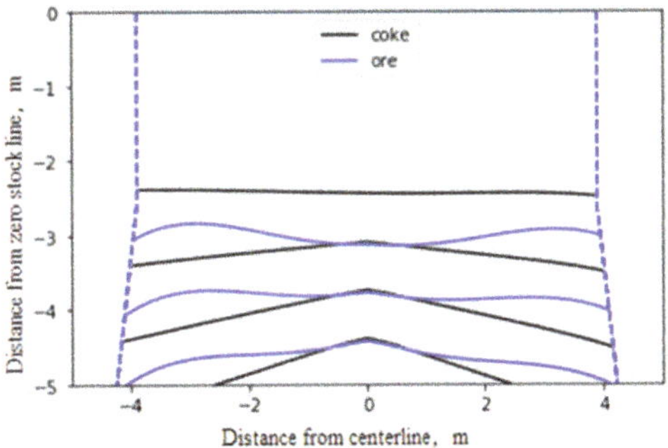

Figure 12. Burden distribution with layer by layer.

2.3. Parameters of Blast Furnace and Assumptions of Calculation

In order to combine the mathematical model with the radar data for a blast furnace, the relevant parameters of the charging system are listed in Tables 1–3.

Table 1. Parameters of the bell-less top blast furnace.

Property	Value
Throat diameter (mm)	8300
Throat height (mm)	2600
Shaft angle (°)	84.15
D_c [1] (mm)	4010
D_b [2] (mm)	1030
D_z [3] (mm)	4601

[1] Distance from throttle to chute suspension point. [2] Distance from the chute suspension point to the chute bottom plate. [3] Distance from the chute suspension point to the zero line.

Table 2. Parameters of rotating chute.

Property	Value
The length of the chute (mm)	3890
Chute speed $(r \cdot s^{-1})$	0.133
Coefficient of coke friction	0.758
Coefficient of ore friction	0.638
Velocity attenuation coefficient of coke	0.70
Velocity attenuation coefficient of ore	0.71

Table 3. Physical parameters of raw material.

Property	Ore	Coke
Bulk density (kg/m^3)	1800	550
Repose angle (°)	31.5	32.5

Considering the influence of the charging parameters, burden distribution properties, and practical experience of blast furnace operators, the mathematical model combining the radar data was derived on the basis of the following assumptions [5,23]:

(1) Velocity of particles after collision with chute can be described by an attenuation factor without bouncing of particles in the chute.

(2) The chute rotates around the centerline of the blast furnace at any inclination angles with the revolution speed of 8 ring/min.

(3) There is no size distribution of particles in the raw material. The drag force of particles in the air after the chute and Coriolis force can be ignored.

(4) Burden is distributed in three dimensions, is uniform around the circumference, and is symmetrical around the centerline of the blast furnace.

(5) Burden keeps an alternate layer structure during the descent.

3. Application of the Combined Model and Results

3.1. Mathematical Model Test

The trajectory of the burden flow model was used to calculate the material flow path at different inclination angles of the chute as shown in Figure 13. With the increase of the inclination angle of the chute, the trajectory moves to the periphery. The drop point for a large inclination angle of the chute is farther away from the center line of the blast furnace. When the inclination angle of the chute is less than 15°, the chute dumps the materials directly to the center of the furnace (called center-charged burden), which cannot be observed in Figure 13. Figure 14 shows the radial velocity of the burden descent calculated by the burden decent velocity model (Equations (19)–(21)). The descent velocity increases with the increase of the distance from the centerline.

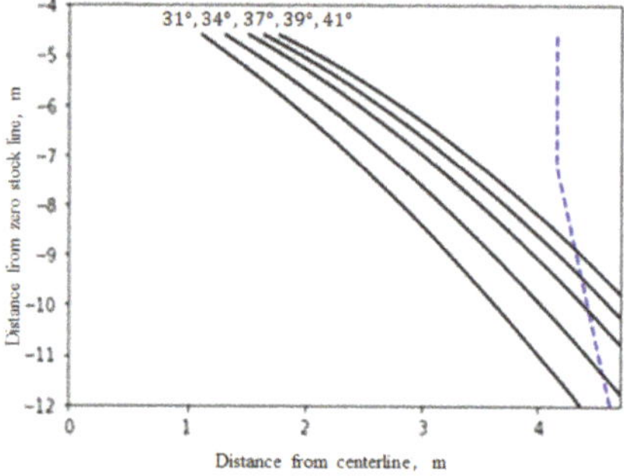

Figure 13. Main trajectories of coke flow.

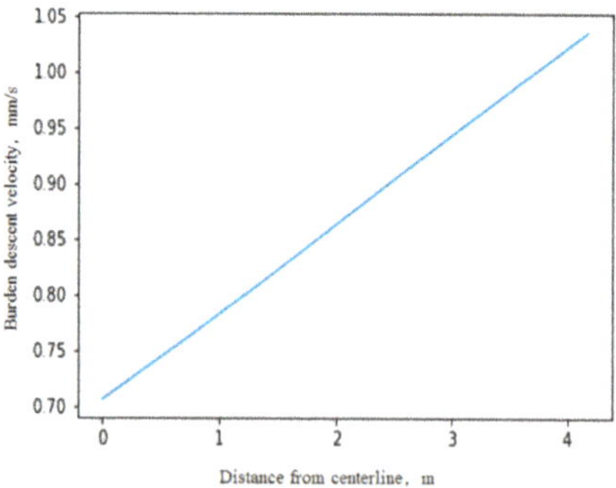

Figure 14. Burden descent velocity from the mathematical model.

Based on the parameters of Tables 1–3 and charge matrix of Table 4, the burden distribution of a multi-ring charging program was calculated by the burden profile model. The results are shown in Figure 15. Blue, green, and red lines express the initial material surface, and the ore and coke surface, respectively. An ore profile with a single ring is shown in Figure 15a. After the full burden matrix, the burden distribution of a batch of ore and coke can be calculated as shown in Figure 15b. The structure of multi-batch burden layers (two coke and two ore layers) was calculated by iterative calculation, as shown in Figure 15c. From the latter, the same type of material has a similar shape and the apexes of the burden profile move toward the periphery a little during the descent due to the effect of the shaft angle of the furnace.

Table 4. Charging matrix of a blast furnace.

Chute Angle (°)	46	44	41.5	39	36.5	15
Coke charging ring number		2	2	2	2	2
Ore charging ring number	3	3	2	2		

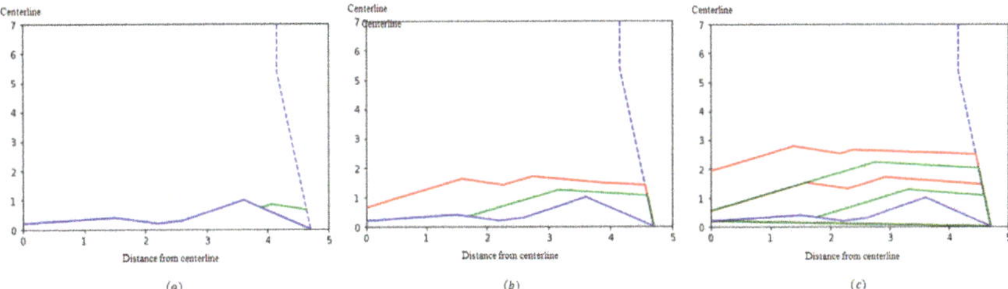

Figure 15. Burden distribution by a multi-ring charging program based on Tables 1–4: (**a**) 1st ring of ore; (**b**) a full burden distribution of an ore and a coke layer with multi-ring; (**c**) 2 coke and 2 ore layers. Blue line: initial material surface. Blue dash-dotted line: wall. Green line: ore layer. Red line: coke layer.

In order to test the mathematical model, a number of cases are listed in Table 5. A comparison of the burden distributions for different charging matrixes is shown in Figure 16. Figure 16a shows burden profiles with central coke when the inclination angle of the chute is 12° and the number of coke rings is 5. Figure 16b shows the burden profile with exactly the same parameters as in Figure 16a but for four coke rings. Comparing Figure 16a,b, only one ring of coke moves to an angle of 27°, which means the central coke becomes thinner and the thickness of the coke layer at an angle of 27° becomes bigger. With another ring of coke moving to 27° (Figure 16c), the thickness of the central coke becomes much thinner and the layer becomes much thicker than in Figure 16a. Figure 16d shows the burden distribution without the central coke and three rings moved to an angle of 20°. Compared to Figure 16c, the central coke has disappeared and the layer thickness at an angle of 20° becomes bigger than in Figure 16d. a comparison of Figure 16d,e shows that only a ring of coke moves from an angle of 20° to 27°. Therefore, the only difference between them is that the coke thicknesses at these two angles are somewhat different. Figure 16f shows the burden distribution with one ring less of coke at an angle of 27° compared to (e). Comparing Figure 16f,g, the burden distribution without two rings at an angle of 20° in the latter yields a thin layer at an angle of 20°. In Figure 16h, every inclination angle of the chute decreased by 1° except 27° for coke. Therefore, the ore layers move toward the center. Figure 17 shows the effect of the ore batch on the burden distribution (a. with ore batch of 63 t and b. with ore batch of 55 t). When the ore batch decreased from (a) to (b), the ore layer thickness became smaller.

Table 5. Charging programs with different matrixes to test the mathematical model.

Figure No.	Inclination Angle of the Chute (°)	41	39	37	34	31	27	20	12
a	Ore	2	3	3	2	1			
	Coke	3	3	3	2	2			5
b	Ore	2	3	3	2	1			
	Coke	3	3	3	2	2	1		4
c	Ore	2	3	3	2	1			
	Coke	3	3	3	2	2	2		3
d	Ore	2	3	3	2				
	Coke	3	3	3	2		2	3	
e	Ore	2	3	3	2	1			
	Coke	3	3	3	2	2	3	2	
f	Ore	2	3	3	2	1			
	Coke	3	3	3	2	2	2	2	
g	Ore	2	3	3	2	1			
	Coke	3	3	3	2	2	2		
Inclination angle of the Chute (°)		40	38	36	33	30	27		
h	Ore	2	3	3	2	1			
	Coke	3	3	3	2	2	2		

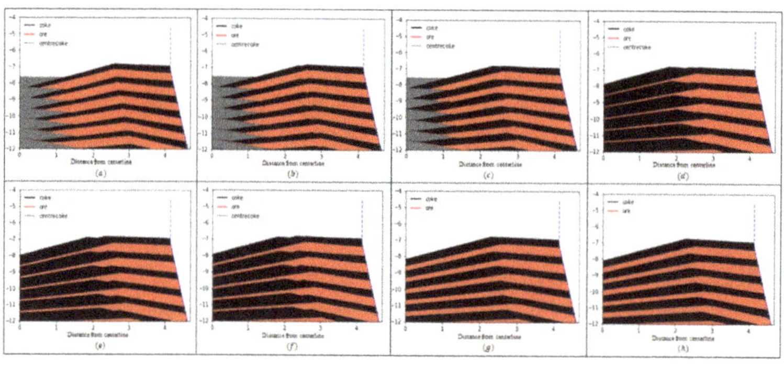

Figure 16. (a–h) are the comparison of burden distributions corresponding to the charging matrix of (a–h) in Table 5.

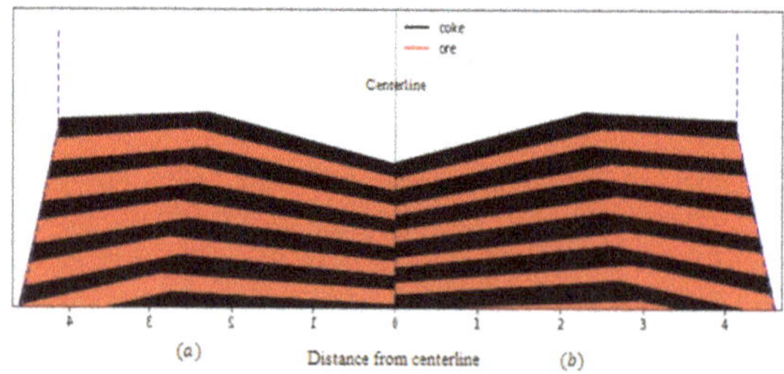

Figure 17. Comparison of the burden distribution with different ore batches: (a) Ore batch of 63 t; (b) Ore batch of 55 t.

Figures 16 and 17 shows the test of the sensitivity of the mathematical model. When the number of burden rings increased, the corresponding thickness of the burden layer increased. When the inclination angle of the chute changed, the structure of the burden layers changed accordingly. According to the conservation of volume, the burden distribution changes with the change of the burden batch. In short, changes in the inclination angle of the chute, ring of charging, and ore batch will cause corresponding changes of the layers predicted by the mathematical model.

The burden distribution with the charging matrix of Table 5a is shown in Figure 18a. The ratio of ore to coke (Equation (23)) is defined by the last two layers and is shown in Figure 18b. The ratio is zero at the center due to the central-coke layers and has a highest value at r = 1.3 m at the inclination angle of the chute of 31°.

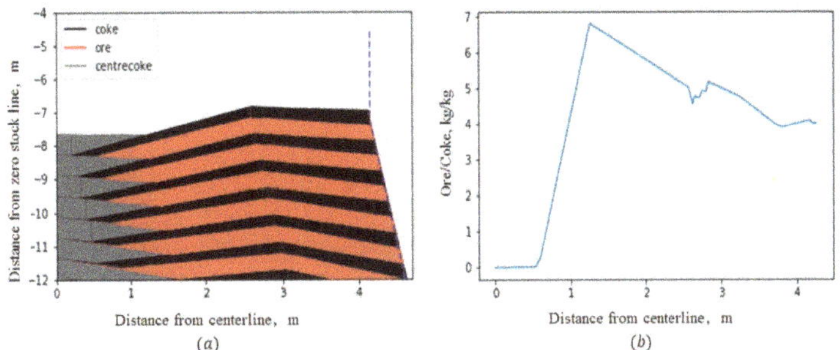

Figure 18. (**a**) Burden distribution; (**b**) ratio of ore to coke from (**a**) case.

3.2. Combination of the Mathematical Model and Radar Data

Radar data can work together with the mathematical model to support and guide the operation of blast furnace charging. Therefore, it is necessary to compile them into a visual interface. Based on the charging parameters and radar data from a plant in East China, a 2D simulation software of blast furnace charging was developed. The model was programed in Python and the visualization code was provided by JavaScript to give a friendly web interface.

A cloud map drawn by radar scanning data is shown in Figure 19a. Radar scanning data includes many points in an interval. After noise removal and feature extraction in the interval, 20 relatively stable points (green curve) were obtained. These points were used to calculate the burden distribution by the burden distribution model. The calculated burden layers' structure is shown in Figure 19b.

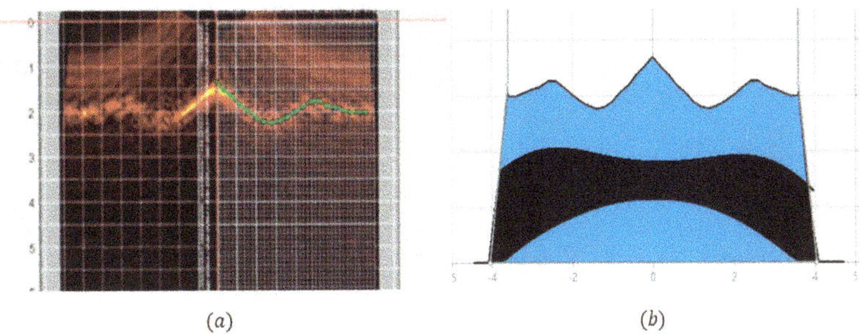

Figure 19. Burden profile: (**a**) Burden profile from radar data; (**b**) Burden profile from the mathematical model combined with radar data.

In order to display the radar data and the results of the mathematical model, we designed a user-friendly operator interface. Its main functions are shown in Figure 20. After logging into the software (Figure 20a), the left column is the function menu of the software. A column is designed to enter the parameter settings of the furnace in Figure 20b and includes blast furnace parameters, chute parameters, raw material properties parameters, charging matrix, and so on. The burden distribution display is the main page of the software and is drawn out by Echarts after radar data processing in Figure 20c. After noise removal and feature extraction, the radar data combined with the mathematical model illustrate the structure of the material layers. The burden descent velocity and ratio of ore to coke are also drawn in the main interface. In addition, the system parameters of the radar are also monitored, such as the chip temperature, nitrogen temperature and nitrogen flow rate, valve pressure in radar system, and liquid flow rate.

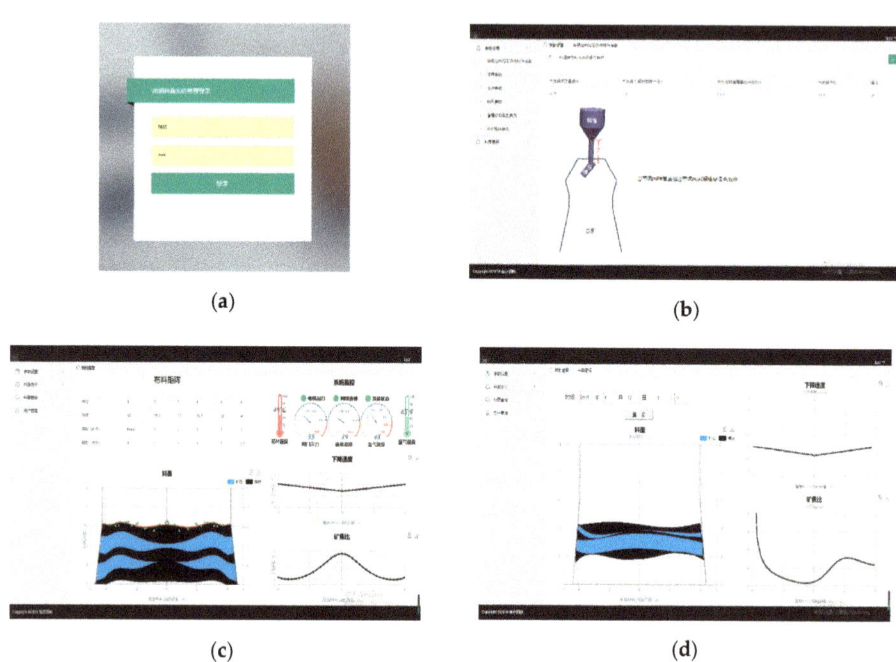

Figure 20. Visualization of our software by the combination of the mathematical model and radar data: (**a**) Software login interface; (**b**) Parameter setting interface; (**c**) Burden distribution display and radar data monitor; (**d**) User management interface.

In order to guarantee the security of the data and analyze the data of different users and different furnaces, a user management system was designed for user administration, as shown in Figure 20d.

4. Conclusions

The charging of the blast furnace directly affects the burden distribution in the throat, which influences the gas distribution in the shaft of the furnace. Adjusting the charging can improve the distribution of gas flow, increase the gas utilization efficiency, reduce energy consumption, and prolong the life of the furnace.

In this paper, with the help of computer technology, a mathematical model of the charging system was developed, composed of the trajectory of burden flow, burden profile, burden layer structure, and burden distribution evaluation. Serial cases were used to test the mathematical model's sensitivity. After noise removal and feature extraction, radar data of the burden profile was combined with the mathematical model to improve the accuracy of the model. A 2D view was obtained by combining the mathematical model and the radar data to visualize the charging, burden distribution, radar data, mathematical model, and relative equipment state. Based on the data from a blast furnace, the software was found to be more consistent than the present state-of-the-art tools used at the plant and ran smoothly to help the operation of the furnace.

Author Contributions: Conceptualization, M.L. and Y.Y.; methodology, M.L., H.W. and Y.Y.; software, M.L.; validation, Y.Y.; formal analysis, H.W. and Y.Y.; investigation, M.L. and Y.Y.; resources, G.X. and Y.Y.; data curation, Y.G. and G.X.; writing—original draft preparation, M.L. and Y.G.; writing—review and editing, H.W. and Y.Y.; visualization, M.L.; supervision, G.X. and Y.Y.; project administration, G.X. and Y.Y.; funding acquisition, Y.Y. All authors have read and agree to the published version of the manuscript.

Funding: We gratefully acknowledge financial support from The Program for Professor of Special Appointment (Eastern Scholar) at Shanghai Institutions of Higher Learning (No.TP2015039), National Natural Science Foundation of China (No.51974182), National 111 project, Grant/Award No.17002 and Jinheng information technology in Nanjing.

Conflicts of Interest: The authors declare no conflict of interest.

References

1. Zhang, L.L.; An, G.; Zhang, Z.G.; Wang, S.T.; Qiu, X.X. Application of making infrares video on blast furnace roof. *Hebei Metall.* **2007**, *6*, 35–37.
2. Venkatesh, V.; Akhil, G.; Liang, G.; Rangarajan, V. Finite Element Based Physical Chemical Modeling of Corrosion in Magnesium Alloys. *Metals* **2017**, *7*, 83.
3. Nishio, H.; Ariyama, T.; Sato, M.; Nakatani, G.; Maki, A.; Saito, N. Development of a Simulation Model Burden Distribution. *(Study Burden Distribution Bell-Less Top-II)*. 1983, 23, p. b319. Available online: https://www.researchgate.net/publication/298277111_DEVELOPMENT_OF_A_SIMULATION_MODEL_OF_BURDEN_DISTRIBUTION_STUDY_ON_BURDEN_DISTRIBUTION_IN_BELL-LESS_TOP_-_II (accessed on 8 December 2019).
4. Jung, S.K.; Chung, W.S. Improvement of gas flow through analyzing discharge behavior in the bunker used in blast furnace. *ISIJ Int.* **2001**, *41*, 1324–1330. [CrossRef]
5. Radhakrishnan, V.R.; Ram, K.M. Mathematical model for predictive control of the bell-less top charging system of a blast furnace. *J. Process Contr.* **2001**, *11*, 565–586. [CrossRef]
6. Saxén, H.; Hinnelä, J. Model for burden distribution tracking in the blast furnace. *Miner. Process. Extr. Metall. Rev.* **2004**, *25*, 1–27. [CrossRef]
7. Hinnelä, J.; Saxén, H.; Pettersson, F. Modeling of the Blast Furnace Burden Distribution by Evolving Neural Networks. *Ind. Eng. Chem. Res.* **2003**, *42*, 2314–2323. [CrossRef]
8. Nag, S.; Koranne, V.M. Development of material trajectory simulation model for blast furnace compact bell-less top. *Ironmak. Steelmak.* **2009**, *36*, 371–378. [CrossRef]
9. Park, J.I.; Baek, U.H.; Jang, K.S.; Oh, H.S.; Han, J.W. Development of the Burden Distribution and Gas Flow Model in the Blast Furnace Shaft. *ISIJ Int.* **2011**, *51*, 1617–1623. [CrossRef]
10. Nag, S.; Gupta, A.; Paul, S.; Gavel, D.J.; Aich, B. Prediction of Heap Shape in Blast Furnace Burden Distribution. *ISIJ Int.* **2014**, *54*, 1517–1520. [CrossRef]
11. Shi, P.Y.; Zhou, P.; Fu, D.; Zhou, C.Q. Mathematical model for burden distribution in blast furnace. *Ironmak. Steelmak.* **2016**, *43*, 74–81. [CrossRef]
12. Xiao, G.C. Application of rotary radar technology in blast furnace of nanjing steel. In Proceedings of the Workshop on Experience Analysis of Production Organization in Limited Season and Staggered Season of Blast Furnace in 2018, Tai'an, Shandong, China, 18 April 2018; pp. 124–128.
13. Zhang, S.; Wei, K.D.; Zhang, H.G.; Yin, Y.X.; Chen, X.Z. 3D Surface Reconstrution of Blast Furnace Based on Phased Array Radar Data. In Proceedings of the 11th China Steel Annual Conference, Beijing, China, 21 November 2017; p. 6. Available online: http://cpfd.cnki.com.cn/Article/CPFDTOTAL-ZGJS201711002143.htm (accessed on 8 December 2019).
14. Liu, D.X.; Li, X.L.; Lu, S.T.; Ding, D.W.; Chen, X.Z. Application of Fuzzy Neural Network in Burden Surface Clustering. In Proceedings of the 2012 24th Chinese Control and Decision Conference (CCDC), Taiyuan, China, 23 May 2012; pp. 1094–1099. Available online: https://ieeexplore.ieee.org/document/6244174?tp=&arnumber=6244174&contentType=Conference%20Publications&refinements%3D4291944822%26sortType%3Ddesc_p_Publication_Year%26ranges%3D2012_2012_p_Publication_Year%26pageNumber%3D107%26rowsPerPage%3D100= (accessed on 8 December 2019).
15. Gao, Z.K. Innovation and Practices of Blast Furnace Visualization and Simulation Technology. *China Metall.* **2013**, *23*, 8–14.

16. Zhu, Q.; Lu, C.L.; Yin, Y.X.; Chen, X.Z. Burden Distribution Calculation of Bell-Less Top of Blast Furnace Based on Multi-Radar Data. *J. Iron Steel Res. Int.* **2013**, *20*, 33–37. [CrossRef]
17. Li, X.L.; Liu, D.X.; Jia, C.; Chen, X.Z. Multi-model control of blast furnace burden surface based on fuzzy SVM. *Neurocomputing* **2015**, *55*, 1146–1156. [CrossRef]
18. Tian, J.; Tanaka, A.; Hou, Q.; Chen, X. Radar Detection-based Modeling in a Blast Furnace: A Prediction Model of Burden Surface Shape after Charging. *ISIJ Int.* **2018**, *58*, 1999–2008. [CrossRef]
19. Tian, J.; Tanaka, A.; Hou, Q.; Chen, X. Radar Detection-Based Modeling in a Blast Furnace: A Prediction Model of Burden Surface Descent Speed. *Metals* **2019**, *9*, 609. [CrossRef]
20. Miao, L.L.; Chen, X.Z.; Bai, Z.L.; Huang, Y.Q.; Hou, Q.W.; Yin, Y.X. Blast furnace line shape measurement fusion and compensation algorithm based on radar. *J. Univ. Sci. Technol. Beijing* **2014**, *36*, 82–88.
21. Li, H.; Saxén, H.; Liu, W.; Shao, L.; Zou, Z. Model-based analysis of factors affecting the burden layer structure in the blast furnace shaft. *Metals* **2019**, *9*, 1003. [CrossRef]
22. Wang, P. Measurement and analysis of burden flow trajectory and width in bell-less top with two concentric vertical hoppers. *Iron Steel* **2003**, *38*, 10–14.
23. Yu, Y.W. *Research and Development on the Model for Bell-Less Charging of Blast Furnace*; Chongqin University: Chongqin, China, 2008.
24. Nishida, T.; Hachiya, S.; Tanaka, K.; Satoh, K.-I.; Inaba, S.-I.; Okimoto, K.-I. Application of the results of a burden distribution experiment on a bell-less scale model to blast furnace operation. *Kobe Steel Eng. Rep. Res. Dev.* **1982**, *32*, 73–76.

Article

Numerical Study of Binary Trickle Flow of Liquid Iron and Molten Slag in Coke Bed by Smoothed Particle Hydrodynamics

Shungo Natsui [1],*, Kazui Tonya [2], Hiroshi Nogami [1], Tatsuya Kikuchi [2], Ryosuke O. Suzuki [2], Ko-ichiro Ohno [3], Sohei Sukenaga [1], Tatsuya Kon [4], Shingo Ishihara [1] and Shigeru Ueda [1]

[1] Institute of Multidisciplinary Research for Advanced Materials, Tohoku University, Katahira 2-1-1, Aoba-ku, Sendai, Miyagi 980-8577, Japan; nogami@tohoku.ac.jp (H.N.); sohei.sukenaga.d3@tohoku.ac.jp (S.S.); ishihara@tohoku.ac.jp (S.I.); tie@tohoku.ac.jp (S.U.)

[2] Faculty of Engineering, Hokkaido University, N13-W8, Kita-ku, Sapporo, Hokkaido 060-8628, Japan; tonya_913@frontier.hokudai.ac.jp (K.T.); kiku@eng.hokudai.ac.jp (T.K.); rsuzuki@eng.hokudai.ac.jp (R.O.S.)

[3] Department of Materials Science and Engineering, Faculty of Engineering, Kyushu University, Motooka 744, Nishi-ku, Fukuoka 819-0395, Japan; ohno@zaiko.kyushu-u.ac.jp

[4] Magnetic Powder Metallurgy Research Center, National Institute of Advanced Industrial Science and Technology, 2266-98 Anagahora, Shimo-Shidami, Moriyama-ku, Nagoya, Aichi 463-8560, Japan; t-kon@aist.go.jp

* Correspondence: natsui@tohoku.ac.jp

Received: 16 January 2020; Accepted: 12 February 2020; Published: 14 February 2020

Abstract: In the bottom region of blast furnaces during the ironmaking process, the liquid iron and molten slag drip into the coke bed by the action of gravity. In this study, a practical multi-interfacial smoothed particle hydrodynamics (SPH) simulation is carried out to track the complex liquid transient dripping behavior involving two immiscible phases in the coke bed. Numerical simulations were performed for different conditions corresponding to different values of wettability force between molten slag and cokes. The predicted dripping velocity changes and interfacial shape were investigated. The relaxation of the surface force of liquid iron plays a significant role in the dripping rate; i.e., the molten slag on the cokes acts as a lubricant against liquid iron flow. If the attractive force between the coke and slag is smaller than the gravitational force, the slag then drops together with the liquid iron. When the attractive force between the coke and slag becomes dominant, the iron-slag interface will be preferentially detached. These results indicate that transient interface morphology is formed by the balance between the momentum of the melt and the force acting on each interface.

Keywords: ironmaking blast furnace; coke bed; trickle flow; molten slag; liquid iron; SPH

1. Introduction

An efficient trickle flow of high-temperature melts in coke beds is necessary in ironmaking blast furnaces, as it ensures a smooth continuous process, which is a prerequisite for high productivity in operations that use lower amounts of reducing agents. The trickle flow is driven by the gravitational force that acts on the liquid iron and molten slag (two immiscible liquids), and the dripping behavior is balanced by viscous and interfacial forces. When the volume of the high-strength coke, which acts as a spacer in the lower part of the blast furnace, is reduced, the resistance to the melt passage increases, which can cause instability. It is, therefore, essential to maintain a desired trickle flow through the coke bed to prevent clogging problems that may occur if the pressure drop in the bed becomes excessive or shows large variations. This approach emphasizes the need to understand the dripping behavior of the two liquids in the coke bed.

Over the years, liquid flow in blast furnaces has been investigated experimentally and mathematically. Several studies focused on the volume fraction of the liquid in the packed bed, called "liquid hold-up", and the effect of various in-furnace conditions, including the physical properties of the liquid and the wettability of coke. The relationship between hold-up and dimensionless number has been determined by applying an experimental system using room temperature media [1–3] or actual melt [4–6]. Since the hold-up of CaO-SiO_2-Al_2O_3-based molten slag (>1773 K) is significantly affected by the wettability of coke [7], several experimental studies have been conducted on the static contact angle between molten slag and "carbonaceous material" [8–10]. The molten slag wetting behavior relatively depends on the carbonaceous material substrate, i.e., not only the equilibrium contact angles but the periods to reach a stable contact angle vary considerably. If basicity (CaO/SiO_2 mass %) ≤ 1, the decrease in the contact angle with time as the basicity decreases is generally attributed to the formation of interface product SiC [11], but the change in the state of the Fe-Si-C system is still unclear because this change depends on the Si form (which may exist in a gaseous form). In investigating coke-slag wettability, the differences in small amounts of components (such as FeO and MgO) in slag and the differences in the specific properties (such as crystal structure and void structure) of carbonaceous substrates should be considered.

Computational fluid dynamics (CFD) is useful for understanding this phenomenon. The advantage of this approach is the precise tracking of time-varying fluid interfaces, which is difficult to observe experimentally. Most recent CFD studies demonstrate that, with regard to the wettability of the packed bed, the geometry of the bed has a significant effect on the interface dynamics, leading to the co-existence of complex interfaces; thus, the change in wettability leads to various interfacial movements that are unidentified under static conditions. Molten slag flow in a packed coke bed was investigated by using the three-dimensional combined discrete element method (DEM) and CFD [12–14]. A series of two-dimensional [15,16] and three-dimensional [17] high-resolution direct numerical simulations were also conducted with the aim of understanding the pore-scale fluid dynamics. Although the two-liquid flow in packed beds has received attention but usually focused on room temperature media [18,19], there is no guarantee that the information available from the water-oil-rock system can be applied to the slag-iron-coke system in blast furnaces. For example, the density difference between iron and slag (≈ 4000 kg/m^3) is considerably higher than that of water and oil (≈ 200 kg/m^3) [20]. It is essential to evaluate the interfacial force in an unsteady condition under the condition where the momentum in the gravity direction appears. CFD studies have been conducted for such cases. The phase field modeling for the behavior of molten slag confirmed that the increment of the interfacial tension between liquid iron and molten slag was the driving force of the molten slag separation from the melt [21]. Using the volume of fluid simulation, the driving force of the liquid iron penetration into the coke bed was proposed due to the total energy reduction by extending the area covered with the liquid iron [22]. On the other hand, the fully Lagrangian approach, which can track moving calculation points, has been adopted for solving multi-fluid physics problems during high-temperature metallurgical processes with complicated interfaces [23,24]. The smoothed particle hydrodynamics (SPH) method discretizes a continuous fluid phase by moving particles and is suitable for analyzing interfacial flow, even for numerous dispersed phases. This method can track the dripping flow [25], the movement of both the gas and the liquid phase [26], fluid flow in bed structures having different shapes (coupled with multi-sphere DEM) [27,28], and solid particle penetration into the liquid iron bath [29,30] directly. Recently, with the increase in computer capacity, the SPH method has been applied to very complicated interface problems [31]. For convenience, surface forces are converted into body forces in the SPH method, but recent expansions in computer capacity have alleviated this weakness. In this study, the trickle flow of a liquid iron and molten slag was simulated in the lower part of the blast furnace, and the effect of the wettability of molten slag and coke was investigated.

2. Methods

2.1. Basic Formulation of SPH

From the basic idea of SPH, it is evident that fluid motion is represented using a set of particle motion equations. A kernel function is introduced as an integral interpolation to solve a differential equation. In other words, the formulation is based on an interpolation scheme that allows the estimation of a vector or scalar function f at position $\mathbf{r}$ in terms of the values of the function at the discretization points.

$$f(\mathbf{r}) \cong \int f(\mathbf{r}')W(\mathbf{r},h)dV \tag{1}$$

In Equation (1), $\mathbf{r}$ denotes arbitrary coordinates, $\mathbf{r}'$ denotes particle positions, V denotes the volume, W is the smoothing kernel function, and h is the radius of influence. The summation of function f can be replaced with a summation over particles only within the distance h from $\mathbf{r}_i$ owing to the compactness; thus, $W(\mathbf{r}_{ij},h) = 0$ when $|\mathbf{r}_{ij}| > h$. The kernel must possess a form symmetrical to $|\mathbf{r}_{ij}| = 0$. Here, i is the particle index, and j is the index of the neighboring particle around i. The kernel has at least a continuous first derivative and must satisfy the normalization condition as $\int W(\mathbf{r}_{ij},h)\,d\mathbf{r} = 1$. Within the $h \to 0$ limit, the kernel is required to reduce to a Dirac delta function $\delta(\mathbf{r}_{ij})$. Wendland's kernel can be applied to prevent having various kernel artifacts in a multiphase system [32]:

$$W(\mathbf{r}_{ij},h) = \frac{21}{16\pi h^3}\begin{cases} \left(1-\frac{q}{2}\right)^4(2q+1), & q < 2 \\ 0, & q \geq 2 \end{cases} \tag{2}$$

where $|\mathbf{r}_{ij}|_0$ is the interparticle distance corresponding to the initial conditions, $q = |\mathbf{r}_{ij}|/h$, and it is assumed that $h = 1.05|\mathbf{r}_{ij}|_0$ [27]. The gradient form of Equation (1) can be represented by using the divergence theorem as follows:

$$\nabla f(\mathbf{r}) \cong -\int f(\mathbf{r}')\nabla W(\mathbf{r},h)dV \tag{3}$$

When applying this approximation to dispersed fluids, the discontinuities in the density distribution of the fluids become significant, which increases numerical errors. Nonuniform distribution or insufficiency of particles in the regions near the interface results in a significant fluctuation of pressure. The moving least squares (MLS) method is useful for approximating the function to solve this problem [33], which is related to the numerical fluctuations in the pressure at the nearby interface. The pressure is a function of the local density, and thus, the smooth density field of a bulk phase leads to a continuous pressure distribution. The MLS method improves the mass-area-density consistency and filters small-scale pressure oscillations, as described briefly in the following section.

2.2. Density Approximation

The SPH formulation can be transformed into a particle-based format to express the mass-area-density consistency process. The density of the particles was expressed in terms of the sum of the kernel functions of the N particles present within the radius of influence as follows:

$$\rho_i = \sum_{j=1}^{N} m_j W(\mathbf{r}_{ij},h) \tag{4}$$

where the subscripts i and j denote the particle indices, m is the mass, and ρ is the local density around the particle. Further, m_j is the mass of particle j. Therefore, the kernel function around particle i can be discretized using the following equation, which is derived from Equations (1)–(3):

$$f_i(\mathbf{r}) = \sum_{j=1}^{N} \frac{m_j}{\rho_j} f(\mathbf{r}_j) W(\mathbf{r}_{ij}, h) \tag{5}$$

The gradient of f_i can be represented as expressed in Equation (6):

$$\nabla f_i(\mathbf{r}) = -\sum_{j=1}^{N} \frac{m_j}{\rho_j} f(\mathbf{r}_j) \nabla W(\mathbf{r}_{ij}, h) \tag{6}$$

MLS involves a first-order consistent gradient approximation, which allows pressure smoothing, and its first derivative values are obtained using the method for the homogenous bulk phase mentioned above. The method of the least square interplant with constraint condition (CLS) represents an improved scheme, leading to a more accurate approximation than the MLS method around the sampling points [33,34]. In the CLS method, the particles can be made to directly represent a physical quantity by extending the MLS method in the one-dimensional error space for multiple dimensions. In the three-dimensional space, the CLS method approximates the values of various physical parameters around the particles based on the following equation: $\langle \rho \rangle_i = a_0 + a_1(x - x_i) + a_2(y - y_i) + a_3(z - z_i)$, where, x, y, and z are the coordinates of the sampling points and a_0, a_1, a_2, and a_3 are the undetermined coefficients. If $x = x_i$, $y = y_i$, and $z = z_i$, then $\langle \rho_i \rangle = a_0$ at particle i. Refer to a previous report for parameters determining the procedure [26].

2.3. Fluid Motion Equation and Discretization

The governing equations for a weakly compressible viscous flow are based on the relationship between the velocity of sound and the flow density under adiabatic conditions, as well as the Navier-Stokes Equations:

$$\left(\frac{Dp}{D\rho} \right)_S = c^2 \tag{7}$$

$$\rho \frac{D\mathbf{v}}{Dt} = -\nabla p + \mu \nabla^2 \mathbf{v} + \rho + \mathbf{F}_s \tag{8}$$

where $\mathbf{v}$ is the fluid velocity, p is the pressure, c is the velocity of sound, μ is the viscosity, and $\mathbf{F}_s$ is the interfacial force. Subsequently, Equation (8) can be formulated for each particle as follows:

$$m_i \frac{D\mathbf{v}_i}{Dt} = -\sum_{j=1}^{N} \left(\langle p \rangle_i V_i^2 + \langle p \rangle_j V_j^2 + \Pi_{ij} \right) \nabla W_{ij} + \sum_{j=1}^{N} \frac{2\mu_i \mu_j}{\mu_i + \mu_j} \left(V_i^2 + V_j^2 \right) \frac{\mathbf{r}_{ij}}{|\mathbf{r}_{ij}|^2} \mathbf{v}_{ij} \nabla W_{ij} + m_i \mathbf{g} + \langle \mathbf{F}_s \rangle_i \tag{9}$$

where Π is the artificial viscosity term, which is usually added to the pressure gradient term to help in diffusing sharp variations in the flow and dissipate the energy of the high-frequency term [35]. To determine the time derivative of pressure from Equation (7), Tait's equation of state can generally be used [28]:

$$\langle p \rangle_i = \frac{c^2 \rho_0}{\gamma} \left\{ \left(\frac{\rho_i}{\rho_0} \right)^\gamma - 1 \right\} \tag{10}$$

where γ $(= 7.0)$ is the adiabatic exponent and ρ_0 is the true density value of the material. Considering the balance of the time step and the incompressible behavior of the artificial compressible fluid, an optimal value for c must exist.

2.4. Interfacial Force Model

Considering the interfacial force $\mathbf{F}_s$, the interparticle potential force is defined using the space derivative of potential $E\left(\left|\mathbf{r}_{ij}\right|\right)$. $\mathbf{F}_s$ is localized at the liquid interface by applying it to the liquid elements in the transition region of the interface. The force per unit area $\langle\mathbf{F}_s\rangle_i$ is then converted into force per unit volume using the expression [36]:

$$\langle\mathbf{F}_s\rangle_i = -2\sigma_i\left|\mathbf{r}_{ij}\right|_0^2\left(\sum_{j=1}^{N} E\left(\left|\mathbf{r}_{ij}\right|\right)\right)^{-1} \cdot \sum_{j=1}^{N} \frac{\partial E\left(\left|\mathbf{r}_{ij}\right|\right)}{\partial \mathbf{r}} \frac{\mathbf{r}_{ij}}{\left|\mathbf{r}_{ij}\right|} \tag{11}$$

where σ_i is the surface tension or interfacial tension of particle i. The Fowkes hypothesis is considered in calculating the interfacial force on the multiphase boundary [37]. The Fowkes hypothesis explains that, in a system in which two immiscible liquid phases (liquid iron and molten slag) are in contact, the elements present at the two-phase interface are subject to forces. At the interface between liquid iron and molten slag, liquid iron interface elements receive the attractive force σ_m equivalent to the "surface tension" of liquid iron and the dispersion force σ_D from molten slag. The force acting on the interface elements of the molten slag can be described similarly. Hence, the interfacial tension σ_{ms} is expressed as follows:

$$\sigma_{ms} = \sigma_m + \sigma_s - 2\sigma_D \tag{12}$$

This simple hypothesis indicates that the unknown dispersion force and interfacial tension can be calculated explicitly by applying the surface tension as the input and the interfacial tension of the two liquid phases in contact as the conditions. An immiscible blend of liquid iron and molten slag contacting the coke plate is considered, as illustrated in Figure 1. In terms of the tension balance on the solid-gas-liquid triple line in which liquid iron, coke, and gas are in contact with one another, the surface tension σ_m of the liquid iron, surface tension σ_c of the solid phase, and the solid-liquid interfacial tension σ_{mc} are assumed to be balanced by the contact angle θ_m. In other words, Young's equation reflects a horizontal balance among the interfacial tensions: $\sigma_m \cos \theta_m + \sigma_{mc} = \sigma_c$. Here, the unknown solid surface tension and the solid-liquid interfacial tension are eliminated from Young's and Fowkes' equations to obtain the following equation:

$$\cos \theta_m = 2\frac{\sigma_D}{\sigma_m} - 1 \tag{13}$$

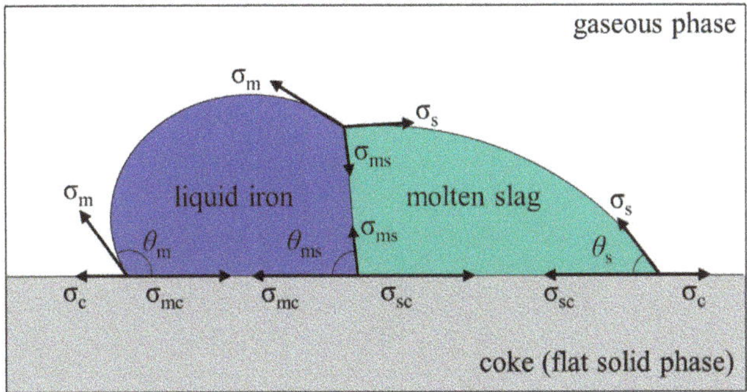

Figure 1. Two liquid droplets in contact on a flat, solid surface. These forces balance one another.

Equation (13) indicates that θ_m is determined by the surface tension of the liquid phase and the dispersion force acting between the different phases. The dispersion force is explicitly defined by this equation, and the static contact angle can be calculated using the potential interparticle model.

Furthermore, considering other triple lines, such as that existing between the molten slag, the solid, and the gas, and that between the two liquid phases and the solid, the following equation is obtained:

$$\cos \theta_{ms} = \frac{\sigma_m}{\sigma_{ms}} \cos \theta_m - \frac{\sigma_s}{\sigma_{ms}} \cos \theta_s \tag{14}$$

In Equation (14), θ_s is the contact angle between the molten slag and solid plate, and θ_{ms} is the contact angle between the two liquid phases and the solid plate. This equation indicates that the liquid iron-molten slag-solid contact angle θ_{ms} is represented by the contact angles θ_m and θ_s.

2.5. Physical Properties of Liquid Iron and Molten Slag

The condition of the lower part of the blast furnace was examined to determine the physical properties of liquid iron (ρ_m, μ_m, and σ_m) and molten slag (ρ_s, μ_s, and σ_s). Assuming that the molten iron at the lower part of the blast furnace has a sufficiently low oxygen concentration and is in a carbon-saturated state, the molten iron can be regarded as chemically stable at 1773 K. Hence, it can be assumed as follows: ρ_m = 6800 kg/m^3, μ_m = 0.01 Pa s, and σ_m = 1.25 N/m [38]. The contact angle between carbon-saturated iron and coke is reported to exceed 120°. However, the physical properties of molten slag vary over a wide range. The typical slag composition in the dripping zone depends on the CaO-SiO$_2$-Al$_2$O$_3$-MgO system at 1773 K; the basicity (CaO/SiO$_2$ mass %) ranges from 0.7–2.0 and decreases as the slag descends in the furnace. When the slag contains MgO, the reduction of MgO may proceed preferentially over that of SiO$_2$, thereby influencing the wetting behavior of the molten slag on the coke substrate [8]. In this study, the liquid phase composition was considered as the simplest 40 mol% CaO-40 mol% SiO$_2$-20 mol% Al$_2$O$_3$-0 mol% MgO slag, in which the physical property data were assumed in a single phase at a constant temperature of 1773 K. Since the physical properties of molten slag exhibit large deviations among the reported values, typical datasets from the same research group are adopted [39,40]: ρ_s = 2574 kg/m^3, μ_s = 0.53 Pa s, and σ_s = 0.49 N/m. According to a previous study, there is no agreement on the contact angle between the CaO-SiO$_2$-Al$_2$O$_3$-based molten slag and the carbonaceous material. On one hand, the contact angle decreases when CaO/SiO$_2$ $\leqq$ 1 [11,41]; on the other hand, the contact angle remains constant at approximately 160° in the case of graphite, regardless of the basicity [10]. These differences are thought to be due to the different compositions of the carbonaceous material. The change in these contact angles might be due to the reduction reaction of SiO$_2$ between the carbonaceous material and molten slag, as well as the formation of SiC as the interfacial product because an initial contact angle of 45° between SiC and the slag was reported [11]. Therefore, in this study, the contact angle between the molten slag and coke was defined by taking θ_m = 160° for low-reactivity interface and θ_m = 30° for high-reactivity interface [8,10].

2.6. Calculation Condition

Detailed packed-bed-structure digital data consisting of multiple coke samples was constructed to simulate the lower part of the blast furnace. Assuming that the coke distribution was just above the raceway, representative coke samples with an average equivalent spherical volume diameter D = 0.0247 m was selected [42]. By using a three-dimensional scanning technique [26], 100 pieces of representative coke three-dimensional surface shape dataset were numerically obtained. About 300,000 surface points on each coke sample were obtained with a minimum resolution of 0.43 mm. The obtained coordinates were converted to standard triangulated language, and the surface shape was polygonal with a triangular mesh. The natural packed structure comprising these particles can be obtained by a DEM-based scheme. The basic format of the DEM is to track spherical particles. Multi-sphere (MS) DEM is a method utilizing a DEM contact force model that is expanded to handle the motion of freely shaped solids. It arrays spherical particles and expresses complex shapes to enable intuitive mounting. The position and rotation angles of each coke sample were determined by using a pseudorandom number, and the packed bed structure was determined by the MS-DEM simulation of a box-type container with 0.12-m sides, similar as a previous report [43]. Next, two immiscible liquids with a 0.12-m width and 0.02-m height were placed immediately above the packed bed to simulate the liquid iron-molten-slag trickle flow in the coke-packed bed. Position of liquid iron and molten slag was determined by using a pseudorandom number. The volume ratio was set as 8:2 to achieve the conditions similar to that of the actual operation. In SPH simulations for multi-phase flow that include a gas, the pressure differential becomes large between liquid and gas, and, consequently, the pressure gradient becomes excessive, thus making convergent calculation difficult. Therefore, in this research, a gas phase is assumed to exist in spaces where particles do not exist. The calculation domain is depicted in Figure 2. The well-known Courant–Friedrichs–Lewy condition was applied to determine dt. In this study, a calculation particle diameter d_p of 1.00 mm was adopted as a constant value in all calculation processes. Thus, the following values were determined: $dt = 1.0 \times 10^{-5}$ s, the analysis time is 10 s, and the number of total particles is 1,402,280. All the programs were coded by the author. Each computer code was written in Fortran 90/95 and compiled and executed by an Intel Fortran compiler on a Windows system. The CPU used in this work was Intel Core i7-7820 × (3.6 GHz, 8 cores). A basic parallelized algorithm was applied using a multi-core processor and OpenMP. Simulation time was 1 week or more.

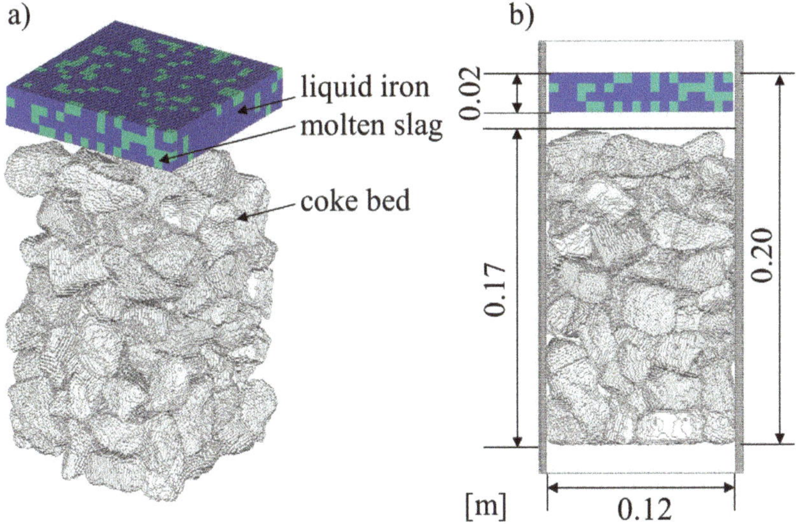

Figure 2. Schematic diagram of initial conditions: (**a**) three-dimensional view and (**b**) horizontal view of packed structure and liquid iron and molten slag.

3. Results

Figure 3 shows the temporal change in the three-dimensional distributions of the liquid iron (blue color) and molten slag (green color) as a result of the model calculation. The packed bed structure formed by irregularly shaped cokes is geometrically complex, and the liquid iron flow through the continuous void drops in the form of strings or droplets. From the viewpoint of localized flow behavior, several regions do not receive any flow, and it is assumed that liquid iron flows only through the limited region. However, the flow behavior of liquid iron is influenced by the presence of slag with different coke wettability, as shown in Figure 3b,c. As shown in Figure 3b, for a coke surface with poor wettability containing molten slag, the molten slag drops along with the liquid iron. However, the coke surface with good wettability shown in Figure 3c retains molten slag at the upper part of the packed bed owing to a high attractive force acting against the molten slag.

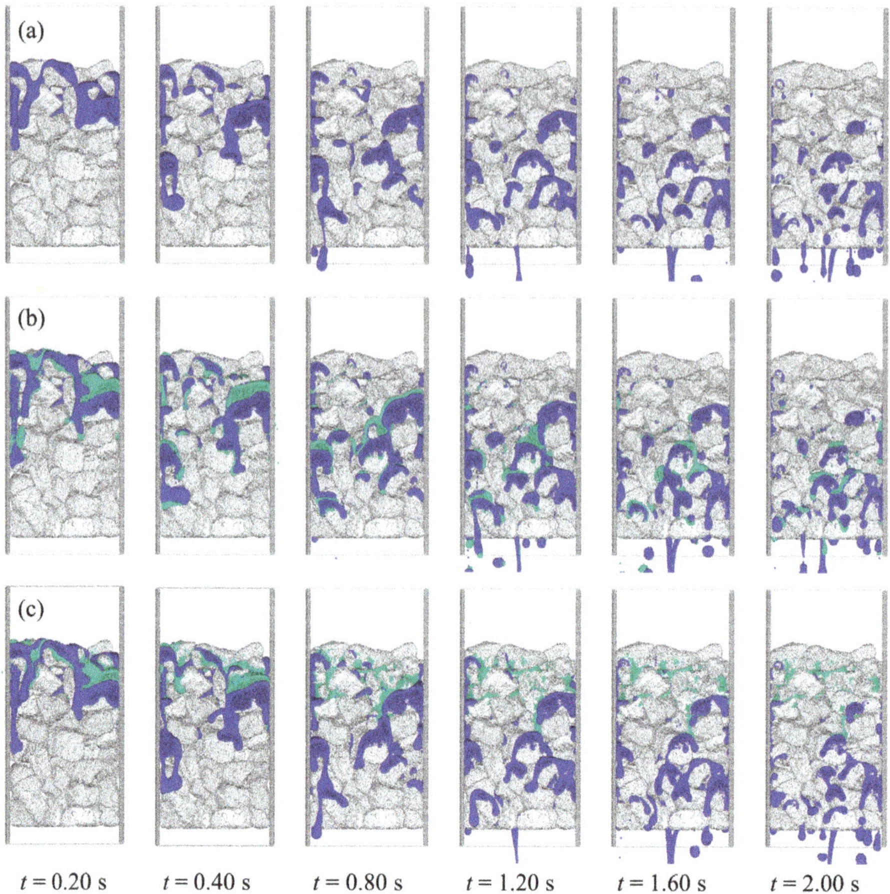

$t = 0.20$ s $t = 0.40$ s $t = 0.80$ s $t = 1.20$ s $t = 1.60$ s $t = 2.00$ s

Figure 3. Liquid iron and molten slag distributions on vertical cross-sections of a coke bed. (**a**) Without slag, (**b**) $\theta s = 160°$, and (**c**) $\theta s = 30°$.

Figure 4 shows the volume distribution of liquid iron and molten slag in the height direction at t = 0.20 and 1.20 s. In Figure 4a, liquid iron almost has the same distribution at t = 0.20 s, irrespective of the presence of slag and the change in θs. Conversely, at t = 1.20 s, the liquid iron with slag exists at a lower height compared to "without slag" iron. In the region shown in (a-1), liquid iron is concentrated regardless of θs, and in the region shown in (a-2), liquid iron is concentrated when θs = 30°. In Figure 4b, molten slag exists at a lower height when θs = 160° than that at 30°. The attractive force due to the wettability of the coke surface prevents slag from dripping in the direction of gravity. This change in wettability affects the flow form of the liquid iron shown in (a-1) and (a-2). It is necessary to determine the time change of each melt velocity to consider the dripping mechanism leading to this result. The average velocity in the gravitational direction per unit time is given, considering the temporal change in the center of gravity of each phase, as follows [44]:

$$\mathbf{v}_z = \frac{1}{V\Delta t} \int (\mathbf{r}_{i,t+\Delta t} - \mathbf{r}_{i,t}) dV \tag{15}$$

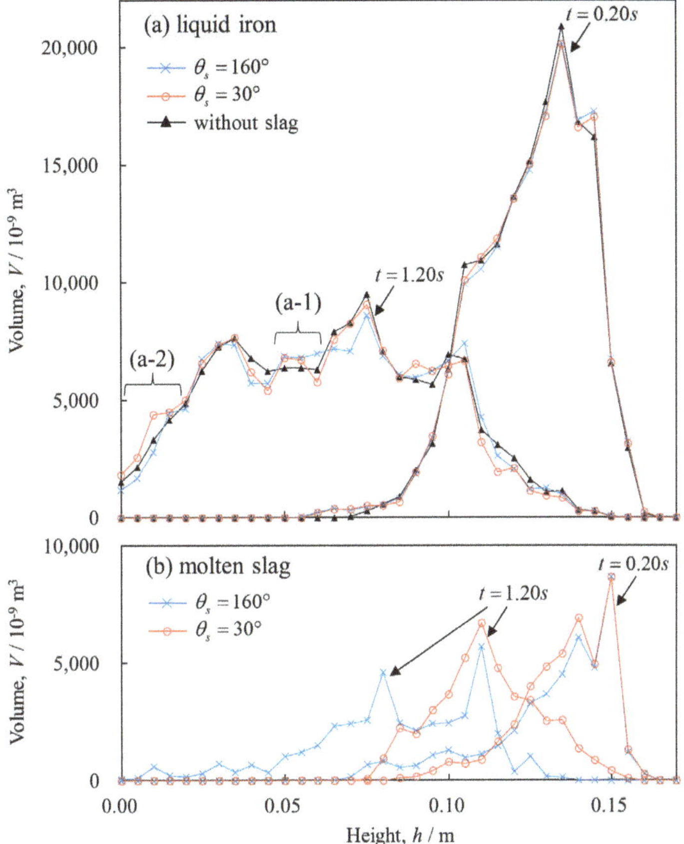

Figure 4. Dripping profiles of liquid iron and molten slag. Calculation domain of coke bed was divided into 34 control volumes in the height direction ($\Delta z = 0.005$ m), and the profiles were derived by counting the number of liquid particles existing in each control volume as time passes.

Figure 5 displays the average velocity of each melt. At $t > 0.5$ s, the liquid iron almost reaches a steady state and almost corresponds to the flow at the lower part of the blast furnace presented by Sugiyama et al. [45] and also appears to follow the Darcy-type equation. From a detailed perspective, however, the liquid iron tends to increase the dripping rate because of the presence of slag. Moreover, the velocity of liquid iron is affected by the wettability between the slag and coke. The angle $\theta s = 30°$ yields the maximum liquid-iron velocity within range (A); in contrast, $\theta s = 160°$ represents the maximum velocity in region (B). This reversal occurs at $t = 0.31$ s. As observed from Figure 4, the liquid iron dripping rate is higher than that of molten slag. Due to the difference in density between the two melts, the gravitational force of liquid iron was 2.7 times higher than that of molten slag. As molten slag has a high viscosity and low density, it appears that it prevents liquid iron from descending. However, when $\theta s = 30°$, the coke bed surface is wet by slag immediately and facilitates the sliding of the liquid iron. With time, this "lubrication" by molten slag is lost, because the liquid iron moves below the molten slag. When $\theta s = 160°$, the effect of wetting becomes weak, and the molten slag drips along with the liquid iron. Thus, the "lubrication" between the molten slag and liquid iron remains on the coke surface. Since the molten slag drops slowly, owing to the 0.37 times density and 53 times viscosity coefficient of the liquid iron, these observations appear as slight differences.

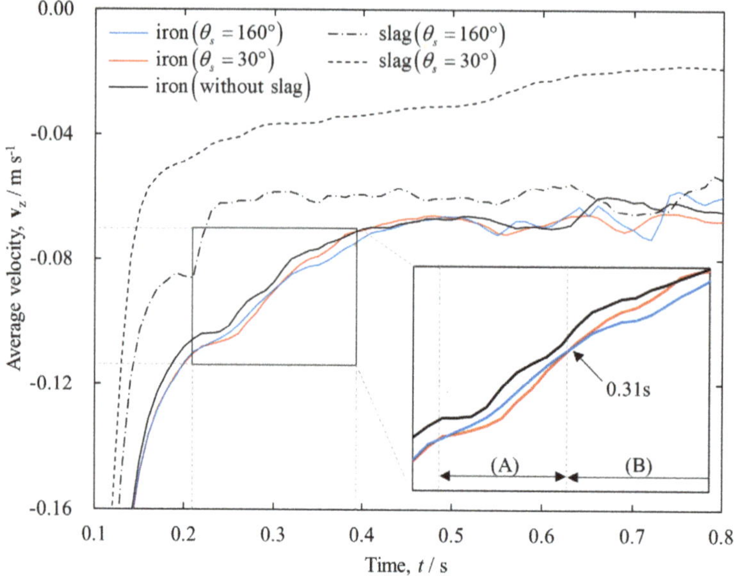

Figure 5. Time variations in the mean dripping velocity. These profiles were obtained using a space integral by considering the center of gravity of all the droplets each time.

The variation in the wettability between the coke and molten slag modifies the interfacial area between the four-phase iron-slag-coke-gas by the following mechanism. When some droplets are dispersed in the system under isothermal conditions, the Helmholtz free energy, $F = \sum \sigma_{ij} A_i$, of the system increases. Since the potential energy corresponding to the flow of the liquid iron droplets through the packed bed is equal for each case, the kinetic energy corresponding to the secondary droplet formation may be the same for all the conditions. However, the total energy in the system is not conserved due to the viscous damping from this calculation, but it is useful to clarify the effect of wettability between the coke and slag on the interfacial area of each phase. As our interest is focused on the effect of wettability between the slag and coke on the interfacial energy of each phase, the interfacial area for each phase should be estimated. The main advantage of the SPH method is its rapid prediction of the interface area A from its initial condition A_0 and the interface-judged particle number n.

$$A(t) \cong \frac{n(t)}{n_0} A_0 \tag{16}$$

A_0 is geometrically determined from the initial conditions. In this study, the free surface of the liquid iron and slag are 0.04992 and 0.01248 m^2, respectively, and the initial iron-slag interface area is 0.02110 m^2. See Section 2 and a previous report [46] for the counting procedure of n. Figure 6 depicts the time change of the estimated interfacial area of each phase. At $t = 0.1$ s, the melt penetrates the coke bed, and both interfacial areas significantly vary simultaneously. In Figure 6a, for "without slag" liquid iron, the initial iron-gas surface area is significantly different from that of "with slag", but the area becomes almost equal to other conditions at $t = 0.1$ s. The case of $\theta s = 160°$ sometimes indicates a larger iron-gas surface area than that of "without slag", because the iron slides down first and forms a contact interface at a lower height. Beyond $t = 0.1$ s, "without slag" iron has the lowest iron-gas surface area. This behavior can be explained as follows. As shown in Figure 6b, since the liquid is not affected by molten slag in "without slag" liquid iron, the iron-coke interfacial area significantly increases compared to that of "with slag" at $t > 0.1$ s; in other words, the iron-coke interface decreases with the presence of the molten slag. From Figure 6c, although the iron-slag interface area has a larger value at $\theta s = 160°$

than at $\theta_s = 30°$, the slag-coke interface area exhibits the opposite trend, as observed in Figure 6d. Since the coke surface with good wettability is covered with slag, it will likely increase the contact area between the free surface of the slag and iron-slag in a static state. In the case of $\theta_s = 160°$, the slag drops with the liquid iron due to gravity. This dripping occurs because the attractive force acting between the coke and slag is less than the gravitational force. However, for $\theta_s = 30°$, the attractive force acting between the coke-slag interface dominates, and the iron-slag interface is preferentially detached. Here, the effect of momentum in the direction of gravity appears. Since a "quasi-stable" interface is formed by the balance between the melt momentum and the force acting on each interface, the state predicted by the equilibrium theory is not always achieved during dripping [47]. This result indicates that static hold-up of the molten slag may promote the smooth dripping of liquid iron, as variations in the wettability between the coke and slag transiently change the flow field of the liquid iron. It emphasizes the distinctive interfacial features that can depend on interactions with the wettability of nonuniform packed cokes and the transient behavior of two melts. In the future, the calculation region of this model can be expanded and the model may be applied to continuous dripping, and the application of the model can also be extended to the entire bottom of the blast furnace.

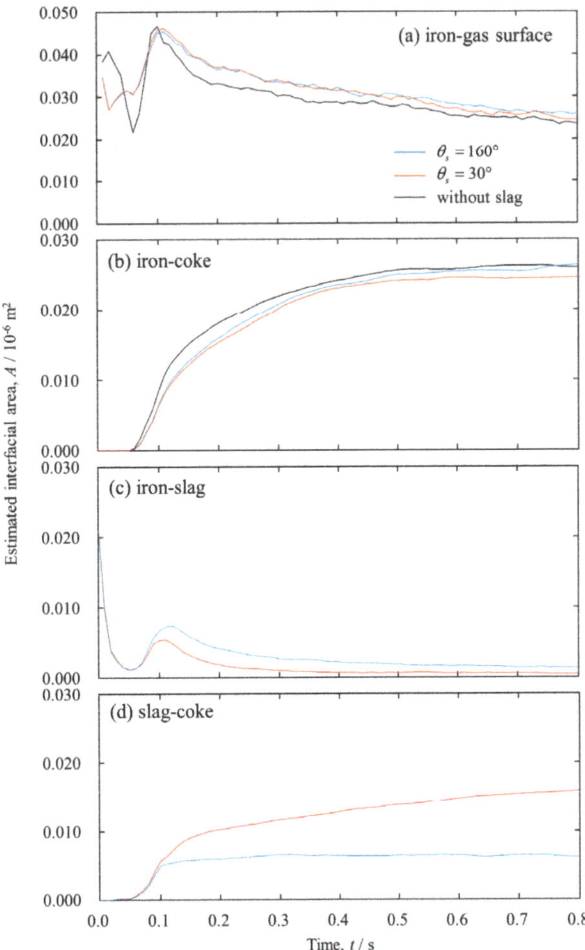

Figure 6. Time change of estimated interfacial area between each phase.

4. Conclusions

Numerical simulations using a multi-phase SPH method were performed on two immiscible high-temperature melts dripping through a coke-packed bed during an ironmaking blast furnace process, namely, a binary trickle flow in which liquid iron and molten slag dripped simultaneously. The advantage of this method is the direct estimation of the transient three-dimensional behavior of multi-phase flow based on body forces considering the force acting between different phases, including complex dispersed phases. The present detailed analysis helps in the rationalization of the unexplained transient behavior of liquid iron and molten slag presented by previous experiments. In particular, the effect of coke surface wettability for molten slag on binary trickle flow was investigated.

The flow behavior of liquid iron is influenced by the presence of slag in coke surfaces with different coke wettability. If the coke surface has poor wettability and contains molten slag, the molten slag drops along the molten iron. However, a coke surface with good wettability retains molten slag owing to a higher attractive force acting between the coke and slag rather than an interfacial force of attraction acting between the iron and slag. Although the molten slag dripping rate is lower than that of the liquid iron because of high viscosity and low density, the coke bed surface is wet by slag immediately and facilitates the sliding of the liquid iron. These results demonstrate that the static hold-up of molten slag may promote the smooth dripping of liquid iron. As predicted by novel findings presented in this study, when the low reducing agent ratio operation is realized, increases in the amount of liquid iron and molten slag passing through a unit volume of coke bed finally influence the amount of stagnant melt behavior in the coke bed through a mechanism in which the descent velocity of each melt depends on coke-slag wettability. Through this analysis, the mechanism underlying the simultaneous trickle flow of liquid iron-molten slag behaviors, as described in conventional research, can be explained from a unified point of view.

Author Contributions: Data curation, S.N., H.N., T.K. (Tatsuya Kikuchi), S.I., and S.U.; funding acquisition, S.N. and K.-i.O.; investigation, K.-i.O. and S.S.; methodology, S.N. and T.K. (Tatsuya Kon); software, S.N. and T.K. (Tatsuya Kon); validation, T.K. (Tatsuya Kon); visualization, T.K. (Tatsuya Kon); writing—original draft, S.N.; and writing—review and editing, S.N., H.N., K.T. and R.O.S. All authors have read and agreed to the published version of the manuscript.

Funding: This study was funded by the Steel Foundation for Environmental Protection Technology (SEPT), Japan, grant number C-41-52, and supported by the Japan Society for the Promotion of Science (JSPS) KAKENHI, grant number 15H04168.

Acknowledgments: The authors thank Shinsuke Taya, Kentaro Baba, Azuma Hirai, Akihisa Ito, and Miho Hayasaka for their support with the construction of the computing environment at Tohoku University, IMRAM.

References

1. Fukutake, T.; Rajakumar, V. Liquid holdups and abnormal flow phenomena of the gas-liquid counter-current flow in packed beds under simulating conditions of the flow in the dropping zone of a blast furnace. *Tetsu-to-Hagané* **1980**, *66*, 1937–1946. [CrossRef]
2. Chew, S.J.; Zulli, P.; Yu, A. Modelling of liquid flow in the blast furnace. Application in a comprehensive blast furnace model. *ISIJ Int.* **2001**, *41*, 1122–1130. [CrossRef]
3. Kawabata, H.; Shinmyou, K.; Harada, T.; Usui, T. Influence of channeling factor on liquid hold-ups in an initially unsoaked bed. *ISIJ Int.* **2005**, *45*, 1474–1481. [CrossRef]
4. Husslage, W.M.; Bakker, T.; Steeghs, A.G.S.; Reuter, M.A.; Heerema, R.H. Flow of molten slag and iron at 1500 °C to 1600 °C through packed coke beds. *Metall. Mater. Trans. B* **2005**, *36*, 765–776. [CrossRef]
5. Jang, D.; Shin, M.; Oh, J.S.; Kim, H.S.; Yi, S.H.; Lee, J. Static holdup of liquid slag in carbonaceous beds. *ISIJ Int.* **2014**, *54*, 1251–1255. [CrossRef]
6. Wang, G.; Liu, Y.; Zhou, Z.; Wang, J.; Xue, Q. Static holdup of liquid slag in simulated packed coke bed under oxygen blast furnace ironmaking conditions. *JOM* **2018**, *70*, 29–33. [CrossRef]
7. Ohgusu, H.; Sassa, Y.; Tomita, Y.; Tanaka, K.; Hasegawa, M. Main factors affecting static holdup of molten slag in coke-packed bed. *Tetsu-to-Hagané* **1992**, *78*, 1164–1170. [CrossRef]

8. Oh, J.S.; Lee, J. Composition-dependent reactive wetting of molten slag on coke substrates. *J. Mater. Sci.* **2016**, *51*, 1813–1819. [CrossRef]

9. White, J.F.; Lee, J.; Hessling, O.; Glaser, B. Reactions between liquid CaO-SiO$_2$ slags and graphite substrates. *Metall. Mater. Trans. B* **2017**, *48*, 506–515. [CrossRef]

10. Saito, K.; Ohno, K.I.; Miki, T.; Sasaki, Y.; Hino, M. Behavior of ironmaking slag permeation to carbonaceous material layer. *ISIJ Int.* **2006**, *46*, 1783–1790. [CrossRef]

11. Kang, T.W.; Gupta, S.; Saha-Chaudhury, N.; Sahajwalla, V. Wetting and interfacial reaction investigations of coke/slag systems and associated liquid permeability of blast furnaces. *ISIJ Int.* **2005**, *45*, 1526–1535. [CrossRef]

12. Geleta, D.D.; Siddiqui, M.I.H.; Lee, J. Characterization of Slag Flow in Fixed Packed Bed of Coke Particles. *Metall. Mater. Trans. B* **2019**. [CrossRef]

13. Vångö, M.; Pirker, S.; Lichtenegger, T. Unresolved CFD-DEM modeling of multiphase flow in densely packed particle beds. *Appl. Math. Model.* **2018**, *56*, 501–516. [CrossRef]

14. Jiao, L.; Kuang, S.; Yu, A.; Li, Y.; Mao, X.; Xu, H. Three-dimensional modeling of an ironmaking blast furnace with a layered cohesive zone. *Metall. Mater. Trans. B* **2020**, *51*, 258–275. [CrossRef]

15. Rabbani, H.S.; Joekar-Niasar, V.; Pak, T.; Shokri, N. New insights on the complex dynamics of two-phase flow in porous media under intermediate-wet conditions. *Sci. Rep.* **2017**, *7*, 4584. [CrossRef] [PubMed]

16. Rabbani, H.S.; Or, D.; Liu, Y.; Lai, C.Y.; Lu, N.B.; Datta, S.S.; Stone, H.A.; Shokri, N. Suppressing viscous fingering in structured porous media. *Proc. Natl. Acad. Sci. USA* **2018**, *115*, 4833–4838. [CrossRef]

17. Natsui, S.; Ishihara, S.; Kon, T.; Ohno, K.; Nogami, H. Detailed Modelling of Packed-bed Gas Clogging Due to Thermal-softening of Iron Ore by Eulerian-Lagrangian Approach. *Chem. Eng. J..* in press. [CrossRef]

18. Alpak, F.O.; Berg, S.; Zacharoudiou, I. Prediction of fluid topology and relative permeability in imbibition in sandstone rock by direct numerical simulation. *Adv. Water Resour.* **2018**, *122*, 49–59. [CrossRef]

19. Tartakovsky, A.M.; Panchenko, A. Pairwise force smoothed particle hydrodynamics model for multiphase flow: Surface tension and contact line dynamics. *J. Comput. Phys.* **2016**, *305*, 1119–1146. [CrossRef]

20. Shao, L.; Saxén, H. A simulation study of two-liquid flow in the taphole of the blast furnace. *ISIJ Int.* **2013**, *53*, 988–994. [CrossRef]

21. Kim, H.S.; Kim, J.G.; Sasaki, Y. The role of molten slag in iron melting process for the direct contact carburization: Wetting and separation. *ISIJ Int.* **2010**, *50*, 1099–1106. [CrossRef]

22. Jeong, I.H.; Kim, H.S.; Sasaki, Y. Trickle flow behaviors of liquid iron and molten slag in the lower part of blast furnace. *ISIJ Int.* **2013**, *53*, 2090–2098. [CrossRef]

23. Ghosh, S.; Viswanathan, N.N.; Ballal, N.B. Flow phenomena in the dripping zone of blast furnace—A review. *Steel Res. Int.* **2017**, *88*, 1600440. [CrossRef]

24. Kon, T.; Natsui, S.; Ueda, S.; Nogami, H. Analysis of effect of packed bed structure on liquid flow in packed bed using moving particle semi-implicit method. *ISIJ Int.* **2015**, *55*, 1284–1290. [CrossRef]

25. Natsui, S.; Kikuchi, T.; Suzuki, R.O.; Kon, T.; Ueda, S.; Nogami, H. Characterization of liquid trickle flow in poor-wetting packed bed. *ISIJ Int.* **2015**, *55*, 1259–1266. [CrossRef]

26. Natsui, S.; Nashimoto, R.; Takai, H.; Kumagai, T.; Kikuchi, T.; Suzuki, R.O. SPH simulations of the behavior of the interface between two immiscible liquid stirred by the movement of a gas bubble. *Chem. Eng. Sci.* **2016**, *141*, 342–355. [CrossRef]

27. Natsui, S.; Sawada, A.; Kikuchi, T.; Suzuki, R.O. Holdup characteristics of melt in coke beds of different shapes. *ISIJ Int.* **2018**, *58*, 1742–1744. [CrossRef]

28. Natsui, S.; Nashimoto, R.; Kikuchi, T.; Suzuki, R.O.; Takai, H.; Ohno, K.; Sukenaga, S. Capturing the non-spherical shape of granular media and its trickle flow characteristics using fully-Lagrangian method. *AIChE J.* **2017**, *63*, 2257–2271. [CrossRef]

29. Nakano, M.; Ito, K. Three dimensional simulation of lime particle penetration into molten iron bath using smoothed particle hydrodynamics. *ISIJ Int.* **2016**, *56*, 1537–1542. [CrossRef]

30. Tsuboi, M.; Ito, K. Cold model experiment and numerical simulation of flow characteristics of multi-phase slag. *ISIJ Int.* **2017**, *57*, 1191–1196. [CrossRef]

31. Hosono, N.; Karato, S.I.; Makino, J.; Saitoh, T.R. Terrestrial magma ocean origin of the moon. *Nat. Geosci.* **2019**, *12*, 418–423. [CrossRef]

32. Wendland, H. Piecewise polynomial, positive definite and compactly supported radial functions of minimal degree. *Adv. Comput. Math.* **1995**, *4*, 389–396. [CrossRef]

33. Colagrossi, A.; Landrini, M. Numerical simulation of interfacial flows by smoothed particle hydrodynamics. *J. Comput. Phys.* **2003**, *191*, 448–475. [CrossRef]

34. Lancaster, P.; Salkauskas, K. Surfaces generated by moving least squares methods. *Math. Comput.* **1981**, *37*, 141–158. [CrossRef]

35. Monaghan, J.J. An introduction to SPH. *Comput. Phys. Commun.* **1988**, *48*, 89–96. [CrossRef]

36. Kondo, M.; Koshizuka, S.; Suzuki, K.; Takimoto, M. Surface Tension Model Using Inter-particle Force in Particle Method. In Proceedings of the ASME/JSME 2007 5th Joint Fluids Engineering Conference, San Diego, CA, USA, 30 July 2007; American Society of Mechanical Engineers Digital Collection. pp. 93–98.

37. Fowkes, F.M. Attractive forces at interfaces. *Ind. Eng. Chem.* **1964**, *56*, 40–52. [CrossRef]

38. Keverian, J.; Taylor, H.F. Effects of gaseous and solid addition on surface tension and contact angle (on graphite) if various iron-carbon alloys. *Trans. AFS* **1957**, *65*, 212–221.

39. Machin, J.S.; Yee, T.B. Viscosity studies of system CaO-MgO-Al_2O_3-SiO_2: II, CaO-Al_2O_3-SiO_2. *J. Am. Ceram. Soc.* **1948**, *31*, 200–204. [CrossRef]

40. Mukai, K.; Ishikawa, T. Surface tension measurements on liquid slags in CaO-SiO_2, CaO-Al_2O_3 and CaO-Al_2O_3-SiO_2 systems by a pendant drop method. *J. Jpn. Inst. Met.* **1981**, *45*, 147. [CrossRef]

41. Mehta, A.S.; Sahajwalla, V. Influence of composition of slag and carbonaceous materials on the wettability at the slag/carbon interface during pulverised coal injection in a blast furnace. *Scand. J. Metall.* **2000**, *29*, 17–29. [CrossRef]

42. Watakabe, S.; Takeda, K.; Igawa, K. Coke properties and operational conditions of blast furnace to prevent coke degradation in the raceway. *Tetsu-to-Hagané* **2002**, *88*, 8–15. [CrossRef]

43. Natsui, N.; Sawada, A.; Terui, T.; Kashihara, Y.; Kikuchi, T.; Suzuki, R.O. Evaluation of coke degradation effect on flow characteristics in packed bed using 3D scanning for rotational mechanical strength test and solid-liquid-gas three-phase dynamic model analysis. *Tetsu-to-Hagané* **2018**, *104*, 347–357. [CrossRef]

44. Natsui, S.; Ohno, K.I.; Sukenaga, S.; Kikuchi, T.; Suzuki, R.O. Detailed modeling of melt dripping in coke bed by DEM-SPH. *ISIJ Int.* **2018**, *58*, 282–291. [CrossRef]

45. Sugiyama, T.; Nakagawa, T.; Sibaike, H.; Oda, Y. Analysis on liquid flow in the dripping zone of blast furnace. *Tetsu-to-Hagané* **1987**, *73*, 2044–2051. [CrossRef]

46. Natsui, S.; Nashimoto, R.; Kumagai, T.; Kikuchi, T.; Suzuki, R.O. An SPH study of molten matte–slag dispersion. *Metall. Mater. Trans. B* **2017**, *48*, 1792–1806. [CrossRef]

47. Kon, T.; Sukenaga, S.; Ueda, S. Dynamic wettability of liquids on gasified metallurgical cokes. *ISIJ Int.* **2017**, *57*, 1166–1172. [CrossRef]

Review

Dead-Man Behavior in the Blast Furnace Hearth—A Brief Review

Lei Shao [1,2], Qilin Xiao [1,2], Chengbo Zhang [1,2], Zongshu Zou [1,2] and Henrik Saxén [3,*]

1   School of Metallurgy, Northeastern University, Shenyang 110819, China; shaolei@mail.neu.edu.cn (L.S.); xiaoqilin1997@163.com (Q.X.); zhangcb654@163.com (C.Z.); zouzs@mail.neu.edu.cn (Z.Z.)
2   State Key Laboratory of Rolling and Automation, Northeastern University, Shenyang 110819, China
3   Process and Systems Engineering Laboratory, Åbo Akademi University, 20500 Åbo, Finland
*   Correspondence: hsaxen@abo.fi; Tel.: +358-40-5443301

Received: 5 October 2020; Accepted: 19 October 2020; Published: 22 October 2020

Abstract: The blast furnace campaign length is today usually restricted by the hearth life, which is strongly related to the drainage and behavior of the coke bed in the hearth, usually referred to as the dead man. Because the hearth is inaccessible and the conditions are complex, knowledge and understanding of the state of the dead man are still limited compared to other parts of the blast furnace process. Since a number of publications have studied different aspects of the dead man in the literature, the purpose of the current review is to compile the findings and knowledge in a comprehensive document. We mainly focus on contributions with respect to the dead man state, and those assessing its influence on the hearth performance in terms of liquid flow patterns, lining wear and drainage behavior. A set of common modeling approaches in this specific furnace area is also briefly presented. The aim of the review is also to deepen the understanding and stimulate further research on open questions related to the dead man in the blast furnace hearth.

Keywords: blast furnace hearth; dead man; iron and slag flow; lining wear; hearth drainage

1. Introduction

The blast furnace (BF) route still remains the dominant one in the production of liquid iron, which is the primary raw material for large-scale steelmaking. In recent years, the trend has been to construct larger furnaces and close small and inefficient ones. Along with the growth of furnace size, more liquid iron and slag are also stored in the BF hearth, and problems of draining and wear are encountered more frequently in the practical operation. Furthermore, due to tougher global competition, the furnace campaign lengths should be extended, and the furnace body should withstand operation under production rates that vary with the market conditions. It is today commonly recognized that drainage and wear problems in the hearth region play the key role in determining the BF campaign life [1].

Both practical and theoretical investigations have revealed that most of the draining and wear problems in the hearths of large furnaces are related to the state and behavior of the porous coke beds that fill them, i.e., the dead man. In order to fully utilize the potential of large BFs in terms of high productivity and lower unit costs of production, it is of considerable significance to understand the governing mechanisms of the dead man behavior in the BF hearth. However, knowledge and information about the dead man are still severely limited, mainly due to the fact that the hearth is the most inaccessible part of the BF, lacking direct measurements. Nevertheless, a number of publications related to or addressing different aspects of the dead man can be found in the literature. The purpose of the current brief review is to summarize these results in a comprehensive document, where the main emphasis is put on presenting contributions shedding light on the dead man state and assessing its influence on BF hearth performance, with respect to liquid flow patterns, hearth lining wear and drainage behavior. The basic modeling approaches used in the analysis are also briefly treated.

2. Investigations and Modeling of the Dead Man

The internal state of the BF, as schematically illustrated in Figure 1, has been gradually revealed by extensive dissection investigations [2,3], which confirmed that there exists a stagnant column of coke particles in the lower part of the furnace. The stagnant column, where coke particles move downwards with a highly reduced velocity, was called "dead man" because it was earlier assumed to exert negligible influence on the whole ironmaking process [4]. However, this assumption was later been found to be incorrect. As a matter of fact, it is nowadays commonly believed that the dead-man state significantly affects the gas and liquid flows in the BF lower part, which, in turn, determine the temperature distribution within the hearth, the liquid drainage, as well as wear of the hearth lining [5]. In addition, the dead man appears to be rather "active" since it is usually claimed to have an average porosity of $\varepsilon = 0.3 - 0.5$ [2], and can be renewed in periods varying between a few days and some weeks.

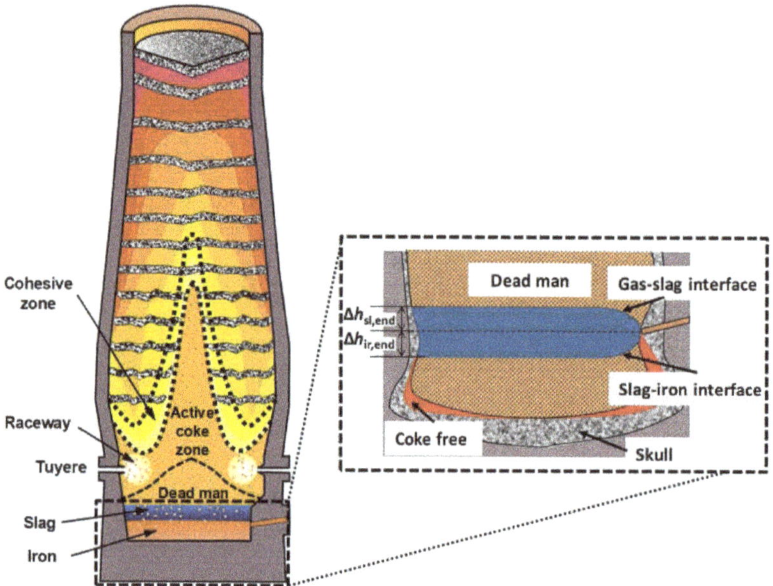

Figure 1. Schematic vertical cross-section of the ironmaking blast furnace and its hearth.

2.1. Structure and Renewal of the Dead Man

As depicted in Figure 1, the dead man is located under the active coke zone beneath the cohesive zone. The upper part of the dead man, which is in the region between the raceways, is cone-shaped with a rounded top. The inclination from the apex of the dead man to the raceway plane is usually claimed to be associated with the repose angle of the coke particles that are charged [6], but also other factors (e.g., charging pattern [7], gas and liquid flow rates) may influence its state. Traditionally, the solid flow has been studied via small-scale experiments, but along with the emergence of more efficient software and hardware, it has recently become feasible to study the dead man formation and the flow of coke in the hearth via the Discrete Element Method (DEM).

The lower part of the dead man is in the region of the hearth where liquid iron and slag dripping from the cohesive zone accumulate before they are tapped out intermittently. Thus, the lower part of the dead man is submerged in a bath of liquid iron and slag, and is thus subjected to a buoyancy force that depends on how deeply the coke is submerged in the two liquids. Because of this, it is not straightforward to deduce the bottom shape and position of the dead man, especially taking into account the dynamic changes in the liquid levels during the tap cycle, as well as other operation variables that may affect or interact with the buoyancy force. In principle, the bottom shape and position of the dead man can be estimated by balancing the forces acting on the dead man, whereby the buoyancy force (often) varies with the liquid levels. A majority of the studies on the estimation of the dead man bottom shape and position are summarized in a separate subsection. In this context, it should be mentioned that some authors use the term dead man only referring to its lower part (i.e., the region below the tuyere level), while others use the broader definitions used above (cf. Figure 1).

In the upper part beneath the active coke zone, the renewal of the dead man is relatively fast due to the short distance to the raceways where coke is intensively consumed. It has been clarified that there exists a small quasi-stagnant region in the center of the active coke zone where the coke particles descend slowly towards the raceways. Therefore, the dead-man porosity (and permeability) can be improved by feeding high-quality coke into the BF center [8–10]. It was also reported that the dead man can be lifted with a sufficient buoyancy force, and the coke right beneath the tuyere level can be "pushed" into the raceways. Renewal thus occurs as the "old" particles are forced to go out of the dead man and "new" particles enter to fill the voids through the upper surface of the dead man [11–15]. In the lower part, especially below the taphole level, it is, however, difficult for the coke to flow upwards into the raceways. Various other renewal mechanisms have been presented and the prevailing ones include FeO reduction, carbon solution loss and carbonization of liquid iron [16–20]. Data from a radioactive tracer test indicated that coke in the peripheral region within the hearth is consumed in 2–3 days, due to a more intensive flow of liquid iron that dissolves coke carbon. This number can be deduced from a simple carbon balance of the hearth, as follows: Consider a large BF with a hearth diameter of 14 m and a daily production rate of 9000 tons, where the hearth coke (85% carbon) occupies about 1000 m^3, with a dead man voidage of $\varepsilon = 0.35$. If the iron that enters the hearth has a carbon content of 2.5% and a content of 4.5% at tapping, an average renewal rate of about 3 days is obtained. However, in the core region of the dead man, the coke is much more gradually dissolved, which is often claimed to occur within a few weeks [3]. It is expected that the dead man in the hearth is heterogeneous with respect to permeability, since the renewal processes are all strongly dominated by liquid flow and heat transfer, which are usually non-uniform in the hearth [21]. Another aspect that indicates a non-uniform permeability distribution is that smaller coke particles may move up and down with the hearth liquids in the voids formed between larger coke particles [5].

The permeability of the dead man can be estimated qualitatively by studying the residence time distribution of tracers injected through a tuyere, or via a probe inserted into the dead man at the tuyere level. The tracer particles dissolve in the iron and/or the slag and are measured in the runner afterwards. By analyzing the response of the tracer in the outflow, information about the flow pattern inside the hearth can be obtained. If the tracer particles are injected at different locations along the radius using a probe, the residence time can be used to evaluate the permeability of different zones of the hearth coke. In some industrial trials with radioactive coke particles, it was found that the core of the dead man in the hearth was very impermeable [22,23].

2.2. Floating State of the Dead Man

The bottom shape and position of the dead man depend on the liquid levels in the hearth and on the force acting on the bed from above. In a normal tap cycle, the liquid levels vary with varying outflow rates of iron and slag, due to the intermittent tapping, as the taphole is eroded during tapping, and because of the "competition" between iron and slag flow in the taphole [24]. Mainly based on balance equations of mass and force, the liquid levels in the BF hearth with two different dead man floating states were estimated and are shown in Figure 2, where the corresponding filtered electromotive force (emf) signals measured at the hearth shell are also depicted [25]. By comparison of simulated liquid levels and emf, Brännbacka [26] found that the iron level corresponds better to the emf signal. Comparing the maxima of the iron and slag level with the emf, this observation can also be confirmed in Figure 2. However, it should be mentioned that the variation in true liquid levels is very hard to measure, even though some indirect measuring methods have been proposed [27–29].

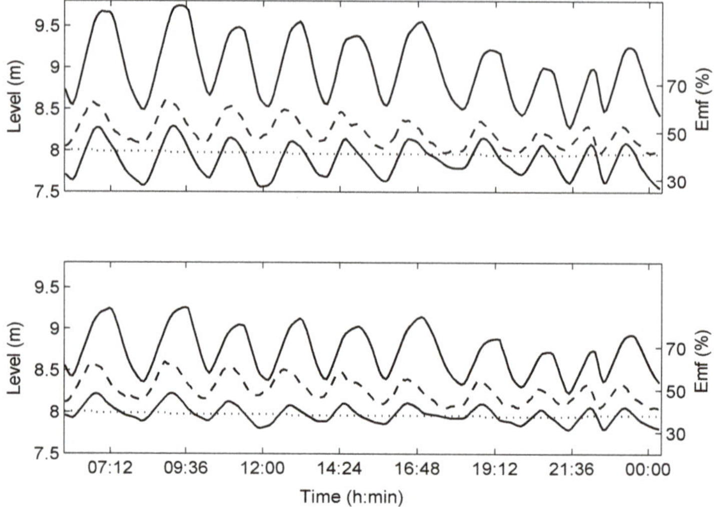

Figure 2. Estimated slag (upper solid curves) and iron (lower solid curves) levels in two BFs with different dead man floating states: sitting (upper panel) and floating (lower panel). Scaled emf signals (dashed curves) and the taphole levels (dotted lines) are also depicted. Reproduced with permission [25].

In an operating furnace, the dead man state in the hearth cannot be measured directly or monitored, owing to the high temperatures, wear, and extremely hostile environment. The dynamic behavior of the dead man has therefore mainly been investigated by utilizing scale models and/or mathematical models under simplified conditions. By visual inspection, it was found that the dead man moves vertically in a cyclic manner during a tap cycle [11,30]. The dead man sits completely on the hearth bottom (i.e., fully fills the hearth) when the liquid levels measured from the hearth bottom are low, while it floats to some degree of height or fully if the liquid levels are high, thus creating a free passage (i.e., coke-free zone) for the liquid flow. Since the iron, with a density of about 2.5 times that of slag, exerts a stronger buoyancy, it is generally considered that the distance between the hearth bottom and the inner end of the taphole ("sump depth") is decisive for dead man floating [31]. In the experimental runs with a pilot model where air was blown in through several tuyeres located in the sidewall and particles were discharged near the tuyeres (in order to mimic the coke consumption by combustion), the dead-man bottom was demonstrated to assume a profile with higher floating levels near the sidewall. This bottom shape has been confirmed in a set of quenched furnaces and two examples are presented in [32,33].

A sophisticated mathematical model [34] also taking into account BF hearth geometry and operation parameters categorized the floating state of the dead man into four different groups: (A) completely floating with a flat bottom, (B) completely floating, but floating higher near the wall, (C) partly floating at the wall, and (D) completely sitting. The results are shown in Figure 3a, where the hearth depth is defined as the distance between hearth bottom and the taphole level. It can be seen that the floating state of the dead man depends strongly on the hearth ("sump") depth and liquid level. The corresponding conditions of some Japanese furnaces were also examined by the mathematical model. As elucidated in Figure 3, the prevailing conditions (i.e., types B and C) of Japanese furnaces appear left of the vertical dashed line, while the aforementioned floating type A (i.e., completely floating with a flat bottom profile) appears right of the dashed vertical line, and is thus in conflict with the prevailing conditions.

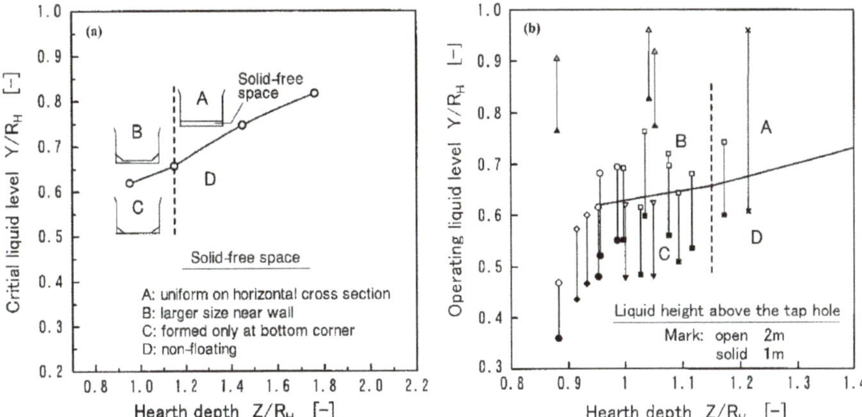

Figure 3. Influences of liquid level and hearth depth on the dead-man state, where Y, Z and R_H refer to the liquid level from hearth bottom, hearth depth, and hearth radius, respectively. (**a**) Critical liquid level as a function of hearth depth; (**b**) operating liquid levels of Japanese blast furnaces. Reproduced with permission [34].

2.3. Modeling of Dead Man State

The dead-man state can be estimated by conducting a balance between the buoyancy of iron and slag in the hearth and the force pressing down on the dead man. The buoyancy force, which is a function of liquid level and dead man porosity, is relatively straightforwardly expressed. Nevertheless, the vertical force pressing down on the dead man is more complicated, since it is related to a set of furnace operating conditions, e.g., raceway length, gas drag and burden weight, as well as liquid holdup above the hearth liquids [34]. The BF lower part is schematically illustrated in Figure 4, where the dead man is divided into two particular regions based on the raceway length/gas drag intensity, i.e., a central region and a region under raceways [35].

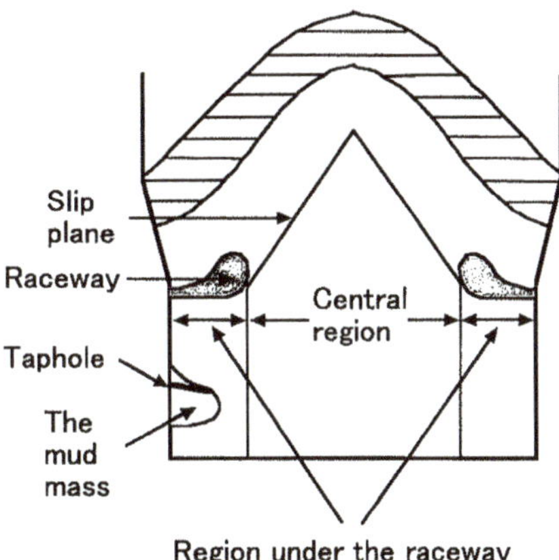

Figure 4. Schematic sketch representing the lower part of the blast furnace. Reproduced with permission [35].

It should be stressed that both the buoyancy force and the downward-acting force are often expressed per unit area, i.e., in the form of pressure. The downward-acting stress at the tuyere level (cf. Figure 4), which was investigated by conducting both experimental runs and numerical calculations [34], is depicted in Figure 5. As can be seen in the figure, the stress is highly reduced in the region where the raceway is located. This can be explained by the drag of the upward-flowing gas from the raceway, which compensates for a portion of the burden weight above the tuyere level. In the region under the raceways, consequently, the dead man could float higher if the buoyancy force is sufficient.

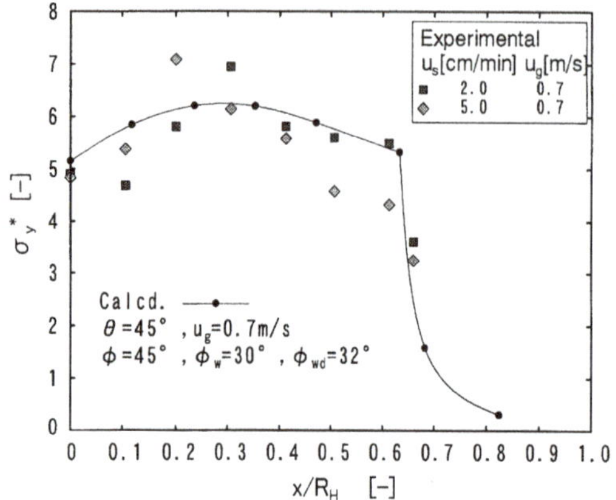

Figure 5. Lateral distribution of the downward-acting stress at the tuyere level. Reproduced with permission [34].

As the radial distribution of the downward-acting force per area, p_d, may vary with the operating conditions, Brännbacka et al. [25,26,36–38] suggested the simple but flexible parameterized expression

$$p_d = \begin{cases} \bar{p}_d & \text{if } r \leq r_0 \\ \bar{p}_d - a\left(\frac{r-r_0}{R}\right)^n & \text{if } r > r_0 \end{cases} \tag{1}$$

where $\bar{p}_d$, r and r_0 are the overall downward pressure, radial position as well as radius of the central region where the downward-acting pressure is unaffected by the raceways, respectively. R is the hearth radius and a is a scaling factor, while n is a parameter that influences the arising shape of the dead man bottom under the raceways. The magnitude of the overall force can be obtained by calculating the burden weight as reduced by the lifting force of the gas drag and the friction of the furnace wall. The vertical position of the dead man bottom deduced from the force balance is

$$z_{dm} = \begin{cases} z_{sl} - \dfrac{p_d}{\rho_{sl}g(1-\varepsilon)} & \text{if } 0 \leq p_d \leq p_{b,sl}^{max} \\ z_{ir} + \dfrac{\rho_{sl}}{\rho_{ir}}(z_{sl} - z_{ir}) - \dfrac{p_d}{\rho_{ir}g(1-\varepsilon)} & \text{if } p_{b,sl}^{max} < p_d \leq p_{b,sl}^{max} + p_{b,ir}^{max} \\ z_{hb} & \text{if } p_d > p_{b,sl}^{max} + p_{b,ir}^{max} \end{cases} \tag{2}$$

with

$$p_{b,sl}^{max} = \rho_{sl}g(1-\varepsilon)(z_{sl} - z_{ir}); \quad p_{b,ir}^{max} = \rho_{ir}g(1-\varepsilon)(z_{ir} - z_{hb}) \tag{3}$$

where ρ_{ir}, ρ_{sl}, g and z_{hb} are the densities of liquid iron and slag, gravitational acceleration and the vertical position of the hearth bottom, respectively, while ε, z_{ir} and z_{sl} are the dead-man porosity and the levels of liquid iron and slag, respectively.

If the parameters of Equation (1) are given, the dead man bottom profile can be calculated based on the quantities of hearth liquids and an average dead man porosity. Figure 6 shows the estimated evolution of the iron and slag levels and the bottom shape of the dead man during a tap cycle in a BF, where the inner hearth profile was estimated by solving an inverse heat transfer problem [25]. The corresponding three-dimensional coke-free zones beneath the dead man are depicted in Figure 7, where the black bar indicates the location of the taphole.

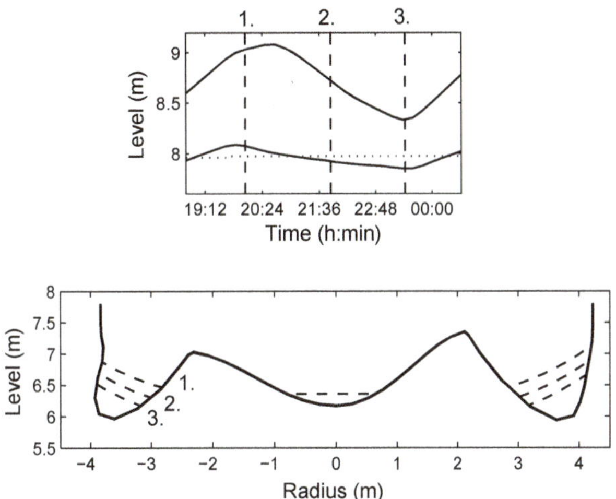

Figure 6. Evolution of the iron and slag levels (upper panel) and the dead-man bottom profile (lower panel) during a tap cycle in an eroded blast furnace. Reproduced with permission [25].

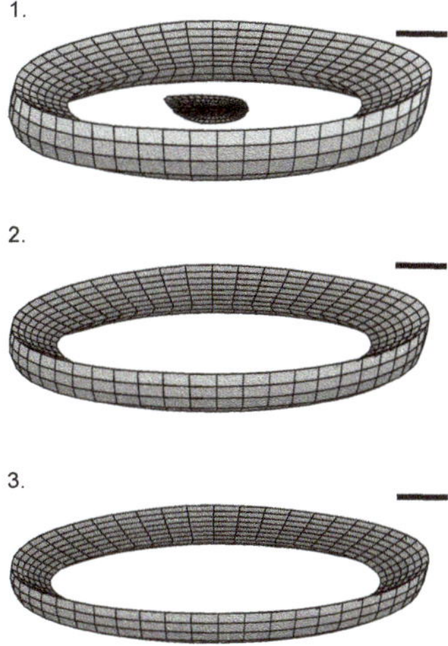

Figure 7. Three-dimensional illustration of the coke-free zones in cases 1–3 of Figure 6. The black horizontal bar marks the location of the taphole. Reproduced with permission [25].

2.4. Effect of Dead Man State on Hearth Performance

2.4.1. Liquid Flow Pattern and Flow-Induced Shear Stress

In the hearth, the floating state of the dead man plays a key role in determining the lining wear and the pattern of liquid flow [31]. Through dissections of quenched furnaces and based on observations at the campaign end when the hearth is relined, the wear profile of the hearth lining has been investigated. The profiles reported in such investigations usually indicate an elephant-foot-shaped profile with severe erosion of the lining in the lower periphery of the hearth. It has been recognized that an elephant-foot-shaped profile is caused by the intensive circumferential flow of hot metal that can occur when the permeability in the dead man's core deteriorates and/or the dead man floats partly, forming a coke-free zone ("gutter") at the hearth corner [39,40]. A bowl-shaped profile, where the lining in the middle of the hearth bottom is excessively eroded, has also been reported. This pattern could be the result if the dead man floats completely, or if the porosity of the dead man is fairly uniform and it occupies the whole hearth. The latter can be expected for hearth designs where the sump depth is large.

The pattern of liquid flow in the BF hearth has been investigated by using both physical and numerical models. Physical studies utilizing scale models have usually considered only steady-state iron flow through a heterogeneous dead man with zones of different permeability. The modeling results can still give insightful information concerning the liquid flow close to the hearth bottom, where the lining erosion is mainly attributed to iron flow.

A number of computational fluid dynamics (CFD) models, focusing on the phenomena in the BF hearth and considering liquid flow and/or heat transfer, have also been built in the past. Usually, the dead man is taken as a fixed packed bed, and Darcy's/Ergun's equation can be applied. The influences of dead man properties, such as packing structure and floating state, on the liquid flow paths and distribution of temperature in the hearth lining have been thoroughly evaluated [41–50]. Figure 8 (based on unpublished results using the model outlined in [48]) illustrates the general streamlines of hot metal in one half of the hearth. These results indicate that a partly floating dead man leads to an intensive circumferential flow, thus exerting a strong heat load on the lining at the hearth corner. However, some simulation results have implied that the distribution of temperature at the hearth bottom is less sensitive to the dead man's properties, since the local heat transfer is controlled by the high thermal resistance of the hearth lining refractories (i.e., ceramic pad) [47].

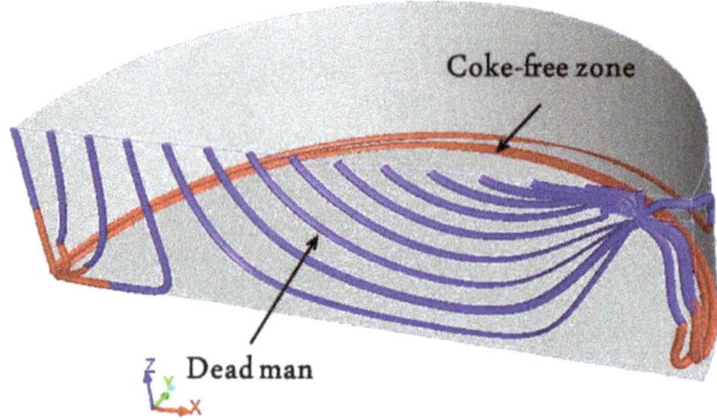

Figure 8. General streamlines of iron flow in a hearth with a partially floating dead man.

The possible hearth lining wear mechanisms include chemical reactions between the lining materials and molten liquids, abrasion and friction caused by coke particles in the hearth, as well as thermo-mechanical stress and flow-induced shear stress. The last one that is caused by the near-wall flow field can lead to lining erosion alone, and a combination of it with other mechanisms could give rise to more wear, eventually resulting in severe hearth damage. Thus, it is imperative to understand the shear stress in terms of its generation and distribution so that appropriate precautions can be taken to reduce the shear stress in order to prolong the campaign life of the hearth. The flow-induced shear stress has been studied mainly using CFD models, since no direct measurements of the BF hearth state variables exist [48–53]. The contours of the shear stress on the hearth bottom under different dead man states, calculated in [48,49], are depicted in Figure 9. It can be seen that with a sitting dead man, the high shear stress zone emerges in the interior of the hearth bottom, particularly below the taphole entrance. Nevertheless, the zone moves to the peripheral region when the dead man partly floats, with a coke-free zone emerging at the hearth corner. In addition, a fully floating dead man state could mitigate the hearth bottom shear stress to some extent, since the shear stress distributes quite uniformly. This would imply that in furnaces where the dead man has started floating completely, the hearth bottom erosion would not progress much.

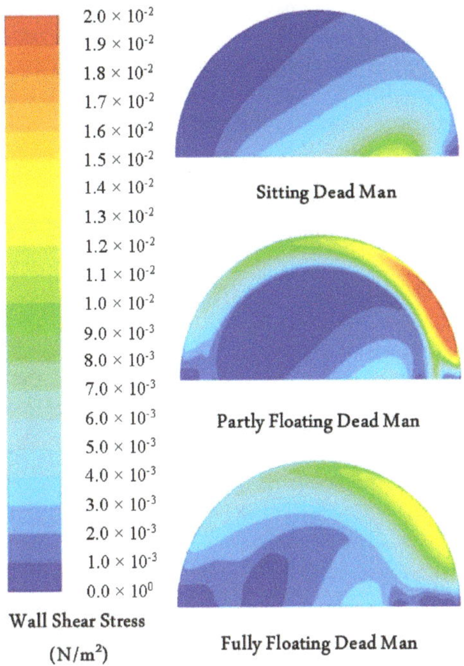

Figure 9. Influences of dead man floating state on the shear stress exerted on the hearth bottom. Reproduced with permission [48].

It was also reported that the high shear stress and heat load in the vicinity of the taphole can be effectively reduced with a longer taphole, since the overall liquid flow is forced to bend towards the center of the dead man, and the circumferential flow caused by a partly floating dead man, or a dead man with a blocked core, can be restrained [52]. This is actually the main reason why a long taphole is usually a prerequisite for protecting the hearth lining near the taphole from severe erosion. A long taphole is often associated with a high carbon content of the liquid iron, which supports the above hypothesis. Furthermore, it has been demonstrated [35] that to achieve a longer taphole, the injected mud must be in good contact with the dead man. Thus, if the dead man floats excessively at the wall, the taphole becomes short, and severe sidewall erosion may follow. In summary, the dead man state is strongly associated with hearth lining erosion, as discussed in the following subsection.

2.4.2. Hearth Wear Profile

The internal geometry of the scale models and the computational domains that represent the hearth profile have usually been simple and regular in physical experiments and in CFD simulations. However, as discussed above, the inner hearth profile may assume different states, partly as a result of the dead-man state. When the hearth lining is cooled at the cold face, the local temperature and velocity of liquid iron in the vicinity of the hot face could become insufficient to keep the iron in liquid form. As a result, the iron may gradually solidify, forming a skull layer on the hot surface of the remaining lining. Therefore, as is often seen by observing the evolution of thermocouples embedded in the hearth lining, the internal geometry of the hearth varies during the campaign [54].

In view of the inherent coupling between liquid flow and heat transfer, an intensive circumferential flow leading to severe erosion is usually linked with an increased heat load on the hearth lining materials. Temperatures measured by thermocouples embedded in the hearth lining can therefore be utilized to predict the hearth wear profile. In practice, the 1150 °C isotherm is often regarded as the internal liquid–solid interface in the hearth. In order to estimate this isotherm, mathematical models where an inverse problem of heat transfer is solved have been proposed [31,54]. It should be stressed that 1150 °C is the lowest temperature at which carbon-saturated iron is present in liquid form, and consequently only heat conduction is solved in this kind of hearth wear profile model. A basic algorithm for estimating the hearth wear profile is outlined in Figure 10.

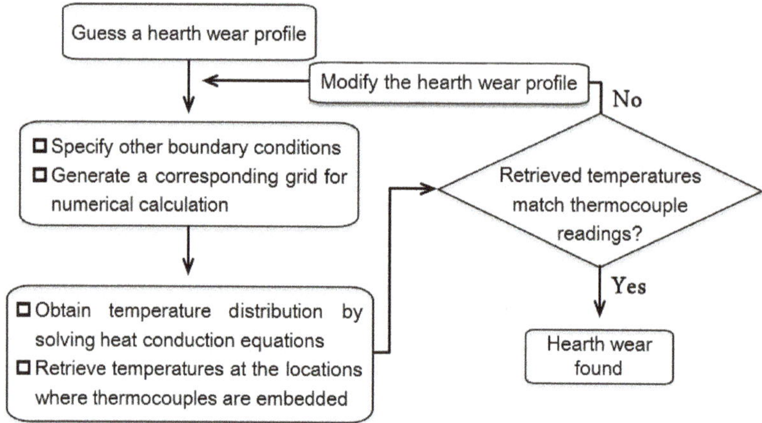

Figure 10. Basic algorithm of a hearth wear profile estimation model.

As a rule, wear models detect the inner profile of the intact lining by matching the most severe erosion experienced during the campaign. If later points correspond to less severe erosion, this is taken as an indication of the formation of a skull layer. It is still a complicated task to accurately identify the remaining sound lining and the skull thickness, particularly for cases wherein the historical thermocouple readings are not available for the whole furnace campaign. A systematic and two-dimensional approach is needed in order to estimate the progress of erosion, and also the formation of skull, during the campaign [31,54,55]. The erosion and skull lines estimated with such an approach for two different BFs (unpublished results using the model outlined in [55]) are presented in Figure 11. It can be seen that the two furnaces have different but characteristic wear profiles, i.e., bowl-shaped and elephant-foot-shaped profiles, which are likely to reflect the state of the iron flow in or below the dead man during the campaigns.

Hearth wear profile estimation models are commonly implemented as monitoring and diagnosis tools to aid the operation of BFs. By analyzing the estimated hearth wear profile, indications may be obtained pertaining to the need to adjust the operation towards conditions less prone to cause erosion, e.g., by reducing the production rate or by blanking tuyeres. The latter means both reducing the local inflow of iron "from above" and suppressing a possible local floating of the dead man. On the other hand, it may be argued that such blanking may instead promote liquid flow, as the liquid holdup decreases at a lower local bosh gas flow. It seems that the efficiency of such actions depends on the state of the furnace when the change is implemented.

The estimated erosion profile can be used to support the interpretation of the liquid flow pattern and dead man floating state that are intimately related to hearth erosion. For instance, a too strong peripheral flow or too low carbon content of the hot metal may be counteracted by the center charging of strong and large coke, which (in the long run) will promote a more uniform flow of iron through the dead man, enhancing the contact of it with the hearth wall. Still, it should be kept in mind that the accuracy of erosion models depends on the validity of the modeling assumptions. In particular, brittle lining layers of low conductivity may lead to inaccurate estimates of the progress of the erosion profile, and the occurrence of such should be detected by other means [56,57].

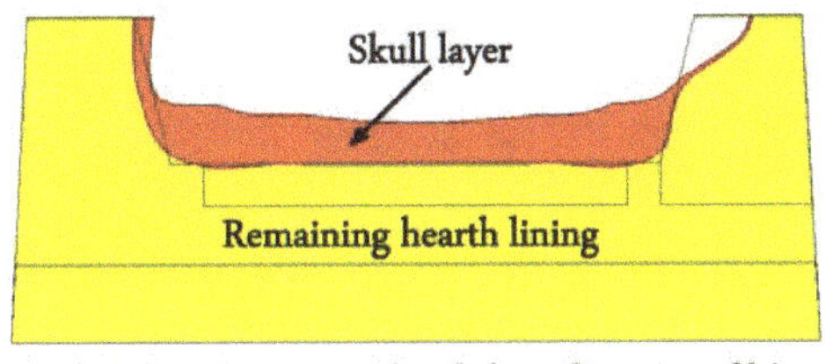

(a) Blast furnace A (bowl shaped wear profile)

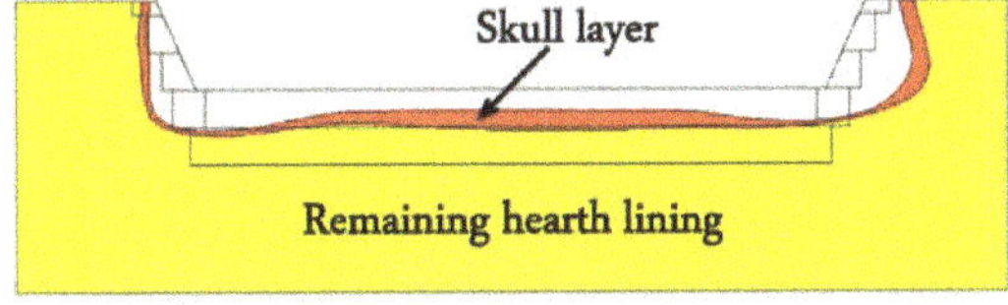

(b) Blast furnace B (elephant-foot shaped wear profile)

Figure 11. Erosion and skull lines in two industrial furnaces estimated based on inverse heat transfer calculation.

2.4.3. Drainage Behavior

In the BF hearth, the void space is occupied by the immiscible liquid iron and slag. As a result of gravity, the rivulets of molten iron and slag from the cohesive zone [58] will eventually separate into two different layers, i.e., the upper slag layer and the lower iron layer. Two interfaces, i.e., gas–slag and slag–iron, are thus formed in the hearth. In practice, a tap starts when one taphole is drilled open. A substantial pressure drop in the vicinity of the taphole can be formed as the high-viscosity slag is driven to flow through the dead man towards the taphole. As a result, the gas–slag interface is tilted down locally near the taphole [24,59–62]. Thus, the overall gas–slag interface is above the taphole level at the moment when gas bursts out and the tap is terminated. This is actually the reason why some amount of slag still remains in the hearth at the tap end, and a "dry hearth" is impossible in practice.

It was earlier commonly assumed that the slag–iron interface is horizontal at the level of the taphole when the slag phase approaches the taphole during tapping. Based on this assumption, extensive physical experiments were carried out by Fukutake and Okabe in order to estimate the tap end slag residual ratio [63–65]. By analyzing the experimental results mainly using the theories of fluid dynamics, the authors proposed a dimensionless flow-out coefficient that was found to correlate monotonically with the slag residual ratio defined. The flow-out coefficient is strongly affected by the state of the dead man, including its voidage and coke particle size. Later, the aforementioned assumption of a horizontal slag–iron interface was found to be incorrect, since both practical observations and physical experiments with two immiscible liquids indicated that the lower phase (liquid iron) can be pumped up from some level below the taphole during the period when slag and iron are drained out simultaneously [59,66,67]. In the BF hearth, a substantial pressure drop is induced in the taphole vicinity when the high-viscosity slag flows through the dead man. This large pressure drop is sufficient to compensate for the flow resistance of the low-viscosity iron, and to overcome the gravitational force when iron is drained up from some level below the taphole. The iron and slag levels at the end of a tap in the BF hearth are sketched in Figure 12, where the hearth internal profile is idealized.

The flow-out coefficient [63–65] was later modified by other authors [68,69] to take into account the non-horizontal slag–iron interface as it is depicted in Figure 12, the coke-free zone beneath the dead man, the varying drainage rates of iron and slag due to taphole erosion, and the continuous production of the liquids. Both observations in the practical operation of the hearth and CFD-based simulations [70,71] have shown that an increase in the slag residual ratio is often attributed to a decrease in dead man permeability, or an increase in draining rate or in slag viscosity. It has also been found that a coke-free zone beneath the dead man directly affects hearth drainage only if it extends close to or above the taphole level [32,68,72]. However, even a partial floating of the dead man may have implications for the drainage via the accumulation and depletion of liquid iron in the coke-free zone during the tap cycle, which affects the liquid levels [25,37,38] and therefore the pressure-loss terms of iron and slag in front of the taphole [24].

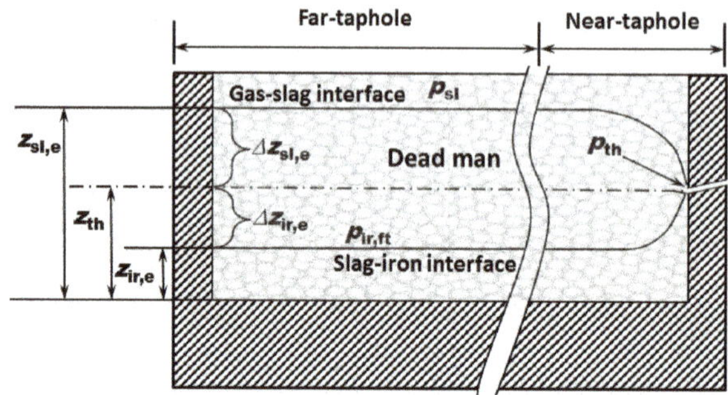

Figure 12. Schematic of the gas–slag and slag–iron interfaces at the end of a tap, where $z_{sl,e}$, z_{th} and $z_{ir,e}$ refer to the vertical distances from the overall gas-slag interface, centerline of taphole, and slag-iron interface to the hearth bottom, while p_{sl}, p_{th} and $p_{ir,ft}$ refer to the static pressure at the overall gas–slag interface, in front of the taphole, and at the overall slag–iron interface, respectively.

With reference to Figure 12, the asymptotic relation between the tap end iron and slag levels that are related to the dead man state can be derived based on a simplified pressure balance [66,67]:

$$\frac{\Delta z_{ir,e}}{\Delta z_{sl,e}} = \frac{\rho_{sl}}{\rho_{ir} - \rho_{sl}} \tag{4}$$

where z and ρ are the vertical distance and liquid density, and where subscripts ir, sl and e denote iron, slag and tap end, respectively.

Equation (4) has been adopted as an asymptotic relation in some hearth drainage studies to validate the calculated results [25,73]. It should, however, be noted that the end points of individual taps may depart considerably from the above relation, simply because the "initial" volume of slag is not sufficient: if too little slag is drained, the duration of the tap is not long enough for the liquids to reach this asymptotic state. The occurrence of local liquid levels in a BF with an impermeable dead man further complicates the general interpretation [62,74–76].

3. Concluding Remarks

Over the years, several aspects regarding the BF dead man have been studied by means of dissection investigations, physical experiments, and theoretical and numerical calculations. Today, the importance of the dead man state and its influence on the performance of the BF hearth have been commonly recognized, even though direct (long-term) measurements of the pertaining state variables are still impossible.

The structure and renewal mechanisms of the dead man have been clarified. It has been demonstrated that the dead man is heterogeneous in terms of its permeability distribution, and that the permeability can be improved in practice by charging high-quality coke into the BF center. The dead man bottom shape and position strongly depend on the balance between the buoyancy force exerted by the in-hearth liquid iron and slag, and the downward-acting force that is reduced towards the furnace wall due to the drag of the upward-flowing gas from the raceways and wall friction. In general, the dead man sits completely on the hearth bottom when the levels of the in-hearth liquids are low. If the liquid levels are high, however, the dead-man bottom assumes a profile where it floats higher at the hearth corner.

Observed hearth lining wear profiles, i.e., elephant-foot-shaped and bowl-shaped, are intimately related to the dead man floating state and its permeability distribution. The lining profile can be estimated utilizing hearth wear models, whereby an inverse problem of heat conduction is solved to predict the position of the 1150 °C isotherm. Such models are today used in several BFs as monitoring and diagnosis tools. The estimated profile can be used to assist the interpretation of the in-hearth liquid flow pattern and the floating state of the dead man. By analyzing the estimated hearth wear profile, indications may also be obtained pertaining to the need to change the operation towards conditions less prone to cause erosion, including lowering of the production rate, blanking of tuyeres in regions with strong local hearth wear, or modifying the tapping strategy.

The drainage of the BF hearth is complicated and a dry tap is impossible because some amount of slag always remains in the hearth at the end of a normal tap. It has been shown that the slag residual ratio at the tap end can be correlated with the flow-out coefficient. In practice, an increase in the slag residual ratio is often attributed to a decrease in dead man permeability, or an increase in draining rate or in slag viscosity. The complexity of the hearth drainage behavior is basically due to the multiphase flow of immiscible fluids (gas, slag and iron), the existence of the dead man and the erosion of the taphole. As a tap proceeds, both the gas–slag and slag–iron interfaces gradually tilt towards the taphole. Therefore, the overall slag and iron levels are located above and below the taphole at the tap-end, respectively. This interface titling phenomenon has been investigated, and an absolute asymptotic limit has been derived and can be applied to validate the related modeling results. However, much work is still required in order to understand the effects of, e.g., local permeability changes in the dead man or local dead man motion on the drainage patterns from individual tapholes. A deeper understanding of the drainage can be the basis of a better design of the tapping operation, which can be influenced by the duration of the inter-cast period, the drill diameter and the taphole angle. The role of the taphole length, e.g., how this variable can be controlled and how it affects the dead-man state, are also factors that should be studied and clarified in the future.

Author Contributions: Conceptualization, L.S. and H.S.; literature review, L.S., Q.X., C.Z., Z.Z. and H.S., original draft preparation, L.S.; writing—review and editing, L.S., Z.Z. and H.S.; funding acquisition, L.S. and Z.Z. All authors have read and agreed to the published version of the manuscript.

Funding: This research was partially funded by the National Science Foundation of China, grant number 51604068. The authors gratefully acknowledge the support.

Conflicts of Interest: The authors declare no conflict of interest. The funders had no role in the design of the study or interpretation of the results.

References

1. Liu, Z.J.; Zhang, J.L.; Zuo, H.B.; Yang, T.J. Recent progress on long service life design of Chinese blast furnace hearth. *ISIJ Int.* **2012**, *52*, 1713–1723. [CrossRef]
2. Kanbara, K.; Hagiwara, T.; Shigemi, A.; Kondo, S.; Kanayama, Y.; Wakabayashi, K.; Hiramoto, N. Dissection of blast furnaces and their internal state. *Trans. Iron Steel Inst. Jpn.* **1977**, *17*, 371–380. [CrossRef]
3. Omori, Y. *Blast Furnace Phenomena and Modelling*; Elsevier: London, UK, 1987.
4. Nishio, H.; Wenzel, W.; Gudenau, H.W. Significance of the dead (man) zone in the blast furnace. *Stahl Eisen.* **1977**, *97*, 867–875.
5. Nightingale, R.J. The Development and Application of Hearth Voidage Estimation and Deadman Cleanliness Index for the Control of Blast Furnace Hearth Operation. Ph.D. Thesis, University of Wollongong, New South Wales, Australia, January 2000.
6. Raipala, K. On Hearth Phenomena and Hot Metal Carbon Content in Blast Furnace. Ph.D. Thesis, Helsinki University of Technology, Helsinki, Finland, November 2003.
7. Ichida, M.; Nishihara, K.; Tamura, K.; Sugata, M.; Ono, H. Influence of ore/coke distribution on descending and melting behavior of burden in the blast furnace. *ISIJ Int.* **1991**, *31*, 505–514. [CrossRef]
8. Takahashi, H.; Komatsu, N. Cold model study on burden behaviour in the lower part of blast furnace. *ISIJ Int.* **1993**, *33*, 655–663. [CrossRef]

9.   Takahashi, H.; Tanno, M.; Katayama, J. Burden descending behavior with renewal of deadman in a two dimensional cold model of blast furnace. *ISIJ Int.* **1996**, *36*, 1354–1359. [CrossRef]

10.   Nogami, H.; Toda, K.; Pintowantoro, S.; Yagi, J.I. Cold-model experiments on deadman renewal rate due to sink-float motion of hearth coke bed. *ISIJ Int.* **2004**, *44*, 2127–2133. [CrossRef]

11.   Shibata, K.; Kimura, Y.; Shimizu, M.; Inaba, S. Dynamics of dead-man coke and hot metal flow in a blast furnace hearth. *ISIJ Int.* **1990**, *30*, 208–215. [CrossRef]

12.   Shinohara, K.; Saitoh, J. Mechanism of solids segregation over a two-dimensional dead man in a blast furnace. *ISIJ Int.* **1993**, *33*, 672–680. [CrossRef]

13.   Zhang, S.J.; Yu, A.B.; Zulli, P.; Wright, B.; Austin, P. Numerical simulation of solids flow in a blast furnace. *Appl. Math. Modell.* **2002**, *26*, 141–154. [CrossRef]

14.   Kawai, H.; Takahashi, H. Solid behavior in shaft and deadman in a cold model of blast furnace with floating-sinking motion of hearth packed bed studied by experimental and numerical DEM analyses. *ISIJ Int.* **2004**, *44*, 1140–1149. [CrossRef]

15.   Bambauer, F.; Wirtz, S.; Scherer, V.; Bartusch, H. Transient DEM-CFD simulation of solid and fluid flow in a three dimensional blast furnace model. *Powder Technol.* **2018**, *334*, 53–64. [CrossRef]

16.   Sunahara, K.; Inada, T.; Iwanaga, Y. Size degradation of deadman coke by reaction with molten FeO in blast furnace. *ISIJ Int.* **1993**, *33*, 275–283. [CrossRef]

17.   Kasai, A.; Kiguchi, J.; Kamijo, T.; Shimizu, M. Degradation of coke by molten iron oxide in the cohesive zone and dripping zone of a blast furnace. *Tetsu-to-Hagané* **1998**, *84*, 9–13. [CrossRef]

18.   Sun, H. Analysis of reaction rate between solid carbon and molten iron by mathematical models. *ISIJ Int.* **2005**, *45*, 1482–1488. [CrossRef]

19.   Li, K.; Zhang, J.; Liu, Y.; Barati, M.; Liu, Z.; Zhong, J.; Yang, T. Graphitization of coke and its interaction with slag in the hearth of a blast furnace. *Metall. Mater. Trans. B* **2016**, *47*, 811–813. [CrossRef]

20.   Post, J.R. Simulation of the Inhomogeneous Deadman of an Ironmaking Blast Furnace. Ph.D. Thesis, Delft University of Technology, Delft, The Netherlands, November 2019.

21.   Sugiyama, T. Experimental and numerical analysis on the movement and the accumulation of powder in the deadman and the dripping zone of blast furnace. *Tetsu-to-Hagané* **1996**, *82*, 29–34. [CrossRef]

22.   Raipala, K. Deadman and hearth phenomena in the blast furnace. *Scand. J. Metall.* **2000**, *29*, 39–46. [CrossRef]

23.   Negro, P.; Petit, C.; Urvoy, A.; Sert, D.; Pierret, H. Characterization of the permeability of the blast furnace lower part. *Revue Métallurgie* **2001**, *98*, 521–531. [CrossRef]

24.   Roche, M.; Helle, M.; van der Stel, J.; Louwerse, G.; Storm, J.; Saxén, H. Drainage model of multi-taphole blast furnace. *Metall. Mater. Trans. B* **2020**, *51*, 1731–1749. [CrossRef]

25.   Brännbacka, J.; Saxén, H. Novel model for estimation of liquid levels in the blast furnace hearth. *Chem. Eng. Sci.* **2004**, *59*, 3423–3432. [CrossRef]

26.   Brännbacka, J. Model Analysis of Dead-Man Floating State and Liquid Levels in the Blast Furnace Hearth. Ph.D. Thesis, Åbo Akademi University, Turku, Finland, October 2004.

27.   Desai, D. Analysis of blast furnace hearth drainage based on the measurement of liquid pressure inside of the hearth. *Iron Steelmak.* **1992**, *19*, 52–57.

28.   Danloy, G.; Stolz, C.; Crahay, J.; Dubois, P. Measurement of iron and slag levels in blast furnace hearth. In Proceedings of the 58th Ironmaking Conference, Chicago, IL, USA, 21–24 March 1999; pp. 89–98.

29.   Havelange, O.; Danloy, G.; Franssen, C. The dead man, floating or not? *Revue Métallurgie* **2004**, *101*, 195–201. [CrossRef]

30.   Shinotake, A.; Ichida, M.; Ootsuka, H.; Kurita, Y. Bottom shape of blast furnace deadman and its floating/sinking behavior by 3-dimensional model experiment. *Tetsu-to-Hagané* **2003**, *89*, 573–580. [CrossRef]

31.   Andreev, K.; Louwerse, G.; Peeters, T.; van der Stel, J. Blast furnace campaign extension by fundamental understanding of hearth processes. *Ironmak. Steelmak.* **2017**, *44*, 81–91. [CrossRef]

32.   Nouchi, T.; Yasui, M.; Takeda, K. Effects of particle free space on hearth drainage efficiency. *ISIJ Int.* **2003**, *43*, 175–180. [CrossRef]

33.   Inada, T.; Kasai, A.; Nakano, K.; Komatsu, S.; Ogawa, A. Dissection investigation of blast furnace hearth—Kokura No. 2 Blast Furnace (2nd campaign). *ISIJ Int.* **2009**, *49*, 470–478. [CrossRef]

34.   Takahashi, H.; Kawai, H.; Suzuki, Y. Analysis of stress and buoyancy for solids flow in the lower part of a blast furnace. *Chem. Eng. Sci.* **2002**, *57*, 215–226. [CrossRef]

35. Tsuchiya, N.; Fukutake, T.; Yamauchi, Y.; Matsumoto, T. In-furnace conditions as prerequisites for proper use and design of mud to control blast furnace taphole length. *ISIJ Int.* **1998**, *38*, 116–125. [CrossRef]
36. Brännbacka, J.; Saxén, H. Modeling the liquid levels in the blast furnace hearth. *ISIJ Int.* **2001**, *41*, 1131–1138. [CrossRef]
37. Brännbacka, J.; Saxén, H. Model analysis of the operation of the blast furnace hearth with a sitting and floating dead man. *ISIJ Int.* **2003**, *43*, 1519–1527. [CrossRef]
38. Brännbacka, J.; Torrkulla, J.; Saxén, H. Simple simulation model of blast furnace hearth. *Ironmak. Steelmak.* **2005**, *32*, 479–486. [CrossRef]
39. Takeda, K.; Watakabe, S.; Sawa, Y.; Itaya, H.; Kawai, T.; Matsumoto, T. Prevention of hearth brick wear by forming a stable solidified layer. *Ironmak. Steelmak.* **2000**, *27*, 79–84.
40. Li, Y.; Cheng, S.; Zhang, P.; Zhou, S. Sensitive influence of floating state of blast furnace deadman on molten iron flow and hearth erosion. *ISIJ Int.* **2015**, *55*, 2332–2341. [CrossRef]
41. Ohno, J.; Tachimori, M.; Nakamura, M.; Hara, Y. Influence of hot metal flow on the heat transfer in a blast furnace hearth. *Tetsu-to-Hagané* **1985**, *71*, 34–40. [CrossRef]
42. Preuer, A.; Winter, J.; Hiebler, H. Computation of the erosion in the hearth of a blast furnace. *Steel Res.* **1992**, *63*, 147–151. [CrossRef]
43. Preuer, A.; Winter, J. Numerical simulation of refractory erosion caused by carbon dissolution in blast furnace. *Revue Métallurgie* **1993**, *90*, 955–964. [CrossRef]
44. Inada, T.; Takatani, K.; Miyahara, M.; Wakabayasi, S.; Yamamoto, T.; Kasai, A.; Takata, K. Development and application of an advanced numerical analysis technology for blast furnace hearth. In Proceedings of the 58th Ironmaking Conference, Chicago, IL, USA, 21–24 March 1999; pp. 633–639.
45. Panjkovic, V.; Truelove, J.S.; Zulli, P. Numerical modelling of iron flow and heat transfer in blast furnace hearth. *Ironmak. Steelmak.* **2002**, *29*, 390–400. [CrossRef]
46. Guo, B.Y.; Yu, A.B.; Zulli, P.; Maldonado, D. CFD modelling and analysis of the flow, heat transfer and mass transfer in a blast furnace hearth. *Steel Res. Int.* **2011**, *82*, 579–586. [CrossRef]
47. Guo, B.Y.; Maldonado, D.; Zulli, P.; Yu, A.B. CFD modelling of liquid metal flow and heat transfer in blast furnace hearth. *ISIJ Int.* **2008**, *48*, 1676–1685. [CrossRef]
48. Shao, L.; Saxén, H. Numerical prediction of iron flow and bottom erosion in the blast furnace hearth. *Steel Res. Int.* **2012**, *83*, 878–885. [CrossRef]
49. Shao, L. Model-Based Estimation of Liquid Flows in the Blast Furnace Hearth and Taphole. Ph.D. Thesis, Åbo Akademi University, Turku, Finland, September 2013.
50. Cheng, W.T.; Huang, E.N.; Du, S.W. Numerical analysis on transient thermal flow of the blast furnace hearth in tapping process through CFD. *Int. Commun. Heat Mass Transf.* **2014**, *57*, 13–21. [CrossRef]
51. Vats, A.K.; Dash, S.K. Flow induced stress distribution on wall of blast furnace hearth. *Ironmak. Steelmak.* **2000**, *27*, 123–128. [CrossRef]
52. Dash, S.K.; Ajmani, S.K.; Kumar, A.; Sandhu, H.S. Optimum taphole length and flow induced stresses. *Ironmak. Steelmak.* **2001**, *28*, 110–116. [CrossRef]
53. Dash, S.K.; Jha, D.N.; Ajmani, S.K.; Upadhyaya, A. Optimisation of taphole angle to minimise flow induced wall shear stress on the hearth. *Ironmak. Steelmak.* **2004**, *31*, 207–215. [CrossRef]
54. Torrkulla, J.; Saxén, H. Model of the state of the blast furnace hearth. *ISIJ Int.* **2000**, *40*, 438–447. [CrossRef]
55. Brännbacka, J.; Saxén, H. Model for fast computation of blast furnace hearth erosion and buildup profiles. *Ind. Eng. Chem. Res.* **2008**, *47*, 7793–7801. [CrossRef]
56. Shinotake, A.; Nakamura, H.; Yadoumaru, N.; Morizane, Y.; Meguro, M. Investigation of blast-furnace hearth sidewall erosion by core sample analysis and consideration of campaign operation. *ISIJ Int.* **2003**, *43*, 321–330. [CrossRef]
57. Kaymak, Y.; Bartusch, H.; Hauck, T.; Mernitz, J.; Rausch, H.; Lin, R. Multiphysics model of the hearth lining state. *Steel Res. Int.* **2020**, 200055. [CrossRef]
58. Wang, G.X.; Chew, S.J.; Yu, A.B.; Zulli, P. Model Study of Liquid Flow in the Blast Furnace Lower Zone. *ISIJ Int.* **1997**, *37*, 573–582. [CrossRef]
59. Tanzil, W.B.; Zulli, P.; Burgess, J.M.; Pinczewski, W.V. Experimental model study of the physical mechanisms governing blast furnace hearth drainage. *Trans. Iron Steel Inst. Jpn.* **1984**, *24*, 197–205. [CrossRef]
60. Shao, L.; Saxén, H. Pressure drop in the blast furnace hearth with a sitting deadman. *ISIJ Int.* **2011**, *51*, 1014–1016. [CrossRef]

61. Shao, L.; Saxén, H. A simulation study of two-liquid flow in the taphole of the blast furnace. *ISIJ Int.* **2013**, *53*, 988–994. [CrossRef]

62. Saxén, H. Model of draining of the blast furnace hearth with an impermeable zone. *Metall. Mater. Trans. B* **2015**, *46*, 421–431. [CrossRef]

63. Fukutake, T.; Okabe, K. Investigation of slag flow in the blast furnace hearth based on the fluid dynamics and of relation between residual slag amount and tapping out conditions. *Tetsu-to-Hagané* **1974**, *60*, 607–621. [CrossRef]

64. Fukutake, T.; Okabe, K. Experimental studies of slag flow in the blast furnace hearth during tapping operation. *Trans. Iron Steel Inst. Jpn.* **1976**, *16*, 309–316. [CrossRef]

65. Fukutake, T.; Okabe, K. Influences of slag tapping conditions on the amount of residual slag in blast furnace hearth. *Trans. Iron Steel Inst. Jpn.* **1976**, *16*, 317–323. [CrossRef]

66. Tanzil, W.B.U. Blast Furnace Hearth Drainage. Ph.D. Thesis, University of New South Wales, Sydney, Australia, 1985.

67. Tanzil, W.B.U.; Pinczewski, W.V. Blast furnace hearth drainage: Physical mechanisms. *Chem. Eng. Sci.* **1987**, *42*, 2557–2568. [CrossRef]

68. Zulli, P. Blast Furnace Hearth Drainage with and without Coke-Free Layer. Ph.D. Thesis, University of New South Wales, Sydney, Australia, 1991.

69. Bean, I. Blast Furnace Hearth Drainage. Improvement of the Residual-Flowout Correlation. Ph.D. Thesis, University of New South Wales, Sydney, Australia, September 2008.

70. Nishioka, K.; Maeda, T.; Shimizu, M. A three-dimensional mathematical modelling of drainage behavior in blast furnace hearth. *ISIJ Int.* **2005**, *45*, 669–676. [CrossRef]

71. Nishioka, K.; Maeda, T.; Shimizu, M. Effect of various in-furnace conditions on blast furnace hearth drainage. *ISIJ Int.* **2005**, *45*, 1496–1505. [CrossRef]

72. Nouchi, T.; Sato, M.; Takeda, K.; Ariyama, T. Effects of operation condition and casting strategy on drainage efficiency of the blast furnace hearth. *ISIJ Int.* **2005**, *45*, 1515–1520. [CrossRef]

73. Shao, L.; Saxén, H. A simulation study of blast furnace hearth drainage using a two-phase flow model of the taphole. *ISIJ Int.* **2011**, *51*, 228–235. [CrossRef]

74. Iida, M.; Ogura, K.; Hakone, T. Analysis of drainage rate variation of molten iron and slag from blast furnace during tapping. *ISIJ Int.* **2008**, *48*, 412–419. [CrossRef]

75. Iida, M.; Ogura, K.; Hakone, T. Numerical study on metal/slag drainage rate deviation during blast furnace tapping. *ISIJ Int.* **2009**, *49*, 1123–1132. [CrossRef]

76. Roche, M.; Helle, M.; van der Stel, J.; Louwerse, G.; Shao, L.; Saxén, H. Off-line model of blast furnace liquid levels. *ISIJ Int.* **2018**, *58*, 2236–2245. [CrossRef]

Publisher's Note: MDPI stays neutral with regard to jurisdictional claims in published maps and institutional affiliations.

Article

Evolution and Physical Characteristics of a Raceway Based on a Transient Eulerian Multiphase Flow Model

Xing Peng, Jingsong Wang *, Haibin Zuo and Qingguo Xue

State Key Laboratory of Advanced Metallurgy, University of Science and Technology Beijing, Beijing 100083, China; pengxing_hunan@sina.com (X.P.); zuohaibin@ustb.edu.cn (H.Z.); xueqingguo@ustb.edu.cn (Q.X.)
* Correspondence: wangjingsong@ustb.edu.cn; Tel.: +86-010-82375181

Received: 10 September 2020; Accepted: 17 October 2020; Published: 20 October 2020

Abstract: In industrial processes, a semi-cavity area formed by airflow wherein the particles circulate is called a "raceway". In a blast furnace, the role of the raceway is particularly important. To understand and predict the evolution and physical characteristics of the raceway, a three-dimensional transient Eulerian multiphase flow model in a packed particle bed was developed. In the model, it was assumed that the gas and solid (particle) phases constitute an interpenetrating continuum. The gas-phase turbulence was described as a k–ε dispersed model. The gas-phase stress was considered in terms of the effective viscosity of the gas. The solid-phase constitutive relationship was expressed in terms of solid stress. It was found that the evolution process of the raceway can be divided into three stages: (1) rapid expansion, (2) slow contraction, and (3) gradual stabilization. When the blast velocity was increased from 150 m/s to 300 m/s, the surface area of the raceway increased from 0.194 m^2 to 1.644 m^2. The depth and height of the raceway increased considerably with velocity, while the width slightly increased.

Keywords: raceway evolution; raceway size; flow pattern; Eulerian multiphase flow

---

1. Introduction

In a blast furnace (BF), the raceway is formed by airflow wherein the particles circulate. The combustion of coke and injected fuels in the raceway supplies gas and heat for the critical endothermic reduction of iron ores and for iron smelting [1]. Therefore, the raceway characteristics directly affect the primary distribution of gas and heat inside the BF. Some previous studies have used empirical size characteristics of the raceway to predict the combustion of pulverized coal and the gas flow distribution, which may considerably differ from those of the actual BF raceway [2–4]. The raceway depth directly affects the burnout rate of pulverized coal and determines the airflow distribution in the center of the blast furnace. The flow pattern will determine the strength of gas–solid mixing and the rate of coke consumption, thereby further affecting the smelting efficiency of the blast furnace. Therefore, it is necessary to understand the evolution process and physical characteristics of the raceway.

Investigations of the BF raceway phenomenon and its characteristics can be carried out via three methods: theoretical analysis, experimental testing, and numerical modeling. In theoretical analyses, some studies analyzed the raceway size on the basis of the force balance of the raceway boundary in different spatial dimensions [5–8]. The phenomenon of raceway hysteresis was explained, together with the effects of chemical reactions, blast velocity, material layer porosity, particle diameter, and other factors. However, this method treats the raceway as a circle or a sphere and disregards the force between the particles. Thus, it can be considered a relatively inaccurate method.

In experimental testing, the microwave reflection method was used to study the formation and depth of the BF raceway during production [9]. The effects of tuyere diameter, air volume, and coal injection on the depth of the raceway zone were investigated. In contrast, considering that the complex environment, in terms of high temperature and pressure in the raceway, implies significant difficulties for direct research, most researchers used cold models to study the formation and physical characteristics of the raceway [10–16]. However, it was challenging for the researchers to experimentally obtain dynamic information and accurately measure the raceway characteristics in three-dimensional (3D) space through experimental testing.

With the advancement of computers, numerical modeling has become a more popular method. A combined computational fluid dynamics and discrete element method (CFD-DEM) was developed [17–23]. The effects of different variables on the raceway were investigated. Nonetheless, previous CFD-DEM-based studies generally used two-dimensional (2D) or pseudo 3D models and small sizes with certain divergences from actual conditions. Hilton et al. [24] and Lichtenegger et al. [25] used the CFD-DEM method to investigate the effect of particle properties on the evolution of the raceway in 3D packed beds. However, these previously reported approaches were computationally expensive. Also, these approaches did not facilitate quantitative analysis of the raceway or the investigation of raceway physical characteristics.

However, the gas–solid flow model based on CFD can achieve high efficiency at low computational cost. The shape and size of the raceway was studied in a 2D state using a transient or steady model based on CFD [26–28]. Rangarajan et al. [29] extensively studied the influence of the operating conditions on raceway properties using a two-fluid model. Based on CFD modeling technology, research on coupling fuel combustion and raceway formation has been carried out, and a lot of information about combustion and gas distribution has been obtained [30–36]. However, no details on constitutive relations, the surface area of the raceway, or the evolution of the raceway penetration depth in a short time interval can be found in these articles.

In this study, we developed an industrial-scale blast furnace 3D slot model based on a transient Eulerian multiphase flow model (EMFM). The influence of the chemical reaction in the BF on the raceway characteristics is mainly reflected in the change in gas flow [26]. For simplicity, we did not set the combustion reaction or heat transfer, but we set the initial bed solid packing fraction to be less than the maximum volume fraction as an approximate replacement. The evolution process and flow pattern of the raceway are revealed. The depth, height, width, and surface area of the raceway were predicted, providing detailed information and theoretical guidance for the process of gas injection into packed beds in industrial processes.

2. Model Description

The model assumes that the gas phase and the solid (particle) phase constitute an interpenetrating continuum. The different phases appear in the same calculated cell and are characterized by the volume fraction, α_i, of each phase i (gas, solid). The gas-phase turbulence was described as a k–ε dispersed model and the gas-phase stress was considered in terms of effective viscosity. An advanced constitutive relation was adopted to describe solid stress.

2.1. Conservation Equations

In the process of gas–solid flow, both the gas and particle flows satisfy the conservation of mass and momentum. Given that there is no mass exchange between the solid particles and the gas phase, they are independent of each other. The mass conservation equation for phase i can be expressed as

$$\frac{\partial(\alpha_i \rho_i)}{\partial t} + \nabla \cdot (\alpha_i \rho_i U_i) = 0, \tag{1}$$

$$\Sigma \alpha_i = 1. \tag{2}$$

The momentum conservation equation for phase i can be written as

$$\frac{\partial(\alpha_i \rho_i U_i)}{\partial t} + \nabla \cdot (\alpha_i \rho_i U_i U_i) = \nabla \cdot \tau_i + \alpha_i \rho_i g + S. \tag{3}$$

The source term, S, is generated by the momentum transfer between the gas and solid phases and is expressed as

$$S = \beta(U_j - U_i), \quad j \neq i. \tag{4}$$

For $\alpha_g > 0.8$, coefficient β is based on the drag force of the fluid acting on a single particle, and for $\alpha_g \leq 0.8$, β is described by Ergun's equation [37]. Thus, β can be expressed as

$$\beta = \begin{cases} \frac{3}{4} C_D \frac{\alpha_s \alpha_g \rho_g |U_s - U_g|}{d_s} \alpha_g^{-2.65} & \alpha_g > 0.8 \\[2mm] 150 \frac{\alpha_s^2 \mu_g}{\alpha_g d_s^2} + 1.75 \frac{\rho_g \alpha_s |U_s - U_g|}{d_s} & \alpha_g \leq 0.8 \end{cases} \tag{5}$$

where d_s is the solid (particle) diameter; the drag coefficient, C_D, is given by

$$C_D = \begin{cases} \frac{24}{\alpha_g Re}\left[1 + 0.15(\alpha_g Re)^{0.687}\right] & Re \leq 1000 \\[2mm] 0.44 & Re > 1000 \end{cases} \tag{6}$$

where Re is the particle Reynolds number and can be expressed as

$$Re = \frac{\rho_g d_s |U_s - U_g|}{\mu_g}. \tag{7}$$

2.2. Constitutive Relations

The gas-phase constitutive equation is characterized by the effective viscosity of the gas. The solid-phase constitutive relationship is expressed in terms of solid stress. Tables 1 and 2 summarize a detailed description of the constitutive relations [37–42].

Table 1. Gas constitutive relations.

Item	Formula
Gas stress	$\tau_g = -P_g I + \mu_{eff,g}(\nabla U_g + (\nabla U_g)^T) - \frac{2}{3}(\mu_{eff,g}(\nabla \cdot U_g)I + \rho_g k_g)$
Gas effective viscosity	$\mu_{eff,g} = \mu_g + \mu_{t,g}$
Gas turbulent viscosity	$\mu_{t,g} = \rho_g C_\mu \frac{k_g^2}{\varepsilon_g}$ $(C_\mu = 0.09)$

Table 2. Solid constitutive relations.

Item	Formula
Solid stress	$\tau_s = (-P_s + \xi_s \nabla \cdot U_s)I + \mu_s\{(\nabla U_s + \nabla U_s^T) - \frac{2}{3}\nabla U_s I\}$
Solid pressure	$P_s = \alpha_s \rho_s \Theta + 2\rho_s(1 + e)\alpha_s^2 g_0 \Theta$
Diffusion coefficient	$k_s = \frac{150 \rho_s d_s \sqrt{\Theta\pi}}{384(1+e)g_0}[1 + \frac{6}{5}g_0\alpha_s(1 + e)]^2 + 2\alpha_s^2 \rho_s d_s g_0(1 + e)(\frac{\Theta}{\pi})^{1/2}$
Particle collisional dissipation of energy	$\gamma_s = 3(1 - e^2)g_0\rho_s\alpha_s^2\Theta(\frac{4}{d_s}(\frac{\Theta}{\pi})^{1/2} - \nabla \cdot U_s)$
Solid radial distribution function	$g_0 = \frac{3}{5}[1 - [\frac{\alpha_s}{\alpha_{s,\,max}}]^{1/3}]^{-1}$
Solid bulk viscosity	$\xi_s = \frac{4}{3}\alpha_s^2 \rho_s d_s g_0(1 + e)(\frac{\Theta}{\pi})^{1/2}$
Solids shear viscosity	$\mu_s = \mu_{s,kin} + \mu_{s,col} + \mu_{s,fr}$
Solid kinetic viscosity	$\mu_{s,kin} = \frac{10\rho_s d_s \sqrt{\Theta\pi}}{96(1+e)g_0}[1 + \frac{4}{5}g_0\alpha_s(1 + e)]^2$
Solid collisional viscosity	$\mu_{s,col} = \frac{4}{5}\alpha_s^2 \rho_s d_s g_0(1 + e)(\frac{\Theta}{\pi})^{1/2}$
Solid frictional viscosity	$\mu_{s,fr} = \frac{P_{friction}\sin\phi}{2\sqrt{I_{2D}}}$
Frictional pressure	$P_{friction} = \begin{cases} Fr\frac{(\alpha_s - \alpha_{s,min})^2}{(\alpha_{s,max} - \alpha_s)^5}, Fr = 0.1\alpha_s, & \alpha_s \geq 0.5 \\ 0 & \alpha_s < 0.5 \end{cases}$

2.3. Turbulence Equations

Turbulence predictions were obtained from a k–ε dispersed model. The transport equations were expressed as follows:

$$\frac{\partial}{\partial t}\left(\alpha_g \rho_g k_g\right) + \nabla\cdot\left(\alpha_g \rho_g U_g k_g\right) = \nabla\cdot\left(\alpha_g \frac{\mu_{t,g}}{\sigma_k}\nabla k_g\right) + \alpha_g G_{k,g} - \alpha_g \rho_g \varepsilon_g + \alpha_g \rho_g \Pi_{k_g} \tag{8}$$

$$\frac{\partial}{\partial t}\left(\alpha_g \rho_g \varepsilon_g\right) + \nabla\cdot\left(\alpha_g \rho_g U_g \varepsilon_g\right) = \nabla\cdot\left(\alpha_g \frac{\mu_{t,g}}{\sigma_\varepsilon}\nabla \varepsilon_g\right) + \alpha_g \frac{\varepsilon_g}{k_g}\left(C_{1\varepsilon} G_{k,g} - C_{2\varepsilon} \rho_g \varepsilon_g\right) + \alpha_g \rho_g \Pi_{\varepsilon_g} \tag{9}$$

where Π_{k_g} and Π_{ε_g} are source terms that can be included to model the influence of the dispersed phases on the continuous phase. The constants for the k–ε model were taken as $\sigma_k = 1.00$, $\sigma_\varepsilon = 1.30$, $C_{1\varepsilon} = 1.44$, and $C_{2\varepsilon} = 1.92$ [26].

2.4. Geometry and Operating Conditions

To save computing resources, a slot model of the lower part of the BF was derived. Figure 1 depicts the computational domain. The geometric model covers the iron slag surface to the furnace bosh, with the deadman removed. Its size is based on a small steel plant BF. The EMFM equations were calculated using ANSYS–FLUENT 17.2. The phase-coupled SIMPLE (PC-SIMPLE) algorithm was used for the coupling between pressure and velocity. The second-order upwind style was used in the discretization scheme.

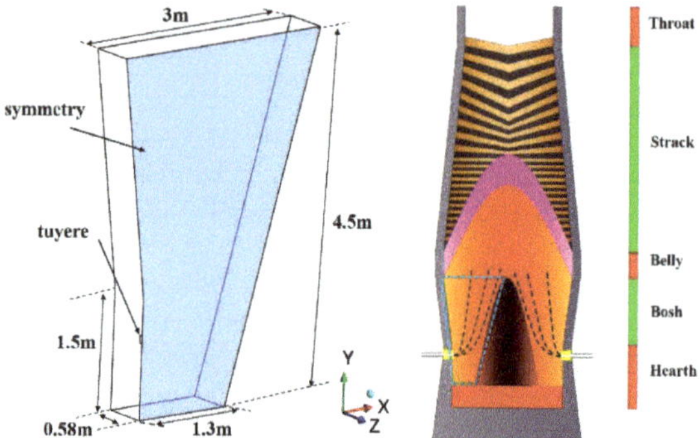

Figure 1. Blast furnace (BF) schematic and geometry model of the calculation domain.

Tables 3 and 4 list the simulation parameters and computational conditions, respectively. Injection angles of 5° were associated with the negative direction of the y axis. The chemical reaction, the polydispersity of the particles, and the liquid phase were not considered during the flow process. Therefore, to replace the effects of the above factors, and in combination with the actual charge void distribution in the lower part of the BF, the solid volume fraction was set to 0.6, which is less than the maximum limiting volume fraction ($\alpha_{s,max}$). Considering the pressure of the upper layer of the BF, the outlet pressure was set to 303,000 Pa.

Table 3. Simulation parameters.

Parameters	Value
Number of calculation units	109,516
Time Step	0.0001 s
Particle density	1000 kg/m^3
Angle of internal friction	30°
Tuyere equivalent diameter	0.113 m
Initial solid volume fraction	0.6
Solid packing limit	0.7
Friction packing limit	0.61
Initial bed particle height	4 m
Outlet pressure	303,000 Pa

Table 4. Computational conditions.

Case	Blast Velocity (m/s)	Injection Angle	Particle Diameter (m)
1	150	5°	0.01
2	200	5°	0.01
3	250	5°	0.01
4	300	5°	0.01

2.5. Grid and Time Step Independence

Table 5 shows the raceway size after stabilization under different grids and time steps. Further refinement of the grid in either direction did not change the raceway size by more than 2%, which verifies the independence of the computational domain grid. The simulation result did not change by more than 1% by further reducing the time step. This demonstrates the reliability of the numerical model.

Table 5. Raceway size of different numbers of grid cells and time steps.

Number of Grid Cells	Time Step (s)	Depth (mm)	Height (mm)	Width (mm)	Deviation
109,516	0.0001	631	458	264	-
300,672	0.0001	640	461	267	<2%
109,516	0.00005	637	462	266	<1%

3. Results and Discussion

3.1. Raceway Evolution Characteristics

Raceway evolution is an important phenomenon, particularly reblowing, which occurs after a temporary wind break in an ironmaking BF. To accurately analyze the evolution process and physical characteristics of the raceway, the boundary of the raceway was previously defined by the values of isostatic stress and solid or gas volume fractions [24–27]. In this study, when the solid volume fraction was less than 0.5, the frictional pressure was 0, and the solid motion was mainly affected by collision. Therefore, the boundary of the raceway was defined as a solid volume fraction of 0.5.

As depicted in Figure 2, at an injection velocity of 150 m/s, the penetration depth of the raceway reached a peak at 1 s, at a value of 0.783 m, and it stabilized at 9 s, at a value of 0.386 m. At an injection velocity of 200 m/s, the penetration depth reached 0.982 m at 1.1 s and then decreased to 0.460 m at 25 s. At 250 m/s, the penetration depth increased to a peak of 1.143 m at 1.5 s and achieved a stable value of 0.631 m at 40 s. At 300 m/s, the penetration depth reached a peak of 1.327 m at 1.7 s and stabilized at 1.109 m at 47.5 s. The evolution process of the raceway can be divided into three stages: (1) rapid expansion, (2) slow contraction, and (3) gradual stabilization. In Stage 1, the penetration depth of the raceway increases rapidly in the early stages of gas injection. The higher the injection velocity, the faster the increase in the penetration depth. Then, in Stage 2, as the particles descend and congregate, the penetration depth decreases slowly after reaching the peak. In Stage 3, the raceway stabilizes.

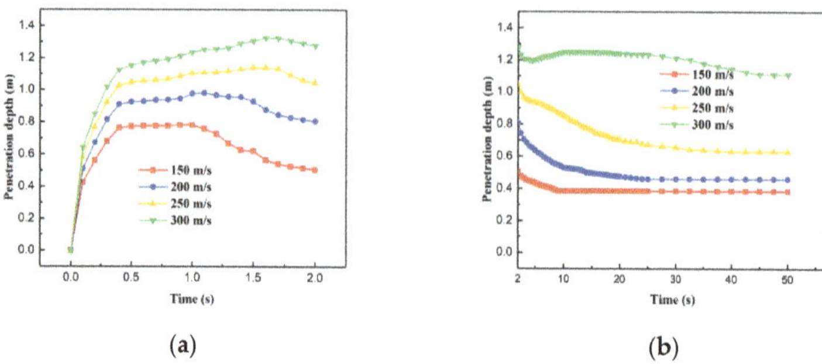

Figure 2. Evolution of penetration depth in the raceway: (**a**) 0–2 s and (**b**) 2–50 s.

A typical blast velocity is close to 250 m/s in a small BF tuyere. Figure 3 depicts the evolution of the raceway. When air was injected through the tuyere, the expansion of the depth of the raceway was more obvious. When the peak was reached, the height changes of the raceway were more obvious. Finally, the raceway stabilized at 40 s. This trend was due to the fact that when the solid phase interacted with the gas phase, the initial solid volume fraction changed toward the maximum limiting volume fraction and eventually stabilized. This created a particle circulation zone attributed to the balance between the drag of the blast and the gravity and pressure of the particles.

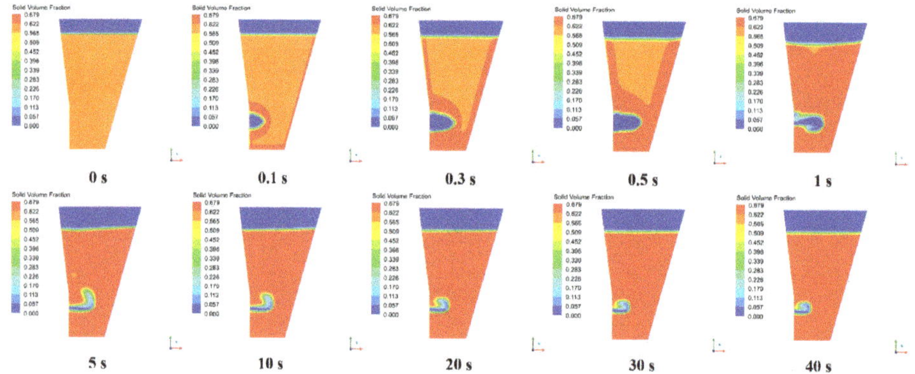

Figure 3. Time evolution of the solid volume fraction for Case 3.

3.2. Raceway Size Characteristics

Different BF operating conditions led to different raceway physical characteristics, which were mainly reflected by their size. Figure 4 shows that the shape of the raceway after it stabilizes is an upturned bag at high blast velocity. In an actual BF, this provides enough space for the combustion of pulverized coal and coke, thereby improving production efficiency.

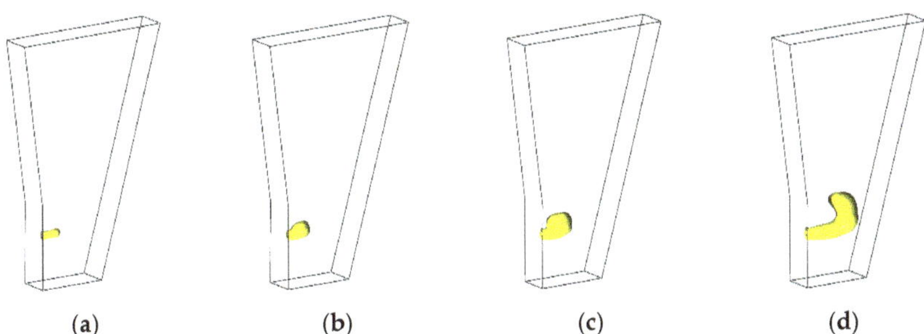

(a)   (b)   (c)   (d)

Figure 4. Raceway shapes after they stabilize at different blast velocities: (**a**) 150 m/s; (**b**) 200 m/s; (**c**) 250 m/s; (**d**) 300 m/s.

The blast velocity increase was obviously beneficial for increasing the depth, height, and surface area of the raceway, while the width was slightly increased, as depicted in Figure 5. The size of the raceway was not linearly related to the blast velocity. When the blast velocity was increased from 150 m/s to 300 m/s, the surface area of the raceway increased from 0.194 m^2 to 1.644 m^2, and the depth increased from 0.386 m to 1.109 m. This was due to the increased gas kinetic energy because of the increased blast volume and velocity. Therefore, increasing the blast velocity is very effective for increasing the depth of the raceway in order to develop the central gas flow in an actual BF.

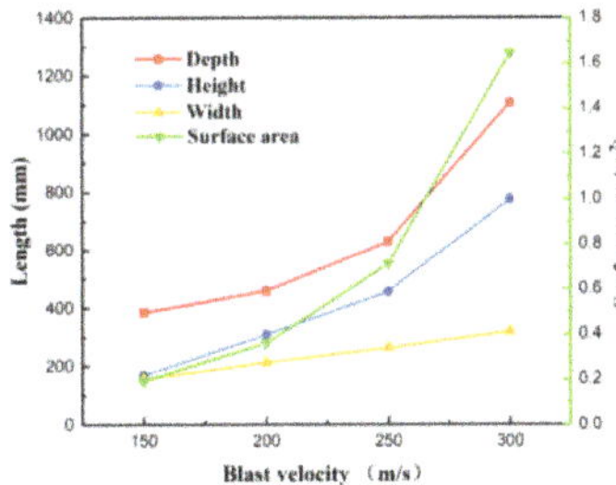

Figure 5. Effect of blast velocity on the raceway size.

3.3. Pressure Distribution

Figure 6a shows that the gas pressure was high in the raceway and decreased as it approached the outlet of the particle bed. In contrast, the solid granular pressure was considerably low in the raceway and at the boundary of the raceway. It is noteworthy that the solid granular pressure reached a local peak at the boundary of the raceway, where gas injection resistance was the highest, as depicted in Figure 6b.

As demonstrated in Figure 7, the gas pressure remained relatively stable up to 0.4 m from the front end of the tuyere because there were fewer particles and low resistance. At a distance equal to or greater than 0.4 m, the air pressure rapidly increased because the gas was subjected to increased particle resistance after deep penetration into the particle bed, and the pressure decreased because the gas velocity decreased and there was further particle resistance. The solid granular pressure in the raceway is close to 0. Near the boundary of the raceway, due to the interaction gas and solid, the solid granular pressure changes drastically, increasing first and then decreasing. However, it slowly increases in the end because the solid were constricted by the wall.

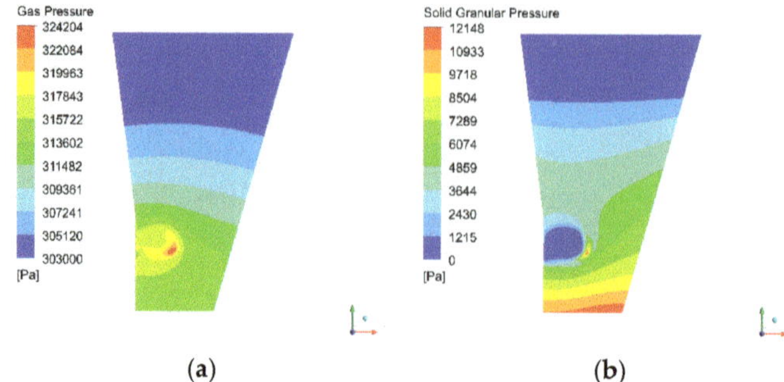

Figure 6. (**a**) Gas pressure and (**b**) solid granular pressure for Case 3 in the symmetry plane at 40 s.

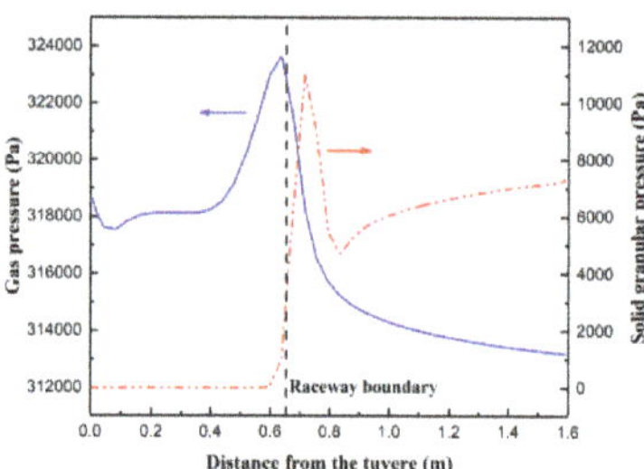

Figure 7. Gas pressure and solid granular pressure for Case 3 on the axis of the tuyere at 40 s.

3.4. Flow Pattern

As depicted in Figures 8a and 9a, the gas in the raceway can be divided into a jet zone and an anti-clockwise flow zone. However, the gas flowed into the particle bed from the boundary of the raceway and did not form a large circulation area. This was due to the injection of high-speed gas into the tuyere, which limited gas circulation in the jet zone. Additionally, the gas had a weak anti-clockwise circulation flow at the edge of the tuyere. This is inconsistent with previous results in which the gas studied according to the CFD-DEM model was divided into anti-clockwise or clockwise circulation or a plume-like flow [21].

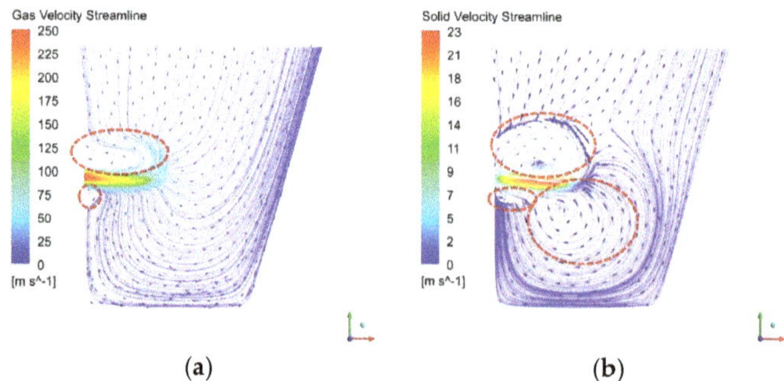

Figure 8. The symmetry plane of Case 3 at 40 s: (**a**) gas velocity streamline; (**b**) solid velocity streamline.

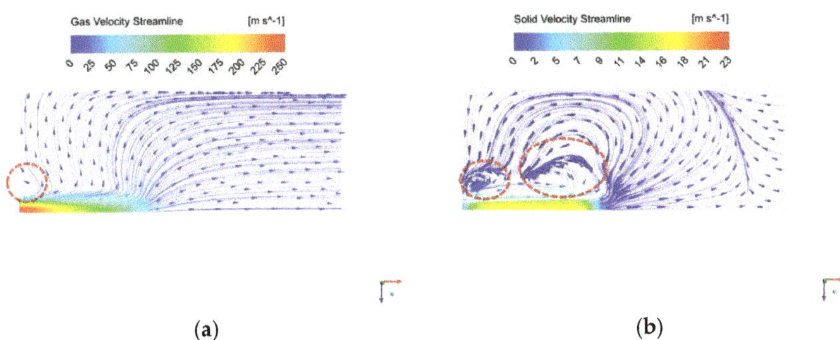

(a)   (b)

Figure 9. The tuyere level plane of Case 3 at 40 s: (**a**) gas velocity streamline; (**b**) solid velocity streamline.

The particles were clearly circulating anti-clockwise in the raceway, as depicted in Figure 8b. This was mainly due to the higher resistance of the solid particles along the axis of the tuyere and the lower pressure on the upper part of the particle bed. Below the tuyere, two clockwise particle circulation zones were formed, but the movement speed was considerably low. This was because the gas was affected by the solid resistance and the forces on the bottom and the wall. This reduced the gas flow velocity in the lower part of the packed bed, resulting in lower resistance. The source of particles in the solid jet area mainly derived from the upper part of the raceway particles, which also caused the upper particles in the particle bed to move downwards. In the horizontal direction, although there were also two anti-clockwise circulating flows to provide particles for the raceway, as depicted in Figure 9b, their velocities were extremely low. Therefore, the BF raceway was not a single-cycle flow as previously reported [21], but an extremely complex multi-cycle flow with gas–solid interaction. The circulation pattern near the tuyere may have a negative impact on the life of the tuyere.

The gas was injected from the tuyere along the axial direction of the tuyere. Owing to the resistance of the solid particles, the gas velocity rapidly decreased until it reached a value of 0.443 m/s at the wall surface, as depicted in Figure 10. The drag of the gas affected the particles. The particle velocity increased rapidly and maintained a relatively stable value in the middle part of the raceway. However, near the boundary of the raceway ($\alpha_s \to 0.5$), the particle collision viscosity increased because of the increased particle volume fraction, which considerably reduced the particle velocity. External to the raceway boundary and with an increase in the particle friction viscosity and a decrease in the gas drag, the particle velocity was reduced to a value close to 0.

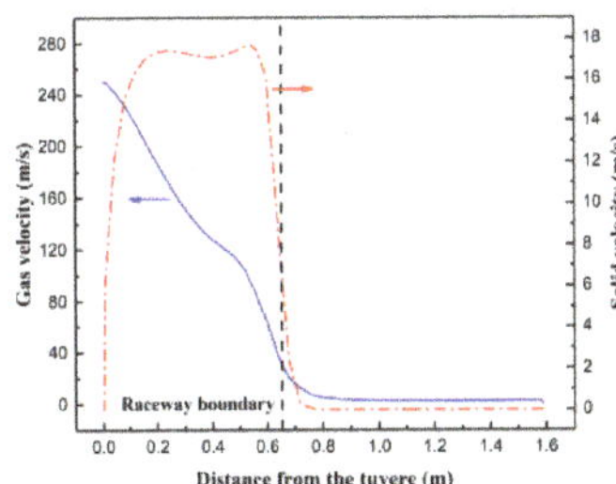

Figure 10. Gas and solid velocity along the axis of the tuyere for Case 3 at 40 s.

4. Conclusions

A 3D transient EMFM was developed to study the evolution and physical characteristics of the raceway in the packed particle bed of an ironmaking BF. The constitutive relation of the gas and solid phases was comprehensively considered in the model. The main conclusions of this study are as follows:

(1) The evolution process of the raceway can be divided into three stages: rapid expansion, slow contraction, and gradual stabilization. The shape of the raceway was that of an upturned bag at high blast velocity.

(2) The blast velocity had a significant effect on the size of the raceway. As the velocity increased, the depth, height, and surface area of the raceway considerably increased, while the width slightly increased.

(3) The gas pressure in the raceway was higher than that of the particle bed, while the solid granular pressure was lower. The raceway did not exhibit a single-cycle flow pattern, but exhibited a complex multiphase and multi-cycle flow pattern.

Author Contributions: Investigation, X.P., H.Z., and J.W.; methodology, X.P. and H.Z.; resources, Q.X.; data curation, X.P.; Formal analysis, X.P. and J.W.; writing—original draft preparation, X.P.; writing—review and editing, X.P. and J.W.; project administration, Q.X. All authors have read and agreed to the published version of the manuscript.

Funding: This research was funded by the National Natural Science Foundation of China, grant number U1960205, and supported by the National Key Research and Development Program, grant number 2016YFB0601304.

Acknowledgments: The authors gratefully acknowledge the financial support provided by the National Natural Science Foundation of China (No. U1960205) and the National Key Research and Development Program (No. 2016YFB0601304).

Conflicts of Interest: The authors declare no conflict of interest.

Notation

Symbol	Meaning
α_i	i phase volume fraction
ρ_i	i phase density, kg/m^3
U_i	i phase velocity, m/s
τ_i	i phase stress–strain tensor, Pa
P_i	i phase pressure, Pa
g	Gravity acceleration, m/s^2
S	Source term
β	Momentum exchange coefficient
C_D	Drag coefficient
d_s	Solid-phase diameter, m
μ_g	Gas-phase viscosity, Pa·s
Re	Reynolds number
I	Unit stress tensor
$\mu_{eff,g}$	Gas effective viscosity, Pa·s
k_g	Gas turbulent kinetic energy, Pa·s
$\mu_{t,g}$	Gas turbulent viscosity, Pa·s
ε_g	Gas turbulent dissipation rate
g_0	Solid radial distribution function
e	Coefficient of restitution for particle collisions
Θ	Granular pseudo-temperature
k_s	Diffusion coefficient
γ_s	Particle collisional dissipation of energy
$\alpha_{s,min}$	Friction packing limit
$\alpha_{s,\,max}$	Packing limit
ξ_s	Solid bulk viscosity, Pa·s
μ_s	Solids shear viscosity, Pa·s
$\mu_{s,kin}$	Solid kinetic viscosity, Pa·s
$\mu_{s,col}$	Solid collisional viscosity, Pa·s
$\mu_{s,fr}$	Solid frictional viscosity, Pa·s
$P_{friction}$	Frictional pressure, Pa

ϕ	Angle of internal friction
I_{2D}	Second invariant of the deviatoric stress tensor
Fr	Froude number
$G_{k,g}$	Gas-phase turbulent kinetic energy
Π_{k_g}	Turbulent kinetic energy source term
Π_{ε_g}	Turbulent dissipation rate source term

References

1. Burgess, J.M. Fuel combustion in the blast furnace raceway zone. *Prog. Energy Combust. Sci.* **1985**, *11*, 61–82. [CrossRef]
2. Zhou, Z.; Xue, Q.; Li, C.; Wang, G.; She, X.; Wang, J. Coal flow and combustion characteristics under oxygen enrichment way of oxygen-coal double lance. *Appl. Therm. Eng.* **2017**, *123*, 1096–1105. [CrossRef]
3. Liu, Y.; Shen, Y. Three-dimensional modelling of charcoal combustion in an industrial scale blast furnace. *Fuel* **2019**, *258*, 116088. [CrossRef]
4. Zhou, Z.; Yi, Q.; Wang, R.; Wang, G.; Ma, C. Numerical investigation on coal combustion in ultralow CO_2 blast furnace: Effect of oxygen temperature. *Processes* **2020**, *8*, 877. [CrossRef]
5. Gupta, G.S.; Rajneesh, S.; Rudolph, V.; Singh, V.; Sarkar, S.; Litster, J.D. Mechanics of raceway hysteresis in a packed bed. *Metall. Mater. Trans. B Process Metall. Mater. Process. Sci.* **2005**, *36*, 755–764. [CrossRef]
6. Gupta, G.S.; Rudolph, V. Comparison of blast furnace raceway size with theory. *ISIJ Int.* **2006**, *46*, 195–201. [CrossRef]
7. Guo, J.; Cheng, S.; Zhao, H.; Pan, H.; Du, P.; Teng, Z. A mechanism model for raceway formation and variation in a blast furnace. *Metall. Mater. Trans. B Process Metall. Mater. Process. Sci.* **2013**, *44*, 487–494. [CrossRef]
8. Li, Y.L.; Cheng, S.S.; Zhang, P.; Guo, J. Development of 3-D mathematical model of raceway size in blast furnace. *Ironmak. Steelmak.* **2016**, *43*, 308–315. [CrossRef]
9. Matsui, Y.; Yamaguchi, Y.; Sawayama, M.; Kitano, S.; Nagai, N.; Imai, T. Analyses on blast furnace raceway formation by micro wave reflection gunned through tuyere. *ISIJ Int.* **2005**, *45*, 1432–1438. [CrossRef]
10. Flint, P.J.; Burgess, J.M. A fundamental study of raceway size in two dimensions. *Metall. Trans. B* **1992**, *23*, 267–283. [CrossRef]
11. Sarkar, S.; Gupta, G.S.; Litster, J.D.; Rudolph, V.; White, E.T.; Choudhary, S.K. A cold model study of raceway hysteresis. *Metall. Mater. Trans. B Process Metall. Mater. Process. Sci.* **2003**, *34*, 183–191. [CrossRef]
12. Sastry, G.S.S.R.K.; Gupta, G.S.; Lahiri, A.K. Cold model study of raceway under mixed particle conditions. *Ironmak. Steelmak.* **2003**, *30*, 61–65. [CrossRef]
13. Sastry, G.S.S.R.K.; Gupta, G.S.; Lahiri, A.K. Void formation and breaking in a packed bed. *ISIJ Int.* **2003**, *43*, 153–160. [CrossRef]
14. Rajneesh, S.; Sarkar, S.; Gupta, G.S. Prediction of raceway size in blast furnace from two dimensional experimental correlations. *ISIJ Int.* **2004**, *44*, 1298–1307. [CrossRef]
15. Mojamdar, V.; Gupta, G.S.; Puthukkudi, A. Raceway formation in a moving bed. *ISIJ Int.* **2018**, *58*, 1396–1401. [CrossRef]
16. Lu, Y.; Jiang, Z.; Zhang, X.; Liu, S.; Wang, J.; Zhang, X. Determination of void boundary in a packed bed by laser attenuation measurement. *Particuology* **2020**, *51*, 72–79. [CrossRef]
17. Xu, B.H.; Yu, A.B.; Chew, S.J.; Zulli, P. Numerical simulation of the gas-solid flow in a bed with lateral gas blasting. *Powder Technol.* **2000**, *109*, 13–26. [CrossRef]
18. Feng, Y.-Q.; Pinson, D.; Yu, A.-B.; Chew, S.J.; Zulli, P. Numerical Study of Gas-Solid Flow in the Raceway of a Blast Furnace. *Steel Res. Int.* **2003**, *74*, 523–530. [CrossRef]
19. Umekage, T.; Yuu, S.; Kadowaki, M. Numerical simulation of blast furnace raceway depth and height, and effect of wall cohesive matter on gas and coke particle flows. *ISIJ Int.* **2005**, *45*, 1416–1425. [CrossRef]
20. Sarkar, S.; Gupta, G.S.; Kitamura, S.Y. Prediction of raceway shape and size. *ISIJ Int.* **2007**, *47*, 1738–1744. [CrossRef]
21. Miao, Z.; Zhou, Z.; Yu, A.B.; Shen, Y. CFD-DEM simulation of raceway formation in an ironmaking blast furnace. *Powder Technol.* **2017**, *314*, 542–549. [CrossRef]

22. Cui, J.; Hou, Q.; Shen, Y. CFD-DEM study of coke combustion in the raceway cavity of an ironmaking blast furnace. *Powder Technol.* **2020**, *362*, 539–549. [CrossRef]
23. Lu, Y.; Liu, S.; Zhang, X.; Jiang, Z.E.D. A Probabilistic Statistical Method for the Determination of Void Morphology with CFD-DEM Approach. *Energies* **2020**, *13*, 4041. [CrossRef]
24. Hilton, J.E.; Cleary, P.W. Raceway formation in laterally gas-driven particle beds. *Chem. Eng. Sci.* **2012**, *80*, 306–316. [CrossRef]
25. Lichtenegger, T.; Pirker, S. CFD-DEM modeling of strongly polydisperse particulate systems. *Powder Technol.* **2018**, *325*, 698–711. [CrossRef]
26. Mondal, S.S.; Som, S.K.; Dash, S.K. Numerical predictions on the influences of the air blast velocity, initial bed porosity and bed height on the shape and size of raceway zone in a blast furnace. *J. Phys. D Appl. Phys.* **2005**, *38*, 1301–1307. [CrossRef]
27. Singh, V.; Gupta, G.S.; Sarkar, S. Study of gas cavity size hysteresis in a packed bed using DEM. *Chem. Eng. Sci.* **2007**, *62*, 6102–6111. [CrossRef]
28. Sun, Y.; Chen, R.; Zhang, Z.; Wu, G.; Zhang, H.; Li, L.; Liu, Y.; Li, X.; Huang, Y. Numerical simulation of the raceway zone in melter gasifier of COREX process. *Processes* **2019**, *7*, 867. [CrossRef]
29. Rangarajan, D.; Shiozawa, T.; Shen, Y.; Curtis, J.S.; Yu, A. Influence of operating parameters on raceway properties in a model blast furnace using a two-fluid model. *Ind. Eng. Chem. Res.* **2014**, *53*, 4983–4990. [CrossRef]
30. Gu, M.; Zhang, M.; Selvarasu, N.K.C.; Zhao, Y.C.Q.Z. Numerical Analysis of Pulverized Coal Combustion inside Tuyere and Raceway. *Steel Res. Int.* **2008**, *79*, 17–24. [CrossRef]
31. Fu, D.; Zheng, D.; Zhou, C.Q.; D'Alessio, J.; Ferron, K.J.; Zhao, Y. Parametric studies on pci performances. ASME/JSME 2011 8th Therm. *Eng. Jt. Conf. AJTEC* **2011**. [CrossRef]
32. Okosun, T.; Street, S.J.; Zhao, J.; Wu, B.; Zhou, C.Q. Influence of conveyance methods for pulverised coal injection in a blast furnace. *Ironmak. Steelmak.* **2017**, *44*, 513–525. [CrossRef]
33. Fu, D.; Tang, G.; Zhao, Y.; D'Alessio, J.; Zhou, C.Q. Integration of Tuyere, Raceway and Shaft Models for Predicting Blast Furnace Process. *JOM* **2018**, *70*, 951–957. [CrossRef]
34. Wu, D.; Zhou, P.; Zhou, C.Q. Evaluation of pulverized coal utilization in a blast furnace by numerical simulation and grey relational analysis. *Appl. Energy* **2019**, *250*, 1686–1695. [CrossRef]
35. Okosun, T.; Silaen, A.K.; Zhou, C.Q. Review on Computational Modeling and Visualization of the Ironmaking Blast Furnace at Purdue University Northwest. *Steel Res. Int.* **2019**, *90*, 1–16. [CrossRef]
36. Zhuo, Y.; Shen, Y. Three-dimensional transient modelling of coal and coke co-combustion in the dynamic raceway of ironmaking blast furnaces. *Appl. Energy* **2020**, *261*, 114456. [CrossRef]
37. Gidaspow, D. *Multiphase Flow and Fluidization: Continuum and Kinetic Theory Descriptions*; Academic Press: New York, NY, USA, 1994; 467p.
38. Ding, J.; Gidaspow, D. A bubbling fluidization model using kinetic theory of granular flow. *AIChE J.* **1990**, *36*, 523–538. [CrossRef]
39. Lun, C.K.K.; Savage, S.B.; Jeffrey, D.J.; Chepurniy, N. Kinetic theories for granular flow: Inelastic particles in Couette flow and slightly inelastic particles in a general flowfield. *J. Fluid Mech.* **1984**, *140*, 223–256. [CrossRef]
40. Schaeffer, D.G. Instability in the evolution equations describing incompressible granular flow. *J. Differ. Equ.* **1987**, *66*, 19–50. [CrossRef]
41. Ocone, R.; Sundaresan, S.; Jackson, R. Gas-Particle flow in a duct of arbitrary inclination with particle-particle interactions. *AIChE J.* **1993**, *39*, 1261–1271. [CrossRef]
42. Johnson, P.C.; Jackson, R. Frictional–collisional constitutive relations for granular materials, with application to plane shearing. *J. Fluid Mech.* **1987**, *176*, 67–93. [CrossRef]

Publisher's Note: MDPI stays neutral with regard to jurisdictional claims in published maps and institutional affiliations.

Article

On the Impacts of Pre-Heated Natural Gas Injection in Blast Furnaces

Tyamo Okosun [1], Samuel Nielson [1], John D'Alessio [2], Shamik Ray [2], Stuart Street [3] and Chenn Zhou [1,*]

[1]   Center for Innovation through Visualization and Simulation (CIVS) and Steel Manufacturing Simulation and Visualization Consortium (SMSVC), Purdue University Northwest, Hammond, IN 46323, USA; tokosun@pnw.edu (T.O.); snielson@pnw.edu (S.N.)

[2]   Stelco Inc., Hamilton, ON L8L 8K5, Canada; John.DAlessio@stelco.com (J.D.); shamik.ray@stelco.com (S.R.)

[3]   AK Steel Corp., West Chester, OH 45069, USA; stuart.street@aksteel.com

*   Correspondence: czhou@pnw.edu; Tel.: +1-219-989-2665

Received: 29 May 2020; Accepted: 28 June 2020; Published: 1 July 2020

Abstract: During recent years, there has been great interest in exploring the potential for high-rate natural gas (NG) injection in North American blast furnaces (BFs) due to the fuel's relatively low cost, operational advantages, and reduced carbon footprint. However, it is well documented that increasing NG injection rates results in declining raceway flame temperatures (a quenching effect on the furnace, so to speak), with the end result of a functional limit on the maximum injection rate that can be used while maintaining stable operation. Computational fluid dynamics (CFD) models of the BF raceway and shaft regions developed by Purdue University Northwest's (PNW) Center for Innovation through Visualization and Simulation (CIVS) have been applied to simulate multi-phase reacting flow in industry blast furnaces with the aim of exploring the use of pre-heated NG as a method of widening the BF operating window. Simulations predicted that pre-heated NG injection could increase the flow of sensible heat into the BF and promote complete gas combustion through increased injection velocity and improved turbulent mixing. Modeling also indicated that the quenching effects of a 15% increase in NG injection rate could be countered by a 300K NG pre-heat. This scenario maintained furnace raceway flame temperatures and top gas temperatures at levels similar to those observed in baseline (stable) operation, while reducing coke rate by 6.3%.

Keywords: blast furnace; natural gas; fuel injection; computational fluid dynamics; numerical simulation; combustion; RAFT

---

1. Introduction

Worldwide, auxiliary fuel injection serves as one of the key technologies by which blast furnace (BF) coke consumption rates are reduced. As with many other industries, iron and steelmaking must also contend with a growing focus on carbon emissions and their impacts, resulting in tighter regulations on emissions or the implementation of carbon taxes by various governments. Through the reduction of coke consumption and by the introduction of some hydrogen to the reduction process via fuels such as natural gas, injected fuels are one of the key levers by which operators can reduce carbon footprint and possibly make the blast furnace more efficient. In addition, these injected fuels typically present a significant economic incentive for their greater use, especially in the case of natural gas (NG) in North America. However, no modification can be undertaken in isolation, and changes to any variable will inevitably require an understanding of how they may impact conditions within the furnace.

With high NG injection rates becoming more and more common in addition to, and sometimes in place of, pulverized coal injection (PCI) in North American BFs, it is increasingly important for operators to understand what impacts NG injection can have on conditions in the furnace and how any disadvantages can be mitigated. Of course, in comparison to PCI, NG injection has some obvious benefits, including reduced carbon footprint and the lack of potential for fines buildup from ash or unburned char. However, the most critical limitation of NG injection results from the production of H_2O (and to a lesser extent CO_2) from NG combustion. Within the raceway and packed bed, these species will participate in endothermic reactions with the coke bed, consuming heat and causing a drop in reducing gas temperatures. This phenomenon can be easily observed by examining the raceway adiabatic flame temperature (RAFT), calculated theoretically using heat and mass balance modeling based upon the 'Rist Diagram' or via approaches such as the AISI (American Iron and Steel Institute) formulation [1,2]. This is the temperature of the furnace gases after all reactions have taken place and the only remaining species are CO, H_2, and N_2.

While the minimum stable RAFT level will vary between furnaces, recent publications indicate that the majority of North American BFs operate somewhere above 2020 K [1,3]. A falling RAFT can result in furnace instability and reduced productivity, and generally, operators will mitigate such scenarios by increasing the level of oxygen enrichment in the hot blast. Increased O_2 levels in the blast, however, are known to cause a decline in furnace top temperatures, potentially leading to condensation of moisture in or near the uptakes and damage to the furnace. While increasing NG injection rates can help to bring the top temperature back up [3], the negative impact of O_2 enrichment on top temperatures is stronger than the positive impacts of NG. The competing factors of low top temperatures due to O_2 enrichment and low raceway flame temperatures due to NG injection create a functional constraint on the maximum level of NG injection in the BF, with the highest reported levels occurring at around 150–160 kg/mthm (kg per tonne of hot metal produced by the furnace) [1,3].

It is clear then that a method by which flame temperatures could be increased without relying heavily on O_2 enrichment could make possible higher rates of NG injection. The approach which appears most obvious here would be to increase the hot blast temperature, as that would supply additional sensible heat to the furnace to counter the low flame temperatures without impacting gas chemistry. However, as many furnaces are already maximizing their available hot blast temperature, this method may not be an option. One may then wish to turn to alternative methods of introducing sensible heat to the furnace. Since NG is typically injected at ambient temperature, the injected gas flow presents the most desirable location for this additional heat, with the added benefit of increasing the gas velocity, which may slightly enhance tuyere exit velocity, as well as improve turbulent mixing with the blast and enhance gas combustion.

Pre-heating NG by a reasonable amount (300–400 K) before injection into the furnace is a concept that has been previously noted in a very limited set of publications and trials. The idea has been explored conceptually by researchers working with heat and mass balance models and lab-scale experimentation [3], and at least one industrial facility has reported attempts at studying and implementing NG pre-heating in operational BFs [4,5]. The aforementioned attempt at implementation was conducted at the OJSC LMZ "Svobodny Sokol" and OJSC "Severstal" BFs, with NG temperature increases of 170K and 75K, respectively, achieved by channeling NG flow through a heat recovery apparatus around the outside of the blowpipe. While gas pre-heating was not applied in isolation, the increase in NG temperature led to a reduction in coke consumption rate in both trialed scenarios, and an increase in the NG coke replacement ratio (the ratio of how much coke can be removed from the furnace burden for a fixed amount of fuel injected into the tuyere) from 1.2:1 to 1.42:1 was observed. Table 1 details additional operating conditions under which the technique was implemented for both furnaces.

Table 1. Impacts of pre-heated NG injection at OJSC LMZ "Svobodny Sokol" and OJSC "Severstal". Reproduced with permission from T. Okosun, AISTech 2019; published by AIST, 2019.

	Svobodny Sokol		Severstal	
	Baseline	**Gas Pre-Heat**	**Baseline**	**Gas Pre-Heat**
Production (t/day)	1334	1391	3303	3207
NG Temp. (K)	303	473	303	378
Blast Temp. (K)	1323	1361	1454	1455
Coke rate (kg/mthm)	497	480.4	420.2	411.1
Production % change	+4.15%		−2.9%	
Coke rate % change	−3.3%		−2.2%	

The implementation of such gas pre-heating in the field, whether achieved via heat recovery systems or some other method, would require some expenditure for design and installation. With this in mind, it becomes important to develop a more complete understanding just how much value there is in enabling this additional lever for BF control, and what potential efficiency benefits NG pre-heating might present to BF operation in a range of different scenarios. Industry experience can serve as a judge of feasibility for methods of implementation, but focused research is often needed to understand the specific impacts of untested changes in parameters such as NG pre-heating on the BF. Simulation modeling presents an excellent approach for such research, allowing for scientific-based predictions of how changes to input conditions can impact the chemical reactions, temperatures, flow patterns, and other phenomena inside the BF.

Given its comparable speed and low expense when compared to test rigs or other experimental methods, computational fluid dynamics (CFD) modeling has become a key first step in determining the impacts of new operating parameters on multi-phase reacting flow systems. In particular, CFD modeling of the BF has been applied to great effect at the tuyere level by a wide range of researchers exploring methods of optimization for operation. Such modeling has expanded from simplified one-dimensional approaches to full-scale three-dimensional multi-phase flow models of specific regions of the blast furnace [6–14]. In particular, the Steel Manufacturing Simulation and Visualization Consortium (SMSVC) at Purdue University Northwest's (PNW) Center for Innovation through Visualization and Simulation (CIVS) has performed a range of simulation studies on BF performance using a combination of commercial CFD codes and in-house solvers focusing on combustion, tuyere region design, and more [15–22].

This paper details current simulation research on high-rate NG injection in North American BFs, and the impacts of pre-heated NG injection. CFD models of two BFs of similar scale were conducted as part of this research, with one furnace utilizing co-injection of NG and PC, and the other operating on a pure NG injection basis. Modeling was conducted to determine the impacts of gas pre-heating in isolation and in conjunction with other modifications to operation. Observed benefits included increased analogue flame temperatures and a corresponding reduction in furnace coke rate with NG pre-heating. Additionally, CFD modeling indicated that a combined increase in NG injection rate and NG pre-heat could maintain top gas and flame temperatures at baseline levels while achieving a lower coke consumption rate.

2. Methods, Geometry, and Boundary Conditions

2.1. Computational Modeling Methodology

Accurately modeling the multiphase flow physics, chemical reactions, and heat/mass transfer occurring within the blast furnace requires a delicate balance of detailed modeling techniques and reasonable assumptions in order to produce results in a reasonable time frame. This research focuses on the flow, reactions, and heat transfer in the raceway and shaft regions of the BF, and a combination of CFD models are coupled together to predict phenomena. These models have been discussed in far greater detail in previous publications [15–25], so in the interests of brevity, this paper will present a summary of the approach.

The primary assumption made in this modeling approach is based upon the difference in time-scales between the flow of gases and solids in the furnace. Gases ascend through the shaft very quickly in comparison to the descent speed of the solid burden material, allowing for the burden packed bed to be treated as a steady-state porous media that can participate in heat transfer and chemical reactions, as well as influence gas flow. Dispersed solids, such as the pulverized coal in the raceway region, are treated as interpenetrating continuum in the model, with both gas and dispersed phases having their own corresponding conservation equations. Interphase momentum and mass exchange occurs between the particle and gas phases. For this modeling approach, the flow of liquid slag and hot metal are not directly simulated. The presence of liquid below the cohesive zone and its impact on gas flow are accounted for by decreasing the porosity of the packed bed, resulting in a corresponding increase in gas pressure resistance.

Additionally, the most effective way to reduce the computational cost of a BF simulation is to assume that flow input conditions through each tuyere—this would include wind rate, auxiliary fuel injection rates, hot blast temperature, and other parameters—are axisymmetric around the entire furnace. With this assumption made, it is possible to simulate a single tuyere, corresponding raceway cavity, and similarly axisymmetric region of the BF shaft, greatly reducing the total computational cost of a simulation. For a generic BF, the 3D domain of the simulation would include the blowpipe upstream of the tuyere, any fuel injection lances, the tuyere, the raceway boundary and coke bed, and boundaries representing the location of the furnace walls and deadman.

As previously mentioned, the approach in this research utilizes multiple sub-models coupled together to predict conditions in the BF. Specifically, these models are targeted at two major reaction zones, the raceway and the shaft. Gas flow and injected fuel input conditions are specified for the raceway model, which is used to calculate the combustion and other reactions occurring within the raceway. The corresponding species, temperature, and mass flow distribution of reducing gases leaving the raceway region are then mapped into the lower boundary of the BF shaft model, which can then be used to predict macro-level output parameters such as top gas temperature, cohesive zone shape and location, gas utilization, and furnace coke rate.

For easier adjustment of tuyere, blowpipe, and injection lance design and placement, gas combustion, solids combustion, fluid flow, and heat transfer in the blowpipe and tuyere zones are conducted using a Eulerian-based model in ANSYS Fluent® (v19.2, ANSYS, Canonsburg, PA, USA, 2019), a commercial CFD package. The steady-state Navier–Stokes equations are applied to model conditions in this region using the semi-implicit method for pressure linked equations (SIMPLE) scheme with second-order upwind discretization of the transport equations. Turbulence modeling is handled with the standard k-ε turbulence model [26], a common and robust choice for efficient simulation of multiphase flow. Gas phase reactions— including CO, H_2, CH_4, and coal volatile combustion—are governed by the eddy-dissipation-concept model. Radiation heat transfer is governed in all raceway region simulations by the discrete ordinates (DO) radiation model.

The results of this simulation are then mapped to a two-component approach for simulating the coke bed and raceway cavity downstream of the tuyere. Once the incoming conditions from the tuyere are determined, the size and shape of the raceway in the coke bed is simulated using a transient Eulerian multi-fluid model, conducted again with a second-order application of the SIMPLE scheme in ANSYS Fluent®. The coke bed is treated as a fluidized granular continuous phase, and the gas flow forms a raceway cavity, which can then be frozen. The air phase volume fraction is then used to define the porosity of the coke bed in the raceway, and an in-house CFD solver detailed in previous publications [27–29] is applied to predict chemical reactions, gas flow, and heat transfer in a steady-state simulation of flow through the coke bed in the raceway region. In addition to combustion of injected fuels, the coke bed can react with oxygen in the hot blast to generate combustion products and heat, and if sufficient thermal energy is present, reactions of CO_2 and H_2O with carbon in the coke bed are also possible. Table 2 below contains the kinetic constants for gas-phase reactions of CO, H_2, and CH_4, as well as coke reactions and coal moisture evaporation and devolatilization.

Table 2. Key reaction mechanisms and kinetics for the computational fluid dynamics (CFD) raceway combustion model [30,31].

Reaction	A (1/s)	B (m³/(kg.s))	Activation Energy E (J/mol)
$CH_4 + 2O_2 \rightarrow CO_2 + 2H_2O$	N/A	1.6×10^{10}	1.081×10^5
$2CO + O_2 \rightarrow 2CO_2$	N/A	7.0×10^4	6.651×10^4
$2H_2 + O_2 \rightarrow 2H_2O$	N/A	5.4×10^2	1.255×10^5
Coal Moisture Evaporation	N/A	8.32×10^5	4.228×10^4
Coal Devol. Reaction 1	3.7×10^5	N/A	7.366×10^4
Coal Devol. Reaction 2	1.46×10^{13}	N/A	2.511×10^5
$C + O_2 \rightarrow CO_2$	1.225×10^3	N/A	9.977×10^4
$2C + O_2 \rightarrow 2CO$	1.813×10^3	N/A	1.089×10^5
$C + CO_2 \rightarrow 2CO$	7.351×10^3	N/A	1.380×10^5
$C + H_2O \rightarrow CO + H_2$	1.650×10^5	N/A	1.420×10^5

Combustion reactions of solids in the raceway result in the generation of additional gaseous mass. In addition, the significant variations in temperature can result in changes to gas density. These updated values are mapped back into the raceway formation step in the commercial CFD code in the form of cell-specific gas density values and a source term for gas mass, and the simulation of the raceway shape is re-run. This generates a new coke bed porosity distribution (updated raceway shape), and the combustion model can then be repeated as well. This iterative process will converge to an unchanging raceway shape, at which point the final combustion simulation results can be post-processed and mapped onto the CFD shaft model as input conditions.

Using the reducing gas flow rates, species, and temperature inputs from the raceway sub-model, the blast furnace shaft model simulates reducing gas flow through the burden layers in the BF, iron ore reduction and melting, and related phenomena. Computational domains for the BF shaft stretch from the furnace bosh, just above the raceway region, to the furnace top. As mentioned earlier, the flow is assumed to be steady state once the burden distribution is defined—either with a burden distribution sub-model or based on imported burden profiles from industry. This burden profile is used to define the porosity of the packed bed in the simulation, with coke and ore layers each having their own corresponding porosity and resistance to gas flow depending on particle size.

In this packed bed, fluid flow, chemical reactions, and heat and mass transfer between the gas and burden are all simulated by the CFD solver. Species included in the reaction models are CO, CO_2, H_2, H_2O, and N_2 in the gas phase and Fe_2O_3, Fe_3O_4, FeO, Fe, C, CaO, and MgO in the solid phase. Reactions are primarily gas–solid, though some phase transitions are included, such as moisture evaporation from the burden near the top of the furnace. The most common reactions are, of course, reduction of iron oxides by CO and H_2, coke gasification, and carbonate flux decomposition. The primary reactions included in the BF shaft model are listed in Table 3.

Finally, the cohesive zone (CZ) is defined between the iron ore pellet softening and iron liquidus temperatures (1473–1673 K), and in this region, the porosity of the iron ore layers is decreased to almost zero, rendering these areas impermeable to gas flow and generating a layered CZ. The approach allows gas flow to pass through the coke slits while being blocked by the cohesive ore layers. Additional details on the BF shaft model, including information on the burden distribution sub-model, can be found in previous publications [24,32–34].

Table 3. Reaction mechanisms used in the CFD shaft model.

Reaction	No.	Chemical Equation
Indirect reduction of iron oxide by CO	R1	$3Fe_2O_3(s) + CO(g) \rightarrow 2Fe_3O_4 + CO_2(g)$
	R2	$Fe_3O_4 + CO(g) \rightarrow 3FeO(s) + CO_2(g)$
	R3	$FeO(s) + CO(g) \rightarrow Fe(s) + CO_2(g)$
Indirect reduction of iron oxide by H_2	R4	$3Fe_2O_3(s) + H_2(g) \rightarrow 2Fe_3O_4 + H_2O\ (g)$
	R5	$Fe_3O_4 + H_2(g) \rightarrow 3FeO(s) + H_2O(g)$
	R6	$FeO(s) + H_2(g) \rightarrow Fe(s) + H_2O(g)$
Boudouard reaction	R7	$C(s) + CO_2(g) \rightarrow 2CO(g)$
Water gas reaction	R8	$C(s) + H_2O(g) \rightarrow CO(g) + H_2(g)$
Flux decomposition	R9	$MeCO_3(s) \rightarrow MeO(s) + CO_2(g)\ (Me = Ca, Mg)$
Water gas shift reaction	R10	$H_2(g) + CO_2(g) \rightleftharpoons H_2O(g) + CO(g)$
Direct reduction of liquid FeO	R11	$C(s) + FeO(l) \rightarrow Fe(l) + CO(g)$

2.2. Blast Furnace Geometry and Simulation Boundary Conditions

Research was conducted via simulation of two different blast furnaces of average size and production for North America located at separate North American steel manufacturing facilities. The geometry for the tuyere and blowpipe regions of both furnaces are unique to their different operations, with one furnace utilizing co-injection of pulverized coal and natural gas and the second using only NG injection. These two operations present ideal test beds for exploring the impacts of injected fuel adjustments, with the co-injection case allowing for comparisons of how a co-injection furnace may react during a loss-of-PCI scenario and the NG-injection furnace highlighting the impacts of NG pre-heating on operation.

The area beyond the tuyere nose (containing the raceway and coke bed) for a single tuyere was modeled for both furnaces using computational grids. These domains were approximately two meters high and were bounded by the furnace wall and tuyere at the outside and a fixed boundary representing the deadman towards the furnace center. Symmetry conditions were applied on the bounding walls to either side of the raceway. The combination of the specific physical geometry of the tuyere and furnace with the momentum and reactions from the hot blast generated a unique raceway cavity in the coke bed. As mentioned in the previous section, the raceway geometry was defined by applying the calculated void fraction in the raceway formation model to fixed spatial variations of the coke bed porosity. The geometry for the tuyere region of both industry blast furnaces and a generic raceway region geometry are shown in Figure 1.

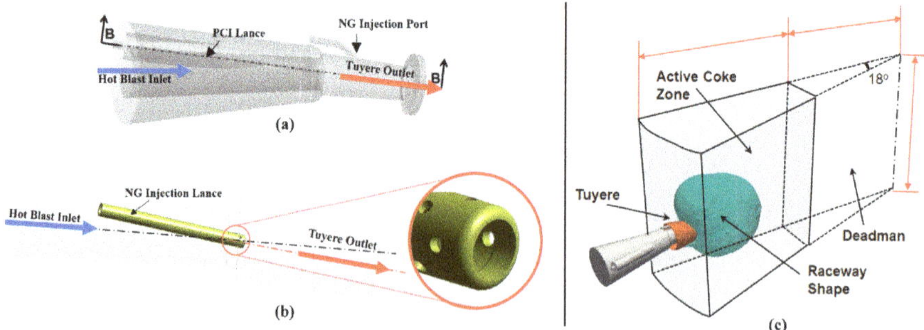

Figure 1. (**a**) Blowpipe and tuyere region simulation domain for co-injection blast furnace (BF); (**b**) Blowpipe and tuyere region simulation domain for natural gas (NG)-only BF; and (**c**) Generic raceway region geometry with tuyere and blowpipe position included.

Additionally, simulations of the shaft region of the NG-only blast furnace were conducted. This geometry was far simpler than the raceway, with the furnace walls defining an axisymmetric cylindrical region with a radius that varied depending on height. The burden charge profile for ore and coke layers was provided by the industry partner, with the ore charge weight varying depending on the natural gas injection rate to adjust furnace production and corresponding coke rate in real-world operation.

Boundary conditions were defined for each individual region simulated. In the tuyere/blowpipe regions, the hot blast inflow, NG inflow, and PC carrier gas inflow were defined as mass flow inlets in the CFD solver. At each inlet, fluid temperature, species distribution, and mass flow rate were defined based on furnace operating conditions. PC injection was handled with the discrete phase method (DPM), and the total mass flow influx of PC particles was defined once again based on industry operating conditions. A pressure outlet boundary condition was used for the tuyere exit, with an estimated emissivity and blackbody temperature used to account for radiation heat flux entering the tuyere from the high-temperature coke bed in the raceway.

Mass flow and species distribution for the gas phase were mapped onto the inlet for the raceway region CFD simulation. The DPM concentration at the tuyere outlet plane was mapped as a scalar value to define the volume fraction of the dispersed particle phase for tracking PC movement in the raceway. Similar mass flow inlet and pressure outlet flow boundary conditions were applied in this region, with symmetry conditions applied on the boundaries located on either side of the raceway envelope. Mass flow rates and species distributions were also mapped from the outlet of the raceway region to the inlet of shaft model, and once again, a pressure outlet was used for the out-flow condition at the top of the furnace.

Environmental heat losses were governed by refractory and steel shell thermal conductivity in the blowpipe, with assumed natural convection at average ambient temperatures of ~305 K. In the water-cooled copper tuyere, assumptions were based on typical industry conditions and expectations, with a heat transfer coefficient of ~3000 W/(m^2 K) and a freestream temperature of 300 K for cooling water applied. For the furnace walls in the raceway and shaft regions, a basic constant heat flux assumption was applied based on expectations of heat losses to cooling water provided by industry partners.

3. Results

3.1. Co-Injection Blast Furnace

The impact of various parameters and design modifications on the operation of the co-injection furnace simulated in this research were documented in previous publications [15–17]. This work focuses in particular on the effects of a significant increase in natural gas injection rate in comparison to standard operating conditions, perhaps in a loss-of-PCI scenario while attempting to maintain production rate. Modeling of this furnace was conducted at typical operating conditions, with a wind rate of ~200,000 Nm^3/h, an oxygen enrichment of 34%, a blast temperature of 1408 K, a PCI rate of 85 kg/mthm, and a NG injection rate of 65 kg/mthm. The pulverized coal particle size distribution was provided by industry partners, with an average particle size of 0.046 mm.

While attention was given to the impacts of design and operational changes on conditions in the tuyere region in previous publications [15–17], in this research, the focus remained on changes to gas temperature and species distribution in the raceway region. Using the CFD model, it is possible to generate a numerical value that can be easily compared to the raceway adiabatic flame temperature (RAFT) by taking a mass-weighted average of the gas temperature in all computational cells in the domain with less than 0.5% H_2O, CO_2, and O_2 content by volume. This corresponds to the theoretical definition of RAFT as the temperature of all gases once they have been reduced from fuel and oxidizer into CO and H_2 (also including inert N_2). From this point forward, the CFD-generated value will be referred to as the flame temperature analogue (FTA).

In the baseline scenario, simultaneous injection of NG and PC led to spatial variations in temperature distribution. Coke combustion occurred where unreacted oxygen encountered the coke bed, while the initial NG flame occurred rapidly near the outlet of the tuyere, followed by a sharp reduction in gas temperature in the regions where NG combustion products (particularly H_2O) engage in endothermic reactions with coke. PC combustion in recirculation zones within the raceway itself generated larger high temperature regions, which can be observed in Figure 2. The predicted baseline FTA was 2244 K, a difference of 2.1% compared with an expected RAFT of 2293 K.

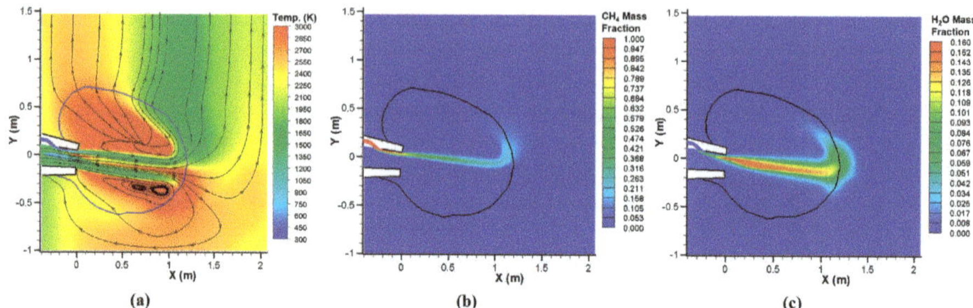

Figure 2. (**a**) Gas temperature contours and flow streamlines for the baseline co-injection BF raceway (raceway boundary shown in blue); (**b**) Contours of CH_4 mass fraction; and (**c**) Contours of H_2O mass fraction.

Operational rules-of-thumb based on industry experience have documented the impacts of increasing or decreasing many tuyere-level parameters on the furnace. Correspondingly, this research aimed to ascertain whether the CFD models used in this study would show similar trends, and if so, to simulate the impacts of a complete loss-of-PCI scenario in which furnace operators attempt to replace all injected fuel with natural gas. Comparisons were conducted with two scenarios in which the ratio between NG and PC injection rates were shifted and all other parameters were held constant. The ratio in the baseline case was 1.31:1 (PC to NG). In Case #1, the ratio was increased to 2.26:1, and in Case #2, the ratio was decreased to 0.87:1. In line with expectations, Case #1 (with a lower NG injection rate) resulted in a high predicted FTA of 2271 K and Case #2 (with a higher NG injection rate) had a lower predicted FTA of 2228 K.

Case #3 took high-rate natural gas operation to an extreme, cutting PCI and increasing the NG injection rate from the baseline value of 65 kg/mthm and increasing it to 150 kg/mthm. The NG in this case was supplied through an NG injection lance. This level of NG injection is generally held as the maximum sustainable for stable operation due to the constraints detailed in Section 1 of this paper (falling RAFT, low top temperatures due to increased O_2 enrichment). Aside from the modification to injection, operating conditions in Case #3 were maintained at baseline levels. Figure 3 shows the distribution of gas temperature, CO_2, and H_2O in the raceway region. Combustion was immediate and was distributed throughout the majority of the tuyere jet, however, gas temperature fell rapidly upon contact with the coke bed as CO_2 and H_2O underwent the aforementioned endothermic reactions with coke. The FTA in this scenario dropped by more than 11% to 1988 K, a value just below the aforementioned minimum RAFT for North American BFs of 2020 K. It should be noted that the increased injection rate and delivery of NG through a lance resulted in NG pushing away from the tuyere center plane, and it was only once the gas had begun to recirculate in the raceway that higher concentrations were observed on the center plane of the raceway as seen in Figure 3b.

The results from this scenario indicated that the CFD modeling techniques applied were able to predict the decline in gas temperatures resulting from high levels of NG injection. Based upon this, the next stage of research aimed to determine the effectiveness of NG pre-heating as a method to counter the decline in predicted FTA and maintain reducing gas temperatures in the furnace. This would serve as a potential method by which a loss-of-PCI scenario could be quickly adapted to in the field, allowing for significantly higher NG injection rates with a much more manageable decline in BF reducing gas temperatures. As previously discussed, NG pre-heating may present another lever for operators beyond adjustments to O_2 enrichment, wind rate, and hot blast temperature.

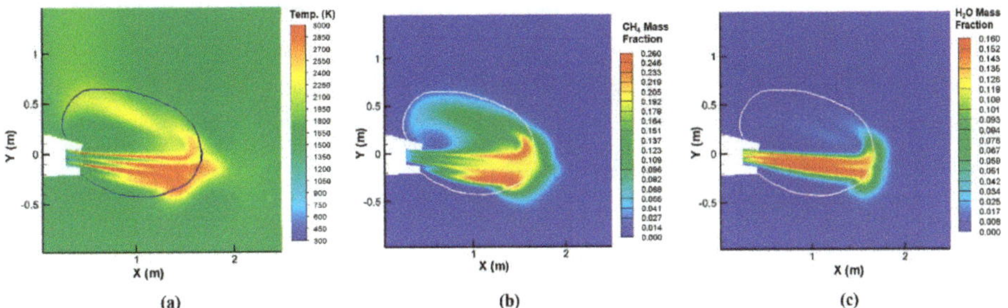

Figure 3. (**a**) Gas temperature contours and flow streamlines for Case #3 in the co-injection BF raceway (raceway boundary shown in blue); (**b**) Contours of CO_2 mass fraction; and (**c**) Contours of H_2O mass fraction.

3.2. Natural Gas Injection Blast Furnace

Following the loss-of-PCI study, an investigation of the specific impacts of NG pre-heating throughout the furnace (both in the raceway and shaft regions) was conducted using the NG injection industry blast furnace. While the NG-only BF was roughly 30% larger than the co-injection BF and operated at a slightly higher production rate, CFD modeling of the BF as a trial for this portion of the research allowed for isolation of gas pre-heating impacts, helping to clearly delineate any benefits and drawbacks associated with the approach. The baseline operating conditions for the NG-only BF included a wind rate of 270,000 Nm^3/h, an oxygen enrichment of 29%, a hot blast temperature of 1448 K, and an NG injection rate of 95 kg/mthm, among other parameters. Natural gas was injected into the tuyere through a lance with multiple ports around the edges near the tip and a single central port for primary gas flow, as shown in Figure 1b.

The FTA predicted for baseline operating conditions at the NG-only BF was 2187 K, which compares favorably to the industry RAFT expectation of 2169 K (0.74% difference). When observing conditions within the NG-only furnace raceway, it can be seen that the location and angle of the NG injection lance has a significant impact on the temperature and gas species distributions in the raceway region. Figure 4 shows contours of gas temperature, CH_4 mass fraction, and H_2O mass fraction. In this case, the NG plume remained to the left (looking from the outside of the furnace in, as if through a peep sight) of the tuyere center plane, as the injection lance was inserted from the right-hand side and pushed gas to the opposite side of the tuyere. This resulted in higher concentrations of NG and NG combustion products on the left-hand side the raceway, leading to a left–right asymmetry in gas temperatures as the CO_2 and H_2O from NG combustion experienced the expected endothermic reactions with coke.

The distributions of gas temperature, mass flow rate, and species distribution were exported from the upper outlet boundary of the raceway region CFD model and were then transferred as importable inlet conditions for the shaft region CFD model. Other required operating conditions for the BF shaft model included the burden distribution—provided by industry collaborators for the baseline scenario and shifts in burden charge weight for increased or decreased NG injection rates—and by-weight moisture content in the charged iron ore and coke layers. The moisture content was of particular interest, as in the wet and cold winter months in some regions of North America, ore moisture content can rise significantly, resulting in impacts on top gas water vapor content and temperature when charged moisture evaporates. For the baseline case, these values were fixed at 2.5% by weight for the ore layers and 6% by weight for the coke layers. Contours of gas temperature and species distribution in the shaft region are shown in Figure 5. Also included are demarcations indicating the location of the burden layers in the furnace and the location of the cohesive zone.

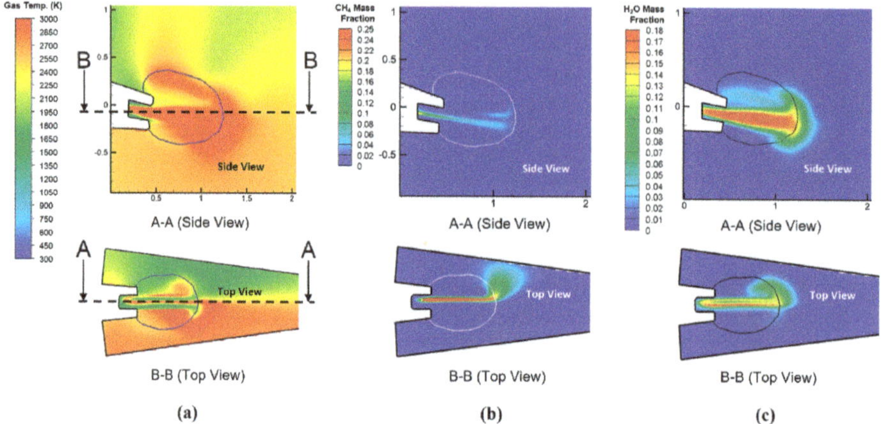

Figure 4. Side view (top) and top view (bottom) of contours of (**a**) temperature; (**b**) CH$_4$ mass fraction contours; and (**c**) H$_2$O mass fraction in the raceway region for the NG-only BF baseline case (raceway boundary shown in blue or white).

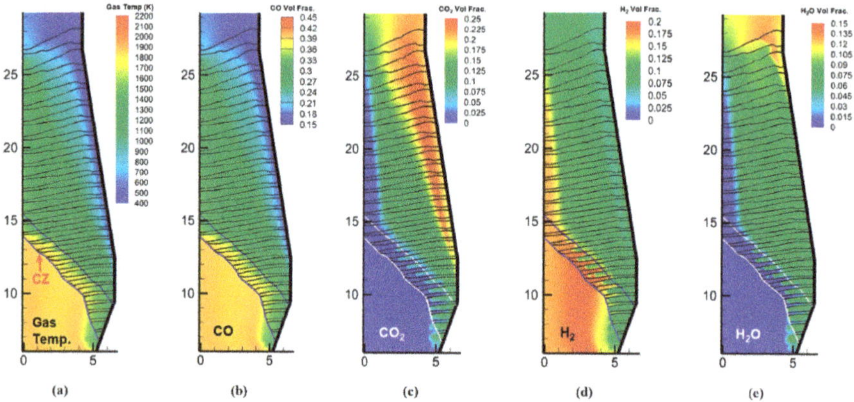

Figure 5. Cross-section views of contours in the shaft region showing (**a**) gas temperature contours; (**b**) CO volume fraction; (**c**) CO$_2$ volume fraction; (**d**) H$_2$ volume fraction; and (**e**) H$_2$O volume fraction.

While the CFD models applied in this research have been extensively validated in previous publications [21,25,26,30–32], additional validation against macro-level parameters from the industry operation of this blast furnace was conducted to build confidence in the ability of the included simulations to predict operating conditions for this particular blast furnace. In particular, CFD predictions of coke rate, reducing gas utilization, and average top gas temperature were compared against industrial values for operation corresponding to the scenario investigated in the baseline case. Top gas temperature and CO and H$_2$ utilization values were determined based on averaged readings from a top gas analyzer during standard operation matching the conditions of the baseline case. These comparisons are detailed in Table 4.

Table 4. Comparison of CFD results against industry data for NG-only BF baseline operation.

Parameter	CFD Prediction	Industrial
Coke Rate	392 kg/mthm	~390 kg/mthm
CO Utilization	50.8%	~50%
H_2 Utilization	51.5%	~50%
Avg. Top Gas Temperature	403 K	~385 K

The impacts of NG pre-heating on the NG injection BF were first explored with six scenarios beyond the baseline case. These cases, selected based on industry feedback, raised the incoming temperature of injected NG from 300 K to 600 K in 100 K increments. The upper temperature boundary was set at 600 K for this study to avoid scenarios with the potential for NG cracking and soot formation in the transport piping, a practical limitation which might well lead to excessive maintenance and higher costs after implementation. Every other operating condition (hot blast temperature, wind rate, oxygen enrichment, NG injection rate) was held constant for the first set of simulation cases to isolate the impacts of NG pre-heating on phenomena within the raceway region.

In the tuyere and blowpipe region, the impacts of NG pre-heating were primarily observable in the increased velocity of the incoming NG plume and the increased average temperature. Comparing a scenario in which NG was preheated to 600 K to the baseline case, gas temperatures at the outlet were 1.4% higher and the average gas velocity was 1.8% higher. Additionally, the higher NG injection velocity pushed the combusting gas plume slightly closer to the tuyere wall, which may present reliability concerns during operation unless the lance is retracted, or a larger diameter lance is used. A 6.3% increase in NG combustion before the tuyere exit was also observed, most likely due to the reduced density and increased velocity of NG, which results in improved turbulent mixing. The average turbulent kinetic energy predicted at the tuyere outlet under standard conditions was 326 m^2/s^2, while with a 300 K NG pre-heat this value rose to 414 m^2/s^2, a 27% increase. A direct comparison between these two cases is shown in Figure 6.

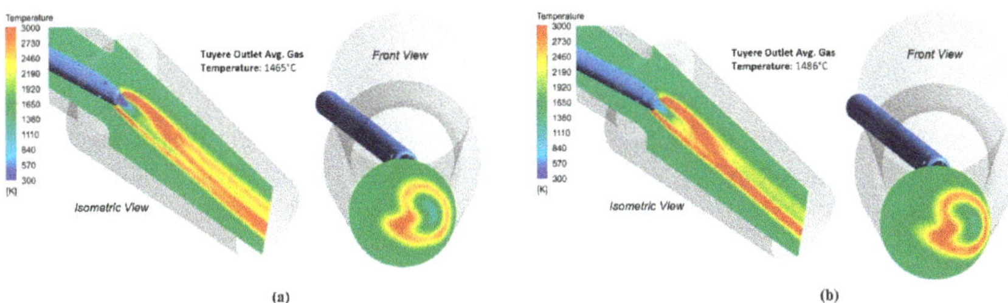

Figure 6. Isometric and front views of gas temperature contours on the tuyere center plane (left) and tuyere outlet (right) for (**a**) the baseline case; and (**b**) the 300 K NG pre-heat scenario.

Changes to species distribution in the raceway region were minimal, as might be expected given that incoming mass flow rates of fuel and oxygen were held constant. Temperature distributions were also similar between cases, with the largest observable variations occurring in the predicted FTA value (which is itself the first parameter targeted by NG pre-heating). Similarly, in the shaft region, the impacts of NG pre-heating are easiest to observe by directly comparing the changes in top gas temperature and coke rate to the corresponding values from the baseline case. Table 5 details the results from the NG pre-heating simulations at the baseline natural gas injection rate.

Table 5. Impacts of NG pre-heating on BF operating conditions in the raceway and shaft regions. FTA: flame temperature analogue.

NG Temp. (K)	FTA (K)	FTA Increase from Base (K)	Top Gas Temp. (K)	Coke Rate (kg/mthm)	Coke Rate Decrease from Base (kg/mthm)
300 (Base)	2187	-	403	392	-
400	2207	20 K (0.9%)	399	389	3
500	2221	34 K (1.6%)	393	387	5
600	2239	52 K (2.4%)	390	385	7

It is clear that NG pre-heating had an observable impact on the predicted FTA value, with each 100 K increase in NG temperature resulting in a 17 K increase in predicted FTA on average. Additionally, the BF coke rate declined by up to 7 kg/mthm at the maximum level of NG pre-heating, indicating the potential for improvements to operational efficiency with this approach.

During the course of this research, the potential of NG pre-heating as a lever to address declining top gas temperatures during winter months in North America due to high burden moisture content was also explored. The CFD shaft model was applied to determine the impact of ore moisture, with results indicating that each 0.5% increase in ore moisture content by weight resulted in a 5 K decline in top gas temperature. This agrees with measurements and experience from both industrial partner facilities modeled in this research. CFD modeling of the impacts of NG pre-heating indicated that each 100 K increase in NG temperature resulted in a 4.5 K drop in average top gas temperature. While this aligns well with blast furnace operational rules-of-thumb (parameters that increase RAFT such as hot blast temperature or oxygen enrichment typically also result in a decline in top gas temperatures) it also means that NG pre-heating alone is unlikely to serve as a direct counter to the low top temperatures observed in high burden moisture content scenarios.

Having established the potential benefits and drawbacks of NG pre-heating in isolation, the next step was to determine whether pre-heating could be applied together with the modification of other variable parameters to widen the BF operating window. For instance, does pre-heating allow for a potential solution to moisture-generated low top temperatures if combined with increased NG injection rates (which typically increase top gas temperature)? Could pre-heating serve as a tool that might enable operators to push higher NG injection rates in the case of the aforementioned loss-of-PCI scenario at a co-injection BF? Or might it be possible to push beyond the typical limit of 150 kg/mthm of NG injection by combining the increased injection rate with NG pre-heating? The next stage of research aimed to answer these questions with a parametric study involving a range of NG injection rates, NG pre-heating levels, and oxygen enrichment levels.

Eight different NG injection rates were simulated for this study. In addition to the baseline NG injection rate of 95 kg/mthm, scenarios were modeled at NG injection rates of 85, 105, 110, 115, 120, 130, 140, and 150 kg/mthm. For each of these new injection rates, the ratio between ore and coke in the burden charge was altered based on industry operation guidelines for the furnace. As the injection rate rose, the ore charge weight was increased while maintaining the same burden distribution, leading to thicker ore layers and a higher potential production rate. NG pre-heating was also tested for each of these injection rates, so each of these eight additional injection rate scenarios had four sub-cases at the same NG temperatures (300 K, 400 K, 500 K, and 600 K) as the baseline case.

While results in the tuyere region are not different enough between cases to merit a full review here, it is important to note that the increased NG injection rates were achieved by simply increasing the mass flow rate of NG into the lance. At the upper end of the injection rate range (140–150 kg/mthm), this results in very high NG velocities exiting the lance, especially when combined with NG pre-heating. A realistic implementation of such high NG injection rates with pre-heating would require a larger diameter injection lance to manage NG injection velocity and avoid potential impingement on the inner surface of the tuyere.

Focusing on the raceway region, the predicted FTA fell quickly with increasing NG injection rate, as expected. Additionally, the FTA increasesd with NG pre-heating for all simulated NG injection rates, save for the 150 kg/mthm scenario. Figure 7 details the predicted FTA results for the full range of cases, with the declining slope of the FTA vs. NG pre-heat as the NG rate (in kg/mthm) increased. These slopes represent the "pre-heating efficiency" of NG pre-heating in raising the predicted FTA (unit increase in FTA per unit increase of NG temperature).

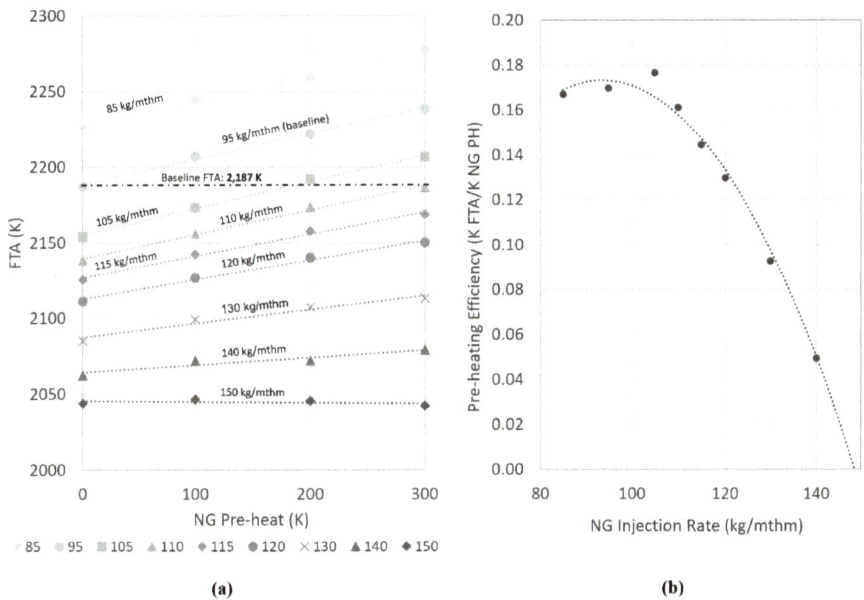

(a)

(b)

Figure 7. (**a**) Predicted FTA vs. NG pre-heat; and (**b**) NG pre-heating efficiency vs. NG injection rate.

Included in Figure 7 is a marker noting the location of the baseline FTA value that allows for easy comparison with other cases. For instance, at an NG injection rate of 105 kg/mthm (10 kg higher than the baseline), an equivalent FTA to the baseline case could be achieved with an NG pre-heat level of 200 K (NG temperature of 500 K). At 110 kg/mthm, this FTA could be achieved with a pre-heat of a little over 300 K (NG temperature of 600 K). If it is assumed that the furnace remains stable and matches productivity at a given reducing gas temperature with a fixed burden distribution, these comparisons seem to indicate the potential for operators to push higher NG injection rates through the use of pre-heated NG, with relatively minor impacts on furnace operation otherwise.

It is worth noting here that from an NG injection rate of 85 kg/mthm to an injection rate of 105–110 kg/mthm, it appears the NG "pre-heating efficiency" increased to a maximum of a 0.177 K increase in FTA for each 1 K increase in NG temperature. When the NG injection rate was increased further, the "pre-heating efficiency" began to decline. There are multiple potential causes for this decline. First, the increase in NG injection rate for these cases was not accompanied by a corresponding increase in hot blast O_2 enrichment. The NG fuel rate exceeded the available oxygen for combustion around the 110 kg/mthm mark, and pre-heating at injection rates beyond this point likely contributed to NG decomposition. Additionally, the increased injection rate reduced gas residence time inside the raceway and drove the NG plume closer to the tuyere wall due to higher gas velocity, potentially hindering combustion. The combination of these factors likely led to the results observed in the 150 kg/mthm case, in which NG pre-heating actually decreased the predicted FTA.

In addition to the baseline oxygen enrichment level of ~29%, two additional scenarios were added to each case at oxygen enrichment levels of 32% and 35%. In all cases, wind rates were held constant at baseline levels and the volume fraction of oxygen in the wind was adjusted. Figures 8 and 9 show the predicted FTA values for the 32% and 35% oxygen enrichment case sets respectively. As expected, FTA values were inflated for these cases. Additionally, the availability of additional oxygen resulted in improved "pre-heating efficiency" at the higher NG injection rates compared to the baseline oxygen enrichment scenario, though in each set of cases, most benefits from pre-heating had essentially vanished by the 150 kg/mthm injection rate. It is likely that some combination of additional measures such as increased oxygen enrichment, wind rate, and NG pre-heating would be necessary for efficient operation at these high injection rates for this particular furnace.

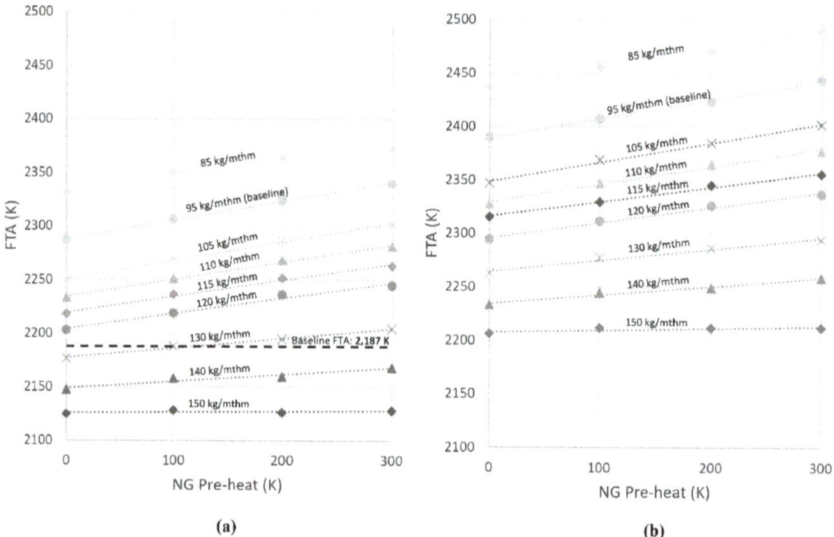

Figure 8. Predicted FTA vs. NG pre-heat for (**a**) the 32% O_2 enrichment scenario; and (**b**) the 35% O_2 enrichment scenario.

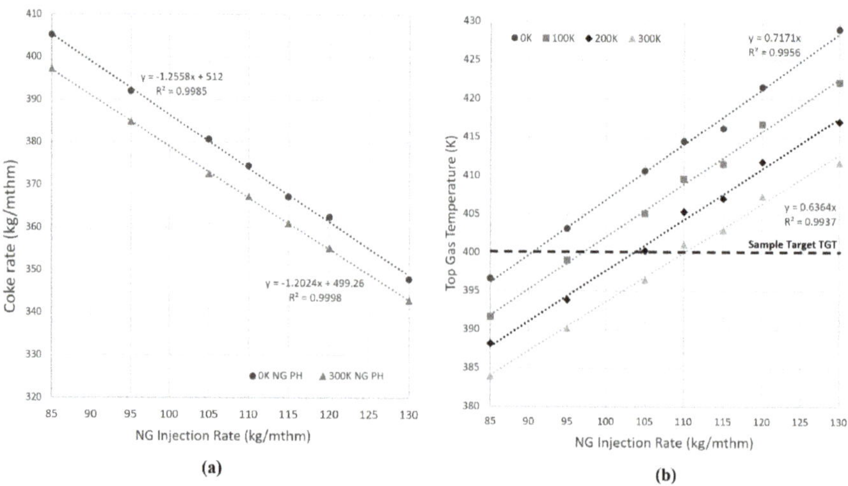

Figure 9. (**a**) Predicted BF coke rate for 0 K and 300 K NG pre-heat vs. NG injection rate; and (**b**) Predicted average top gas temperature for all NG pre-heat levels vs. NG injection rate.

For the shaft region, simulations focused on the baseline oxygen enrichment scenario with increases to NG injection rate that would require no adjustments to the existing infrastructure at the industry partner facility. These cases aimed to determine the combined impacts of NG pre-heating and increased NG injection rates on the furnace top gas temperature and coke rate. Figure 9 shows the predicted impacts of NG injection rates (from 85 to 130 kg/mthm) on furnace coke rate and top gas temperature.

While changes in both coke rate and top gas temperature due to NG injection rate remained linear through the range of 85–130 kg/mthm, two trends appeared from the impacts of NG pre-heating. First, it appears that the coke rate savings observed from pre-heating decline at the higher NG injection rates, likely due to the aforementioned limiting factors in the raceway. Second, the impact of pre-heating on top gas temperature appears to become more significant as the injection rate rises, with a larger drop in top gas temperature observed for each 100 K pre-heat at the higher end of the injection rate range. The increased decomposition of NG in the raceway in lieu of gas combustion will result in reducing gas compositions with a balance shifted slightly towards H_2 (as carbon from the NG will deposit in the solid phase and require CO_2 or O_2 to convert it into reducing gas), and modeling predicts that this shift promotes endothermic hydrogen reduction reactions. Specifically, simulations predicted that the increase from an NG temperature of 300 K to 600 K would result in a 1.8% increase in wustite reduction by H_2 in the shaft for the 130 kg/mthm NG injection rate scenario. In comparison, the same increase in NG temperature results in only a 0.77% increase in the wustite-H_2 reduction reaction for the 85 kg/mthm NG injection rate scenario. Since wustite reduction by H_2 consumes ~25 kJ/mol, while wustite reduction by CO releases 16 kJ/mol, this shift towards H_2 reduction (together with the reduction in available heat due to NG cracking rather than combusting) seems to explain the increasing detrimental impact on top gas temperatures. Further research in this area could prove illuminating when exploring high natural gas injection rates and their impacts on the furnace.

Overall, the modeling predicts that the NG injection rate will have a far more significant impact on predicted coke rate than NG pre-heating, which is to be expected given the additional shifts in gas composition and reduction reactions under higher NG injection conditions. For top gas temperature, the impacts between the two parameters are more similar, and it is in this region that the greatest benefits of combining NG pre-heating and increased NG injection rates can be observed. Selecting a sample acceptable top gas temperature value of 400 K (125 °C) and moving across the range of NG injection rates, it is clear that there are numerous combinations of NG injection rate and NG pre-heat level that provide an average top gas temperature close to matching the criterion (95 kg/mthm NG injection rate and 100 K pre-heat, 105 kg/mthm NG injection rate and 200 K pre-heat, 110 kg/mthm NG injection rate and 300 K pre-heat).

4. Discussion

4.1. Potential Impacts and Benefits of NG Pre-Heating

Both in isolation and in concert with modest increases in NG injection rate, pre-heating of NG before injection into the BF appears to present an additional method by which operators could improve the operating efficiency of their furnaces and widen the window of stable operating conditions. CFD modeling results indicated that there were observable benefits to blast furnace coke rates from pre-heating NG up to a temperature of 600 K, though this did result in a decline in top gas temperatures, similar to the expected impacts from increased hot blast temperature or other parameters that result in a boost to FTA (RAFT). While the predicted 7 kg/mthm decrease in BF coke rate in this scenario was a desirable outcome, the corresponding decline of 13 K in top gas temperature may lead to top gas temperatures lower than 373 K (100 °C) and the possibility of condensation of water vapor in the uptakes. However, by combining higher NG temperatures with an increased NG injection rate, significant benefits to the BF coke rate can be achieved while maintaining constant levels for both FTA and top gas temperature (TGT). A set of such cases are outlined in Table 6. Scenario #1 has the standard operating conditions from the baseline case, with an NG pre-heat of 300 K. Scenarios #2 and #3 apply combinations of NG pre-heating with increased NG injection rates (using the levels at which pre-heating was most effective for this BF) to maintain the FTA and TGT at or near baseline levels while allowing for the furnace to operate at a lower coke rate.

Table 6. Potential operating condition modifications and corresponding economic impacts. TGT: top gas temperature.

Case	NG Rate (kg/mthm)	NG Pre-heat (K)	FTA (K)	TGT (K)	Coke Rate (kg/mthm)
Baseline	95	0	2187	403	392
Scenario #1	95	300	2238	390	385
Scenario #2	105	200	2192	400	375
Scenario #3	110	300	2186	401	367

The CFD simulations conducted for Scenario #3, using a 110 kg/mthm NG injection rate and a 300 K NG pre-heat, predicted an FTA of 2186 K (1 K lower than baseline) and a TGT of 401 K (2 K lower than baseline). However, in this case, the BF coke rate decreased from 392 kg/mthm in the baseline to 367 kg/mthm (25 kg/mthm lower), presenting the potential for significant savings. Of course, it is necessary to consider the potential costs and safety concerns involved in pre-heating the NG, as the effort would likely be greatly dependent on the exact method of NG pre-heating.

It is worth returning here to the concept of using high-rate NG injection to compensate for a loss-of-PCI situation at a co-injection BF. Clearly, NG pre-heating presents a net benefit to predicted FTA. While not enough to entirely counter the decline seen in the NG-only operation case at the co-injection BF, applying pre-heating at higher NG injection rates (up to the 120–130 kg/mthm mark) would likely enable furnace operators to maintain stable operation in a loss-of-PCI scenario. In the NG-only furnace, balancing NG injection rate with various levels of NG pre-heating allows for the furnace FTA and TGT to remain at baseline levels. Similarly, it is expected that this strategy could be applied for furnaces with lower total levels of injectants (such that acceptable RAFTs could be achieved), or furnaces with the flexibility to supply additional oxygen through enrichment or wind to further supplement the pre-heating impacts.

4.2. Conceptual Methods for Achieving NG Pre-Heating

Several potential methods could be used to achieve an increase in NG temperature from ambient (~300 K or 25 °C) temperature to the range of pre-heating explored in this study (~600 K or 325 °C). From a practical standpoint, it would likely be best to apply a pre-heating method that does not require any additional heat sources or fuel use in order to maximize the operating expense reduction benefits highlighted in the previous section. It is clear that a single capital investment to develop and implement an NG pre-heating system with minimal ongoing costs would certainly be more attractive to BF operators than the alternative. The previously reported exploration of NG pre-heating at two Russian BFs [4,5] approached the concept from this angle, electing to route the incoming ambient NG through cooling flanges around the exterior of the blowpipe and elbow in an effort to capture heat that would otherwise be lost to the environment from the hot blast and direct it back into the BF.

Some preliminary CFD modeling of similar concepts was undertaken for the industry BF studied in this research. Some sample designs of a blowpipe-based heat exchanger or heat recovery system were explored, including in-refractory channels and an exterior piping wrap similar to the concepts shared in previous publications. A visualization of these potential designs is shown in Figure 10.

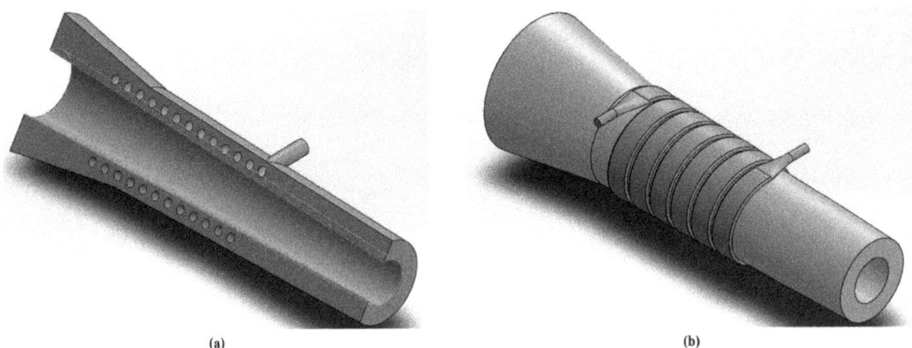

Figure 10. Conceptual designs for (**a**) the blowpipe channel NG pre-heater; and (**b**) the exterior wrap NG pre-heater.

Preliminary CFD simulations of convection heat transfer using the blowpipe channel NG pre-heater were conducted using baseline operating conditions (hot blast temperature, wind rate, and natural gas injection rate). These simple trials predicted a NG pre-heat of 170 K (NG temperature of 470 K) by the end of the channel, though the exact impact of this heat exchanger design on the hot blast temperature flowing through the blowpipe has yet to be determined. A potential loss in hot blast temperature would certainly be a significant detractor from such a design implementation and might serve to encourage the exploration of alternative approaches.

In addition, these designs present significant complexity in a region of the BF that can already have a great deal of design constraints, so it may be beneficial to explore methods for pre-heating elsewhere. A heat recovery system based around the exhaust or ambient heat losses from the BF hot stoves may present an acceptable solution, and future work may focus on determining the viability of such a system in terms of achievable NG pre-heat and specific impacts on BF coke rate, top gas temperature, and FTA.

4.3. Industrial Perspective on Feasibility of Implementation

The computational simulations conducted in this research indicate that pre-heating of natural gas prior to injection into the blast furnace could offer process benefits. However, modeling does not consider the effort required for industrial implementation and the associated difficulties that such an effort might present.

Several routes to pre-heating exist, recuperative (including waste heat), on-demand, and in-situ partial combustion; each route having a certain level of technical maturity, capability, controllability, as well as CAPEX and OPEX characteristics that are plant unique. Conceptually, most of the identified heating technologies would heat gas sourced singly from the main furnace distribution line, i.e., all the injected natural gas would be passed through a single recuperative heater, while the heating routes illustrated in Figure 10 would heat individually closer to the point of consumption.

Implementation for the single source is somewhat easier, although potentially more expensive, as much of the construction and commissioning could be completed while the furnace remains online. However, pre-heating via the blowpipe route would require an outage of suitable length to change all the blowpipes, or a planned approach to replace at regular intervals, which could take several years and leave the furnace with unbalanced pre-heating in the meantime.

Industrial consideration also needs to be given for material selection (pipes, valves, and seals need to be capable of operating at higher temperatures) and process control (which may have varying feasibility depending on the route chose, i.e., blowpipe heating). In addition, modified operating and maintenance practices would need to be developed, and safety protocols would also require careful review. One example would be RAFT control, which is widely adopted as a key variable in ironmaking.

Typically, front-line operators use a simple empirical model such as the AISI RAFT model equation to determine real-time RAFT values. As currently defined, this model does not account for pre-heating of natural gas. Similarly, the well-known operating rules-of-thumb would need expansion to include the effect of pre-heating temperature on RAFT, top heat, fuel rate, and productivity.

5. Conclusions

CFD modeling techniques were applied to ascertain the predicted impacts of varied NG injection rates and pre-heating levels on operation at two different industrial BFs. Simulations indicated that there is a clear possibility to utilize NG pre-heating in conjunction with increased NG injection rates to provide an avenue for improved BF operating efficiency (reduced coke rates while maintaining stable raceway flame and top gas temperatures). Impacts of NG pre-heating in isolation appear linear, though the slope changes significantly with NG injection rate, and the benefits disappear quickly when pushing towards injection rates above 140 kg/mthm. The application of NG pre-heating also appears to have a similar negative impact on top gas temperatures as that of other parameters that increase furnace RAFT, such as hot blast temperature or oxygen enrichment. However, this impact is small enough that it can be offset by the increased NG injection rates made possible under these operating conditions.

As noted in the discussion section, achieving NG pre-heating could be approached from a multitude of different avenues, including pre-combustion or an upstream recuperative heater. Some initial simulation testing of a conceptual system for heating NG near the BF blowpipe indicated that sufficient heating could be provided to the incoming NG to achieve the predicted benefits explored in this research. However, any industrial implementation would doubtless need to account for the control mechanisms, CAPEX considerations, and other limitations that would figure into such a system in a real-world environment. The difficulties and costs would need to be weighed against the potential benefits, with an eye towards furnace life-span and the potential operational flexibility and return on investment that an NG pre-heating system would enable.

Author Contributions: Conceptualization, T.O., J.D., and S.S.; Data curation, S.N.; Formal analysis, T.O. and S.N.; Investigation, T.O., S.N., J.D., S.R., and S.S.; Methodology, T.O. and C.Z.; Project administration, C.Z.; Resources, T.O. and C.Z.; Software, T.O. and S.N.; Supervision, T.O. and C.Z.; Validation, T.O., S.N., J.D., S.R., and S.S.; Visualization, T.O. and S.N.; Writing—original draft, T.O. and S.S.; Writing—review and editing, T.O. All authors have read and agreed to the published version of the manuscript.

Funding: This research received no external funding.

Acknowledgments: The authors would like to thank the agencies and companies involved in supporting previous and current development of blast furnace research at PNW, including AK Steel, AISI, AIST, ArcelorMittal, the U.S. Dept. of Energy, the Indiana 21st Century Technology and Development Fund, Praxair, U.S. Steel, Stelco, Union Gas, and all members of the Steel Manufacturing Simulation and Visualization Consortium (SMSVC). In particular, the authors would like to acknowledge Megha Jampani of Linde plc for her contributions to conceptual development. The support of many individual industry collaborators, along with CIVS staff and students, are also appreciated.

Conflicts of Interest: The authors declare no conflict of interest.

References

1. Pistorius, P.C.; Gibson, J.; Jampani, M. Natural Gas Utilization in Blast Furnace Ironmaking: Tuyère Injection, Shaft Injection and Prereduction. In *Applications of Process Engineering Principles in Materials Processing, Energy and Environmental Technologies*; Springer: Cham, Switzerland, 2017; pp. 283–292.

2. Peacey, J.G.; Davenport, W.G. *The Iron Blast Furnace: Theory and Practice*; Pergamon Press: Oxford, UK, 1979; pp. 44–57.

3. Geerdes, M.; Chigneau, R.; Kurunov, I.; Lingiardi, O.; Ricketts, J. *Modern Blast Furnace Ironmaking: An Introduction*, 2nd ed.; IOS Press: Amsterdam, The Netherlands, 2015; pp. 59–75.

4. Feshchenko, S.A.; Pleshkov, V.I.; Lizunov, B.N.; Lapshin, A.A.; Soveiko, K.N.; Loginov, V.N.; Vasil'ev, L.E. Making Blast-Furnace Smelting More Efficient through the Injection of Heated Natural Gas. *Metallurgist* **2007**, *51*, 605–611. [CrossRef]

5. Feshchenko, S.A.; Pleshkov, V.I.; Loginov, V.N.; Kurunov, I.F. Synergetic Effect of Natural Gas Pre-heating Prior to its Injection into a Blast Furnace. In Proceedings of the AISTech 2008, Pittsburgh, PA, USA, 5–8 May 2008; p. 6.

6. Zhen, M.; Zhou, Z.; Yu, A.B.; Shen, Y. CFD-DEM Simulation of Raceway Formation in an Ironmaking Blast Furnace. *Powder Technol.* **2017**, *314*, 542–549.

7. Yeh, C.P.; Du, S.W.; Tsai, C.H.; Yang, R.J. Numerical Analysis of Flow and Combustion Behavior in Tuyere and Raceway of Blast Furnace Fueled with Pulverized Coal and Recycled Top Gas. *Energy* **2012**, *42*, 233–240. [CrossRef]

8. Babich, A.; Senk, D.; Gudenau, H.W. An Outline of the Process. In *Ironmaking*; Verlag Stahleisen GmbH: Dusseldorf, Germany, 2016; pp. 180–185.

9. Vuokila, A.; Mattila, O.; Keiski, R.L.; Muurinen, E. CFD Study on the Heavy Oil Lance Positioning in the Blast Furnace Tuyere to Improve Combustion. *ISIJ Int.* **2017**, *57*, 1911–1920. [CrossRef]

10. Austin, P.R.; Nogami, H.; Yagi, J. A Mathematical Model of Four Phase Motion and Heat Transfer in the Blast Furnace. *ISIJ Int.* **1997**, *37*, 458–467. [CrossRef]

11. Natsui, S.; Ueda, S.; Nogami, H.; Kano, J.; Inoue, R.; Ariyama, T. Analysis of Non-Uniform Gas Flow in Blast Furnace Based on DEM-CFD Combined Model. *Steel Res. Int.* **2011**, *82*, 964–971. [CrossRef]

12. Kon, T.; Natsui, S.; Matsuhashi, S.; Ueda, S.; Inoue, R.; Ariyama, T. Influence of Cohesive Zone Thickness on Gas Flow in Blast Furnace Analyzed by DEM-CFD Model Considering Low Coke Operation. *Steel Res. Int.* **2013**, *84*, 1146–1156. [CrossRef]

13. Majeski, A.; Runstedtler, A.; D'Alessio, J.; Macfadyen, N. Injection of Pulverized Coal and Natural Gas into Blast Furnaces for Iron-making: Lance Positioning and Design. *ISIJ Int.* **2015**, *55*, 1377–1383. [CrossRef]

14. Dong, X.F.; Yu, A.B.; Chew, S.J.; Zulli, P. Modeling of Blast Furnace with Layered Cohesive Zone. *Metall. Mater. Trans. B* **2010**, *41*, 330–349. [CrossRef]

15. Fu, D.; Zheng, D.; Zhou, C.Q.; D'Alessio, J.; Ferron, K.J.; Zhao, Y. Parametric Studies on PCI Performances. In Proceedings of the ASME/JSME 2011 8th Thermal Engineering Joint Conference, Paper No. AJTEC2011-44608, Honolulu, HI, USA, 13 March 2011.

16. Chen, Y.; Fu, D.; Zhou, C.Q. Numerical Simulation of the Co-Injection of Natural Gas and Pulverized Coal in Blast Furnace. In Proceedings of the AISTech 2013, Pittsburgh, PA, USA, 6–9 May 2013; pp. 573–580.

17. Silaen, A.K.; Okosun, T.; Chen, Y.; Wu, B.; Zhao, J.; Zhao, Y.; D'Alessio, J.; Capo, J.; Zhou, C.Q. Investigation of High Rate Natural Gas Injection through Various Lance Designs in a Blast Furnace. In Proceedings of the AISTech 2015, Cleveland, OH, USA, 4–7 May 2015; pp. 1536–1549.

18. Okosun, T.; Street, S.; Chen, Y.; Zhao, J.; Wu, B.; Zhou, C.Q. Investigation of Co-Injection of Natural Gas and Pulverized Coal in a Blast Furnace. In Proceedings of the AISTech 2015, Cleveland, OH, USA, 4–7 May 2015; pp. 1581–1594.

19. Okosun, T.; Street, S.J.; Zhao, J.; Wu, B.; Zhou, C.Q. Investigation of Dual Lance Designs for Pulverized Coal and Natural Gas Co-Injection. In Proceedings of the AISTech 2016, Pittsburgh, PA, USA, 16–19 May 2016; pp. 581–594.

20. Okosun, T.; Street, S.J.; Zhao, J.; Wu, B.; Zhou, C.Q. Influence of Conveyance Methods for Pulverized Coal Injection in a Blast Furnace. *Ironmak. Steelmak.* **2017**, *44*, 513–525. [CrossRef]

21. Okosun, T.; Liu, X.; Silaen, A.K.; Barker, D.; Dybzinksi, D.P.; Zhou, C.Q. Effects of Blast Furnace Auxiliary Fuel Injection Conditions and Design Parameters on Combustion Characteristics and Injection Lance Wear. In Proceedings of the AISTech 2017, Nashville, TN, USA, 8–11 May 2017; p. 11.

22. Okosun, T.; Nielson, S.; D'Alessio, J.; Klaas, M.; Street, S.J.; Zhou, C.Q. Investigation of High-Rate and Pre-heated Natural Gas Injection in the Blast Furnace. In Proceedings of the AISTech 2019, Pittsburgh, PA, USA, 6–9 May 2019; p. 15.

23. Okosun, T.; Silaen, A.K.; Zhou, C.Q. Review on Computational Modeling and Visualization of the Ironmaking Blast Furnace at Purdue University Northwest. *Steel Res. Int.* **2019**, *90*, 1900046. [CrossRef]

24. Fu, D. Numerical Simulation of Ironmaking Blast Furnace Shaft. Ph.D. Thesis, Purdue University, West Lafayette, IN, USA, May 2014.

25. Okosun, T. Numerical Simulation of Combustion in the Ironmaking Blast Furnace Raceway. Ph.D. Thesis, Purdue University, West Lafayette, IN, USA, May 2018.
26. Launder, B.; Spalding, D. *Lectures in Mathematical Models of Turbulence*; Academic Press: New York, NY, USA, 1972.
27. Gu, M.; Zhang, M.; Selvarasu, N.K.C.; Zhao, Y.; Zhou, C.Q. Numerical Analysis of Pulverized Coal Combustion inside Tuyere and Raceway. *Steel Res. Int.* **2008**, *79*, 17–24. [CrossRef]
28. Gu, M.; Chen, G.; Zhang, M.; Huang, D.; Chaubal, P.; Zhou, C.Q. Three-dimensional Simulation of the Pulverized Coal Combustion inside Blast Furnace Tuyere. *Appl. Math. Model.* **2010**, *34*, 3536–3546. [CrossRef]
29. Huang, D.; Tian, F.; Chen, N.; Zhou, C.Q. A Comprehensive Simulation of the Raceway Formation and Combustions. In Proceedings of the AISTech 2009, St. Louis, MO, USA, 4–7 May 2009.
30. Zhou, L. *Combustion Theory and Chemical Fluid Dynamics*; Science Press: Beijing, China, 1986.
31. Peters, N. Premixed Burning in Diffusion Flames. *Int. J. Heat Mass Transf.* **1979**, *22*, 691–703. [CrossRef]
32. Fu, D.; Huang, F.; Tian, F.; Zhou, C.Q. Burden Descending and Redistribution in a Blast Furnace. In Proceedings of the AISTech 2010, Pittsburgh, PA, USA, 3–6 May 2010.
33. Zhou, C.Q. Minimization of Blast Furnace Fuel Rate by Optimizing Burden and Gas Distribution. In *Final Technical Report to U.S. Department of Energy (DOE)*; Purdue University: West Lafayette, IN, USA, 2012.
34. Fu, D.; Chen, Y.; Rahman, M.d.T.; Zhou, C.Q. Prediction of the Cohesive Zone in a Blast Furnace. In Proceedings of the AISTech 2011, Indianapolis, IN, USA, 2–5 May 2011.

Article

Numerical Analysis on Velocity and Temperature of the Fluid in a Blast Furnace Main Trough

Yao Ge [1], Meng Li [1], Han Wei [1], Dong Liang [2], Xuebin Wang [2] and Yaowei Yu [1,*]

[1] State Key Laboratory of Advanced Special Steel, Shanghai Key Laboratory of Advanced Ferrometallurgy, School of Materials Science and Engineering, Shanghai University, Shanghai 102100, China; ge_geyao@hotmail.com (Y.G.); L_limeng@163.com (M.L.); weihan@shu.edu.cn (H.W.)

[2] LaiSteel Research and Technology Center, LaiSteel, Jinan 250000, China; ironlaoliang@163.com (D.L.); erli2000@126.com (X.W.)

* Correspondence: yaowei.yu@hotmail.com

Received: 29 December 2019; Accepted: 18 February 2020; Published: 22 February 2020

Abstract: The main trough is a part of the blast furnace process for hot metal and molten slag transportation from the tap hole to the torpedo, and mechanical erosion of the trough is an important reason for a short life of a campaign. This article employed OpenFoam code to numerically study and analyze velocity, temperature and wall shear stress of the fluids in the main trough during a full tapping process. In the code, a three-dimensional transient mass, momentum and energy conservation equations, including the standard k-ε turbulence model, were developed for the fluid in the trough. Temperature distribution in refractory is solved by the Fourier equation through conjugate heat transfer with the fluid in the trough. Change velocities of the fluid during the full tapping process are exactly described by a parabolic equation. The investigation results show that there are strong turbulences at the area of hot metal's falling position and the turbulences have influence on velocity, temperature and wall shear stress of the fluid. With the increase of the angle of the tap hole, the wall shear stress increases. Mechanical erosion of the trough has the smallest value and the campaign of the main trough is estimated to expand over 5 days at the tap hole angle of 7°.

Keywords: main trough; transient fluid of hot metal and molten slag; wall shear stress; conjugate heat transfer; refractory

---

1. Introduction

The main trough of the blast furnace is a drainage channel for molten iron and slag. In tapping of a 3000 m^3 blast furnace, 4 to 7 tons per minute of molten slag and hot metal with 1773 K flows into the main trough from a tap hole. Tapping time changes from 70 to 120 min and tapping number is around 15 every day [1]. Then, with gravity force, molten slag moves to a skimmer on the top of hot metal and is separated into a slag trough by the skimmer. Due to a harsh working environment, the main trough of the 3000 m^3 blast furnace has 9 to 10 campaigns per year and 45 tons of casting material (with a price of \$857 per ton) is needed for every campaign. Each campaign runs for about 35 days and needs five minor maintenances. Each minor maintenance consumes 3 tons of ramming material (with a price of \$823 per ton). The cost of the blast furnace main trough is around \$0.52 million per year excluding manpower, time and environmental cost [2]. The maintenance cost of the main trough is very expensive. Therefore, the internal state of the main trough should be known, and the erosion mechanism of refractory materials must be understood by the operators and the managers of the blast furnace. Erosional factors of the blast furnace trough include [3]: (1) Mechanical (physical) erosion of fluid flows of molten slag and hot metal, (2) chemical reaction erosion between refractory and the fluid, and (3) thermal stress erosion of intermittent tapping. The main one is the first (mechanical erosion), which is proven by the fact that the erosional extent of an iron storage trough is quite little in the new generation of huge blast furnaces.

In order to reduce the erosion of the main trough, scientists and engineers have done a lot of works to understand the inner situation of the main trough. There are two approaches to study the mechanical erosion. One is the hydraulic model experiment with a tracer. Locations and extent of the physical erosion are predicted through analyzing the range and the depth of the tracer color [4]. The other is a numerical method of Computational Fluid Dynamics (CFD) to analyze fluid properties, such as velocity, temperature, pressure drop, viscosity and thermal stress. The hydraulic model experiment has inherent defects, such as high-cost, high-labor and limit-specific results of experiments. Therefore, many scientists choose the numerical method to investigate their work.

Luo et al. [5] applied Ansys commercial software (Fluent) to analyze velocity distribution of molten slag and hot metal in a main trough. The results show that the fluid's velocities in the center of the trough are faster than ones near the wall and depend on the shape of the trough. Dash et al. [6] studied the fluid and turbulent kinetic energy in a main trough by the numerical analysis and investigated the effect of the slope of the main trough on the velocity distribution. Luomala et al. [7] used CFD and a 1/4 scaled-down hydraulic model with a laser Doppler velocimeter to study the properties of fluid in the main trough and the effect of the dam height on velocity distribution. Duan et al. [3] calculated the temperature distribution of a main trough using a three-dimensional (3D) model considering natural convection and forced convection and proposed that a new main trough be designed based on the gradient arrangement of the bricks. Wang et al. [8] combined the turbulent model and the volume of fraction (VOF) to develop a 3D fluid model of a main trough and studied the effects of the tap hole stream velocity and the trough geometry on the fluid flow. Chang et al. [9] used a momentum conservation equation and VOF to analyze a main trough flow velocity and wall shear stress, and proposed a method to reduce the refractory wear of the blast furnace. The above literature only concentrates on the flow properties (velocity, pressure, viscosity and so on), temperature distribution in the trough and studies the influence of the trough structure on the fluid. However, the effect of the hot metal trajectory leaving the tap hole on the velocity and the temperature of a trough and the refractory erosion during tapping are not reported. Therefore, this investigation will focus on the effect of the hot metal trajectory.

In this paper, OpenFoam is used to solve the transient Navier–Stocks equations including the mass, momentum and energy conservation equations. In Section 2, the solved issue will be addressed. Then, a mathematical model, boundary conditions and solution of the mathematical model are presented in detail. In Section 2, calculation results are presented and discussed. For example, the velocity, the temperature and the wall shear stress of the main trough are analyzed at different tapping moments. Furthermore, the shear stress under different tap hole angles is analyzed and temperature in the refractory is studied by conjugate heat transfer between the refractory and the fluid. In the last section, the conclusions from the work are summarized.

2. Problem Formulation

2.1. Physical Model

According to the shape of a blast furnace trough from a steel plant in China, a physical model is established in Figure 1. During tapping, molten slag and hot metal are regarded as a mixed continuous and incompressible fluid flowing out from the tap hole, and then fall down into the main trough. The mixture fluid keeps a constant level in the main trough and around 300 mm from the upper surface in the calculation. It is separated by the skimmer, and then hot metal flows into a torpedo. Falling position of the mixture fluid trajectory (FPMFT) in the trough defines the inlet of the model. It moves from 4 m away from the origin of the coordinates in the beginning to the tap hole direction in the tapping process. According to References [5,10], Table 1 lists the physical properties of the mixture fluid and the refractory in the study. Chemical reaction between the refractory and the mixture fluid is neglected in simulations. The diameter and the angle of the tap hole is 60 mm and 10 degrees in the simulations, respectively.

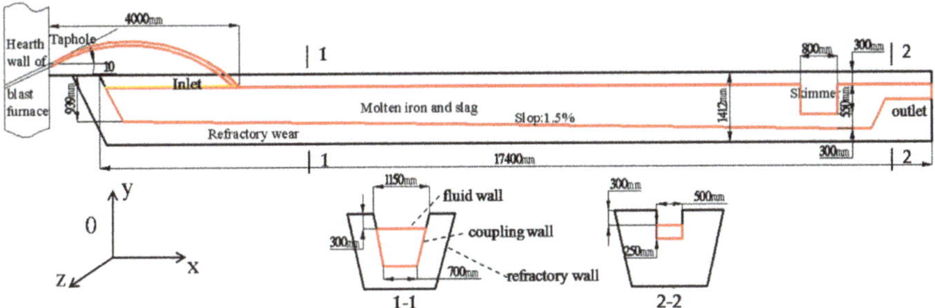

Figure 1. Schematic diagram of the main trough in front view: 1-1 cross-section view at the main trough; 2-2 cross-section view at the outlet.

Table 1. Physical properties of the mixture and the refractory.

Property	Value
Temperature of the inlet (K)	1773
Temperature of the main trough (K)	1583
Temperature of the refractory (K)	1273
Thermal conductivity of the fluid (kg·m·s^{-3}·K^{-1})	33
Thermal conductivity of the refractory (kg·m·s^{-3}·K^{-1})	0.16
Density of iron (kg·m^{-3})	6900
Density of slag (kg·m^{-3})	2600
Viscosity of iron (kg·m^{-1}·s^{-1})	0.0045
Viscosity of slag (kg·m^{-1}·s^{-1})	0.25

Where temperature and viscosity are given in constant values for calculation speed, and the effect of them on the simulation will be focused on later.

2.2. Mathematical Model

2.2.1. Mathematical Model of Molten Slag and Hot Metal

The governing equations of the mixture fluid include a mass conservation equation, a momentum equation based on Reynolds-averaged one, and an energy conservation equation. The fluid in the study was an incompressible Newtonian fluid and its volume expansion ratio $\frac{\partial u_i}{\partial x_i}$ is zero. In order to maintain the conservation of the mixture, the mass conservation equation must be met [11], as:

$$\frac{\partial u_i}{\partial x_i} = 0, \tag{1}$$

where, x_i and u_i express the coordinates of space points (m) and the velocity component at point x_i of the time t coordinate (m·s^{-1}), respectively. x_1, x_2 and x_3 define the three directions of x, y and z, respectively.

The viscous stress tensor P and the deformation rate tensor S of Newtonian fluid have a linear and isotropic function relationship [12]. The Newtonian fluid constitutive equation is substituted into the dynamic equation to obtain the momentum conservation equation of the incompressible Newtonian fluid [11], as:

$$\frac{\partial u_i}{\partial t} + u_j \frac{\partial u_i}{\partial x_j} = -\frac{1}{\rho} \frac{\partial p}{\partial x_i} + v \frac{\partial}{\partial x_j}\left(\frac{\partial u_i}{\partial x_j}\right), \tag{2}$$

where, p, ρ, μ and ν express pressure of the fluid (kg·m^{-1}·s^{-2}), viscosity of the fluid (kg·m^{-3}), kinetic viscosity (kg·m^{-1}·s^{-1}) and kinematic viscosity (m^2·s^{-1}), respectively.

In this study, the standard k-ε turbulence model is used in the simulation. The momentum conservation equation is a time average one to obtain the Reynolds-averaged N-S equation [13]:

$$\frac{\partial \langle u_i \rangle}{\partial t} + \langle u_j \rangle \frac{\partial \langle u_i \rangle}{\partial x_j} + \frac{\partial \langle u_i' u_j' \rangle}{\partial x_j} = -\frac{1}{\rho} \frac{\partial \langle p \rangle}{\partial x_i} + \nu \frac{\partial}{\partial x_j}\left(\frac{\partial \langle u_i \rangle}{\partial x_j}\right), \tag{3}$$

where, u_i' and u_j' define pulse values of the velocity (m·s^{-1}), respectively. $\langle u_i \rangle$ is time average, and $u_i = \langle u_i \rangle + u_i'$.

Reynolds stress tensor term, $-\langle u_i' u_j' \rangle$, is added into Equation (3). This makes the equations disable to close and introduces a turbulence model. According to the Boussinesq hypothesis, the expression of Reynolds [13] stress is:

$$-\langle u_i' u_j' \rangle = \nu_t \left[\frac{\partial}{\partial x_j} \langle u_i \rangle + \frac{\partial}{\partial x_i} u_j\right] - \frac{2}{3} \delta_{ij} k, \tag{4}$$

where, δ_{ij} is Kronecker delta and $\delta_{ij} = \begin{cases} 1, & i = j \\ 0, & i \neq j \end{cases}$. k is turbulent energy (m^2·s^{-2}).

Due to high velocity at the inlet, the mixture fluid in the main trough has a high Reynolds number. The standard k–ε turbulence model has a few empirical constants for this condition. In the standard k–ε model, the turbulent energy k and the turbulent dissipation rate ε are associated with the turbulence ν_t, the formula [13] is as follows:

$$\nu_t = C_\mu \frac{k^2}{\varepsilon}, \tag{5}$$

where, C_μ expresses the empirical constant and a value of 0.09 is used in the simulation.

k and ε are solved in an incompressible fluid using the following two equations [13]:

$$\frac{\partial(\rho k)}{\partial t} + \langle u_i \rangle \frac{\partial(\rho k)}{\partial x_i} = \frac{\partial}{\partial x_i}\left[\left(\mu + \frac{\nu_t}{\sigma_k}\right)\frac{\partial k}{\partial x_j}\right] + G_k - \rho\varepsilon, \tag{6}$$

$$\frac{\partial(\rho \varepsilon)}{\partial t} + \langle u_i \rangle \frac{\partial(\rho \varepsilon)}{\partial x_i} = \frac{\partial}{\partial x_i}\left[\left(\mu + \frac{\nu_t}{\sigma_\varepsilon}\right)\frac{\partial \varepsilon}{\partial x_i}\right] + \frac{C_{1\varepsilon}\varepsilon}{k}G_k - C_{2\varepsilon}\rho\frac{\varepsilon^2}{k}, \tag{7}$$

where, $C_{1\varepsilon}$, $C_{2\varepsilon}$, σ_k and σ_ε express 1.44, 1.92, 1.0 and 1.3, respectively. G_k defines the increase in turbulent kinetic energy caused by the average velocity gradient and is calculated as follows:

$$G_k = \nu_t \left(\frac{\partial \langle u_i \rangle}{\partial u_j} + \frac{\partial \langle u_j \rangle}{\partial u_i}\right)\frac{\partial \langle u_i \rangle}{\partial x_j}, \tag{8}$$

The above eight equations jointly solve the velocity and the pressure of the mixture fluid region and the energy conservation equation is expressed [11] by:

$$\frac{\partial T}{\partial t} + u_i\left(\frac{\partial T}{\partial x_i}\right) = \frac{\partial}{\partial x_i}\frac{\lambda}{\rho C_p}\frac{\partial T}{\partial x_i}, \tag{9}$$

where, λ and C_p express the fluid heat transfer coefficient (W·m^{-1}·K^{-1}) and the specific heat capacity of fluid (J·m^{-1}·s^{-1}), respectively.

The energy conservation equation is also a time average one. Equation (9) is added to the Reynolds heat conduction term ($\langle u_i' T' \rangle$) after time-average, and it becomes:

$$\frac{D\langle u_i' T' \rangle}{Dt} = \frac{\partial}{\partial x_j}\left[C_T \frac{k^2}{\varepsilon}\frac{\partial \langle u_i' T' \rangle}{\partial x_j} + a\frac{\partial \langle u_i' T' \rangle}{\partial x_j}\right] - \left(\langle u_i' u_j' \rangle\frac{\partial \langle T \rangle}{\partial x_j} + \langle u_j' T' \rangle\frac{\partial \langle u_i \rangle}{\partial x_j}\right) - C_{T1}\frac{\varepsilon}{k}\langle u_i' T' \rangle - C_{T2}\frac{\partial \langle u_i \rangle}{\partial x_j}\langle u_j' T' \rangle, \tag{10}$$

where, C_T, C_{T1} and C_{T2} are empirical coefficients. The values of them are 0.07, 3.2 and 0.5 in the simulation, respectively.

2.2.2. Mathematical Model of Refractory

Heat transfer between the mixture fluid and the refractory is coupled to each other. For the refractory heat transfer calculations, only the Fourier equation [14] is solved:

$$\frac{\partial \langle T \rangle}{\partial t} = \frac{\partial}{\partial x_i} \frac{\lambda}{\rho C_p} \frac{\partial \langle T \rangle}{\partial x_i}, \tag{11}$$

where, λ defines the solid heat transfer coefficient (W m^{-1}·K^{-1}). C_p expresses the specific heat of the refractory (Al$_2$O$_3$–SiC–C) and is 0.628 kJ·kg^{-1}·K^{-1}.

2.3. Boundary Conditions (cf. Figure 2)

(1) Inlet boundary conditions. Due to the decrease of the pressure in the furnace during tapping, mass flow of the mixture flow from the tap hole decreases and FPMFT moves to the tap hole direction. Therefore, boundary condition at the inlet is velocity type and it can change in the direction and in the magnitude at the same time. A parabolic Equation (12) is used to define the velocity. Its maximum magnitude is 6.635 m·s^{-1} and is estimated from the FPMFT at the beginning of the tapping. Thermal and pressure boundary conditions at the inlet are constants of 1773 K and zero gradient, respectively.

$$y = x \cdot \tan \alpha - \frac{g}{2u_0^2 \cos^2 \alpha} \cdot x^2, \tag{12}$$

where, α is the inclined angle of the tap hole (°), u_0 defines the velocity of the mixture fluid stream (m·s^{-1}) and changes with the time, x and y are the coordinates of FPMFT (m, m) and g expresses the acceleration of gravity (m·s^{-2}).

(2) Outlet boundary conditions. The outlet of the main trough defines pressure type and is expressed as zero. Thermal and velocity boundary conditions are constants of 1583 K and zero gradient, respectively.

(3) Wall boundary conditions. Boundary condition for free surface of the main trough is no slip. The temperature and pressure are constants of 1583 K and zero gradient at the walls, respectively.

(4) Interaction wall boundary conditions (see yellow surface in Figure 2) for the mixture fluid and the refractory. Temperature boundary condition is a conjugate heat transfer. Velocity and pressure are constants of zero and zero gradient at the interaction wall, respectively.

(5) Refractory wall boundary conditions. Since the refractory only needs to solve the Fourier's equation, there is only a temperature boundary condition with a constant of 1273 K.

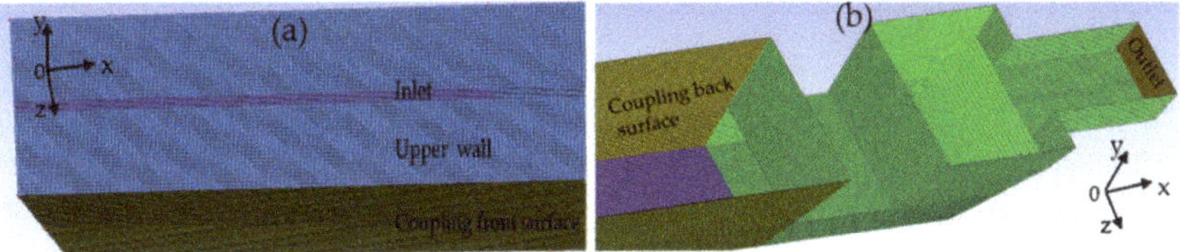

Figure 2. Computational grid: (**a**) the grids at the inlet and the upper wall, (**b**) the grids at the outlet.

2.4. Numerical Procedure

(1) Pre-process. A 3D modeling software is employed to draw a geometry. Hexahedron structure meshes of the geometry are created by the Integrated Computer Engineering and Manufacturing

(ICEM) and are shown in Figure 2. There are 6,433,548 grids, including 1,477,213 for the mixture fluid and 4,956,335 for the refractory.

(2) Solution. The mixture fluid in the main trough is solved by Equations (1), (3) and (9), and the solid in the refractory is solved by Equation (11). For transient simulation, these equations are discretized in time and a time step of 0.001 s is used. In every time step, the simulation is solved by the semi-implicit method for pressure-linked equations (SIMPLE) algorithm as a steady state. Then, the pressure implicit with splitting of operator (PISO) is employed to calculate the transient discretization until the last time step.

(3) Post processing. Paraview and Tecplot software is used to visualize the simulations. Locations of 5 faces and 6 lines in the main trough are shown in Figure 3a, b to analyze the results. A center plane (red) is the central cross-section of the main trough. The distance between the center plane and plane 1 (blue) and one between plane 1 and 2 (orange) are both 0.2 m. Plane 3 (green) defines the back surface of the mixture fluid in the main trough. Plane 4 (deep blue) is perpendicular to other planes with a horizontal distance of 3 m from the origin of the coordinates and also includes some parts in the refractory. Except for line 2, other lines are located on the center plane. Line 1 expresses the intersection of the central plane and the bottom surface of the main trough. The distance between line 1 and 3, one between line 3 and 4, one between line 4 and 5 and one between line 5 and 6 are 0.02, 0.2, 0.2 and 0.2 m, respectively. The intersection of the front surface (cf. Figure 2a) and the bottom surface of the main trough defines line 2.

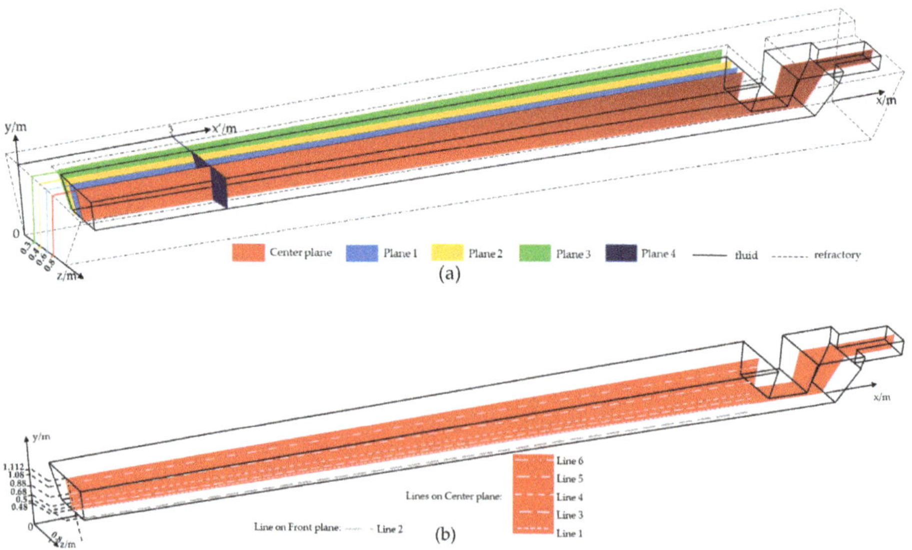

Figure 3. Locations of 5 faces and 6 lines in the main trough for post-processing analysis: (**a**) planes in the full main trough, (**b**) lines in the mixture fluid of the main trough.

3. Results and Discussion

3.1. Residual Errors in Simulation

The residuals include internal and external, two parts in the simulation. When transient state calculations happen in OpenFoam, the number of iterations in each time step is controlled by the internal residuals. The external residual is the difference between the calculated results at adjacent time steps. When the convergence criterion is reached, the calculation moves to the next time simulation. The smaller the residual errors are, the better convergence the calculations have. Figure 4 shows residual error values of pressure, velocity and temperature at the initial moment of calculation. It can be observed that with the increase of the calculation time, the residuals errors gradually decrease. All of them meet the convergence criteria and the calculation results are reliable.

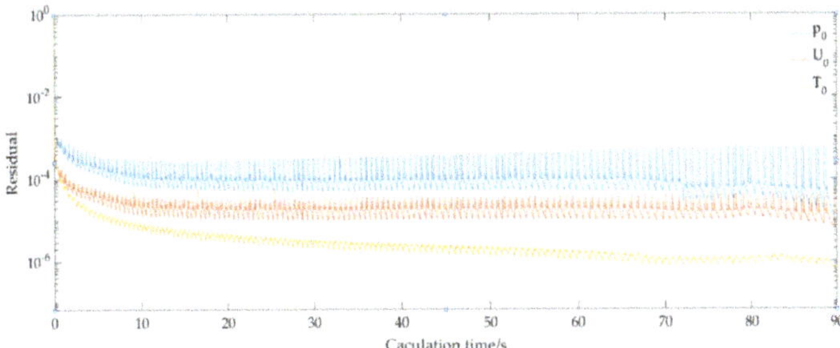

Figure 4. Monitoring residual error curves of pressure, velocity and temperature in the simulation.

3.2. Velocity Distribution of the Mixture Fluid in the Main Trough

Tapping lasts around 90 min in practice and it is not necessary to simulate the full process. Because, except for the beginning of 1 to 2 min and the end of 1 to 2 min, molten slag and hot metal keep a constant mass flow and flow out from the tap holes. Simulation should include the following three steps: the start moment, constant state and the end. Therefore, the total calculation is 90 s and includes the three stages. Furthermore, time step is 0.001 s and the courant number is smaller than 1.

The calculation results at 5, 30, 55 and 80 s are used to analyze the velocity distribution of the mixture fluid at the initial, early intermedia, late intermedia and the end of the tapping. Figure 5 shows the velocity vector of the center plane (cf. Figure 3a) along the main trough. The FPMFT is 2.7 m from the origin of the coordinates in the main trough and downstream of the FPMFT is strongly influenced by the tap hole flow at 5 s. Therefore, a counter clockwise turbulence is observed near 4.5 m, but the impact of the turbulence on the upstream of the FPMFT is quite weak.

The FPMFT moves to the tap hole direction during tapping. The turbulence of FPMFT's downstream is fully developed and its influence range is obviously increased at 30 s. It decreases a lot at 50 s and disappears at 80 s. Comparing four figures in Figure 5, the angle of the mixture fluid at the moving inlet (cf. Equation (12)) changes hugely, and the mixture fluid flow hits the bottom of the main trough. Therefore, more hot metal and molten slag mechanically erode the bottom wall of the refractory.

Figure 5 shows that the erosion near the FPMFT is more serious and the velocity distribution at the FPMFT of 3 m is studied. Figure 6 shows the velocity vector at 3 m from the origin of the coordinates in the main trough at 5, 30, 55 and 80 s. With the increase of the time, the velocity on plane 4 is also significantly reduced. The velocity at the center of plane 4 is large in the period of 2 to 2.6 m/s at 5 s. At 30 s, obvious turbulence is observed to form in the main trough and velocity at the bottom of the main trough is the largest (around 0.6 m/s). Therefore, physical erosion at the bottom of the main trough is more possible and serious. At 55 s, the mixture fluid velocity in the lower side walls of the main trough is larger than other locations. At 80 s, two "donuts" flows appear on the cross-section, but velocity magnitude of the flow is quite small (maximum 0.05 m/s). Therefore, the mixture fluid flow has little effect on the main trough.

Comparing maximum velocities (red arrows) in Figures 5 and 6, velocities of the mixture fluid from the inlet become smaller and smaller during the tapping, which means that mechanical erosion mainly happens in the beginning of the tapping.

In summary, the mixture fluid flow of the main trough is significantly affected by the fluid from the inlet, and a strong turbulence is formed at the downstream of FPMFT and the turbulent area also expands toward the skimmer. The turbulent area near the initial moment of FPHMT exists for a long time. The velocity at the bottom and the lower side walls of the main trough is bigger than others.

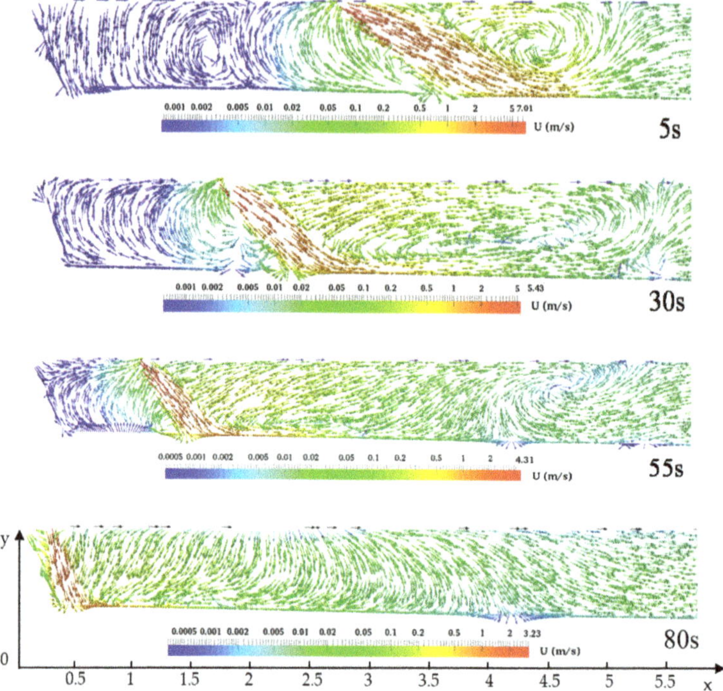

Figure 5. Velocity distribution on the center plane (cf. Figure 3a) at 5, 30, 55 and 80 s.

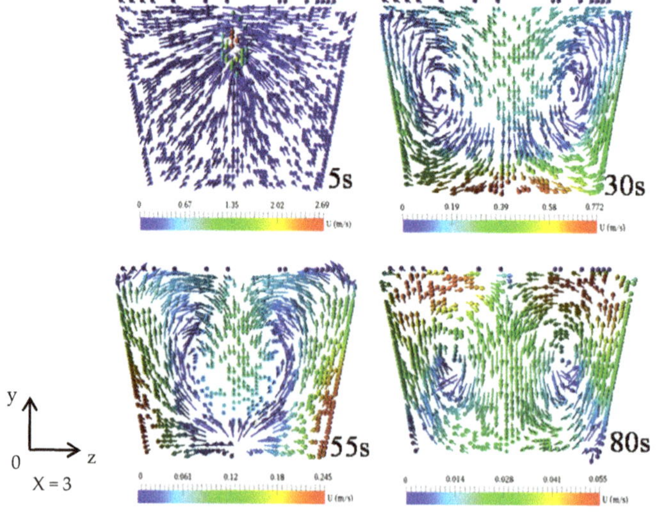

Figure 6. Velocity distribution of the mixture fluid on plane 4 (cf. Figure 3a) at 5, 30, 55 and 80 s.

3.3. Wall Shear Stress

In order to quantitatively study and describe mechanical erosion, wall shear stress is chosen and expressed by:

$$\tau = -\mu \frac{\vec{u}}{\vec{n}},\tag{13}$$

where, $\vec{n}$ and μ indicate the normal vector (m) and the mixed viscosity of molten iron and slag (kg·m^{-1}·s^{-1}), respectively. $\frac{\vec{u}}{\vec{n}}$ is change rate of velocity perpendicular to the direction of the fluid movement.

Figure 7 shows wall shear stress distribution of the refractory on line 1 and line 2 at 5 s. Wall shear stress reaches the maximum at 4 m, and its value at the bottom (line 1) of the main trough is significantly larger than the side wall (line 2). The FPMFT at 2.7 m and the velocity at 4 m is strongly influenced by the mixture fluid flow from the inlet at the moment. Due to effect of the skimmer, the wall shear stress increases a little at 15 m but there is no difference between the side wall and the bottom wall.

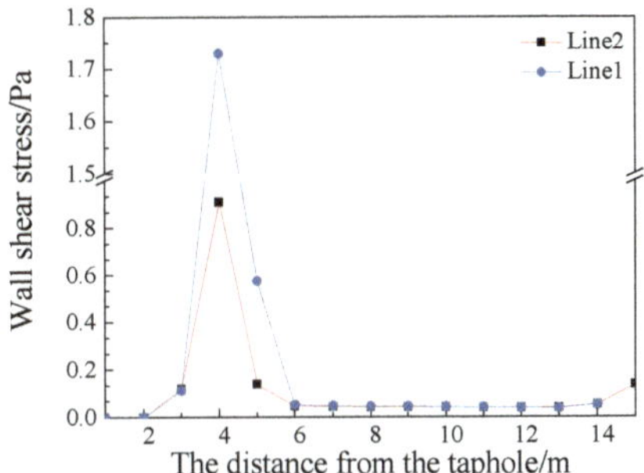

Figure 7. Wall shear stress on the line 1 (blue curve, cf. Figure 3b) and line 2 (red curve, cf. Figure 3b).

3.4. Temperature Distribution of the Main Trough

Figure 8 shows the temperature distribution of the central plane (cf. Figure 3a) at 5, 30, 55 and 80 s. The temperature distribution matches the velocity distribution in the previous section. The downstream of the FPMFT has a significant increase of the temperature due to the mixture fluid from the inlet with high energy, while the temperature of the upstream mixture changes little.

Figure 9 shows the temperature distribution of the mixture fluid on plane 4 (cf. Figure 3a) at 5, 30 and 55 s. At 5 s, the mixture fluid flows from the tap hole and a high temperature is concentrated on the center of the cross-section. At 30 s, the bottom of the main trough forms a higher temperature region than others. The temperature at the others is around 1605 K and is much higher than that at 5 s. At 55 s, the falling position of the mixture fluid is located at 1.2 m and the high-temperature zone is located at the side and the bottom walls of the main trough.

Figure 10 show temperature changes of four lines during tapping ((a) line 3, (b) line 4, (c) line 5 and (d) line 6, cf. Figure 3b). Temperature varies greatly on the main trough direction, especially when the mixture fluid is close to the FPMFT. Due to movement of the FPMFT, temperature varies greatly before 3 m. There is a constant temperature zone of 1610 k from 3 to 6 m. The largest fluctuation happens near the bottom of the mixture fluid (Figure 10a), which means the bottom refractory suffers the highest frequent changes of thermal stress and is highly probable to erode.

Figure 11 shows the temperature distribution of the mixture fluid from the central plane to plane 3 of the main trough (cf. Figure 3a). Temperature near FPMFT is obviously higher than others, and temperature at the downstream of FPMFT is apparently higher than the initial boundary set 1573 K due to the conduction and the convection of heat transfer. Furthermore, the temperature gradually decreases from the central plane trough (z = 0.8) to the plane 3 (z = 0.3). Due to a low thermal conductivity of the refractory and the low initial temperature, the temperature gradient in the refractory is really small. However, the temperature at the interaction boundary wall of the refractory and the mixture fluid is exactly the same as each other.

In summary, the temperature of the mixture fluid increases significantly during tapping. From the center of the mixture fluid to the refractory, the temperature gradually decreases, but the temperature distribution is consistent with the velocity field. In the vertical direction, temperature increases with the increase of the height. In the horizontal direction, the temperature of the mixture fluid near FPMFT changes greatly.

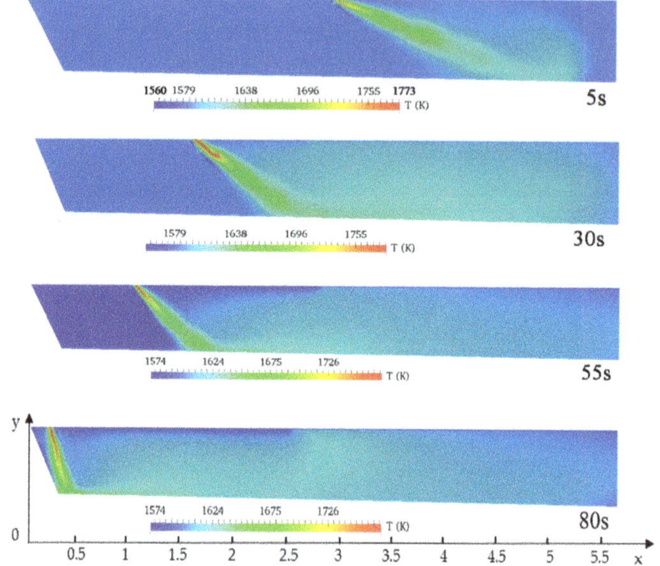

Figure 8. Temperature distribution of the center plane (cf. Figure 3a) at 5, 30, 55 and 80 s.

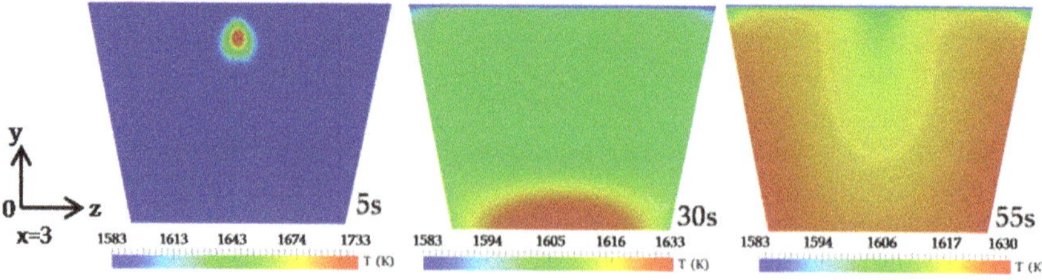

Figure 9. Temperature distribution of the mixture fluid on plane 4 (cf. Figure 3a) at 5, 30 and 55 s.

3.5. Influence of Tap hole Angles on the Flow of the Main Trough

Intermittent tapping of a blast furnace must be punched before the tapping and be plugged after the tapping. The angle of the tap hole can change during tapping. Therefore, it is important to understand the influence of the angle of the tap hole to the temperature, velocity and shear stress of the mixture flow in the main trough. Since velocity and temperature distributions are quite similar in Figures 5–11, they are not analyzed any more.

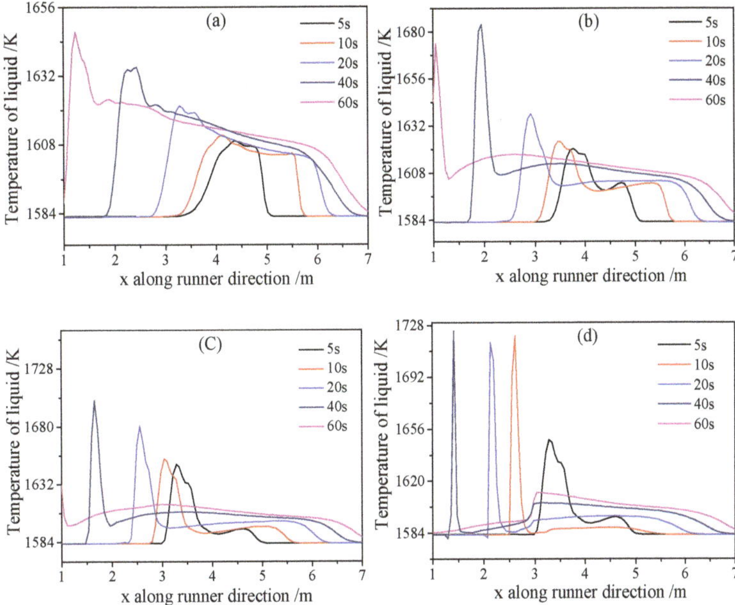

Figure 10. Temperature fluctuations of four lines during tapping ((**a**) line 3, (**b**) line 4, (**c**) line 5 and (**d**) line 6, cf. Figure 3b).

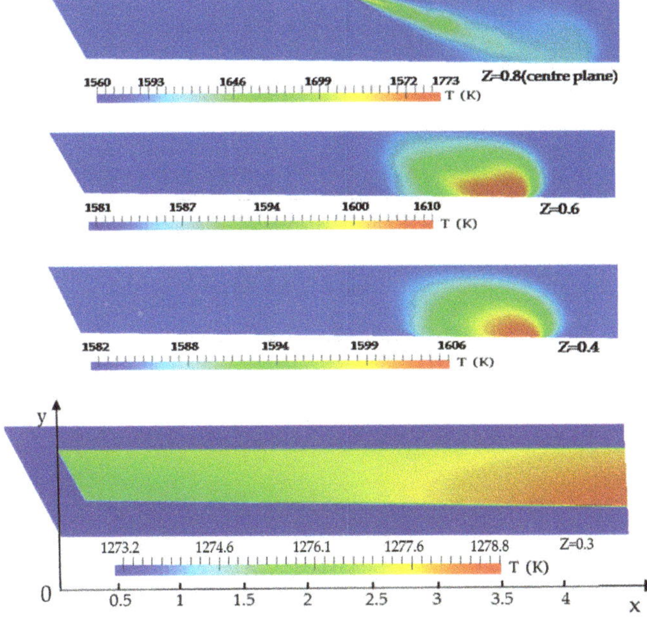

Figure 11. Temperature distributions from the central plane to plane 3 (cf. Figure 3a).

Generally speaking, the angle of the tap hole changes between 7 to 12°. Therefore, two values in the interval were selected to highlight the effect of the angle. As shown in Figure 12, with the increase of the tap hole angle, the wall shear stress increases from 0.73 to 0.87 because molten slag and hot metal gets a higher velocity when they fall into the main trough from a bigger angle of the tap hole. The little increase of wall shear stress happens at 15 m due to the fact that the fluid velocity becomes bigger at the skimmer. When the tap hole angle is 7°, mechanical erosion of the trough has the smallest value. The maximum of wall shear stress is reduced by 16% and the campaign of the main trough is estimated to expand over 5 days, comparing with the tap hole angle of 10° and 7°.

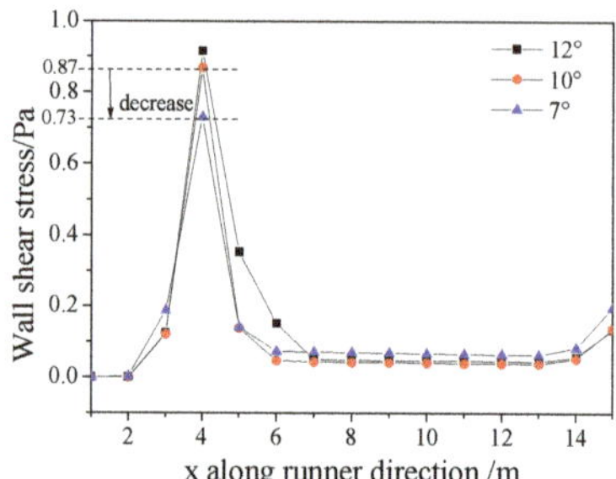

Figure 12. Wall shear stress distribution at different angles of the tap hole on line 2 (cf. Figure 3b).

Figure 13 shows temperature changes of line 3 during tapping with different tap hole angles, (a) 7°, (b) 10° and (c) 12°. With the increase of the angle of the tap hole, change of temperature during tapping is little. But as the location of FPMFT moves toward the direction of the tap hole, the maximum temperature of it increases. During tapping, reasonable adjustment of the tap hole angle can reduce the extent of mechanical erosion in some areas, which is helpful to expand the sieve campaign of the main trough.

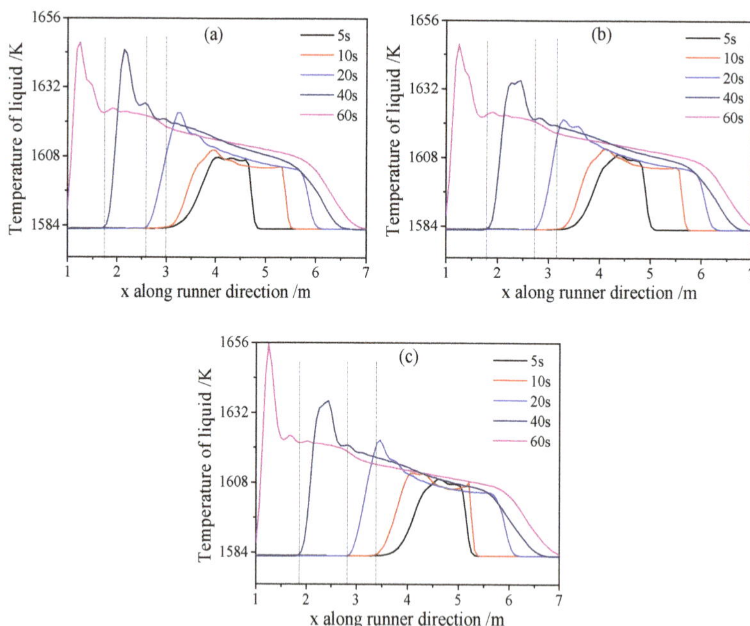

Figure 13. Temperature fluctuations on line 3 (cf. Figure 3b) during tapping with different tap hole angles, (**a**) 7°, (**b**) 10° and (**c**) 12°.

4. Conclusions

In this study, OpenFoam was employed to analyze the velocity, the temperature, the wall shear stress in the main trough of a blast furnace and the influence of different tap hole angles on the main trough. Some conclusions are highlighted as follows.

(1)   Velocity of the mixture fluid at the center of the main trough is generally larger than that at the wall. Velocity of the FPMFT at the downstream is larger than that at the upstream. However, due to strong turbulence at the downstream, the lower front wall also has larger velocity.

(2)   Maximum of shear stress occurs at the downstream of the FPMFT, and the shear stress on the bottom wall is bigger than that on the front wall.

(3)   Due to velocity increase of the FPMFT, the shear stress on the wall of the main trough increases with the increase of the tap hole angle. When the tap hole angle is 7°, mechanical erosion of the trough has the smallest value and the campaign of the trough expands over 5 days.

Author Contributions: Y.G. is responsible for writing the article and simulation calculation of blast furnace main trough. M.L. is responsible for the guidance of code writing in simulation. H.W. is responsible for the language and format of the article. D.L. and X.W. provide blast furnace main trough data involved in the simulation process. Y.Y. guide the overall direction of the article, provide simulation technical guidance and simulation equipment. All authors have read and agreed to the published version of the manuscript.

Funding: We gratefully acknowledge financial support from the Program for Professor of Special Appointment (Eastern Scholar) at Shanghai Institutions of Higher Learning (No.TP2015039), National Natural Science Foundation of China (No.51974182), National 111 project, Grant/Award No. 17002 and Laiwu Steel. The simulations and analyses were completed using OpenFoam open source.

Conflicts of Interest: The authors declare no conflict of interest.

References

1.   Lv, C.Y. Research on Al2O3-SiC-C Castables for Blast Furnace Trough Strengthened by Sialon In-situ Synthesized. Master's Thesis, Wuhan University of Technology, Wuhan, China, 10 May 2004.

2.   Yu, H.L. Application of high-strength hot-feeding in main trough of Shougang blast furnace. In Proceedings of the Collection of China Iron Society's 2004 National Ironmaking Technology and Annual Ironmaking Conference, Chinese Metal Society, Beijing, China, 15 June 2004.

3.   Duan, G.J.; Li, X.J. Study on Main Iron Runner Structyre and Its Cooling Mode. *Ironmaking* **2011**, *30*, 13.

4.   Sun, Y.C. Study of Hydraulic Model Experiment on Flowing of Hot Metal in Iron Runner. Master's Thesis, Laoning institute of Science and Technology, Anshan, China, 2006.

5.   Luo, X.G.; Zuo, H.B. Numerical simulation of the flow field in the blast furnace taphole. *J. Wuhan Univ. Sci. Technol. (Nat. Sci. Ed.)* **2014**, *37*, 5.

6.   Dash, S.K.; Ajmani, S.K. A fluid dynamic analysis of the blast furnace trough at Tata Steel. *Proc. Camme* **1996**, *10*, 26–36.

7.   Luomala, M.J.; Paananen, T.T. Modelling of fluid flows in the blast furnace trough. *Steel Res. Int.* **2001**, *72*, 6. [CrossRef]

8.   Wang, L.; Pan, C.N. Numerical Analysis on Flow Behavior of Molten Iron and Slag in Main Trough of Blast Furnace during Tapping Process. *Adv. Numer. Anal.* **2017**, *2017*, 11. [CrossRef]

9.   Chang, C.M.; Lin, Y.S. Numerical Analysis on the Refractory Wear of the Blast Furnace Main Trough. *Adv. Sci. Technol.* **2014**, *92*, 294–300. [CrossRef]

10.   Jimbomand, I.; Cramb, A.W. The Density of Liquid Iron-carbon Alloys. *Metall. Mater. Trans. B* **1993**, *24*, 5–10.

11.   Zhang, Z.X.; Cui, G.X. *Fluid mechanics*, 3rd ed.; Shi, L., Zhao, C.M., Eds.; Tsinghua University Press: Beijing, China, 2004; pp. 287–289.

12.   Zhang, J.J.; Yan, D.F. Approximate Calculation of Shear Rate of Turbulent Flow of Newtonian Fluid in a Circular Tube. *J. Univ. Pet. (Sci. Technol. Sect.)* **2002**, *4*, 78–80.

13. Yabushita, K.; Yamashita, M. An analytic solution of projectile motion with the quadratic resistance law using the homotopy analysis method. *Phys. A Math. Theor.* **2007**, *40*, 8403–8416. [CrossRef]
14. Wang, J.W. Numerical Prediction of Dangerous Regions in A 3D Turbine Blade with Multilayer-Structure TBCs by A Fluid-Solid Coupling Method. Master's Thesis, Xiangtan University, Xiangtan, China, 2016.

Article

Principal Component Analysis of Blast Furnace Drainage Patterns

Mauricio Roche *, Mikko Helle and Henrik Saxén

Thermal and Flow Engineering Lab, Åbo Akademi University, 20500 Turku, Finland
* Correspondence: mroche@abo.fi

Received: 4 July 2019; Accepted: 5 August 2019; Published: 7 August 2019

Abstract: Monitoring and control of the blast furnace hearth is critical to achieve the required production levels and adequate process operation, as well as to extend the campaign length. Because of the complexity of the draining, the outflows of iron and slag may progress in different ways during tapping in large blast furnaces. To categorize the hearth draining behavior, principal component analysis (PCA) was applied to two extensive sets of process data from an operating blast furnace with three tapholes in order to develop an interpretation of the outflow patterns. Representing the complex outflow patterns in low dimensions made it possible to study and illustrate the time evolution of the drainage, as well as to detect similarities and differences in the performance of the tapholes. The model was used to explain the observations of other variables and factors that are known to be affected by, or affect, the state of the hearth, such as stoppages, liquid levels, and tap duration.

Keywords: hearth; drainage; PCA; analysis tool; pattern; tapholes

1. Introduction

The blast furnace represents the most relevant process on the main route for ore-based production of iron in the steelmaking industry. An important part of the process is its lowest region, better known as the hearth, where molten iron and slag are collected before they are drained ("tapped") out through the tapholes. In large-size furnaces, this drainage procedure is carried out by opening alternating tapholes from where both liquid phases are tapped out simultaneously. Different tapping rates may occur, partly because of differences in the production or internal conditions in the hearth, leading to variation of the levels of the liquid phases. The outflow rates are anticipated to be lower in the beginning of the tapping, increasing as the drainage proceeds owing to erosion of the taphole. The erosion may be further induced by the arrival of newly produced hotter liquids that have not experienced heat loss to the hearth lining. The goal is to drain sufficient amounts according to observations and estimations to keep the liquid levels low in the furnace [1,2].

In large-scale operation, productivity and efficient operation of the casthouse might be jeopardized by the variation of operating factors. Some of the most common concerns are liquid-level fluctuations caused by production/extraction imbalance, and erosion of the hearth lining. Because of the complexity of the process and besides the lack of direct measurements, the understanding and control of the tapping procedure represent both difficult and essential tasks in order to ensure an efficient operation.

Earlier studies of alternating tapping practice presented differences in taphole length, amounts of liquids extracted, and hot metal temperature between the operating tapholes [1,3,4]. Even though some correlations among production variables and erosion of both lining and taphole were established, little is known about what induces such variations and how. Computational fluid dynamics (CFD) simulations were undertaken with the goal to clarify the relation between production variables, implicitly or explicitly suggesting that there may be considerable spatial distribution of the variables or differences depending on local conditions at the tapholes [5–8].

Many previous efforts have been made to shed light on the hearth fluid dynamics and the phenomena that occur during and between the taps. In contrast to earlier notions, the fact that iron can be drained to levels below the taphole was clarified in the 1980s by Tanzil et al. [9], who made small-scale experiments supported by computational analysis. The investigators showed that the flow of the viscous slag phase leads to a lower pressure in front of the taphole, which makes it possible to elevate iron from levels well below the taphole, thus leading to simultaneous outflow of the two liquid phases. Nouchi et al. [3] presented experimental work that associated slag ratio (mass ratio of slag to iron) with liquid levels and tap duration, as well as how the time corresponding to a maximum of the slag ratio shifted with changes in the conditions. Furthermore, the results showed that at constant conditions, the maximum slag ratio increases if low-permeability zones are present in the hearth. On the other hand, Nishioka et al. [10] pointed out that only taphole mud quality or erosion of the operating taphole could not explain the duration of the tapping and the fluctuations in the drainage rate. Even though some causal factors have been identified or suggested, little is reported about the occurrence or frequency of such fluctuations. Iida et al. [2] reached similar conclusions to Nishioka et al. [10], also suggesting that coke particle size and configuration in front of the taphole play important roles for the liquid outflow rates. These results indicate that local conditions at the taphole might cause such imbalance. Furthermore, in later efforts [1], the authors developed a mathematical model that led them to conclude that discrepancies between operating tapholes in terms of liquid drainage rate and tap duration were the result of local zones of low permeability. According to their model, the metal fraction depended on the vertical level of the iron–slag interface compared with the taphole level.

Several studies in the literature have presented models of hearth drainage [1,2,10–13]. A recent study by the present authors proposed a simple offline model simulating the liquid level fluctuations in a hearth with intermittent tapping [14]. By applying different conditions to each taphole, it was demonstrated that some outflow patterns and slag delays observed in a blast furnace could be mimicked. Thus, the model makes it possible to theoretically evaluate the role of different parameters. Furthermore, an online liquid-level model was also proposed based on similar simplifications and assumptions, but using estimates of the production and outflow rates [4]. The model considers a division of the hearth into pools with individual liquid levels. Its application indicated non-uniform drainage among the tapholes, as well as periods of accumulation and depletion of both iron and slag. However, certain drainage behaviors were observed that could not be explained by the models, so a deeper understanding of the hearth drainage behavior is needed. An analysis of the recurrence of outflow patterns and transition from one to another could provide some understanding of the blast furnace hearth state, even though irregularities frequently seen in the data make it difficult to identify and categorize the observations in an appropriate way. It is thus clear that a data-driven analysis of the liquids outflow patterns could be useful. The present paper addresses this very problem using principal component analysis to compress the information to make it easier to understand and illustrate.

2. Principal Component Analysis

The main purpose of principal component analysis (PCA) is to reduce the dimensionality of multivariate data. The method also provides guidelines on the number of components needed to represent the data in question with sufficient accuracy.

Consider a sample data Y of mean-centered measurements with n observations on q variables [15,16]. First, the linear combination $t_1 = Yp_1$ is found that accounts for the maximum variance subject to $|p_1| = 1$. This represents the first principal component. The second component is a combination defined by $t_2 = Yp_2$ that has the next greatest variance subject to $|p_2| = 1$ with the conditions that it is uncorrelated with, and orthogonal to, the first component t_1. The following components are similarly determined. The sample principal component loading vectors p_i are the eigenvectors of the covariance matrix of Y. PCA decomposes the observation matrix Y as

$$Y = TP^T = \sum_{i=1}^{q} t_i P_i^T. \tag{1}$$

If a considerable part of the variation is represented by the contribution of a smaller number of components, these components can replace the original data.

The present paper aims to identify and classify iron and slag outflow patterns from a reference blast furnace by applying PCA to two data sets. The data representation in lower-dimensional components is presented in the following section in order to identify and visualize possible trends in the drainage patterns. Some interpretations of the results will also be presented with respect to hearth dynamics and tapping procedure.

3. Method and Data Sets

The data analyzed consists of two data sets, Data set 1 and 2, of three and six months worth of measurements of liquid outflow rates, respectively, from a large blast furnace with three tapholes, TH-1, TH-2, and TH-3. Between the end of Data set 1 and the beginning of Data set 2, 15 months of operation passed. The iron outflow rate was obtained from torpedo weighing, while the slag outflow rate was estimated based on measurements from the granulation unit. Both signals were available as one-minute averages. Considering the recorded times from the individual taps, the liquid outflow rate for each tapping was calculated, resulting in a total of 887 consecutive ones for Data set 1 and 1729 for Data set 2. From here on, individual tapping samples will be referred as taps, while tapping will refer to the drainage process.

3.1. Pre-Processing of the Data Set

After visual inspection of the samples, systematic abrupt changes of the iron outflow rate (>5 tonne/min in a couple of minutes) were noticed, which did not agree with the overall trend. Such deviations were found to be caused by torpedo changes, so they were filtered out of the samples. The outflow samples were subjected to additional pre-processing conditions in order to discard samples that, owing to error in the measurements or records, could potentially corrupt the classification. Among these conditions were the following:

- outflow near zero for any of the phases;
- taps shorter than 40 min that may occur as a result of disturbances, stoppages, or contingencies;
- any of the phases shows an outflow rate above 2% of the production rate at all times (i.e., also during the intercast period, where there should be no liquid outflows); and
- the total outflow abruptly goes near zero as a result of measurement errors or possibly taphole clogging.

After excluding such samples, 776 taps remained in Data set 1 and 1463 taps in Data set 2, accounting for 87% and 86% of the original data sets, respectively. Nonetheless, the excluded samples were taken into account when estimating the slag delay (see below).

It was found to be appropriate to use a threshold for the outflow rate, set at 2% of the production rate, to indicate when the flow of the liquid in question (iron or slag) has started during the tap. This condition was needed as liquids from a tap often progressed in the runner to the torpedo and the granulation units for some minutes after the tapping ended, and so, based on this information, the moments when iron and slag started flowing out were noted, and the difference between these two times is called "slag delay" in what follows. Note that a negative slag delay means that slag starts flowing out first. The slag delay has an important meaning as it reflects the vertical location of the iron–slag interface in the hearth at the moment when the taphole is opened; therefore, it is an important feature to identify with the present method. Figure 1 illustrates two cases where different locations of the iron–slag interface determine which liquid phase will be "seen" first, that is, if the slag delay will be positive or negative. The figure depicts the state of the interfaces just before the tapping ends (solid lines), just after the tap has ended and the interfaces have become horizontal (dotted lines) and the state just before the next tapping starts (dashed lines). In case (a), the right taphole has drained more slag and less iron, so the iron–slag interface is not significantly below the left taphole as the tapping ends, and while the tapholes are plugged, it rises above the left taphole. This will result in a positive slag delay in the following tap from the left taphole. As this taphole will start with iron-only flow, slag keeps on accumulating in the hearth, elevating the slag–gas interface. The duration of the tap is long and the iron–slag interface will have time to descend well below the taphole, eventually leading to the state depicted in the right panel, yielding a negative slag delay for the right taphole, as shown in case (b).

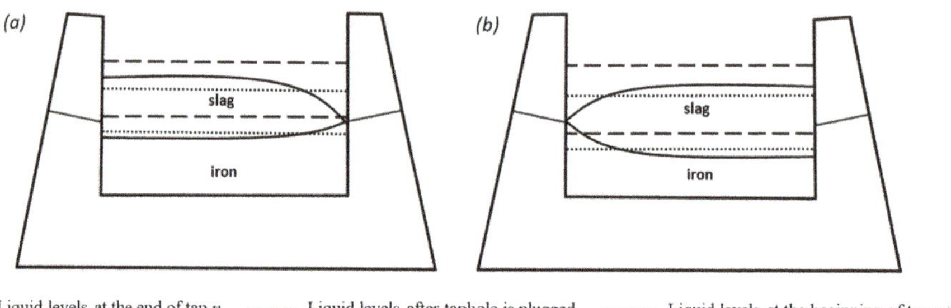

Figure 1. Illustration of two cases of slag delay: (**a**) positive slag delay (for the left taphole), (**b**) negative slag delay (for the right taphole).

PCA is a scale-dependent method. Thus, if the data matrix has variables of considerably bigger magnitude than others, these will have dominant weights in the sub-dimension. Therefore, the outflows were normalized by calculating the slag share, defined as

$$S_{\text{slag}}(i) = \frac{m_{\text{slag}}(i)}{m_{\text{iron}}(i) + m_{\text{slag}}(i)}, \tag{2}$$

where i denotes the time, while $m_{\text{iron}}(i)$ and $m_{\text{slag}}(i)$ are the corresponding mass flow rates of iron and slag from the furnace for the same time. The share of slag was considered an interesting variable, because work by other investigators has demonstrated that the hearth state may affect the distribution of the two liquids in the taps from multi-taphole furnaces [2,3,10]. Furthermore, in order to normalize the tap duration (defined as the time elapsed between the moment when the first liquid starts flowing out till the last liquid stops flowing out of the taphole), S_{slag} was down-sampled to 10 average points, each representing one tenth of the duration of the tap. Thus, all taps were represented by ten values of the slag share, which facilitated a comparison between the results of different taps despite differences in tap duration. This down-sampling was found to yield an acceptable representation of the variable without losing much significant or relevant information. Thus, the vector length of the samples to be categorized was of 10. The method aims to identify outflow patterns within the production levels of the BF; therefore, both slag fraction and tapping duration were required to results pattern-dependent. Figure 2 presents a few examples from Data set 2 that depict some typical liquid outflow patterns, illustrating the phase that is first observed—hence, the start of the tap—and how the tap durations may vary. The top panels of Figure 3 show the corresponding slag shares calculated by Equation (2) for the same four taps. The lower panels of the figure show, using solid squares, the corresponding 10-point normalized curves. As seen from the four examples, the evolution of the share of slag in the outflowing liquid mixture may show quite different characteristics.

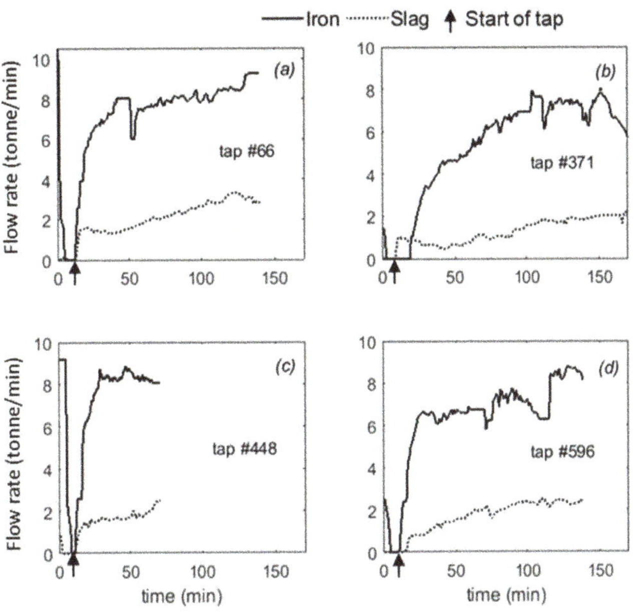

Figure 2. Examples of liquid outflow patterns from the Data set 2: (**a**) zero slag delay and 135 min tap duration, (**b**) negative slag delay and 170 min tap duration, (**c**) zero slag delay and 70 min tap duration, and (**d**) positive slag delay and 140 min tap duration.

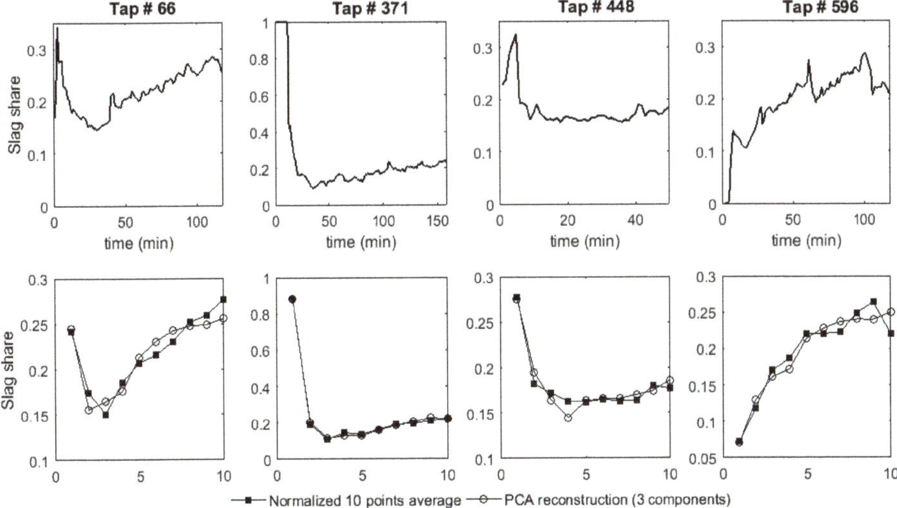

Figure 3. Slag share for the taps presented in Figure 2. Upper panels: one-minute values obtained from the measurements. Lower panels: information discretized into 10 points (solid squares) and principal component analysis (PCA) reconstruction (open circles).

A preliminary study of the normalized outflow patterns showed differences between data sets and tapholes. Figure 4 shows the slag share for the 10 average points or variables of the data sets after preprocessing. Data set 1 is presented in Figure 4a, while Data set 2 is presented for comparison in two equivalent consecutive subsets of 720 and 763 taps in Figure 4b,c, respectively. The median curves for each taphole are depicted, as well as the median for the whole data set or subset and their corresponding 25th and 75th percentiles. The median for TH-1 in Data set 1 is seen to be similar to the 25th percentile of this dataset, while for TH-2, it is similar to the 75th percentile, particularly for variables 1 and 8–10. The median of TH-3 is, on the other hand, comparable to the global median. The median for the first subset of Data set 2 shows higher slag shares for all variables compared with Data Set 1, particularly for variable 1, where the median of TH-1 falls below the global median, while median of TH-2 falls above, and for TH-3 at the same level. Taphole median values are found above and below the global median in different orders for the other nine variables. In the second subset of Data set 2 (cf. Figure 4c), variable 1 shows a decrease from 0.35 to 0.23, with TH-1 and TH-2 medians below the global median and TH-3 above. Similar median curves are observed for variables 2–10 as in the first subset in Figure 4b.

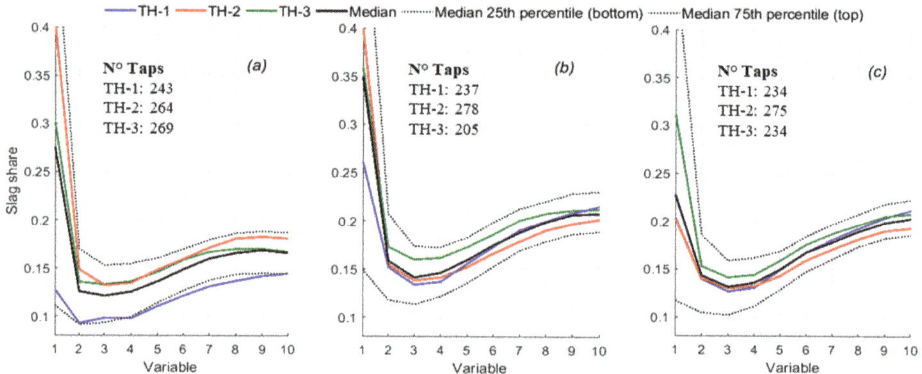

Figure 4. Overall median curves of Data set 1 and 2 and median by taphole (TH), (**a**) Data set 1, (**b**) first half of Data set 2, (**c**) second half of Data set 2.

In summary, interesting variability is observed within and between the data sets, and between the different tapholes. This motivates the effort to identify the outflow patterns that may occur during certain periods and how they persist or develop during the operation of the blast furnace. Next, a systematic PCA-based analysis of the outflow patterns of the two data sets is presented.

3.2. Post-Processing of the Results

Matlab's PCA algorithm was applied separately to the two discretized data sets. After subtracting the mean value corresponding to each variable, the algorithm computes by singular value decomposition the coefficient matrix corresponding to each variable per component. The 10-dimensional vector is represented by 10 components yielding a 10×10 coefficient matrix. The algorithm also computes the scores matrix corresponding to each observation per component, that is, there are as many individual score vectors as samples. On the basis of Equation (1) and adding the mean value of each variable, the observations can be reproduced. In this way, the variability of each component or dimension can be studied. Figure 5 presents the different pattern ranges that the first four principal components represent in the two data sets.

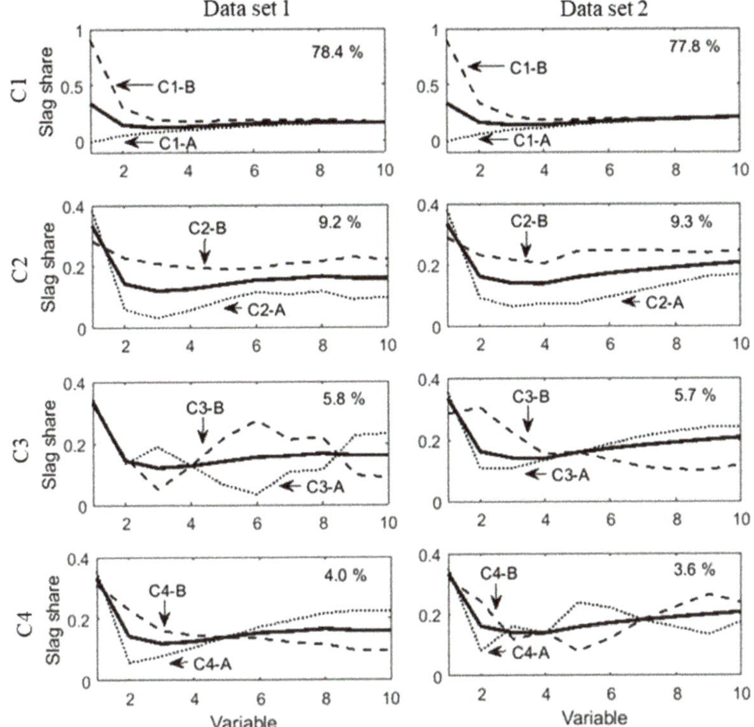

Figure 5. Slag share ranges represented by the principal component (C1 … C4) for Data set 1 (left panels) and Data set 2 (right panels). Solid lines represent the *mean pattern*, dotted lines the lowest values (labelled A), and dashed lines the highest (labelled B). The percentage of variance explained by each component is reported in the right top part of each subpanel.

After studying the results, it can be hypothesized that the first component (C1, top panels) largely reflects the slag delay expressed in terms of slag share in the initial part of the tapping, where zero corresponds to iron-first and one to slag-first drainage. The second component (C2) captures mainly the level of the slag share (excluding variable 1), falling in the range of 0.1 to 0.3 for both data sets (corresponding to a slag ratio of 110–430 kg/t_{hm}, where subscript hm denotes hot metal). The third and fourth components (C3 and C4) represent changes in the share of slag during tapping. Even though the data sets were treated separately, C1 and C2 display a similar meaningful representation of the data in the principal component space. C4 in Data set 1 and C3 in Data set 2 show similar pattern ranges as well. As seen in Figure 4, the outflow patterns of the three tapholes display different trends from variable 2 to 10, and this trend can be captured by the slopes that C4 in Data set 1 and C3 in Data set 2 describe. It can be concluded that these slopes reflect how the slag share develops during the tapping when both phases flow out. Figure 5 also indicates by solid lines the pattern corresponding to the mean values for each data set, which are seen to be very similar. In the principal component space, this pattern corresponds to the interception of the components, which is referred to as the *mean pattern* in what follows. The dotted lines correspond to lowest values in the principal component space (denoted by A) and the dashed lines correspond to highest values (denoted by B).

The algorithm also reports the percentage of the total variance explained by each component. In Data set 1, the first four components explain 78.4%, 9.2%, 5.8%, and 4.0%, respectively, while in Data set 2, they explain 77.8%, 9.3%, 5.7%, and 3.6%, respectively. Considering the meaningful information that the components provide, C1, C2, and C4 were selected for Data set 1, which together explain 91.6%

of the data variation. To provide an equal representation of the data sets and to allow for a comparison, C1, C2, and C3 were selected for Data set 2, together explaining 92.8% of the variation. An accuracy exceeding 90% was considered sufficient to represent the data sets and consequently enough to make a meaningful analysis of the results.

4. Results

The method as a batch algorithm organizes all samples in the component dimensions regardless of the taphole operated. In order to process the results, moving averages of the selected principal components for the latest 20 taps (corresponding to approximately two days of operation) were considered. The average was calculated for each taphole, taking into account the taps that occurred within the latest 20-tap window. Thus, this post-processing criterion yields a quasi-time evolution. It is important to stress that the main objective of the work is to reveal similar or different behavior of the drainage from the different tapholes to gain understanding of the variability seen during operation. The left panel of Figure 6 shows a lay-out of the three tapholes in the furnace. The right top and bottom panels present the periods during which the tapholes were operated in Data set 1 and Data set 2, respectively. Each taphole is typically operated about three weeks for the blast furnace in question. This schedule results in the alternating operation of two tapholes at the time, with some overlap (where all three tapholes are used) during the transition periods. The figure also indicates the (numbered) periods that are analyzed in the following subsections.

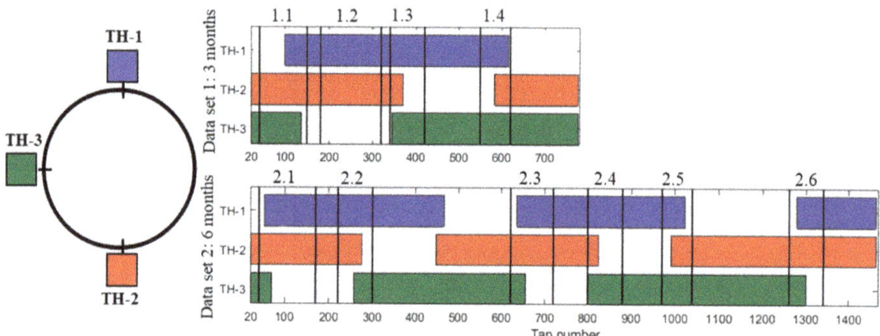

Figure 6. Taphole location in the reference furnace (**left**) and operating periods of the tapholes for the Data set 1 (**right top**) and Data set 2 (**right bottom**). The sub-periods studied in following sections are also indicated.

It should be noted that the representation, because of the data compression and averaging, does not provide detailed information about single days of operation. However, the comparison between tapholes at any given point could indicate agreement or disagreement between the outflow patterns. We will present some snapshots (cf. Figures 7–10) of the motion of the averages of the principal components as time evolves. The corresponding components of the tapholes are depicted by blue line and diamonds (TH-1), red line and squares (TH-2), and green line and circles (TH-3). For the sake of clarity, the two most important components (C1 and C2) are plotted versus each other in a "phase plot" for shorter sub-periods (cf. Figures 7 and 9).

The positions of the markers on each curve indicate the average value of the components corresponding to each tapholes for the last 20 observations. For instance, a blue diamond labeled "70" indicates the average location in the principal component space of taps from TH-1 from taps 50 to 70 regardless of taphole. As the tapholes are operated alternatively, the same number label will appear next to the markers on the curves of the operating tapholes, indicating the pattern progression. For clarity, arrows show the direction of the evolution, and labels "start" and "end" are used to indicate the first and last points of operation of a taphole if it occurs in the period. On the axes of each component, the limits of the component (lower by A, upper by B, cf. Figure 5) are also indicated. In addition to the phase plot of C1 and C2, similar averages of an additional principal component, C4 for Data set 1 (cf. Figure 8) and C3 for Data set 2 (cf. Figure 10), are plotted in a separate panel for each data set along with the average tap duration of the operating tapholes.

4.1. Data Set 1

Figure 7 presents C1 versus C2 for Data set 1, while Figure 8 depicts the evolution of C4 and the tap duration. Four periods are studied from Data set 1, referred to as Periods 1.1–1.4 (cf. Figure 6). They range from 7 to 16 days of operations and include segments of operation where one taphole is replaced by another, as well as segments where a taphole pair operates.

4.1.1. Period 1.1

Figure 7a shows the evolution of C1 versus C2 for Period 1.1 ranging from tap 40 to 150 and corresponding to 15 days of operation. Initially, TH-2 and TH-3 operate with stable outflow patterns with little variation for taps 40–70. TH-2 shows short negative slag delay (positive C1) where slag flows out first, while TH-3 shows a short positive slag delay (negative C1) where iron flows out first. This slight imbalance can be explained by the reasoning presented in Figure 1. By tap 70, both tapholes exhibit outflow patterns resembling the mean. However, at the time when TH-1 starts operating (after tap 100), TH-3 has shifted towards a higher slag share (positive C2). As TH-1 initially drains a relatively high share of iron, the system reacts to compensate for such an imbalance with TH-3. The balance before this moment and the imbalance that follows are also seen in C4 in the top panel of Figure 8. By tap 130, the drainage reaches a more stable distribution, with TH-1 showing a positive slag delay and TH-2 and TH-3 a negative slag delay. After TH-3 ends its operation, TH-1 and TH-2 show little variation, with TH-1 draining more iron than TH-2. One-day stoppages occur at taps 74 and 141, but do not affect the drainage patterns.

Figure 8, in turn, shows that C4 values higher than the mean occur when the tap duration is short. As seen in Figure 5, high C4 values correspond to a negative slope of slag share. Thus, if the slag share is high at the beginning of the tapping and, as a consequence, the descent of the slag–gas interface is rapid, a short tap duration can be expected because the moment when the slag–gas interface bends down to the taphole will end the tap [9] (cf. Figure 1). This holds true for TH-1 and TH-2 beyond Period 1.1.

4.1.2. Period 1.2

Figure 7b shows C1 versus C2 for taps 180–320, corresponding to the 16 days of operation in Period 1.2 that starts two days after the end of Period 1.1. Period 1.2, where TH-1 and TH-2 operate alternately, is particularly interesting because the tapholes are opposite (cf. Figure 6), so possible differences in the in-hearth conditions should be discerned during the period. The results indeed suggest drainage imbalance: at tap 180, TH-1 and TH-2 display similar patterns to those seen at the end of Period 1.1, that is, positive slag delay for TH-1 and negative for TH-2. For taps 180–210, the slag share decreases in TH-1, while it increases in TH-2 and the outflow patterns remain fundamentally different, but from tap 210 onward, the imbalance gradually disappears and the patterns become more similar. By tap 255, the two tapholes show patterns where the two liquids flow out simultaneously. TH-2 still drains a higher share of iron than TH-1, but the differences gradually become smaller, even though TH-1 still drains iron first while TH-2 drains slag first. As for C4, in Period 1.2 (like the previous period), a negative correlation is seen with the tap duration (cf. Figure 8). The taphole showing a larger C4 value changes from one to the other, in pace with the taphole showing the shortest tap duration. In the end of this period, all principal components considered approach the mean values for both tapholes, but the strong imbalance lasted for about 100 taps of operation of the taphole pair, reflecting the slow dynamics of the system.

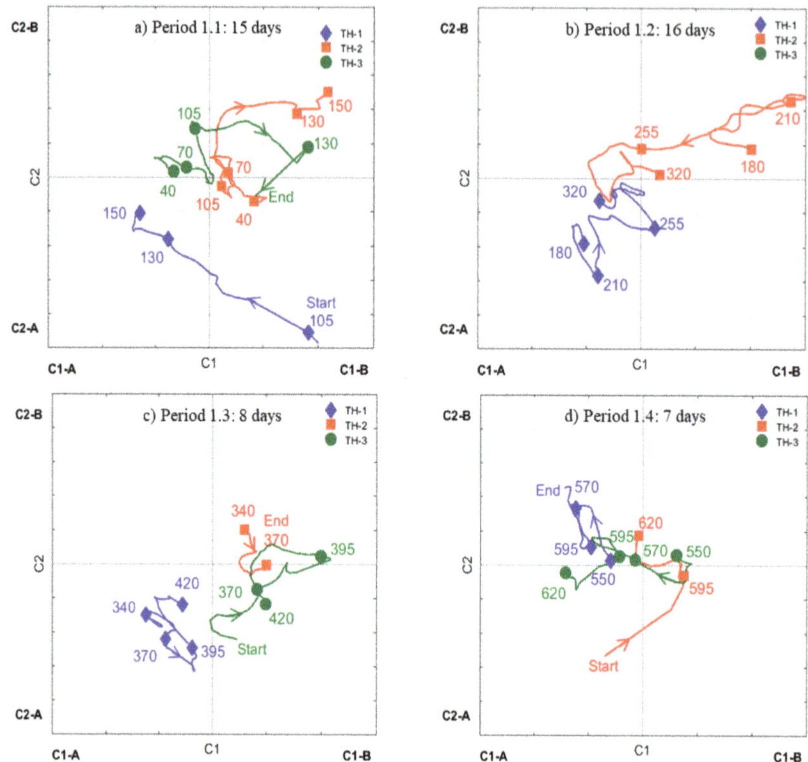

Figure 7. First vs. second principal components progression (C1, C2) corresponding to Data set 1 at different tap labels. (**a**) Period 1.1, (**b**) Period 1.2, (**c**) Period 1.3, (**d**) Period 1.4.

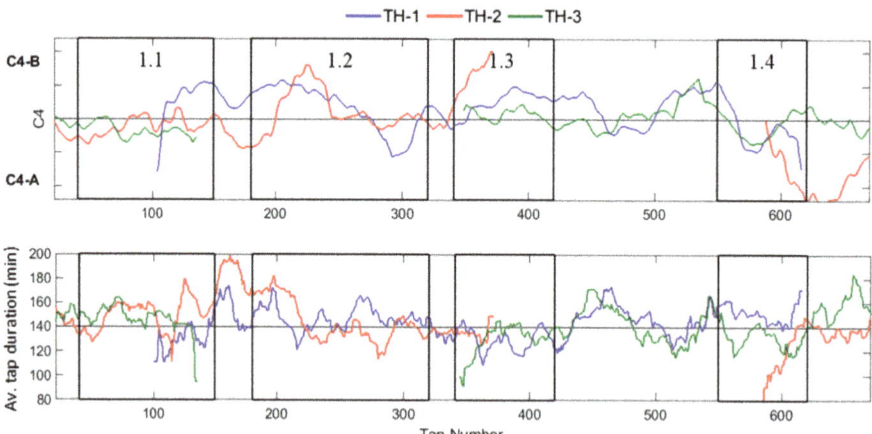

Figure 8. Fourth principal component (C4, **top**) and average tap duration (**bottom**) corresponding to the four periods of Data set 1. Blue line (TH-1), red line (TH-2), and green line (TH-3).

4.1.3. Period 1.3

Figure 7c shows the period between taps 340 and 420, starting two days after Period 1.2 ended. It covers eight days of operation, where TH-2 ends while TH-3 starts its operation. At tap 340, TH-1 and TH-2 show outflow patterns of similar characteristics to those in the previous period: TH-1 with lower slag share and iron-first flow and TH-2 with higher slag share and slag-first flow. In this drainage state, the "incoming" taphole TH-3 starts with a simultaneous drainage of both liquid phases and with a slightly higher iron share than what TH-1 exhibited initially in Period 1.1. This suggests that at the moment when TH-3 starts its operation, the iron–slag interface is at the taphole level. For taps 370–395, TH-3 adopts the same performance that TH-2 showed before its operation ended. However, by tap 420, both TH-1 and TH-3 drain a higher iron share, but a positive and negative slag delay, respectively. This drainage state should trigger an imbalance in the systems as more iron is drained than produced, but may reflect a temporal state (with low slag rate). Yet, the internal conditions still seem to allow for a rather uniform liquid flow within the hearth. Figure 8 shows that all three tapholes yield medium or high C4 values and short or average tap durations.

4.1.4. Period 1.4

The results of the final period of Data set 1, starting six days after Period 1.3 has ended and lasting for seven days ranging from tap 550 to 620, are shown in Figure 7d. At tap 550, both TH-2 and TH-3 show a slag share close to the average (0.2) and short positive and negative slag delays, respectively. As expected, the slag share increased after Period 1.3, and the increase continues in TH-1. TH-3 shows little variation in the share of slag, but shifts to a positive slag delay around tap 595. By contrast, the slag share for TH-2, which is taken into operation, increases rapidly. The tap duration for TH-1 is considerable longer than for TH-3 (cf. Figure 8), suggesting that the system reaches a balanced drainage state mainly through changes in the outflows of TH-1.

4.2. Data Set 2

Figure 9 presents the C1 versus C2 for Data set 2 and Figure 10 the results for C3 in the top panel and tap duration in the bottom panel. Six periods are studied (cf. Figure 6), with period lengths ranging from 9 to 16 days of operation, including segments where one taphole starts its operation and another ends it.

4.2.1. Period 2.1

Figure 9a shows the evolution of C1 versus C2 for Period 2.1, taps 40–170, corresponding to 16 days of operation of Data set 2. TH-1 starts operating around tap 50 with a long positive slag delay, but with a higher drained share of slag compared with the other tapholes. Figure 10 confirms these features by showing a low C3 value, that is, a positive slope of the slag share: as iron drains first, the slag outflow rate is small initially, but increases as the iron level descends and the taphole erodes. As expected, the tap duration is long; an initial high iron share indicates a high iron level in the hearth and a longer time will elapse until the slag–gas interface has descended sufficiently to end the tapping. However, as the operation continues, the slag share steadily decreases in TH-1. Inspection of the individual outflows from this taphole during the period revealed that the iron flow of some taps was interrupted or highly irregular. This suggests an unsuccessful drainage of iron from the taphole, even though iron was initially the first phase to flow out. For TH-3, the outflow patterns fall close to the mean and vary little during the last 30 taps of this taphole. As for TH-2—taphole that operates throughout the whole period—fluctuations between positive and negative slag delays are observed, but with low variation in C2 and an overall slag share close to 0.2. With time, both TH-1 and TH-2 converge around tap 170 to similar patterns with short negative slag delays. The tap duration decreases for both tapholes, which may indicate a poor slag drainage. C3 in Figure 10 also indicates that from tap 170 onwards, TH-1 shows a negative slope, that is, the iron share increases as the tapping progresses. This is a significant pattern shift from the drainage in the earlier parts of the period.

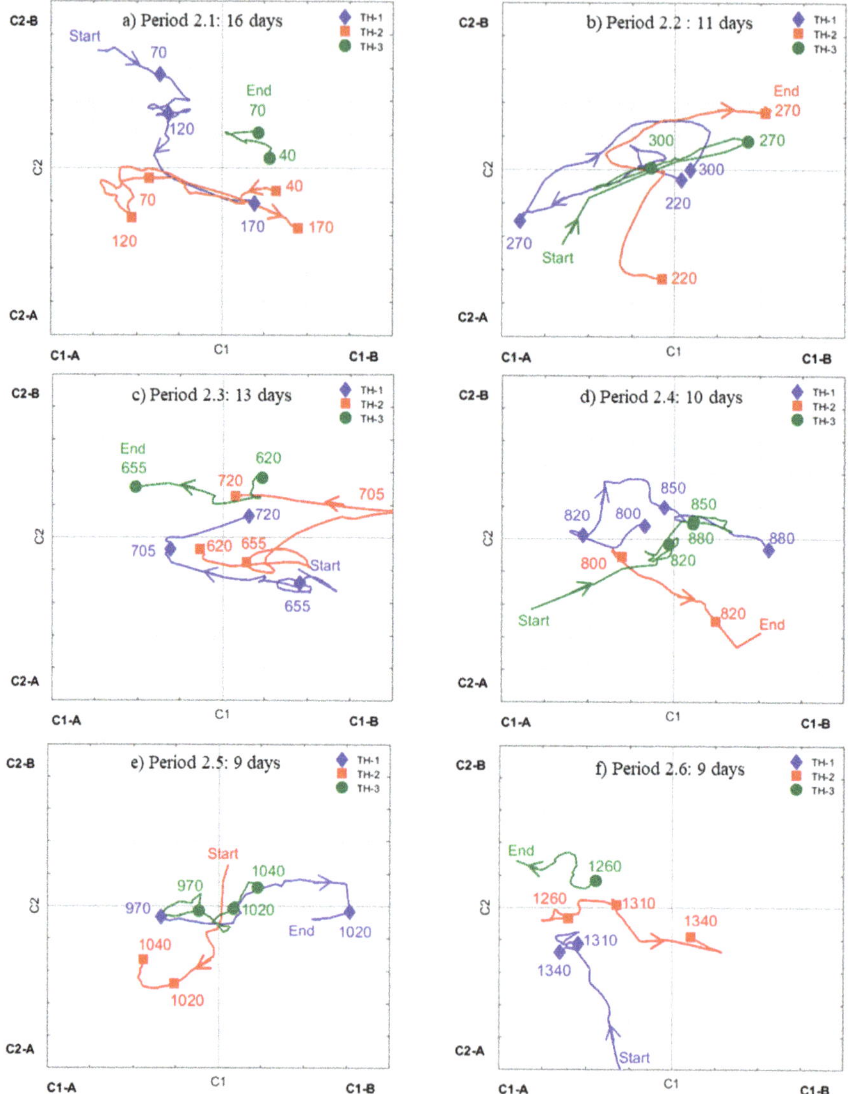

Figure 9. First vs. second principal components progression (C1, C2) corresponding to Data set 2 at different tap labels. (**a**) Period 2.1, (**b**) Period (2.2), (**c**) Period 2.3, (**d**) Period 2.4, (**e**) Period 2.5, (**f**) Period 2.6.

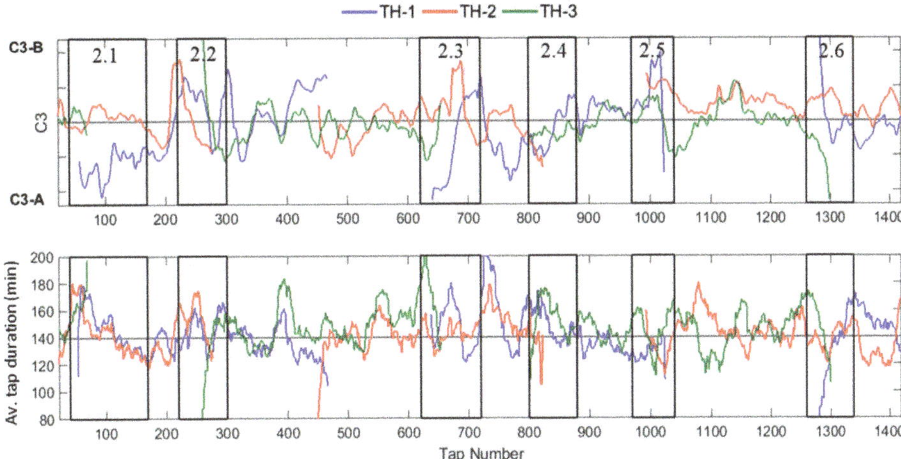

Figure 10. Third principal component (C3, **top**) and average tap duration (**bottom**) corresponding to the six periods of Data set 2. Blue line (TH-1), red line (TH-2), and green line (TH-3).

4.2.2. Period 2.2

Period 2.2 (Figure 9b), with taps 220–300, covering 11 days of operation, shows another type of evolution of the principal components. TH-1 operates throughout the period, TH-2 up to tap 270, while TH-3 is taken into operation some taps earlier. TH-1 shows a cyclical evolution in the (C1, C2) plane, where C1 first decreases and then increases back to the level of the starting point, while the changes in C2 are smaller. Thus, the main differences are in the slag delay, which changes from zero to clearly positive, and then back again. By contrast, for TH-2, taps 220–270, which represent the last 50 taps with this taphole operating, the main change is seen in C2: the operation shifts from a low slag share towards a high one with no slag delay, finally drifting to a state with a negative slag delay. A descent of the iron–slag interface in the hearth at TH-2 could lead to an increase of the slag share in the tapping and eventually also to a negative slag delay. By tap 270, TH-1 drains enough iron to lower the iron–slag interface to yield a negative slag delay (cf. Figure 1b) for TH-2, as it is ending its operation, and for TH-3, a few taps after its operation starts. Such behavior is discussed at some length in the work of [17]. C3 for TH-2 (Figure 10), in turn, indicates that the slope of the slag share increases, which can also be explained by the descent of the iron–slag interface. As TH-3 is taken into operation, the taps initially show a low slag share, but the outflow patterns rapidly change to become similar to those of TH-2 (around tap 270). Compared with Period 2.1, the tap duration for TH-1 is shorter as C2 decreases and C3 increases, even though C1 stays low.

Summarizing the findings, the changes in the hearth during the period reflect a strong imbalance in the liquid volumes tapped from the tapholes. Measures in the casthouse taken at the beginning and at the end of the period, for example, different tapping sequences or the use of different drill diameters, may trigger such an imbalance. Nonetheless, local conditions in certain hearth regions or taphole conditions cannot be ruled out as influencing factors. More information about specific taps should be analyzed in order to explain the reason for the imbalance observed.

In the six-week segment between Periods 2.2 and 2.3 (not shown), TH-1 shows some variation of the slag delay, but the overall slag share stays close to the mean. TH-3 initially shows a negative slag delay and then drifts towards the mean pattern, and to a positive slag delay as TH-2 is taken into operation. This "new" taphole starts its operation with a long positive slag delay and high iron share to converge to the states of the other two tapholes, suggesting that liquid levels are in balance in the different regions of the hearth. The changes in slopes described by C3 in the top panel of Figure 10 may reflect the system's response to maintain such a balance. Like in previous periods, a positive slope (lower C3) implies longer tap duration. For taps 500–620, the outflows from TH-2 and TH-3 show less variation in the three principal components, despite a six-day stoppage after tap 603.

4.2.3. Period 2.3

Figure 9c presents the evolution of (C1, C2) for the period that spans from taps 620 to 720, corresponding to 13 days of operation. This period occurs after a six-day stoppage, where operation restarts with tap 604. Thus, tap 620 in the (C1, C2) plane represents the average location of the 16 taps after operation was resumed. TH-3 ends its operation as TH-1 is taken into use. At tap 620, the principal components for TH-2 suggest drainage close to the mean pattern, while the flow from TH-3 has a higher slag share. Thus, it is expected that TH-1, when introduced (replacing TH-3), will count with a higher iron share, as is also seen in C2. TH-1 initially shows a high iron share and negative slag delay, to later show higher a slag share. Also, the slag delay changes from negative to positive and then negative again. The behavior of TH-2 is almost opposite, indicating that the end levels of the iron–slag interface show a mirror relation (noting that TH-1 and TH-2 are opposite tapholes). Finally, the outflows of TH-1 and TH-2 converge to similar patterns by tap 720. C3 indicates a progressive change in opposite directions of TH-1 and TH-2 before tap 700 to then reach a more balance distribution around the mean. A negative correlation between C3 and tap duration is again observed.

4.2.4. Period 2.4

Period 2.4 represents seven days of quite stable and uniform operation (Figure 9d, taps 800–880), where TH-1 operates throughout the period, TH-2 ends, and TH-3 starts its operation in the beginning of the period. TH-1 and TH-2 initially display a slightly positive slag delay and a mean slag share of 0.2, and as TH-3 is introduced, a positive slag delay is also expected. However, local conditions or operating actions result in a high iron share for TH-3, inducing imbalance as explained in Figure 1. At tap 820, TH-3 drains according to the mean pattern without slag delay, and the high iron share of TH-2 is counteracted by a shift in outflow of TH-1 to a higher slag share. The variations in C3 for all tapholes are small (Figure 10). After TH-2 is taken out of operation, the other tapholes show stable outflows, but TH-1 drifts towards a state with negative slag delay.

4.2.5. Periods 2.5 and 2.6

Period 2.5 (Figure 2e) represents a nine-day segment (taps 970–1040) that holds both positive and negative slag delays, but the slag share varies little. It can be argued that similar drainage conditions prevail at the tapholes with a uniform distribution of the liquids in the hearth. However, at the point where TH-1 is substituted by TH-2 (before tap 1020), its iron drainage has become poor: TH-1 experiences problematic taps much more often than TH-2 and TH-3. As a result, TH-2 drifts to a higher iron share and positive slag delay by tap 1020. Even though TH-2 drains enough iron, the tap duration is short, suggesting that slag is also poorly drained. As a consequence, both TH-2 and TH-3 show a higher slag share in the outflows by tap 1040. The imbalance of the drainage system is also seen in C3, but for taps 1050–1250, a more stable operation is noticed. This can be compared to similar segments following periods with a larger variation, for example, taps 350–600 and taps 800–100.

Finally, Period 2.6 is studied, which represents the nine-day segment with taps 1260–1340. During this period, TH-1 starts operating, while TH-3 ends. As also seen in Period 2.4, when both operating tapholes show iron-first tapping (i.e., positive slag delay), the "incoming" taphole will also do so. The initial imbalance in the share of slag drained between TH-1 and TH-2 vanishes by tap 1310, when TH-3 has already been taken out of operation. TH-1 maintains a long positive slag delay until tap 1340, while TH-2 shifts to a negative one, but towards the end of Data set 2, the outflows of the two tapholes fall close to the mean.

5. Results of PCA and Liquid Level Model

The large hearth diameter in multi-taphole blast furnaces may give rise to zones of different coke-bed permeability, which leads to differences in the local conditions and liquid levels [18,19]. A mathematical model was developed by the present authors [14] and applied to study the effect of different parameters and variables on the liquid levels and drainage process. The hearth was modelled as two interconnected pools of liquids, where communication factors can control the flow from one taphole zone to the other. Figure 1 illustrates the states of a hearth with good communication between the zones, resulting in uniform liquid levels (except the in/declining parts in the vicinity of the tapholes). A sensitivity analysis of the model was conducted to evaluate the effect of parameters on the tap duration and slag delay. Therefore, some findings from the earlier study can be associated with the PCA results presented here.

For instance, the model shows that poor iron communication increases the tap duration and slag delay (Periods 2.1, 2.3, and 2.4), while poor slag communication does not change the tap duration, but results in a negative slag delay (Periods 1.3 and 2.3). Asymmetric cases, where the pool sizes are different, give rise to conditions resembling those observed during periods when one taphole ends its operation and another is taken into use. Roche et al. [14] demonstrated that the asymmetric cases are characterized by more negative slag delay (for one taphole) and longer tap duration (for the other). This is observed in periods where the "incoming" taphole shifts from an initially positive slag delay to a slight negative one after some taps (Periods 1.3, 2.1, 2.2, and 2.4).

Changes in the conditions in front of the taphole due to the interaction between the injected mud and the coke bed [5] could potentially alter the internal level of the taphole with respect to the slag–iron interface, causing differences in the slag delay and tap duration between operating tapholes. The final declivity of the gas–slag interface is affected by the slag viscosity and coke diameter [20–22], which may vary with time and affect the tap duration, and, therefore, the liquid levels, triggering some imbalance.

Regardless of the conditions in the hearth, the outflow from one taphole will affect the outflow of the other. A sensitivity study of the outflow parameters by the model suggests that the system is prompted to changes in slag delay and tap duration when different outflow patterns are imposed on one taphole. A slower initial drainage rate of iron from one taphole yields a positive slag delay and longer tap duration for the other, which, in turn, leads to a negative slag delay for the first taphole because of low liquids levels (cf. Figure 1). This behavior is seen in several periods where the outflow patterns are found in opposite regions of the (C1, C2) plane (Periods 1.1, 1.2, 1.3, 2.4, and 2.6). On the other hand, a slow initial drainage rate of slag from one taphole yields slightly negative slag delays in both tapholes and longer tapping. This seems to be the case with tapholes ending their operation when compared with the operating one (Periods 1.3, 2.1, and 2.5). As observed in the results, the outflow patterns diverge as a result of a triggering factor, but often converge after some period of operation towards the mean pattern as the systems reaches balance again (as seen in Periods 1.2, 1.4, 2.2, and 2.4). A constant monitoring of the variables during operation, for example, iron production, liquid outflows, and slag delay, is a key to keep such imbalances within reasonable limits.

6. Summary of the Analysis

For the taps in the two data sets analyzed, TH-1 entered into operation four times, while TH-2 and TH-3 entered three times. Even though the data sets analyzed were quite extensive, the number of taphole changes was still limited, and it is thus difficult to draw general conclusions about the draining conditions of the taphole that is brought into operation. However, it can still be deduced that if the taphole that enters into operation is located farther off (here, TH-1 or TH-2) and the liquids are not in balance, it likely leads to more imbalance in the drainage. This is particular notable for TH-1 in Periods 1.1, 2.1, and 2.6. This effect can be caused by smaller random effects, for example, in the conditions in the coke bed at the taphole inlet that accentuate the characteristics of the pattern when a taphole starts operating. Thus, an existing drainage imbalance sets the incoming taphole in an extreme initial state. On the other hand, when the drainage of the system is balanced, the operation of the incoming taphole is not very different, even though it often initially shows some discrepancy in terms of (typically positive) slag delay and slag share. This suggests that inactivity of a taphole gradually leads to different conditions in the hearth in the vicinity of it. In the reference BF, the taphole length was found to be initially considerably longer for the fresh taphole compared with the ones that had been operating. As for the end of the operating period of a taphole pair, more extreme drainage patterns were often observed. In the steel works in question, the incoming and outgoing tapholes were operated alternately for a few taps, most likely to induce a smooth transition, but disturbances were still seen. This also supports the hypothesis that different local conditions in the hearth may arise, such as different liquid levels and differences in the hearth-coke permeability. The model-based analysis of the liquid levels referred to in Section 5 indicated that the effective pool of the hearth that is drained may grow during tapping [14]. Such distribution of volumes would not only lead to different outflow patterns, but could also induce zones with different permeability with impact on lining and taphole erosion.

Also, the results of C4 for Data set 1 and C3 for Data set 2 are interesting because of the cyclical trend seen for all tapholes. These principal components showed negative correlation with the tap duration in both data sets and for all three tapholes. As mentioned earlier, with an initial higher slag share in the outflow, the slag–gas interface descends more rapidly. Additionally, some variation of the slag share as the tapping evolves is expected as a result of taphole erosion. Another explanation is that the fluid dynamics of the system induces such behavior, for example, the fact that iron has to be elevated from levels below the taphole.

It should be stressed that the two-phase outflow, the unknown conditions of the coke bed and the taphole entrance, and the taphole erosion make it difficult to explain the dynamics of the system. In general, TH-3 showed a more uniform distribution of the samples around the mean pattern, most likely owing to its location in the hearth. When operating alternately with either of the other two tapholes, similar hearth conditions are expected because of their location. Further, TH-1 and TH-2 often showed opposite behavior in order to balance the drainage.

The liquid level model supporting the findings [14] has proposed some system parameters that affect the outflow patterns of iron and slag. In light of the information gained in the present study, there may be a need to re-evaluate the system with the model. A challenge is that every tapping seems to show individual features, but the results presented in this paper still reveal some general drainage patterns of the hearth, and interrelations between the drainage of two alternating tapholes. In particular, it would be interesting to understand the transition paths between different common or particular outflow patterns, which could be useful when an online model is developed. In order to present more conclusive interpretations, a larger number of samples is necessary, where each taphole is represented by a sufficient number of operating periods. Despite these limitations, the PCA-based tool must be considered efficient in illustrating and condensing the information from individual samples into a comprehensible and compact form. The tool can be used along with other information from the BF to analyze how the system reacts after stoppages of different duration, an increase or decrease in the production rate, interchanges of tapholes, changes of drill diameters, and other specific measures taken in the casthouse.

7. Conclusions

Two data sets, corresponding to three and six months of operation of a blast furnace, were analyzed with respect to the drainage of the furnace hearth. The blast furnace studied has three tapholes (TH-1, TH-2, and TH-3) and the liquid outflow rates were calculated based on weighing of the torpedos (hot metal) and signals from the granulation unit (slag). After inspecting individual outflow rates throughout the data, certain outflow patterns were detected and some of them were found to recur. To simplify the interpretation, the share of slag in the outflow was taken as the variable of primary concern. After normalizing the tap duration, the evolution of the share of slag during the tapping was down-sampled to ten variables to filter out some noise. To further compress the information and to make it possible to visualize the time evolution of the drainage patterns, principal component analysis (PCA) was applied to the data sets. The results of the compression were analyzed and the first two principal components were found to describe two relevant tapping features: the first principal component (C1) basically describes the slag delay (i.e., the length of the period of one-phase flow in the beginning of the tapping), and the second principal component (C2) describes the mean or end slag share for the tapping. To better analyze the results, one additional component was considered, which expresses how the slag share develops as both iron and slag are being drained. This component was found to be the fourth component (C4) in Data set 1 and the third component (C3) in Data set 2. For the sake of clarity, the results were post-processed into moving averages over the latest 20 taps. The results presented in the paper focus mainly on detecting diverging or converging behavior of the outflows of the operating tapholes based on the evolution of the principal components. The data sets were divided into different periods. TH-1 was found to display the most erratic patterns, followed by TH-2, while TH-3 exhibited less fluctuation. This is logical, as TH-3 is located between TH-1 and TH-2. Often, taphole changes were followed by draining imbalance, where the outflows from all three tapholes occasionally showed large fluctuations, which were seen as extreme values in the principal component space. After some time, the system typically stabilized, leading to drainage that corresponded to the mean outflow pattern. Even though the reason for the disturbances often remained unknown, the general behavior of the system could partly be explained by arguments based on fluid dynamics. To generalize the findings, a longer data set may be needed, because random events seem to have a considerable effect on the dynamics. Nonetheless, the PCA-based tool developed has proven to be useful in capturing the evolution of hearth drainage in a few key indices, which may be studied and analyzed in an attempt to gain further understanding of the complex industrial system at hand. To assess the usefulness of the PCA-based approach, it is proposed that the model should be implemented in the plant and the evolution of the (main) principal components could be studied and correlated with other BF variables. This will allow for gathering further experience on what affects the lower-dimensional representation of the liquids outflows and how the information could be utilized in the casthouse operation. In conjunction with further analysis, the accuracy of the estimated liquid outflow rates should also be evaluated, and, in particular, the quality of the estimated slag outflow rate.

Author Contributions: conceptualization, M.R. and H.S.; methodology, M.R.; software, M.R.; formal analysis, M.R., M.H., and H.S.; data curation, M.H.; writing—original draft preparation, M.R.; writing—review and editing, M.R., M.H., and H.S.; supervision, M.H. and H.S.; funding acquisition, H.S.

Funding: The research leading to these results had received funding from the European Union's Research Fund for Coal and Steel (RFCS) research program under grant agreement no. RFSR-CT-2015-00001, and this support is gratefully acknowledged.

Acknowledgments: The authors are grateful for the valuable questions, comments, and suggestions by the anonymous reviewers, which helped us to improve quality of the manuscript.

Conflicts of Interest: The authors declare no conflict of interest.

References

1. Iida, M.; Ogura, K.; Hakone, T. Numerical Study on Metal/Slag Drainage Rate Deviation during Blast Furnace Tapping. *ISIJ Int.* **2009**, *49*, 1123–1132. [CrossRef]
2. Iida, M.; Ogura, K.; Hakone, T. Analysis of Drainage Rate Variation of Molten Iron and Slag from Blast Furnace during Tapping. *ISIJ Int.* **2008**, *48*, 412–419. [CrossRef]
3. Nouchi, T.; Sato, M.; Takeda, K.; Ariyama, T. Effects of Operation Condition and Casting Strategy on Drainage Efficiency of the Blast Furnace Hearth. *ISIJ Int.* **2005**, *45*, 1515–1520. [CrossRef]
4. Roche, M.; Helle, M.; van der Stel, J.; Louwerse, G.; Shao, L.; Saxén, H. On-Line Estimation of Liquid Levels in the Blast Furnace Hearth. *Steel Res. Int.* **2019**, *90*, 1800420. [CrossRef]
5. Tsuchiya, N.; Fukutake, T.; Yamauchi, Y.; Matsumoto, T. In-furnace Conditions as Prerequisites for Proper Use and Design of Mud to Control Blast Furnace Taphole Length. *ISIJ Int.* **1998**, *38*, 116–125. [CrossRef]

6. Zhao, Y.; Fu, D.; Lherbier, L.W.; Chen, Y.; Zhou, C.Q.; Grindey, J.G. Investigation of Skull Formation in a Blast Furnace Hearth. *Steel Res. Int.* **2014**, *85*, 891–901. [CrossRef]

7. Takatani, K.; Inada, T.; Takata, K. Mathematical Model for Transient Erosion Process of Blast Furnace Hearth. *ISIJ Int.* **2001**, *41*, 1139–1145. [CrossRef]

8. Panjkovic, V.; Truelove, J.S.; Zulli, P. Numerical modelling of iron flow and heat transfer in blast furnace hearth. *Ironmak. Steelmak.* **2002**, *29*, 390–400. [CrossRef]

9. Tanzil, W.B.U.; Zulli, P.; Burgess, J.M.; Pinczewski, W.V. Experimental Model Study of the Physical Mechanisms Governing Blast Furnace Hearth Drainage. *Trans. Iron Steel Inst. Jpn.* **1984**, *24*, 197–205. [CrossRef]

10. Nishioka, K.; Maeda, T.; Shimizu, M. A Three-dimensional Mathematical Modelling of Drainage Behavior in Blast Furnace Hearth. *ISIJ Int.* **2005**, *45*, 669–676. [CrossRef]

11. Agrawal, A.; Kor, S.C.; Nandy, U.; Choudhary, A.R.; Tripathi, V.R. Real-time blast furnace hearth liquid level monitoring system. *Ironmak. Steelmak.* **2016**, *43*, 550–558. [CrossRef]

12. Saxén, H. Model of Draining of the Blast Furnace Hearth with an Impermeable Zone. *Metall. Mater. Trans. B* **2015**, *46*, 421–431. [CrossRef]

13. Saxén, H.; Brännbacka, J. Dynamic model of liquid levels in the blast furnace hearth. *Scand. J. Metall.* **2005**, *34*, 116–121. [CrossRef]

14. Roche, M.; Helle, M.; van der Stel, J.; Louwerse, G.; Shao, L.; Saxén, H. Off-line Model of Blast Furnace Liquid Levels. *ISIJ Int.* **2018**, *58*, 2236–2245. [CrossRef]

15. Bartholomew, D.J. Principal Components Analysis. *Int. Encycl. Educ. (Third Ed.)* **2010**, 374–377. [CrossRef]

16. Kourti, T. 4.02 Multivariate Statistical Process Control and Process Control, Using Latent Variables. *Compr. Chemom.* **2009**, 21–54. [CrossRef]

17. Helle, M.; Roche, M.; Saxén, H. On-line estimation of liquid levels and local drainage characteristics in the blast furnace. In Proceedings of the SteelSim, Toronto, ON, Canada, 13–15 August 2019. paper 24.

18. Hu, X.; Sundqvist Ökvist, L.; Ölund, M. Materials Properties and Liquid Flow in the Hearth of the Experimental Blast Furnace. *Metals* **2019**, *9*, 527. [CrossRef]

19. Nishioka, K.; Maeda, T.; Shimizu, M. Effect of Various In-furnace Conditions on Blast Furnace Hearth Drainage. *ISIJ Int.* **2005**, *45*, 1496–1505. [CrossRef]

20. Fukutake, T.; Okabe, K. Experimental Studies of Slag Flow in the Blast Furnace Hearth During Tapping Operation. *Trans. Iron Steel Inst. Jpn.* **1976**, *16*, 309–316.

21. Fukutake, T.; Okabe, K. Influences of Slag Tapping Conditions on the Amount of Residual Slag in the Blast Furnace Hearth. *Trans. Iron Steel Inst. Jpn.* **1976**, *16*, 317–323.

22. Zulli, P. Blast Furnace Hearth Drainage with and without a Coke Free Layer. Doctoral Dissertation, Faculty of Engineering, UNSW, Sydney, Australia, 1991.

Article

Parametric Dimensional Analysis on a C-H$_2$ Smelting Reduction Furnace with Double-Row Side Nozzles

Jinyin Xie, Bo Wang and Jieyu Zhang *

State Key Laboratory of Advanced Special Steel, Shanghai Key Laboratory of Advanced Ferrometallurgy, School of Materials Science and Engineering, Shanghai University, Shanghai 200444, China; xjy_1215@shu.edu.cn (J.X.); bowang@shu.edu.cn (B.W.)
* Correspondence: zhangjieyu@shu.edu.cn

Received: 12 November 2019; Accepted: 14 January 2020; Published: 21 January 2020

Abstract: Higher requirements for steel smelting technology have been put forward based on the increasing awareness of energy conservation and environmental protection. In the field of iron making, carbon reduction processes are often used. In this study, molten iron was smelted by designing a C-H$_2$ smelting reduction method. Although previous researchers have studied this through a large number of physical and numerical simulations, they have not yet refined general laws from the perspective of dimensional analysis. In this paper, a double-row side blow hydraulics simulation was carried out in the C-H$_2$ smelting reduction furnace, and an entire list of dimensionless groups of input and output parameters was proposed based on its hydraulics simulation data. The expressions between the dimensionless group of mixing time and dimensionless groups such as Capillary number (Ca) and Lagrange group (La$_1$) were obtained by multiple linear regression based on the experimental research results and data analysis. By verifying the calculated and experimental values of the dimensionless group of mixing time, it can be seen that both have a good positive correlation. This study provides a better methodology for controlling key parameters and lays the foundation for the optimal design of the process parameters for the C-H$_2$ smelting reduction furnace.

Keywords: C-H$_2$ smelting reduction furnace; double-row side nozzles; dimensional analysis; mixing time; multiple linear regression

---

1. Introduction

The steel industry is one of the most important raw material industries. The iron-making process in the early steel process is mainly done through a blast furnace [1]. With changes in the times, a series of drawbacks of the blast furnace have gradually emerged. One is the heavy dependence on coke resources [2], and the second is the increasing pressure on environmental protection. With the rapid depletion of natural resources and awareness regarding environmental protection, the smelting reduction process came into being. This significantly reduces energy consumption and investment by recycling the waste heat and eliminating sintering and coking plants [3]. Among the variants of this process, representative ones are COREX [4], jointly developed by VAI and Korf, FINEX [5], jointly developed by POSCO Steel and VAI Engineering, HISMELT [6], jointly developed by CRA Australia and Kloekner, and DIOS [7], jointly developed by the Japan Iron and Steel Federation to organize eight domestic steel companies.

In the field of smelting reduction, many researchers have done considerable research on physical simulation and mathematical simulation. Zou et al. [8,9] proposed a two-step three segment iron bath smelting reduction process with a thick slag layer, which was developed for the gradient separation of the oxidation zone and the reduction zone by a thick slag layer with physical simulation. Srishilan et al. [10] proposed a predictive thermochemical model of the COREX process which enables the rapid computation of process parameters. The model helps in predicting the variations in process parameters with respect to the degree of metallization and post-combustion ratio for the given raw material conditions. A series of recent representative research results have also been obtained in smelting reduction and hydrocarbon reduction. In the field of plasma reduction, for example, Behera et al. [11] succeeded in the smelting reduction of iron ore (hematite) in thermal hydrogen plasma. Mandal et al. [12] described a new concept for maintaining an inert atmosphere with a high temperature of about 1973 K (1700 °C) inside the furnace during smelting reduction. In terms of direct reduction, Shim et al. [13] reported that the direct reduction rate in a melter–gasifier was roughly drawn as the product of the CO_2 content in the ascending gas and the reaction rate constant of coal with CO_2, and the way to minimize the direct reduction ratio was discussed with that diagram. You et al. [14] used Sn-bearing complex iron ore via reduction with mixed H_2/CO gas to prepare Sn-enriched direct reduced iron (DRI). In the behavior of reduction, Park et al. [15,16] investigated the high-temperature behavior of a magnetite–coke composite pellet fluxed with dolomite by a customized thermogravimetric analyzer (TGA) at 1573 K (1300 °C) and the influence of different coal types on the reduction of the composite pellet. At the same time, previous research has proposed a new generation of C-H_2 smelting reduction furnace (cf. Figure 1).

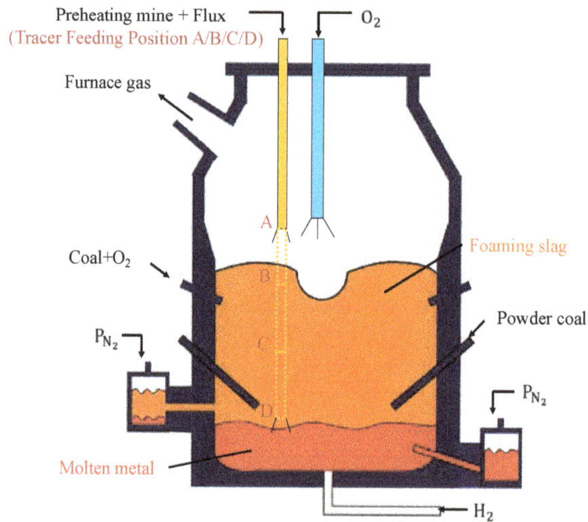

Figure 1. A new generation of C-H_2 iron bath smelting reduction process configuration.

The theory of similarity is the theory of experimentation. For the formulas of complex phenomena that cannot be solved by mathematical analysis, the similarity theory provides an experimental solution. The similar first theorem specifies the physical quantities that need to be measured during the experiment, as well as the conditions that the model experiment should follow. The π theorem is a universal dimensional analysis method which was proposed by E. Buckinghan in 1915, so it is also known as the Buckingham Theorem. The Buckingham Theorem is described as F(x1, x2, ... , xn) = 0, if there are n variables that are mutual functions for a physical phenomenon. If these variables contain m basic quantities, they can be arranged into the functional relation $\varphi(\pi 1, \pi 2, ... , \pi n \leq m)$ of (n ≤ m) dimensionless numbers, and then n physical quantities can be combined into (n ≤ m) dimensionless π numbers. In this paper, the dimensional analysis is based on the Buckingham theorem, which specifies how to organize the experimental results. The experimental results need to be organized into a functional relationship between dimensionless groups. Finally, the dimensionless groups and the coefficients of similarity are determined according to the experimental results. The similarity theory is used to solve the problem in various fields. Zhen et al. [17] focuses on providing a quantitative methodology on how each parameter affects the structural response of the subsurface tension leg platform (STLP), which will facilitate establishing the unique design dimensionless groups as regards to STLP. Vatankhah et al. [18] predicted the rate of discharge through different side holes in irrigation and hydraulic engineering. Sharp-crested side triangular orifices were studied experimentally and analytically, and several models were derived for the discharge coefficient based on Buckingham's theorem of dimensional analysis. Meng et al. [19] used the π theorem, an improved dimensional analysis method, and the dimensionless quantity, which can effectively reflect the relationship between the non-sinusoidal vibration parameters of a continuous casting mold, was given. The dimensionless function correlation formula, which can objectively describe the actual phenomenon, was obtained.

Side blowing is one of the key processes of smelting reduction. Fuel and enriched oxygen are injected into the furnace by a side nozzle to provide the heat required for the reduction reaction and the melt is stirred at the same time. By designing and optimizing the parameters such as the arrangement, the angle, and the flow rate of the side nozzle, this provides better reaction conditions for melting in the furnace and prolongs the service life of the side nozzle, and finally achieves the combination with the top and bottom blowing. For side blowing, many scholars had conducted relevant research in this field. For example, Feng [20] outlined the production overview of the side-blown melting reduction furnace, introduced the construct design of main components, and listed the practice and summarized the characteristics of the side-blown reduction furnace. Due to a large number of side nozzle parameters and furnace parameters, the experimental results were not regular on the surface, so it was necessary to sort out the data through dimensional analysis and provide an analytical formula, so as to obtain the quantitative law. Wang et al. [21] investigated the penetration behavior of immersion side-blowing gas flow in a slag lade water model by photography and the dimensional analysis method, in order to provide a theoretical base for improving the reaction speed between the gas–liquid interface and the oxygen gun jet distribution. The maximum penetration depth calculated by the empirical formula is in agreement with the measured data from the experiments.

Although many studies [22–24] have been carried out on the C-H_2 smelting reduction furnace through physical simulation and mathematical simulation, the experimental data and results have not been quantitatively analyzed. This paper is aimed at providing a quantitative method for examining how each parameter affects the mixing time in the C-H_2 smelting reduction furnace, and proposes a dimensionless input and output parameter based on the Buckingham theory, a complete list of derived dimensionless groups. This is helpful for establishing a single design standard for C-H_2 smelting reduction furnaces. This study provides a means to understand critical parameters better and lays the foundation for the optimal design of the side blowing parameters of the C-H_2 smelting reduction furnace. The conclusions obtained are also widely applicable to the engineering design and design analysis of the smelting reduction furnace.

2. Experimental Setup and Methods

In this paper, a comprehensive experimental study on the flow characteristics of a C-H_2 smelting reduction furnace was carried out. Based on the similarity principle, the smelting reduction iron-making process under the high temperature conditions in the prototype was studied by hydraulic simulation at room temperature in this experiment. The schematic diagram of the C-H_2 smelting reduction model apparatus is shown in Figure 2. The model was simulated by the scale ratio of 1:1 to the prototype, in which the molten iron of the prototype was 200 kg. The experiment was carried out in a cylindrical transparent plexiglass furnace with a diameter of about 0.4 m and a height of 1.68 m. In the experiment, the molten iron was simulated by water, high vacuum oil was used to simulate the slag, and oxygen-enriched air was blown on the top and bottom and side nozzle to simulate the flux injection and the bottom blowing hydrogen. The volume ratio of water to high vacuum oil in the model was 1:2 [25,26], in which the water phase was 0.246 m and the height of the oil phase was 0.492 m. The physical parameters of the experiment are shown in Table 1. In the prototype, the combined top, bottom, and side blowing of the C-H_2 smelting reduction process occurs. This paper is the first phase of the project, aiming to carry out the physical simulation and dimensional analysis of the single-side blowing of double-row side nozzles. The side blowing nozzle was divided into the upper side nozzle and the lower side nozzle, and the diameter of the side nozzle was 0.004 m. The upper side nozzle was 0.574 m from the bottom and was located at 1/3 above the slag phase. The lower side nozzle was 0.492 m from the bottom and was located in the middle of the slag layer. The prototype and dimensions of the water model are shown in Figure 3a,b, respectively.

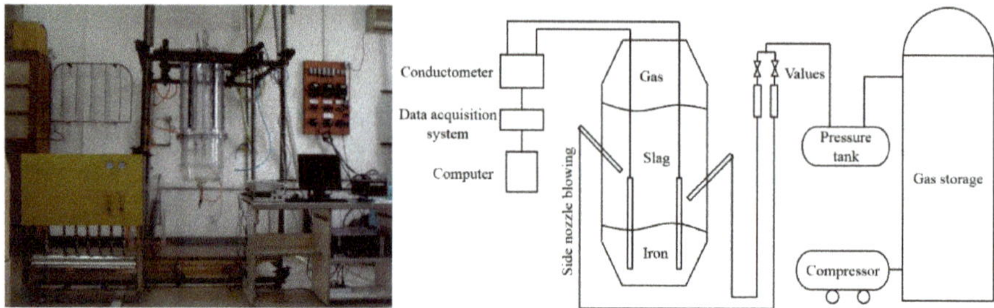

Figure 2. Schematic diagram of the C-H$_2$ smelting reduction model apparatus.

Table 1. Physical parameters of the experiment.

	Density (kg/m^3)	Kinematic Viscosity (m^2/s)
Iron	7020	9×10^{-5}
Slag	3000	1.33×10^{-4}
Air	1.205	1.506×10^{-5}
High vacuum oil (25 °C)	860	8.75×10^{-5}
Water	1000	1×10^{-6}

In this study, several different influencing factors were set. The effects of various factors and mixing time were obtained by orthogonal tests. By sorting out the experimental data, it was organized to be a functional relationship between the dimensionless groups. The first factor was the tracer feeding position. In the prototype, the tracer feeding port was added to the preheating mine and flux. It is of considerable significance to investigate the feeding position of raw materials for the mixing effect in the molten pool. Position A was at 2 cm above the slag interface (0.758 m from the bottom). Position B was located at 2 cm below the slag interface (0.718 m from the bottom). Position C was at the center of the slag (0.492 m from the bottom). Position D was located at 2 cm above the slag and the molten interface (0.266 m from the bottom), as shown in Figure 4a.

Similarly, several other factors were considered separately. These included the relative angle between the upper side nozzle and the lower side nozzle (60°, asymmetrical side blowing; 120°, asymmetrical side blowing; 180°, symmetric side blowing) (cf. Figure 3c), the horizontal angle of the upper side nozzle and the lower side nozzle (−15°, 0°, 15°) (cf. Figure 3d), the insertion depth of the side nozzle, and the flow rate of the side nozzle.

The inside of the molten pool was strongly agitated and disturbed in the combined upper side nozzle and lower side nozzle. Therefore, the efficiency of mass transfer and heat transfer was increased, and the rate of chemical reaction was also increased. In order to study the mixing phenomenon in the C-H$_2$ smelting reduction bath, the mixing time was regarded as an important index. The mixing time [27] was defined as the period required for an instantaneous tracer concentration to settle within ±5% deviation around the final tracer concentration in the C-H$_2$ smelting reduction reactor bath. This definition is referred to as the 95% mixing time. In the C-H$_2$ smelting reduction bath, the mixing time was measured by the conductivity of three electrodes, which was 0.05 m away from the bottom of the bath. Figure 4b is the position and angle of the sensor and tracer. In the prototype, the feed port was used for feed preheating ore and flux. In this experiment, in order to simulate the effect of different raw material positions on the mixing time of the molten pool, a saturated Sodium chloride (NaCl) aqueous solution (75 mL) was fed from the intermediate to the C-H$_2$ smelting reduction furnace. The conductivity of water was measured by three DSS-IIA conductivity meters and recorded automatically by a computer software recorder. For each physical simulation test site in each mode of operation, the measurements were taken at least three times and the arithmetic mean of the average residence time was obtained. Through the orthogonal test and analysis, the relationship between the average residence time and various experimental parameters can be obtained. These results would eventually be organized into a functional relationship between the dimensionless groups.

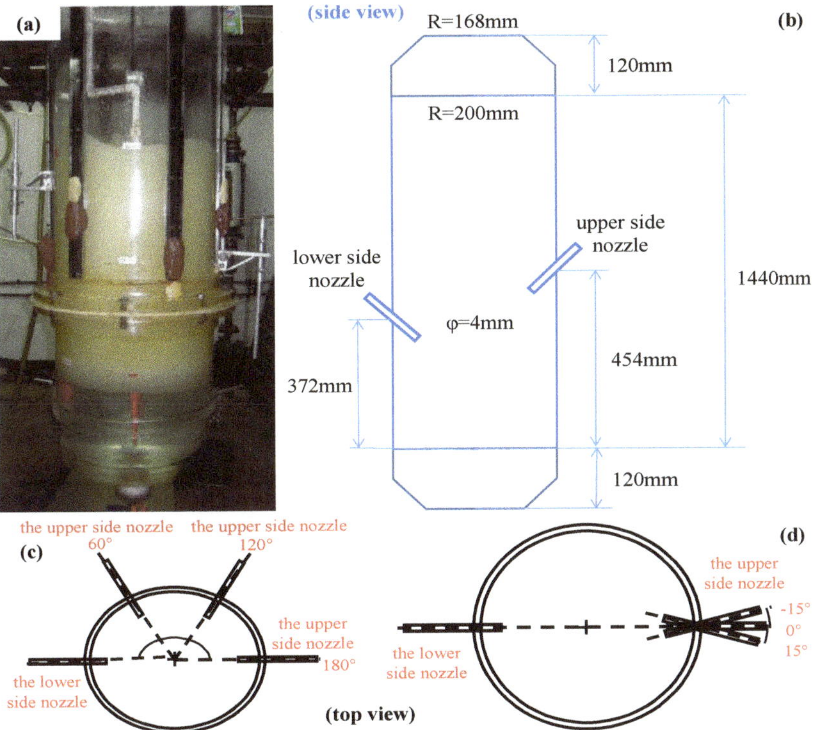

Figure 3. Prototype (**a**) and dimensions (**b**) of water model, the relative angle between the upper side nozzle and the lower side nozzle (**c**), the horizontal angle of the upper side nozzle and the lower side nozzle (**d**).

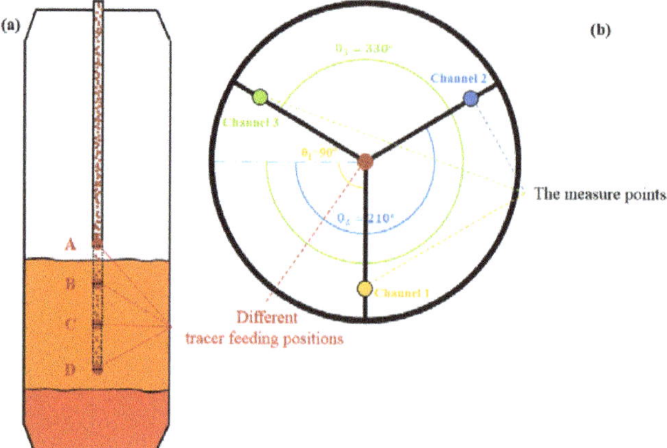

Figure 4. Different tracer feeding positions (**a**), and the position and angle of the sensor and tracer (**b**).

3. Dimensional Analysis

The purpose of this study was to establish a relationship between the mixing time and other valid variables. Their dimensions were considered through dimensional analysis. In this method, the first step is to select the appropriate initial parameters, including the input parameters and output parameters of the mixing time. The second step is to group these initial parameters into a dimensionless group and organize the new relationships between the various parameters. In particular, it is important to select the initial parameters precisely, since there is a need for a unique relationship between the chosen parameters [28,29].

Several factors may affect the value of mixing time. The variables that may affect the mixing time are density ρ, kinematic viscosity υ, surface tension σ, stirring energy ε [30], etc. In this study, 20 initial parameters were selected to identify the dimensionless groups of the parameters which can quantify the mixing time, as shown in Table 2. Based on this, 17 corresponding dimensionless groups were obtained from the selected initial parameters, according to the Vaschy–Buckingham theorem.

Table 2. Selected parameters for the dimensional analysis of the mixing time.

Category	Variable	Symbol	Unit	M	L	t
	Density	ρ	kg·m^{-3}	1	−3	0
	Kinematic viscosity	υ	m^2·s^{-1}	0	2	−1
Physical	Surface tension	σ	kg·s^{-2}	1	0	−2
Parameters	Mixing time	τ	s	0	0	1
	Acceleration of gravity	g	m·s^{-2}	0	1	−2
	Stirring energy	ε	kg·m^2·s^{-3}	1	2	−3
	Height of furnace	H_f	m	0	1	0
	Diameter of furnace	D_f	m	0	1	0
	Height of high vacuum oil	H_{oil}	m	0	1	0
Model Size	Height of water	H_w	m	0	1	0
	Height of upper side nozzle	H_{us}	m	0	1	0
	Height of lower side nozzle	H_{ls}	m	0	1	0
	Diameter of side nozzle	D_s	m	0	1	0

Table 2. *Cont.*

Category	Variable	Symbol	Unit	M	L	t
Tracer Position	Height of tracer	H_i	m	0	1	0
	Distance from tracer to the center of the circle	R_i	m	0	1	0
Measuring Point	Height of measuring point	H_m	m	0	1	0
	Distance from measuring point to the center of the circle	R_m	m	0	1	0
Working Condition	Insertion depth of side nozzle	h_s	m	0	1	0
	Flow velocity of side nozzle	V_s	m·s^{-1}	0	1	−1
	Flow rate of single side nozzle	q_s	m^3·s^{-1}	0	3	−1

In Table 2, M, L, t are three basic dimensions, where M stands for the mass dimension, kg, L for the length dimension, m, and t for the time dimension, s.

The mixing time can be represented by the following functional relationships:

$$\pi = F\left(\rho, \upsilon, \sigma, \tau, g, \varepsilon, H_f, D_f, H_{oil}, H_w, H_{us}, H_{ls}, D_s, H_i, R_i, H_m, R_m, h_s, V_s, q_s\right) \tag{1}$$

Meanwhile, since many factors are constant in this study, such as the height of the furnace, the height of the high vacuum oil, the height of the side nozzle, the diameter of the side nozzle and so on, the functional relationships can be further simplified to

$$\pi = F(\rho, \upsilon, \sigma, \tau, g, \varepsilon, D_s, H_i, h_s, V_s, q_s) \tag{2}$$

The equation can be expressed in the following dimensionless equation:

$$\pi = \varphi(\pi_1, \pi_2, \pi_3, \pi_4, \pi_5, \pi_6, \pi_7, \pi_8) \tag{3}$$

where $\pi_1 \sim \pi_8$ is a dimensionless group, and φ is a functional symbol. After substituting in each variable, it gives

$$\pi = \rho^a \upsilon^b \sigma^c \tau^d g^e \varepsilon^f D_s^g H_i^h h_s^i V_s^j q_s^k \tag{4}$$

$$M^0 L^0 t^0 = \left(ML^{-3}\right)^a \left(L^2 t^{-1}\right)^b \left(Mt^{-2}\right)^c (t)^d \left(Lt^{-2}\right)^e \left(ML^2 t^{-3}\right)^f (L)^g (L)^h (L)^i \left(Lt^{-1}\right)^j \left(L^3 t^{-1}\right)^k \tag{5}$$

By replacing the other variables with ρ, D_s and V_s, we get

$$\pi = \rho^{(-c-f)} \upsilon^b \sigma^c \tau^d g^e \varepsilon^f D_s^{(-b-c-d+e-2f-h-i-2k)} H_i^h h_s^i V_s^{(-b-2c+d-2e-3f-k)} q_s^k \tag{6}$$

$$\pi = \left(\frac{\upsilon}{D_s V_s}\right)^b \left(\frac{\sigma}{\rho D_s V_s^2}\right)^c \left(\frac{\tau V_s}{D_s}\right)^d \left(\frac{g D_s}{V_s^2}\right)^e \left(\frac{\varepsilon}{\rho D_s^2 V_s^3}\right)^f \left(\frac{H_i}{D_s}\right)^h \left(\frac{h_s}{D_s}\right)^i \left(\frac{q_s}{D_s^2 V_s}\right)^k \tag{7}$$

Since the side-blown nozzle is located in the high vacuum oil, the kinematic viscosity here is selected as the high vacuum oil kinematic viscosity, υ_{oil}, m^2·s^{-1}. The tracer feeding position and the insertion position of the side nozzle are closely related to the total height of the water phase and oil phase, the D_S in the tracer feeding position number and the insertion depth number of side nozzle are replaced by the total height of the water phase and oil phase H_{oil+w}, which can be converted into the following equation:

$$\pi = \left(\frac{\upsilon_{oil}}{D_s V_s}\right)^b \left(\frac{\sigma}{\rho D_s V_s^2}\right)^c \left(\frac{\tau V_s}{D_s}\right)^d \left(\frac{g D_s}{V_s^2}\right)^e \left(\frac{\varepsilon}{\rho D_s^2 V_s^3}\right)^f \left(\frac{H_i}{H_{oil+w}}\right)^h \left(\frac{h_s}{H_{oil+w}}\right)^i \left(\frac{q_s}{D_s^2 V_s}\right)^k \tag{8}$$

In this work, the angle between the upper side nozzle and the lower side nozzle is also a relatively important factor, including the relative angle and horizontal angle, so the angle factor should be taken into account when studying the dimensionless groups. The momentum dimensionless group is studied based on different angles. δ is introduced here, that is,

$$\delta = \left(\frac{\pi}{4}\right)^{\frac{1}{2}} \times \left(2 \times (V_s \times \cos\alpha)^2\right)^{\frac{1}{4}} \times \left(\frac{D_f}{(Q_s)^{\frac{1}{2}}}\right) \tag{9}$$

where α is combined the the relative angle between the upper side nozzle and the lower side nozzle and the horizontal angle of the upper side nozzle and the lower side nozzle. This parameter synthetically considers the influence of flow velocity with various angles, V_s is flow velocity of the side nozzle, $m{\cdot}s^{-1}$. D_f is the diameter of the furnace, m, and Q_s is the volumetric flow rate, $m^3{\cdot}s^{-1}$.

Through the above dimensionless group derivation, a series of dimensionless groups can be obtained as follows in Table 3.

Table 3. Expression of dimensionless groups based on Equations (8) and (9).

Dimensionless Groups	Expressions
π_1	$\dfrac{v_{oil}}{D_s V_s}$
π_2	$\dfrac{\sigma}{\rho D_s V_s^2}$
π_3	$\dfrac{\tau V_s}{D_s}$
π_4	$\dfrac{g D_s}{V_s^2}$
π_5	$\dfrac{\varepsilon}{\rho D_s^2 V_s^3}$
π_6	δ
π_7	$\dfrac{q_s}{D_s^2 V_s}$
π_8	$\dfrac{H_i}{H_{oil+w}}$
π_9	$\dfrac{h_s}{H_{oil+w}}$

The original dimensionless groups mentioned above can be substituted into Equation (10) to obtain

$$\pi = \varphi\left(\frac{v_{oil}}{D_s V_s}, \frac{\sigma}{\rho D_s V_s^2}, \frac{\tau V_s}{D_s}, \frac{g D_s}{V_s^2}, \frac{\varepsilon}{\rho D_s^2 V_s^3}, \delta, \frac{H_i}{H_{oil+w}}, \frac{h_s}{H_{oil+w}}, \frac{q_s}{D_s^2 V_s}\right) \tag{10}$$

Extracting the τ from Equation (10) and rearranging, we get

$$\tau = \frac{D_s}{V_s}\left(\frac{v_{oil}}{D_s V_s}, \frac{\sigma}{\rho D_s V_s^2}, \frac{g D_s}{V_s^2}, \frac{\varepsilon}{\rho D_s^2 V_s^3}, \delta, \frac{H_i}{H_{oil+w}}, \frac{h_s}{H_{oil+w}}, \frac{q_s}{D_s^2 V_s}\right) \tag{11}$$

where v_{oil} is the kinematic viscosity of the high vacuum oil, $m^2{\cdot}s^{-1}$, σ is the surface tension of the high vacuum oil, $kg{\cdot}s^{-2}$, g is the acceleration of gravity, $m{\cdot}s^{-2}$, ε is the stirring energy of the side nozzle, $kg{\cdot}m^2{\cdot}s^{-3}$, δ is the dimensionless groups of momentum, q_s is the flow rate of the single side nozzle, $m^3{\cdot}s^{-1}$, h_s is the insertion depth of the side nozzle, m, H_{oil+w} is the total height of the high vacuum oil and water, m, D_s is the diameter of the single side nozzle, m, and V_s is the flow velocity of the side nozzle, $m{\cdot}s^{-1}$.

The dimensionless groups related to Equation (11) based on a physical chemistry handbook are shown in Table 4:

Table 4. The dimensionless groups associated with Equation (11) based on the physical chemistry handbook [31].

Symbol	Name	Expression
Re	Reynolds number	$\dfrac{D_s V_s}{\upsilon_{oil}}$
Ca	Capillary number	$\dfrac{\rho \upsilon_{oil} V_s}{\sigma}$
La$_1$	Lagrange group	$\dfrac{\varepsilon \sigma_{oil}}{\rho D_s^3 V_s^4}$
H$_{o1}$	Homochronous number	$\dfrac{\tau V_s}{D_s}$
K$_F$	Capillarity-buoyancy number	$\dfrac{g \rho^4 \upsilon_{oil}^4}{\rho \sigma^3}$
Z	Ohnesorge number	$\dfrac{\rho \upsilon_{oil}}{(\rho \sigma D_s)^{\frac{1}{2}}}$
δ	Diameter group	$\left(\dfrac{\pi}{4}\right)^{\frac{1}{2}} \times \left(2 \times (V_s \times \cos\alpha)^2\right)^{\frac{1}{4}} \times \left(\dfrac{D_f}{(Q_s)^{\frac{1}{2}}}\right)$
j$_M$	J-factor	$\left(\dfrac{q_s \upsilon_{oil}^8}{D_s^{10} V_s^9}\right)^{\frac{1}{3}}$
N$_{sh}$	Sherwood number	$\dfrac{V_s D_s^2}{q_s}$
N$_{sc}$	Schmidt number	$\dfrac{\upsilon_{oil} D_s}{q_s}$

Combining Tables 3 and 4, the dimensionless groups in Equation (11) can be sorted into Table 5.

Table 5. The dimensionless groups in Equation (11).

Dimensionless Groups	Expression
$\pi_1 = \dfrac{\upsilon_{oil}}{D_s V_s}$	$\dfrac{1}{\mathrm{Re}}$
$\pi_2 = \dfrac{\sigma}{\rho D_s V_s^2}$	$\dfrac{1}{\mathrm{Ca \times Re}}$
$\pi_3 = \dfrac{\tau V_s}{D_s}$	H_{o1}
$\pi_4 = \dfrac{g D_s}{V_s^2}$	$\dfrac{\mathrm{K_F}}{\mathrm{Z}^2} \times \dfrac{1}{\mathrm{Ca}}$
$\pi_5 = \dfrac{\varepsilon}{\rho D_s^2 V_s^3}$	$\mathrm{La}_1 \times \mathrm{Re}$
$\pi_6 = \delta$	δ
$\pi_7 = \dfrac{q_s}{D_s^2 V_s}$	$\dfrac{\mathrm{j_M \times N_{sh}}}{\mathrm{N}_{sc}^{\frac{5}{3}}} \times \mathrm{Re}$

Based on the fact that $\frac{\mathrm{K_F}}{\mathrm{Z}^2}$ is constant in this condition, it will be removed here. At the same time, N_{jm} is named in order to simplify the expression of $\frac{\mathrm{j_M \times N_{sh}}}{\mathrm{N}_{sc}^{\frac{5}{3}}}$, that is $N_{jm} = \frac{\mathrm{j_M \times N_{sh}}}{\mathrm{N}_{sc}^{\frac{5}{3}}}$. According to Tables 3–5 and the actual working conditions of this study, the equation of dimensionless groups can be expressed as follows:

$$\tau = \frac{D_s}{V_s}\left(\mathrm{Re}^a, \mathrm{Ca}^b, \mathrm{La}_1{}^c, \delta^d, \mathrm{N}_{jm}^e, \left(\frac{H_i}{H_{oil+w}}\right)^f, \left(\frac{h_s}{H_{oil+w}}\right)^g\right) \tag{12}$$

As can be seen from Table 4, $\frac{\mathrm{Re}}{\mathrm{Ca}} = \frac{\sigma D_s}{\rho \upsilon_{oil}^2}$ is a constant. Since Ca is the dimensionless group representing the relative effect of viscous drag forces versus surface tension forces acting across an interface between a liquid and a gas, or between two immiscible liquids, so the Capillary number (Ca) is retained here, thus the following dimensionless groups equation can be further obtained:

$$\tau = \frac{D_s}{V_s}\left(\mathrm{Ca}^a, \mathrm{La}_1{}^b, \delta^c, \mathrm{N}_{jm}^d, \left(\frac{H_i}{H_{oil+w}}\right)^e, \left(\frac{h_s}{H_{oil+w}}\right)^f\right) \tag{13}$$

where a, b, c, d, e, and f are the empirical coefficient, to be derived from the experimental data.

Through the analysis of the data obtained in the hydraulic simulation experiment, the influence of the dimensionless quantities and the mixing time was obtained.

4. Results and Discussions

In this study, multiple linear regression was used to fit the dimensionless groups equation in order to obtain many empirical coefficients in Equation (13). Comparing the calculated value of the fitted dimensionless groups equation with the water simulation experiment value allowed us to verify the accuracy. The multiple linear regression equation of dimensionless groups can be rewritten into the following mode:

$$\lg\frac{\tau V_s}{D_s} = a_1\lg Ca + a_2\lg La_1 + a_3\lg\delta + a_4\lg N_{jm} + a_5\lg\left(\frac{H_i}{H_{oil+w}}\right) + a_6\lg\left(\frac{h_s}{H_{oil+w}}\right) \tag{14}$$

4.1. The Effects of Ca and La₁

After the parameters such as the injection velocity of the upper side nozzle and the lower side nozzle are respectively substituted into Equation (14), the data of multiple linear regression are shown in the Table 6:

Table 6. Value and standard error of lg Ca and lg La₁ by multiple linear regression for the upper side nozzle and lower side nozzle.

	Y Axis	X Axis	Value	Standard Error
Upper side nozzle	$\lg\tau_{exp}V_{us}/D_s$	"lgCa"	−0.9564	0.0684
		"lgLa₁"	−0.6496	0.0006
Lower side nozzle	$\lg\tau_{exp}V_{ls}/D_s$	"lgCa"	−0.7403	0.0647
		"lgLa₁"	−0.6685	0.0058

From Table 6, the equation of the mixing time of the upper side nozzle and the lower side nozzle and the similar number of Ca, La₁ can be obtained, respectively:

$$\tau = \frac{D_s}{V_{us}}\left(Ca^{-0.96}La_1^{-0.65}\right) \tag{15}$$

$$\tau = \frac{D_s}{V_{ls}}\left(Ca^{-0.74}La_1^{-0.67}\right) \tag{16}$$

where V_{us} is the injection velocity of the upper side nozzle, m·s^{-1}, V_{ls} is the injection velocity of the lower side nozzle, m·s^{-1}.

Figure 5a,b shows the relationship between the experimental value and the calculated value of the dimensionless group of the mixing time of the upper side nozzle and the lower side nozzle, respectively, made according to the Equations (15) and (16). As can be seen from Figure 5a,b, for both the upper side nozzle and the lower side nozzle, the experimental values have a considerable linear relationship with the calculated values. At the same time, to further verify the comparison, Figure 6a–f shows the relation diagram of the experimental value and the calculated value of the mixing time dimensionless groups in three monitoring points τ_1, τ_2 and τ_3 of the upper and lower side nozzle, respectively. It can also be seen that it has a higher consistency with the τ (cf. Figure 5).

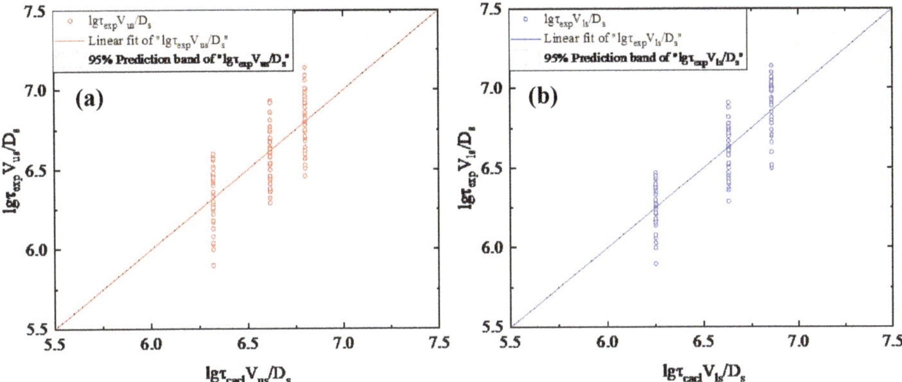

Figure 5. Comparison of experimental $\lg\tau V_s/D_s$ with calculated ones for upper side nozzle (**a**) and lower side nozzle (**b**), respectively, using proposed Equations (15) and (16).

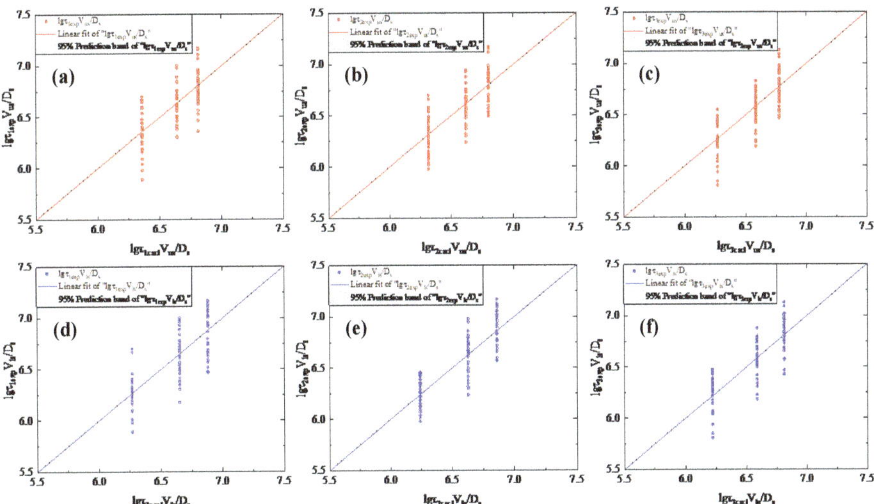

Figure 6. Comparison of experimental $\lg\tau V_s/D_s$ with calculated ones for upper side nozzle (**a**–**c**) and lower side nozzle (**d**–**f**), respectively, using τ_1, τ_2, τ_3.

4.2. The Effects of Ca, La$_1$ and N$_{jm}$

After studying the effect of Ca and La$_1$, the N$_{jm}$ dimensionless group is added here for the purpose of investigating the relationship between the dimensionless group of the mixing time and the dimensionless groups such as density, dynamic viscosity, stirring energy, flow rate, and so on based on the variable of velocity. The following table contains the multiple linear regression data based on the three dimensionless numbers of upper side nozzle and lower side nozzle, respectively:

From Table 7, the equations between the dimensionless group of the mixing time and the three dimensionless groups of Ca, La$_1$ and N$_{jm}$ for the upper and lower side nozzles can be obtained, respectively, as shown below:

$$\tau = \frac{D_s}{V_{us}}\left(\mathrm{Ca}^{-1.90}\mathrm{La_1}^{-2.61}\mathrm{N}_{jm}^{5.01}\right) \tag{17}$$

$$\tau = \frac{D_s}{V_{ls}}\left(Ca^{-1.51}La_1^{-2.27}N_{jm}^{4.11}\right) \tag{18}$$

Table 7. Value and standard error of lg Ca, lg La$_1$, and lgN$_{jm}$ by multiple linear regression for upper side nozzle and low side nozzle.

	Y Axis	X Axis	Value	Standard Error
Upper side nozzle	$\lg\tau_{\exp}V_{us}/D_s$	"lgCa"	−1.8971	1.3603
		"lgLa$_1$"	−2.6093	2.8302
		"lgN$_{jm}$"	5.0122	7.2387
Lower side nozzle	$\lg\tau_{\exp}V_{ls}/D_s$	"lgCa"	−1.5112	1.2876
		"lgLa$_1$"	−2.2745	2.679
		"lgN$_{jm}$"	−4.1078	6.8521

Similarly, by drawing the experimental values and calculated values for the dimensionless group of mixing time, we can see from Figure 7a,b that they have a good linear correlation, and the comparison chart (cf. Figure 8a–f) of τ_1, τ_2, τ_3 is also shown a high consistency with τ (cf. Figure 7), respectively.

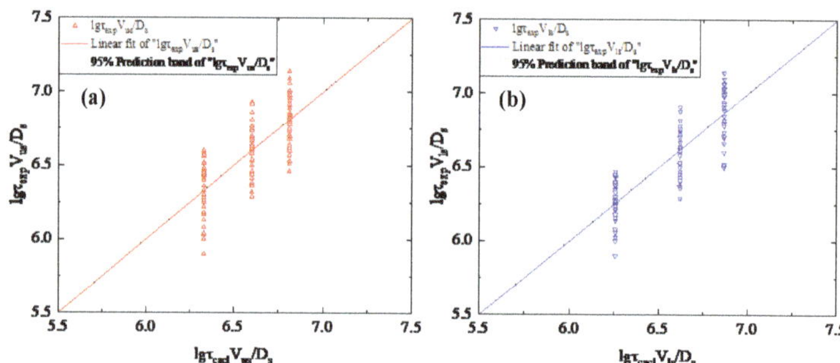

Figure 7. Comparison of experimental $\lg\tau V_s/D_s$ with calculated ones for upper side nozzle (**a**) and lower side nozzle (**b**), respectively, using proposed Equations (17) and (18).

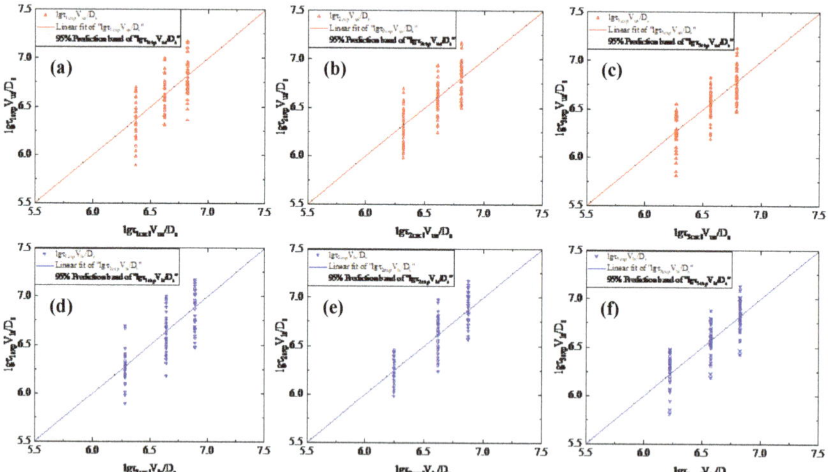

Figure 8. Comparison of experimental $\lg\tau V_s/D_s$ with calculated ones for upper side nozzle (**a–c**) and lower side nozzle (**d–f**) respectively using τ_1, τ_2, τ_3.

4.3. The Effects of Ca, La₁, Nⱼₘ, and δ

The above three dimensionless groups are mainly based on the variable factor of injection velocity. Nevertheless, the variables of the angles are also important variable parameters, such as the relative angle of the upper and lower side nozzle, the horizontal angle of the upper side nozzle, the horizontal angle of the lower side nozzle, etc. and dimensionless group δ is the group of the related angle. Therefore, the dimensionless group δ is introduced here to study the expression of the dimensionless group of mixing time under the action of these four dimensionless groups. The following Table 8 shows multiple linear regression data for the experimental values of four dimensionless groups to the mixing time:

Table 8. Value and standard error of lg Ca, lg La₁, lg Nⱼₘ and lg δ by multiple linear regression for the upper side nozzle and lower side nozzle.

	Y Axis	X Axis	Value	Standard Error
Upper side nozzle	$\lg\tau_{\exp} V_{us}/D_s$	"lgCa"	−1.8802	1.3685
		"lgLa₁"	−2.7131	2.8806
		"lgNⱼₘ"	5.3305	7.4090
		"lgδ"	0.1140	0.5086
Lower side nozzle	$\lg\tau_{\exp} V_{ls}/D_s$	"lgCa"	−1.8503	1.3312
		"lgLa₁"	−2.6127	2.7000
		"lgNⱼₘ"	4.7726	6.8838
		"lgδ"	−0.4049	0.4034

Based on the data in Table 8, the following equations of the mixing time dimensionless group of the upper side nozzle and the mixing time dimensionless group of the lower side nozzle can be obtained, respectively:

$$\tau = \frac{D_s}{V_{us}}\left(Ca^{-1.88}La_1^{-2.71}N_{jm}^{5.33}\delta^{0.11}\right) \tag{19}$$

$$\tau = \frac{D_s}{V_{ls}}\left(Ca^{-1.85}La_1^{-2.61}N_{jm}^{4.77}\delta^{-0.40}\right) \tag{20}$$

Finally, by comparing the experimental values of the dimensionless group of mixing time with the calculated value of the dimensionless group, it can be seen that the experimental value of the dimensionless group has a good linear relationship with the calculated value of the dimensionless group (cf. Figure 9a,b), and it can also be seen from Figure 10a–f that there is a high consistency between the figure τ_1, τ_2, τ_3 (cf. Figure 10) and τ (cf. Figure 9).

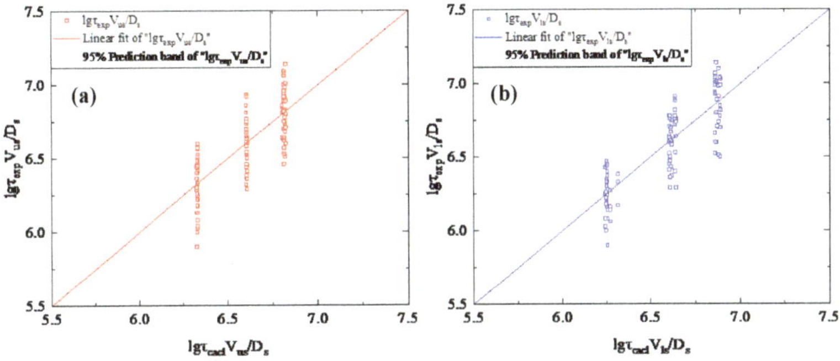

Figure 9. Comparison of experimental $\lg\tau V_s/D_s$ with calculated ones for upper side nozzle (**a**) and lower side nozzle (**b**), respectively, using proposed Equations (19) and (20).

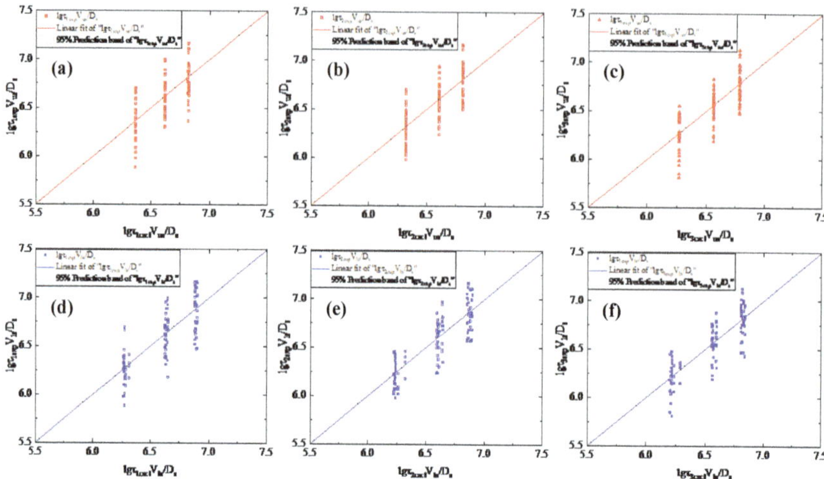

Figure 10. Comparison of experimental $\lg\tau V_s/D_s$ with calculated ones for upper side nozzle (**a–c**) and lower side nozzle (**d–f**), respectively, using τ_1, τ_2, τ_3.

4.4. The Effects of Ca, La₁, N_{jm}, δ, H_i and h_i

The position of the tracer feeding position and the insertion depth of the side nozzle are also important variable parameters. The expression of the dimensionless group of mixing time was studied after adding these two dimensionless groups. On the basis of the above dimensionless groups, the dimensionless group related to the tracer feeding position and the dimensionless group related to the insertion depth of the side nozzle was added unceasingly into the equation of mixing time dimensionless group. After substituting the parameters such as the flow velocity of upper side nozzle and the insertion depth of upper side nozzle into the Equation (14), multiple linear regression was performed through the origin, and the results are shown in in Table 9:

Table 9. Value and standard error of lg Ca, lg La₁, lg N_{jm}, lgδ, lgH_i/(H_{oil+w}), and lg(h_s/H_{oil+w}) by multiple linear regression for upper side nozzle and lower side nozzle.

	Y Axis	X Axis	Value	Standard Error
		"lgCa"	−1.8704	1.2735
		"lgLa₁"	−2.7167	2.6804
Upper side nozzle	$\lg\tau_{exp}V_{us}/D_s$	"lgN_{jm}"	5.3508	6.8942
		"lgδ"	0.1180	0.4733
		"lgH_i/(H_{oil+w})"	−0.3671	0.0872
		"lg(h_{us}/H_{oil+w})"	0.0103	0.0168
		"lgCa"	−1.8389	1.2257
		"lgLa₁"	−2.6102	2.4860
Lower side nozzle	$\lg\tau_{exp}V_{ls}/D_s$	"lgN_{jm}"	4.7765	6.3382
		"lgδ"	−0.4019	0.3715
		"lgH_i/(H_{oil+w})"	−0.3632	0.0808
		"lg(h_{ls}/H_{oil+w})"	0.0108	0.0157

As can be seen from Table 9, the equation of the dimensionless group of the mixing time of the upper side nozzle is as follows in Equation (21):

$$\tau = \frac{D_s}{V_{us}}\left(\mathrm{Ca}^{-1.87}\mathrm{La_1}^{-2.72}\mathrm{N}_{jm}^{5.35}\delta^{0.12}\left(\frac{H_i}{H_{oil+w}}\right)^{-0.36}\left(\frac{h_{us}}{H_{oil+w}}\right)^{0.01}\right) \tag{21}$$

$$\tau = \frac{D_s}{V_{ls}}(\mathrm{Ca}^{-1.84}\mathrm{La}_1{}^{-2.61}\mathrm{N}_{jm}{}^{4.77}\delta^{-0.40}\left(\frac{H_i}{H_{oil+w}}\right)^{-0.36}\left(\frac{h_{ls}}{H_{oil+w}}\right)^{0.01}) \qquad (22)$$

Similarly to the upper side nozzle, Table 9 can also be obtained after the injection velocity and insertion depth of the lower side nozzle are replaced in the Equation (14) by multiple linear regression. According to the data in Table 9, the equation of the dimensionless group of mixing time of lower side nozzle can be obtained in Equation (22).

By comparing the dimensionless group calculated by the dimensionless Equations (21) and (22) with the dimensionless groups measured in the experiment, the correctness of the above dimensionless group can be verified. Figure 11a shows the relationship between the experimental value and the calculated value of the mixing time dimensionless group for the upper side nozzle, while Figure 11b shows the relationship between the experimental value and the calculated value of the mixing time dimensionless group for the lower side nozzle. It can be seen from Figure 11a,b that the fitting effect of a dimensionless group of mixing time is good. At the same time, in order to further verify its accuracy, Figure 12a–f shows the fitting curves of the three mixing time monitoring points at the respective operating conditions of the upper side nozzle and the lower side nozzle, respectively. It can also be seen that the distribution of the showcases 1, 2, and 3 has a relatively high degree of fitness. Therefore, it can be concluded that the equation of the mixing time dimensionless group of the upper side nozzle and the equation of the mixing time dimensionless group of the lower side nozzle have practical reference value.

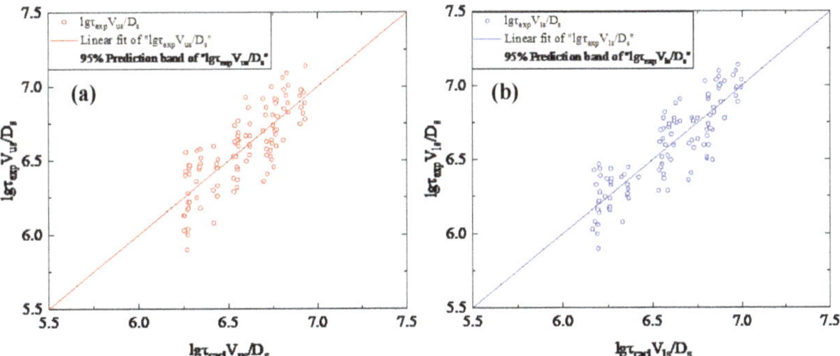

Figure 11. Comparison of experimental $\lg\tau V_s/D_s$ with calculated ones for upper side nozzle (**a**) and lower side nozzle (**b**), respectively, using proposed Equations (21) and (22).

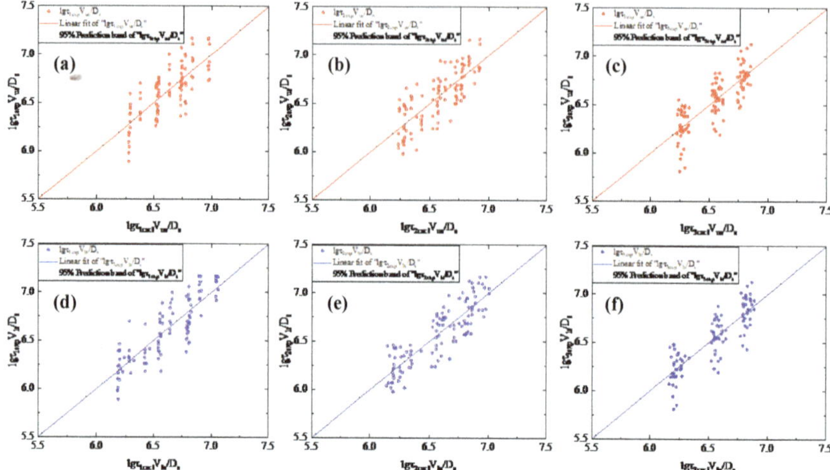

Figure 12. Comparison of experimental $\lg \tau V_s/D_s$ with calculated ones for upper side nozzle (**a–c**) and lower side nozzle (**d–f**), respectively, using τ_1, τ_2, τ_3.

5. Conclusions

Based on the experimental research and data analysis, the dimensionless groups of mixing time and kinetic viscosity, the surface tension, the tracer feeding position, the insertion depth of the side nozzle, etc., were obtained. In this paper, the expressions between the dimensionless group of mixing time and dimensionless groups such as Ca and La_1 were obtained by multiple linear regression. It can be seen from the expressions that the indexes of the dimensionless groups have a higher identity when they have more than three dimensionless groups. By verifying the calculated and experimental values of the dimensionless group of mixing time, it can be seen that both have a good positive correlation. At the same time, it can also be seen from the comparison between the calculated values of τ_1, τ_2, τ_3 and the experimental values that they are in good agreement with the corresponding τ, which indicates that the fitting expressions have higher reliability. Because density, surface tension, and other parameters of the medium have not been changed in this study, Equations (15) and (16) are more suitable for the study of the side nozzle velocity and related angle. Equations (21) and (22) will be of great significance when the density, viscosity, surface tension, and furnace diameter of the medium are changed in further work. This conclusion will better provide help for the control of key parameters, help to establish the design standard of C-H_2 smelting reduction furnaces, and lay a foundation for the optimization of side nozzle parameters of C-H_2 smelting reduction furnaces.

Author Contributions: Conceptualization, J.X. and J.Z.; Data curation, J.X. and J.Z.; Formal analysis, J.X.; Funding acquisition, J.Z.; Investigation, J.X. and J.Z.; Methodology, J.X., B.W., and J.Z.; Project administration, J.Z.; Resources, J.X., B.W., and J.Z.; Supervision, J.Z.; Validation, J.X., B.W., and J.Z.; Writing—original draft, J.X.; Writing—review and editing, J.X., B.W., and J.Z. All authors have read and agreed to the published version of the manuscript.

Funding: This study was funded by the National Science and Technology Support Program for development of smelting reduction iron smelting process based on hydrogen metallurgy (2006BAE03A12).

Acknowledgments: The authors gratefully acknowledge the resources partially provided by the State Key Laboratory of Advanced Special Steel, Shanghai University of Materials Science and Engineering.

Conflicts of Interest: The authors declare no conflict of interest.

References

1. Tylecote, R.; Hua, J. *History of Metallurgical Development in the World*, 1st ed.; Beijing Science and Technology Literature Publishing House: Beijing, China, 1985; pp. 576–582.

2. Chen, J.; Lin, W.; Zhao, J. *Non-Coking Coal Metallurgy Technology*, 1st ed.; Chemical Industry Press: Beijing, China, 2007; pp. 25–34.

3. Mohsenzadeh, F.; Payab, H.; Abedi, Z.; Mohammad, A. Reduction of CO_2 emissions and energy consumption by improving equipment in direct reduction ironmaking plant. *Clean Technol. Environ. Policy* **2019**, *21*, 847–860. [CrossRef]

4. Ali, H.; Marlene, A.; Lynn, P. Alternative emerging ironmaking technologies for energy-efficiency and carbon dioxide emissions reduction: A technical review. *Renew. Sustain. Energy Rev.* **2014**, *33*, 645–658.

5. You, Y.; Li, Y.; Luo, Z.; Lia, H.; Zou, Z.; Yang, R. Investigating the effect of particle shape on the charging process in melter gasifiers in COREX. *Powder Technol.* **2019**, *351*, 305–313. [CrossRef]

6. Xu, C.; Cang, D. A brief overview of low CO_2 emission technologies for iron and steel making. *J. Iron Steel Res. Int.* **2010**, *17*, 1–7. [CrossRef]

7. Guo, Z.; Xie, Y.; Wang, D. Evaluation of sulfur distribution in the process of coal-based smelting reduction ironmaking. *Energy Convers. Manag.* **1997**, *38*, 1413–1419.

8. He, Y.; Li, C.; Wei, G.; Zou, Z. Static model of iron bath smelting reduction process with thick slag layer. *J. Northeast. Univ. Nat. Sci.* **2015**, *36*, 651–654.

9. Li, C.; He, Y.; Li, Q.; Zou, Z. Physical simulation of fluid mixing in a smelting reduction iron-bath with thick slag layer. *J. Northeast. Univ. Nat. Sci.* **2014**, *35*, 1266–1269.

10. Srishilan, C.; Shukla, A. Static thermochemical model of COREX melter gasifier. *Metall. Mater. Trans. B* **2018**, *49*, 388–398. [CrossRef]

11. Behera, P.; Bhoi, B.; Paramguru, R.; Mukherjee, P.; Mishra, B. Hydrogen plasma smelting reduction of Fe_2O_3. *Metall. Mater. Trans. B* **2018**, *50*, 262–270. [CrossRef]

12. Mandal, A.; Sinha, O. Recovery of multi-metallic components from bottom ash by smelting reduction under plasma environment. *Metall. Mater. Trans. B* **2016**, *47*, 19–22. [CrossRef]

13. Shim, Y.; Jung, S. Conditions for minimizing direct reduction in smelting reduction iron making. *ISIJ Int.* **2018**, *58*, 274–281. [CrossRef]

14. You, Z.; Li, G.; Wen, P.; Peng, Z.; Zhang, Y.; Jiang, T. Reduction of Sn-bearing iron concentrate with mixed H_2/CO gas for preparation of Sn-enriched direct reduced iron. *Metall. Mater. Trans. B* **2017**, *48*, 1486–1493. [CrossRef]

15. Park, H.; Sohn, I.; Tsalapatis, J.; Sahajwalla, V. Reduction behavior of dolomite-fluxed magnetite: Coke composite pellets at 1573 K (1300 °C). *Metall. Mater. Trans. B* **2018**, *49*, 1109–1118. [CrossRef]

16. Park, H.; Sohn, I.; Freislich, M.; Sahajwalla, V. Investigation on the reduction behavior of coal composite pellet at temperatures between 1373 and 1573 K. *Steel Res. Int.* **2017**, *88*, 1–13. [CrossRef]

17. Zhen, X.; Wu, J.; Huang, Y.; Han, Y.; Yao, J. Parametric dimensional analysis on the structural response of an innovative subsurface tension leg platform in ultra-deep water. *China Ocean Eng.* **2018**, *32*, 482–489. [CrossRef]

18. Vatankhah, A.; Mirnia, H. Predicting discharge coefficient of triangular side orifice under free fow conditions. *J. Irrig. Drain. Eng.* **2018**, *144*, 04018030. [CrossRef]

19. Meng, X.; Zhu, M. Non-sinusoidal vibration parameter control of continuous casting mold based on dimensional analysis. *Metal Ind. Autom.* **2006**, *S2*, 142–144.

20. Feng, S. Design of side-blown melting reduction furnace and its application. *China Nonferrous Metall.* **2015**, *3*, 19–21.

21. Wang, M.; Zhang, L.; Yu, S.; Tu, G.; Sui, Z. Penetration behaviour of immersion side-blowing airstream in molten bath. *Nonferrous Met.* **2006**, *58*, 50–52.

22. Wang, Z.; Lei, M.; Zhang, J.; Zheng, S.; Wang, B.; Hong, X. Experimental study on fluid flow characteristics in the smelting reduction furnace by water model. *Chin. J. Process Eng.* **2009**, *9*, 36–40.

23. Lei, M.; Zhang, J.; Wang, Z.; Wang, B.; Zheng, S.; Hong, X. Effect of side and top lance blowing on fluid flow in smelting reduction furnace. *J. Chin. Rare Earth Soc.* **2008**, *26*, 247–251.

24. Yin, D.; Feng, K.; Cheng, W.; Wang, B.; Mao, J.; Xie, J.; Zhang, J.; Zheng, S.; Hong, X. Experimental study on side-blowing nozzles in smelting reduction furnace with water model. *Chin. J. Process Eng.* **2010**, *10*, 83–87.

25. Meng, S.; Zhang, H. Simulation and analysis of molten reduction process in iron bath. *Angang Technol.* **1995**, *11*, 15–18.

26. Dong, L.; Liu, Q. Present study status on the final reduction of iron bath smelting reduction. *J. Iron Steel Res.* **1999**, *11*, 65–69.

27. Iguchi, M.; Nakamura, K.; Tsujino, R. Mixing time and fluid flow phenomena in liquids of varying kinematic viscosities agitated by bottom gas injection. *Metall. Mater. Trans. B* **1998**, *29*, 569–575. [CrossRef]

28. Palmer, A. *Dimensional Analysis and Intelligent Experimentation*, 1st ed.; World Scientific Publishing: Singapore, 2008; pp. 38–49.

29. Quéau, L.; Kimiaei, M.; Randolph, M. Dimensionless groups governing response of steel catenary risers. *Ocean Eng.* **2013**, *74*, 247–259. [CrossRef]

30. Terrazas, M.; Conejo, A. Effect of nozzle diameter on mixing time during bottom-gas injection in metallurgical ladles. *Metall. Mater. Trans. B* **2015**, *46*, 711–718. [CrossRef]

31. Robert, C.; Melvin, J.; William, H. *Handbook of Chemistry and Physics*, 66th ed.; CRC Press: Boca Raton, FL, USA, 1985; pp. 314–324.

Article

Model Study on Burden Distribution in COREX Melter Gasifier

Haifeng Li [1,2,3,*], **Zongshu Zou** [1,2,3,*], **Zhiguo Luo** [1,3], **Lei Shao** [1,3] and **Wenhui Liu** [1,3]

1   Key Laboratory for Ecological Metallurgy of Multi-metallic Mineral (Ministry of Education), Northeastern University, Shenyang 110819, Liaoning, China; luozg@smm.neu.edu.cn (Z.L.); shaolei@smm.neu.edu.cn (L.S.); knight5300@163.com (W.L.)
2   State Key Laboratory of Rolling and Automation, Northeastern University, Shenyang 110819, Liaoning, China
3   School of Metallurgy, Northeastern University, Shenyang 110819, Liaoning, China
*   Correspondence: lihf@smm.neu.edu.cn (H.L.); zouzs@mail.neu.edu.cn (Z.Z.)

Received: 21 October 2019; Accepted: 26 November 2019; Published: 1 December 2019

Abstract: COREX is one of the commercialized smelting reduction ironmaking processes. It mainly includes two reactors, i.e., a (reduction) shaft furnace (SF) and a melter gasifier (MG). In comparison with the conventional blast furnace (BF), the COREX MG is not only equipped with a more complicated top charging system consisting of one gimbal distributor for coal and eight flap distributors for direct reduction iron (DRI), but also the growth mechanism of its burden pile is in a developing phase, rather than that in a fully-developed phase in a BF. Since the distribution of charged burden plays a crucial role in determining the gas flow and thus in achieving a stable operation, it is of considerable importance to investigate the burden distribution influenced by the charging system of COREX MG. In the present work, a mathematical model is developed for predicting the burden distribution in terms of burden layer structure and radial ore/coal ratio within the COREX MG. Based on the burden pile width measured in the previous physical experiments at different ring radii on a horizontal flat surface, a new growth mechanism of burden pile is proposed. The validity of the model is demonstrated by comparing the simulated burden layer structure with the corresponding results obtained by physical experiments. Furthermore, the usefulness of the mathematical model is illustrated by performing a set of simulation cases under various charging matrixes. It is hoped that the model can be used as a what-if tool in practice for the COREX operator to gain a better understanding of burden distribution in the COREX MG.

Keywords: COREX melter gasifier; mixed charging; burden layer structure; burden pile width

---

1. Introduction

Steel is the world's most popular construction material due to its durability, processability, and cost cheapness. However, producing steel brings high energy consumption and CO_2 emissions, especially in ironmaking process. In order to minimize the energy consumption and CO_2 emissions of the ironmaking process, some alternative liquid iron production technologies to blast furnace (BF), such as the COREX process and the FINEX process, have been developed [1]. The COREX process is a smelting-reduction process developed by Siemens Voest-Alpine Industrieanlagenbau Gmbh & Co. (VAI) in the 1970s, for cost-efficient and environment-friendly production of hot metal from iron ore and non-coking coal. Eight COREX units in the world have been put into use and successfully commercialized in different areas, e.g., South Africa, India, and China; therein, two of them are the latest generation of COREX with a capacity of 1.5 million tons of liquid iron per year and were built in China at the Baosteel Luojing steel plant.

In the COREX process [2], all the metallurgical reactions take place in two separate process reactors, the upper shaft furnace (SF) for the iron ore pre-reduction and the lower melter gasifier (MG) for final reduction and smelting. A schematic process flow sheet is shown in Figure 1. Iron ore (lump ore, pellets, or a mixture thereof) is charged into the upper SF where the burden is reduced to direct reduced iron (DRI) by the reduction gas arising from the lower MG. Discharge screws convey the DRI from the SF into the MG where final reduction and melting take place in addition to all other metallurgical reactions. In comparison with a conventional BF, COREX MG is equipped with a more complicated top charging system consisting of one gimbal distributor for coal and eight flap distributors for DRI.

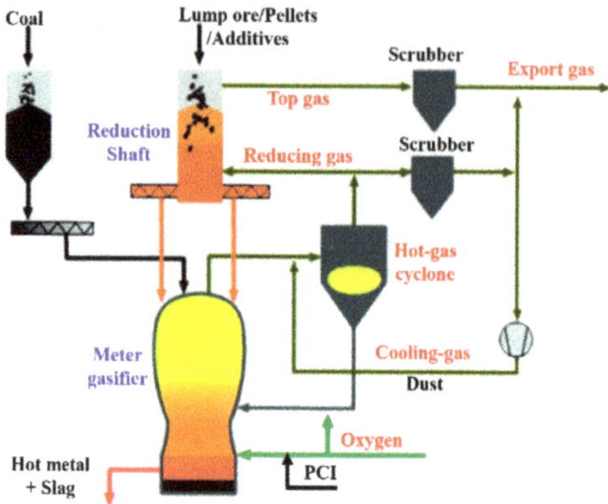

Figure 1. Schematic flow sheet of COREX process.

The BF concept is used, but the BF is virtually split into two parts at the cohesive zone interface (cf. Figure 2) in a COREX process. Compared with the conventional BF route, non-coking coal can be directly used for ore reduction and smelting in a COREX process, which eliminates the need for coke making units. The use of lump ore or pellets also dispenses with the need of sinter plants. Since coking and sintering plants are not required for the COREX process, substantial cost savings of up to 20% can be achieved in the production of hot metal, of a grade similar to that of the blast furnace [3].

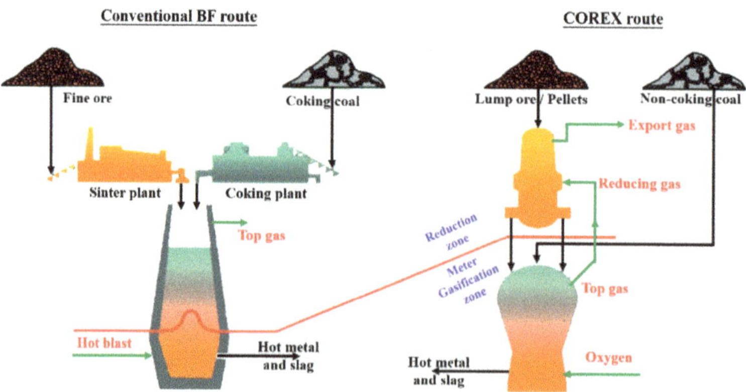

Figure 2. Comparison of concepts between blast furnace (BF) route and COREX route.

Since the distribution of charged burden plays a crucial role in determining the gas flow and thus in achieving a stable operation in BF and SF, it is of considerable importance to investigate the burden distribution influenced by the complicated charging system of COREX MG. The top charging system of COREX MG consists of one coal-gimbal distributor and eight DRI-flap distributors. The gimbal distributor distributes the coal to different radial positions in the furnace by adjusting the angle of the chute. The DRI-flap distributors can charge DRI into the furnace by changing the angle of the flap. The burden distribution is affected by many factors, such as loading equipment and charging patterns. However, as a closed high-temperature reactor, it is difficult to observe directly the burden surface profile and internal structure in the COREX MG. Therefore, it is necessary to establish a mathematical model to predict the profile of the burden surface and the internal structure of burden column.

Many studies [4–7] have established mathematical models of BF and SF charging with chute distributors, and compared with physical experiments [8] or data from industrial onsite experiments [9], and the rationality and accuracy of the models have been verified. However, the mathematical model studies on the charging process in COREX MG remain scarce. In recent years, more and more detailed research works have been carried out on the flow trajectory of the DRI-flap distributor and the coal-gimbal distributor by some research groups [10–16]. Although many physical experiments and numerical simulations have been conducted, sophisticated forecasting software or mathematical models similar to the fast evaluation model of BF charging [17–19] for predicting the burden surface profile and internal layer structure have not been formed.

The authors' experimental studies on COREX MG charging [11,12] has shown that, due to the high free space (low stockline) and small flow rate of burden charging in the COREX MG, the formation of burden pile is in a developing phase with growing pile angles, while that in a BF is commonly recognized as in a fully-developed phase with stable pile angles. The experimental measurement showed also that the pile width was practically the same as the width of burden flow arriving at the burden surface. This makes the formation of burden pile in COREX MG being different from that in a BF. Therefore, this work will focus on the formation process of burden pile when certain material reaches the burden surface, and thereby a thorough understanding of the charging process as well as burden pile evolution can be expected, especially for the mechanism of developing growing with developing pile angles.

2. Mathematical Model

The charging system of MG involves one coal-gimbal distributor and eight DRI-flap distributors. Reasonable charging processes are completed by various combinations of the two sets of equipment. The coal-gimbal distributor consists of consecutively a coal hopper, a down-comer, and the rotating chute; while a DRI-flap distributor consists of consecutively a vertical pipe, an inclined-pipe, and the angle-adjustable flap.

2.1. The Coal-Gimbal Distributor

The mathematical model for the coal-gimbal distributor is similar to the chute distributor of a modern BF with the bell-less charging system (cf. Figure 3). In the mathematical model, the charging process is divided into four consecutive fundamental steps: the free falling of particles from the hopper onto the chute, the sliding of particles along chute, the free falling of particles from the chute tip, and the formation of the burden pile. Detailed treatments of the first three fundamental steps can be found in a previous publication [18,19] of the authors. In the fourth step, compared with the charging process of BF, the burden has a longer descent distance and a wider flow width in the COREX charging process, so the formation mechanism of burden pile is quite different.

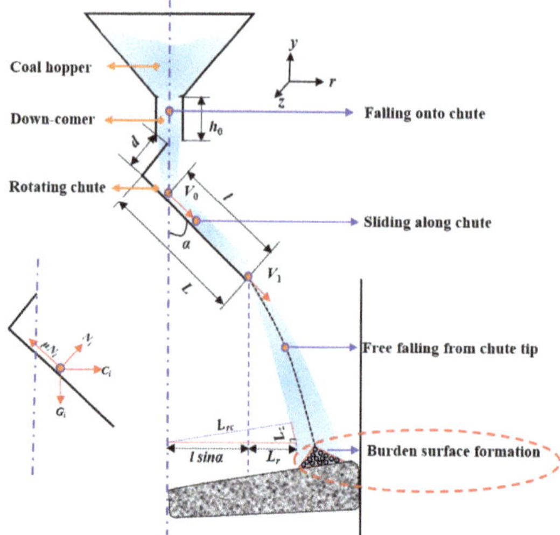

Figure 3. Schematic illustration of coal-gimbal distributor.

2.2. The DRI-Flap Distributor

The pathway of DRI particles in the DRI-flap distributor is shown in Figure 4. The mass of the particle is m. At the bottom of the vertical pipe, the particle attains the velocity of V_1 (m/s) normal to the inclined pipe and it attains the velocity of V_2 (m/s) at the end of the inclined pipe, then the particle collides with DRI-flap and attains the bouncing velocity of V_3 (m/s). The friction coefficient between particle and inclined-pipe is μ, and the length of inclined pipe is l_1. The angle between the inclined pipe and the horizontal direction is α_1, and that between V_3 and the vertical direction is α_2.

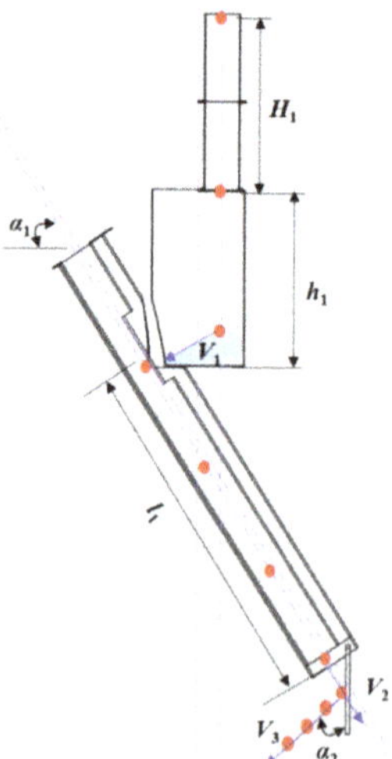

Figure 4. Pathway of direct reduction iron (DRI) in DRI-flap distributor.

DRI particles stored in the upper hopper fall onto the bottom of vertical pipe. This motion is simplified by assuming the flow of a bulk stream similar to that of an individual particle, which is stationary when leaving the upper hopper and gains an exit velocity V_1 along the direction normal to the inclined pipe.

$$V_1 = k_{1,\text{DRI}} \cos \alpha_1 \sqrt{2g(H_1 + h_1)} \tag{1}$$

where a correction coefficient, $k_{1,\text{DRI}}$, is introduced to take into account the imperfect elastic collision of the falling particles.

Since the moving direction of DRI particles from the vertical pipe is normal to the inclined pipe, the initial particle velocity along the inclined pipe is zero. Therefore, the velocity at the end of the inclined pipe is obtained as

$$V_2 = \sqrt{2g(\cos \alpha_1 - \mu \sin \alpha_1)l_1} \tag{2}$$

The velocity V_3 after colliding with the flap is

$$V_3 = k_{2,\text{DRI}} V_2 \tag{3}$$

where a correction coefficient, $k_{2,\text{DRI}}$, is introduced to take into account the imperfect elastic collision between the particles and flap.

Upon leaving the flap, the velocity of the bulk stream is decomposed into two components in the vertical direction ($V_3 \sin\alpha_2$) and the horizontal direction ($V_3 \cos\alpha_2$). After this, the bulk stream will form a parabolic motion similar to the free falling of a particle from a chute tip.

2.3. Formulation of Mathematical Model

The burden structure model in COREX MG is formulated as follows.

(1) Whenever the coal-gimbal distributor or the DRI-flap distributor is used, the burden pile is formed in a ring at certain radius and uniformly distributed in the circumferential direction. It is assumed that the two-dimensional section of the burden profile is a triangle, the shape of the triangle is determined by the width of the burden flow and the volume of the burden batch together, and the position of the triangle in radius direction is determined by the chute angle or flap angle.

(2) As mentioned above, the formation of COREX burden pile is in a developing phase, therefore, the width (W) of the burden pile is considered to be equal to the width of the burden flow arriving at the burden surface. The measured results for the width (W) of burden piles, the locations of the pile's left (L) and right (R) edges for coal, and DRI for different rings in physical experiment are listed in Table 1 [15].

Table 1. Measured results of the width of burden piles for coal and DRI in physical experiment (Units: mm).

Parameters	R0.5	R1.0	R1.5	R2.0	R2.5	R3.0	R3.5	R4.0	R4.5	R5.0
W_{coal}	-	220	240	265	280	300	320	350	365	-
L_{coal}	-	37	94	151	208	264	320	376	431	-
R_{coal}	-	257	334	412	490	567	644	721	797	-
W_{DRI}	-	-	-	300	-	330	-	350	-	370
L_{DRI}	-	-	-	134	-	237	-	389	-	563
R_{DRI}	-	-	-	437	-	563	-	738	-	935

Through regression analysis of the experimental data, the relationships between burden pile width and ring radius were obtained as shown by Equations (4) and (5) for coal and DRI respectively.

$$W_{\text{coal}} = 41.667R_i + 177.92 \tag{4}$$

$$W_{\text{DRI}} = 23R_i + 257 \tag{5}$$

(3) The impinging effect and the mixing between sequential layers (rings) are ignored. Based on observations from physical experiments, it was found that the larger and lighter coal particles more easily to roll than the ore particles if they charged on an inclined surface. Therefore, a modification coefficient (greater than 1) is introduced to take it into account.

2.4. Result of Stable Initial Burden Profile

Compared with BF, the COREX MG charging process has the characteristics of smaller volume of burden dumps (or batches) and lower height of stock level (13~14 m in MG, and 1~2 m in BF). Previous physical experiments [10–12] and numerical simulation [13,14] results indicate that, in MG, it is difficult to form a stable surface with certain internal and external repose angles similar to those in the BF [15,16]. In other words, the mechanism of burden pile formation is different in the two reactors. To solve this problem, a new method based on burden pile width is proposed to calculate the growing process of the pile. The model can be used to characterize the evolution of burden surface profile and internal layer structure under various charging matrixes.

The published physical experiment [12] and numerical simulation [20] studies of the mixed charging in MG show that the stable burden profile has a high center and a low edge with the descent of the whole burden column. By regression analysis of the physical experiment and numerical simulation data, an equation for the stable burden profile in the COREX MG can be obtained, as shown in Figure 5. Such a stable burden profile is then simplified into a line of three segments, that is, the horizontal center and edge segments and the inclined middle segment.

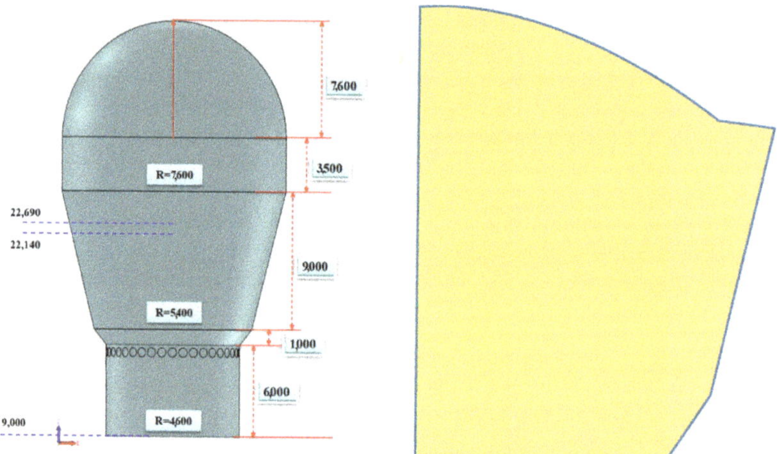

Figure 5. Stable initial burden profile.

2.5. Growing Mechanism of Burden Surface

The characterization of the growing of burden surface mainly requires the determination of the radius locations of the left and right edges of the pile according to the width of burden flow measured from previous physical experiments [12] under various angles of coal-gimbal chute and DRI-flap. Since the stable initial burden profile is composed of three segments, the growing mechanisms at different locations are more complicated than that on a horizontal flat profile. In this model, there are mainly five pile patterns as shown in Figure 6. In the figure, L and R represent the left and right ends of the burden pile, respectively, on a horizontal plane. OM represents the radius of the pile ring which is determined by the angles of coal-gimbal chute and DRI-flap. Point O is the intersection point between the old burden surface and the vertical line at pile ring radius, and point M can be obtained according the dump volume. The critical width of burden pile ($W_{critical}$) is the width when the pile attains stable repose angles, and is determined by the height of stock line, the angle of chute or flap, and the physical property parameters including size distribution and rolling friction coefficient. L, R, O and $W_{critical}$ are measured on a horizontal flat burden surface by previous physical experiment under various angles of the coal-gimbal chute and DRI-flap, and these data has just been listed in the above Section 2.3.

To characterize the formation of such complicated piles, the growing of a pile is calculated by iteration with a small volume step till the dump volume. In this way, the growth mechanism of burden pile on the horizontal section is shown in Figure 6a, including an early developing phase and a later developed phase. In the developing phase, both the inner and outer angles of the pile increase before the pile reaches the developed phase, where the pile undergoes a parallel growing mechanism similar to that in a BF. The horizontal distance LR is the width of burden flow arriving at the original burden surface, while OM represents the radial location of the burden ring.

The growing mechanism of a burden pile sitting on the turning point from a horizontal segment to a declining one is shown in from Figure 6b, where both points L and M are on the horizontal segment and point R is on the declining segment. In such cases, the point L is set as the left end of the burden flow. The right end of the first-step burden pile is the first intersection R_0, between the old burden surface and the line segment of MR. When the right point R_0 reaches R, the outer angle starts increasing until it reaches the developed phase, and then the growing mechanism becomes parallel in growth with the outer angle, similar to that in a BF. The evolution of the inner angle follows a similar procedure as the outer angle.

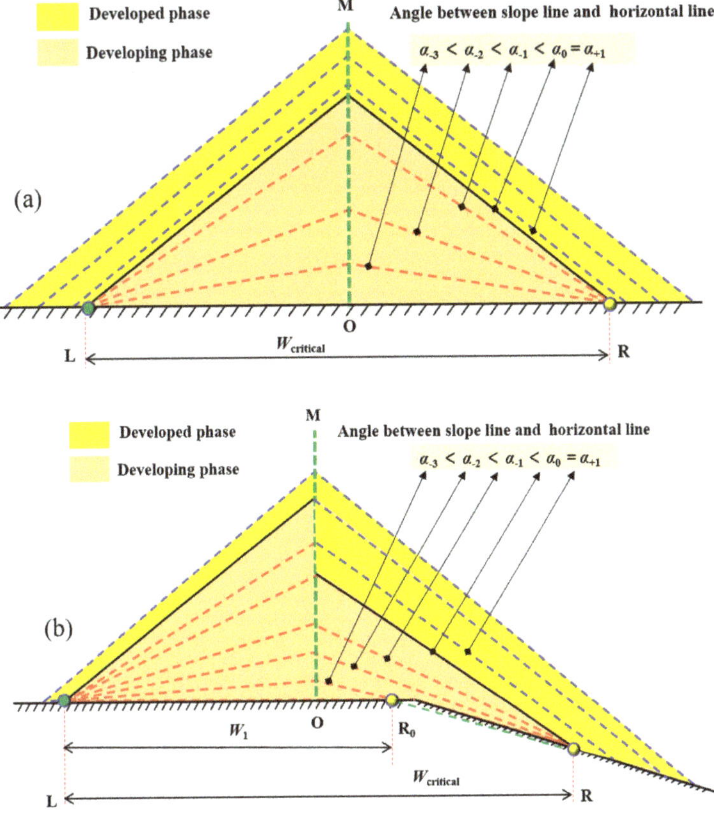

Figure 6. *Cont.*

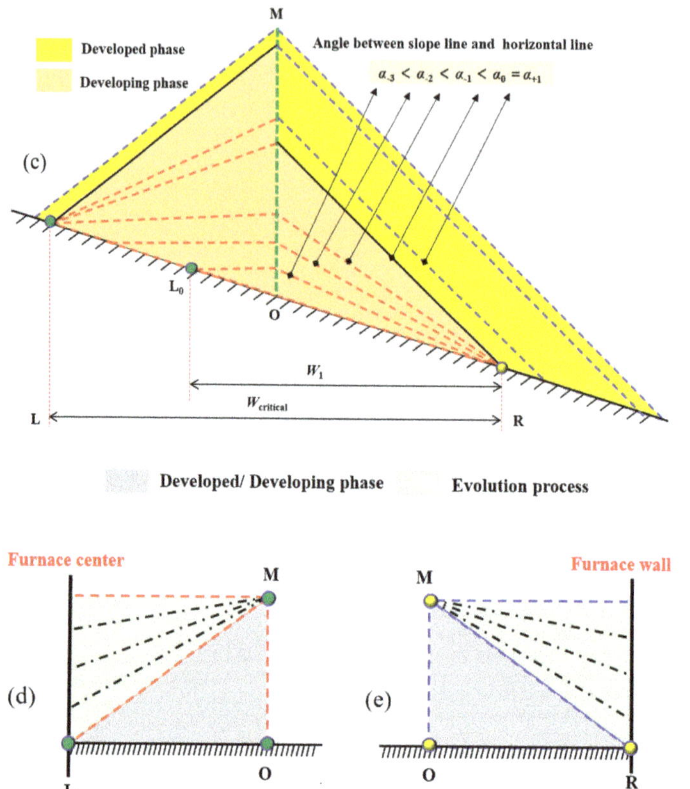

Figure 6. Growing mechanism of burden surface. (**a**) on the horizontal section; (**b**) from a horizontal segment to a declining one; (**c**) on the horizontal section; (**d**) near the furnace center section; and (**e**) near the furnace wall section.

When a pile is developed on the declining segment, its growing mechanism is shown in Figure 6c, where all the three points L, R, and M are on the declining segment. In this case, point R is set as the right end of the burden flow. The left point of first-step burden pile is the first intersection L_0 between the old burden profile and the horizon line. The inner angle increases until it reaches the developed phase and then turns to parallel growth with inner angle similar to that in the BF.

The special cases where one side of the pile reaches the furnace center and the furnace wall are considered as shown in Figure 6d,e. Regardless of whether the inner angle or the outer angle is in the developed or developing phases, as long as one side of the pile encounters an obstacle (furnace center or furnace wall), the inner or outer angle will decrease until it reaches the horizontal level, that is, a flat platform is formed by the center or wall.

3. Results

3.1. Verification of the Model

To show the feasibility and effectiveness of the proposed mathematical model, a comparison between the model prediction and the results of a physical experiment is conducted. Two important parameters, the radial coal/ore ratio and burden surface profile, are compared and analyzed. Table 2 shows the mixed charging pattern of Case 1 used in the physical experiment, where the time period of charging is 300 s (the same as the physical experiment), and the volumes of coal and DRI in one charging period (including two coal dumps and two DRI dumps) are 14.5 m^3 and 8.6 m^3 respectively.

Table 2. Data of charging pattern in Case 1.

Parameters \ Rings	R0.5	R1.0	R1.5	R2.0	R2.5	R3.0	R3.5	R4.0	R4.5	R5.0
Relative thickness of coal	-	-	-	0.5	0.6	0.7	0.6	0.6	0.5	-
Volume of coal/m^3	-	-	-	0.639	0.958	1.341	1.341	1.533	1.437	-
Relative thickness of DRI	-	-	-	-	-	0.5	0.8	1	1	-
Volume of DRI/m^3	-	-	-	-	-	0.504	0.941	1.344	1.512	

According to the similarity principle, the scale ratio of the physical model to the actual furnace is 1:7.5, so the width of burden pile in the actual process should be expanded by 7.5 times. Therefore, the results for burden piles of coal and DRI in the mathematical model are listed in Table 3, including the inner end and outer end positions of the consecutive piles.

Table 3. Results for the burden piles of coal and DRI in mathematical model of Case 1 (Units: mm).

Parameters	R0.5	R1.0	R1.5	R2.0	R2.5	R3.0	R3.5	R4.0	R4.5	R5.0
L_{coal}	0.00	276.98	704.83	1131.57	1556.92	1980.52	2401.81	2819.98	3233.79	3641.21
R_{coal}	1490.65	1923.88	2507.98	3090.97	3672.58	4252.43	4829.96	5404.39	5974.45	6538.12
L_{DRI}	0.00	0.00	0.00	1003.81	1373.31	1777.01	2285.95	2916.84	3588.18	4224.70
R_{DRI}	2013.75	2100.00	2186.25	3276.31	3732.06	4222.01	4817.20	5534.34	6291.93	7014.70

The burden surface profile of Case 1 obtained by the mathematical model is shown in Figure 7. As can be seen from Figure 7a, the height of the burden surface varies from 12.9 m to 14.5 m, and the burden surface is higher in the middle area and lower in the edge area. Figure 7b is an enlarged part view of the burden surface. By comparing with the result of the physical experiment (cf. Figure 7c), it can be seen that the profile calculated by the model is similar to the result of the physical experiment, showing that the mathematical model based on the burden flow width has a certain accuracy for predicting the burden profile. Both the results of the physical experiment and mathematical model show that the height of burden surface near the furnace center is lower than that in the middle region, and it then decreases along the radial direction but slightly rises near the wall. In this case, because the coal and DRI particles were charged from 2.0 m and 3.0 m outwards respectively, less particles reach the center and more particles were charged to the wall region. But the descent rate of the wall region is higher than that of the center region, and thus a lower burden bed profile was formed.

To further validate the mathematical model, the radial distribution of radial ore/coal mass ratio is compared. Figure 8 shows a comparison of the ore/coal mass ratio in the radial direction between model prediction and experimental measurement. In the physical experiment, eight samples (about 2 kg each) are collected from the layer formed in the last charging cycle at the locations of every 100 mm from center to wall. The ore/coal mass ratio is calculated based on the mass of coal and DRI particles of each sample.

Both the model prediction and experimental measurement in Figure 8 show that the ore/coal mass ratio increases first and then decreases, attaining a maximum at a radial position of about 5 m for the present charging matrix. The charging region of coal is in the radial range of 2~4.5 m, and that of DRI is in the range of 3~4.5 m. Therefore, the volume of coal distributed before the radial distance of 2 m is more than that of ore, and the mass ratio of ore/coal is close to zero. In the radial range of 2.0~4.5 m, the increase rate of coal is lower than that of DRI (cf. Table 1), so the mass ratio of ore/coal is increased. However, due to that larger size and lighter mass, coal particles tend to roll and easily be pushed to the descending region near the wall, the ore/coal mass ratio is decreased near the furnace wall.

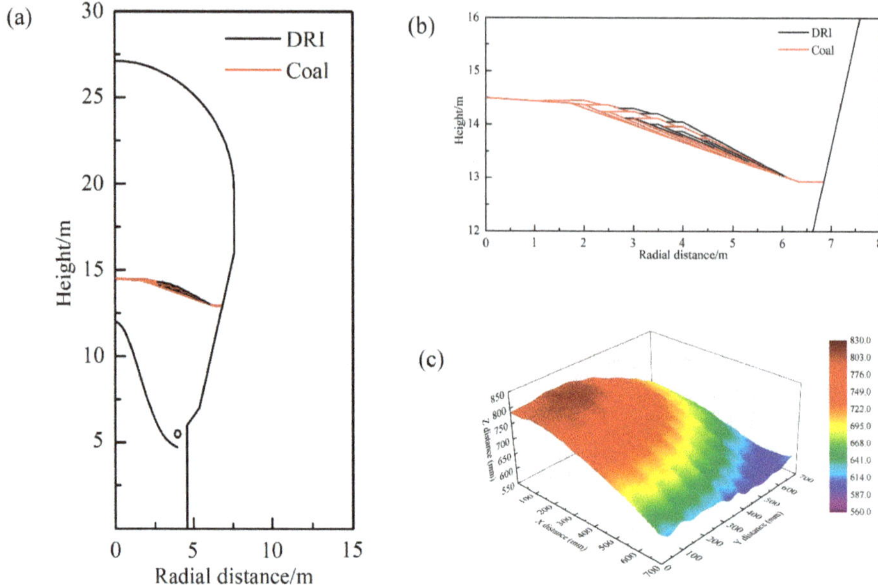

Figure 7. Comparison of burden profile between model prediction and experimental measurement. (**a**) Overall schematic of model prediction; (**b**) partially enlarged view of model prediction; and (**c**) burden profile of experimental measurement.

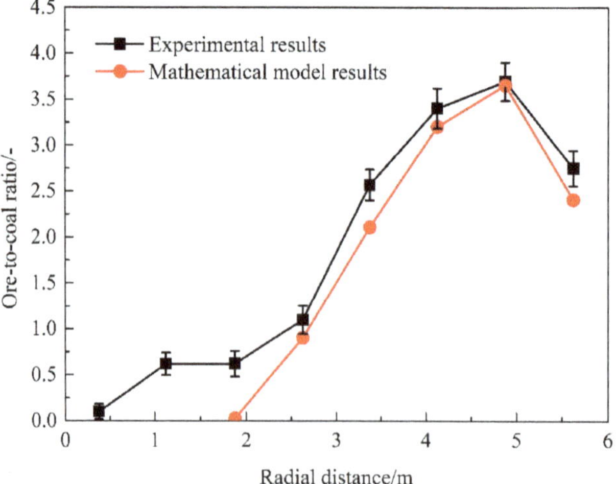

Figure 8. Comparison of ore/coal ratio distribution between model prediction and experimental measurement.

Comparing the results of the mathematical model and physical experiment, it can be found that the trend is basically the same, but the value of ore/coal mass ratio calculated by the mathematical model is smaller than that measured in the physical experiment, which is mainly caused by the shape of DRI-flap in the physical experiment. In a previous work [12], it was observed that DRI particles were unevenly distributed in the circumferential direction due to the effect of DRI-flap shape, and the relative quantity of DRI distributed in the area between two adjacent flaps is somewhat more than that in the other areas. The sampling in the physical experiment is exactly in the points between the two DRI-flaps, and thus the ore quantity is somewhat large. In the mathematical model, however, it is assumed that the DRI is evenly distributed in the circumferential direction. In other words, the circumferential segregation is not considered, so the overall ore/coal ratio is somewhat smaller than the experimental value.

Since the burden distribution is determined according to its flow width (measured at a horizontal level), and do not consider the rolling of the particles at a horizontal level in the mathematical model, no material reaches the central area of the furnace (especially DRI). This is somewhat different from the results of the physical experiments. The model should thus be further improved in future studies to make it more consistent with the results of physical experiments. Overall, although the model calculation results are slightly different from the physical experiment results, the overall trend is consistent, which further proves that the mathematical model can effectively predict the mass ratio of ore/coal and the burden profile.

3.2. Application of the Model

The charging matrix determines the surface profile and internal structure of the burden column in the furnace. In this section, three different charging matrixes are used to study the effect of charging pattern on the burden distribution. Detailed data of the charging matrixes are listed in Table 4. The on-site charging pattern can be classified into three categories, including inner coal and outer ore (e.g., Case 1#), outer coal and inner ore (e.g., Case 2#), and co-location of coal ore (e.g., Case 3#).

Table 4. Data of charging matrixes.

Case NO.	Materials Type	Relative Thickness (–) at Different Rings Locations (m)								
		R1.0	R1.5	R2.0	R2.5	R3.0	R3.5	R4.0	R4.5	R5.0
1#	Coal	-	-	0.5	0.6	0.7	0.6	0.6	0.5	-
	DRI	-	-	-	-	0.5	0.8	1.0	1.0	-
2#	Coal	-	-	-	0.5	0.8	0.7	0.6	0.15	-
	DRI	-	-	1.0	0.8	0.8	0.7	0.6	0.2	-
3#	Coal	-	-	0.7	0.7	0.8	0.9	0.8	-	-
	DRI	-	-	1.0	0.8	0.8	0.8	0.75	-	-

The evolution of burden profiles under various charging matrixes is shown in Figure 9, where the respective initial stable burden surfaces are determined according to the physical experiment as described above. Taking Case 1 as an example, one charging period including two coal dumps and two DRI dumps, that is, the first coal dump is charged from center to wall, followed by the first DRI charging from center to wall, and then the second coal dump is charged from wall to center, followed by the second DRI charging from wall to center. As can be seen from the figure, the burden surface profile is higher in the center and lower near the wall. As the charging position of Case 3 is farther from the wall than that of Case 1 and Case 2, the original burden surface does not have a flat platform near by the wall. Therefore, the burden profile of Case 3 is relatively flat in the radial direction, and more material gets to the wall region, while the middle area thickness of the burden pile in Case 1 and Case 2 is relatively greater than that in Case 3.

The results of the ore/coal mass ratio distribution for different charging matrixes are shown in Figure 10. It can be seen from the figure that the charging pattern has a great influence on the radial ore/coal ratio distribution.

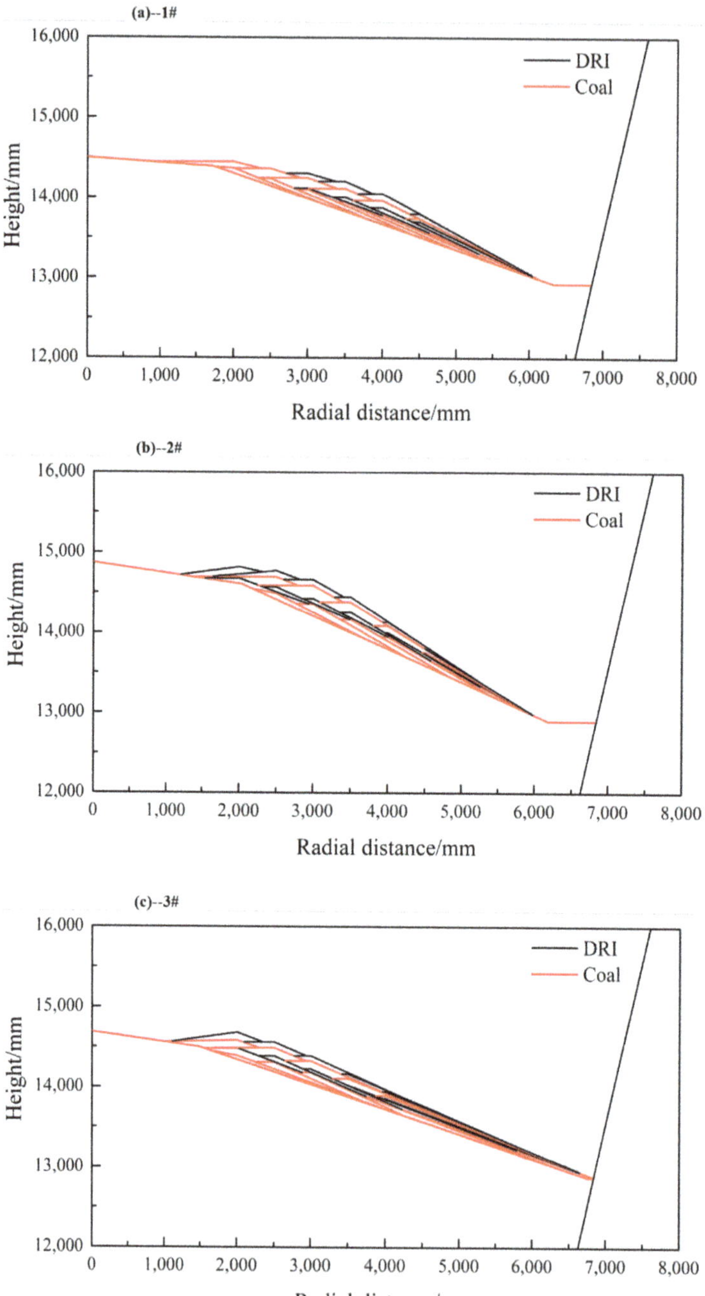

Figure 9. Evolution of burden surface profiles under different charging patterns. (**a**) Burden surface profile in Case 1; (**b**) Burden surface profile in Case 2; and (**c**) Burden surface profile in Case 3.

For Case 1, the charging region of coal is in the range of 2.0~4.5 m, and that of DRI is in the range of 3.0~4.5 m. Since the starting position of coal charging is closer to the furnace center than DRI, there is an ore-free region in the central area. The ore/coal ratio is zero before approximately 2.0 m, and then increases rapidly along the radial direction till $R \approx 4.75$ m, and finally decreases in the wall region.

For Case 2, the charging region of coal is in the range of 2.5~4.5 m and that of DRI is in the range of 2.0~4.5 m. Therefore, the volume of ore distributed before the radial distance of 1.0 m (the DRI left ends of *R*2.0 in Table 3) is more than that of coal, and the ore/coal ratio is larger in this region than Case 1. In the region of 2.5~3.0 m, the relative coal increase rate is greater than DRI, so the ore/coal ratio decreases along radial direction. However, due to larger size and lighter mass, the coal particles tend to roll and are easily pushed to the descending region near the wall, so the ore/coal ratio is decreased near the furnace wall.

For Case 3, the charging regions of both coal and ore are in the same range of 2.0~4.0 m. Since the starting positions of coal and ore charging are the same, the ore/coal ratio distribution in the radial direction is relatively uniform in Case 3. In Case 1 and Case 2, the ore/coal ratio is relatively higher in the wall area, so the ore content or the coal load is larger, which may lead to a poor permeability of the burden bed.

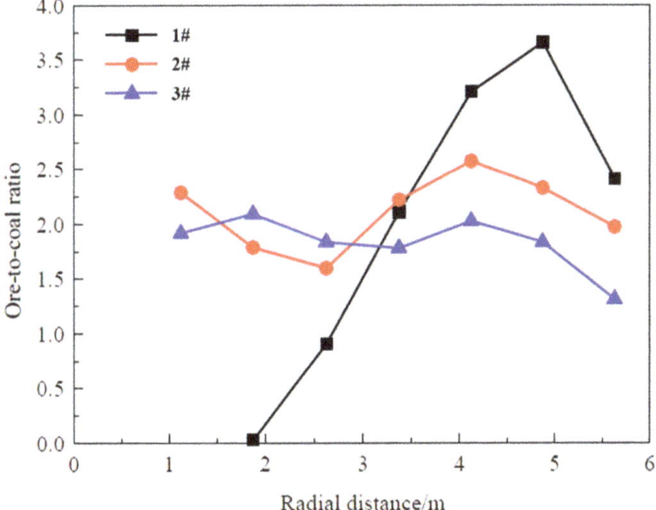

Figure 10. Calculated radial ore/coal ratio distribution with different charging patterns.

4. Conclusions and Future Perspectives

Based on the previous physical experiment, a new approach was proposed to characterize the growing mechanism of burden surface with the width of burden flow arriving at the burden surface in the COREX MG. The validity of the model is demonstrated by comparing the simulated burden layer structure with the corresponding results obtained by physical experiments. The usefulness of the mathematical model is illustrated by performing a set of simulation cases under various charging matrixes. The main conclusions are as follows.

(1) A mathematical model for characterizing the layer structure has been established based on the burden flow width. Compared with physical experiment, the model prediction is reasonably reliable, and the model can be used to predict the burden distribution with the complicated charging system of gimbal and flaps.

(2) The model can be used to predict the radial distribution of ore/coal mass ratio with different charging matrixes. It can be seen that the charging matrix has a great influence on the radial ore/coal ratio distribution. The co-location charging of coal and ore results in the most uniform distribution of the ore/coal ratio in the radial direction.

In the future, the mathematical model will be further improved by considering the influences of coal pushing, mixing between layers, and burden column descending. It is hoped that the model can be used as a what-if tool in practice for the COREX operator to gain a better understanding of burden distribution in the COREX MG and to supply boundary conditions for a mathematical model of the COREX MG to be developed.

Author Contributions: Conceptualization, H.L., Z.L., and Z.Z.; data curation, H.L., Z.L., and W.L.; funding acquisition, Z.Z.; investigation, W.L. and L.S.; methodology, W.L.; project administration, Z.Z.; resources, Z.Z.; software, H.L. and W.L.; supervision, Z.Z.; writing—original draft, H.L.; writing—review and editing, Z.Z. and L.S.

Funding: This research was funded by the National Science Foundation of China grant number 51604068, 51574064 And the APC was funded by China Scholarship Council (CSC No. 201706085021).

Acknowledgments: Financial supports from the National Science Foundation of China (Grants 51604068, 51574064) and China Scholarship Council (CSC No. 201706085021) are gratefully acknowledged.

Conflicts of Interest: The authors declare no conflict of interest.

References

1. Hasanbeigi, A.; Arens, M.; Price, L. Alternative emerging ironmaking technologies for energy-efficiency and carbon dioxide emissions reduction: A technical review. *Renew. Sustain. Energy Rev.* **2014**, *33*, 645–658. [CrossRef]
2. Srishilan, C.; Shukla, A.K. Static thermochemical model of COREX melter gasifier. *Metall. Mater. Trans. B* **2018**, *49*, 388–398. [CrossRef]
3. Song, J.Y.; Jiang, Z.Y.; Bao, C.; Xu, A.J. Comparison of Energy Consumption and CO_2 Emission for Three Steel Production Routes—Integrated Steel Plant Equipped with Blast Furnace, Oxygen Blast Furnace or COREX. *Metals* **2019**, *9*, 364. [CrossRef]
4. Xu, J.; Wu, S.L.; Kou, M.Y.; Zhang, L.H.; Yu, X.B. Circumferential burden distribution behaviors at bell-less top blast furnace with parallel type hoppers. *Appl. Math. Model.* **2011**, *35*, 1439–1455. [CrossRef]
5. Ren, T.Z.; Jin, X.; Ben, H.Y.; Yu, C.Z. Burden distribution for bell-less top with two parallel hoppers. *J. Iron Steel Res. Int.* **2006**, *13*, 14–17. [CrossRef]
6. Du, P.Y.; Cheng, S.S.; Teng, Z.J. Research of snakelike deviation in the burden distribution of a parallel-hopper bell-less top. *J. Univ. Sci. Technol. Beijing* **2011**, *33*, 479–485. (In Chinese)
7. Shi, L.; Zhao, G.; Li, M.; Ma, X. A model for burden distribution and gas flow distribution of bell-less top blast furnace with parallel hoppers. *Appl. Math. Model.* **2016**, *40*, 10254–10273. [CrossRef]
8. Park, J.I.; Jung, H.J.; Jo, M.K.; Oh, H.S.; Han, J.W. Mathematical modeling of the burden distribution in the blast furnace shaft. *Met. Mater. Int.* **2011**, *17*, 485–496. [CrossRef]
9. Guo, H.W.; Zhang, J.L.; Chen, L.K.; Che, Y.M.; Yang, T.Y. Simulation of charging system in bell-less BF. *Chin. J. Process. Eng.* **2009**, *9*, 415–419. (In Chinese)
10. Chen, L.S.; Luo, Z.G.; You, Y.; Zou, Z.S. Effects of Flap Angles on the Charging Procedure of Flap Distributors. *J. Northeast. Univ. Nat. Sci.* **2013**, *34*, 971–974. (In Chinese)
11. You, Y.; Luo, Z.G.; Hou, Q.F.; Li, H.F.; Zhou, H.; Chen, R.; Zou, Z.S. Experimental Study of Burden Distribution in the COREX Melter Gasifier. *Steel Res. Int.* **2017**, *88*, 1700025. [CrossRef]
12. Luo, Z.G.; You, Y.; Li, H.F.; Zhou, H.; Zou, Z.S. Experimental study on charging process in the COREX Melter Gasifier. *Metall. Mater. Trans. B* **2018**, *49*, 1740–1749. [CrossRef]
13. Li, H.F.; You, Y.; Zou, Z.S.; Cai, J.J. Numerical simulation on the charging process of new DRI-Flap distributor. *J. Northeast. Univ. Nat. Sci.* **2016**, *37*, 800–804. (In Chinese)
14. You, Y.; Luo, Z.G.; Li, H.F.; Zou, Z.S.; Yang, R.Y. Effects of the shape and inclination angle of DRI-flaps on DRI distribution in COREX Melter Gasifiers. *Powder Technol.* **2018**, *339*, 854–862. [CrossRef]
15. Li, H.F.; You, Y.; Zhou, H.; Luo, Z.G.; Zou, Z.S. Study on burden pile profile prediction model for COREX-3000 Melter Gasifier. *J. Chongqing Univ.* **2015**, *38*, 39–44. (In Chinese)
16. Zhang, H.M. CFD-DEM Modeling of Multiphase Flow in a FINEX Melter Gasifier. Ph.D. Thesis, University New South Wales, Sydney, Australia, 2015; pp. 95–144.
17. Saxén, H.; Hinnelä, J. Model for Burden Distribution Tracking in the Blast Furnace. *Miner. Process. Extr. Metall. Rev.* **2004**, *25*, 1–27. [CrossRef]

18. Mitra, T.; Saxén, H. Model for Fast Evaluation of Charging Programs in the Blast Furnace. *Metall. Mater. Trans. B* **2014**, *45*, 2382–2394. [CrossRef]
19. Li, H.F.; Saxén, H.; Liu, W.Q.; Shao, L.; Zou, Z.S. Model-based analysis of factors affecting the burden layer structure in the blast furnace shaft. *Metals* **2019**, *9*, 1003. [CrossRef]
20. You, Y.; Luo, Z.G.; Zou, Z.S.; Yang, R.Y. Numerical study on mixed charging process and gas-solid flow in COREX Melter Gasifier. *Powder Technol.* **2019**. [CrossRef]

Article

Numerical Simulation of the Raceway Zone in Melter Gasifier of COREX Process

Ye Sun [1,2,*], Ren Chen [1,2], Zuoliang Zhang [1,2], Guoxi Wu [1,2], Huishu Zhang [1,2], Lingling Li [1,2], Yan Liu [1,2], Xiaoliang Li [1,2] and Yan Huang [1,2]

[1] Liaoning Key Laboratory of Optimization and Utilization of Non-associated Low-grade Iron Ore in Benxi, Liaoning Institute of Science and Technology, Benxi 117004, China; toughold@163.com (R.C.); zhang231167@lnist.edu.cn (Z.Z.); wgxlnkjxy@163.com (G.W.); huishuzhang@163.com (H.Z.); linglinglidlut@163.com (L.L.); lkyliuyan@163.com (Y.L.); lixiaoliang521@foxmail.com (X.L.); 0303yanyan@163.com (Y.H.)

[2] School of Metallurgy Engineering, Liaoning Institute of Science and Technology, Benxi 117004, China

* Correspondence: sunye0412@sina.com;

Received: 28 September 2019; Accepted: 13 November 2019; Published: 20 November 2019

Abstract: The physical and chemical processes in the raceway zone of the COREX melter–gasifier express are similar to those inside the blast furnace. Based on the research achievements on blast furnaces, the unsteady numerical simulation of a gas-solid two-phase in the raceway was carried out by using computational fluid software. The formation process of the raceway in the COREX melter–gasifier was simulated. The shape and size of the raceway were obtained. Then, the effect of gas flow on the depth and height of the raceway was analyzed in this paper.

Keywords: COREX; raceway zone; numerical simulation; gas flow

1. Introduction

COREX is the world's first commercially established and industrially proven smelting-reduction process [1,2]. It is a two-stage process that involves pre-reduction in a shaft furnace, followed by final reduction and separation in a melter gasifier [3–5]. The melter gasifier is the key reactor of the COREX process. Lateral injection of high-speed gas into the packed bed in a melter gasifier can cause the formation of granular circulation regions within the bed. These are commonly called 'raceways' due to the distinct shape of the path taken by entrained particles within the bed [6,7]. The raceway plays a critical role in providing energy and reducing agents for the successful and stable operation of the melter gasifier. Exploring the characteristics of the raceway in the COREX melter gasifier is beneficial for the design and optimization of the chemical process.

There are different approaches for the investigation of raceway phenomena in a packed bed. One method is cold model simulation. Two-dimensional or pseudo three-dimensional models were used to study the size of the raceway through photography [8–11]. However, due to the rather intensive particle-fluid interactions and high temperature environment, there are some deviations between the experimental results and the practice dates. The temperature measurement method is also a good way to explore the properties of the raceway zone. However, no matter the direct measurement method or non-contact measurement method, the size and volume of the raceway cannot be accurately obtained [12–14]. To overcome these difficulties in experimental and measurement studies, numerical simulations of the raceway have becomes more and more popular. For example, Sarkar et al. studied the raceway boundary using a continuum two-fluid model [15]. Frank et al. developed a computational fluid dynamics (CFD) model for describing the pressure field, temperature field, gas composition field, and so on [16]. Shen et al. established a series of CFD models to study coal combustion in the raceway zone [17–20]. Recently, a coupled CFD-DEM model was also developed to determine particle and gas flow in the raceway in a blast furnace [21–23]. All these studies are useful for understanding the characteristics of the raceway in a blast furnace, whereas the difference between a blast furnace and a COREX process makes it impossible for the latter to draw experience directly from the blast furnace. Indeed, the raceway boundary and gas–solid flow behaviours of the raceway in a COREX melter gasifier were investigated by Sun et al. [24,25]. However, under pure oxygen injection in a COREX melter gasifier, the formation process of the raceway needs further research.

In this paper, a CFD model is developed to study the raceway zone in the melter gasifier of the COREX process. The raceway formation process is first discussed, and the influence of the amount of the blowing gas on raceway size is also studied. The findings of this work will be useful for the design, control, and optimization of COREX melter gasifier process operation.

2. Mathematical Model

2.1. The Basic Assumptions of This Paper

(1)   In each location of the flow field, the particulate phase coexists with the gas phase and both penetrate each other, with each phase having its own velocity, temperature, and volume fraction, but the particles of each size group have the same velocity and temperature.
(2)   Each particle phase (size group) has a continuous distribution of velocity, temperature, and volume fraction in space.
(3)   Each particle phase and gas phase, in addition to quality, momentum, and energy interactions, also has its own turbulence.
(4)   The initial size distribution is used to distinguish the particle groups.
(5)   For dense particle suspensions, particle collision can cause additional particle viscosity, diffusion, and heat conduction. A two-fluid model is used in the study (also called Eulerian model).

2.2. Volume Fraction

The Ansys-Fluent 14.5 commercial software was used in the study. The volume fraction represents the volume percentage of each phase, and each phase satisfies the law of conservation of mass and momentum. The volume fraction of the q phase (V_q) is defined as

$$V_q = \int_V \alpha_q dV \tag{1}$$

where $\sum_{q=1}^{n} \alpha_q = 1$.

The effective density of the q phase is

$$\widehat{\rho}_q = \alpha_q \rho_q \tag{2}$$

where ρ_q is the physical density of the q phase.

2.3. Mass Conservation Equation

The mass conservation equation of the q phase is

$$\frac{\partial}{\partial t}\left(\alpha_q \rho_q\right) + \nabla \cdot \left(\alpha_q \rho_q \vec{v}_q\right) = \sum_{p=1}^{n}\left(\dot{m}_{pq} - \dot{m}_{qp}\right) + S_q \tag{3}$$

where $\vec{v}_q$, $\dot{m}_{pq}$, and S_q are the velocity of the q phase, the mass transfer from the p phase to q phase, and the source phase, respectively.

2.4. Momentum Conservation Equation

The momentum conservation equation of the q phase is

$$\frac{\partial}{\partial t}\left(\alpha_q \rho_q \vec{v}_q\right) + \nabla \cdot \left(\alpha_q \rho_q \vec{v}_q \vec{v}_q\right) = -\alpha_q \nabla p + \nabla \cdot \overline{\overline{\tau}}_q + \alpha_q \rho_q \vec{g}$$
$$+ \sum_{p=1}^{n}\left(\vec{R}_{pq} + \dot{m}_{pq}\vec{v}_{pq} - \dot{m}_{qp}\vec{v}_{qp}\right) + \left(\vec{F}_q + \vec{F}_{lift,q} + \vec{F}_{vm,q}\right) \tag{4}$$

where τ_q is the pressure strain tensor of the q phase; $\vec{F}_q$, $\vec{F}_{lift,q}$, and $\vec{F}_{vm,q}$ are the volume force, lift force, and virtual mass force of the q phase, respectively; and $\vec{R}_{pq}$ is an interaction term.

2.5. Conditions

Using the melter–gasifier as the prototype, an unsteady simulation is performed from the computational domain consisting of the upper coke bed region of the slag layer on the packed bed region. The calculation domain size is shown in Figure 1. The bottom edge is 4690 mm, the height is 4230 mm, and the furnace wall inclination angle is 30°, which is close to the actual size, and the tuyere spray gun is 1200 mm. Compared with the whole model, the insertion depth of the tuyere spray gun is negligible. For the sake of the simplicity of the model calculation, the inlet of the tuyere spray gun is located at the wall. The average mesh size is 40 mm in the present simulation. They are mostly structured meshes. We performed a sensitivity study of the mesh size with an average size of 100 mm, 80 mm, 60 mm, 40 mm, 30 mm, and 20 mm. The difference in the depth of the raceway (gas velocity is 200 m/s) between 60 mm and 40 mm is 2.9%, while that between 40 mm and 30 mm is within 0.5%. This suggests that a mesh size of 40 mm is reasonable and confirms the mesh independence.

In the simulation, a no-slip condition is applied to walls. The pressure–velocity decoupling is done with the PISO algorithm. The tuyere zone is the velocity inlet. An explicit scheme is used to describe the shape of the interface. The time step is 0.001 s. The numerical solution is considered to be convergent when the residual errors of the variables are less than 10^{-5}. The main parameters used in the model are shown in Table 1.

Table 1. Parameters used in this model.

Project	Values	Unit
Gun inlet diameter	30	mm
Coke bed porosity	0.60	-
Max focal bed porosity	0.63	-
Bed height	2230	mm
Type of blowing gas	Air	-
Air density	1.205	kg/m^3
Air viscosity	1.76×10^{-5}	Pa·s
Injecting gas velocity	50–250	m/s
Coke diameter	40	mm
Coke density	600	kg/m^3
Gas injection inclination	4	degree (°)
Operating pressure	3.5	atm

The unsteady simulation initialization conditions are shown in Figure 2. The lower part is the coke layer, and the porosity is 0.4; the upper part is gas.

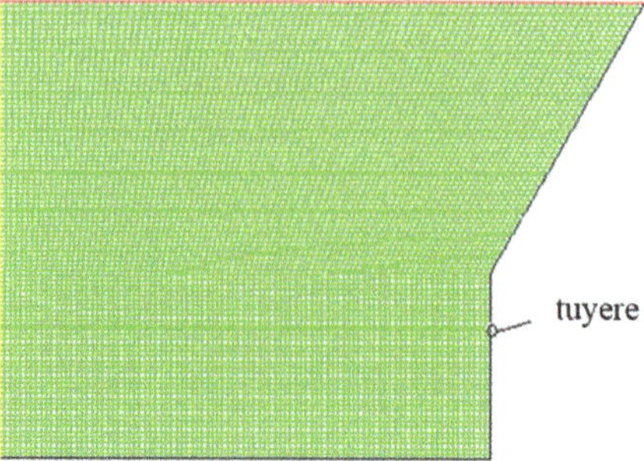

Figure 1. Schematic diagram of the computational domain for the tuyere zone.

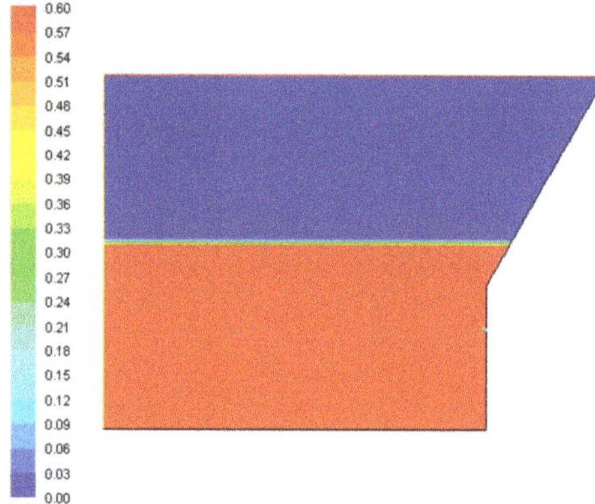

Figure 2. Initialization condition of the coke volume fraction in the model.

3. Results

3.1. Unsteady Simulation of the Raceway Formation Process

Taking the typical COREX melter–gasifier nozzle injection operation parameters as an example, the gas velocity is set to 200 m/s; the spray inclination angle is 4°, and the parameters of the shape of the raceway zone and the volume fraction of the gas–solid particles at different times are tracked. The gas volume fraction at different time periods is shown in Figure 3.

It can be seen from Figure 3 that the volume fraction of the gas phase changes significantly over time. At the beginning of the injection, the gas develops radially toward the center. Then, the gas moves straight to the center, and the upper and lower sides develop slowly. Over time, the expansion of the radial direction begins to become slow and finally stagnate, at which point the gas begins to extend upwards in the axial direction, slowly forming an arc-shaped cavity region. The upper portion of the cavity region exhibits a semicircular or semi-elliptical shape, which lasts for a long period of time. Finally, the gas forming bubble-like sphere begins to drift upward when the gas pressure in the cavity accumulates to a certain extent.

Combining the above theory, it can be determined from the change of the gas phase volume fraction, and from the start of the injection to the formation of the raceway zone, that the gas first develops in the radial direction and reaches a certain degree before starting to develop upward. The reason for this development is that the porosity of the coke bed is continuously reduced under the compression of the gas; the gap between the coke particles becomes increasingly smaller, and the resistance of the gas to the depth development is increased. When compressed to a certain limit (simulated by setting the coke porosity limit to 0.37), the resistance of the gas in the radial direction is so large that the gas cannot expand forward, and the gas grows upwards.

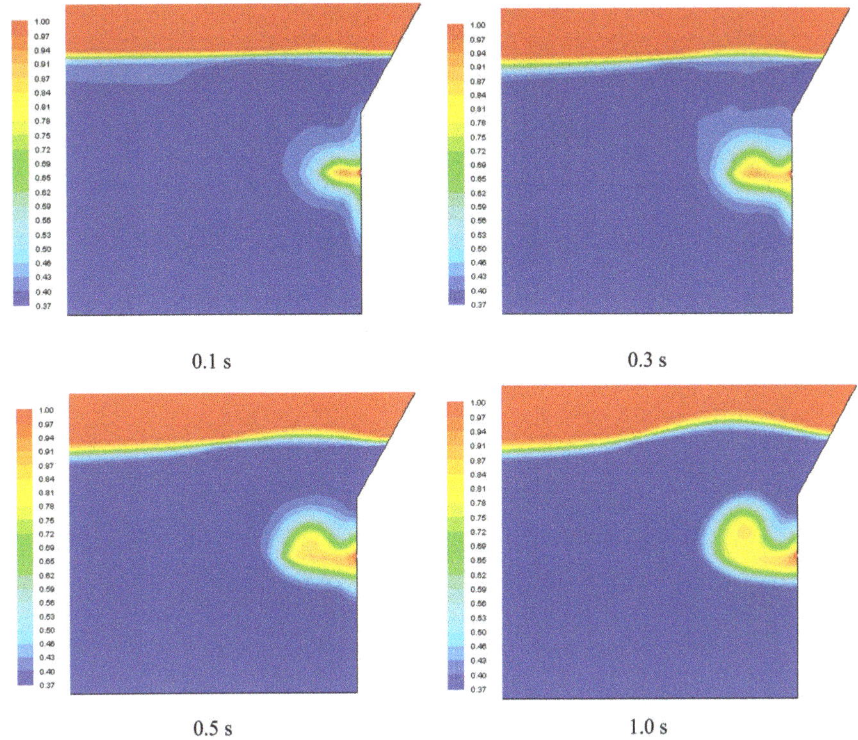

Figure 3. Variation of gas phase volume fraction with time.

Figure 4 shows the depth variation of the radial development of the gas over time. It can be seen from the figure that, as the blowing time changes, the depth of the raceway zone first increases rapidly, and after 0.4 s, it begins to stabilize and finally reaches a maximum of about 800 mm.

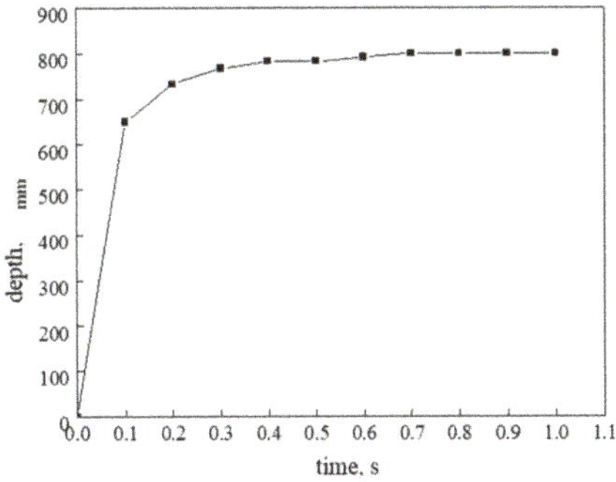

Figure 4. Variation of raceway depth with blowing time.

Figure 5 shows the raceway behaviour in our previous experimental study [26]. The result indicates that particles rotate in front of the tuyere, and a stable cavity can be observed. Comparing the present numerical simulation and the previous physical simulation, the calculation result is consistent with the results of the physical experiment. The shape of the raceway shows similar features. These agreements verify the applicability of the present model for investigating the characteristics of the raceway in a melter gasifier under different conditions.

Figure 5. Raceway behaviour in the experiment.

3.2. Effect of the Velocity of the Blowing Gas on the Cavity Size of the Raceway

According to the previous analysis, the shape and boundary of the raceway zone are generally stable from the time when the depth reaches the maximum and the time when the air mass begins to drift away from the raceway zone. On this basis, the effects of the different gas kinetic energies of blasting (i.e., airflow velocity) on the shape and size of the raceway zone are considered. Figure 6 shows the volume fraction diagram of the gas phase when the injection angle is four degrees and the gas velocities are 150 m/s, 200 m/s, 250 m/s, and 300 m/s, respectively.

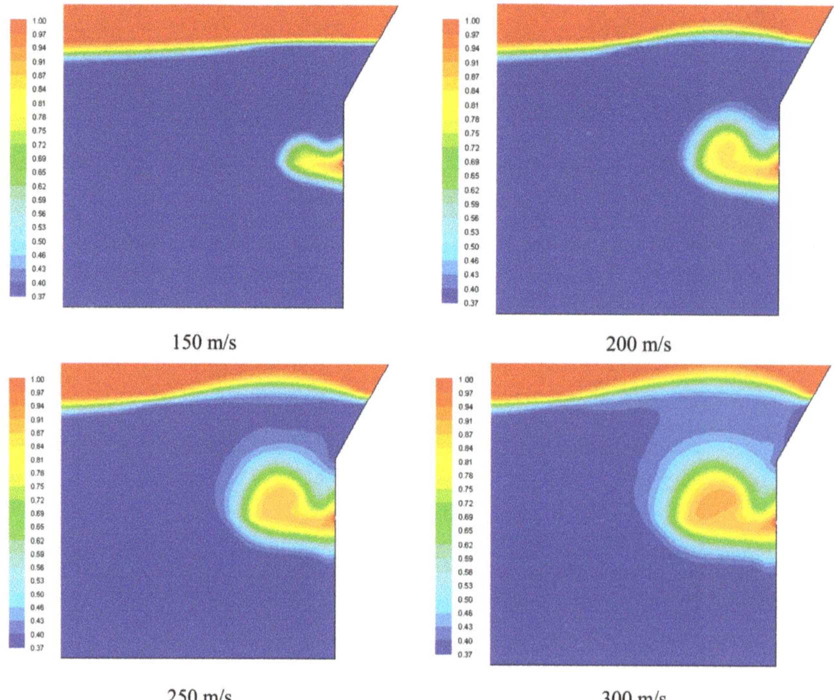

Figure 6. Gas phase volume fraction fractions at different blowing velocities.

Combined with the velocity cloud map and the gas phase volume fraction map, it can be seen that the greater the gas injection speed, the deeper the depth of the raceway, and the larger the cavity volume when it is stabilized, which can also be seen from its depth and height. Figure 7 shows the depth variation of the raceway under different gas velocity conditions.

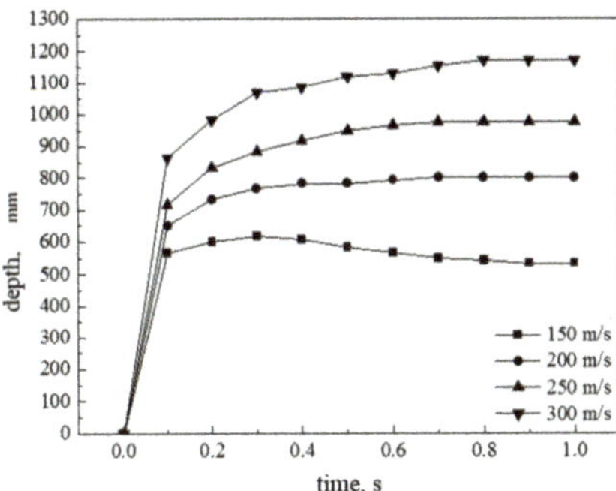

Figure 7. Variation of raceway depth with time at different blowing velocities.

Under different gas velocities, the penetration depth is different. With an increase in gas velocity, the depth of the coke layer, through which the gas flows, increases. As can be seen from Figure 7, the cavity depth changes in the four blowing situations are basically the same—that is, both increase from the beginning and finally stabilize. The variation of the depth of the raceway also directly reflects the variation of the volume of the cavity in the raceway. When the raceway zone is stable at 150 m/s, the depth is about 500 mm, at 200 m/s, the depth is about 800 mm, at 250 m/s, the depth is about 950 mm, and at 300 m/s, the depth is about 1200 mm.

The depth and height values of the raceway zone in the stable period at each blowing speed are plotted, as shown in Figure 8. This makes it easier to visually observe how the two change with the injection speed. The main reason for this trend is that the velocity of the blowing gas increases—that is, the kinetic energy of the blast increases, the amount of gas injected per unit time increases, the gas pressure in the cavity increases, and the ability to compress the coke particles also increases. When the gas pressure and the coke layer resistance are balanced, the cavity volume is also large.

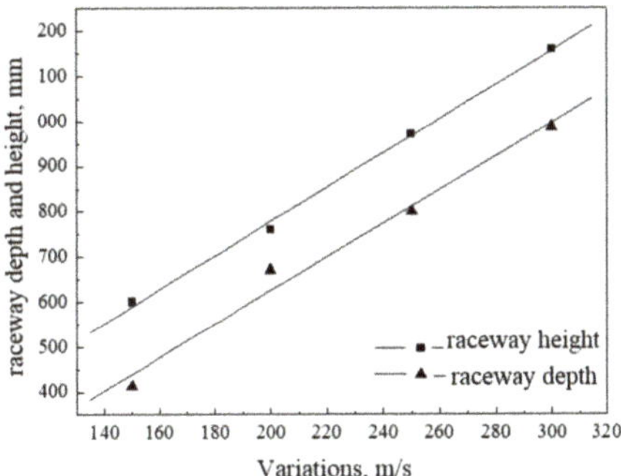

Figure 8. Variations of raceway depth and height with blowing velocity.

4. Conclusions

The formation process of the tuyere of the COREX melter–gasifier was simulated by Euler gas–solid two-phase flow theory. The shape and size of the relatively stable period of the raceway were obtained. The influence of the jet velocity on the depth and height of the raceway is analyzed. The main results are as follows.

(1)   As the gas continues to inject, the cavity first grows deep into the furnace. After reaching a certain depth, the cavity begins to develop upwards, and the cavity volume increases. The time taken from the start of the gas to the formation of the stable raceway shape is short, and then the depth and height of the raceway and the volume of the cavity are stable for a long period of time.

(2)   Under the condition that the normal blowing speed of the COREX melter–gasifier is 250 m/s and the blowing angle is 4°, the depth of the raceway is about 950 mm, and the shape of the raceway is approximately semi-elliptical.

(3)   As the velocity of the tuyere gas injection increases, the depth and height of the raceway increase, and the volume of the cavity in the raceway zone increases as it stabilizes, but the shape of the raceway does not change significantly.

Author Contributions: Conceptualization, Y.S. and R.C.; methodology, Y.S. and Z.Z.; software, G.W. and Y.S.; formal analysis, H.Z. and Y.L.; investigation, X.L. and Y.H.; writing, Y.S. and R.C.; supervision, G.W. and L.L.; funding acquisition, Y.S. and R.C.

Funding: This research was funded by the Natural Science Foundation of Liaoning Province (No.20170540476), Liaoning Province Doctor Startup Fund (No.20170520079). Also, special thanks to the project, which is sponsored by the 'Liaoning BaiQianWan Talents Program'.

Conflicts of Interest: The authors declare no conflict of interest.

References

1. Anameric, B.; Kawatra, S.K. Direct iron smelting reduction processes. *Miner. Process. Extr. Metall. Rev.* **2008**, *30*, 1–51. [CrossRef]
2. Qu, Y.X.; Zou, Z.S.; Xiao, Y.P. A Comprehensive Static Model for COREX Process. *ISIJ Int.* **2012**, *52*, 2186–2193. [CrossRef]
3. Zhou, H.; Wu, S.L.; Kou, M.Y.; Luo, Z.G.; He, W.; Zou, Z.S.; Shen, Y.S. Discrete Particle Simulation of Solid Flow in a Large-Scale Reduction Shaft Furnace with Center Gas Supply Device. *ISIJ Int.* **2018**, *58*, 422–430. [CrossRef]
4. Kumar, P.P.; Gupta, P.K.; Ranjan, M. Operating experiences with Corex and blast furnace at JSW Steel Ltd. *Ironmak. Steelmak.* **2008**, *35*, 260–263. [CrossRef]
5. Zhou, H.; Wu, S.L.; Kou, M.Y.; Yao, S.; Shen, Y.S. Analysis of Coke Oven Gas Injection from Dome in COREX Melter Gasifier for Adjusting Dome Temperature. *Metals* **2018**, *8*, 921. [CrossRef]
6. Di, Z.X.; Luo, Z.G.; Zou, Z.S. Fractal study on raceway boundary. *J. Iron Steel Res. Int.* **2011**, *18*, 16–19. [CrossRef]
7. Gupta, G.S.; Rudolph, V. Comparison of blast furnace raceway size with theory. *ISIJ Int.* **2006**, *46*, 195–201. [CrossRef]
8. Hatano, M.; Fukuda, M.; Takeuchi, M. An experimental study of the formation of raceway using a cold model. *Tetsu Hagane* **1976**, *1*, 25–32. [CrossRef]
9. Sastry, G.K.; Gupta, G.S.; Lahiri, A.K. Void formation and breaking in a packed bed. *ISIJ Int.* **2003**, *43*, 153–160. [CrossRef]
10. Rajneesh, S.; Sarkar, S.; Gupta, G.S. Prediction of raceway size in blast furnace from two dimensional experimental correlations. *ISIJ Int.* **2004**, *44*, 1298–1307. [CrossRef]
11. Hiroshi, T.; Nobuyuki, K. Cold model study on burden behaviour in the lower part of blast furnace. *ISIJ Int.* **1993**, *33*, 655–663.
12. Zhou, D.D.; Cheng, S.S.; Zhang, R.X.; Li, Y.; Chen, T. Uniformity and Activity of Blast Furnace Hearth by Monitoring Flame Temperature of Raceway Zone. *ISIJ Int.* **2017**, *57*, 1509–1516. [CrossRef]
13. Li, Y.; Cheng, S.S.; Zhang, R.X.; Chen, T. Reconstruction of Three-dimensional Temperature Distribution with Radiative Image by Monte Carlo Method in Blast Furnace Raceway. *ISIJ Int.* **2017**, *57*, 2141–2147. [CrossRef]
14. Zhou, D.D.; Cheng, S.S. Measurement study of the PCI process on the temperature distribution in raceway zone of blast furnace by using digital imaging techniques. *Energy* **2019**, *174*, 814–822. [CrossRef]
15. Sarkar, S.; Gupta, G.S.; Kitamura, S.Y. Prediction of Raceway Shape and Size. *ISIJ Int.* **2007**, *47*, 1738–1744. [CrossRef]
16. Frank, H.D.; Tian, F.G.; Chen, N.W. A comprehensive simulation of the raceway formation and combustions. *AISTech 2009 Proc.* **2009**, *1*, 333–344.
17. Du, S.W.; Chen, W.H. Numerical prediction and practical improvement of pulverized coal combustion in blast furnace. *Int. Commun. Heat Mass Transf.* **2006**, *33*, 327–334. [CrossRef]

18. Gu, M.Y.; Zhang, M.C.; Selvarasu, N.C. Numerical analysis of pulverized coal combustion inside tuyere and raceway. *Steel Res. Int.* **2008**, *79*, 17–24. [CrossRef]

19. Shen, Y.S.; Shiozawa, T.; Austin, P.; Yu, A.B. Modelling of injecting a ternary coal blend into a model ironmaking blast furnace. *Miner. Eng.* **2016**, *90*, 89–95. [CrossRef]

20. Shen, Y.S.; Yu, A.B. Model study of the effect of bird's nest on transport phenomena in the raceway of an ironmaking blast furnace. *Miner. Eng.* **2014**, *63*, 91–99. [CrossRef]

21. Hou, Q.F.; E, D.Y.; Yu, A.B. Discrete Particle Modeling of Lateral Jets into a Packed Bed and Micromechanical Analysis of the Stability of Raceways. *AIChE* **2016**, *62*, 4240–4250. [CrossRef]

22. Hilton, J.E.; Cleary, P.W. Raceway formation in laterally gas-driven particle beds. *Chem. Eng. Sci.* **2012**, *80*, 306–316. [CrossRef]

23. Wei, G.C.; Zhang, H.; An, X.Z.; Xiong, B.; Jiang, S.Q. CFD-DEM study on heat transfer characteristics and microstructure of the blast furnace raceway with ellipsoidal particles. *Powder Technol.* **2019**, *346*, 350–362. [CrossRef]

24. Sun, J.J.; Luo, Z.G.; Di, Z.X.; Zhang, T.; Zhou, H.; Zou, Z.S. Definition of Raceway Boundary Using Fractal Theory. *J. Iron Steel Res. Int.* **2015**, *22*, 36–41. [CrossRef]

25. Sun, J.J.; Luo, Z.G.; Zou, Z.S. Numerical simulation of raceway phenomena in a COREX melter gasifier. *Powder Technol.* **2015**, *281*, 159–166. [CrossRef]

26. Sun, Y.; Luo, Z.G.; Zou, Z.S.; Liu, H.H. Determining raceway boundary by image processing via high-speed video camera. *J. Northeast. Univ. (Nat. Sci.)* **2009**, *30*, 1458–1461.

Article

Computational Approaches for Studying Slag–Matte Interactions in the Flash Smelting Furnace (FSF) Settler

Jani-Petteri Jylhä, Nadir Ali Khan and Ari Jokilaakso *

Department of Chemical and Metallurgical Engineering, Aalto University, Kemistintie 1, P.O. Box 16100, FI-00076 Aalto, Finland; jani-petteri.jylha@aalto.fi (J.-P.J.); nadir.khan@aalto.fi (N.A.K.)

* Correspondence: ari.jokilaakso@aalto.fi; Tel.: +358-50-3138-885

Received: 12 March 2020; Accepted: 15 April 2020; Published: 22 April 2020

Abstract: Computational methods have become reliable tools in many disciplines for research and industrial design. There are, however, an ever-increasing number of details waiting to be included in the models and software, including, e.g., chemical reactions and many physical phenomena, such as particle and droplet behavior and their interactions. The dominant method for copper production, flash smelting, has been extensively investigated, but the settler part of the furnace containing molten high temperature melts termed slag and matte, still lacks a computational modeling tool. In this paper, two commercial modeling software programs have been used for simulating slag–matte interactions in the settler, the target being first to develop a robust computational fluid dynamics (CFD) model and, second, to apply a new approach for molten droplet behavior in a continuum. The latter is based on CFD coupled with the discrete element method (DEM), which was originally developed for modeling solid particle–particle interactions and movement, and is applied here for individual droplets for the first time. The results suggest distinct settling flow phenomena and the significance of droplet coalescence for settling velocity and efficiency. The computing capacity requirement for both approaches is the main limiting factor preventing full-scale geometry modeling with detailed droplet interactions.

Keywords: computational fluid dynamics; CFD–DEM; coalescence; settling; funneling flow

1. Introduction

In the flash smelting (FS) process, a mixture of sulfide-based concentrate and flux is continuously fed to the reaction shaft through a concentrate burner. Additional recycled materials, such as copper scrap and waste electrical and electronic equipment (WEEE) scrap can be used in the flash smelting feed. With the solid feed, oxygen enriched air is also blown through the burner. The air is used to create the exothermic reaction of sulfide oxidation, which creates the energy needed to melt the feed. This forms molten slag and matte phases as separate layers in the settler, with the lighter slag layer on top of the matte layer. The main functions of the slag are to protect the matte from oxygen, collect impurities such as Zn, Co, Ni, Sb, As, and Mo [1–3], and thermally insulate the matte to minimize energy losses. The slag and matte are tapped through tapping holes in the furnace wall. Unlike the feed, the tapping takes place at regular intervals, which causes variations in the thickness of the melt layers. Besides the melts, SO_2 gas is also formed. The gas exits through the uptake shaft carrying dust that is collected in gas cleaning, and the gas is then used in sulfuric acid production. In addition, the thermal energy of the gas is recovered and used for heating input gases and possibly for the local community. The collected dust is then circulated back to the process [4].

The slag and matte layers are not stagnant in this continuous process. More material constantly descends from the reaction shaft and matte droplets settle through the slag layer, creating flows in the matte and slag layers. Xia et al. studied the slag and matte flows in a FS settler [5–7]. They found that tapping of the slag and the matte creates complex turbulent flows in the settler. The flows are not uniform: the effect of the tapping flow is reduced as the distance to the tapping hole increases. Furthermore, the region affected by the tapping decreases as the flow velocity is reduced.

However, the area under the reaction shaft has strong flows as the majority of the droplets will be descending relatively directly from the concentrate burner. Similar trajectories have been reported in a study by Zhou et al. [8]. Studies by Khan and Jokilaakso, and Jylhä and Jokilaakso show a funneling effect created by drag flows in the slag [9,10]. These drag flows were caused by settling matte droplets.

Mechanical copper losses in the FS process can be traced to copper as flue dust or matte entrained in the slag. The settling of matte droplets through the slag phase in the FS furnace (FSF) as shown in Figure 1, is an important phenomenon and determines the overall copper yield in the smelting unit process. The copper losses during the settling process are due, in addition to the mechanical entrainment of the copper matte droplets, to the chemical dissolution of copper in the slag phase. This not only reveals the loss of matte droplets during the settling process, but also how quick the settling process is. The entrainment of matte droplets has been studied by many researchers [11–17] and it has been concluded that mostly droplets ≤100 μm in size are trapped in the slag phase and are ultimately carried away in the slag phase through the slag outlet.

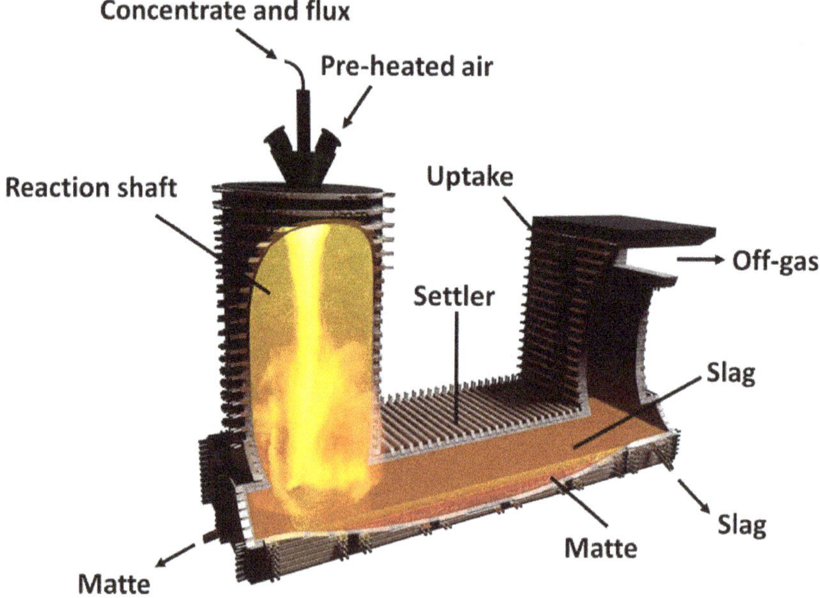

Figure 1. Illustration of the Outotec flash smelting furnace.

The entrained copper losses to slag are due to, for example, solid spinels, small size, or droplets being lifted from the surface of the matte layer [18–20]. Spinels may be attached to a droplet or surround the droplet completely [21]. The attached spinel affects the settling of the droplet: the bulk density of the droplet–spinel entity is lower than that of a droplet without an attached spinel, but the size of the entity increases. Thus, the attached spinel may either increase or decrease the settling velocity of the droplet–spinel entity. However, as the droplet–spinel has better wetting with the slag than the matte, and has lower density than the matte, the droplet–spinel entity cannot enter the matte layer, and thus, droplets with spinel may be removed with the slag in tapping. Additionally, turbulence can lift parts of the surface layer of the matte to the slag [18]. Furthermore, the distance between the tapping hole and the slag–matte interface affects copper losses. In a study by Jiménez et al. [19], matte droplets were found to form a dispersion layer above the matte surface. With too short a distance between the matte surface and the tapping hole, droplets from the dispersion layer may be sucked upwards and removed through the tapping hole. According to their study, at least 20 cm distance is needed to minimize the quantity of droplets sucked from the dispersion layer.

To limit mechanical copper losses, several parameters should be optimized [22]. The density difference between the slag and the matte should be as high as possible, while the viscosity of the slag should be as low as possible. Also, the matte droplets should be as large as possible. These three factors decrease the residence time of the matte droplets in the slag. The density difference and the slag viscosity are greatly affected by slag chemistry and variations in solid content, oxygen levels, and temperature in the slag. The droplet size is affected by coalescence and reactions, which are also affected by the properties of the slag and droplets themselves.

Mathematical and computational modeling of pyrometallurgical processes and furnaces started decades ago, as there was an increasing need for better understanding of the phenomena occurring inside them. A significant increase in modeling development has been seen since the introduction of commercially available computational fluid dynamics (CFD) software. The development of the Outotec flash smelting furnace (FSF) and process modeling has been reviewed in a recent review paper [23]. The aim of the modeling studies reported here is to find a feasible computational method by using commercial simulation software packages for gaining a better understanding of the fluid dynamics of the settling matte droplets in the slag layer. After validating the fluid dynamics behavior with physical model results, additional droplet–droplet or droplet–spinel interactions and chemical reactions between matte droplets and slag will be included in the simulation. However, as the geometries and phenomena in the models are computationally extremely intensive, the additional features have to be included one by one as a longer-term target. Finally, the models and knowledge developed will be used to help find ways to reduce copper losses and develop processes and their operational efficiency in a similar manner as has been done with the FS reaction shaft models [23]. In this paper, the fluid dynamics modeling results are presented, and the results obtained with the two software packages are compared. The settling flow behavior of the molten matte phase has been revealed, and the results from statistical (or traditional) CFD modeling and from the new CFD–DEM coupling approach are consistent, suggesting a channeling or funneling kind of flow pattern.

The software used in this study was ANSYS® Fluent 19.2 and EDEM® 2019.1 with EDEM-Fluent coupling v2.2 provided by DEM Solutions Ltd., Edinburgh, Scotland, UK [24].

2. Methods

Information or research reports on the breakup or coalescence of matte droplets in the FS settler in the literature are scarce. Therefore, this study focused on this part of the process so that the entraining and settling behavior of matte droplets could be investigated thoroughly.

2.1. Population Balance Model

The high-volume fraction of the dispersed matte phase during the settling process in the FS settler was described by the population balance model (Eulerian–Eulerian approach). The dispersed phase model (DPM, Eulerian–Lagrange approach) method has limitations when the volume fraction is above 10%, since computational time becomes expensive. Several methods are available for droplet classifications using the population balance model. These include size class formation and the moment method. Size class formation divides the system into several size groups. Therefore, it is a more accurate way to estimate the system. Size classes are further classified into the homogeneous and inhomogeneous methods. The homogeneous method classifies the different size groups into a single velocity field, which means it treats all the size groups as one phase and, therefore, the results from this method are not as accurate. In contrast, the inhomogeneous method defines a velocity profile for each size group, so this method is more feasible for an accurate representation of a system that contains different size groups. The inhomogeneous method has been used for an oil, water and air system [25], and it accurately predicted the system. Therefore, this method was preferred for this study as well. However, defining the different size groups with the inhomogeneous method is not enough, since the model equations need to be parametrized to truly depict the scenario population balance, especially for coalescence efficiency and frequency, to capture the correct coalescence rate. Nonetheless, since it is not easy to obtain experimental data under the harsh conditions of the settler, these methods may give a good estimate of what is happening inside the settler.

The population balance model distributes different droplet sizes into various size groups. These groups represent a range of droplet sizes and the volume fraction of each droplet size is estimated through mathematical models. The equation representing the population balance model (PBE) in a control volume is shown in Equation (1) [26].

$$\frac{\partial f(\mathbf{x},\ \xi,\ t)}{\partial t} + \nabla \cdot (V(\mathbf{x},\ \xi,\ t) f(\mathbf{x},\ \xi,\ t)) = S(\mathbf{x},\ \xi,\ t) \tag{1}$$

where f represents the density function with parameters: x, ξ, and t, where x represents the physical coordinates, and ξ represents the internal coordinates of the particle, for example, diameter. Internal coordinates could be either a scalar or a vector depending on how many properties of the droplet or particle you wish to include. Since in this work the focus is on the diameter of the droplet, ε is considered as a scalar. Finally, t represents the time, V represents velocity, and S an external source term.

If the death and birth of particles are not considered, then the number of particles in a control volume is constant, and

$$\frac{dN}{dx} = \frac{\partial f(\mathbf{x},\ \xi,\ t)}{\partial t} + \nabla \cdot (V(\mathbf{x},\ \xi,\ t) f(\mathbf{x},\ \xi,\ t)) = S(\mathbf{x},\ \xi,\ t) = 0 \tag{2}$$

Two different schemes, the class method and the moment method, are available for estimating the particle size distribution (PSD) in the domain. The class method divides the PSD into various discrete size groups and each size group is represented by the PBE. For some applications, the multiple size groups (MUSIG) model has been developed, which uses a homogeneous and inhomogeneous approach to solve the PSD field.

Inhomogeneous particle size distribution was used for this work since homogeneous PSD schemes use a single velocity field for all particle sizes groups and therefore, may not be an accurate representation of the velocity fields for different size groups. In contrast, inhomogeneous PSD uses different velocity fields for the size groups [26].

2.2. Coalescence Model

The Luo coalescence model [27] over-predicts the coalescence frequency and requires adjustment of coefficients in the equation using sensitivity analysis and validation with experimental data [28]. Therefore, the turbulence model was used for detailed analysis. However, a few results from the Luo model are presented as well for comparison. The turbulence model is governed by the following empirical equations.

The droplet collision rate is determined by the local shear within the eddy in the suspension mixture and is represented by the following equation [29]

$$a\left(L_i,\ L_j\right) = \varsigma T \sqrt{\frac{8\pi}{15}} \gamma \frac{\left(L_i + L_j\right)^3}{8} \tag{3}$$

where $\dot{Y}$ is the shear rate of the droplet and ςT is the capture efficiency coefficient of turbulent collision. Finally, the aggregation or coalescence rate is determined by the equation presented below [30]. The Brownian effect is negligible in turbulent flows and on droplet scale, so it was excluded.

$$a\left(L_i, L_j\right) = \varsigma T 2^{\frac{3}{2}} \sqrt{\pi} \frac{\left(L_i + L_j\right)^2}{4} \sqrt{\left(U_i^2 + U_j^2\right)} \tag{4}$$

2.3. Coupled CFD–DEM

Matte settling through the slag was also investigated with CFD coupled to a discrete element method (DEM). With the CFD–DEM method, individual droplets can be simulated by approximating them as soft spheres that are affected by slag flow. In the coupling, CFD is used to calculate the slag flow and drag forces for each droplet, while DEM is used to solve the contact forces and movement of the droplets. Additional models can be used to take more complex phenomena, such as coalescence, into account. CFD and DEM are not computed simultaneously; instead, first CFD solves one time step and transfers the drag forces to DEM, which then calculates until the CFD time step is reached and returns the droplet locations to CFD. Generally, DEM calculates several time steps between each CFD time step as the CFD time steps are longer. The basic principle of CFD–DEM simulation is illustrated in Figure 2.

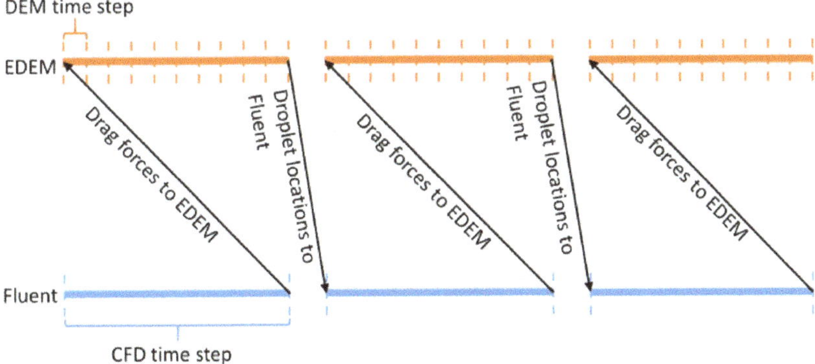

Figure 2. Illustration of the computational fluid dynamics–discrete element method (CFD–DEM) calculation process.

In CFD–DEM simulation, coalescence was calculated using the bubble coalescence probability by Wang et al. [31], which was calculated for every droplet–droplet contact. If the probability was at least 50%, the droplets were considered to have coalesced.

2.4. Geometry Dimensions and Materials

As CFD–DEM is a computationally demanding method, the FS settler model had to be scaled down for studying the settling phenomenon. The high computational demand is caused by using two methods in parallel and solving interactions of a very large number of individual elements. The geometry for the slag was a cube with 20 cm sides, as presented in Figure 3. As both slag and matte descend from the reaction shaft, the droplet and slag inlet was set at the center of the top face of the cube, while the tapping hole was depicted with a small tube on one side. The inlet and tapping hole had diameters of 15 cm and 1.5 cm, respectively. The slag flow rate (70 t/h) was taken from the literature [6] and scaled down with the reaction shaft diameter (4.5 m) reported in the literature [32]. The matte droplet feed rate was set as 40% of the slag feed rate and varied in size according to a normal distribution. The values of the slag and matte parameters used in the simulation are listed in Table 1.

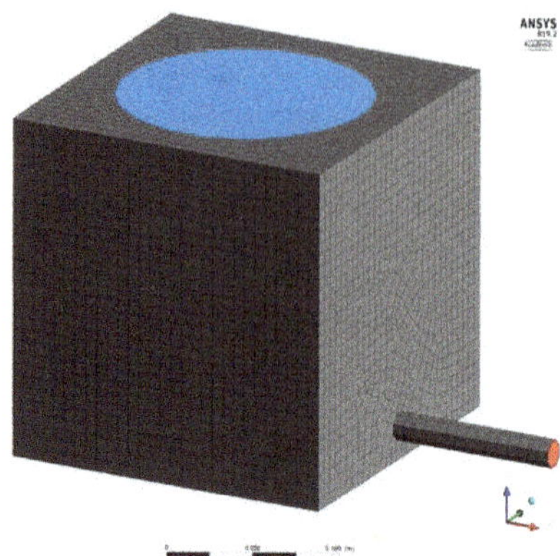

Figure 3. Geometry and mesh of the CFD model for CFD–DEM simulation. Inlet (blue) on the top and tapping hole (red) on the right side.

Table 1. Values of the slag and matte parameters.

Parameter	Slag	Matte
Feed rate (kg/s)	0.022	0.0086
ρ (kg/m^3)	3150	5100
Viscosity (kg/ms)	0.45	-
Mean diameter (μm)	-	500
Standard deviation	-	0.1

An additional user-defined model was used to simulate coalescence of the matte droplets. Coalescence was approximated by deleting the coalescing droplets and creating a new one in their center of mass with a mass equal to the sum of the deleted ones. Due to the limitations of the model, coalescence was instantaneous. The maximum size for coalesced droplets was limited to 2 mm, as larger droplets caused the software to use all the available memory due to the coalescence affected too many droplets at once.

This scaled-down settler geometry with similar slag and matte physical properties was modeled with CFD software as well. However, in addition to the previous research in this regard (single particle size of 500 μm) [10] and the CFD–DEM modeling part, in this CFD modeling the particle size distribution (PSD) scheme was also applied. The final selection and distribution of different droplets present in the initial mixture (PSD) were taken from the work of Jun 2018 [33], which is presented in Table 2.

Table 2. Size distribution of the droplets at the inlet.

Matte Droplets Size (μm)	Mass %	Volume Fraction in Mixture
500	2	0.006
300	67	0.201
150	18	0.054
100	4	0.012
75	1	0.003
60	2	0.006
50	6	0.018

Parallel with the development of these computational models, experimental investigations were ongoing on laboratory scale. The kinetics of slag–matte reactions [34,35] at temperatures prevailing in the FSF was studied in a vertical tube furnace in air and argon (oxygen-free) atmospheres simulating the settler area reactions between matte droplets and partly and nonreacted feed particles. Additionally, the main impurity elements (Sb, As, Bi) and pieces of waste electric and electronic equipment (WEEE) [36] were used in the experiments.

3. Results

3.1. CFD Simulation

Coalescence was studied using the Luo coalescence model [27] and the turbulence coalescence model [30]; however, the Luo model over-predicted the results. For example, the coalescence rate was quite high and almost all the smaller-sized droplets were coalesced into large size droplets, and therefore, the concentration of large droplets was high in the scaled-down settler. The results after 20 s are presented in Table 3. The results were obtained at a plane passing through the center of the scaled-down settler at x = 0.

Table 3. Results for Luo [27] coalescence model.

Droplet Size (μm)	Volume Fraction
1300	
900	
500	

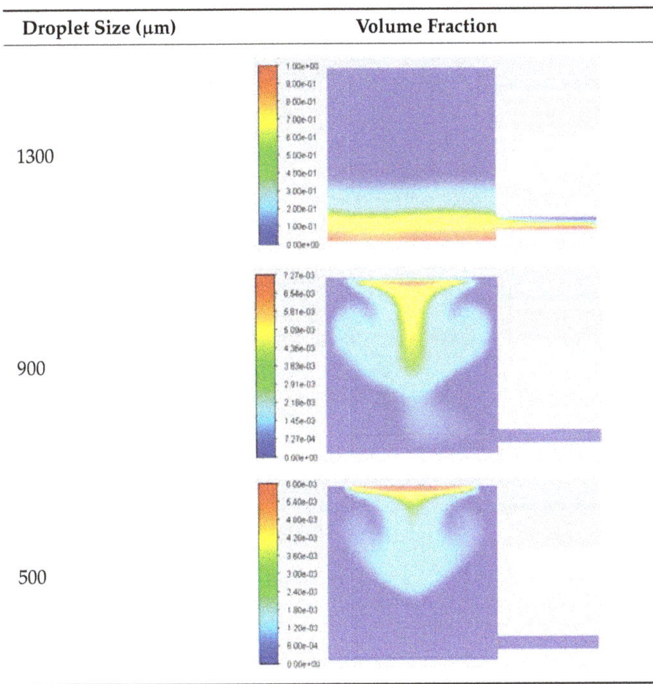

From the above results it can be concluded that all the smaller size droplets had coalesced into large size droplets and were not present in the settler after 20 s, which reveals that the coalescence rate was unrealistic. However, even though the coalescence rate might not be accurate, the model shows similar flowing behavior as depicted in the CFD–DEM method below. Thus, it can be concluded that the flow is aligned more towards the center of the settler forming a conical shape. The turbulence model was also used in lieu of the Luo model to obtain a comparison, and the results are presented in Table 4.

Table 4. Volume Fraction Contours for 500 µm Droplets at 15, 17, 18, and 20 s.

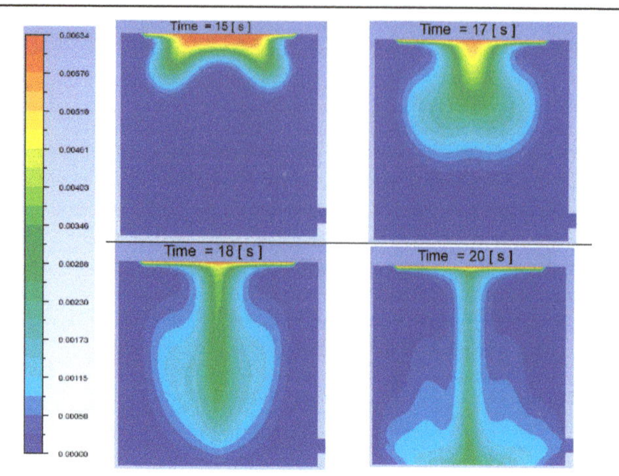

The volume fraction contours for the small-sized settler, taken at a plane in the center of the settler at x = 0.1 for 500 µm droplets, are presented in Table 4. It can be deduced that coalescence mostly occurred near the top of the slag surface, and as the particles started to settle, they took their own path. Second, the volume fraction was decreasing towards the larger size droplets, which shows that a small number of droplets bigger than 900 µm were present in the settler. Almost all the droplets reached the bottom of the settler within 20 s. In the future, it would be good to focus the study of droplet coalescence on sizes between 500–1000 µm.

The velocity vector profiles for the small settler model with 500 µm droplets are presented in Table 5. The velocity vector profiles confirm the formation of a funnel-shaped flow. After entering the slag layer, the flow channels the droplets towards the center of the settler.

Table 5. Velocity Vector Contours for 500 µm Droplets at 15, 17, 18, and 20 s.

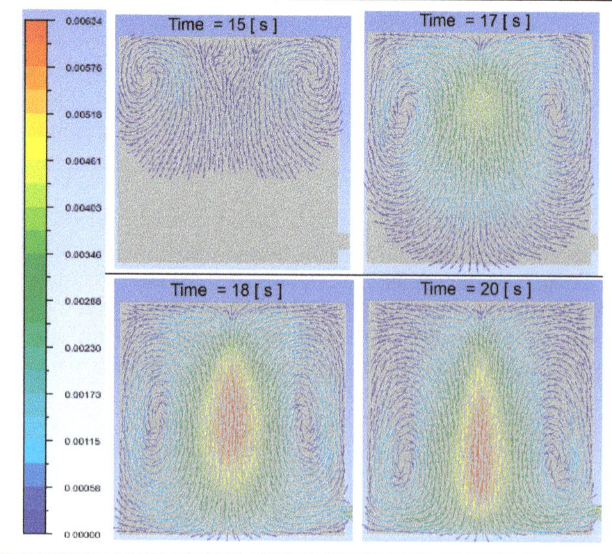

The volume fraction contours for the full-sized settler taken at the center plane of the settler (x = 0.5) containing the settler inlet and slag outlet are presented in Table 6. These volume fraction contours reveal that most of the coalescence occurred at the surface of the slag. However, a small number of droplets also coalesced during the descending route because of the turbulence underneath the inlet. This chaotic behavior was created because of the reverse flows caused by vortices and slag movement in the upward direction due to the density difference. This phenomenon is presented in the velocity vector profiles in Table 7.

Table 6. Volume fraction contours for 500 μm droplets.

Time (Sec.)	
70	
75	
80	

Table 7. Velocity Vector Contours for 500 μm Droplets (m/s).

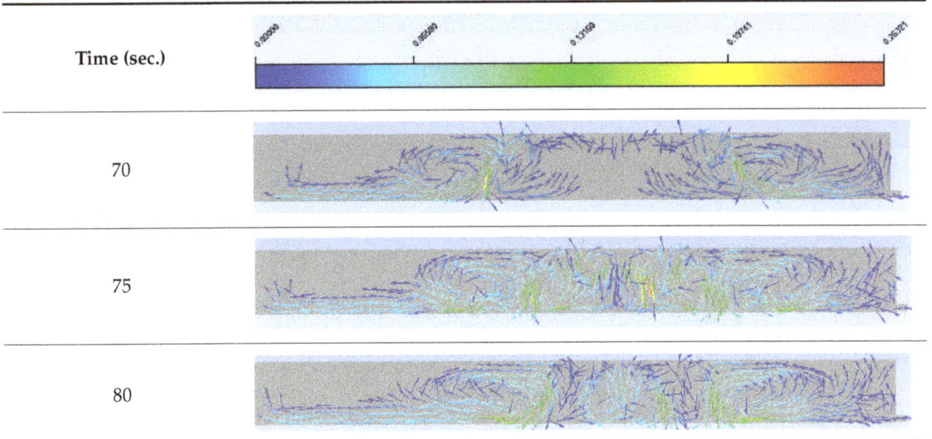

Time (sec.)	
70	
75	
80	

Finally, when we compare the volume fraction contours of the smaller-sized settler presented in Table 4, and the volume fraction contours of the full-size settler presented in Table 6, we can observe a significant difference. This is because in the full-size settler instead of a single funnel-shaped flow, different locally channeled flows are formed, which is exhibited in the profile. The same phenomenon is depicted in the velocity vector profiles in Tables 5 and 7.

The velocity vector profiles of 500 μm droplets are presented at a plane of the settler center (x = 0.5 m). This vector profile exhibits the development of 500 μm droplets and settling behavior. As the matte and slag mixture enters the settler, the matte droplets start settling, and after separation the slag circulates upwards. This creates a turbulent flow and vortices underneath the inlet illustrated by the vector profiles, which also confirm the formation of the local channeling flows aligned to their centers.

Compared to the 900 µm droplets, 1300 µm droplets are present in the settler in fairly low numbers, which is revealed by the number fraction of the different-sized matte droplets plot in Figure 4 and the volume fraction distribution in Figure 5. Figure 5 shows the volume fraction distribution of different-sized droplets present in the settler at different time intervals. Figure 4 also illustrates that a higher number of small droplets are present in the system, which suggests that their coalescence rate is low, and thus, the coalescence rate is low in general. The dominant droplet size is smaller than 300 µm.

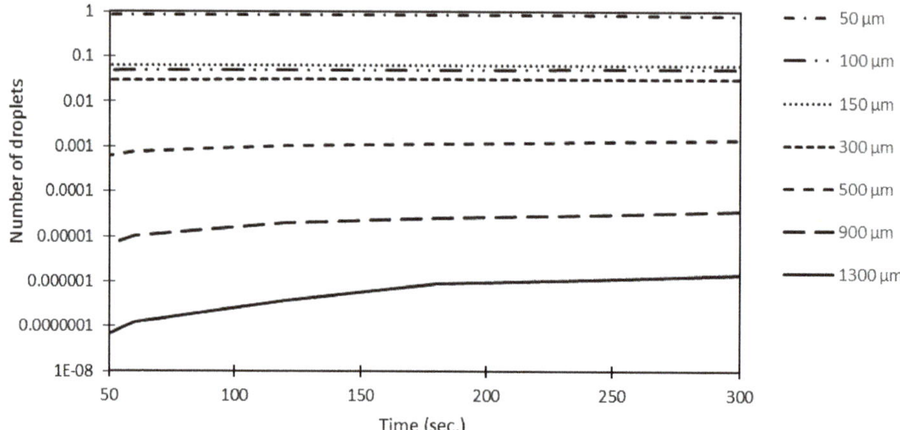

Figure 4. Number fraction distribution of different-sized droplets inside the settler.

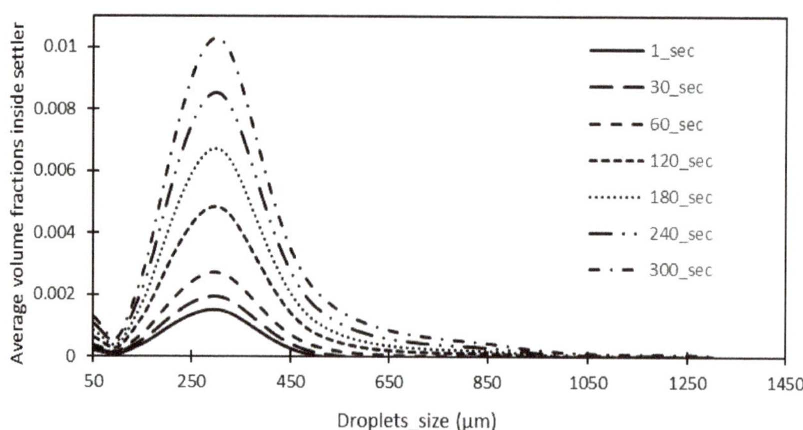

Figure 5. Volume fraction distribution curves.

The volume fraction distribution curves presented in Figure 5 show that most of the droplets are between 200 µm to 400 µm. The peak of these curves is usually achieved at around 300 µm. This is also an indication that the transition to large droplets above 400 µm is significantly low.

3.2. Coupled CFD–DEM Simulation

The CFD–DEM simulation revealed a strong funneling effect in the settling droplet cloud. The effect was caused by the drag of the settling matte, pulling droplets towards the centerline of the cloud, and subsequently increasing the bulk density as the droplets became more closely packed. This caused an increased settling velocity which, in turn, increased the drag, and thus, created a self-sustaining funneling effect. However, the CFD–DEM simulation did not show the development of two light matte streams that can be seen at 15 s in Table 4. The development of the effect is shown in Figure 6.

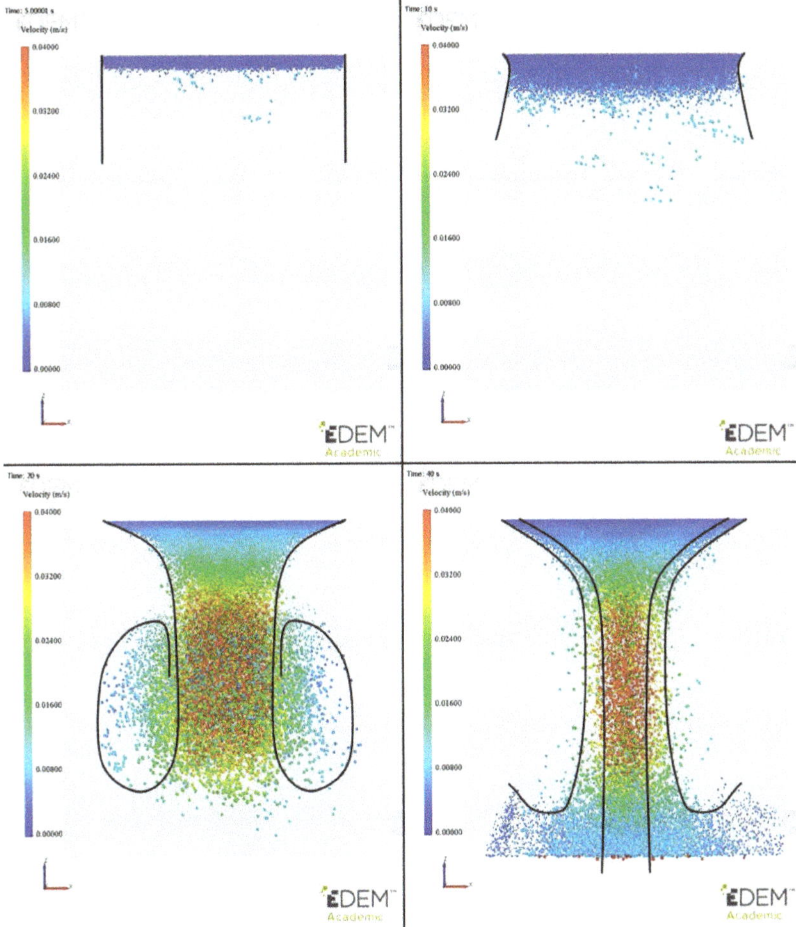

Figure 6. Formation of the funneling effect in the slag at 5 s, 10 s, 20 s, and 40 s.

The funneling effect increased the settling efficiency of the matte by increasing the settling velocity, as presented in Figure 7. Also shown is a comparison to a calculated velocity. Stokes's law was used to calculate the settling velocity for an average droplet size for every 0.5 s, while the function calculator in Ansys CFD-Post was used to calculate the average downward slag flow velocity. These were summed to obtain the calculated velocity, which was found to match well with the results from the simulation. The average velocity decreased as the cloud reached the slag–matte interface. The decrease was caused by the slag scattering droplets on the outer sides of the bottom cloud. The scattering was caused by downward flowing slag being deflected by the matte. As the droplets were scattered, some of the smaller droplets were pushed sideways and slightly upwards, prolonging the residence time of the droplets. Some of these droplets were removed through the tapping hole. Also, some of the droplets could have been trapped by the slag flows, preventing them from settling onto the matte layer and leading to copper losses.

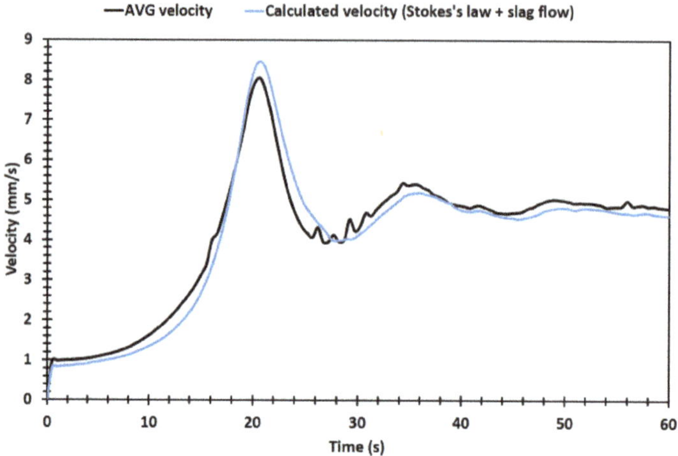

Figure 7. Average velocity of droplets and calculated velocity.

In addition to the funneling effect, droplet size has a significant effect on settling. The droplet size increases as two droplets collide and coalesce. The droplet size can be seen to increase rapidly near the surface of the slag as drag decelerates the falling droplets. The average diameter of the droplets in the whole slag layer at 60 s was 627 μm. As the droplets started to settle and coalesce, the size increased rapidly, averaging around 1280 μm below a depth of 50 mm, as presented in Figure 8. However, the size distribution and the figure show that a significant number of the droplets coalesced very near the slag surface.

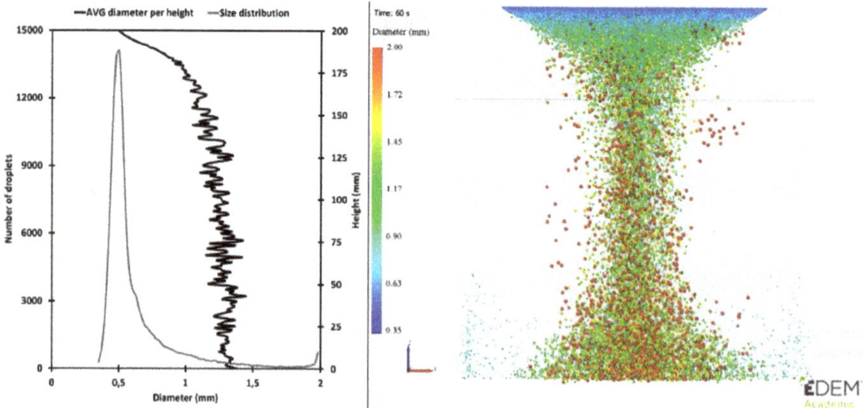

Figure 8. Droplet sizes in the slag layer at 60 s. The 50 mm settling distance is marked with a line.

The total feed rate was around 25,000 droplets per second. The cloud penetrated the slag layer in around 23 s, during which time over 500,000 droplets were injected in the slag. However, the droplet numbers in the simulation are significantly lower due to the high coalescence rate; 72% of all droplets in the simulation had coalesced. As can be seen in Figure 9, the droplet number peaks at around 135,000 droplets at the 12 second mark, but then begins to decrease. The decrease can be accounted for by the start of the funnel formation, which pulls the droplets closer together and thus enhances coalescence. However, the number of coalesced droplets does not show a similar peak during the funnel formation period, which can be explained by the increasing droplet diameter as the coalesced droplets further coalesce with each other. The average diameter stabilized soon after the funnel effect had fully formed. Also, the number of droplets in the simulation stabilized at around 124,000 after 30 s.

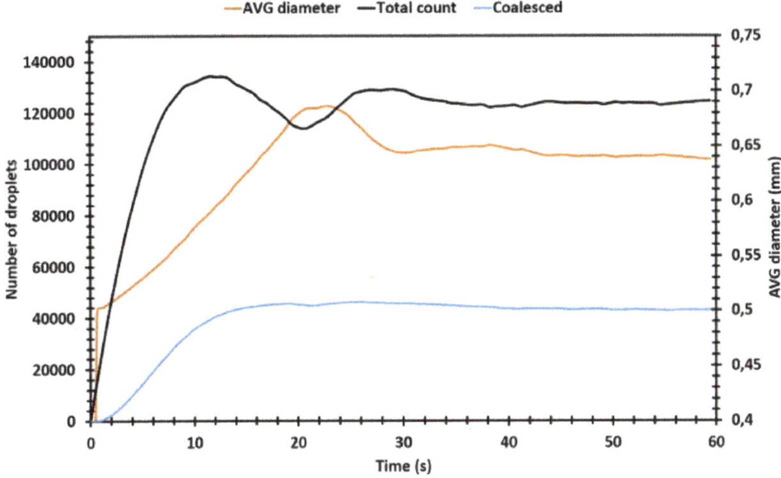

Figure 9. Droplet count and diameter development during the simulation.

4. Discussion

In the literature, [5,7] research regarding matte droplets settling through the slag phase was conducted at steady state using the Eulerian–Eulerian approach. In addition, the Eulerian–Lagrangian approach has been used to study the paths of individual droplets during settling and the effect of the slag tapping position and velocity on settling [6]. The conclusion drawn in these studies was that mostly droplets of 100 μm and below remained suspended in the slag phase, and that a higher tapping velocity and tap position close to the inlet disturbed the settling and entrapped the matte droplets. Similar conclusions were later drawn when using the Eulerian–Eulerian approach and transition state conditions while studying 100 μm, 300 μm, and 500 μm sized droplets individually [10]. The current investigation continued by adding the coalescence of droplets. Furthermore, instead of mono-sized droplets at the inlet, a mixture of different-sized droplets as reported in literature [33] were used for the inlet conditions, and the results from the full-scale settler were consistent with those reported earlier [6,7,10].

As expected, large droplets settled quickly and droplets of 500 μm and above mostly settled within 90 s. However, small droplets below 500 μm were still settling after 90 s, especially the 100 μm and 50 μm droplets, which were mostly dispersed inside the slag. Additionally, the formation of a vortex under the inlet and chaotic flows due to opposing currents of matte droplets trying to settle [7], and the lighter slag phase moving upwards were likewise observed in this work. Nonetheless, the coalescence rate of large droplets was relatively low, as indicated by the low volume fraction of 900 μm and 1300 μm sized droplets. In the future work, more droplets with a wider size distribution will be used as the inlet condition to get closer to the real situation.

The results in the down-scaled model and industrial scale geometry suggested a funneling settling of the matte phase. However, in the industrial scale simulation, there were several funnels whereas in the small model, only one central funnel formed. Intuitively, it is quite obvious that the large inlet area of the industrial settler would not produce only one single channel for the whole matte phase, but rather, several localized "nucleation" sites from where the funneled settling would start.

The settling of individual matte droplets was simulated using CFD–DEM coupling. The funneling effect was also formed in the slag layer as the drag pulled the droplets towards the centerline of the settling cluster. The funneling effect increased the settling velocity of the droplets, and thus, increased the settling efficiency.

A user-defined model for coalescence was also included in the CFD–DEM simulation. The behavior of the droplets, and consequently, coalescence is practically impossible to observe in a real FS process due to the extreme environment in the furnace. However, the effects of the coalescence in the simulations seemed reasonable: the settling velocity of the droplets increased due to the increasing size of the droplets. Also, coalesced droplets created a relatively large size distribution in the slag layer. Compared to the Eulerian–Eulerian method, the coalescence rate was higher in the CFD–DEM simulation. This difference could have been caused by different methods for solving the droplet–droplet contact or coalescence criteria. Nonetheless, some of the smaller droplets did not coalesce and never settled, and consequently, they were entrained in the slag and would have caused copper losses.

Due to computational instabilities, the maximum droplet size had to be limited to 2 mm. This could have had some effect on settling, as some droplets could have grown more, leading to a smaller number of droplets with a larger average size. They would cause stronger drag, and thus strengthen the funneling effect. However, the number of such large droplets would most likely be relatively low, limiting their effect. Also, very large droplets could break into smaller ones due to inertial forces. Furthermore, a significant majority of the droplets would most likely be much smaller.

The funneling effect and coalescence together formed a kind of feedback loop. Coalescence increased the droplet size which increased drag, and subsequently strengthened the funneling effect. The effect caused the droplets to move closer to each other, which consequently increased the coalescence rate. Both phenomena increased the settling velocity. The results were in good agreement with the values calculated by combining the average slag flow velocity and Stokes's law velocity of an average-sized droplet.

The slag flow deflecting from the matte surface created a kind of dispersion layer above the slag–matte interface, as some of the smaller droplets were pushed sideways with the slag. This kind of layer has also been reported in the literature [20]. Some droplets from the dispersion layer were sucked by the tapping hole, leading to copper losses. Such a situation can be seen in Figure 6; however, the effect in the simulation is likely over-emphasized by the small geometry as the tapping hole is much closer to the matte–slag interface than it would be in a real settler.

The use of well-established CFD modeling with the commercial and widely used tool ANSYS Fluent, together with a coupled CFD–DEM modeling with EDEM software was the first computational attempt to understand the settling of matte droplets in the FS settler. The results of the scaled-down geometry obtained with both computational approaches suggest that droplet coalescence plays a crucial role in the resulting flow pattern. The matte droplets entering the slag surface layer coalesce rapidly causing accelerated settling which, in turn, drags the droplets towards a central path or channel down to the underlaying matte layer. In the full-scale geometry, in contrast, the flow of matte droplets, although quickly coalescing, seems to follow a more complicated settling pattern. This is believed to be due to the more complicated flow of the continuous slag phase induced by the input material flow.

A comparison of the two computational approaches for the matte settling phenomena can be seen in Figure 10, where the small-scale model results from the CFD and CFD–DEM models are shown. Based on the computational results of the small-scale model of the matte droplets settling through the slag layer, it seems to be fair to conclude that the two approaches used, CFD and coupled CFD–DEM, both predict the settling pattern to be channeled or funneled to a central "trail" of coalescing droplets. However, these results have to be validated by a physical model before this can be confirmed.

The relation of the funneling effect and size of the settling cloud requires further study. In a real FS furnace, the area corresponding to the inlet in the model is significantly larger. The diameter of the reaction shaft can be, for example, 4.5 m compared to the 15 cm of the scaled-down model in this study. Additionally, the inlet area would not be as well defined as in the computational study. Instead, there would be a gradient in the feed density. Further research is needed to evaluate the effect of these factors on the formation and strength of the funneling effect. Moreover, the small geometry of the model may affect the ease of forming the funneling effect as walls close to the cloud may create stronger turbulence in the slag. Thus, larger model geometry should also be studied and the funneling effect experimentally validated with a physical water–oil model. The results of the validation will be presented in a future article.

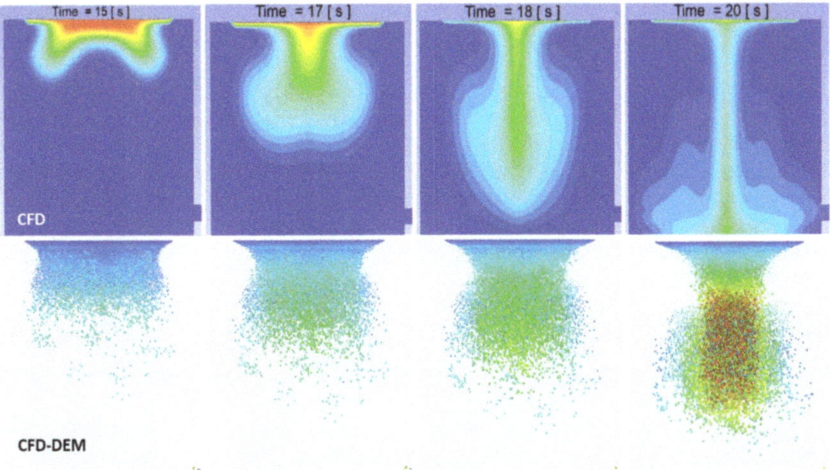

Figure 10. A comparison of the CFD and CFD–DEM results for matte settling in the small-scale settler model, revealing a similar funneling flow pattern at 15, 17, 18, and 20 s. Upper row: CFD, lower row: CFD–DEM.

Further development is needed to improve the accuracy of the coalescence model. An improved coalescence model could decrease memory issues (especially with the CFD–DEM computation), if the droplet size limit were increased. In addition, simulating the formation and effect of spinel particles should be considered. DEM is capable of simulating spinel particles too, but their attachment to droplets and formation require an additional model. The spinel model would also be supplemented by a reaction kinetics model. However, development of these two models would require further development in the software used as well. Additionally, a more powerful computer is required until full-scale settler simulations with CFD–DEM can be considered as even small-scale simulations take a long time. For example, the small cube simulation took over a month of computing and the larger model [9] took over two months with Intel Xeon E3-1230 v5 @ 3.4 GHz.

5. Conclusions

A comparison of two computational approaches, CFD and CFD–DEM, was carried out for the matte settling phenomena in settler geometries. The results of the small-scale model of matte droplets settling through a slag layer suggest that the settling flow pattern is channeled or funneled as a central trail of coalescing droplets. For a sounder conclusion, these results still have to be validated by a physical model. Nevertheless, as a conclusion, the applicability of the coupled CFD–DEM software equal to CFD software for the liquid–liquid system used in this study can be confirmed.

One conclusion regarding the settling pattern of matte droplets in the industrial scale settler geometry, is that it is plausible that the droplets coalesce and form centralized funnels instead of settling uniformly across the whole inlet area. This phenomenon is beneficial for the efficient separation of the slag and matte phases, and thus, for the recovery of the metal. Consistently with industrial reality, the smallest droplets were very slow to settle, and consequently they require a very long time or a very thin slag layer to reach the matte layer before the slag leaves the settler through the tapping hole. One option could be to operate with continuous tapping and a thin slag layer.

The studied case clearly indicated that the current computing power is still the limiting step in computer modeling of a full-scale industrial settler with realistic droplet coalescence calculation. Although there are some similarities in the settling pattern in the small- and full-scale models, they are too different for conclusions to be made based on the small-scale model. This poses a challenge for the validation of the full-scale model results as there are no possibilities for direct observations of the flow phenomena in the molten slag layer inside the settler. So far, the only option is to compare copper losses from the CFD model with industrial data [10].

Author Contributions: Conceptualization, all; methodology, J.-P.J. and N.A.K.; analysis, all; investigation, J.-P.J. and N.A.K.; resources, A.J.; writing the manuscript, J.-P.J. and N.A.K.; review and editing, A.J.; supervision, A.J.; funding acquisition, A.J. All authors have read and agreed to the published version of the manuscript.

Funding: The funding from the Steel and Metal Producers' Fund in Finland, and from the Aalto University School of Chemical Engineering is greatly acknowledged.

Conflicts of Interest: The authors declare no conflict of interest. The funders had no role in the design of the study; in the collection, analyses, or interpretation of data; in the writing of the manuscript, or in the decision to publish the results.

References

1. Arslan, C.; Arslan, F. Recovery of copper, cobalt, and zinc from copper smelter and converter slags. *Hydrometallurgy* **2002**, *67*, 1–7. [CrossRef]
2. Sarfo, P.; Das, A.; Wyss, G.; Young, C. Recovery of metal values from copper slag and reuse of residual secondary slag. *Waste Manag.* **2017**, *70*, 272–281. [CrossRef] [PubMed]
3. Kambham, K.; Sangameswaran, S.; Datar, S.R.; Kura, B. Copper slag: Optimization of productivity and consumption for cleaner production in dry abrasive blasting. *J. Clean. Prod.* **2007**, *15*, 2007. [CrossRef]
4. Kojo, I.V.; Storch, H. Copper production with Outokumpu Flash Smelting: An Update. In *International Symposium on Sulfide Smelting, Proceedings of the Sohn International Symposium: Advanced Processing of Metals and Materials, San Diego, California, 27–31 August 2006*; TMS: Warrendale, PA, USA, 2006; Volume 8.
5. Xia, J.L.; Ahokainen, T.; Kankaanpää, T.; Järvi, J. Numerical Modeling of Flow and Heat Transfer in the Settler of a Copper Flash Smelting. In Proceedings of the 6th International Symposium on Heat Transfer, Beijing, China, 15–19 June 2004.
6. Xia, J.L.; Ahokainen, T.; Kankaanpää, T.; Järvi, J.; Taskinen, P. Numerical modelling of copper droplet setting behavior in the settler of a flash smelting furnace. In Proceedings of the European Metallurgical Conference, EMC 2005, Dresden, Germany, 18–21 September 2005.
7. Xia, J.L.; Ahokainen, T.; Kankaanpää, T.; Järvi, J.; Taskinen, P. Flow and heat transfer performance of slag and matte in the settler of a copper flash smelting furnace. *Steel Res. Int.* **2007**, *78*, 155–159. [CrossRef]
8. Zhou, J.; Zhou, J.; Chen, Z.; Mao, Y. Influence Analysis of Air Flow Momentum on Concentrate Dispersion and Combustion in Copper Flash Smelting Furnace by CFD Simulation. *JOM* **2014**, *66*, 1629–1637. [CrossRef]
9. Jylhä, J.-P.; Jokilaakso, A. CFD-DEM modelling of matte droplet behaviour in a flash smelting settler. In Proceedings of the 58th Conference of Metallurgists Hosting Copper 2019, Vancouver, BC, Canada, 18–21 August 2019.
10. Khan, N.A.; Jokilaakso, A. Dynamic Modelling of Molten Slag-Matte Interactions in an Industrial Flash Smelting Furnace Settler. In Proceedings of the Extraction 2018, Ottawa, ON, Canada, 26–29 August 2018.
11. Niemi, T. *Particle Size Distribution in CFD Simulation of Gas-Particle Flows*; Aalto University: Espoo, Finland, 2012.
12. Warczok, A.; Utigard, A.T. Settling of copper drops in molten slags. *Met. Materi. Trans. B* **1995**, *26*, 1165–1173. [CrossRef]
13. Yu, D.; Mambakkam, V.; Rivera, H.A.; Li, D.; Chattopadhyay, K. Spent Potlining (SPL): A myriad of opportunities. In *Aluminium International Today*; Quartz Business Media Ltd.: Redhill, Surrey, UK, 2015.

14. Elliott, F.J.; Mounier, M. Surface and Interfacial Tensions in Copper Matte-Slag Systems, 1200 °C. *Can. J. Met. Mat. Sci.* **2013**, *21*, 415–428. [CrossRef]

15. Phelan, D. The Modelling of Matte Droplet Coalescence in the Vanykov Process. Master's thesis, University of Wollongong, Wollongong, Australia, 1999.

16. Cheng, X.; Cui, Z.; Contreras, L.; Chen, M.; Nguyen, A.; Zhao, B. Introduction of Matte Droplets in Copper Smelting Slag. In Proceedings of the 8th International Symposium on High-Temperature Metallurgical Processing, Gewerbestrasse, Switzerland, 9 February 2017; pp. 385–394.

17. Iwamasa, P.K.; Fruehan, R.J. Separation of Metal Droplets from Slag. *ISIJ Int.* **1996**, *36*, 1319–1327. [CrossRef]

18. Miettinen, E. Thermal Conductivity and Characteristics of Copper Flash Smelting Flue Dust Accretions. Ph.D. Thesis, Helsinki University of Technology, Espoo, Finland, 2008.

19. De Wilde, E.; Bellemans, I.; Campforts, M.; Guo, M.; Blanpain, B.; Moelans, N.; Verbeken, K. Investigation of High-Temperature Slag/Copper/Spinel Interactions. *Metall. Mater. Trans. B Process Metall. Mater. Process. Sci.* **2016**, *47*, 3421–3434. [CrossRef]

20. Jiménez, F.; Ríos, G.; Martinez, J.; Fernández-Caliani, J. Speciation of copper in flash, converter and slag-cleaning furnace slags. In Proceedings of the Copper 2013, Santiago, Chile, 1–4 December 2013.

21. De Wilde, E.; Bellemans, I.; Zheng, L.; Campforts, M.; Guo, M.; Blanpain, B.; Moelans, N.; Verbeken, K. Origin and sedimentation of Cu-droplets sticking to spinel solids in pyrometallurgical slags. *Mater. Sci. Tech.* **2016**, *32*, 1911–1924. [CrossRef]

22. Bellemans, I.; De Wilde, E.; Moelans, N.; Verbeken, K. Metal losses in pyrometallurgical operations—A review. *Adv. Colloid Interface Sci.* **2018**, *255*, 47–63. [CrossRef] [PubMed]

23. Taskinen, P.; Jokilaakso, A.; Lindberg, D.; Xia, J. Modelling copper smelting–the flash smelting plant, process and equipment. *Min. Proc. Ext. Met.* **2019**. [CrossRef]

24. DEM Solutions Ltd. *EDEM 2019.1 User Guide*; DEM Solutions Ltd.: Edinburgh, UK, 2019.

25. Fagerlund, O.K.; Jalkanen, H. Microscale Simulation of Settler Processes in Copper Matte Smelting. *Met. Materi. Trans. B* **2000**, *31*, 439–451.

26. De Wilde, E. Methodology Development and Experimental Determination of the Origin of Sticking Copper Droplets in Pyrometallurgical Slags. Ph.D. Thesis, KU Leuven, Leuven, Belgium, 2015.

27. Luo, H.; Svendsen, H.F. Theoretical model for drop and bubble breakup in turbulent dispersions. *Am. Inst. Chem. Eng. J.* **1996**, *42*, 1225–1233. [CrossRef]

28. Zhou, J.; Chen, Z.; Zhou, P.; Yu, J.; Liu, A. Numerical simulation of flow characteristics in settler of flash furnace. *Trans. Nonf. Met. Soc. China* **2012**, *22*, 1517–1525. [CrossRef]

29. Ping, Z.; Ping, J.Y.; Rong, H.C.; Chi, M. Settling mechanism and influencing factors on matte droplets in settler slag of copper flash smelting furnace. *Chin. J. Nonf. Met.* **2006**, *16*, 2032–2037.

30. Kamp, A.M.; Chesters, A.K.; Colin, C.; Fabre, J. Bubble coalescence in turbulent flows: A mechanistic model for turbulence-induced coalescence applied to microgravity bubbly pipe flow. *Int. J. Multip. Flow* **2001**, *27*, 1363–1396. [CrossRef]

31. Wang, T.; Wang, J.; Jin, Y. Theoretical prediction of flow regime transition in bubble columns by the population balance model. *Chem. Eng. Sci.* **2005**, *60*, 6199–6209. [CrossRef]

32. Davenport, W.G.; Partelpoeg, E.H. *Flash Smelting: Analysis, Control and Optimization*, 1st ed.; Pergamon Press: New York, NJ, USA, 1987; pp. 22–28.

33. Jun, Z.; Zhuo, C. Smelting Mechanism in the Reaction Shaft of a Commercial Copper Flash Furnace. In Proceedings of the Extraction 2018, Ottawa, ON, Canada, 26–29 August 2018.

34. Guntoro, P.I.; Jokilaakso, A.; Hellstén, N.; Taskinen, P. Copper matte—Slag reaction sequences and separation processes in matte smelting. *J. Min. Metall. B* **2018**, *54*, 301–311. [CrossRef]

35. Wan, X.; Shen, L.; Jokilaakso, A.; Eriç, H.; Taskinen, P. Experimental approach to matte-slag reactions in the flash smelting process. *Min. Proc. Ext. Met. Rev* **2020**, in press. [CrossRef]

36. Wan, X.; Fellman, J.; Jokilaakso, A.; Klemettinen, L.; Marjakoski, M. Behavior of waste printed circuit board (WPCB) materials in the copper matte smelting process. *Metals* **2018**, *8*, 887. [CrossRef]

 *processes*

Article

Simulation Study and Industrial Application of Enhanced Arsenic Removal by Regulating the Proportion of Concentrates in the SKS Copper Smelting Process

Qinmeng Wang *, Qiongqiong Wang, Qinghua Tian * and Xueyi Guo *

School of Metallurgy and Environment, Central South University, Changsha 410083, China; qqwang0112@163.com

* Correspondence: qmwang@csu.edu.cn (Q.W.); qinghua@csu.edu.cn (Q.T.); xyguo@csu.edu.cn (X.G.);
 Tel./Fax: +86-731-8887-6255 (Q.W.); +86-731-8887-7863 (Q.T.); +86-731-8887-6089 (X.G.)

Received: 24 February 2020; Accepted: 20 March 2020; Published: 26 March 2020

Abstract: Arsenic removal is a crucial issue in all copper smelters. Based on the Fangyuan 1[#] smelter, the effects of major elements (Cu, Fe and S) in sulfide concentrates on arsenic removal in the SKS copper smelting process were studied in this paper. The results show that Cu, Fe and S in concentrates have a significant influence on the oxygen/sulfur potential of smelting systems, and also affect the efficiency of arsenic removal. By regulating the proportion of the major elements in sulfide concentrates, the concentrate composition was changed from its original proportions (Cu 24.4%, Fe 26.8%, S 28.7%, and other 20%) to optimized proportions (Cu 19%, Fe 32%, S 29%, and other 20%). The distribution of arsenic among three phases in the original production process (gas 82.01%, slag 12.08%, matte 5.91%) was improved to obtain an optimal result (gas 94.37%, slag 3.45%, matte 2.18%). More arsenic was removed into the gas phase, and the mass fraction of arsenic in matte was reduced from 0.07% to 0.02%. The findings were applied to actual production processes in several other copper smelters, such as the Hengbang copper smelter, Yuguang smelter and Fangyuan 2[#] smelter. Therefore, the optimized result obtained in this work could provide direct guidance for actual production.

Keywords: arsenic removal; copper smelting; SKS; Shuikoushan process; oxygen bottom blown

1. Introduction

Arsenic is an element that is toxic for the environment and people's health [1–4], especially As_2O_3. In the smelting process for copper sulfide ores, arsenic is dispersed into the fly ash, smelting slag and copper matte [5–7]. Due to the increasing complexity of copper concentrate, arsenic control has been listed as an important issue in copper smelters [8]. The SKS (Shuikoushan) copper smelting process is very adaptable to complex concentrates and it has high arsenic removal efficiency [9–13]. In recent years, the SKS copper smelting process has become a popular research topic [14–20]. Arsenic removal in the SKS smelting process is affected by several factors, such as the composition of concentrate, matte grade, oxygen concentration in air blown into the furnace, oxygen/ore ratio, smelting temperature, the Fe/SiO_2 ratio in slag, and so on. In our previous study, the effects of matte grade, oxygen/ore ratio, oxygen concentration in air blown into the furnace, smelting temperature and ratio of Fe/SiO_2 in slag were investigated [21].

However, the composition of concentrate, especially the major elements (Cu, Fe and S), could affect the oxygen/sulfur potential of the smelting system, and further affect the removal of arsenic. Therefore, in this work, the content of the major elements (Cu, Fe and S) in sulfide concentrate was adjusted, and the removal of arsenic from the matte to gas phase in the SKS copper smelting process was investigated through the commercial simulation software SKSSIM [22]. This work will help us to better understand the SKS copper smelting process.

2. Research Methodology

The study was carried out by SKSSIM simulation software, combined with actual production in the Fangyuan 1# smelter in Dongying, China.

SKSSIM is an efficient simulation software for the SKS process, and it is based on the SKS smelting mechanism model [22] and the theory of Gibbs free energy minimization [23]. In the mechanism model, the SKS furnace is divided into seven functional layers from top to bottom and three functional regions along the length direction [21–24]. The particle swarm optimization algorithm, C# computer programming language, and Microsoft Visual Studio were used to develop the SKSSIM software. The development process (including activity coefficient, Gibbs free energy including activity coefficient, Gibbs free energy, phase entrainment coefficient, model verification and modification) has been presented in detail in our previous work [23]. SKSSIM has been successfully validated by the actual production process in Fangyuan 1# smelter [21,23,24]. Therefore, SKSSIM is a convenient method to carry out this study.

3. Results and Discussion

3.1. Arsenic Distribution in the Actual Production Process and by SKSSIM

The initial conditions and operation parameters of the SKS process are given in our previous work [21]. The calculated results are compared with the actual industrial production data in Tables 1 and 2.

Table 1. Chemical composition of matte and slag.

Composition (wt %)		As	Cu	Fe	S	SiO$_2$	Others
Actual production	matte	0.07	70.77	5.52	20.22	0.51	2.91
	slag	0.08	3.16	42.58	0.86	25.24	28.08
By SKSSIM	matte	0.07	70.31	4.80	20.38	0.82	3.62
	slag	0.07	2.93	42.07	0.73	25.18	29.02

Table 2. Arsenic fractional distribution among gas, matte and slag phases.

Distribution (wt %)	Gas	Slag	Matte
Actual production	82.01	12.08	5.91
By SKSSIM	82.71	11.06	6.23

3.2. Effect of Cu Content in Concentrate on Arsenic Distribution

Figure 1 shows the effect of Cu content in concentrate on the distribution of arsenic among the gas, slag and matte phases. With an increase in Cu content, there is an increase in the proportion of arsenic in both the matte and slag phases, and the fractional distribution in the gas phase decreases. At a low Cu content (such as 18%), around 95 pct of the arsenic reports to the gas phase. As the Cu content increases to 27%, about 20% of the arsenic reports to the matte phase, and only 55% reports to the gas phase. Under the fixed total volume of oxygen blown into the SKS furnace and ratio of oxygen/ore (total volume of oxygen/total mass of dry mixed concentrates), the matte grade increases as the Cu content increases. The activity coefficient of As in matte decreases with an increase in matte grade, and higher-grade matte has a higher affinity for arsenic; hence it reduces the activity and vapor pressure of arsenic [7,25].

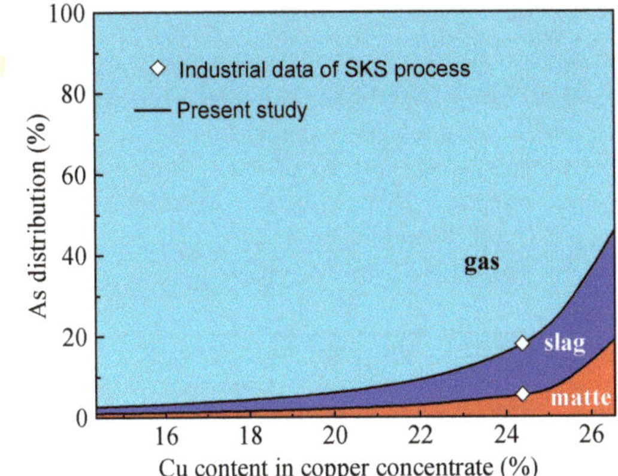

Figure 1. Comparison between the actual production and simulation results of the effect of Cu content in concentrate on the distribution of arsenic among the gas, slag and matte phases.

Therefore, a higher initial Cu content in the concentrate results in more arsenic in the matte phase, and the removal of arsenic from the smelting system to the gas phase is not very efficient.

After arsenic enters the gas phase, it then goes through the waste heat boiler, electrostatic precipitator and flue gas scrubber with the SO_2 off-gas. Most arsenic compounds in off-gas condense into the solid phase and are collected in dust by the waste heat boiler and electrostatic precipitator. In the flue gas scrubber, some of the arsenic enters the waste acid. Arsenic can be detoxified or recovered from the aforementioned dust and waste acid.

3.3. Effect of Fe Content in Concentrate on Arsenic Distribution

In Figure 2, the effect of the initial Fe content in the concentrate on the calculated distribution of arsenic among the gas, slag and matte phases in the SKS process are presented. With the increase in Fe content, the proportion of arsenic reporting to both the matte and slag phases decreases, and the fractional distribution in the gas phase increases. However, the change trend is not as obvious as that caused by the initial Cu content in the concentrate. When the Fe content in the concentrate is varied from 17% to 37%, the distribution of arsenic among the gas, slag and matte phases varies from 75% to 85%, from 15% to 10%, and from 10% to 5%, respectively. Therefore, the change in Fe content in the initial concentrate has a relatively tiny influence on the distribution of arsenic.

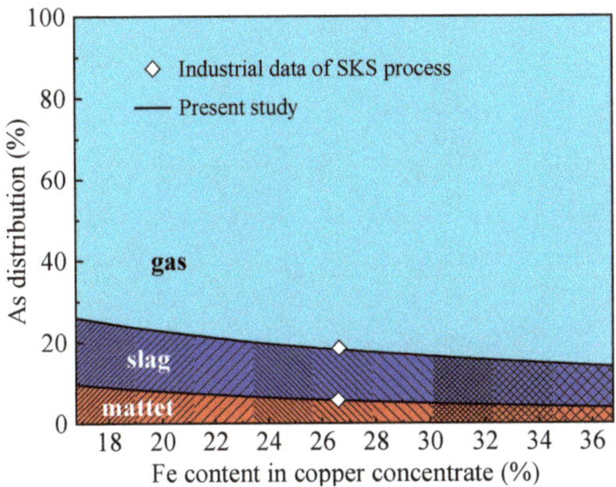

Figure 2. Comparison between the actual production and simulation results of the effect of Fe content in concentrate on the distribution of arsenic among the gas, slag and matte.

3.4. Effect of S Content in Concentrate on Arsenic Distribution

In Figure 3, as the S content in the initial concentrate increases, the proportion of arsenic reporting to the matte and slag phases decreases sharply, and the arsenic reporting to the gas phase increases. Therefore, the change in S content has a relatively significant influence on the distribution of arsenic. As the S content in the feed increases, the oxygen potential in the smelting system decreases and the sulfur potential increases, causing higher partial pressure of S_2 and hence, a higher degree of volatilization of arsenic as AsS [7,24]. Therefore, high S content in the initial concentrate is beneficial for removing arsenic to the gas phase from the matte phase.

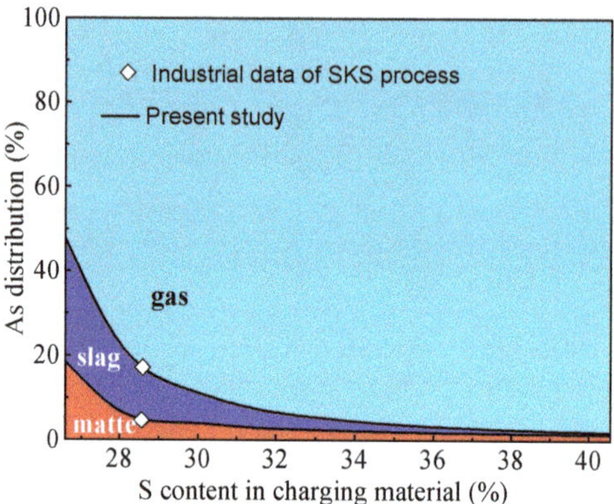

Figure 3. Comparison between the actual production and simulation results of the effect of S content in concentrate on the distribution of arsenic among the gas, slag and matte.

3.5. Effect of Arsenic Content in Concentrate on Arsenic Distribution

Figure 4 shows the effect of the initial content of arsenic in copper concentrate on the deportment of arsenic to the phases. In China, the maximum allowed content of arsenic in copper concentrates imported from abroad is 0.5%. In general, the arsenic content in feed is 0.2%–0.7% in the copper smelting process. As shown in Figure 4, in the range studied, the proportion of arsenic reporting to both the gas and slag phases increases slightly with an increase in the arsenic content in copper concentrate. The volatilization of arsenic increases as a result of the arsenic being transferred to the gas phase as AsS, AsO, and As_2.

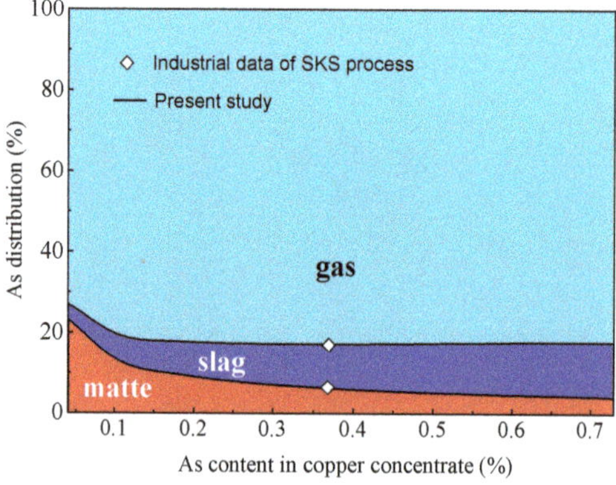

Figure 4. Comparison between the actual production and simulation results of the effect of arsenic content in copper concentrate on the distribution of arsenic among the gas, slag and matte.

3.6. Interactive Effect of Cu and S Content in Concentrate on Arsenic Distribution

As shown in Figures 1–3, it is obvious that the Cu and S elements in concentrates have a more significant influence than Fe on the distribution of arsenic among the gas, slag and matte phases. Therefore, it is necessary to discuss the interactive effect of Cu and S content on arsenic distribution.

3.6.1. Interactive Effect on Arsenic Distribution

In this work, the symbol % represents the proportion of arsenic distribution among the gas, slag, and matte phases, and the symbol wt % represents the mass fraction of arsenic in different phases.

The fractional distribution of arsenic in the gas phase with the change in the initial content of Cu and S in concentrate is given in Figure 5a. The mass fraction of arsenic in gas is also shown in Figure 5b. The variation trends in the two figures are basically consistent. As both the Cu and S content in concentrate increases, both the fractional distribution and mass fraction of arsenic in the gas phase decrease. With Cu 18% and S 28%, around 95% of the arsenic enters the gas phase, and the mass fraction of arsenic in the gas phase is nearly 0.6%. However, as the Cu and S content in the concentrate increase to 24% and 34%, respectively, only 74% of the arsenic reports to the gas phase, and the mass fraction of arsenic in the gas phase is around 0.41%. Therefore, low initial content of Cu and S in concentrate helps to eliminate arsenic from the smelting system to the gas phase. However, these results contradict the result of Figure 3, that is, as the S content in the initial concentrate increases, the proportion of arsenic reporting to the gas phase increases. This phenomenon may be explained because Cu in concentrate has a stronger impact on arsenic distribution than S. Cu, S and Fe exist in concentrate in the form of $CuFeS_2$, CuS, Cu_2S, FeS_2, etc., the content of the elements have close relevance, and the interactive relationship to arsenic distribution is complex.

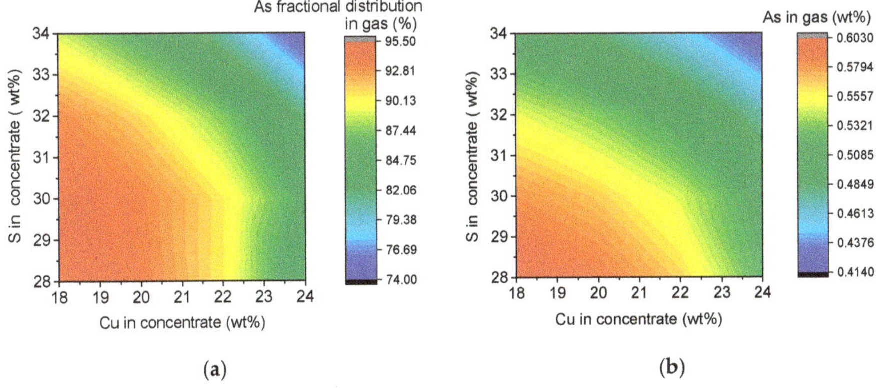

Figure 5. Arsenic distribution in the gas phase: (**a**) fractional distribution; (**b**) mass fraction.

The fractional distribution and mass fraction of arsenic in the slag phase with the change in the initial content of Cu and S in the concentrate are presented in Figure 6a,b. The variation trend in Figure 6 is contrary to Figure 5. As the Cu and S content in concentrate increases, the fractional distribution and mass fraction of arsenic in the slag phase also increase. With Cu of 18% and S of 28%, only around 2.1% arsenic enters the slag phase, and the mass fraction of arsenic in the slag phase is about 0.012%. However, as the Cu and S content in concentrate increases to 24% and 34%, respectively, about 24% arsenic reports to the slag phase, and the mass fraction of arsenic in the slag phase is nearly 0.18%. Therefore, a high initial content of Cu and S in concentrate helps to eliminate arsenic from the smelting system to the slag phase.

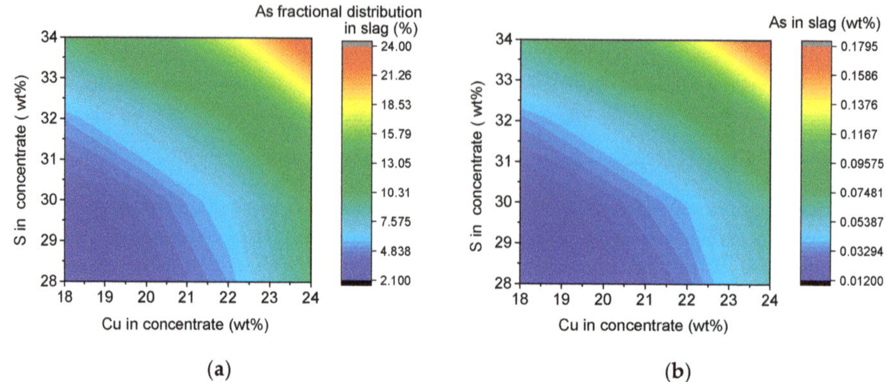

(a)

(b)

Figure 6. Arsenic distribution in the slag phase: (**a**) fractional distribution; (**b**) mass fraction.

In the matte phase, the fractional distribution and mass fraction of arsenic in relation to the change in the initial content of Cu and S in concentrate are presented in Figure 7a,b. The variation trend in the matte phase is different from the gas and slag phases. As the S content in concentrate increases and Cu decreases, both the fractional distribution and mass fraction of arsenic decrease in the matte phase. With Cu 24% and S 28% in concentrate, about 6.3% arsenic reports to the matte phase, and mass fraction of arsenic in the matte phase is nearly 0.073%. However, with Cu 18% and S 34% in concentrate, about 1.1% of the arsenic enters the matte phase, and the mass fraction of arsenic in the matte phase is nearly 0.012%.

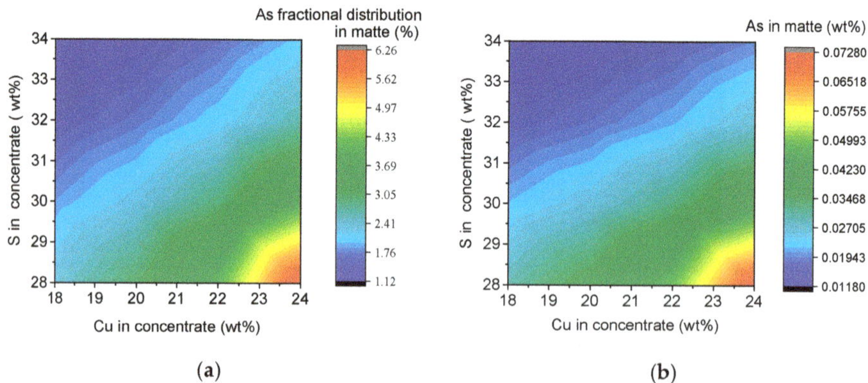

(a)

(b)

Figure 7. Arsenic distribution in the matte phase: (**a**) fractional distribution; (**b**) mass fraction.

In the copper smelting process, it is necessary to eliminate arsenic from the matte to gas or slag phase as much as possible. However, arsenic removal is just one parameter; if the concentrate composition is changed, many other important parameters need to be considered, such as smelting temperature, matte grade, copper loss to slag, slag type (Fe/SiO$_2$), Fe$_3$O$_4$ content in slag, and so on.

3.6.2. Interactive Effects on Other Parameters of the SKS Process

In order to ensure normal production and avoid big changes in other major process parameters (smelting temperature, copper loss to slag, matte grade, etc.), the impact of major elements (Cu, Fe and S) in sulfide concentrate on other parameters (matte grade, slag type Fe/SiO$_2$, smelting temperature, copper loss to slag, Fe$_3$O$_4$ content in slag, S content in slag) was evaluated.

Matte grade, smelting temperature and slag type are the main factors considered in the copper smelting process. Matte grade with the change in the initial content of Cu and S in concentrate is given in Figure 8a. As the S content in concentrate decreases and Cu increases, matte grade increases. With Cu 24% and S 28% in concentrate, the matte grade increases to about 70%. With Cu 18% and S 34% in concentrate, the matte grade drops to around 50%. Therefore, as the total oxygen blown into the SKS furnace is fixed, to produce high grade matte, the initial content of Cu in concentrate should be increased, and the initial content of S in concentrate should be reduced. However, in the Hengbang copper smelter (Yantai, China), low grade matte (around 50% [13]) was chosen, and more arsenic was removed to the gas phase.

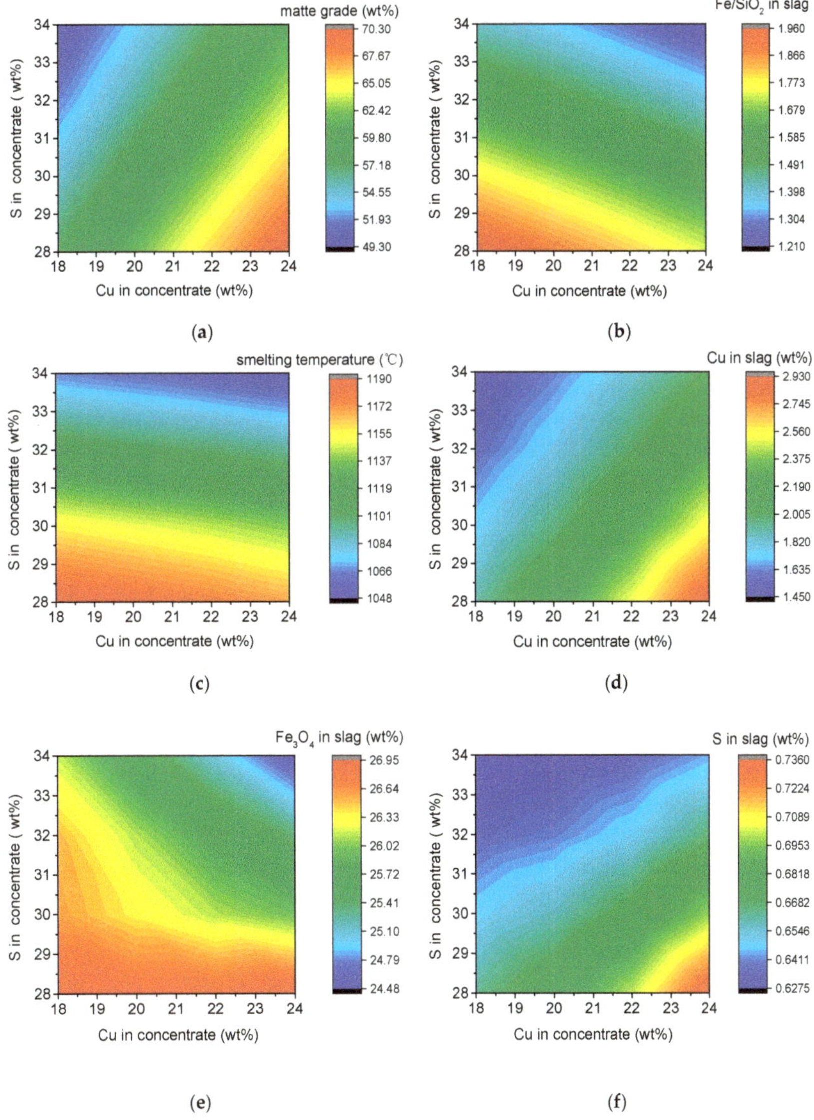

Figure 8. Interactive effects on other parameters of the SKS process: (**a**) matte grade; (**b**) slag type Fe/SiO$_2$; (**c**) smelting temperature; (**d**) copper loss to slag; (**e**) Fe$_3$O$_4$ content in slag; (**f**) S content in slag.

Slag type could affect the copper loss to slag and the capacity to remove impurities. Slag type Fe/SiO_2 with the change in the initial content of Cu and S in concentrate is given in Figure 8b. As the Cu and S content in concentrate both increase, slag type Fe/SiO_2 decreases. With Cu 18% and S 28%, Fe/SiO_2 is about 1.9. However, as Cu and S content in concentrate increase to 24% and 34%, respectively, Fe/SiO_2 is nearly 1.2. In the SKS process, Fe/SiO_2 should be around 1.7.

Smelting temperature with the change in the initial content of Cu and S in concentrate is given in Figure 8c. As both the Cu and S content in concentrate increase, the smelting temperature decreases. With Cu 18% and S 28%, Fe/SiO_2 is about 1200 °C. However, as Cu and S content in concentrate increases to 24% and 34%, respectively, Fe/SiO_2 is nearly 1050 °C. In the actual industrial production of the SKS copper smelting process, the smelting temperature should be above 1170 °C.

Copper loss to slag is an important factor in production. The mass fraction of copper in slag with the change in the initial content of Cu and S in concentrate is given in Figure 8d. As S content in concentrate decreases and Cu increases, the copper loss to slag increases. With Cu 24% and S 28% in concentrate, the mass fraction of copper in slag increases to about 2.9%. However, with Cu 18% and S 34% in concentrate, the mass fraction of copper in slag drops to around 1.5%. In industrial production, the mass fraction of copper in slag should be controlled below 2.5% for the high direct recovery rate of copper.

In slag, Fe_3O_4 can increase the viscosity of slag and further increase the copper loss to slag, thus Fe_3O_4 content in slag is a very important indicator. In Figure 8e, Fe_3O_4 content in slag with the change in the initial content of Cu and S in concentrate is given. As both the Cu and S content in concentrate increase, the Fe_3O_4 content in slag decreases. With Cu 18% and S 28%, the Fe_3O_4 content in slag is about 27%. However, when the Cu and S contents in concentrate increase to 24% and 34%, respectively, the Fe_3O_4 content in slag is nearly 24.5%.

The mass fraction of sulfur in slag with the change in the initial content of Cu and S in concentrate is given in Figure 8f. As S content in concentrate decreases and Cu increases, the S content in slag increases. With Cu 24% and S 28% in concentrate, the S content in slag increases to about 0.74%. However, with Cu 18% and S 34% in concentrate, the S content in slag drops to around 0.63%. The S element exists in slag mainly in the form of Cu_2S and FeS by the mechanical entrainment of matte.

In short, under normal production or little change of other major process parameters (smelting temperature, copper loss to slag, matte grade, copper loss to slag, slag type Fe/SiO_2, Fe_3O_4 content in slag, etc.), the proportions of ore should be optimized to remove more arsenic from the matte to the gas or slag phase in the actual industrial production.

3.7. Industrial Application and Verification

The commercial SKS smelting furnace in the Fangyuan 1$^{\#}$ smelter (Dongying, China) is shown in Figure 9. In the Fangyuan 1$^{\#}$ smelter, gold concentrates bearing arsenic are treated with copper sulfide concentrates by using the SKS smelting process. As more and more gold concentrates bearing arsenic are added to copper sulfide concentrates, the Cu content in the final mixed concentrate decreases gradually. Therefore, the composition of the final mixed concentrate should be optimized.

The impurity element As could be greatly eliminated from the matte to gas phase by adjusting the concentrate composition. In this work, the oxygen rate was 12,072 $Nm^3 \cdot h^{-1}$, the oxygen concentration in oxygen-enriched air was 73%, the feed rate of dry concentrate was 66 $t \cdot h^{-1}$, and the ratio of oxygen/ore was 183 $Nm^3 \cdot t^{-1}$. The matte grade, smelting temperature, slag type Fe/SiO_2 were around 70%, 1200 °C and 1.7, respectively.

Through optimization, as shown in Figure 10, the concentrate composition was changed from its original proportions (Cu 24.4%, Fe 26.8%, S 28.7%, and other 20%) to optimized proportions (Cu 19%, Fe 32%, S 29%, and other 20%), and the distribution of arsenic among three phases was changed from the original (gas 82.01%, slag 12.08%, matte 5.91%) to optimized results (gas 94.37%, slag 3.45%, matte 2.18%). The mass fraction of arsenic in matte was reduced from 0.07% to 0.02%.

Figure 9. Commercial SKS smelting furnace in the Fangyuan 1[#] phase smelter.

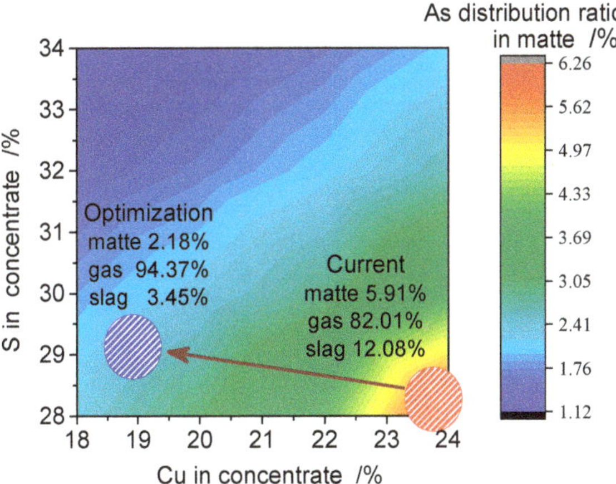

Figure 10. Optimization of ore proportions for arsenic removal in the Fangyuan 1[#] smelter.

The research results and change rules were also confirmed by actual production at several other copper smelters, such as the Hengbang smelter, Yuguang smelter and Fangyuan 2[#] smelter. By decreasing the Cu content in concentrate, the arsenic removal ratio to gas increases, and residual ratios in slag and matte decrease. The change tendency is shown clearly in Table 3 and Figure 11.

Table 3. Main composition of concentrate and distribution ratios of arsenic.

Name of Smelter	Composition of Concentrates (wt %)		Distribution Ratios of Arsenic (%)		
	Cu	Other	Gas	Slag	Matte
Yantai Hengbang [13]	16.10	83.90	88.35	7.51	4.14
Fangyuan 2[#]	20.46	79.54	72.50	20.37	7.13
Jiyuan Yuguang	22.67	77.33	55.89	36.08	8.03

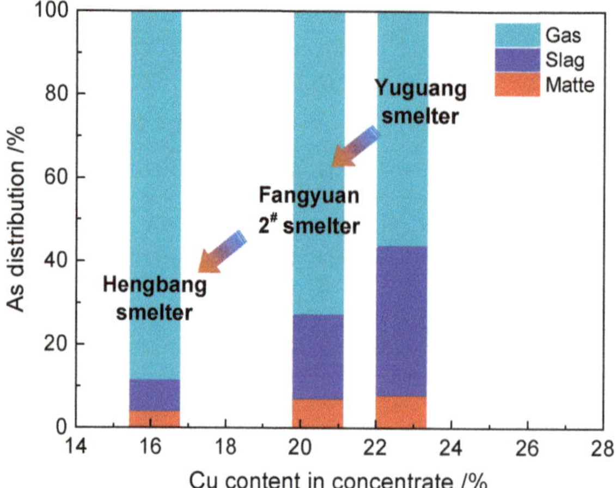

Figure 11. Arsenic distribution in different copper smelters.

Therefore, the optimized result identified in this work provides direct guidance to actual production, and can reduce the load of removing impurities in the conversion, refining and electrolytic processes in SKS copper smelters.

4. Conclusions

The main elements (Cu, Fe and S) in sulfide concentrates have a significant influence on the arsenic distribution among gas, slag and matte phases. When the Cu content in concentrate increases, the fractional distribution of arsenic in smelting products increases, except in the gas phase. As the Fe content increases, the proportion of arsenic reporting to the gas phase increases, however, there are opposite changes in the matte and slag phases. When the S content in the initial concentrate increases, the fractional distribution of arsenic in both the matte and slag phases decreases sharply, and reporting to the gas phase increases. Through optimization of the ore proportions, the concentrate composition was changed from its initial proportions (Cu 24.4%, Fe 26.8%, S 28.7%, and other 20%) to optimized proportions (Cu 19%, Fe 32%, S 29%, and other 20%), and the distribution of arsenic among three phases was improved from the initial (gas 82.01%, slag 12.08%, matte 5.91%) to an optimized distribution (gas 94.37%, slag 3.45%, matte 2.18%). The mass fraction of arsenic in matte was reduced from 0.07% to 0.02%. The optimized result could provide direct guidance for actual production.

Author Contributions: Q.W. (Qinmeng Wang) carried out the research and wrote the paper. X.G. and Q.T. designed the research. Q.W. (Qiongqiong Wang) and X.G. reviewed and contributed to the final manuscript. All authors have read and agreed to the published version of the manuscript.

Funding: The authors appreciate the financial support from the National Natural Science Foundation of China (No.51904351 and No.51620105013), Innovation-Driven Project of Central South University (No.2020CX028) and Hunan Natural Science Fund for Distinguished Young Scholar (No.2019JJ20031).

Conflicts of Interest: The authors declare no conflict of interest.

References

1. Cheng, R.J.; Zhang, H.; Ni, H.W. Arsenic removal from arsenopyrite-bearing iron ore and arsenic recovery from dust ash by roasting method. *Processes* **2019**, *7*, 754. [CrossRef]
2. Dosmukhamedov, N.; Kaplan, V. Efficient removal of arsenic and antimony during blast furnace smelting of lead-containing materials. *JOM* **2017**, *69*, 381–387. [CrossRef]
3. Zhong, D.P.; Li, L.; Tan, C. Separation of arsenic from the antimony-bearing dust through selective oxidation using CuO. *Metall. Mater. Trans. B* **2017**, *48*, 1308–1314. [CrossRef]

4. Yang, W.C.; Tian, S.Q.; Wu, J.X.; Chai, L.Y.; Liao, Q. Distribution and behavior of arsenic during the reducing-matting smelting process. *JOM* **2017**, *69*, 1077–1083. [CrossRef]

5. Yazawa, A.; Azakami, T. Thermodynamics of removing impurities during copper smelting. *Can. Metall. Q.* **1969**, *8*, 257–261. [CrossRef]

6. Nakazawa, S.; Yazawa, A.; Jorgensen, F.R.A. Simulation of the removal of arsenic during the roasting of copper concentrate. *Metall. Mater. Trans. B* **1999**, *30*, 393–401. [CrossRef]

7. Chen, C.L.; Zhang, L.; Jahanshahi, S. Thermodynamic modeling of arsenic in copper smelting process. *Metall. Mater. Trans. B* **2010**, *41*, 1175–1185. [CrossRef]

8. Swinbourne, D.R.; Kho, T.S. Computational thermodynamics modeling of minor element distributions during copper flash converting. *Metall. Mater. Trans. B* **2012**, *43*, 823–829. [CrossRef]

9. Coursol, P.; Mackey, P.J.; Kapusta, J.P.T.; Valencia, N.C. Energy consumption in copper smelting: A new Asian horse in the race. *JOM* **2015**, *67*, 1066–1074. [CrossRef]

10. Li, W.F.; Zhan, J.; Fan, Y.Q.; Wei, C.; Zhang, C.F.; Hwang, J.Y. Research and industrial application of a process for direct reduction of molten high-lead smelting slag. *JOM* **2017**, *69*, 784–789. [CrossRef]

11. Chen, L.; Hao, Z.D.; Yang, T.Z.; Liu, W.F.; Zhang, D.C.; Zhang, L.; Bin, S.; Bin, W.D. A comparison study of the oxygen-rich side blow furnace and the oxygen-rich bottom blow furnace for liquid high lead slag reduction. *JOM* **2015**, *67*, 1123–1129. [CrossRef]

12. Liu, W.F.; Yang, T.Z.; Zhang, D.C.; Chen, L.; Liu, Y.F. A new pyrometallurgical process for producing antimony white from by-product of lead smelting. *JOM* **2014**, *66*, 1694–1700. [CrossRef]

13. Qu, S.L.; Dong, Z.Q.; Chen, T. Distribution of minor elements in complex copper concentrates in oxygen-enriched bottom blown smelting process. *China Nonferrous Metall.* **2016**, *3*, 22–24. (In Chinese)

14. Liu, H.Q.; Cui, Z.X.; Chen, M.; Zhao, B.J. Phase Equilibrium Study of ZnO-"FeO"-SiO_2 System at Fixed P_{O_2} 10^{-8} atm. *Metall. Mater. Trans. B* **2016**, *47*, 164–177. [CrossRef]

15. Liu, H.Q.; Cui, Z.X.; Chen, M.; Zhao, B.J. Phase Equilibria Study of the ZnO- "FeO"-SiO_2-Al_2O_3 System at P_{O_2} 10^{-8} atm. *Metall. Mater. Trans. B* **2016**, *47*, 1113–1123. [CrossRef]

16. Shui, L.; Cui, Z.X.; Ma, X.D.; Rhamdhani, M.A.; Nguyen, A.V.; Zhao, B.J. Mixing phenomena in a bottom blown copper smelter: A water model study. *Metall. Mater. Trans. B* **2015**, *46*, 1218–1225. [CrossRef]

17. Shui, L.; Cui, Z.X.; Ma, X.D.; Rhamdhani, M.A.; Nguyen, A.V.; Zhao, B.J. Understanding of bath surface wave in bottom blown copper smelting furnace. *Metall. Mater. Trans. B* **2016**, *47*, 135–144. [CrossRef]

18. Chen, M.; Cui, Z.X.; Zhao, B.J. Slag chemistry of bottom blown copper smelting furnace at Dongying Fangyuan. In Proceedings of the 6th International Symposium on High-Temperature Metallurgical Processing, Orlando, FL, USA, 15–19 March 2015; pp. 257–264.

19. Zhang, Z.Y.; Chen, Z.; Yan, H.J.; Liu, F.K.; Liu, L.; Cui, Z.X.; Shen, D.B. Numerical simulation of gas-liquid multi-phase flows in oxygen enriched bottom-blown furnace. *Chin. J. Nonferrous Met.* **2012**, *22*, 1826–1834. (In Chinese)

20. Yan, H.J.; Liu, F.K.; Zhang, Z.Y.; Gao, Q.; Liu, L.; Cui, Z.X.; Shen, D.B. Influence of lance arrangement on bottom-blowing bath smelting process. *Chin. J. Nonferrous Met.* **2012**, *22*, 2393–2400. (In Chinese)

21. Wang, Q.M.; Guo, X.Y.; Tian, Q.H.; Chen, M.; Zhao, B.J. Reaction mechanism and distribution behavior of arsenic in the bottom blown copper smelting process. *Metals* **2017**, *7*, 302. [CrossRef]

22. Wang, Q.M.; Guo, X.Y.; Tian, Q.H. Copper smelting mechanism in oxygen bottom-blown furnace. *Trans. Nonferrous Met. Soc. China* **2017**, *27*, 946–953. [CrossRef]

23. Wang, Q.M.; Guo, X.Y.; Tian, Q.H.; Jiang, T.; Chen, M.; Zhao, B.J. Development and application of SKSSIM simulation software for the oxygen bottom blown copper smelting process. *Metals* **2017**, *7*, 431. [CrossRef]

24. Wang, Q.M.; Guo, X.Y.; Tian, Q.H.; Jiang, T.; Chen, M.; Zhao, B.J. Effects of matte grade on the distribution of minor elements (Pb, Zn, As, Sb, and Bi) in the bottom blown copper smelting process. *Metals* **2017**, *7*, 502. [CrossRef]

25. Chen, C.L.; Jahanshahi, S. Thermodynamics of arsenic in FeO_X-CaO-SiO_2 slags. *Metall. Mater. Trans. B* **2010**, *41*, 1166–1174. [CrossRef]

Article

Quantitative Methods to Support Data Acquisition Modernization within Copper Smelters

Alessandro Navarra [1,*], Ryan Wilson [1], Roberto Parra [2], Norman Toro [3,4], Andrés Ross [5], Jean-Christophe Nave [5] and Phillip J. Mackey [6]

1   Department of Mining and Materials Engineering, McGill University, 3610 University Street, Montreal, QC H3A 0C5, Canada; ryan.wilson@mail.mcgill.ca
2   Department of Metallurgical Engineering, University of Concepción, E. Larenas 285, Concepción 4070371, Chile; rparra@udec.cl
3   Departamento de Ingeniería en Metalurgia y Minas, Universidad Católica del Norte, Av. Angamos 610, Antofagasta 1270709, Chile; ntoro@ucn.cl
4   Faculty of Engineering and Architecture, Universidad Arturo Prat, Almirante Juan José Latorre 2901, Antofagasta 1244260, Chile
5   Department of Mathematics and Statistics, McGill University, 805 Sherbrooke Street West, Montreal, QC H3A 0B9, Canada; andres.ross@mail.mcgill.ca (A.R.); jean-christophe.nave@mcgill.ca (J.-C.N.)
6   P.J. Mackey Technology Inc., Kirkland, QC H9J 1P7, Canada; pjmackey@hotmail.com
*   Correspondence: alessandro.navarra@mcgill.ca

Received: 9 October 2020; Accepted: 13 November 2020; Published: 17 November 2020

Abstract: Sensors and process control systems are essential for process automation and optimization. Many sectors have adapted to the Industry 4.0 paradigm, but copper smelters remain hesitant to implement these technologies without appropriate justification, as many critical functions remain subject to ground operator experience. Recent experiments and industrial trials using radiometric optoelectronic data acquisition, coupled with advanced quantitative methods and expert systems, have successfully distinguished between mineral species in reactive vessels with high classification rates. These experiments demonstrate the increasing potential for the online monitoring of the state of a charge in pyrometallurgical furnaces, allowing data-driven adjustments to critical operational parameters. However, the justification to implement an innovative control system requires a quantitative framework that is conducive to multiphase engineering projects. This paper presents a unified quantitative framework for copper and nickel-copper smelters, which integrates thermochemical modeling into discrete event simulation and is, indeed, able to simulate smelters, with and without a proposed set of sensors, thus quantifying the benefit of these sensors. Sample computations are presented, which are based on the authors' experiences in smelter reengineering projects.

Keywords: Industry 4.0; copper smelter; nickel-copper smelter; radiometric sensors; Peirce-smith converting; matte-slag chemistry; discrete event simulation; adaptive finite differences

---

1. Introduction

Modern metallurgical installations such as steel plants and copper smelters require a range of plant sensors and process control systems to attain their highest efficiency. It is often stated that the levels of so-called "smart automation" today represents the fourth-generation industrial revolution (sometimes also called "Industry 4.0")—following the first, considered as the power generation and mechanical automation (early 1800s), the second as widespread industrialization (early 1900s), the third as electronic automation (starting 1950s), and, now, the fourth [1], benefitting from modern information and communication technology (ICT), as illustrated in Figure 1. Learning from history, it is well-known that if a plant does not remain up to date, taking advantage of modern but proven equipment and controls, then it lags its competitors. This paper discusses the application of techniques, sensors, and mathematical modeling that can support copper smelters in their endeavors to modernize their operations through data acquisition and, thus, remain competitive.

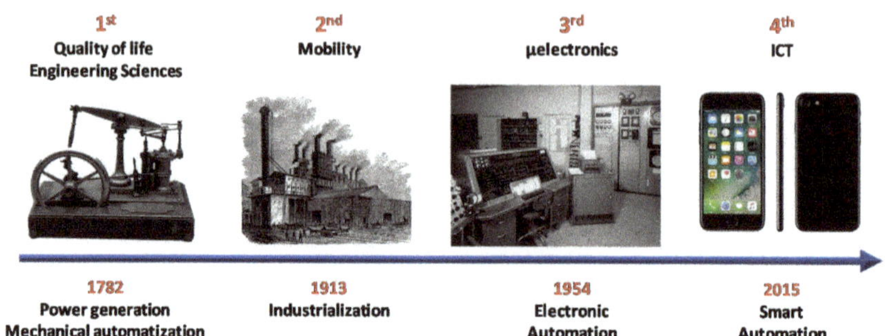

Figure 1. Representation of "Industry 4.0"—the 4th industrial revolution. Adapted from [1]. ICT: information and communication technology.

By way of background, copper smelters [2] are quite complex, with a number of individual high-performance operating units functioning together. Two of the key primary instrumentation and control needs in copper smelting today—and throughout history—are air (or oxygen) measurement and control and furnace temperature. In February 1956, a paper was presented at the AIME Annual Meeting in New York on the fine and well-recognized converter operation at the Noranda smelter (now Glencore Horne) located in Rouyn-Noranda, Quebec, Canada [3].

After describing how the production capacity at the fairly new plant increased over the previous two decades since the start-up (in 1927) by simply installing additional and/or larger units, the author discussed a major effort at further increasing production capacity that included the installation of reliable instruments for measuring both the instantaneous air flow and reporting the accumulated air blown over any given time interval. The significance and benefits of this new instrumentation (higher production rates) were subsequently described; of interest, the converters at this plant reached record or near-record air blowing rates in the industry at this time, reflecting the fact that it was one of the leading plants of the day. Of course, such equipment is mandatory today, coupled with the distributed control system (DCS) and advanced computer models to provide feedback for optimizing the operation.

Furnace temperature measurements at the time were made by an optical pyrometer mounted so that it pointed downward onto the molten bath. While generally reliable, it was often prone to dust and fume build-up affecting a clear sight line; unsmelted flux and other materials on the bath could also lead to low temperature readings. Some thirty years after the above noted paper [3], a new pyrometer instrument operated by sighting directly into the bath through one of the tuyères was described at the Copper 87 conference held in Viña del Mar, Chile [4]. During the discussion, the authors were asked about the cost of the new instrument. An indicative preproduction price range was given when the questioner loudly stated that he could buy "a Mercedes-Benz automobile" for the price mentioned, whereupon the authors responded that if the questioner installed one of these instruments at his plant, the profits he would make would allow him to purchase quite a few Mercedes! The point is that proper selection and use of reliable instruments for automated data acquisition and control are vital for a proper functioning plant to remain competitive, as the cost of sensors are comparatively modest.

In batch processes, such as steelmaking or copper converting, knowledge of the precise endpoint of the operation is extremely important for compositional control and optimization of the subsequent refining operations.

The Boliden's Rönnskär smelter, located at Skelleftehamn, Sweden, operates one of the best converter aisles in the world today, in the authors' opinion. They also pioneered the use of the ingenious Semtech Optical Process Control (OPC) sensor device for precise endpoint determination [5] to detect when the last of the iron was expelled into slag and, subsequently, when the last of the sulfur was expelled into the SO_2 offgas. This instrument spectrometrically measures trace amounts of lead and copper sulfides and oxides in the offgas and signals when the level of sulfides starts to decrease, signaling the end of the batch [6]. The precise endpoint control saves significantly in processing time at Rönnskär in the subsequent anode furnace, thus lowering operating costs and allowing higher throughput.

In spite of the successful implementations of modern sensors at certain smelters such as Rönnskär, there is a general hesitance for copper smelters to adopt sensor arrays, as it is often unclear what the operational implications will be. Indeed, it is unclear what will be the series of changes that will be necessary so that the perceived benefits of the sensors will be manifest, and it is therefore difficult to quantify what will be the true impact of these new sensors. This paper presents a unified quantitative framework for copper and nickel-copper smelters, which integrate thermochemical modeling into discrete event simulations and is intended to assist in smelter reengineering projects that feature innovative sensors in consideration the Industry 4.0 paradigm.

2. Radiometric Sensors for Extractive Pyrometallurgy of Copper

Within copper smelters, the reality is that even the 3.0 industrial revolution (information technology (IT) and automation) still has room for continued development. This current condition limits the ability to integrate Industry 4.0 developments for process optimization. Indeed, the degree of analytical instrumentation usage for the monitoring and control of the smelter processes is limited. As a result, the information available for operational decision-making at many plants is mainly based on static mass and energy balances. In such cases, the operational dynamics continues to depend largely on the experience of ground operators.

Only a handful of advanced measurement instruments have gained industry acceptance for operational monitoring and control in copper pyrometallurgical reactors. The aforementioned Semtech OPC system, which has been on the market for more than 25 years, monitors pyrometallurgical variables by analyzing the emission spectrum of flames emitted by gases during the conversion of copper sulfide mattes within Peirce-Smith (PS) converters. Additionally, the Noranda Pyrometer, configured solely for bath furnaces using blowing tuyères, has been on the market since the 1980s without any change to the original concept and with only some upgrades to the unit's robustness (with regard to assembly, material selection, etc.). By measuring radiation through an analog array, it monitors the temperature of the molten bath by applying Planck's law for two fixed wavelengths. Other instruments include level measurement in a feed bin and furnace melts and equipment vibration monitoring.

One option for the advancement of process monitoring and control systems in copper smelters is to modernize the use of radiometric measurements, since this approach was previously validated with both the OPC system and the Noranda Pyrometer. A series of proposed concepts and results from initial industrial trials are discussed here.

2.1. Reactive Systems

A recent work focused on radiometric optoelectronic sensors that consider a broad spectrum, which includes the visible range up the near infrared range [7–9]. This approach is different from available commercialized sensors, which only analyze a limited selection of wavelengths. The idea here is to analyze both the continuous and discontinuous parts of the spectrum, as well as its dynamics in different time scales. The purpose is to correlate the measured spectral radiation to the operational conditions of the reactors following the emission of radiation from the oxidation reactions. These sensors also measure temperature with better precision than the Noranda Pyrometer by using more sophisticated techniques to select the appropriate (two or more) wavelengths. The sensor thus solves the complex problem of simultaneously measuring emissivity and temperature.

Results from an experiment using a drop tube setup for the flash oxidation of copper concentrates (Figure 2) validate the formation of wustite (FeO), magnetite (Fe_3O_4), and copper oxides (CuO/Cu_2O) as indicators of the concentrate smelting/oxidation/combustion process [7–9]. This information is of special interest to track the physicochemical dynamics of the process in real-time and establish operating criteria for a smelting reactor. Examples of such criteria could include adjustments to concentration/oxygen ratios, oxygen enrichment for the incoming blast, and the quantity of cold charge that will be required to maintain the heat balance.

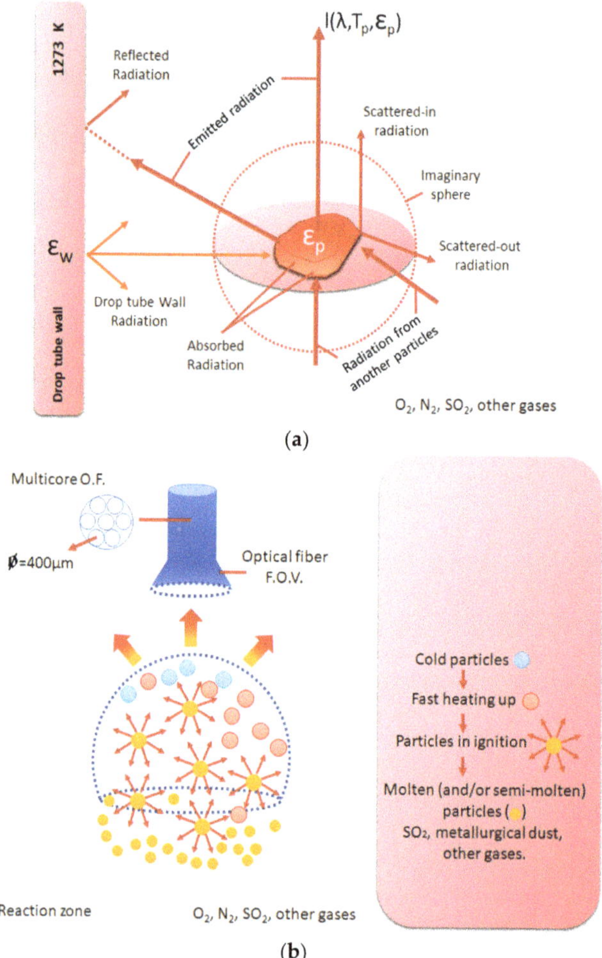

Figure 2. Radiometric measurement scheme and associated radiative processes: (**a**) single-heated particle radiative emission with its surroundings, in which the intensity I of the emitted radiation is a function of wavelength λ, particle temperature T_p and particle emissivity ε_p and (**b**) sensing scheme depicting the different particle states as they fall through the reaction zone (adapted from [8]).

The spectral acquisition system consists of a multicore optical fiber with its own cooling process (Figure 2b), which measures the combustion flame radiation. Figure 2b shows a simple combustion scheme for sulfide particles covering the physical phase changes that a particle can experience inside the reaction zone. However, measuring the spectrum is a complex task. As shown in Figure 2a, there is an ensemble of physical and chemical processes that complicate this measurement. One example is drop tube radiation caused by increased electrical resistance due to the higher temperatures; processing these spectral signals can, however, mitigate the effect of the unwanted radiation [8,9].

Applying specialized algorithms in the treatment of the spectral signals obtained from the experiment, coupled with multivariate data analysis methodologies, allows for the identification and classification of copper and iron sulfide minerals present in the blend [10]. These results are particularly important as they demonstrate that spectral data obtained from the oxidation process can be used to identify the type of charge being treated within a molten bath. The controlled bench-scale laboratory study was carried out on several different types of concentrate. An exploratory analysis of the results using principal component analysis (PCA) applied to the spectral data depicted high correlation features among species with different mineral characteristics but similar elemental compositions. Classification algorithms were tested on the spectral data, and a classification accuracy of 95.3% was achieved using a support vector machine (SVM) classifier with a Gaussian kernel. Initial industrial-scale trials with a prototype have confirmed these results [10].

2.2. Nonreactive Systems

Despite tremendous advances in the development of passive and active photonic sensors, such as hyperspectral imaging (HI) and laser-induced breakdown spectroscopy (LIBS), real-time analytical sensors do not exist at present for the conditions of pyrometallurgical copper processing. To date, there is no commercial instrument capable of online quantification of copper content (% Cu) without contact during tapping operations. Nor can the available sensors discriminate between the phases that are of particular interest to smelter operations. The distinction between matte and slag during the tapping of a smelting furnace can significantly impact copper recovery, yet this function is heavily reliant on the experience of operators involved in tapping.

However, in the last decade, the spectral behavior of pig iron and slag in the ferrous industry has been studied to estimate different variables that allow for improved control of the tapping process [11,12]. These models describe the parameters contained in the iron–slag mixture during blast furnace tapping, such as iron emissivity, casting depth, slag layer thickness and absorption coefficients, and radiometric parameters (e.g., reflectance at the iron-slag interface). The methodology begins by determining a spectral range in which the radiation of the molten phases is comparatively high. This facilitates detection by a silicon charge coupled device (CCD) camera, which is sensitive in the visible spectral range and part of the near-infrared. An optical filter centered at 650 nm was used together with the optics, such that the radiation emitted by the wash was partially filtered in the indicated spectral range.

The results confirm that the difference in emissivity of iron and slag at 650 nm allows for the spatial distinction of these phases. Furthermore, it was identified that the radiation intensity of the molten iron remains practically constant during the process, while that of the slag fluctuates. This fluctuation is due to differences in the thickness of the slag layer, as it absorbs and transmits the radiation coming from the steel to varying degrees. Additionally, the optical system was calibrated with a high-temperature black-body radiator, allowing the temperatures to be estimated at 1500–1600 °C, which is considerably higher than the copper smelter temperatures (1200–1350 °C). The results confirm that, by using sensitive optoelectronic systems in the molten iron spectral emission band, coupled with appropriate spectral models and processing software, it is possible to develop reliable and robust systems at both the laboratory and industrial scales. The authors believe that this approach could be adapted for copper smelter processes and represents a natural pathway for future work.

3. Unifying Framework for Copper and Nickel–Copper Smelter Dynamics

3.1. Overview of Copper and Nickel–Copper Smelter Operations

The copper pyrometallurgy process treats mineral concentrates to produce copper anodes that are, in turn, electrorefined to generate effectively pure (99.99% Cu) end-product cathodes. The concentrates are comprised of copper–iron sulfide minerals with particle sizes of generally less than 150 μm. Elemental compositions typically range from 25% to 30% Cu, 25% to 35% S, and 20% to 40% Fe, with the remainder made up of gangue material (oxides). The smelter targets the selective oxidation of Fe and S in order to retain Cu for the final product. Specifically, the iron is skimmed away as liquid slag and the sulfur is removed as SO_2 offgas, eventually resulting in molten copper [13,14]. This pyrometallurgical technique accounts for approximately 75% of the primary copper production worldwide [15,16], the majority of which is carried out using the conventional approach depicted in Figure 3. In some cases, the incoming concentrates are subject to roasting reactions prior to being fed into the smelting operation; these roasted concentrates are known as calcines.

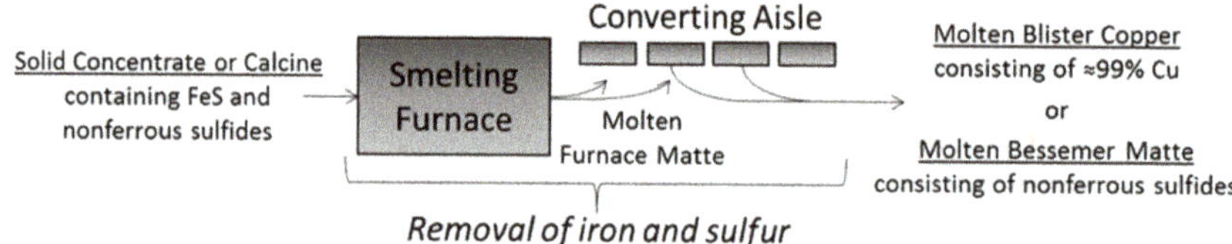

Figure 3. Schematic of conventional copper or nickel–copper sulfide smelter operations. The smelting furnace and converting aisle eliminate iron and sulfur, producing blister copper in the case of copper smelters and Bessemer matte in the case of nickel–copper smelters.

Within a copper smelter, the concentrate or calcine is smelted by exothermic reactions while controlling the oxygen mass balance, which produces an intermediate matte phase composed of copper, iron, and sulfur. This molten matte is then subject to further oxidation, which converts the matte into a molten metal product called blister copper (>98% purity). The subsequent liquid–state refinement (fire-refining) precedes the casting of a Cu anode product (~99.5% purity). The anodes are then transferred to an electrorefinery for a final upgrade to end-product Cu cathodes that are 99.99% pure; the electrorefinery also recovers gold, silver, and other valuable byproducts that are contained in the anodes. Nonetheless, smelting and converting are the central processes in a copper smelter, since they transform the mineral input into an initial metallic output [14].

The smelting operation is a continuous process typically executed in one (or at certain large plants, two) large furnace(s). There are two main smelting technologies that induce similar chemical transformations but differ in the mechanism by which oxygen is delivered to the concentrate. Flash smelting, in which the oxidation of the concentrate takes place in a generally vertically mounted burner, currently accounts for roughly 45% of the world copper smelting capacity [2,15,17]. Within a flash smelting furnace, the refractory-lined reaction shaft is like an industrial scale "drop tube" (Figure 2). On the other hand, bath smelting technologies inject air or oxygen-enriched air into the molten bath either via a top-mounted lance (called top smelting lance or TSL technology) or via submerged tuyères; bath smelting technologies account for roughly 50% of smelting capacity, and this proportion has recently increasing due to new bath smelting technologies introduced in China [15,17].

Peirce-Smith (PS) converting is the most longstanding and widely used technology in conventional copper smelters [14] and is performed in discrete batches. PS converting is indeed a remnant of the second industrial revolution (Figure 1) and was influenced by the 19th century developments of Sir Henry Bessemer in steelmaking [16]. PS batch conversions are often carried out in parallel (Figure 3) and may share a limited set of resources (e.g., oxygen and offgas handling capacities). The continuous-discrete contrast of smelting converting is central to conventional smelter dynamics, in which PS converting can be a major bottleneck in conventional copper smelters [14]. Moreover, PS converting is also a feature of nickel–copper smelters (Figure 3), noting the difference in the final discharge product (Figure 3). Copper PS results in blister copper, whereas nickel–copper PS converting is simply to remove the iron and its associated sulfur; the resulting iron-free matte still contains considerable sulfur and is known as Bessemer matte, in honor of Sir Henry Bessemer.

Smelting furnaces can generally produce, and hold, matte in excess of the converting capacity, which means that the smelting schedule can depend on the converting schedule. Given that smelting and converting are central to the overall smelter operation, all other critical functions at the smelter plant can also be restricted by the converting cycles. Each converting cycle begins when a fresh charge of matte (and possibly some amount of cold charge) is delivered to an empty converter and ends with the final discharge. The matte is subjected to pressurized blowing wherein oxygen-enriched air is blown into the melt, and N_2 and SO_2 are exhausted through a hood mounted on the vessel (Figure 4). The offgas is captured in order to convert SO_2 into sulfuric acid; meanwhile, N_2 acts as a coolant in the process [14]. Copper and nickel–copper smelters both apply the first stage of PS converting, which is called the slag-blow, producing an iron-rich slag that forms atop the denser matte (Figure 4a). This stage may require intermittent pauses in order to skim away slag accumulation and replenish the vessel with fresh matte and cold charge. Once all of the slag is removed (<1% Fe in matte), copper smelters continue blowing the remaining matte; this final stage of converting is known as the copper-blow, as it results in the formation of blister copper that sinks to the bottom of the vessel (Figure 4b). The copper-blow does not produce any more slag and, therefore, does not require intermittent skimming. Nickel–copper smelters, however, only apply the slag-blow (Figure 4a), not the copper-blow. In either context, the cycle is complete when all of the matte is converted to the correct endpoint and discharged (Figure 3).

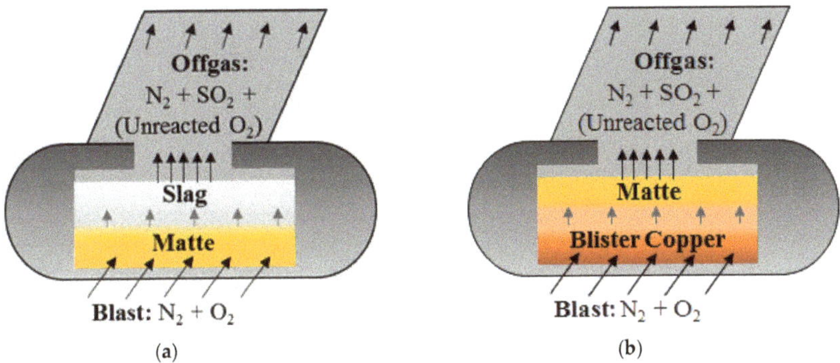

Figure 4. Cross-section of a Pierce-Smith converter, during (**a**) a slag-blow and (**b**) a copper-blow.

In addition to scheduling constraints, various process parameters are also subject to statistical variations, including the chemical composition of incoming plant feed, matte grade, furnace performance, and converter cycle times, among other global factors. It is critical to measure, model, and simulate such uncertainties in order to streamline and support the decision-making processes in the design, development, and operation stages of industrial systems.

3.2. Detailing of Smelter Dynamics Within Discrete Event Simulation

In a previous work, Navarra et al. [18,19] incorporated thermochemical equilibria within a discrete event simulation (DES); it was suggested generally that the hybridization of time-adaptive finite differences (TAFD) and DES is a suitable paradigm for multiphase smelter reengineering projects [20]. Within copper and nickel–copper smelters, thermochemical equilibria determine the iron-speciation of smelting and converting slags, as described in the following section. However, the early phases of a smelter reengineering project can assume fixed molar ratios of iron and oxygen.

Assuming that the smelter feed is composed mainly of iron, sulfur, and copper, a smelter accomplishes the following unbalanced reaction:

$$(Cu, Fe, S)_{(Feed)} + O_{2(Blast)} + Flux \rightarrow (FeO_x, Flux)_{(Slag)} + SO_{2(Offgas)} + Cu_{(Blister)} \tag{1a}$$

In which FeO_x represents a mixture of wustite FeO and magnetite Fe_3O_4, such that $x = 1$ and $x = 1.25$ corresponds to pure wustite and magnetite, respectively. For simplicity, x can be fixed to 1, although typical values can range between 1.0 and 1.1, depending on the nature and quantity of the flux and the monitoring and control of the process itself; in particular, a low level of magnetite in slag is desirable, which is associated with low slag viscosity. In practice, the flux is predominantly silica SiO_2, but certain smelters include varying quantities of CaO and other stable oxides; CaO is especially common in continuous converting [21], which is an alternative to the conventional PS converting [16]. The SO_2 is captured for sulfuric acid production, and the blister copper is subject to fire refining prior to being cast into anodes that undergo electrolytic refining. A similar reaction can describe nickel–copper smelters:

$$(NI, CU, CO, FE, S)_{(FEED)} + O_{2(Blast)} + Flux$$
$$\rightarrow (FeO_x, Flux)_{(Slag)} + SO_{2(Offgas)} + (Ni, Cu, Co, S)_{(Bessemer)} \tag{1b}$$

The subsequent processing of Bessemer matte depends on the given nickel–copper plant. Equation (1a,b) provide more detail to Figure 3 for copper and nickel–copper smelters, respectively.

Depending on the scope and phase of the project, x can be regarded as a single global value for the entire smelter or as distinct values for the smelting furnace(s) and converters. Equation (1a) can thus be rewritten

$$(Cu, Fe, S)_{(Feed)} + O_{2(Blast)} + SFlux + CFlux$$
$$\rightarrow \left(FeO_{x_S}, SFlux\right)_{(SSlag)} + \left(FeO_{x_C}, CFlux\right)_{(CSlag)} + SO_{2(Offgas)} + Cu_{(Blister)} \tag{2}$$

For copper smelters, which decompose the global x into x_S and x_C, characterize the slag of the smelting furnace. A similar decomposition of the global x could be applied to Equation (1b) in the case of nickel–copper smelters.

Moreover, the individual slag-blow segments of PS cycles can each be assigned appropriate x values. Therefore, Equation (2) could be further detailed as:

$$(Cu, Fe, S)_{(Feed)} + O_{2(Blast)} + SFlux + CFlux1 + CFlux2 + \cdots + CFluxn$$
$$\rightarrow$$
$$\left(FeO_{x_S}, SFlux\right)_{(SSlag)} + \left(FeO_{x_{C1}}, CFlux1\right)_{(CSlag1)} + \left(FeO_{x_{C2}}, CFlux2\right)_{(CSlag2)} + \cdots \tag{3}$$
$$+ \left(FeO_{x_{Cn}}, CFluxn\right)_{(CSlagn)} + SO_{2(Offgas)} + Cu_{(Blister)}$$

In which x_S characterizes the slag of the smelting furnace, and, depending on the level of detail, x_{Ci} can characterize the types of converter cycles or can characterize the individual types of slag-blow segments, for $i = 1$ to n. For example, Figure 5a shows an action graph that occurs within a smelter that practices two kinds of converter cycles, long and short; hence, $n = 2$. Figure 5b shows a more detailed representation, which considers 13 kinds of blow segments. (The slag-blow segments are punctuated with charging and skimming actions, although these are not explicitly shown in Figure 5b). For a conventional copper smelter, actions 1–9 describe slag-blow segments (Figure 4a); hence, $n = 9$, and the remaining actions 10–13 represent copper-blow segments (Figure 4b) that complete the cycle as a batch of blister copper is discharged. In the case of a nickel–copper smelter, all of the arcs represent slag-blow segments; hence, $n = 13$, noting that the discharge is the so-called Bessemer matte (Ni, Cu, Co and S) that is described in Figure 3 and Equation (1b).

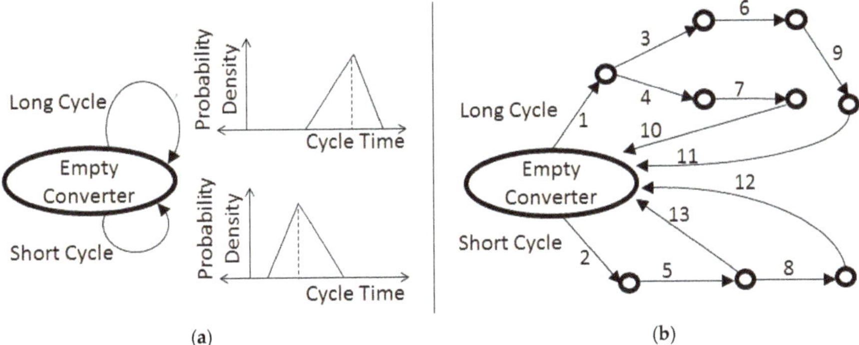

Figure 5. Examples of action graphs that represent Peirce-Smith converting cycles, which consider two types of cycles: long and short. (**a**) The low-detail representation shows the long and short cycles as single actions that are characterized by broad distributions of cycle times, whereas (**b**) a more detailed representation considers individual blow segments, from 1 to 13; each of the segments can be characterized by comparatively narrow time durations (which were omitted from the figure).

The decision to apply one segment versus another (e.g., segment 3 versus 4) would depend on the state of the plant, to the extent that the state variables can monitored with the available sensors. Even if the resulting slag compositions for the different cycles (Figure 5a) or blow segments (Figure 5b) are relatively consistent, it may be unclear how frequently each cycle or segment will be applied, e.g., depending on how often certain plant conditions occur. A global mass balance based on Equation (3) requires an estimation of how often each of the different cycles (Figure 5a) or segments (Figure 5b) are applied; such estimations are the result of DES computations, as described below.

The broad distributions of Figure 5a approximate the combined effects of narrower distributions that would characterize the individual segments of Figure 5b. A proposed technological change within the smelter (e.g., installation of new sensors) may require a rethinking of the slag-blow and the definition of new action graphs. Depending on the project, it may be necessary to further decompose the actions of Figure 5b into sub-actions and sub-sub-actions, possibly including thermochemical modeling [22,23] or computational fluid dynamics [24]. This decomposition may be essential in order to properly simulate the system with and without the technological change, thereby evaluating the benefit of the proposed change. In the case of sensors, it is necessary to simulate how the additional information will be incorporated into the decisions and operational actions of the smelter, thereby computing the value of these better-informed decisions and actions.

In many reengineering projects, the phenomena that occur within the smelter may be less important than the phenomena that occur outside of the smelter. For instance, the DES model of the Hernán Videla Lira (HVL) Smelter developed by Navarra et al. [25] focuses on the smelter-wide response to changing meteorological conditions and has a comparatively simple representation of converter cycles, similar to Figure 5a. The HVL Smelter considers distinct categories of meteorological conditions—normal, unfavorable, and extreme—to describe the potential for the surrounding atmosphere to disperse the SO_2 effluent. If the smelter is running in its normal operational mode when the unfavorable meteorological conditions emerge, there is a so-called "environmental incident". The model of Navarra et al. [25] computes the trade-off between production and environmental risk. Moreover, this model quantifies the improved trade-off that can result from a more accurate array of meteorological sensors.

DES development is a means to extend the static mass balances, to detail the critical phenomena that are driving and/or constraining a particular phase of an engineering project. The simulated events can dynamically affect the mass balances that are detailed throughout the model. Figure 6 makes the distinction between the events that occur outside of the smelter and within the smelter, which constitute the external and internal logistics, respectively. There is a further distinction between the logistical coordination of smelter equipment (furnaces, ladles, cranes, etc.) and the kinetics that occur within the equipment. In general, the state variables that describe the system, and the events that would alter these variables, can be positioned within the concentric ellipses of Figure 6. The incorporation of variables and events within a DES model must be guided by the scope and the phase of the engineering project. For example, it is not recommended to detail individual crane movements, unless the particular project would benefit from a comprehension of this aspect [26]. Likewise, it is not recommended to detail particular equipment breakdown events, unless the project would benefit from a comprehension of this aspect [27].

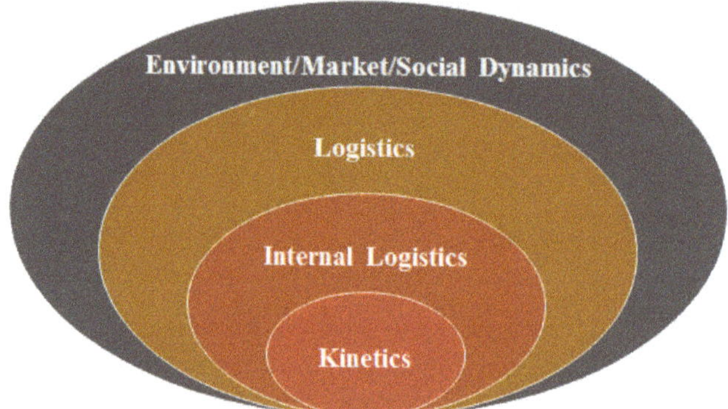

Figure 6. Relationship between smelter kinetics, internal and external smelter logistics, and broader system dynamics.

The computational efficiency of DES is due to adaptive time stepping, as the virtual clock advances from one event to the next without explicitly representing the dynamics that occur between events (Figure 7a). The sequencing of events is governed by the future event list (Figure 7b). Within this scheme, a prolonged activity is represented by a sequence of events, including starting and ending the activity. For example, a basic representation of converter cycles described by Figure 5a may include only two events: the start and end of the cycle. A more detailed representation (e.g., Figure 5b) may include several intervening events to represent individual slag- and copper-blow segments, as well as the intervening skimming, charging, actions of the operators, etc. with the level of details that correspond to the given project. Incidentally, DES applies random number generation to determine the duration and outcome of the activities and is thus a form of Monte Carlo simulation [28]; the distributions and action paths illustrated in Figure 5 can be incorporated into the framework.

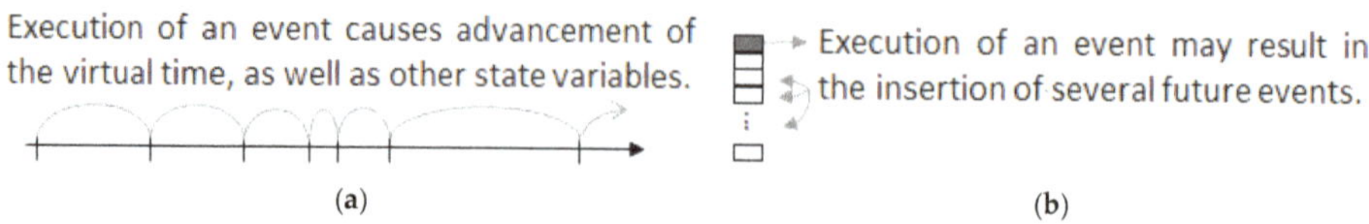

Figure 7. Fundamental components of a discrete event simulation (DES) framework, including (a) a virtual timeline that is subject to discrete steps and (b) a future event list.

Moreover, the simultaneous operation of several converters in unison with other logistical phenomena is integrated into one single future event list (Figure 7b); hence, a system-wide representation. Periods of time with relatively few events are computed relatively quickly, thereby focusing the computational efforts on periods of time that are more heavily packed with activity. This time-adaptive aspect of DES allows the simulation of thousands of operating days within minutes.

A DES framework can include operational criteria that determine the action pathway of converter cycles, allowing the computation of frequency confidence intervals. Following the example of Figure 5a, the average frequency of short cycles may be between 2.8 and 3.2 cycles/day with 95% confidence and that of long cycles may be between 0.9 and 1.2 cycles/day with 95% confidence; this result will allow a mass balance based on Equation (3), given the data about the matte that are charged within each cycle and the corresponding flux and oxygen requirements. In a slightly more detailed representation, the DES framework may include the criteria that would determine the more detailed action paths of Figure 5b.

Standard DES frameworks do not explicitly represent the dynamics that occur between events. However, a linearly dynamic state variable can be represented as a combination of discretely dynamic state variables. For instance, the mass of feed stockpile k may be computed at a time t, as

$$m_k(t) = m_k^{\text{Previous}} + \dot{m}_k^{\text{Previous}} \cdot \left(t - t^{\text{Previous}}\right) \tag{4}$$

in which m_k^{Previous} and $\dot{m}_k^{\text{Previous}}$ are the mass and rate change of k that were computed at the previous event, which occurred at time t^{Previous}. Thus, each feed k would require two discretely dynamic variables (m_k^{Previous} and $\dot{m}_k^{\text{Previous}}$), in addition to the t^{Previous} variable that remembers the time of the previous event. Equation (4) can used in simulations that consider alternating modes of operation that control feed blends in response to imbalances in incoming concentrates [29].

Other linearly continuous variables can be implemented in a manner similar to Equation (4), representing each of these continuous variables as two discrete variables: level and rate (e.g., m_k^{Previous} and $\dot{m}_k^{\text{Previous}}$). Considering the DES representation of time (Figure 7), this constitutes a time-adaptive finite difference (TAFD) scheme. However, a full representation of the continuous dynamics requires the detection of threshold-crossing events, as described in Section 3.4. These threshold-crossing events are especially important in assessing the installation of sensors whose role may be to signal the need for corrective actions precisely when critical thresholds are crossed.

3.3. Slag Iron Speciation and Other Thermochemical Considerations

The iron oxide speciation, i.e., the balance between FeO and Fe_3O_4, can be quantified as the oxygen-to-iron ratio x presented in Equations (1)–(3). Indeed, x represents a degree of freedom that must be resolved in order to complete the mass balance. This degree of freedom can also be expressed as the ratio of ferric to ferrous ions within the slag, $\alpha = Fe^{3+}/Fe^{2+}$, often called the degree of oxidation. The homeomorphic relationship between x and α is given by

$$x = \frac{2 + 3\alpha}{2 + 2\alpha} \tag{5}$$

Equation (1a) can thus be rewritten as

$$(Cu, Fe, S)_{(Feed)} + O_{2(Blast)} + Flux \rightarrow \left(FeO_{\left(\frac{2+3\alpha}{2+2\alpha}\right)}, Flux\right)_{(Slag)} + SO_{2(Offgas)} + Cu_{(Blister)} \tag{6}$$

similar to Equation (1b), in which the minimum $\alpha = 0$ corresponds to pure wustite FeO and $\alpha = 2$ corresponds to pure magnetite Fe_3O_4. Equation (6) can be further detailed in a similar manner as Equations (2) and (3) by assigning appropriate subscripts to α, as in [20].

In the modeling of slag chemistry, α is preferred over x to avoid ambiguity between the reactive oxygen of the blast and the inert oxygen that is strongly bonded within the flux (i.e., within the SiO_2, CaO, etc.). For instance, the role of SiO_2 flux is made more evident by expressing the wustite as a component within a fayalite matrix $FeO \cdot 2SiO_2$; hence, the balance of FeO versus Fe_3O_4 is considered as $FeO \cdot 2SiO_2$ and Fe_3O_4. In practice, SiO_2 is added into the slag in proportions that surpass the stoichiometry of fayalite and may be accompanied by other stable oxides. Under matte-processing conditions, the stable molecules SiO_2, CaO, etc. can be regarded as if they were indivisible atoms. Most notably, the strongly bonded oxygen is not explicitly represented in Equations (1)–(3) and is not taken into account in x; these equations only explicitly consider the blast oxygen. The degree of oxidation α considers only the iron species isolated from any mention of the blast and flux oxygen.

Within Equation (6) and its nickel–copper equivalent, the slag-blow reaction can be isolated and balanced as:

$$FeS_{(Matte)} + \left(\frac{2+3\alpha}{4+4\alpha}\right)O_{2(Blast)} \rightarrow FeO_{\left(\frac{2+3\alpha}{2+2\alpha}\right)(Slag)} + SO_{2(Offgas)} \tag{7}$$

which applies to both the smelting and converting furnaces for both copper and nickel–copper smelters. Indeed, the melting of the feed of Equation (6) results in molten matte that is a mixture of FeS and Cu_2S; in the case of nickel–copper smelters, the matte will also contain nickel and cobalt sulfides [18], but Equation (7) is still correct. The incoming blast includes N_2, as well as O_2 (see Figure 4). As the N_2 passes through the bath and is exhausted into the offgas, along with the SO_2, it carries away sensible heat and is a critical consideration in controlling the bath temperature.

To resolve the degree of freedom α (or equivalently x), the equilibrium between iron oxide species can be expressed as

$$FeS_{(Matte)} + 3Fe_3O_{4(Slag)} \leftrightarrow 10FeO_{(Slag)} + SO_{2(Offgas)} \tag{8}$$

having enthalpy and entropy values ΔH_0 = 622,549 J/mol and ΔS_0 = 342.64 J/mol K, respectively, which can be obtained from HSC Chemistry^{TM}. The corresponding Gibbs free energy balance is

$$\Delta G = \Delta H_0 - T\Delta S_0 + RT \ln\left[\frac{\left(a_{FeO,Slag}\right)^{10} p_{SO2,Offgas}}{a_{FeS,Matte}\left(a_{Fe3O4,Slag}\right)^3}\right] \tag{9}$$

which is set to zero to assume equilibrium. R is the ideal gas constant, T is the bath temperature, and a_{ij} is the activity of species i within phase j. The activity of SO_2 in the offgas is taken to be the partial pressure $p_{SO2,Offgas}$.

Within Equation (9), the activities ($a_{FeS,Matte}$, $a_{FeO,Slag}$, and $a_{Fe3O4,Slag}$) can be re-expressed in terms of α, T, and the operational parameters. The usual parameters include the oxygen enrichment of the blast φ and the silica–iron mass ratio $r = (m_{SiO2,Slag}/m_{Fe,Slag})$, which are considered in Section 4. Empirical measurements relate the activities a_{ij} to their respective mole fractions X_{ij}. In particular, the classic model of Goto [30,31] is validated for smelting and converting, in both the copper and nickel–copper contexts [32], and is the subject of Appendix A.

Iron speciation computations are simpler for smelting furnaces than for converting, since the smelting bath temperature can usually be treated as if it were at a steady state and is approximately uniform and constant. Under this simplification, α can be resolved through an application of Newton's Method [18,19]:

$$\alpha^{(k)} = \alpha^{(k-1)} - \frac{f_G}{\frac{\partial f_G}{\partial \alpha}} \tag{10}$$

or, in case T is not constant,

$$\left(\begin{array}{c} T^{(k)} \\ \alpha^{(k)} \end{array}\right) = \left(\begin{array}{c} T^{(k-1)} \\ \alpha^{(k-1)} \end{array}\right) - \frac{1}{\frac{\partial f_H}{\partial T}\frac{\partial f_G}{\partial \alpha} - \frac{\partial f_H}{\partial \alpha}\frac{\partial f_G}{\partial T}}\left[\begin{array}{cc} \frac{\partial f_G}{\partial \alpha} & -\frac{\partial f_H}{\partial \alpha} \\ -\frac{\partial f_G}{\partial T} & \frac{\partial f_H}{\partial T} \end{array}\right]\left[\begin{array}{c} f_G \\ f_H \end{array}\right] \tag{11}$$

which is a two-variable form of Newton's Method, in which $(T^{(k)}, \alpha^{(k)})$ denote the results of the kth Newton iteration. The righthand sides of Equations (10) and (11) include proxy functions, f_G and f_H, and their derivatives, which are all evaluated at the preceding values $(T^{(k-1)}, \alpha^{(k-1)})$, considering $(T^{(0)}, \alpha^{(0)})$ = (1473 K, 0.15) as typical starting values. The proxy function f_G must be formulated so that

$f_G = 0$ when the Gibbs free energy balance of Equation (9) is satisfied, i.e., when $\Delta G = 0$. Appendix A presents a formulation of f_G that is based on the classic Goto model [30,31]. Likewise, f_H is formulated such that $f_H = 0$ when the heat balance is satisfied [19].

Following the results of Appendix A, it is relatively simple to program Equations (10) and (11) into a simulation platform, thereby relating slag chemistry to the wustite–magnetite balance and, indeed, to the overall mass balance. Depending on the project, Goto's model may be an appropriate starting point, although it does not consider olivine slags [21], nor does it consider the transport of minor elements. In practice, it is preferable to have more wustite than magnetite, since the latter increases the slag viscosity and the entrainment of matte into the slag [21,33]. The modification of slag chemistry through flux additions affects the migration of minor elements [22,23], possibly at the expense of having higher slag viscosity [21].

There is an interest to develop DES platforms that draw upon state-of-the-art thermochemical databases [19] to assist in the retrieval of valuable elements such as gold, silver, and platinum and the handling of deleterious elements such as arsenic, bismuth, and antimony. The authors suspect that the partition of trace elements can be efficiently computed as a function of the main elements a posteriori when Equations (10) and (11) converge. This is an area of future research, and the resulting platforms would support smelter-wide strategies for the processing increasingly problematic feeds. Yet, in reality, for similar apparent conditions like matte grade, temperature, etc., the balance of FeO versus Fe_3O_4 can depend on various parameters, including flux quality, refractory wear, and amount of charge (hence, affecting mixing), and so, an empirical approach to speciation may be more effective.

3.4. Estimation of Threshold Crossing Times Using Time-Adaptive Finite Differences

Time-adaptive finite differences are an important feature of our smelter DES frameworks, allowing the correct placement of threshold-crossing events within the future event list (Figure 8). In the approach of Navarra et al. [18,19], the Newton iterations of Equations (10) and (11) are nested within the Runge-Kutta-Fehlberg (RKF) method, which is a well-known time-adaptive finite difference (TAFD) scheme [34,35]. This method is a combination of finite difference schemes based on fourth and fifth-order Taylor expansions [34]; an attempted timestep is only accepted if the fourth and fifth order estimates of T and α are within an acceptable error tolerance; otherwise, a smaller timestep is attempted. Moreover, each attempted timestep uses the previous results of T and α as a starting point for the Newton iterations, so that convergence is faster, thus making the computation more efficient.

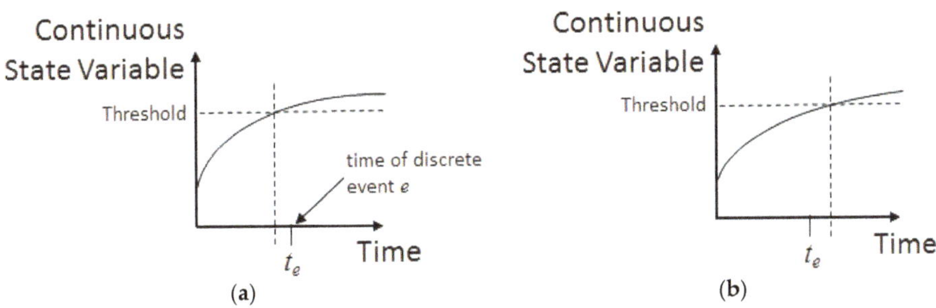

Figure 8. Threshold-crossing event in relation to another event *e*. Time-adaptive finite differences determine if the threshold is crossed (**a**) before *e* or (**b**) after *e*.

The efficiency of DES depends on this time-adaptive aspect to limit the computational effort devoted to dynamics that occur between the discrete events (Figure 7). Alternatives to the approach of Navarra et al. [18,19] include the use of cubic Hermite spline interpolation [36,37] and Richardson extrapolation [38]. These techniques can be adapted to isolate the threshold crossing event of a state variable and improve the accuracy with which we can find the time at which it crosses. In fact, it is important to recognize that both the efficiency and accuracy of the method rely on an accurate time estimate for threshold crossing. In our case, the threshold is first contained in the interval bounded by the two timesteps bracketing an evaluation of the state variable below and above the threshold. This precisely guarantees we have access to values already computed at both of the upper and lower estimates (as a result of the RKF attempt), which includes state variable values, as well as their time derivatives (by direct evaluation if the right-hand side at the state variable values). With this data, we constructed a cubic Hermite spline, which is, therefore, a valid fourth-order interpolation of the solution over this interval. Since the Hermite spline is, in fact, a cubic polynomial, we can compute analytically the crossing time as a simple root-finding problem. We also note that this approach is robust, since the data at the end of the interval provided need not be exact. In fact, if the state variable data has a fourth-order accurate truncation error (e.g., by only using a third order Runge-Kutta integrator) and the derivatives are third-order accurate, the interpolant is guaranteed to be fourth-order accurate [37]. Further, while this approach is tailored to the RKF used here, it may be expanded by considering Hermite quintics if a further need for accuracy is required.

While conventional DES frameworks support basic representations of smelter logistics, a hybridization with time-adaptive finite differences supports a detailed representation of the kinetics that occur within the individual unit operations. In particular, the detailed representation of individual slag-blow segments may require a dynamic simulation of the evolving thermochemical states of matte and slag. Moreover, the DES-TAFD hybridization allows the simulation of several simultaneous converting cycles, in conjunction with the intervening actions involving cranes and ladles, the delivery of a cold charge, and other phenomena. Nonetheless, the explicit representation of slag-blow segments should only be implemented if it is beneficial to the particular engineering project. Beyond the internal dynamics of smelter, a project may require an explicit representation of market-related or environmental phenomena (Figure 6). The hybridization of DES and TAFD is indeed capable of linking detailed representations of diverse aspects throughout the smelter and beyond, whose coordination may be critical to the sustainability of the smelter. Furthermore, aging smelters will not be sustainable unless they can successfully modernize their operational practices, benefitting from sensors and other novel technology.

4. Sample Computations and Context

The sample calculations presented in this section are typical of an aging copper smelter that:

- has been successful for decades in processing reasonably clean feeds;
- is confronted with increasingly challenging feeds that carry excess quantities of arsenic, bismuth, and antimony; and
- is aware of an approach to draw a critical portion of the undesirable elements into the converting slag, which is only effective as the iron in matte approaches zero

Regarding the third point, there may be operator experiences in treating marginal feeds that had a manageable amount of the trace elements; such feeds were treated by driving the individual slag-blow segments to a relatively low iron content (e.g., ~3% Fe), in combination with particular flux compositions. Indeed, a more forceful elimination of the undesirable elements could be induced at an even lower iron content, but this would increase the level of entrained into the slag, under imperfect mixing, as blister copper would be precipitated heterogeneously in certain regions of the vessel. Better endpoint control on the individual segments would allow a more careful advance toward 0% Fe with a limited risk of copper entrainment and could be attained by adapting the endpoint approach at Boliden's Rönnskär Smelter [5], mentioned earlier; this approach depended on customized sensor development [10–12].

From the authors' experience in copper smelter projects, it is incumbent on outside experts to integrate their general understanding with the detailed understanding of in-house experts at the given smelter. There may be issues that were already confronted with some amount of success, but these issues may gradually become critical. For instance, the increasing presence of penalty elements is especially common in custom smelters that receive concentrates from a combination of regional mines, which are themselves confronted with ore blending challenges [29]. The experiences of the smelter should be cross-checked with thermochemical models and a literature review [22,23] in proposing new operational modes that may require a technological upgrade. In effect, this can be a standardization and optimization of approaches that the in-house experts were already considering.

The DES framework described in Sections 3.1–3.4 was implemented using Rockwell Arena[TM] software and replicated the general aspects of conventional smelters while incorporating smelter-specific data. Tables 1–3 contain sample data that are loosely based on published values from [2,5,6,13,39,40]. As stated in Section 3.1, conventional smelters have smelting capacities that normally exceed the downstream converting capacity; indeed, the smelting furnace should not usually produce matte at full capacity, nor should it function in fits and starts. The definition of so-called short and long converter cycles (Figure 6) provides the operational flexibility to set a fixed smelting rate. Moreover, some smelters may have several cycle types under consideration to respond to build-ups of cold charges or compositional imbalances in the feeds that are received from the supplying mines. For simplicity, the current computations consider only two types of cycles.

Table 1. Feed compositions entering the smelting furnace.

Element	Weight%
Copper	20
Iron	39
Sulfur	40
Arsenic	0.4–1.0
Bismuth	0.02–0.20
Antimony	0.02–0.10

Table 2. Examples of smelting furnace operational parameters.

Parameter	Value
Matte holding capacity	900 t
Bath temperature	1275 °C
Blast rate	1100 Nm^3/min
Oxygen enrichment	50 vol%O$_2$
SiO$_2$/Fe in slag	0.7
Matte grade	60 wt% Cu

Table 3. Examples of Peirce–Smith (PS) converting data describing short and long cycles.

	Short Cycle		Long Cycle	
	Duration (h)	* Ladles Added	Duration (h)	* Ladles Added
First slag-blow segment	2.0	3	3.0	4
Second slag-blow segment	0.5 ± 0.2	1	1.0	2
Third slag-blow segment	-	-	0.5 ± 0.2	1
Copper-blow	4.5 ± 0.7	-	5.0 ± 1.2	-

* Assume that ladles are added at the beginning of the segment, each carrying 30 t of matte.

To maintain stable throughput, the blending strategy ensures comparatively stable portions of the main elements (Table 1) but with unavoidable fluctuations in the undesirable trace elements. Additionally, to maintain stable throughput, the smelting parameters are kept constant (Table 2), including the grade of the output furnace matte; given that the matte is a mixture of FeS and Cu_2S, an elemental mass balance determines that a 60% Cu grade corresponds to roughly 24% S and 16% Fe.

Effectively, the smelting furnace blasts 550 Nm^3/min of oxygen into a bath in order to decrease the iron content from 39% to 16% as it burns the FeS. The smelting blast also includes 550 Nm^3/min of nitrogen, which carries away a portion of the heat. The control of the temperature at 1275 °C depends on this nitrogen flow. Given the stable temperature $T = 1275$ °C $= 1548$ K, oxygen enrichment $\varphi = 0.5$, and silica-iron ratio $r = 0.7$, the Newton iterations of Equation (10) are used to obtain the degree of oxidation $\alpha = 0.166$, with a corresponding throughput of 36.928 t/h of matte or equivalently 1.231 ladles/h. At a matte grade of 60% Cu, this corresponds to 22.157 t Cu/h.

Figure 9 shows graphs of the matte content using the data from Tables 1–3. The largest declines correspond to the initiation of long cycles (four ladles = 120 t) and short cycles (three ladles = 90 t), and the other declines correspond to the recharging actions that occur during the cycles at the second and third slag-blow segments. As is common in larger smelters, the offgas handling capacity allows for the execution of two simultaneous converter cycles.

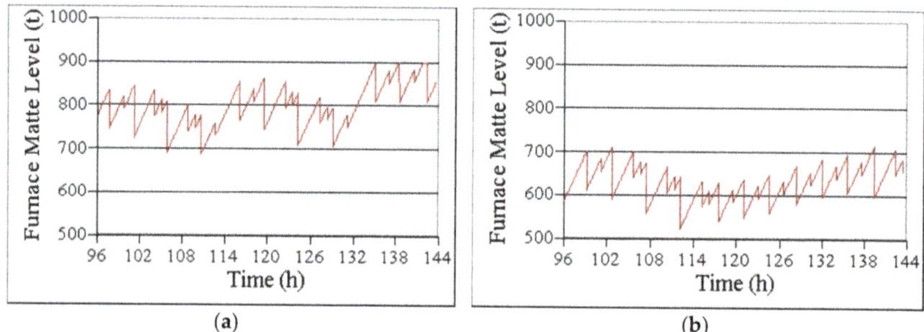

(a)   (b)

Figure 9. Matte levels of the smelting furnace with data from Tables 1–3, in which (**a**) exceeding 800 t triggers two long cycles and (**b**) exceeding 700 triggers three long cycles. The second case is safely below the maximum holding capacity of 900 t.

Within Table 3, the times for the initial slag-blow segments are fixed as operational parameters. However, the durations of the final slag-blow segments depend on the exact nature of the charge (including the cold dope), as well as variable operator behavior and, potentially, other factors. By extension, the copper-blow durations are also variable. For the purposes of Figure 9, these fluctuating values are used to define uniform distributions, e.g., the copper-blow durations are uniformly distributed between 3.8 and 5.2 h for the short cycles and between 3.8 and 6.2 h for the long cycles. Uniform distributions, triangular distributions (Figure 5a), and other simple forms are typical of the early phases of a smelter project. More advanced phases can include more elaborate distributions that are supported by goodness-of-fit testing of the plant data [41].

An analysis of Table 3 reveals that the average conversion rates are 0.571 ladles/h and 0.737 ladles/h for the short and long cycles, respectively. Considering that the smelting furnace produces 1.231 ladles/h that is fed into two simultaneous converter cycles (i.e., 0.615 ladles/h per converter), the balancing of the smelting and converting throughput will require a combination of the long and short cycles. (Incidentally, the long cycle copper-blow is more productive than that of the short cycle; this may be because it is loaded with cold copper scrap, which makes the heat balance less dependent on the nitrogen convection, thus supporting a more intense higher-oxygen blast).

Figure 9 considers a threshold criterion that, whenever the matte level surpasses a critical value, the next several cycles are set to a long cycle. Poor adjustment of this threshold can diminish the overall throughput, as the smelting furnace is deactivated when it reaches the 900-t capacity given in Table 1; this is depicted in Figure 9a, in which the threshold is set to 800 t, which triggers two long cycles, which is insufficient to prevent the approach to 900 t observed at time 140 h. A balanced production is shown in Figure 9b, which applies a threshold of 700 t, which triggers three long cycles. In general, a DES-TAFD hybrid can be adapted to incorporate the operational decision-making of the smelter that is being studied; this flexibility is vital in the context of reengineering projects [20].

Table 4 and Figure 10 describe modified dynamics in which an array of sensors track the approach to 0% Fe and the presence of the undesirable trace elements. The data-driven approach to removing the iron is reflected by the (uniform) variation ascribed to all of the slag-blow segments, whereas the previous configuration (Table 3) fixed predetermined blow times for the initial segments. Moreover, Table 4 follows the action graph of Figure 5b, in which the commitment to an extended cycle is made only after the cycle begins. Indeed, the regular short cycles correspond to 2-5-13, the extended short cycles to 2-5-18-12, the regular long cycles to 1-4-7-10, and the extended long cycles to 1-3-6-9-11. For the sake of Figure 10, it is assumed that 30% of the short cycles are extended and 20% of the long cycles are extended, although different values could be tested and should be driven by actual forecasts of what the incoming feed might be.

Table 4. Example of PS converting data, including short and long cycles that can be extended.

	Regular Short C.		Extended Short C.		Regular Long C.		Extended Long C.	
	Duration (h)	* Ladles Added	Duration (h)	* Ladles Added	Duration (h)	* Ladles Added	Duration (h)	* Ladles Added
First SB segment	2.2 ± 0.2	3	2.2 ± 0.2	3	3.4 ± 0.2	4	3.4 ± 0.3	4
Second SB segment	0.3 ± 0.1	1	0.3 ± 0.1	1	0.7 ± 0.2	2	1.2 ± 0.2	3
Third SB segment	-	-	0.4 ± 0.1	1	0.4 ± 0.1	1	0.3 ± 0.1	1
Fourth SB segment	-	-	-	-	-	-	0.4 ± 0.1	1
Copper-blow	4.5 ± 0.7	-	5.6 ± 0.8	-	5.0 ± 1.2	-	7.0 ± 1.7	-

* Assume that ladles are added at the beginning of the segment, each carrying 30 t of matte.

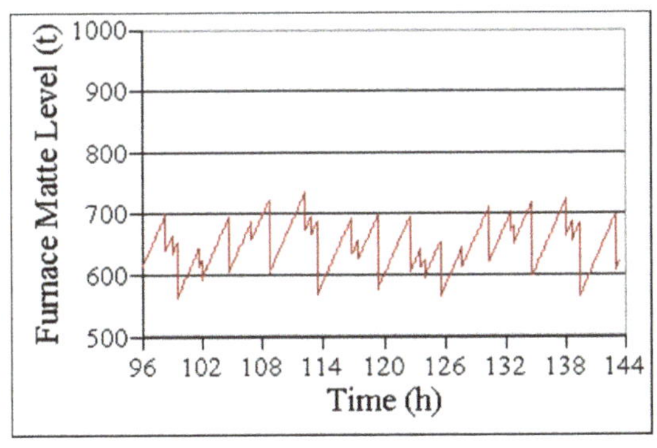

Figure 10. Matte level of smelting furnace with data from Table 1, Table 2, and Table 4, in which exceeding 700 triggers three long cycles, 30% of short cycles are extended, and 20% of long cycles are extended.

A comparison between Figures 9 and 10 shows the effect that responsive converter cycles (Table 4) can have on the matte levels; although the matte levels of Figure 10 remain mostly between 600 and 700 t, the pattern of peaks and dips is far less regular. These fluctuations may be acceptable, considering that it would allow the smelter to handle problematic feeds. Otherwise, the feed rate into the smelting furnace could be controlled in coordination with a variable blast rate, which may be the subject of a following stage of engineering.

To the authors' knowledge, the computations of Figure 10 are the first published instance in which several converters are simultaneously drawing upon a smelting furnace and operational decisions are finalized after the cycles begin, i.e., in response to incoming sensor data. This level of detail is necessary to evaluate the utility of sensors in supporting smelter-wide responses to problematic feeds. The computations described by Table 4 and Figure 10 rely on the representation of individual slag-blow segments (Figure 5b), although other approaches might require the detailed representation of additional aspects within or surrounding the smelter (Figure 6).

5. Conclusions and Future Work

Not only warehouses and manufacturing facilities but metallurgical plants such as a steel plant or copper smelter—which are more difficult to automate—will be transformed in the future with use of new machine capabilities, automation, and improved sensors and controls. Steel is, by far, the major metal produced in the world, representing about 95% by tonnage of all metal production, with the world copper smelter output some 80 to 90 times less than world steel output. This, in part, helps understand that certainly more developments are attained in the iron and steel industry relative to copper. However, the computational framework described in this paper will help close the gap, regarding HI and LIBS and other radiometric sensors, as it enables the implementation of these technologies and justifies their further development within the copper industry. In particular, LIBS is expected to have an increasing importance in handling problematic feeds as existing copper smelters are confronted with increasing amounts of arsenic, bismuth, and antimony [42].

Modern sensors will be vital in supporting smelter-wide responses to increasingly challenging feeds that are being confronted throughout the world, as described in the previous section. It should be noted in particular that smelting and converting operations are central within copper and nickel–copper smelting and are linked to supporting operations throughout the smelter. High-quality and reliable process instrumentation and controls are therefore important in maximizing the global operating efficiency. Additionally, the monitoring of the furnace integrity, refractory wear, preventative maintenance, and plant safety are also key aspects that constantly need attention at the plant. Many high-performance smelting furnaces today include water-cooled copper blocks generally externally mounted on a furnace sidewall to protect the refractory lining at the hot face. The Peirce-Smith converter operates with a converter hood that includes water-cooled panels on the cold face in order to protect the steel wall at the hot face. Recent developments in the instrumentations for detecting and measuring the presence of small levels of water vapor in furnace offgas can signal a water leak and lead to improved furnace monitoring [43]. However, the implementation of such measures often requires quantitative justification.

This paper showed that combining thermochemical equilibrium data with a knowledge of smelting and converting dynamics provides a powerful tool for advancing smelting operations in the form of DES-TAFD hybrid simulations. The specialized use of Newton's Method, Runge-Kutta-Fehlberg, and Hermite interpolation within a DES are, in fact, an advancement within the industrial system analysis, which can be adapted to other industrial contexts, supporting modernization projects that include novel sensors and other technology.

Author Contributions: Conceptualization, N.T.; methodology, A.N.; software, A.N.; validation, A.N., N.T., and A.R.; formal analysis, A.N., A.R., and J.-C.N.; investigation, A.N., R.P., N.T., A.R., and P.J.M.; data curation, A.N.; writing—original draft preparation, A.N., R.W., R.P., A.R., and P.J.M.; writing—review and editing, A.N., R.W., J.N., and P.M.; and funding acquisition, A.N. and R.P. All authors have read and agreed to the published version of the manuscript.

Funding: This research was funded by Conicyt, Anillo Minería ACM 170008, supported by the Chilean government, and NSERC, grant number 2020-04605, supported by the Canadian government.

Conflicts of Interest: The authors declare no conflict of interest.

Appendix A Proxy Function for Gibbs Free Energy Balance Based on Goto's Model

The approach of Navarra et al. [18,19] to obtain a viable proxy function f_G considers that each mole of wustite FeO contains one mole of ferrous, that each mole of magnetite FeO·Fe$_2$O$_3$ contains one ferrous and two ferric, and that all other iron-bearing slag compounds are negligible. It follows that $\alpha = \mathrm{Fe}^{3+}/\mathrm{Fe}^{2+}$ can be taken as

$$\alpha = \frac{2n_{\mathrm{Fe3O4,Slag}}}{n_{\mathrm{FeO,Slag}} + n_{\mathrm{Fe3O4,Slag}}} \tag{A1}$$

in which n_{ij} generally denotes the number of moles of i within phase j.

When setting $\Delta G = 0$, Equation (9) can be reorganized:

$$0 = a_{\mathrm{FeS,Matte}}\left(a_{\mathrm{Fe3O4,Slag}}\right)^3 - \left(a_{\mathrm{FeO,Slag}}\right)^{10} p_{\mathrm{SO2,Offgas}} e^{\left(\frac{\Delta S_0}{R} - \frac{\Delta H_0}{RT}\right)}$$

Using the expressions from Goto [30] and Kemori et al. [31] for the activity coefficients of ($a_{\mathrm{FeS,Matte}}$, $a_{\mathrm{FeO,Slag}}$, and $a_{\mathrm{Fe3O4,Slag}}$), a series of algebraic manipulations were performed by Navarra et al. [18,19] to obtain the following form that explicitly features T and α:

$$0 = \prod_{l=1}^{3}(A_l + B_l\alpha)^{C_l + D_l/T} - \prod_{l=4}^{9}(A_l + B_l\alpha)^{C_l + D_l/T}$$

In which the coefficients (A_l, B_l, C_l, and D_l) are given in Table A1, from which a viable proxy function is obtained:

$$f_G(T,\alpha) = \prod_{l=1}^{3}(A_l + B_l\alpha)^{C_l + D_l/T} - \prod_{l=4}^{9}(A_l + B_l\alpha)^{C_l + D_l/T} \tag{A2}$$

Indeed, $\Delta G = 0$ if and only if $f_G = 0$. Moreover, the partial derivates of f_G can be obtained with respect to α and T, so to complete the Newton iterations described by Equations (10) and (11). To obtain the expression for $\frac{\partial f_G}{\partial T}$, it is helpful to notice that D_l is zero for all factors except for the third and ninth.

$$f_G(T,\alpha) = \left(\prod_{l=1}^{2}(A_l + B_l\alpha)^{C_l}\right)(A_3 + B_3\alpha)^{C_3 + D_3/T} \\ - \left(\prod_{l=4}^{8}(A_l + B_l\alpha)^{C_l}\right)(A_9 + B_9\alpha)^{C_9 + D_9/T} \tag{A3}$$

However, to obtain an expression for $\frac{\partial f_G}{\partial \alpha}$, it is more effective to work directly with Equation (A2).

Within Table A1, φ denotes the volume fraction of oxygen within the blast, which can be related to $p_{\mathrm{SO2,offgas}}$. Additionally, the mole fraction of FeS within the matte is taken to be

$$X_{\mathrm{FeS,Matte}} = \frac{n_{\mathrm{FeS,Matte}}}{n_{\mathrm{FeS,Matte}} + n_{\mathrm{NiS,Matte}} + n_{\mathrm{Cu2S,Matte}} + n_{\mathrm{CoS,Matte}}} \tag{A4}$$

which supports the modeling of nickel–copper smelters, as well as copper smelters in which $n_{\mathrm{NiS,Matte}}$ and $n_{\mathrm{CoS,Matte}}$ are set to zero. Moreover, Table A1 has several instances of the silica-to-iron mole ratio ($n_{\mathrm{SiO2,Slag}}/n_{\mathrm{Fe,Slag}}$), which can be related to the silica-to-iron mass ratio r within the slag:

$$r = \left(\frac{M_{\mathrm{SiO2}}}{M_{\mathrm{Fe}}}\right)\left(\frac{n_{\mathrm{SiO2,Slag}}}{n_{\mathrm{Fe,Slag}}}\right) \tag{A5}$$

in which M_{SiO2} and M_{Fe} are the molar masses of silica and iron, respectively; r is a common operational parameter used to control the flux additions.

Table A1. Coefficients for Equation (A2) (adapted from [19]).

l	A_l	B_l	C_l	D_l
1	1	1	1	0
2	2	-1	10	0
3	$\left(2.44 - 0.4\left(\frac{n_{SiO2,Slag}}{n_{Fe,Slag}}\right)\right)$	$-\left(1.42 - 0.4\left(\frac{n_{SiO2,Slag}}{n_{Fe,Slag}}\right)\right)$	0	15,430
4	$X_{FeS,Matte}\, e^{\frac{\Delta S_0}{R}}$	0	1	0
5	0	1	3	0
6	$\left(\frac{3-\phi}{2\phi}\right)$	$\left(\frac{7-3\phi}{4\phi}\right)$	1	0
7	$\left(1.38 + 12.28\left(\frac{n_{SiO2,Slag}}{n_{Fe,Slag}}\right)\right)$	$\left(56.8 + 12.28\left(\frac{n_{SiO2,Slag}}{n_{Fe,Slag}}\right)\right)$	3	0
8	$2\left(1 + \left(\frac{n_{SiO2,Slag}}{n_{Fe,Slag}}\right)\right)$	$2\left(\frac{n_{SiO2,Slag}}{n_{Fe,Slag}}\right)$	4	0
* 9	$2\left(1 + \left(\frac{n_{SiO2,Slag}}{n_{Fe,Slag}}\right)\right)K$	$2\left(\frac{n_{SiO2,Slag}}{n_{Fe,Slag}}\right)K$	0	15,430

* In which $K = e^{-\Delta H_0/15{,}430R}\left(0.54 + 0.52 X_{FeS,Matte} + 1.4 X_{FeS,Matte}\ln X_{FeS,Matte}\right)^{1458/15{,}430}$.

References

1. Industry 4.0. Available online: www.industrialagilesolutions.com/industry-4-0 (accessed on 16 October 2020).
2. Schlesinger, M.; King, M.; Sole, K.; Davenport, W. Extracting copper from copper-iron-sulfide ores. In *Extractive Metallurgy of Copper*; Elsevier: Oxford, UK, 2011; pp. 2–8.
3. Anderson, J.N. Converter operation at Noranda. In Proceedings of the AIME Annual Meeting, New York, NY, USA, 2–7 July 1956; pp. 133–158.
4. Pelletier, A.; Lucas, J.M.; Mackey, P.J. A new approach to furnace temperature measurement. In Proceedings of the Copper 87 Conference, Viña del Mar, Chile, 30 November–3 December 1987; pp. 489–508.
5. Prietl, T.; Filzwieser, A.; Wallner, S. Productivity increase in a Peirce-Smith converter using the COP KIN and OPC system. In *Converter and Fire Refining Practices, Proceedings of the Annual TMS Meeting, San Francisco, CA, USA, 13–17 February 2005*; The Minerals, Metals, & Materials Society: Warrendale, PA, USA, 2005; pp. 177–190.
6. Ek, M.; Olsson, P. Recent developments on the Peirce-Smith converting process at the Rönnskär Smelter. In *Converter and Fire Refining Practices, Proceedings of the Annual TMS Meeting, San Francisco, CA, USA, 13–17 February 2005*; The Minerals, Metals, & Materials Society: Warrendale, PA, USA, 2005; pp. 19–26.
7. Arias, L.; Torres, S.; Toro, C.; Balladares, E.; Parra, R.; Loeza, C.; Villagrán, C.; Coelho, P. Flash smelting copper concentrates spectral emission measurements. *Sensors* **2018**, *18*, 2009. [CrossRef] [PubMed]
8. Toro, C.; Torres, S.; Parra, V.; Fuentes, R.; Castillo, R.; Díaz, W.; Reyes, G.; Balladares, E.; Parra, R. On the detection of spectral emissions of iron oxides in combustion experiments of pyrite concentrates. *Sensors* **2020**, *20*, 1284. [CrossRef] [PubMed]
9. Marion, M.; Toro, C.; Arias, L.; Balladares, E. Estimation of spectral emissivity and S/Cu ratio from emissions of copper concentrates at the flash smelting process. *IEEE Access* **2019**, *7*, 103346–103353. [CrossRef]
10. Díaz, W.; Toro, C.; Balladares, E.; Parra, V.; Coelho, P.; Reyes, G.; Parra, R. Spectral characterization of copper and iron sulfide combustion: A multivariate data analysis approach for mineral identification on the blend. *Metals* **2019**, *9*, 1017. [CrossRef]
11. Sugiura, M.; Otani, Y.; Nakashima, M. Radiation thermometry for high-temperature liquid stream at blast furnace. In Proceedings of the SICE Annual Conference, Tokyo, Japan, 13–18 September 2011; pp. 472–475.
12. Ugiura, M.; Otani, Y.; Nakashima, M.; Omoto, N. Continuous temperature measurement of liquid iron and slag tapped from a blast furnace. *SICE JCMSI* **2014**, *7*, 147–151.
13. Schlesinger, M.; King, M.; Sole, K.; Davenport, W. Batch converting of copper matte. In *Extractive Metallurgy of Copper*; Elsevier: Oxford, UK, 2011; pp. 127–153.
14. Navarra, A.; Kuan, S.H.; Parra, R.; Davis, B.; Mucciardi, F. Debottlenecking of conventional copper smelters. In Proceedings of the International Conference on Industrial Engineering and Operations Management, Kuala Lumpur, Malaysia, 8–10 March 2016; pp. 2395–2406.
15. Watt, L.; Kapusta, J. The 2019 Copper Smelting Survey. In *Phillip Mackey Honorary Symposium, Proceedings of the Copper 2019 Conference, Vancouver, VA, Canada, 18–21 August 2019*; The Metallurgical Society of CIM: Montreal, QC, Canada, 2019; p. 528496.

16. Price, T.; Harris, C.; Hills, S.; Boyd, W.; Wraith, A. Peirce-Smith converting: Another 100 years? In *International Peirce-Smith Converting Centennial Symposium, Proceedings of the TMS Annual Meeting, San Francisco, CA, USA, 15–19 February 2009*; The Minerals, Metals, & Materials Society: Warrendale, PA, USA, 2009; pp. 181–197.

17. Mackey, P.J. Copper smelting technologies in 2013 and beyond. Plenary paper. In Proceedings of the Copper 2013 Conference, Santiago, Chile, 1–4 December 2013.

18. Navarra, A.; Valenzuela, R.; Cruz, R.; Arancibia, C.; Yañez, R.; Acuña, C. Incorporation of matte-slag thermochemistry into sulphide smelter discrete event simulation. *Can. Metall. Quart.* **2018**, *57*, 70–79. [CrossRef]

19. Navarra, A.; Lemoine, N.; Zaroubi, N.; Marin, T. Semi-discrete dynamics and simulation of Peirce-Smith converting. In Proceedings of the Extraction 2018, Ottawa, ON, Canada, 26–29 August 2018; pp. 273–285.

20. Navarra, A.; Ross, A.; Toro, N.; Ayala, F.; Marin, T. Quantitative methods for copper smelter reengineering projects. In *Phillip Mackey Honorary Symposium, Proceedings of the Copper 2019 Conference, Vancouver, VA, Canada, 18–21 August 2019*; The Metallurgical Society of CIM: Montreal, QC, Canada, 2019; p. 595570.

21. Selivanov, E.; Gulyaeva, R.; Istomin, S.; Belyaev, V.; Tyushnyakov, S.; Bykov, A. Viscosity and thermal properties of slag in the process of autogenous smelting of copper-zinc concentrates. *Miner Process Extr. Metall.* **2015**, *124*, 88–95. [CrossRef]

22. Mackey, P.J. The physical chemistry of copper smelting slags. *Can. Metall. Quart.* **1982**, *21*, 221–260. [CrossRef]

23. Shishin, D.; Hayes, P.C.; Jak, E. Development and applications of thermodynamic database in copper smelter. In *Phillip Mackey Honorary Symposium, Proceedings of the Copper 2019 Conference, Vancouver, Canada, 18–21 August 2019*; The Metallurgical Society of CIM: Montreal, QC, Canada, 2019; p. 594861.

24. Almaraz, A.; López, C.; Arellano, I.; Barrón, M.; Jaramillo, D.; Reyes, F.; Plascencia, G. CFD modelling of fluid Flow in a Peirce-Smith converter with more than one injection point. *Miner Eng.* **2014**, *56*, 102–108. [CrossRef]

25. Navarra, A.; Marambio, H.; Oyarzún, F.; Parra, R.; Mucciardi, F. System dynamics and discrete event simulation of copper smelters. *Miner Metall. Proc.* **2017**, *34*, 96–106. [CrossRef]

26. Coursol, P.; Mackey, P.J.; Morissette, S.; Simard, J.M. Optimization of the Xstrata copper-Horne smelter operation using discrete event simulation. *CIM Bull.* **2009**, *102*, 5–10.

27. Campbell, A.; Reed, M.; Warner, A. Debottlenecking and optimisation of copper smelters leveraging simulation. In *Nickolas Themelis Symposium on Pyrometallury and Process Engineering, Proceedings of the Copper 2013 Conference, Santiago, Chile, 1–4 December 2013*; The Metallurgical Society of CIM: Montreal, QC, Canada, 2013; pp. 1071–1080.

28. Altiok, T.; Melamed, B. Monte Carlo sampling and histories. In *Simulation Modeling and Analysis with Arena*; Elsevier: Oxford, UK, 2007; pp. 15–19.

29. Navarra, A.; Álvarez, M.; Rojas, C.; Menzies, A.; Pax, R.; Waters, K. Concentrator operational modes in response to geological variation. *Miner Eng.* **2019**, *134*, 356–364. [CrossRef]

30. Goto, S. Equilibrium calculations between matte, slag and gaseous phases in copper smelting. In Proceedings of the Annual Meeting of the Institution of Mining and Metallurgy, Brussels, Belgium, 19–23 July 1974; pp. 11–13.

31. Kemori, N.; Kimura, T.; Mori, Y.; Goto, S. An application of Goto's model to a copper flash smelting furnace. In Proceedings of the Annual Meeting of the Institution of Mining and Metallurgy, Brussels, Belgium, 20 November 1987; pp. 21–23.

32. Kyllo, A.; Richards, G. A mathematical model of the nickel converter: Part 1—model development and verification. *Metall. Trans. B* **1991**, *22*, 153–161. [CrossRef]

33. Cheng, X.; Cui, Z.; Contreras, L.; Chen, M.; Nguyen, A.; Zhao, B. Introduction of matte droplets in copper smelting slag. In *International Symposium on High-Temperature Metallurgical Processing, Proceedings of the TMS Annual Meeting, San Diego, CA, USA, 26 February–2 March 2017*; The Minerals, Metals, & Materials Society: Warrendale, PA, USA, 2017; pp. 385–394.

34. Harder, D.W. Runge Kutta Fehlberg. Topic 14.5 of Numerical Analysis for Engineering (University of Waterloo). Available online: Ece.uwaterloo.ca/~{}dwharder/NumericalAnalysis/14IVPs/rkf45/complete.html (accessed on 17 October 2020).

35. Kelton, W.; Sadowski, R.; Swets, N. Continuous and combined discrete/continuous models. In *Simulation with Arena*; McGraw-Hill: New York, NY, USA, 2010; pp. 473–512.

36. Nave, J.-C.; Rosales, R.R.; Seibold, B. A gradient-augmented level set method with an optimally local, coherent advection scheme. *J. Comp. Phys.* **2010**, *229*, 3802–3827. [CrossRef]

37. Kreyszig, E. Spline interpolation. In *Advanced Engineering Mathematics*; Wiley: Hoboken, NJ, USA, 2005; pp. 810–816.

38. Popova, O.A. Using Richardson extrapolation to improve the accuracy of processing and analyzing empirical data. *Meas. Technol.* **2019**, *62*, 111–117. [CrossRef]

39. Navarra, A.; Kapusta, J. Decision-making software for the incremental improvement of Peirce-Smith converters. In *International Peirce-Smith Converting Centennial Symposium, Proceedings of the TMS Annual Meeting, San Francisco, CA, USA, 15–19 February 2009*; The Minerals, Metals, & Materials Society: Warrendale, PA, USA, 2009; pp. 231–250.

40. How to Remove Arsenic, Antimony and Bismuth from Copper. Available online: www.911metallurgist.com/ eliminate-arsenic-antimony-bismuth-copper/ (accessed on 19 October 2020).

41. Altiok, T.; Melamed, B. Goodness-of-fit tests for distributions. In *Simulation Modeling and Analysis with Arena*; Elsevier: Oxford, UK, 2007; pp. 134–137.

42. Yañez, J.; Torres, S.; Sbarbaro, D.; Parra, R.; Saavedra, C. Analytical instrumentation for copper pyrometallurgy: Challenges and opportunities. *IFAC Pap. Online* **2018**, *51*, 251–256.

43. Dennis, P.; Ganguly, S. SAF water leak detection by measurement of gaseous water vapour. In Proceedings of the Twelfth International Ferroalloys Congress, Helsinki, Finland, 6–9 June 2010; pp. 759–768.

Publisher's Note: MDPI stays neutral with regard to jurisdictional claims in published maps and institutional affiliations.

Article

Gaussian Process-Based Hybrid Model for Predicting Oxygen Consumption in the Converter Steelmaking Process

Sheng-Long Jiang *, Xinyue Shen and Zhong Zheng *

College of Materials Science and Engineering, Chongqing University, Chongqing 400044, China; xinyue_shen953@126.com

* Correspondence: sh.l.jiang.jx@gmail.com (S.-L.J.); zhengzh@cqu.edu.cn (Z.Z.);
  Tel.: +86-182-2323-5220 (S.-L.J.); +86-136-3833-1821 (Z.Z.)

Received: 24 April 2019; Accepted: 4 June 2019; Published: 8 June 2019

Abstract: Oxygen is one of the most important energies used in converter steelmaking processes of integrated iron and steel works. Precisely forecasting oxygen consumption before processing can benefit process control and energy optimization. This paper assumes there is a linear relationship between the oxygen consumption and input materials, and random noises are caused by other unmeasurable materials and unobserved reactions. Then, a novel hybrid prediction model integrating multiple linear regression (MLR) and Gaussian process regression (GPR) is introduced. In the hybrid model, the MLR method is developed to figure the global trend of the oxygen consumption, and the GPR method is applied to explore the local fluctuation caused by noise. Additionally, to accelerate the computational speed on the practical data set, a K-means clustering method is devised to respectively train a number of GPR models. The proposed hybrid model is validated with the actual data collected from an integrated iron and steel work in China, and compared with benchmark prediction models including MLR, artificial neural network, support vector machine and standard GPR. The forecasting results indicate that the suggested model is able to not only produce satisfactory point forecasts, but also estimate accurate probabilistic intervals.

Keywords: steelmaking; oxygen consumption; GPR; prediction model

---

1. Introduction

In modern integrated iron and steel works, oxygen is one of the most important energy resources used in various production processes, such as oxygen-rich combustion for ironmaking, converter blowing for steelmaking, and flame cutting for casting [1]. Statistically, about 20% [2] plant-wide electric power is used to produce the oxygen, and more than 50% [2] oxygen is used in the steelmaking process. To precisely monitor for oxygen consumption not only improves the process controlling performance in the steelmaking process, but also benefits making a satisfactory schedule for oxygen production to achieve the goal of energy saving and economic profits [3].

Traditionally, the main task of the steelmaking process is to produce various grades of steel by removing impurities in hot metal, such as excess carbon, silicon, manganese and phosphorus [4]. The primary steelmaking equipment is called the basic oxygen furnace (BOF) also known as Linz–Donawitz (LD) or oxygen converter (as shown in Figure 1). Theoretically, converter steelmaking is a complex process including melting, purifying, and alloying which are carried out at approximately 1600 °C (2900 °F) in melting conditions. First, the hot metal at 1200–1300 °C, an amount of scrap steel and calcined lime are charged into a converter, which produces violent reactions on the surface of the hot metal and slags. Then, oxygen from a lance is blown onto the liquid metal bath surface within typically 15 min, which continuously increases temperature and reduces the carbon-rich hot metal to low-carbon steel between 0–1.5 percent. The slag-gas emulsion is formed during blowing and will decrease in the later period of blowing. Finally, specified chemical compositions and temperatures are reached by initiating numerous chemical reactions in sequence or simultaneously. In addition, the flux of burnt lime, dolomite and other chemical materials are added to further remove impurities and protect the lining of the converter. In fact, because each reaction couples and interplays with others [5], the process of converter steelmaking is very complex, and the condition of oxygen consumption is hard to be monitored in actual environments. However, with the current trend of Industry 4.0, the industrial process tends to be more expensive, which requires higher system reliability and performance. Therefore, it is needed to find an access to forecast the volume of oxygen consumption, which can boost the control performance of steelmaking process and the operational performance of oxygen distribution to a new level.

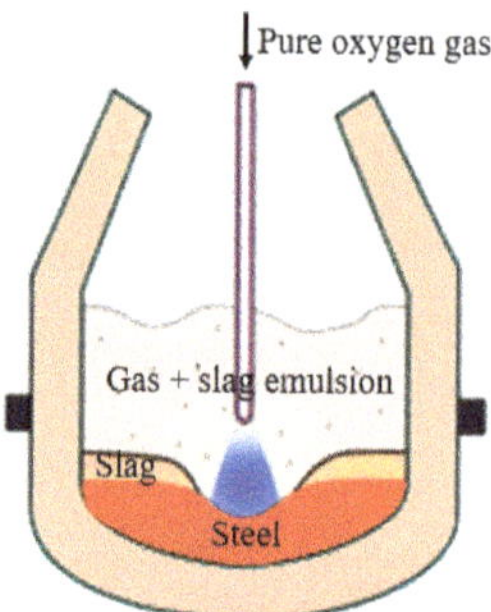

Figure 1. Converter steelmaking process.

With rapid developments of industrial automation and information systems, the industry has established big-data platforms and has collected a vast amount of valuable data [6]. Hence, a considerable number of researchers have used data-driven methodologies to predict these parameters that are hard to directly collect or calculate in the complex process [7,8], such as semiconductor [9], petrochemical [10], and energy process [11]. With concerns on the converter steelmaking process, many efforts have been devoted into predicting the end-point temperature and carbon content [12] in past decades, but the studies on forecasting oxygen consumption have seldom been reported. Since a large number of complex chemical reactions and physical changes in the steelmaking process exist, the exact amount of oxygen required to complete the steelmaking process cannot be calculated directly. Traditionally, these unmeasurable variables were estimated by constructing explicit knowledge models, i.e., material balance, heat balance, kinetics and other theories [13]. However, there are still many unknown factors in the converter steelmaking process, which may cause low precision and unreliability. Therefore, to develop its theoretical model is a high-cost and high-challenge task. However, industrial data produced in the steelmaking process provide a potential way to forecast these unmeasurable variables. These data-driven models complete the prediction task by building the mapping relationship between the input-output data without knowledge about the process. Owing to these advantages, a wide range of data-driven models are applied in the steel industry. Saxén et al. [14] reviewed a variety of prediction models for the silicon content of hot metal produced in blast furnaces. Liu et al. [15] applied a least squares support vector machine (LS-SVM) model to investigate several real-time prediction problems in the converter steelmaking process. Tian and Mao [16] proposed an ensemble extreme learning machine (ELM) to forecast the temperature of liquid steel in the ladle furnace. Han and Liu [17] proposed an ELM with optimized parameters to predict the endpoint carbon content and temperature of liquid steel. Laha et al. [18] compared a number of machine learning models for predicting yield steel in a steelmaking work, and verified that the support vector regression (SVR) is the most powerful one.

However, the data-driven model is a black-box lack of interpretation. Therefore, many researchers paid more attention to the hybrid model based on theory and data. Yang et al. [19] developed a hybrid model to predict the electricity demand in the fused magnesia smelting process, in which its mechanism is formulated by a linear model and the unknown factors are modeled by neural networks. Chen et al. [20] proposed a hybrid model combined with a finite element method and an artificial neural network (ANN) to predict the degree of void closure produced in the cold rolling process. Lei and Su [21] developed an SVR-based hybrid model to forecast the mold level of the continuous casting production process.

The remaining sections of the paper are organized as follows. Section 2 provides the details of the proposed Gaussian process-based hybrid model named HyGPR. The experimental setup and results are analyzed and discussed in Section 3. Finally, Section 4 draws conclusions of this study and identifies several topics for future studies.

2. Methodology

The consumed volume of oxygen during the converter steelmaking process is mainly related with the following four types of factors:

(1) The amount of input materials, such as the carbon, silicon, manganese, phosphorus, sulfur content of hot metal.
(2) The control parameters of blowing, e.g., lance position, and blowing pattern.
(3) The final smelting targets, e.g., oxygen consumption highly depends on the carbon target.
(4) The equipment conditions, e.g., converter lining and internal converter geometry.

In this study, the exact amount of oxygen that will be consumed in the not-started steelmaking process is forecasted. This task is usually realized by static prediction models (before blowing) which obviously differ with dynamic prediction models (during blowing). Hence, the control parameters are not selected in this study. Since the final carbon target of the similar steel grades is very close, prediction models can be trained independently with different carbon target. The equipment conditions cannot be directly observed, so we assume they are constants within a short period. Therefore, only the input materials are considered in our proposed methodology stated in the following subsections.

2.1. Reaction-Based Linear Model

To describe the converter steelmaking process, reaction Equations (1)–(9) are defined where the symbols $[\cdot]$, $(\cdot)$ and $\{\cdot\}$ indicate metal, slag and gas phases, respectively. As shown in (1)–(6) oxidation is the most important chemical reaction mainly carried out on the hot metal, which converts carbon to carbon oxide, silicon to silica, manganese to manganous oxide, phosphorus to phosphate, and sulphur to sulfur dioxide.

$$[C] + \frac{1}{2}\{O_2\} = \{CO\} \tag{1}$$

$$[C] + \{O_2\} = \{CO_2\} \tag{2}$$

$$[Si] + \{O_2\} = (SiO_2) \tag{3}$$

$$[Mn] + \frac{1}{2}\{O_2\} = (MnO) \tag{4}$$

$$2[P] + \frac{5}{2}\{O_2\} = (P_2O_5) \tag{5}$$

$$[S] + \{O_2\} = \{SO_2\} \tag{6}$$

Unfortunately, a little iron is also combined with oxygen in addition to these chemical reactions, and produces Fe_xO_y as follows.

$$2[Fe] + \frac{3}{2}\{O_2\} = (Fe_2O_3) \tag{7}$$

$$[Fe] + \frac{1}{2}\{O_2\} = (FeO) \tag{8}$$

In addition, the liquid slag releases a little oxygen, and acts as an oxidizer to produce some by-products:

$$[S] + (CaO) \rightarrow (CaS) + [O] \tag{9}$$

It should be noted that, a number of other oxygen-related reactions occur in the steelmaking process. For instance, the post reaction $\{CO\} + \frac{1}{2}\{O_2\} \rightarrow \{CO_2\}$ will consume an amount of oxygen, and the C element in the converter lining will also absorb oxygen. Specially, when the phenomenon of rephosphoration and remanganization occur with rising temperatures and low FeO contents in the slag, the reduction of the (P_2O_5) and (MnO) with the solved $[C]$ in the steel droplets in the slag/gas emulsion will release a little oxygen. However, these reactions fail to be directly observed and recorded. Therefore, the consumed and released oxygen during the reaction or blowing are identified as constants or random noises.

We classify these input materials (x) into two sets: The materials consumed oxygen O^+ and the materials released oxygen O^-. To estimate the value of oxygen consumption (y), we assume there is linear relationship between x and:

$$y = f(x) = \sum_{i=1}^{m_1} w_i x_i + \sum_{i=1}^{m_2} w_i(-x_i) + w_0 = \sum_{i=1}^{m_1} w_i x_i - \sum_{i=1}^{m_2} w_i x_i + w_0 \tag{10}$$

where $x = (x_1, \cdots, x_m)$ represents the materials reacting with oxygen, w_i denotes their reactions coefficient, w_0 is a constant term determined by learning from data, m_1 and m_2 respectively represent the size of O^+ and O^-, and $m_1 + m_2 = m$. To determine the values of $w = (w_0, w_1, \cdots, w_m)$, the pre-defined loss function Equation (11) is minimized.

$$J(w) = \frac{1}{2m} \sum_{j=1}^{m} \left(f\left(x^{(j)}\right) - y^{(j)} \right) \tag{11}$$

where $x^{(j)}$ and $y^{(j)}$ respectively represent the input and output values of jth sample data. The loss function $J(w)$ can be solved by the least square or gradient descent least angle method [22].

However, the suggested multiple linear regression (MLR) model is an ideal theoretical model, because the converter steelmaking process in nature is a complex system with multi-component, multi-phase and multi-reaction, the detailed process of each reaction is impossible to be precisely formulated. Additionally, in actual production environments, considerable number factors played in or affected the reactions fail to be observed. Therefore, the MLR model based on these reactions always suffers from low precision and low robustness in the actual production process. To overcome this shortage, this study develops the data-driven prediction model in Section 2.2.

2.2. Gaussian Process Regression with Noise

Gaussian process regression (GPR) [23] is a non-parametric prediction model based on the Gaussian prior distribution. The two main advantages of GPR are the interpretability between the prediction and observations, and the probabilistic sense when some prior models are embedded. In the past decades, theoretical research and real-world application have proved that GPR is a powerful tool for supervised learning applications [24]. Given a dataset $\mathcal{D} = \{X, y\}$, where $X \in \mathbb{R}^{n \times m}$, $y \in \mathbb{R}^{n \times 1}$, n is the sample size, and m is the sample dimension. Assume the regression function f mapping an input vector x to an output value y can be written as:

$$y = f(x) + \epsilon \tag{12}$$

where noise ϵ is the noise with Gaussian distribution $\mathcal{N}\left(0, \sigma_n^2\right)$ and the "signal" term $f(x)$ and noise ϵ are mutually independent. The signal term $f(x)$ is also assumed to be a random variable with Gaussian distribution.

$$f(x) \sim \mathcal{GP}(m(x), k(x, x')) \tag{13}$$

where $m(x) = E(f(x))$ is a mean function which often set to 0, and $k(x, x') = E[(f(x) - m(x))(f(x') - m(x'))]$ is a covariance that illustrates prior assumptions including likely smoothness and patterns in the data. The covariance function k is also identified as the *kernel function* of Gaussian process [25].

Given a collected data set $\mathcal{D} = \{X, y\}$, a predicted signal function f_* should be constructed in order to forecast a new output y^* based on a new input x^*. Once we have determined the mean function and the kernel, the predicted function f_* can be sampled as follows.

$$f_* \sim \mathcal{N}(0, k(x^*, x)) \tag{14}$$

Then, the joint probabilistic distribution of the training outputs y and the predicted function f_* can be written as:

$$\begin{bmatrix} y \\ f_* \end{bmatrix} = \mathcal{N}\left(0, \begin{bmatrix} K(X, X) + \sigma_n^2 I_n & K(X, x_*) \\ K(x_*, X) & k(x_*, x_*) \end{bmatrix}\right) \tag{15}$$

where $K(X, X)$ denotes the covariance matrix between all training inputs, $K(X, x_*)$ represents the covariance matrix between the training inputs and test inputs, $K(x_*, X)$ stands for the covariance matrix between the test inputs and training inputs, $k(x_*, x_*)$ is the covariance between test inputs. I_n is an identity matrix and σ_n^2 is the assumed variance of training samples.

The main task of GPR is to forecast the most likely value of y^* related to x^*. Based on the Bayes' principle, the conditional distribution is concluded [23] as:

$$p\left((f_* | x_*, X, y)\right) \sim \mathcal{N}(m_*, Cov(f_*)) \tag{16}$$

$$m_* = K(x_*, X)\left[K(X, X) + \sigma_n^2 I_n\right]^{-1} y \tag{17}$$

$$Cov(f_*) = k(x_*, x_*) - K(x_*, X)\left[K(X, X) + \sigma_n^2 I_n\right]^{-1} K(x_*, X) \tag{18}$$

Based on these theoretical analysis, the mean and covariance function are the two most important elements in GPR. The kernel function k directly illustrates prior knowledge about the function f, and the combinations between two different kernel functions still can be identified as a kernel [25]. In this paper, we use a composite covariance function with the squared exponential kernel function $k_1(x, x')$ Equation (19) to express smooth trend of the data and the exponential kernel functions $k_2(x, x')$ Equation (20) to illustrate the irregularity of the data.

$$k_1(x, x') = \sigma_{f_1}^2 \exp\left(-\frac{1}{2}(x - x')^T \begin{bmatrix} l_{s1}^2 & & \\ & \ddots & \\ & & l_{sp}^2 \end{bmatrix}^{-1} (x - x')\right) \tag{19}$$

$$k_2(x, x') = \sigma_{f_2}^2 \exp\left(-\sqrt{(x - x') \begin{bmatrix} l_{e1}^2 & & \\ & \ddots & \\ & & l_{ep}^2 \end{bmatrix}^{-1} (x - x')}\right) \tag{20}$$

$$k(x, x') = k_1(x, x') + k_2(x, x') \tag{21}$$

2.3. HyGPR with K-Means Clustering

In this study, the novel hybrid model HyGPR integrating the parametric MLR model and the non-parametric GPR model are constructed to forecast the oxygen consumption in the converter steelmaking process.

$$f(x) = w^T x + g(x) \tag{22}$$

where w^T is the weight vector of the MLR model defined in Equation (10) and $g(x) \sim \mathcal{GP}(0, k(x, x'))$.

Such a hybrid model can bridge the gap between the interpretability of the parametric model MLR and the accuracy of non-parametric model GPR, where MLR is identified as the prior model. Note that the proposed hybrid model can be identified as a special GPR, where the mean function is defined as a linear function and the covariance function is formed as a composite kernel function defined in Equations (19)–(21).

The hyper-parameters of the proposed hybrid model are formed as a vector $\theta = \left\{ w_0, w_1, \ldots, w_m, \sigma_{f_1}^2, l_{s1}^2, \ldots, l_{sm}^2, l_{e1}^2, \ldots, l_{em}^2 \right\}$. To seek the optimal hyper-parameters, we need to maximize the log marginal likelihood [23].

$$\log p(y|\theta) = -\frac{1}{2}(y - m)^T K^{-1}(y - m) - \frac{1}{2}\log|K| - \frac{n}{2}\log 2\pi \tag{23}$$

where $m = w^T X$, and $K_{ij} = k(x_i, x_j)$. When solving this equation, the most challengeable and time-consuming task is finding the inverse matrix K^{-1} with high dimensions.

To apply the proposed HyGPR model in an actual environment, we employ a K-means clustering method to reduce the training sample size (as shown in Figure 2). When training the noise function $g(x)$, we use the K-means clustering method with the same input variables as MLR and GPR to decompose the training set $\mathcal{D} = \{X, y\}$ into Q subsets, and respectively train Q GPR models. When a new input x' arrives, the HyGPR firstly forecasts the value of $f(x')$ using the MLR model, and then predicts the value of $g(x')$ using the qth GPR model selected by the K-means clustering model. With this decomposition manner, the training speed of the GPR is assumed to be accelerated because the dimension of the observed matrix is reduced.

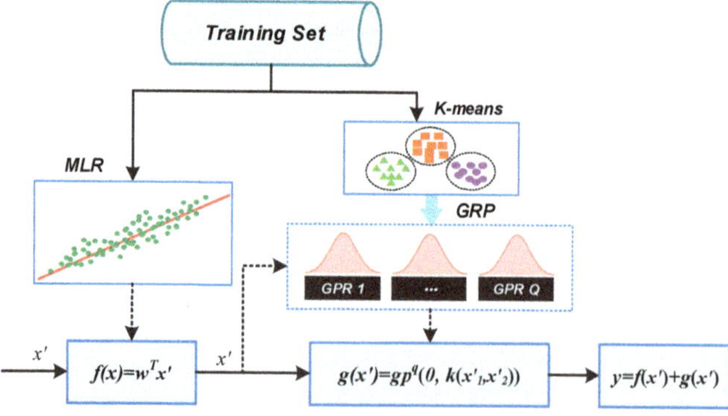

Figure 2. HyGPR model with K-means clustering. MLR: multiple linear regression; GPR: Gaussian process regression.

3. Experiments and Discussion

3.1. Data Set

To test the proposed HyGPR model, we collected the real-world process data of the converter steelmaking process in an integrated iron and steel works situated in the north of China. The data set has 1534 observed samples between 1 April 2018 and 30 June 2018. Figure 3 indicates the distribution of the observed outputs, which is irregular and fluctuates severely. The selected input variables includes:

(1)	The weight of hot metal (Fe).
(2)	The weight of impurity elements, e.g., carbon (C), silicon (Si), manganese (Mn), sulphur (S) and phosphorus (S) which are the products of the weights of hot metal and the element percentages.
(3)	Five additional materials (AM) for steelmaking, of which the real compositions are secreted.

The statistics information such as means, standard deviations, minimum and maximum values, are summarized in Table 1. To evaluate the performance of the proposed model, we divided the dataset into two sets with the handout way: The former 1381 samples (about 90%) for learning HyGPR and second 153 samples (10%) for testing.

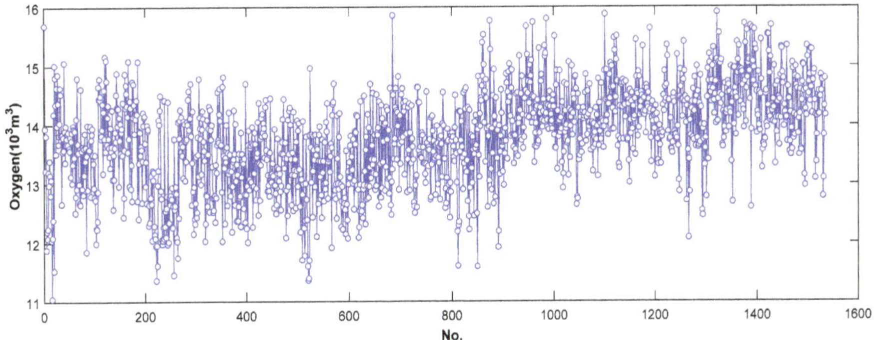

Figure 3. Oxygen consumption of each process in converters.

Table 1. Descriptive statistics of data set.

Parameters	Mean.	Std.	Minimum	Maximum
Fe (t)	276.8450	252.7000	289.8000	21.6711
C (t)	12.1259	0.5061	10.3288	14.0078
Si (t)	0.9477	0.4456	0.0318	2.332
Mn (t)	0.4332	0.2047	0.1252	2.2066
P (t)	0.2431	0.0221	0.1736	0.4119
S (t)	0.0023	0.0022	0	0.0343
AM1 (t)	1.7980	1.4659	0	9.908
AM2 (t)	1.7487	1.6683	0	8.172
AM3 (t)	14.1139	2.5374	5.9610	28.937
AM4 (t)	4.1623	1.3636	0	9.66
AM5 (t)	0.7833	1.6976	0	11.994
O2 (10^3m^3)	13.7823	0.8079	11.0480	15.913

3.2. Evaluation Metrics

In order to compare HyGPR with the other benchmark prediction model, we defined four metrics to assess quantitatively its point and interval forecasting ability. There accuracy metrics for the point prediction including root mean square error (RMSE), mean absolute error (MAE), and mean absolute percentage error (MAPE) were formulated in Equations (24)–(26).

$$\text{RMSE} = \sqrt{\frac{\sum_{i=1}^{N}(y_i - \hat{y}_i)^2}{N}} \tag{24}$$

$$\text{MAE} = \frac{1}{N}\sum_{i=1}^{N}|y_i - \hat{y}_i| \tag{25}$$

$$\text{MAPE} = \frac{1}{N} \sum_{i=1}^{N} \left| \frac{y_i - \hat{y}_i}{y_i} \right| \times 100\% \tag{26}$$

where y_i and $\hat{y}_i$ denote respectively the observed and predicted oxygen consumption in the ith test sample; N is the size of the test sample. Note that the small values of these metrics indicate high prediction accuracy.

The proposed HyGPR model is able to provide not only the forecasting point $\hat{y}_i$ but also the confidence interval $\left[\hat{y}_i^-, \hat{y}_i^+ \right]$ of future oxygen consumption. Therefore, we defined a coverage metric for interval prediction named hit ratio (HRI) in Equation (27) which is applied to calculate the number of test samples fallen into the 95% confidence interval.

$$\text{HRI} = \frac{1}{N} \sum_{i=1}^{N} I_A(\hat{y}_i) \times 100\% \tag{27}$$

where $A = \left\{ \hat{y}_i \middle| \hat{y}_i^- \leq \hat{y}_i \leq \hat{y}_i^+ \right\}$, and $I_A(\hat{y}_i) = \begin{cases} 1, & \hat{y}_i \in A \\ 0, & \hat{y}_i \notin A \end{cases}$.

Additionally, we also used the CPU running time (seconds) to evaluate the learning speed of the tested models.

3.3. Results and Analysis

In this study, all proposed and benchmark models are implemented with the MATLAB 2017 software. Especially, we used the GPML (Gaussian processes for machine learning) toolbox [26] to construct the GPR model and the one in HyGPR, and other compared models were provided by the toolboxes installed in MATLAB. All programs ran on a personal computer with an Intel Core i7-8550U Processor (1.8550GHz) and 16.0GB Memory and installing a Windows 10 operating system.

In the proposed HyGPR model, the cluster count (Q) of K-means is a very important factor that may influence the final prediction performance. To select the most appropriate value of Q, we carried out five group experiments with different clusters. The results of the four accuracy metrics were listed in Table 2 and the forecasting plots were shown in Figure 4.

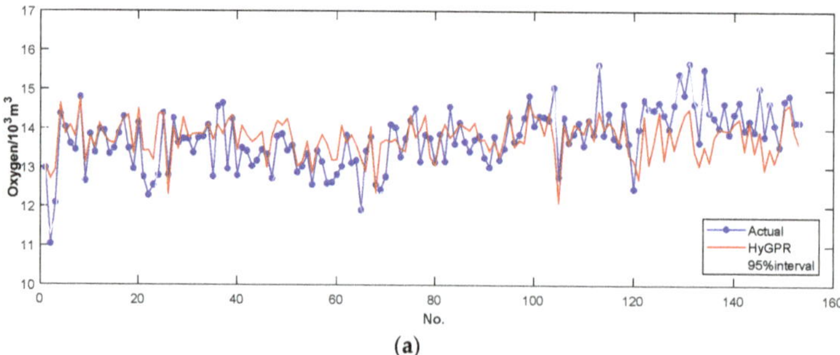

(a)

Figure 4. *Cont.*

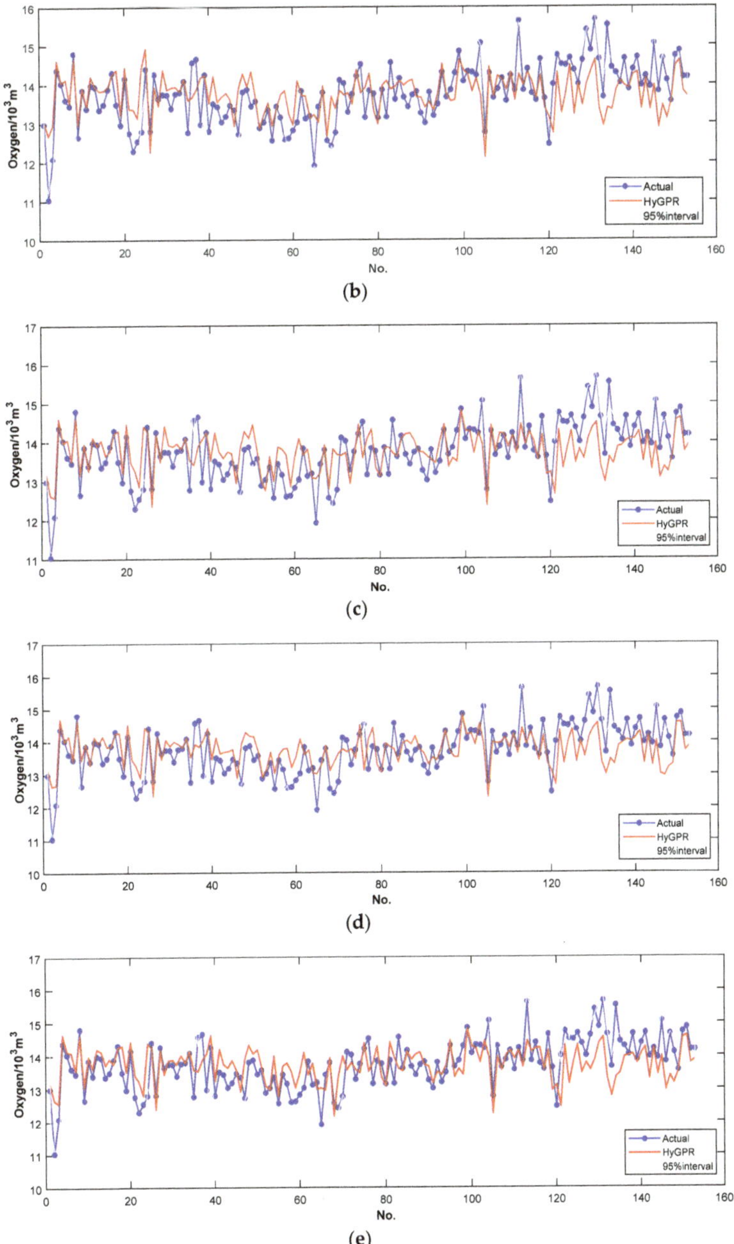

Figure 4. Forecasting results with candidate clusters: (**a**) $Q = 1$; (**b**) $Q = 2$; (**c**) $Q = 3$; (**d**) $Q = 4$; (**e**) $Q = 5$.

According to results listed in Table 2, we found that the RMSE, MAE and MAPE of the HyGPR model with different clusters were approximate, but the CPU time was reduced greatly when $Q > 1$. Since the HyGPR with four clusters could run successfully within the shortest time, we set $Q = 4$ in following computational experiments. In addition, we also found that most of the actual data located in the 95%

confidence interval was provided by HyGPR, which means the proposed model was able to make a probabilistic sense.

Table 2. Experimental results for candidate clusters (The best metrics are highlighted in bold).

Cluster Count	RMSE (10^3m^3)	MAE (10^3m^3)	MAPE (%)	CPU Time (S)
$Q = 1$	0.64	0.49	3.56	62.47
$Q = 2$	0.64	**0.48**	**3.51**	27.62
$Q = 3$	**0.63**	0.49	3.55	19.82
$Q = 4$	**0.63**	**0.48**	3.52	**13.32**
$Q = 5$	0.64	0.53	3.90	15.49

In order to further evaluate the prediction accuracy of the HyGPR model, we compared it with three benchmark models including MLR, ANN [27], and the support vector machine (SVM) [28]. The MLR model was created by the function *fitlm* in MATLAB. The function *fitnet* in MATLAB was adopted to construct ANN with three layers and 10 nodes in the hidden layer, in which the network parameters were optimized by the Levenberg–Marquardt method. The function *fitrsvm* in MATLAB was used to construct the SVM model with the radial basis function (RBF) kernel, in which the hyperparameters were automatically optimized by minimizing the five-fold cross-validation loss function.

To quantitatively select the best one from the testing models, the computational results of the point evaluation metrics were listed in Table 3. It can be observed that the proposed HyGPR model obtained the smallest of RMSE, MAE and MAPE, while the proposed MLR got the worst results of RMSE, MAE and MAPE. The above results focused on the prediction accuracy of the single valued point predictions (as shown in Figure 5).

Table 3. Performance evaluation of compared models on the test set (The best metrics are highlighted in bold).

Model	RMSE (10^3m^3)	MAE (10^3m^3)	MAPE (%)
MLR	0.70	0.54	3.95
SVM	0.69	0.53	3.92
ANN	0.68	0.52	3.84
HyGPR	**0.63**	**0.48**	**3.52**

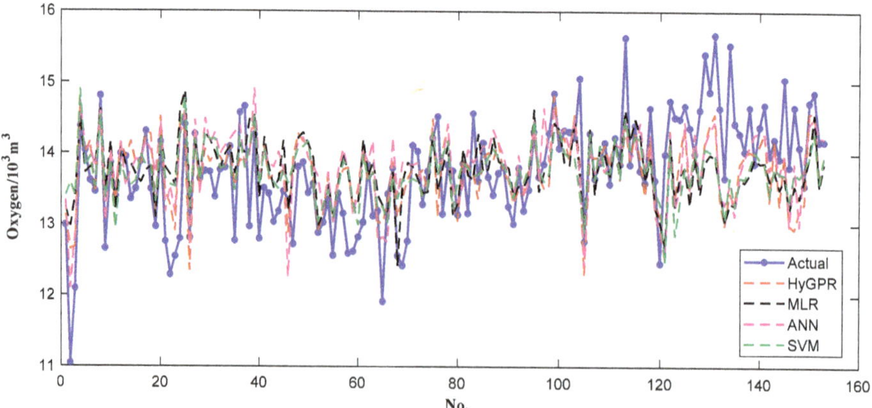

Figure 5. Point forecasting results of the oxygen consumption.

Usually, the MLR, ANN and SVM can only provide point estimations of future oxygen consumption. However, the HyGPR model can provide not only the forecasting point $\hat{y}_i$ but also the confidence interval $\left[\hat{y}_i^-, \hat{y}_i^+\right]$ of future oxygen consumption. To test its HRI within 95% confidence interval of HyGPR, we compared it with the standard GPR with the *squared exponential* kernel function. Table 4 listed the evaluation results of GPR and HyGPR. The HyGPR obtained the same value of RMSE and slightly better value of MAE, MAPE and HRI, but its computing speed is more than five times as fast as the standard GPR. Figure 6 showed the point forecasts and the corresponding 95% confidence intervals. It can also be observed that nearly all of the actual observations fell in the confidence intervals.

Table 4. Accuracy and interval metrics of GPR and HyGPR (The best metrics are highlighted in bold).

Model	RMSE ($10^3 m^3$)	MAE ($10^3 m^3$)	MAPE (%)	HRI (%)	CPU (S)
GPR	0.63	0.49	3.59	92.42%	75.74
HyGPR	0.63	**0.48**	**3.52**	**96.08%**	**14.14**

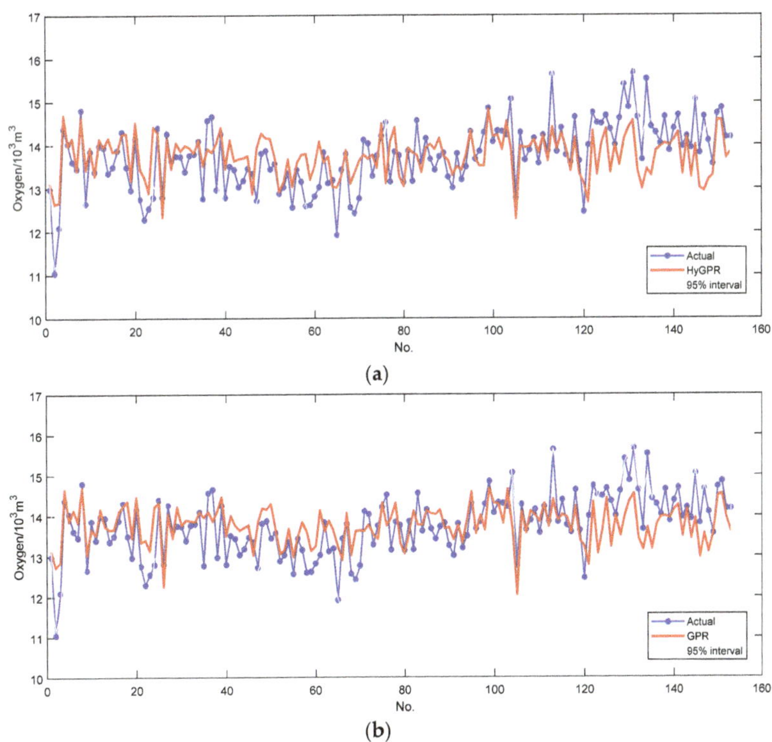

Figure 6. Interval forecasting results of oxygen consumption: (**a**) HyGPR; (**b**) standard GPR.

4. Conclusions

With the increased concerns on the management and optimization of energy systems, it is necessary for modern integrated iron and steel works to develop an accurate and robust model to forecast oxygen consumption. However, it is challengeable to directly forecast oxygen consumption with a simple regression model due to its intermittent and uncertain features. In this study, we introduce a novel hybrid model named HyGPR integrating MLR and GPR. In the proposed prediction model, the MLR model is developed to figure the global trend of the oxygen consumption, and the GPR is applied to explore the local fluctuation caused by noise. Additionally, to overcome the shortcoming of GPR on training speed, a K-means clustering method is applied to decompose the training dataset into a number of subsets. The effectiveness of the HyGPR was verified using the actual process data collected from a large integrated iron and steel works located in the north of China. Afterwards, HyGPR is compared with MLR, ANN, SVM and GPR. The results show that HyGPR can obtain the best point prediction metrics in terms of RMSE, MAE, and MAPE, and the better interval prediction performance in terms of HRI. Furthermore, it runs more than five times faster than the standard GPR. Therefore, it can be concluded that the proposed method is an effective tool to improve the forecasting accuracy and coverage. Moreover, HyGPR runs faster than the standard GPR model due to implementing the decomposition policy.

In future studies, we will investigate the following issues that may be meaningful for industrial application and scientific research:

(1). The online prediction model involving dynamic operating parameters, such as the position of oxygen lance, the pressure of oxygen blowing and the duration of oxygen blowing.

(2). The prediction model to forecast slopping events, which is very important to reduce production costs and environmental impacts.

(3). In this study, we assume that the noise of the steelmaking process is a Gaussian distribution. However, when it comes to the small-sample and high-dimensional data set, the assumption is incorrect. So it needs to develop a non-Gaussian prediction model applied in other environments.

Author Contributions: Methodology and writing, S.-L.J.; Data curation, X.S.; Supervision, Z.Z.

Funding: National Natural Science Foundation of China (No. 51734004, 61873042) and Fundamental Research Funds for the Central Universities (No. 2018CDXYCL0018).

Acknowledgments: We would like to thank the anonymous reviewers and the editors for their constructive and pertinent comments.

Conflicts of Interest: The authors declare no conflict of interest.

References

1. Han, Z.; Zhao, J.; Wang, W.; Liu, Y. A two-stage method for predicting and scheduling energy in an oxygen/nitrogen system of the steel industry. *Control. Eng. Pract.* **2016**, *52*, 35–45. [CrossRef]

2. Fan, J. Research on Production Scheduling and Charge Energy Consumption Prediction of Converter in Steelmaking Works. Master's Thesis, Chongqing University, Chongqing, China, May 2018. (In Chinese)

3. Han, Z.; Zhao, J.; Wang, W. An optimized oxygen system scheduling with electricity cost consideration in steel industry. *IEEE/CAA J. Autom. Sin.* **2017**, *4*, 216–222. [CrossRef]

4. Alam, M.; Naser, J.; Brooks, G. Computational fluid dynamics simulation of supersonic oxygen jet behavior at steelmaking temperature. *Metall. Mater. Trans. B* **2010**, *41*, 636–645. [CrossRef]

5. Ghosh, A. *Secondary Steelmaking: Principles and Applications*; CRC Press: Boca Raton, FL, USA, 2000.

6. Reis, M.; Gins, G. Industrial process monitoring in the big data/industry 4.0 era: From detection, to diagnosis, to prognosis. *Processes* **2017**, *5*, 35. [CrossRef]

7. Kadlec, P.; Gabrys, B.; Strandt, S. Data-driven soft sensors in the process industry. *Comput. Chem. Eng.* **2009**, *33*, 795–814. [CrossRef]

8. Kadlec, P.; Grbić, R.; Gabrys, B. Review of adaptation mechanisms for data-driven soft sensors. *Comput. Chem. Eng.* **2011**, *35*, 1–24. [CrossRef]

9. Duan, Y.; Liu, M.; Dong, M.; Wu, C. A two-stage clustered multi-task learning method for operational optimization in chemical mechanical polishing. *J. Process Control* **2015**, *35*, 169–177. [CrossRef]

10. Li, D. Perspective for smart factory in petrochemical industry. *Comput. Chem. Eng.* **2016**, *91*, 136–148. [CrossRef]

11. Liukkonen, M.; Hälikkä, E.; Hiltunen, T.; Hiltunen, Y. Dynamic soft sensors for NOx emissions in a circulating fluidized bed boiler. *Appl. Energy* **2012**, *97*, 483–490. [CrossRef]

12. Cox, I.J.; Lewis, R.W.; Ransing, R.S.; Laszczewski, H.; Berni, G. Application of neural computing in basic oxygen steelmaking. *J. Mater. Process. Technol.* **2002**, *120*, 310–315. [CrossRef]

13. Zhou, X.; Yang, C.; Gui, W. Modeling and control of nonferrous metallurgical processes on the perspective of global optimization. *Control. Theory Appl.* **2015**, *32*, 1158–1169.

14. Saxén, H.; Gao, C.; Gao, Z. Data-driven time discrete models for dynamic prediction of the hot metal silicon content in the blast furnace—A review. *IEEE Trans. Ind. Inform.* **2013**, *9*, 2213–2225. [CrossRef]

15. Liu, C.; Tang, L.; Liu, J.; Tang, Z. A dynamic analytics method based on multistage modeling for a BOF steelmaking process. *IEEE Trans. Autom. Sci. Eng.* **2018**, 1–13. [CrossRef]

16. Tian, H.X.; Mao, Z.Z. An ensemble ELM based on modified AdaBoost. RT algorithm for predicting the temperature of molten steel in ladle furnace. *IEEE Trans. Autom. Sci. Eng.* **2010**, *7*, 73–80. [CrossRef]

17. Han, M.; Liu, C. Endpoint prediction model for basic oxygen furnace steel-making based on membrane algorithm evolving extreme learning machine. *Appl. Soft Comput.* **2014**, *19*, 430–437. [CrossRef]

18. Laha, D.; Ren, Y.; Suganthan, P.N. Modeling of steelmaking process with effective machine learning techniques. *Expert Syst. Appl.* **2015**, *42*, 4687–4696. [CrossRef]

19. Yang, J.; Chai, T.; Luo, C.; Yu, W. Intelligent Demand Forecasting of Smelting Process Using Data-Driven and Mechanism Model. *IEEE Trans. Ind. Electron.* **2018**. [CrossRef]

20. Chen, J.; Chandrashekhara, K.; Mahimkar, C.; Lekakh, S.N.; Richards, V.L. Void closure prediction in cold rolling using finite element analysis and neural network. *J. Mater. Process. Technol.* **2011**, *211*, 245–255. [CrossRef]

21. Lei, Z.; Su, W. Mold Level Predict of Continuous Casting Using Hybrid EMD-SVR-GA Algorithm. *Processes* **2019**, *7*, 177. [CrossRef]

22. Friedman, J.; Hastie, T.; Tibshirani, R. *The Elements of Statistical Learning*; Springer: Berlin, Germany, 2001.

23. Williams, C.K.; Rasmussen, C.E. *Gaussian Processes for Machine Learning*; MIT Press: Cambridge, MA, USA, 2006.

24. Schulz, E.; Speekenbrink, M.; Krause, A. A tutorial on Gaussian process regression: Modelling, exploring, and exploiting functions. *J. Math. Psychol.* **2018**, *85*, 1–16. [CrossRef]

25. Shawe-Taylor, J.; Cristianini, N. *Kernel Methods for Pattern Analysis*; Cambridge University Press: Cambridge, UK, 2004.

26. Rasmussen, C.E.; Nickisch, H. Gaussian processes for machine learning (GPML) toolbox. *J. Mach. Learn. Res.* **2010**, *11*, 3011–3015.

27. Laha, D. ANN modeling of a steelmaking process. In Proceedings of the International Conference on Swarm, Evolutionary, and Memetic Computing, Chennai, India, 19–21 December 2013; Springer: Cham, Switzerland, 2013.

28. Smola, A.J.; Schölkopf, B. A tutorial on support vector regression. *Stat. Comput.* **2004**, *14*, 199–222. [CrossRef]

Article

Machine Learning-Based Prediction of a BOS Reactor Performance from Operating Parameters

Alireza Rahnama [1,*], Zushu Li [2] and Seetharaman Sridhar [1,3]

[1] AI-GARISMO, 1 Sandover House, 124 Spa Road, London SE16 3FD, UK; sseetharaman@mines.edu or sridhar.seetharaman@ai-garismo.com
[2] WMG, The University of Warwick, Gibbet Hill Road, Coventry CV4 7AL, UK; z.li.19@warwick.ac.uk
[3] George S. Ansell Department of Metallurgical and Materials Engineering, Colorado School of Mines, Golden, CO 80401, USA
* Correspondence: alireza.rahnama@ai-garismo.com

Received: 25 November 2019; Accepted: 13 March 2020; Published: 23 March 2020

Abstract: A machine learning-based analysis was applied to process data obtained from a Basic Oxygen Steelmaking (BOS) pilot plant. The first purpose was to identify correlations between operating parameters and reactor performance, defined as rate of decarburization (dc/dt). Correlation analysis showed, as expected a strong positive correlation between the rate of decarburization (dc/dt) and total oxygen flow. On the other hand, the decarburization rate exhibited a negative correlation with lance height. Less obviously, the decarburization rate, also showed a positive correlation with temperature of the waste gas and CO_2 content in the waste gas. The second purpose was to train the pilot-plant dataset and develop a neural network based regression to predict the decarburization rate. This was used to predict the decarburization rate in a BOS furnace in an actual manufacturing plant based on lance height and total oxygen flow. The performance was satisfactory with a coefficient of determination of 0.98, confirming that the trained model can adequately predict the variation in the decarburization rate (dc/dt) within BOS reactors.

Keywords: machine learning; artificial intelligence; neural network; BOS reactor; steelmaking

---

1. Introduction

The processing of lower grade ores is a topic of particular interest, as fluctuation in raw material cost is a key challenge to sustainability in the steel industry. Raw materials flexibility is to a great extent enabled in the basic oxygen steelmaking (BOS) process, wherein oxidizable impurities, such as phosphorus (P) and silicon (Si), are separated into a slag phase. However, the primary purpose of the BOS process is to convert pig iron into crude steel and, therefore, impurity removal being a secondary function of the converter, has to be balanced by the decarburization process.

As a result, processing parameters have to be balanced, and many different chemical reactions compete for the available oxygen. There are also many processing parameters that are interconnected in complicated ways that cannot be readily predicted. The BOS process is in general advised by static mass- and heat-balanced models, which predict, based on inputs, the resultant end-point. The input parameters include at least the quantities of hot metal, steel scrap, iron ore, fluxes and oxygen to blow, while the resultant end-point is the temperature, weight and composition of steel produced. The process control is supported by various measurements, such as in-blow sampling, end-blow sampling, oxygen flow rate/ total oxygen blown, lance height, waste gas flow rate, waste gas pressure, waste gas composition, etc. The operating information is recorded and stored in a plant manufacturing system normally at the interval of one second. Dynamic process modeling of the BOS process, depending on the researchers, can be based on thermodynamics [1], multizone reactions [2–4], empirical or semi-empirical plant data regression, in combination with computational fluid dynamics (CFD) in some cases [5]. Data-driven modeling techniques have been applied in the dynamic modeling of basic oxygen processes. These include the artificial neural network (ANN) [6–8], fuzzy logic [9]), support vector machine [10,11] and case base reasoning [12]. However, most of these models are limited to the end-point control of carbon and temperature.

The pilot plant work published by Millman et al. [2] ("IMPHOS" or Improved phosphorus refining) provides a set of data on top-blown oxygen steelmaking under conditions similar to an actual BOS plant. The trials were carried out in a six-ton pilot converter at the MEFOS in Sweden and provides data for heats through samples obtained from seven vertical positions in the vessel at 2-min intervals. In addition, 28 process data parameters (including decarburization rate) were recorded every second. The major parameters recorded are the total of materials added as a function of time (hot metal and fluxes); oxygen flow rate/total amount of oxygen blown (total oxygen flow); lance height; waste gas (temperature; flow rate and composition of CO, CO_2, N_2 and O_2) and the data calculated from the above measurements, for example, dc/dt (the decarburization rate).

Using machine learning (ML) approaches to predict material properties is a research field that has been increasing in the last years [13]. Thermodynamic stability of perovskite, transition temperatures of inorganic glasses, glass-forming ability of metallic alloys and interphase precipitation in HSLA (High Strength Low Alloy) steels, among others, are examples of material properties studied by machining learning approaches [14–18]. Machine learning algorithms enable us to design new grades of alloys purely based on past data, either available as an open dataset or reported in the scientific literature [19].

The objective of this study is to carry out a machine learning (ML) analysis on the data generated from the six-ton pilot BOS trials (IMPHOS data) with the purposes of (i) identifying correlations between features and (ii) based on a training data and test data developing a neural network-based regression model (algorithm) to predict the decarburization rate. The tools that we employed in this study were Microsoft Power BI for statistical analysis; Python (Scikit-learn, Matplotlib and Seaborn) for performing correlation analysis and statistical analysis and Microsoft Azure Machine Learning Studio for regression models and their subsequent performance evaluation [20]. The algorithm developed here is used to successfully predict the decarburization rate (dc/dt) in an actual manufacturing plant based on two operating parameters of lance height and total oxygen flow.

2. The Multiphysics of the Basic Oxygen Steelmaking Process

The overall purpose of the BOS process is to refine hot metal (pig iron) to a steel of desired chemistry (C, Si, Mn, P and S) and temperature in a controlled manner in a refractory converter (as shown schematically in Figure 1). In the BOS process wherein C-rich pig iron is converted to steel, pure oxygen is injected though a water-cooled lance into the molten metal bath, and the bath is covered by molten oxides (slag phase). The role of the oxygen is to selectively react with oxidizable elements and primarily remove the C (from >4% to less than 0.1%) as a gas phase but also remove other unwanted impurities, stemming from the iron ore as oxides, that separate and dissolve into the slag phase.

The kinetics of the BOS process is very complex, because it involves multiple simultaneous processes. Simultaneous multiphase interactions, heat and mass transfer, gas-slag-metal chemical reactions in multiple zones and vigorous fluid flow caused by the impingement of the oxygen jet occur in a BOS reactor at high temperatures. In addition, it is a dynamic and transient process, which makes the kinetics involved in a reactor more complex. Direct measurements of temperatures and chemistries are very difficult due to the nature of the process, which involves harsh conditions. That is why many researchers have been trying to address these difficulties through modeling the process.

A multiphysics description of the converter process, with the ultimate aim of predicting carbon content in the melt, will have to involve several sub-models at different scales. These models will have to capture various phenomenon occurring in the gas/metal, metal/slag and metal/slag/gas mixtures, as well as transport processes in the bulk metal and bulk slag. The models include, but are not limited to, the following:

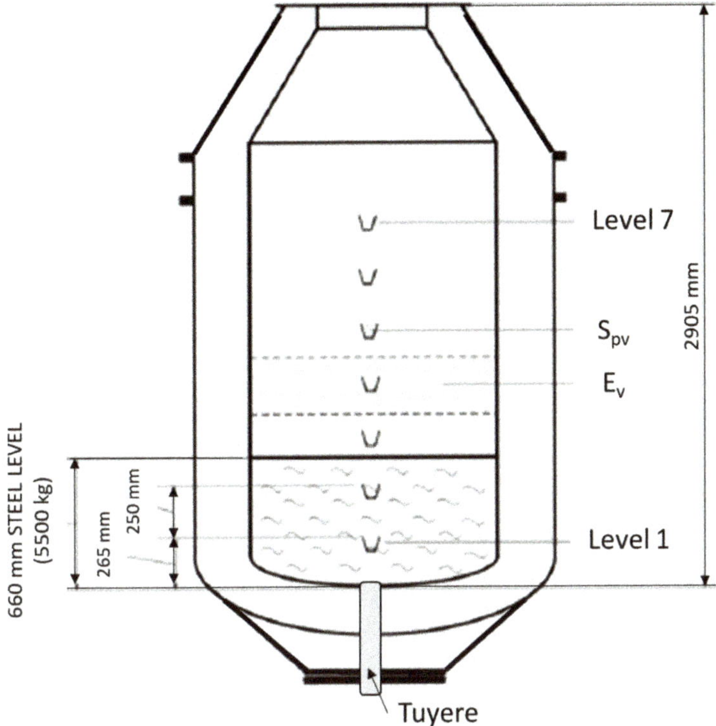

Figure 1. A schematic of the pilot plant BOF converter used in the Improved Phosphorus-refining (IMPHOS) trials with dimensions and indicated location heights of the sampling levels [2].

(i) Multiphase fluid flow in the various multiphase mixtures. Since this is a complex and computational task by itself, it is often simplified to 2D and informed by physical water models [5,21,22].

(ii) Reaction kinetics and competition between the injected oxygen and various dissolved elements, such as C, but also P, Si, Mn, etc. [4,23].

(iii) Microscale interactions between the phases, which include the separation of elements and precipitated phases across interfaces and emulsification [4,24,25].

The decarburization process mainly takes place in two reaction zones: namely, the jet impact zone and the gas-slag-metal emulsion zone [4,24]. At the emulsion zone, a sequence of chemical reactions take place: Firstly, FeO reacts with CO to form CO_2 at the slag-gas interface. Secondly, CO_2 transfers from the slag-gas interface to the gas-steel droplet interface via the gas phase. Subsequently, CO_2 reacts with carbon in an iron droplet at the metal-gas interface, and the formed CO is transferred back to the slag-gas interface via the gas phase. It should be noted that there is disagreement among researchers about the description of the mechanisms and the rate-controlling step(s) under different operating conditions [26]. Some other studies proposed that the reaction rate is controlled by the interfacial area change because of material exchanges during the process. The phase field models developed according to the interfacial instability normally consider only the interfacial tension and fluid flow [24]. The coupling of models across scales is a hard enough (and computationally intensive) problem, which can be simplified and described under simplified process windows. However, the predictive capability under widely ranging process parameters is not possible. Consequently, all the previous studies, although mechanistic in nature, have different levels of simplification to address the complex phenomena in the BOS reactor.

The removal of elements (C, Si, Mn, P and S) from pig iron in the industrial BOS process is mainly controlled by two operating parameters: lance height (i.e., the height of the lance tip above the static steel bath) and oxygen blowing rate (or total oxygen flow/blown). The control strategy can be varying one or both of these two parameters. The lance profile (i.e., lance height as a function of blowing time) is critically important for the efficient production of high-quality steel. For example, a high lance position can help stir and oxidize the slag with a high FeO content, which can accelerate the removal of carbon. A low lance height can substantially increase the impingement of the oxygen jet into the metal bath and splash the metal droplets into the slag layer, which accelerates the formation of the emulsion zone and the removal of carbon and other elements through the interfacial reactions between the massive numbers of metal droplets with emulsified slag [27]. The refining behavior (e.g., the decarburization rate) has a close relationship with the lance profile and oxygen blowing rate (or total oxygen flow). Thus, in this work, we intend to develop a novel algorithm to predict the refining behavior of the BOS reactor (using the decarburization rate as an example) by using machine learning to analyze the massive dataset of operating parameters, including lance height and total oxygen flow.

Using machine learning, we propose a technique that does not employ any simplifying assumptions. On the contrary, the algorithm was trained on a real dataset. In addition, the machine learning algorithm does not require to take into account any of the physics involved in the process. It actually provides a ring road to all the complexities involved in a BOS reactor. In addition, as shown later, the technique is able to predict the decarburization rate precisely. The aim of the current study is, thus, twofold: (1) Firstly, to identify the correlations and trends from a set of processing parameters on carbon removal. (2) Secondly, to try and circumvent the granularity of multiphysics and develop an artificial intelligence-based predictive model for carbon removal.

3. Dataset

A six-ton BOF (Basic Oxygen Furnace) pilot plant converter located at Swerea MEFOS, Sweden, as part of the European commission funded project IMPHOS: Improved Phosphorus refining [2]. In this reactor, operating conditions, such as oxygen blow rate (total oxygen flow), lance height, the quantities of hot metal and fluxes used and off-gas analysis, are recorded. Gas/slag/metal emulsion samples were taken from seven various heights and at 2-min intervals from the start of blowing during a blow via robotic delivery. The sampler lances consist of an inner and outer structure of three mild steel bars, the inner being joined to the sides of inline sample pots and the outer joined to the sides of the disc sample pot lids. The sampler lance is lowered into the converter through an opening in the top (with a slight offset from the oxygen lance), with the lids in the closed position. Once lowered into the position, the outer three bars are retracted, lifting the lids in situ and allowing the sample pots to fill. A schematic of the converter is given in Figure 1 with dimensions. The converter had a lining of magnesia-carbon (fired and fused) bricks. Gas inlets include a single bath agitation tuyere, blowing nitrogen at a rate of $0.5~\text{Nm}^3\text{min}^{-1}$, and a water-cooled oxygen lance blowing oxygen at a norminal rate of $17~\text{Nm}^3\text{min}^{-1}$. During the blow, the following was measured as output data from the reactor: oxygen flow rate/total oxygen flow; lance height and waste gas (temperature; flow rate and composition of CO, CO_2, N_2 and O_2).

There are various models, such as static (mass balance model, thermodynamic model, etc.); empirical (regression based on individual plant data) and dynamic. These models are intended to predict the end-point of steel or the steel/slag compositions as a function of blowing time and to control/optimize BOF steelmaking. In the BOF process, the blown oxygen is used for decarburization; the removal of other impurities of Si, Mn and P; the formation of iron oxides in the slag and post-combustion. A typical example is off-gas information. The industry has been exploring the use of off-gas information for process control. The most important one is to use the off-gas composition to monitor the decarburization rate in the BOS converter. If the decarburization rate can be precisely controlled or predicted, then under the known oxygen-blowing conditions, the amount of oxygen in the slag (iron oxides) and the amount of oxygen to remove other elements, especially for phosphorus, can be predicted. The decarburization rate can be calculated by the off-gas analysis (CO, CO_2, N_2 and O_2). Then, we will link the decarburization rate (calculated from off-gas composition) to the operating parameter(s). This will provide the foundation to predict the decarburization rate (dc/dt) and, as a result, dynamically control the converter operation. The features we are investigating from the dataset are presented in Table 1. The other operating conditions are listed in Table 2, including the chemistries of hot metal; steel and slag; temperature; charges (hot metal, scrap and fluxes) and lance height.

Table 1. The operating parameters and their features obtained from the six-ton pilot plant BOF converter system.

Parameter	Feature
Charge Time (min)	Blowing time (in minutes from O_2 ignition)
dc/dt (kg C/min)	Instant decarburization rate
Oxygen Yield (%)	Instant oxygen yield
Total Oxygen flow rate (Nm^3/min)	Instant oxygen flow rate
Total C removal (kg)	Calculated total C removed from the start of O_2
Total N_2 flow rate (Nm^3/min)	Instant N_2 flow rate
Total O_2 (Nm^3/min)	Total blown oxygen volume
Total N_2 (Nm^3/min)	Total bottom blown nitrogen volume
Total propane (Nm^3/min)	Total blown propane volume
Lance height (mm)	Instant lance height from the point of calculated metal bath level
dC/dt (kg C/s)	Calculated from off-gas composition
dO/dt (kg O/s)	Calculated from off-gas, oxygen for decarburization
dO_s/dt (kg O/s)	Calculated from off-gas composition, oxygen into slag
Temp. waste gas (°C)	Off-gas temperature (measured after water cooling)
CO waste gas (%)	Measured waste gas composition
CO_2 waste gas (%)	Measured waste gas composition
O_2 waste gas (%)	Measured waste gas composition

Note: dC/dt (kg C/s) is calculated from the waste gas composition, which was recorded in the operating system in every second, while dc/dt (kg C/min) is converted from dC/dt (kg C/s). The decarburization rate is generally expressed as dc/dt (kg C/min), and therefore, the target of this study is to predict dc/dt (kg C/min) by a machine-learning (ML)-based algorithm from the operating parameters. It should be pointed out that dC/dt (kg C/s) is still mentioned in this study for the purpose of explanation.

In order to validate the correlation between the dc/dt and the operating parameters (the ML-based methodology, as explained in Section 4, Methods, to predict dc/dt from operating parameters) established from the six-ton BOF converter, production information has been taken from a 330-ton converter at the Tata Steel UK Port Talbot Plant. The operating parameters and the resultant information include oxygen flow rate; total oxygen flow; off-gas analysis (temperature, rate, composition CO, CO_2, N_2 and O_2) and dC/dt (kg C/s) calculated from the off-gas analysis. A total of 1100 data points were obtained at different times for each value corresponding to Table 1, from the 6-ton pilot reactor, which were used for training and test data. In addition, a total of 1200 points corresponding to known conditions in the 330-ton full scale reactor were used as additional test data to investigate if the trained model could predict the decarburization rate in a scaled-up system. It should be noted that differences in the levels/quantities of the parameters exist between the pilot and industrial converters; for example, the heat size (6t for pilot converter and 330t for industrial converter), blowing time (~18 min for the 6t pilot converter and ~20 min for the 330t industrial converter), lance height (110~180 mm for the 6t pilot converter and 2.0~2.6 meters for the 330t industrial converter), O_2 flow rate (14.00~17.30 Nm^3/min for the 6t pilot converter and 598~1025 Nm^3/min for the 330t industrial converter) and total oxygen flow (278 Nm^3 for the 6t pilot converter and 18863 Nm^3 for the 330t industrial converter). However, the parameters and their features recorded in their systems are similar for both conditions, and in this paper, we use the dataset from one heat of both the 6t pilot and the 330t industrial converters, respectively, to develop and demonstrate the machining learning-based model for the prediction of the decarburization rate from operating parameters under different conditions.

Table 2. The operating conditions for both the six-ton (6t) and 330t converters.

6t pilot BOF.	C	Si	Mn	P	S	T (°C)	
Hot metal (%)	3.78~4.25	0.41~0.88	0.39~0.48	0.067~0.095	0.037~0.081	1272~1316	
Steel (%)	0.01~0.41	0~0.15	0.05~0.27	0.008~0.042	0.018~0.035	1669~1772	
	CaO	SiO$_2$	FeO	MnO	Al$_2$O$_3$	MgO	P$_2$O$_5$
Slag (%)	26.2~52.3	6.7~19.2	8.3~30.5	2.6~4.1	0.7~1.6	4.6~17.7	0.73~1.62
330t BOF	C	Si	Mn	P	S	T (°C)	
Hot metal (%)	3.60~5.18	0.30~1.20	0.11~0.52	0.058~0.125	0.025~0.072	1341~1412	
Steel (%)	0.021~0.55	0~0.10	0.03~0.21	0.001~0.057	0.011~0.032	1579~1738	
	CaO	SiO$_2$	FeO	MnO	Al$_2$O$_3$	MgO	P$_2$O$_5$
Slag (%)	36.07~55.35	9.13~19.58	12.15~31.34	2.04~5.62	0.64~4.97	3.61~10.73	0.94~2.38
	6t pilot BOF		330t BOF				
hot metal	4410~5170 kg		269.6~302.5 t				
scrap	500~750 kg		46.0~85.5 t				
lime	250~350 kg		5~25 t				
dolomet	0~30 kg		0~11.5 t				
iron ore	0 kg		0~6.5 t				
total oxygen	263~307 Nm^3		12265~21641 Nm^3				
Lance height	110~180 mm		2.0~2.6 m				

4. Method

While neural networks are mainly known for applications in deep learning, they can be easily used for regression problems and are especially suitable when linear regression does not work. Neural network regression is another supervised learning algorithm available in Microsoft Azure Machine Learning Studio and requires a label column which has to be numerical [20]. Since our label column (dc/dt) was numerical, and the results generated by classical linear regression had low accuracy, we used a neural network regression. Linear regression resulted in a coefficient of determination of 0.45 (Table 3, column 2), which was much lower than that by using neural network regression (0.99) and deemed to be below what is acceptable. The output of a single neuron has a form of $g(\Sigma_j w_{ji} x_j)$, where w_{ji} are the weights and x_j are the input. A continuous activation function, e.g., sigmoid function of a form of $1/(1 + e^{-x})$, was employed. We used "Parameter Range" in the Create trainer mode and subsequently employed the Tune Model Hyperparameters module to iterate over the possible combinations of parameters to achieve the optimal configuration. In "Hidden layer specification" we selected "fully connected case" to create a model that has exactly one hidden layer, and the output layer is fully connected to the hidden layer, and the hidden layer is fully connected to the input layer. Number of nodes in the hidden layer and learning rate were used as the hyperparameters. The optimized number of hidden nodes was calculated to be 100. Another hyperparameter that was tuned was the learning rate. Learning rate is the step taken at each iteration before correction. A large value of the learning rate makes the convergence faster; however, it can result in overshooting the local minima. The tuned learning rate was calculated to be 0.1. Min-max normalizer was selected to linearly rescale each feature to the [0, 1] interval. Rescaling to the [0, 1] interval was carried out by shifting the values of each features, i.e., the minimum value is 0, and then dividing by the new maximum value. The error after each iteration is calculated and "back-propagated" to the network using the chain rule.

Table 3. Evaluation metrics of the predicted values of the dc/dt by using different features in the pilot dataset and using different regression methods.

Evaluated Metrics	Pilot Dataset with All Features [1]	Pilot Dataset with All Features [2]	Pilot Dataset without dc/dt [2]	Pilot Dataset with Total O_2 Flow and Lance Height Only [2]	Industrial Dataset with Total O_2 Flow and Lance Height Only [2]
Mean Absolute Error	0.12	0.029	0.030	0.034	0.25
Root Mean Square Error	0.51	0.043	0.055	0.060	0.62
Relative Absolute Error	0.42	0.005	0.006	0.008	0.04
Relative Square Error	0.48	0.000046	0.00006	0.0001	0.009
Coefficient of Determination	0.45	0.99	0.99	0.97	0.98

[1] Using linear regression and [2] using neural network regression.

5. Results and Discussion

5.1. Machine Learning Predictions of the Decarburization Rate dc/dt

Figure 2 shows the correlation matrix between different features, which was generated by the available dataset. The correlation between two datasets varies between -1 and 1, with 0 implying no correlation. Correlation of +1 implies a perfect positive correlation (e.g., as x increases, so does y), and -1 implies a perfect negative correlation (as x increases, y decreases). Dark blue implies a strong positive correlation, and lighter pink shows a strong negative correlation. For example, there is a strong positive correlation between Total O_2 flow and dO/dt, or there is a strong negative correlation between O_2 waste gas and dc/dt.

Total oxygen flow has nearly perfect positive (>0.9) correlation with dc/dt (kg C/min) or dC/dt (kg C/s), total C removed, dO/dt, dOs/dt (0.71) and waste gas CO_2 composition. The total oxygen flow was mainly used for decarburization escaping from the reactor as waste gas and remaining in the slag (oxidization of Si, Mn, Fe and P elements in the liquid metal). The former is directly linked with parameters dc/dt (or dC/dt), total C removed, dO/dt and waste gas CO_2 composition, while the latter is in the form of dOs/dt. Furthermore, dc/dt (or dC/dt) has nearly perfect positive correlation with total O_2 flow and waste gas CO_2 composition and nearly perfect negative correlation with lance height and waste gas O_2. Except the oxygen from steel scrap and iron ore coolant, the main oxygen comes from the total oxygen blown through the lance that is also the main oxygen source for decarburization. Thus, the dc/dt (or dC/dt) has a nearly perfect correlation with the total O_2 flow. During the pilot plant experiment, O_2 was blown at a fixed flow rate, and the refining performance in the converter was controlled by adjusting the lance height. The decarburization mainly occurred in two zones of a hot spot zone (at the vicinity of the location where the lance releases oxygen to the bath) and gas-slag-metal droplet emulsification zone (where the available area for slag/metal/gas reaction is high). Lower lance height increases the hot spot zone and the amount of metal droplets in the emulsification zone, and the latter increases the decarburization in the gas-slag-metal droplet zone. Therefore, the overall decarburization rate increases with decreasing the lance height, which explains the observed negative correlation. The decarburization rate is calculated from the waste gas composition according to the equation dc/dt $\left(\frac{dc}{dt} = (CO + CO_2) \times waste\ gas\ flow\ rate \times \frac{12}{22.4} \right)$. This explains well the perfect positive correlation between dc/dt (or dC/dt) and waste gas CO_2 concentration. From the above analysis, both the total O_2 flow and the lance height are the controlling parameters for the decarburization, which indicates that the probability of predicting the decarburization rate by the combination of both parameters (see Section 5.3. Prediction of the dc/dt After Excluding Parameters). Finally, dO/dt has nearly perfect positive correlation with the total O_2 flow, dc/dt (or dC/dt) and waste gas CO_2 composition but nearly perfect negative correlation with the lance height and waste gas O_2 content (similar explanation to that of the dc/dt or dC/dt dependence upon the parameters).

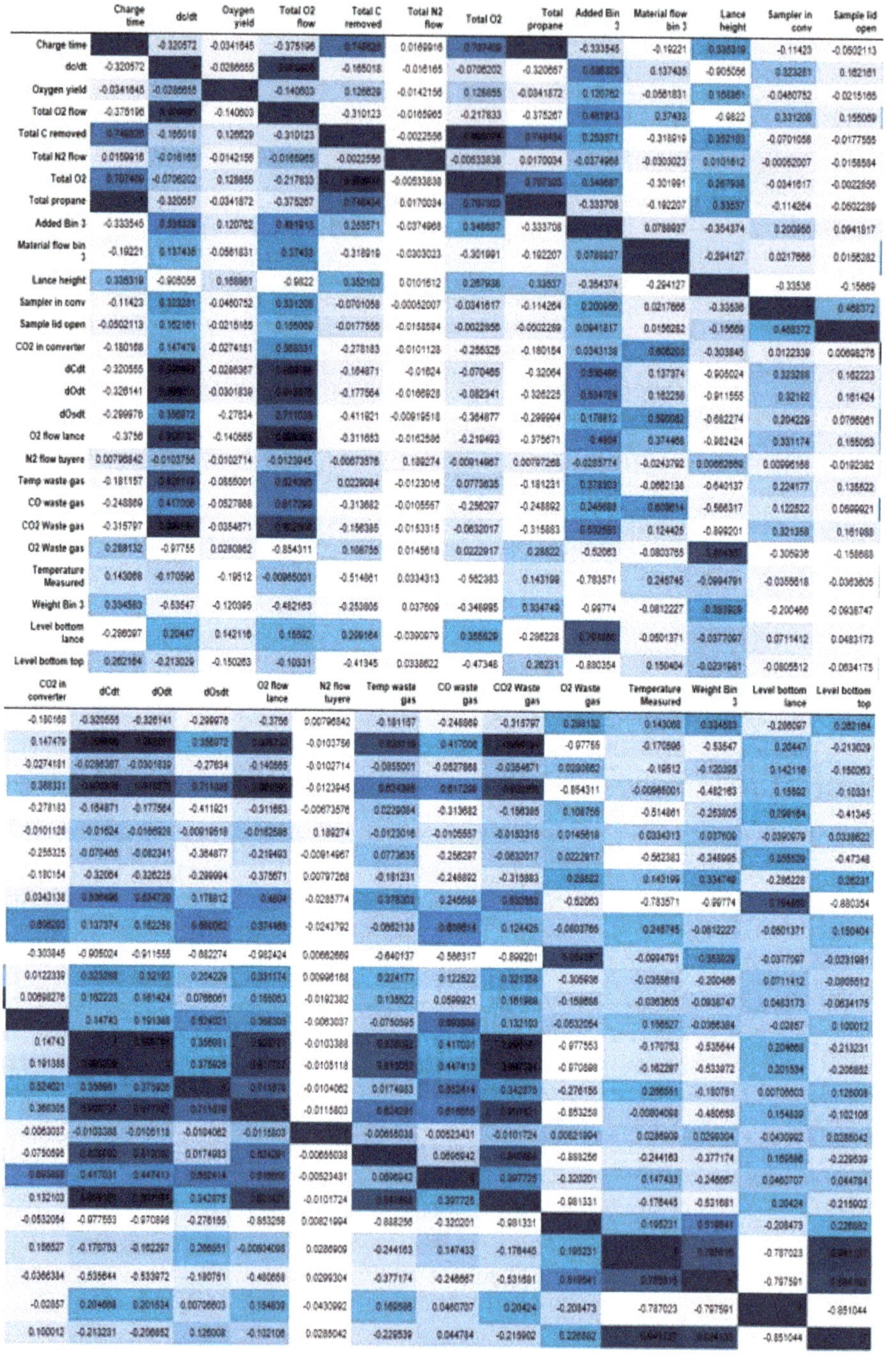

Figure 2. The correlations between different features within the dataset from the six-ton (6t) pilot plant trial. The figure has been vertically split into two parts, and the top and the bottom shown above are actually the left and right parts of the figure, respectively.

5.2. Prediction of dc/dt With All the Features Included in the Dataset

Initially, we used all the features present in the dataset (including dC/dt) to predict the value of the dc/dt. Figure 3 shows a statistical comparison between the actual values of the dc/dt (Figure 3a) and the predicted ones (Figure 3b). The blue bars show the histogram of the values. The green shadowed area shows the cumulative distribution function (CDF) and the blue shadowed area shows the probability density function (PDF). As shown in the figure, the predicted functions closely follow those of the actual values of the dc/dt. Both histograms show a frequency around 700 for a dc/dt value close to 0. In addition, both the histograms of the actual value of dc/dt and that of the predicted value show a bump in frequency around 100 for dc/dt values close to 9.6 and a bump at a frequency about 100 for dc/dt values around 17. As evident from Figure 3, the CDF of the actual values and that of the predicted values have a very similar shape. This confirms that the algorithm successfully captured the statistics of the dataset.

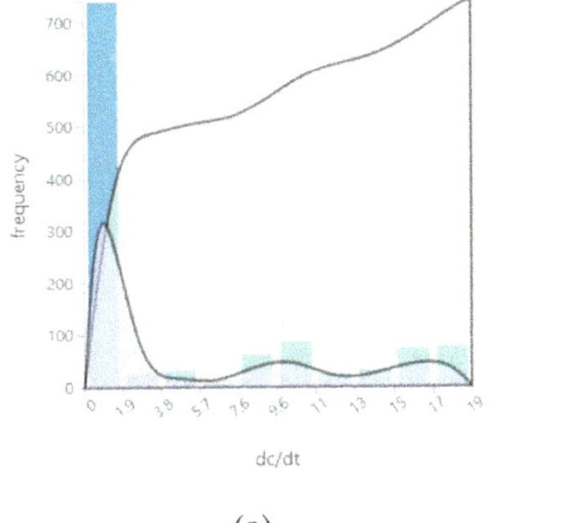

(a)

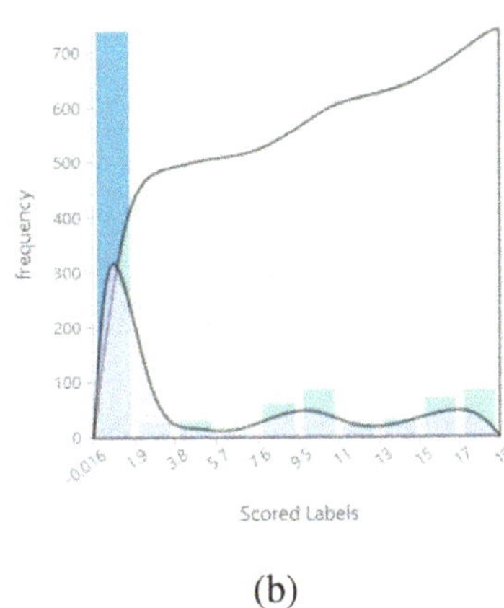

(b)

Figure 3. Histogram (blue), cumulative distribution function (green) and probability density function (purple) for (**a**) the actual values of the dc/dt and (**b**) the predicted values of the dc/dt for the pilot dataset with all the features included.

Table 4 compares the statistical metrics between the actual and predicted values of the dc/dt. Although the statistical metrics for the predicted values are slightly higher, it can be concluded that the machine-learning model successfully captured the statistics of the values of the dc/dt. This is further confirmed by Figure 4, the scatter plot, showing that the neural network regression model could accurately predict the values of the dc/dt. It should be noted that a scatter plot that compares the predicted values of the test set with the "true" values of the target is one the main metrics to evaluate a model performance. As is shown in Figure 4, the scored label as a function of the "true" value of the dc/dt follows a y = x line, meaning the model performed very well.

Table 4. Statistical comparison between the actual values of the dc/dt with the predicted values for the pilot dataset with all the features included.

	Mean	Median	Min	Max	Standard Deviation
Actual Statistics	4.43	0.081	0	19.105	6.42
Predicted Statistics	4.45	0.092	0	19.09	6.43

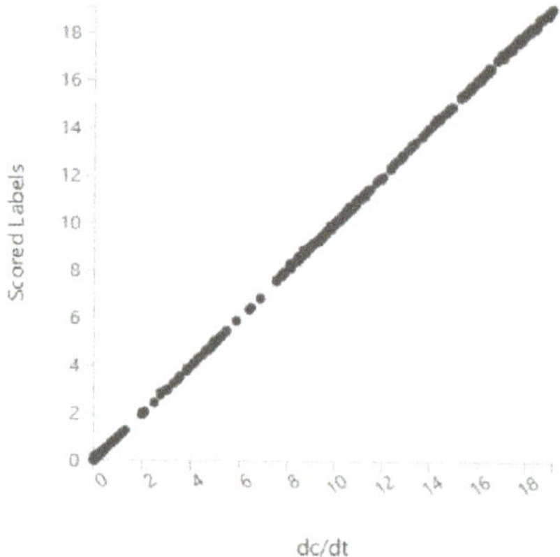

Figure 4. Scatter plot comparing the predicted values of the dc/dt using the neural network method with the actual values of the dc/dt for the pilot dataset with all the features included.

Figure 5 shows the error histogram of the neural network regression model. Errors with a value of 0.000049 had the highest frequency, confirming the excellent performance of the model. Table 3 (column 3) shows the performance metrics of the neural network regression model. The metrics were recorded to be 0.029 for the mean absolute error, 0.043 for the root mean squared error, 0.005 for the relative absolute error and 0.0046 for the relative squared error. The coefficient of determination for the model was calculated to be 0.99, which shows the excellent performance of the neural network regression algorithm.

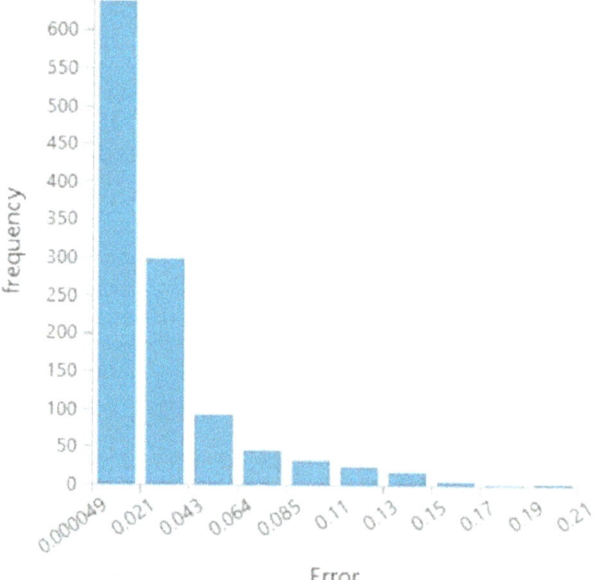

Figure 5. Error histogram of the predicted values of the dc/dt by using all the features included in the pilot plant dataset.

We used the permutation method to measure the feature importance for the prediction of the dc/dt. The feature importance values are shown in Table 5. It was computed that the dC/dt had the highest importance, with a value of 9.06. In second place stands the dO/dt with a value of 0.16.

Table 5. Feature importance for predicting the values of the dc/dt in the pilot dataset with all the features.

dC/dt	dO/dt	Oxygen Yield	CO_2 Waste Gas	CO Waste Gas	Total O_2 Flow	dO_s/dt	Temp. of Waste Gas	Total O_2	O_2 Waste Gas
9.06	0.16	0.07	0.04	0.004	0.002	0.0006	0.0002	0.00014	0.00013

5.3. Prediction of the dc/dt After Excluding Parameters

In order to establish whether the dc/dt could be predicted with reasonable accuracy with fewer parameters, parameters were successively removed. As discussed in the previous section, the dC/dt had a very high prediction power for the values of the dc/dt. However, it was anticipated that the dC/dt might lead to data leakage, as the dC/dt is calculated from the value of the dc/dt. Thus, the feature dC/dt was removed in order to measure the performance of the neural network regression model (Table 3, column 4). It was found that the scored labels (predicted values) have the very similar histogram, CDF and PDF graphs to those of the actual values of the dc/dt. Similar to the previous calculations, the statistical parameters belonging to the predicted values were slightly higher; however, the difference is very low, and it can be thus said that the model was able to capture the statistics of the data when the feature dC/dt was removed.

Then, we attempted to predict the value of the dc/dt using only two features: namely, total O_2 flow and lance height, because these are the two inputs that are controllable in an industrial reactor. The oxygen blown into the converter will be used for the decarburization (dO/dt), oxidization of other elements into the slag (dOs/dt), oxygen in the waste gas, etc.; therefore, in the prediction of the dc/dt, the features of the dO/dt, dOs/dt, etc. are excluded. Figure 6 shows the scatter plot of the dc/dt versus the predicted values. As is evident from the figure, except for some values of the dc/dt where the predicted values were slightly lower, for most of the values, the neural network regression model was able to predict the dc/dt with a good accuracy.

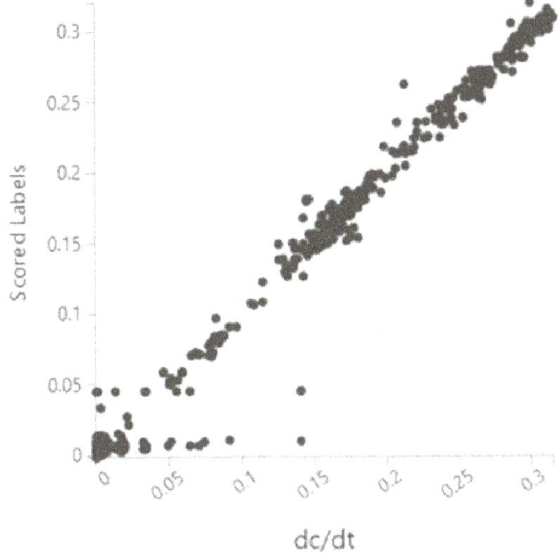

Figure 6. Scatter plot comparing the predicted values of the dc/dt using the neural network method with the actual values of dc/dt by using the two features of total oxygen flow and lance height in the pilot dataset.

Figure 7 shows the error histogram for this prediction. The most frequent error was 0.0000033. Table 3 (column 5) shows the performance metrics for the prediction of the dc/dt using only two features. For this computation, the mean absolute error was calculated to be 0.034, root mean squared error to be 0.06, relative absolute error to be 0.008 and relative squared error to be 0.0001. The coefficient of determination was computed to be 0.97. These performance metrics showed that we were able to successfully predict the value of the dc/dt using only the two variables of total oxygen flow and lance height.

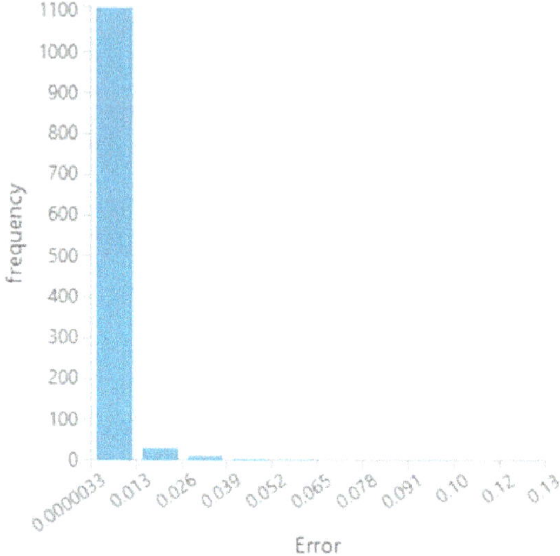

Figure 7. Error histogram of the predicted values of the dc/dt by using the two operating parameters of total oxygen flow and lance height in the pilot plant dataset.

5.4. Prediction of the dc/dt for an Industrial Dataset

We used a dataset that was acquired from an industrial reactor to evaluate the performance of our trained model. Figure 8 shows the comparison between the predicted values and the actual values of the dc/dt. Again, the authors would like to emphasize if the predicted value of the target as a function of the "true" values of the test samples follows the "y = x" line, one can conclude that the model performance is at its best. It is worth mentioning that we only used two features for this prediction: namely, total O_2 flow and lance height. It is always one the goals of any machine-learning development to predict the target with a fewer number of variables/predictors. This is because if the model deals with many predictor variables, then there is a high chance that there are hidden relationships between some of them, leading to redundancy and, even if there is no relationship between any of them, the model can suffer from overfitting when there are a large number of predictor variables. In addition, a model that can predict with a fewer number of predictor variables is more practical due to some considerations, such as data availability, storage, computer resources, time taken for computation, etc. Thus, this is one of the achievements of this study, that the developed model can predict the target by only using two predictor variables.

The scatter plot demonstrates that our trained model could predict the dc/dt precisely. Figure 9 shows the error histogram for this prediction. The most frequent error was 0.000026. The metrics (Table 3, column 6) were recorded to be 0.25, 0.62, 0.04 and 0.009 for the mean absolute error, root mean squared error, relative absolute error and relative squared error, respectively. The coefficient of determination was computed to be 0.98. These values confirm that our trained model can be used at industries to predict and control the variation of the dc/dt in an actual reactor. Figure 8 shows that the machine-learning algorithm can predict the decarburization rate very accurately without employing any simplifications and without taking into account all the reactions, interactions, mass and heat transfers and fluid flows. In fact, all of these parameters are already inherited in the dataset, and an algorithm trained on the real dataset naturally learns the relationships between all of the parameters involved in the process without exactly knowing the physical meanings of them.

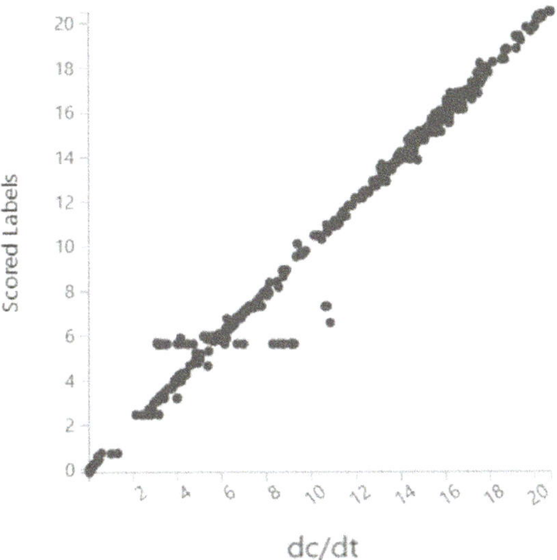

Figure 8. Scatter plot comparing the predicted values of the dc/dt using neural network method with the actual values of the dc/dt for an industrial dataset with the two operating parameters of total oxygen flow and lance height only.

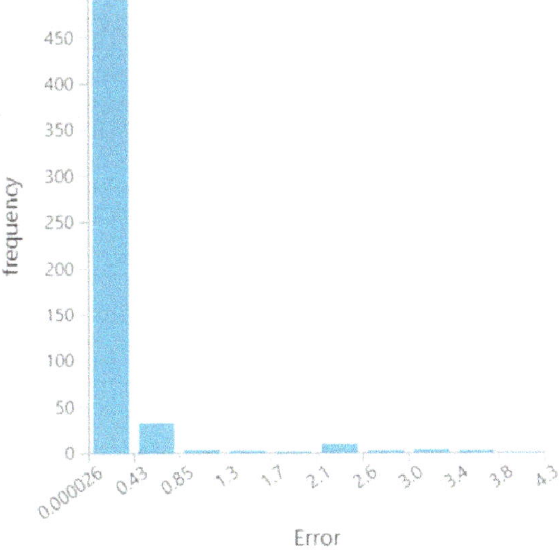

Figure 9. Error histogram of the predicted values of the dc/dt for the industrial dataset with the two operating parameters of total oxygen flow and lance height only.

6. Conclusions

We applied a machine learning-based analysis to a dataset (operating and output data) from a pilot basic oxygen steelmaking (BOS) converter. Correlation analysis showed:

- A strong positive correlation between the rate of decarburization (dc/dt) and total oxygen flow.
- A negative correlation with lance height.
- Less obviously, the decarburization also showed a positive correlation with the temperature of the waste gas, CO_2 content in the waste gas and O_2 in the waste gas.
- The pilot plant dataset was used for training and test data to develop a neural network-based regression to predict the decarburization rate. The developed algorithm was used successfully to predict the decarburization rate in a BOS furnace in an actual manufacturing plant based on the two operating parameters of total oxygen flow and lance height only.
- The performance was satisfactory, with a coefficient of determination of 0.98, confirming that the trained model can adequately predict the variation in the dc/dt within BOS reactors.

The method is easily scalable for industrial applications. In addition, the machine-learning model does not simplify the problem and is able to predict the decarburization rate accurately through learning from the real dataset acquired from BOS pilot plants.

Author Contributions: A.R. and S.S. wrote the machine-learning algorithms and produced the figures. A.R. and S.S. discussed the machine-learning results. Z.L. provided and prepared the datasets. A.R., S.S. and Z.L. discussed the materials case study. All authors wrote and commented on the manuscript and figures. All authors have read and agreed to the published version of the manuscript.

Funding: Z.L. would like to thank the funding from EPSRC (Engineering and Physical Sciences Research Council, UK) under grant number EP/N011368/1(EPSRC Fellowship).

Acknowledgments: The authors are grateful to Johan Eriksson at Swerea MEFOS for the interpretation of the information of the trials carried out in a 6-ton pilot converter at the MEFOS in Sweden. The authors also appreciate Chris Barnes at Tata Steel, UK, who provided the actual industry BOS converter data for the validation of the work.

Conflicts of Interest: The authors declare no conflicts of interest.

References

1. Ogawa, Y.; Yano, M.; Kitamura, S.; Hirata, H. Development of the continuous dephosphorization and decarburization process using BOF. *Steel Res. Int.* **2003**, *74*, 70–76. [CrossRef]
2. Millman, M.S.; Kapilashrami, A.; Bramming, M.; Malmberg, D. *IMPHOS—Improving Phosphorus Refining*; European Commission, Research Fund for Coal and Steel: Luxembourg, 2011; ISSN 1831-9424.
3. Jung, I.H.; Hudon, P.; Van Ende, M.A.; Kim, W.Y. Thermodynamic database for P_2O_5-containing slags and its application to the dephosphorization process. In Proceedings of the Iron & Steel Technology Conference, Indianapolis, IN, USA, 5–8 May 2014; Volume 1, pp. 1257–1268.
4. Rout, B.; Brooks, G.; Rhamdhani, M.A.; Li, Z.; Schrama, F.N.; Sun, J. Dynamic model of basic oxygen steelmaking process based on multi-zone reaction kinetics: Model derivation and validation. *Metall. Mater. Trans. B* **2018**, *49*, 537–557. [CrossRef]
5. Ersson, M.; Hoglund, L.; Tilliander, A. Dynamic coupling of computational fluid dynamics and thermodynamics software: Applied on a top blown converter. *ISIJ Int.* **2008**, *48*, 147–153. [CrossRef]
6. Cox, I.J.; Lewis, R.W.; Ransing, R.S. Application of neural computing in basic oxygen steelmaking. *J. Mater. Process. Technol.* **2002**, *120*, 310–315. [CrossRef]
7. Das, A.; Maiti, J.; Banerjee, R.N. Process control strategies for a steelmaking furnace using ann with Bayesian regularization and anfis. *Expert Syst. Appl.* **2010**, *37*, 1075–1085. [CrossRef]
8. Han, M.; Zhao, Y. Dynamic control model of BOF steelmaking process based on anfis and robust relevance vector machine. *Expert Syst. Appl.* **2011**, *38*, 14786–14798. [CrossRef]
9. Kubat, C.; Takin, H.; Artir, R.; Yilmaz, A. Bofy-fuzzy logic control for the basic oxygen furnace. *Robot. Auton. Syst.* **2004**, *49*, 193–205. [CrossRef]
10. Wang, X.; Han, M.; Wang, J. Applying input variables selection technique on input weighted support vector machine modeling for BOF end point prediction. *Eng. Appl. Artif. Intell.* **2010**, *23*, 1012–1028. [CrossRef]

11. Xu, L. A model of basic oxygen furnace (BOF) endpoint prediction based on spectrum information of the furnace flame with support vector machine (SVM). *Optik* **2011**, *122*, 594–598. [CrossRef]

12. Wang, X.; Xing, J.; Dong, J.; Wang, Z. Data driven based endpoint carbon content real time prediction for BOF steelmaking. In Proceedings of the 36th Chinese Control Conference (CCC), Dalian, China, 26–28 July 2017; pp. 9708–9713.

13. Liu, Y.; Zhao, T.; Ju, W.; Shi, S. Materials discovery and design using machine learning. *J. Mater.* **2017**, *3*, 159–177. [CrossRef]

14. Ward, L.; OKeeffe, S.C.; Stevick, J.; Jelbert, G.R.; Aykol, M.; Wolverton, C. A machine learning approach for engineering bulk metallic glass alloys. *Acta Mater.* **2018**, *159*, 102–111. [CrossRef]

15. Li, W.; Jacobs, R.; Morgan, D. Predicting the thermodynamic stability of perovskite oxides using machine learning models. *Comput. Mater. Sci.* **2018**, *150*, 454–463. [CrossRef]

16. Cassar, D.R.; de Carvalho, A.C.; Zanotto, E.D. Predicting glass transition temperatures using neural networks. *Acta Mater.* **2018**, *159*, 249–256. [CrossRef]

17. Meredig, B.; Agrawal, A.; Kirklin, S.; Saal, J.E.; Doak, J.W.; Thompson, A.; Zhang, K.; Choudhary, A.; Wolverton, C. Combinatorial screening for new materials in unconstrained composition space with machine learning. *Phys. Rev. B* **2014**, *89*, 094104. [CrossRef]

18. Rahnama, A.; Clark, S.; Sridhar, S. Machine learning for predicting occurrence of interphase precipitation in HSLA steels. *Comput. Mater. Sci.* **2018**, *154*, 169–177. [CrossRef]

19. Ramprasad, R.; Batra, R.; Pilania, G.; Mannodi-Kanakkithodi, A.; Kim, C. Machine learning in materials informatics: Recent applications and prospects. *Npj Comput. Mater.* **2017**, *3*, 1–13. [CrossRef]

20. Available online: https://docs.microsoft.com/en-us/azure/machine-learning/studio-module-reference/multiclass-neural-network (accessed on 2 January 2020).

21. Chattopadhyay, K.; Isac, M.; Guthrie, R.I.L. Applications of Computational Fluid Dynamics (CFD) in iron- and steelmaking (part I & part II). *Ironmak. Steelmak.* **2010**, *37*, 554–569.

22. Ek, M.; Shu, Q.F.; van Boggelen, J.; Sichen, D. New approach towards dynamic modelling of dephosphorisation in converter process. *Ironmak. Steelmak.* **2012**, *39*, 77–84. [CrossRef]

23. Oguchi, S.; Robertson, D.G.C.; Deo, B.; Grieveson, P.; Jeffes, J.H.E. Simultaneous dephosphorization and desulphurization of molten pig iron. *Ironmak. Steelmak.* **1984**, *11*, 202–213.

24. Spooner, S.; Rahnama, A.; Warnett, J.M.; Williams, M.A.; Li, Z.; Sridhar, S. Quantifying the Pathway and Predicting Spontaneous Emulsification during Material Exchange in a Two Phase Liquid System. *Sci. Rep.* **2017**, *7*, 14384. [CrossRef]

25. Spooner, S.; Li, Z.; Sridhar, S. Spontaneous emulsification as a function of material exchange. *Sci. Rep.* **2017**, *7*, 5450. [CrossRef] [PubMed]

26. Dogan, N.; Brooks, G.A.; Rhamdhani, M.A. Kinetics of decarburization reaction in oxygen steelmaking process. In *High Temperature Processing Symposium*; Swinburne University of Technology: Melbourne, Australia, 2010; pp. 9–11.

27. The AISE Steel Foundation. *The Making, Shaping and Treating of Steel (Steelmaking and Refining Volume)*; AIST: Pittsburgh, PA, USA, 1998; pp. 475–524.

Article

Dynamic Modeling and Simulation of Basic Oxygen Furnace (BOF) Operation

Daniela Dering [1], Christopher Swartz [1],* and Neslihan Dogan [2]

[1] Department of Chemical Engineering, McMaster Univesity, 1280 Main Street West, Hamilton, ON L8S 4L7, Canada; deringdd@mcmaster.ca
[2] Department of Materials Science and Engineering, McMaster Univesity, 1280 Main Street West, Hamilton, ON L8S 4L7, Canada; dogann@mcmaster.ca
* Correspondence: swartzc@mcmaster.ca

Received: 28 February 2020; Accepted: 12 April 2020; Published: 21 April 2020

Abstract: Basic oxygen furnaces (BOFs) are widely used to produce steel from hot metal. The process typically has limited automation which leads to sub-optimal operation. Economically optimal operation can be potentially achieved by using a dynamic optimization framework to provide operators the best combination of input trajectories. In this paper, a first-principles based dynamic model for the BOF that can be used within the dynamic optimization routine is described. The model extends a previous work by incorporating a model for slag formation and energy balances. In this new version of the mathematical model, the submodel for the decarburization in the emulsion zone is also modified to account for recent findings, and an algebraic equation for the calculation of the calcium oxide saturation in slag is developed. The dynamic model is then used to simulate the operation of two distinct furnaces. It was found that the prediction accuracy of the developed model is significantly superior to its predecessor and the number of process variables that it is able to predict is also higher.

Keywords: dynamic model; simulation; basic oxygen furnace

---

1. Introduction

The Basic Oxygen Furance (BOF) is responsible for approximately 70% of the steel production worldwide [1]. A schematic diagram for the BOF in shown in Figure 1. Scrap metal and hot metal are charged to the BOF, and an oxygen jet at supersonic speed is injected from the top through the lance onto the surface of the metal bath. Some of the species in the metal bath are oxidized and form a less dense slag layer. Flux is added to prevent refractory wearing and to contribute to slag formation. Millions of metal droplets are generated at the impact zone due to the impingement of the supersonic oxygen jet on the liquid metal. Carbon in the metal droplets can react with iron oxide in the slag forming carbon monoxide [2]. Experiments using X-ray fluoroscopy have shown that gas bubbles can be formed within the metal droplet itself [2,3] and not only at the metal droplet-slag interface. Carbon monoxide and carbon dioxide are also formed due to decarburization taking place at the impact zone. The droplet-gas-slag mixture, commonly referred to as emulsion, can occupy most of the furnace volume during the main blow.

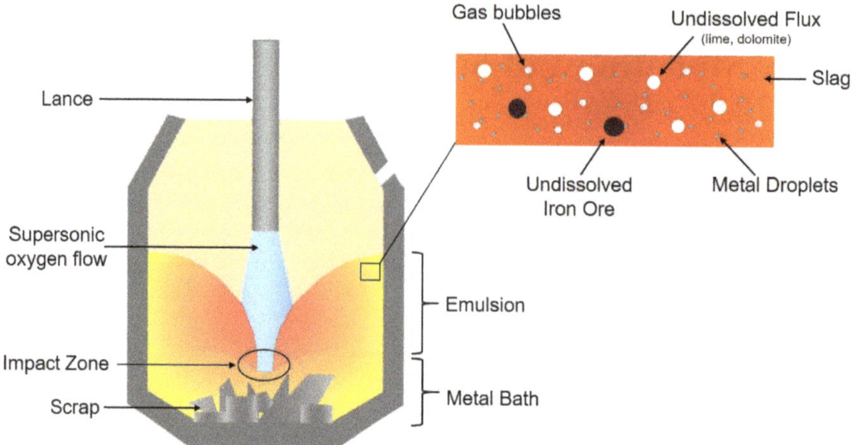

Figure 1. Schematic representation of the Basic Oxygen Furnace (BOF).

The process has limited automation and is highly dependent on operators' knowledge and past experience. Without a framework that can consistently aid the operators with the decision-making process, some of the more complex interactions between the process variables may be overlooked, leading ultimately to sub-optimal operation. Therefore, an optimization framework for the BOF operation that can consistently provide operators with the economically optimal operating conditions could greatly help steelshops reduce production costs. However, in order to have an optimization framework it is necessary to have an accurate enough dynamic model of the physical process.

Several dynamic models [4–9] for the BOF have been developed in recent years. Jalkanen [4] developed the CONSIM 5 simulator for the BOF. All reactions are assumed to take place in one zone and oxygen is partitioned according to its affinity to a certain element. The model accounts for slag formation, energy balance, scrap melting and decarburization.

Sarkar et al. [8] also developed a dynamic model for the BOF. They assumed that the refining reactions take place only in the emulsion zone, between elements dissolved in the droplets and FeO in slag. Oxygen is distributed between the elements in the upper part of the metal bath according to their concentration. The model does not include an energy balance and temperature is assumed to increase linearly during the blow.

Another dynamic model for the basic oxygen furnace was developed by Lytvynyuk et al. [7]. All the supplied oxygen is used to form iron oxide and all other refining reactions take place via reaction with FeO. The rate of the reactions taking place in the metal bath and slag is primarily dependent on the kinetics of mass transfer. For the energy balance, the slag and metal bath are assumed to have the same temperature. The model seems to give excellent prediction of the slag composition, metal bath composition and metal bath temperature at the end of the blow.

A few dynamic models based primarily on empirical relationships have also been developed. Kattenbelt and Roffel [5] used step responses to model the decarburization rate during the main blow. They studied the process response to step changes in lance height, oxygen flow rate and iron ore addition. The model makes extensive use of empirical relationships but provides limited physical insight.

A very comprehensive, first-priniciple dynamic model for the kinetics of decarburization in the BOF was developed by Dogan et al. [6]. Their model accounts for carbon oxidation in two zones: The emulsion and the impact zone (Figure 1). At the impact zone, carbon in the metal bath is assumed to react directly with oxygen and carbon dioxide, and in the emulsion zone carbon in the metal droplets is oxidized by FeO in the slag. The metal bath and slag temperature are assumed to increase linearly with blowing time and the slag composition is required as an input. Rout et al. [9] continued the work of Dogan et al. [6] by modeling the kinetics of manganese, phosphorus and silicon oxidation enabling the model to predict the slag composition.

The current study extends the work of Dogan et al. [6] by incoporating a model for slag formation as well as energy balances. The model for decarburization in the emulsion zone is modified to account for recent fidings [10] and the model for scrap melting is updated based on recent studies [11]. The mathematical model for the BOF is implemented as a system of Differential Algebraic Equations (DAEs) using CasADi [12] with a Python front-end. The resulting framework is used to simulate the Cicutti data [13] as well as 71 heats for the BOF of an industrial collaborator (Plant A).

2. Mathematical Model

2.1. Mass and Energy Balances

The phenomena taking place in the BOF are quite complex, therefore several assumptions are made for the derivation of the mass and energy balances. Flux, iron ore, metal droplets and gas bubbles are considered to be uniformly dispersed in the slag phase. The metal droplets, gas bubbles and slag form an emulsion for which the continuous phase is assumed to be the slag. The oxidation reactions take place mainly at the impact zone (IZ), the interface between the oxygen jet and the metal bath, and there is no resistance to the diffusion of oxides from the metal bath to the slag. A schematic representation of the flow of material in the BOF is shown in Figure 2.

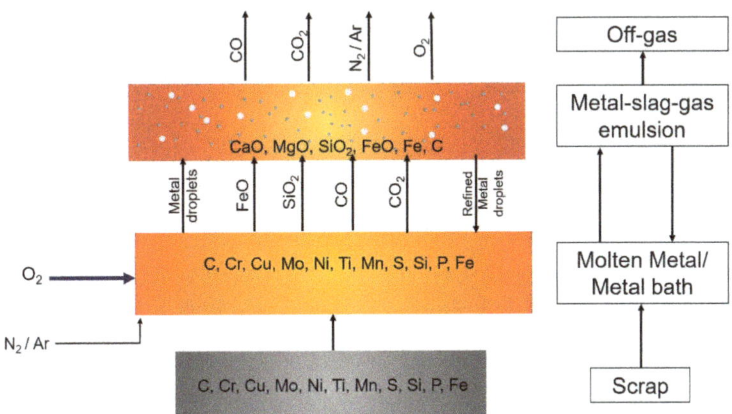

Figure 2. Material flow in the BOF assumed for the current study.

The reactions modelled by the current work are as follows [14], where heats of reaction, ΔH^{rxn}, were obtained from FactSage at 1900 K:

Decarburization:

$$[C] + \frac{1}{2}O_{2(g)} \longrightarrow CO_{(g)} \qquad \Delta H^{rxn} = -118.613 \text{ kJ/mol} \tag{1}$$

$$[C] + CO_{2(g)} \longrightarrow 2\,CO_{(g)} \qquad \Delta H^{rxn} = 160.623 \text{ kJ/mol} \tag{2}$$

$$(FeO) + [C] \longrightarrow [Fe] + CO_{(g)} \qquad \Delta H^{rxn} = 124.644 \text{ kJ/mol} \tag{3}$$

Desiliconization:

$$[Si] + O_{2(g)} \longrightarrow (SiO_2) \qquad \Delta H^{rxn} = -945.172 \text{ kJ/mol} \tag{4}$$

Iron Oxidation:

$$[Fe] + \frac{1}{2}O_{2(g)} \longrightarrow (FeO) \qquad \Delta H^{rxn} = -242.783 \text{ kJ/mol} \tag{5}$$

Post-combustion:

$$CO + \frac{1}{2}O_2 \longrightarrow CO_{2(g)} \qquad \Delta H^{rxn} = -277.927 \text{ kJ/mol} \tag{6}$$

Argon or nitrogen can be injected at the bottom of the BOF to improve mixing. Even though the injected nitrogen can undergo reaction at BOF operating conditions, the amount of nitrogen compounds formed is small and can be neglected. Therefore, it is assumed that the stirring gas leaves the furnace unreacted. At the impact zone, iron, carbon and silicon react with the injected oxygen forming iron oxide, carbon monoxide and silica as shown in Equations (1)–(5). Some of the CO may further react with O_2 forming CO_2 (Equation (6)).

The mass of dissolved flux and oxides is promptly incorporated by the slag. Similarly, the mass of melted scrap is assimilated by the metal bath. All Fe in the iron ore is assumed to be in the form of magnetite (Fe_3O_4) that is reduced to FeO by carbon in the metal bath (Equation (7)).

$$Fe_3O_4 + [C] \longrightarrow (3\,FeO) + CO_{(g)} \qquad \Delta H^{rxn} = 184.660 \text{ kJ/mol} \tag{7}$$

The impact of the oxygen jet on the liquid metal causes millions of metal droplets to be ejected in the emulsion zone. Carbon in the metal droplets can reduce FeO in the slag according to the reaction shown in Equation (3). The refined droplets then fall back to the metal bath.

The gases CO, CO_2, N_2/Ar and unreacted oxygen O_2 form the off-gas stream. Owing to the high temperatures, the residence time for the gases is assumed to be negligible in this work. A mass balance for the metal bath gives:

$$\frac{dW_b}{dt} = \dot{W}_{sc} - \dot{W}_{C,IO} + \dot{W}_{D,e-b} - \dot{W}_{D,b-e} - \dot{W}_{C,b-e} - \dot{W}_{Fe,b-e} - \dot{W}_{Si,b-e} \tag{8}$$

where W_b represents the mass of the metal bath. $\dot{W}_{sc}$ is the scrap melting rate, $\dot{W}_{C,IO}$ is the rate at which carbon in the metal bath is consumed to reduce magnetite in the iron ore to FeO, $\dot{W}_{D,b-e}$ and $\dot{W}_{D,e-b}$ are the mass flow rate of droplets from the metal bath to the emulsion and from the emulsion to the metal bath, respectively. $\dot{W}_{C,b-e}$, $\dot{W}_{Fe,b-e}$, $\dot{W}_{Si,b-e}$ are the mass flow rate of carbon, iron and silicon out of the metal bath and equals the oxidation rate of the respective element at the impact zone. A similar mass balance on the slag–metal–gas emulsion gives:

$$\frac{dW_{sme}}{dt} = \dot{W}_{DissFlux} + \dot{W}_{FeO,IO} + \dot{W}_{FeO,b-e} + \dot{W}_{SiO_2,b-e} + \dot{W}_{D,b-e} - \dot{W}_{D,e-b} - \dot{W}_{CO,EZ} \tag{9}$$

where W_{sme} is the mass of the slag and droplets in the emulsion. $\dot{W}_{DissFlux}$ is the rate of lime and dolomite dissolution, $\dot{W}_{FeO,IO}$ denotes the mass flow rate of iron oxide coming from iron ore, $\dot{W}_{FeO,b-e}$, $\dot{W}_{SiO_2,b-e}$ are the mass flow rate of iron oxide and silicon dioxide into the slag and are equal to the rate of formation of the respective component at the impact zone. $\dot{W}_{CO,EZ}$ is the rate carbon monoxide is formed at the emulsion zone due to droplet decarburization.

Heat is generated in the BOF by the oxidation and post-combustion reactions (Equations (1) and (4)–(6)) and is consumed by scrap melting, to heat the fluxes and iron ore, by decarburization in the emulsion (Equation (3)) and carbon dioxide reduction (Equation (2)). The total heat Q_{gen}^{rxn} generated by the oxidation reactions is given by

$$Q_{gen}^{rxn} = Q_{FeO}^{rxn} + Q_{SiO_2}^{rxn} + Q_{CO}^{rxn} + Q_{CO_2}^{rxn} \tag{10}$$

where Q_{FeO}^{rxn} is the heat from iron oxidation (Equation (5)), $Q_{SiO_2}^{rxn}$ is the heat from silicon oxidation (Equation (4)), Q_{CO}^{rxn} is the net heat generation from decarburization at the impact zone (Equations (1) and (2)), $Q_{CO_2}^{rxn}$ is the heat of CO post-combustion. Q_i^{rxn} is obtained by multiplying the rate of formation of compound i at the impact zone by the respective heat of reaction.

Individual energy balances were derived for the emulsion and metal bath. To simplify the problem, the following assumptions are made:

- The gas hold up in the slag and metal bath is negligible
- The gas exits the furnace at the slag temperature
- The temperature of the slag-metal-emulsion and metal bath is uniform
- A fraction α_l^p of all the heat generated is assumed to be lost to the environment and the remainder is assumed to be absorbed by the slag.

An energy balance for the slag–metal–gas emulsion yields:

$$
\begin{aligned}
(W_s C_{P,s} + W_D C_{P,b})\frac{dT_s}{dt} =& (1 - \alpha_l^p)Q_{gen}^{rxn} - Q_{Fe,EZ}^{rxn} - Q_{D,b-e} \\
& - Q_{FeO,b-e} - Q_{SiO_2,b-e} - Q_{CO,b-e} - Q_{CO2,b-e} \\
& - Q_{IO} - Q_{e-b} - Q_{flux} - Q_{N_2/Ar,b-e} - Q_{O_2,b-e}
\end{aligned}
\tag{11}
$$

with

$$
Q_{i,b-e} = \dot{W}_{i,b-e} C_{P,i}(T_s - T_b) \quad \text{for } i \in \{FeO,\ SiO_2,\ CO,\ CO_2,\ N_2/Ar,\ O_2\}
\tag{12}
$$

In the above, W_s is the mass of slag, W_D is the mass of the droplets in emulsion, $C_{P,s}, C_{P,b}, C_{P,i}$ are the heat capacity of the slag, molten metal and component i, respectively. $Q_{Fe,EZ}$ is the heat consumed by the decarburization reaction in the emulsion, $Q_{D,b-e}$ is the heat required to raise the temperature of the incoming droplets to the temperature of the slag–metal–gas emulsion; $Q_{FeO,b-e}, Q_{SiO_2,b-e}, Q_{CO,b-e}, Q_{CO,b-e}$ and $Q_{CO_2,b-e}$ are the heat required to raise the temperature of the iron, silicon, carbon oxide and carbon dioxide formed at the impact zone to the emulsion temperature, respectively; Q_{e-b} is the heat lost from the emulsion to the metal bath, Q_{flux} and Q_{IO} are the heat consumed by the flux and iron ore additions, $Q_{N_2/Ar,b-e}$ is the heat required to raise the temperature of the stirring gas from the bath to the emulsion temperature, $Q_{O_2,b-e}$ is the heat required to raise the temperature of the non-reacted oxygen from the bath to the emulsion temperature. The rate of heat transfer between the slag and metal bath is given by:

$$
Q_{e-b} = h_{s-b}\pi R^2(T_s - T_b)
\tag{13}
$$

where R is the furnace diameter and h_{s-b} is the heat transfer coefficient, which can be estimated using dynamic data. Applying an energy balance for the metal bath gives:

$$
W_b C_{P,b}\frac{dT_b}{dt} = Q_{e-b} + Q_{D,e-b} - Q_{O_2,in} - Q_{N_2/Ar,in} - Q_{sc}
\tag{14}
$$

where $Q_{D,e-b}$ is the heat transferred by the droplets coming from the emulsion to the metal bath, and Q_{sc} is heat used to melt the scrap, $Q_{O_2,in}$ and $Q_{N_2/Ar,in}$ are the heat consumed to raise the temperature of oxygen and stirring gas to the bath temperature:

$$
Q_{i,in} = F_i C_{P,i}(T_b - T_{i,0}) \quad \text{for } i \in \{N_2/Ar,\ O_2\}
\tag{15}
$$

where F_i, $T_{i,0}$ are the inlet mass flow rate and temperature of gas i.

2.2. The Impact Zone

The impact zone in the BOF is the region where the oxygen jet impinges on the molten metal bath (Figure 1) causing a deformation on the liquid surface that is assumed to have the shape of a paraboloid of height h_c and maximum diameter d_c. The number of impact zones is equivalent to the number of nozzles n_n in the lance as long as there is no coalescence of the oxygen jet. The area of one impact zone is equal to the surface area of the paraboloid:

$$A_{iz} = \frac{d_c \pi}{12 h_c^2} \left[\left(\frac{d_c^2}{4} + 4 h_c^2 \right)^{1.5} - \frac{d_c^3}{8} \right] \tag{16}$$

where d_c and h_c are calculated using the correlations proposed in [15]. For the current work, the decarburization rates previously used by Dogan et al. [6] are modified by a parameter α^p in order to account for the difference between the conditions at which the equation rates were derived and the BOF operating conditions. These parameters can be tuned using process data. Therefore, the decarburization rate via O_2 (Equation (1)) is $-2 r_{O_2,iz}$, where:

$$- r_{O_2,iz} = \alpha^p_{O_2,t} A_{iz} n_n k_{O_2} \ln(1 + P_{O_2}) \tag{17}$$

with:

$$\alpha^p_{O_2,t} = \frac{\alpha^p_{O_2}}{1 + \alpha^p_{Si,C}[\%Si]} \tag{18}$$

The partial pressure in Equation (17) is in atm, and the parameter $\alpha^p_{Si,C}$ in Equation (18) accounts for the inhibiting effect that silicon has on the rate of carbon oxidation [14]. Similarly, the decarburization rate via CO_2 (Equation (2)) is given by:

$$- r_{CO_2,iz} = \alpha^p_{CO_2} A_{iz} n_n k_a P_{CO_2} \tag{19}$$

In the above expressions, k_a and k_{O_2} are the rate coefficients calculated using the correlations found in Dogan et al. [6] and P is the partial pressure calculated as:

$$P_i = \frac{F_i}{\sum_i F_i} P_a \quad \text{for } i \in \{O_2, CO, CO_2, N_2/Ar\} \tag{20}$$

where P_a is the ambient pressure. In Equation (20), F_{O_2} and $F_{N_2/Ar}$ are the inlet molar flow rate of oxygen and bottom stirring gas, and F_{CO} and F_{CO_2} are the molar flux of carbon monoxide and carbon dioxide formed from the decarburization reaction in Equations (1) and (2) and the post-combustion reaction in Equation (6).

When the carbon content of the metal bath $[\%C]$ becomes lower than the critical carbon content C_C, the decarburization rate is given by [6]:

$$- r_{C_c,iz} = \alpha^p_{C_c} k_m \frac{A_{iz} n_n}{V_b} [\%C] \tag{21}$$

where V_b is the volume of liquid metal and k_m is the rate constant calculated using the correlation developed by Kitamura et al. [16].

The desiliconization rate is calculated using the expression found in Rout et al. [17] and modified by a parameter α^p_{Si} to give:

$$- r_{Si,iz} = \alpha^p_{Si} k_m \rho_b ([\%Si] - [\%Si_{eq}]) A_{iz} n_n \tag{22}$$

where ρ_b is the density and $[\%Si]$ is the silicon content of the liquid metal. Rout et al. [9] found that the equilibrium silicon content of the metal bath $([\%Si_{eq}])$ is approximately zero, therefore it is neglected in the calculations. For the current study, it is assumed that the rate of iron oxidation is proportional to the partial pressure of oxygen and is given by:

$$- r_{Fe,iz} = \alpha^p_{Fe} A_{iz} n_n P_{O_2} \tag{23}$$

All the oxygen injected in the system via the lance that is not used for decarburization, desiliconization or iron oxidation, is assumed to be used in the post-combustion of CO.

2.3. Scrap Melting

In the BOF, the heat generated by the oxidation reactions (Equations (1) and (4)–(6)) is much higher than that required to reach the targeted end-point temperature. Therefore, scrap metal is usually added in order to absorb part of the surplus heat via melting. In this section the model used to compute the scrap melting rate $\dot{W}_{sc}$ as well as the heat absorbed Q_{sc} is presented.

Consider a scrap metal plate of half-thickness L, initial temperature T_{sc0} and melting temperature T_m, as shown in Figure 3a.

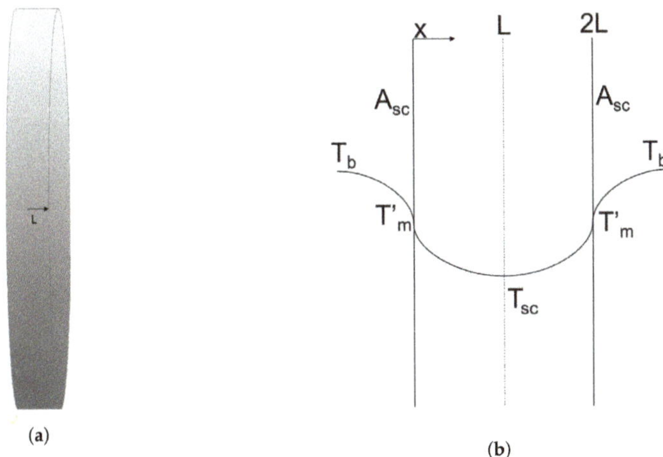

Figure 3. Schematic representation of the melting of scrap. (**a**) Schematic representation of a plate. (**b**) Schematic representation of temperature gradient between hot metal and a cold metal plate.

This plate is then submerged in a metal bath at temperature $T_b > T_m$ and carbon content $[\%C]$. Assuming that heat transfer occurs only in the axial direction and constant physical properties for the plate, a heat balance at the interface between the solid plate and the metal bath gives [11,18–20]:

$$- k_{sc} A_{sc} \frac{\partial T_{sc}(0,t)}{\partial x} - h_{sc} A_{sc}(T_b - T'_m) = \frac{dL}{dt} \rho_{sc}(\Delta H_{sc} + C_{P,sc}(T_b - T'_m)) A_{sc} \quad T_b \geq T_m \quad (24)$$

where k_{sc} is the thermal conductivity of the scrap plate, h_{sc} is the heat transfer coefficient, ρ_{sc} is the scrap density, ΔH_{sc} is the latent heat of melting of the scrap, $C_{P,sc}$ is the specific heat of the scrap, A_{sc} is the interfacial area, x is the distance of a point within the scrap from the interface, t is time and $T_{sc}(x,t)$ is the scrap temperature at a position x. A schematic representation can be found in Figure 3b.

Equation (24) can be solved using the Quasi-Static approach to yield Equations (25)–(27) [21]:

$$\frac{dL}{dt} = \frac{\sum_{n=1}^{\infty} \frac{-A_n}{n\pi}(1 - cos(n\pi))\lambda^2 k_{sc}L\, e^{-\lambda^2 \alpha_{sc} t} - h(T_b - T'_m)}{\rho_{sc}(\Delta H_{sc} + C_{P,sc}(T_b - T'_m)) - \sum_{n=1}^{\infty} \frac{A_n}{n\pi}(1 - cos(n\pi))(\rho_{sc}C_{P,sc} + 2\lambda^2 k_{sc}t)e^{-\lambda^2 \alpha_{sc} t}} \quad (25)$$

$$\lambda = \left(\frac{n\pi}{2L(t)} \right) \quad (26)$$

$$A_n = 2(T_{sc0} - T'_m)\frac{1 - cos(n\pi)}{n\pi} \quad (27)$$

where $\alpha = k_{sc}/\rho_{sc}C_{P_{sc}}$ is the thermal diffusivity. The scrap melting rate $\dot{W}_{sc}$ is then given by:

$$\dot{W}_{sc} = -n_{sc}2\rho_{sc}A_{sc}\frac{dL}{dt} \tag{28}$$

where n_{sc} is the total number of scrap plates.

At the beginning of the BOF operation, the temperature of the molten metal may not be high enough to melt the scrap, in which case the interface temperature T'_m is equal to the bath temperature T_b. The interface temperature is otherwise assumed to correspond to the melting temperature. This is shown in Equation (29) :

$$T'_m = \begin{cases} T_b & T_b < T_m \\ T_m & T_b \geq T_m \end{cases} \tag{29}$$

The melting temperature T_m is dependent on the carbon content C_m at the melting interface and can be calculated using Equation (30) [22]:

$$T_m = \begin{cases} 1810 - 90\%C_m & 0 \leq \%C_m \leq 4.27\% \\ 1425 & \%C_m > 4.27\% \end{cases} \tag{30}$$

Dogan [22] assumed C_m to be equal to the carbon content in the bulk metal [%C]. This assumption reflects, to some extent, what happens in the BOF: Carbon in the metal bath migrates to the scrap surface lowering its melting temperature [21,23]. One consequence of such an assumption is that scrap types of similar physical properties will all melt at the same time, independent of their carbon content. On the other hand, if C_m is assumed to be equal the carbon content of the scrap C_{sc}, low-carbon content scrap types will only melt towards the very end of the blow, which is not observed in practical BOF operations. For the current work, C_m is defined as:

$$C_m = 0.8[\%C] + 0.2C_{sc} \tag{31}$$

The heat transfer coefficient h_{sc} in Equation (25) is calculated according to the correlation proposed by Gaye et al. [24] and given by:

$$h_{sc} = 5000\dot{\epsilon}^{\,0.2} \tag{32}$$

where $\dot{\epsilon}$ is the mixing power due to bottom stirring and top blowing in Wm^{-3}, and h_{sc} is in units of $Wm^{-2}K^{-1}$. The rate Q_{sc} at which heat is absorbed by the scrap is given by the heat conduction term in Equation (24) before melting starts, and by the convective term once melting starts as shown in Equation (33).

$$Q_{sc} = 2n_{sc}A_{sc}\begin{cases} -\frac{\partial T_{sc}(0,t)}{\partial x} & T_b < T_m \\ h(T_b - T'_m) & T_b \geq T_m \end{cases} \tag{33}$$

2.4. Iron Ore Dissolution

Some steel shops also add iron ore before or during the blow to cool down the metal bath and meet the desired end point temperature. Since the mass of iron ore added is usually small compared to the amount of scrap, a rather simple model is used to account for the cooling effect of iron ore.

Iron ore composition can vary greatly according to the region but the predominant element is usually Fe followed by oxygen. For the current model all Fe in the iron ore is assumed to be in the form of magnetite (Fe_3O_4) that is reduced to iron oxide by carbon in the metal bath according to Equation (7). Furthermore, the iron ore melting point T_m is considered to be equal to the FeO melting point (1644 K). Reduction of magnetite and melting of the formed iron oxide are assumed to happen concomitantly.

The iron ore particles are assumed to be spherical and to have uniform temperature, therefore the total heat transfered to the surface of an iron ore particle can be calculated using Equation (34):

$$Q_{conv,ore} = 4\pi r_{ore}^2 h_{ore}(T_s - T_{ore})$$

(34)

where T_{ore} is the iron ore temperature, h_{ore} is the heat transfer coefficient, r_{ore} is the radius and T_s is the slag temperature. It is assumed that a fraction T_{ore}/T_m of the total heat $Q_{conv,ore}$ contributes to melting of the iron ore and the remainder is used for sensible heating. A similar approach was used by MacRosty and Swartz [25] and Bekker et al. [26] to model scrap melting in electric arc furnaces. At the beginning of the process $T_{ore} << T_m$ and most of the heat is used to heat up the iron ore. As T_{ore} approaches T_m, the fraction of the total heat used for melting increases. An energy balance for a single iron ore particle yields:

$$W_{ore} C_{P,ore} \frac{dT_{ore}}{dt} = (1 - T_{ore}/T_m) Q_{conv,ore}$$

(35)

where W_{ore} is the mass of a single iron ore particle and can be obtained from Equation (36):

$$W_{ore} = \frac{4}{3}\pi r_{ore}^3 \rho_{ore}$$

(36)

In the above, ρ_{ore} is the iron ore density and $C_{P,ore}$ is the iron ore heat capacity. A heat balance at the interface of the iron ore particle gives Equation (37) for the rate of dissolution of an iron ore particle:

$$A_{ore}\rho_{ore}(\Delta H_{ore} + Cp_{FeO}(T_s - T_{ore}))\frac{dr_{ore}}{dt} = -(T_{ore}/T_m)Q_{conv,ore}$$

(37)

where ΔH_{ore} is the latent heat of melting of FeO, $C_{P,FeO}$ is the heat capacity of iron oxide and A_{ore} is the area of the interface, here equal to the surface area of a sphere of radius r_{ore}.

The heat consumed to reduce magnetite to iron oxide is given by:

$$Q_{red,ore} = \frac{y_{Fe,ore}}{3M_{Fe}}\dot{W}_{ore,melt}\Delta H^{rxn}$$

(38)

where $y_{Fe,ore}$ is the mass fraction of Fe in the iron ore, ΔH^{rxn} is the heat for the reaction defined in Equation (7), M_{Fe} is the molar mass of iron and $\dot{W}_{ore,melt}$ is the melting rate of iron ore given by:

$$\dot{W}_{ore,melt} = -4\pi \rho_{ore} r_{ore}^2 \frac{dr_{ore}}{dt}$$

(39)

The total heat consumed by iron ore melting and reduction is given by:

$$Q_{IO} = n_{ore}(Q_{conv,ore} + Q_{red,ore})$$

(40)

where n_{ore} is the number of iron ore lumps.

2.5. Flux Dissolution

Similarly to the decarburization rates at the impact zone, flux dissolution is modelled after Dogan et al. [27]. The flux particles are assumed to have a spherical shape, and the rate of flux dissolution is proportional to the rate of change of the particles' radius.

The rate of lime dissolution is given by Equation (41) [27,28], where r_L is the particle radius, $(\%CaO)$ is the concentration of CaO in slag, $(\%CaO_{sat})$ is the saturation concentration of CaO in the slag, k_L is the mass transfer coefficient, ρ_s and ρ_L are the slag and lime density, respectively.

The parameter α_L^p is introduced here to account for deviations between the experimental conditions at which k_L was derived, and the BOF operating conditions.

$$\frac{dr_L}{dt} = \alpha_L^p k_L \frac{\rho_s}{100\rho_L}((\%CaO) - (\%CaO_{sat})) \tag{41}$$

For dolomite, the rate of dissolution is given by [27,29]:

$$\frac{dr_D}{dt} = \begin{cases} \alpha_D^p k_D \frac{\rho_s}{100\rho_D}\left(1 + \frac{M_{MgO}}{M_{CaO}}\right)((\%CaO) - (\%CaO_{sat})) & (\%FeO) < 20\% \\ \alpha_D^p k_D \frac{\rho_s}{100\rho_D}\left(1 + \frac{M_{CaO}}{M_{MgO}}\right)((\%MgO) - (\%MgO_{sat})) & (\%FeO) \geq 20\% \end{cases} \tag{42}$$

where M_{MgO} and M_{CaO} are the molar mass of MgO and CaO, respectively, $(\%MgO)$ is the concentration of MgO in slag, $(\%MgO_{sat})$ is the saturation concentration of CaO in the slag and k_D is the mass transfer coefficient. The rate constants k_L and k_D are calculated using the correlations proposed by Dogan et al. [27] with the parameter used to modify the Reynolds number β set to the nominal value of 1.

Data for the calcium oxide saturation in slag $(\%CaO_{sat})$ for different slag compositions were obtained using the Cell Model [30], a Matlab program that gives the saturation concentration of the individual species in the slag as a function of composition and temperature. Kadrolkar et al. [30] validated the Cell Model results against FactSage^{TM} and ThermoCalc^{TM}. The function 'fitnlm' in Matlab was then used to fit a curve to the generated data, yielding Equation (43).

$$(\%CaO)_{sat} = \frac{3.52 T_s - 4,823.7 e^{-\frac{2.93}{100}(\%SiO_2)} + 12.4(\%FeO) - 9.71(\%MgO) + 17.9(\%CaO)}{100} \tag{43}$$

An R^2 of 0.9 was obtained for the nonlinear regression. The goodness of Equation (43) to predict the calcium oxide saturation in slag was evaluated using the Cicutti data [13]. The calculated values for CaO_{sat} using the Cell Model and Equation (43) are shown in Figure 4, where it can be seen that there is excellent agreement between them.

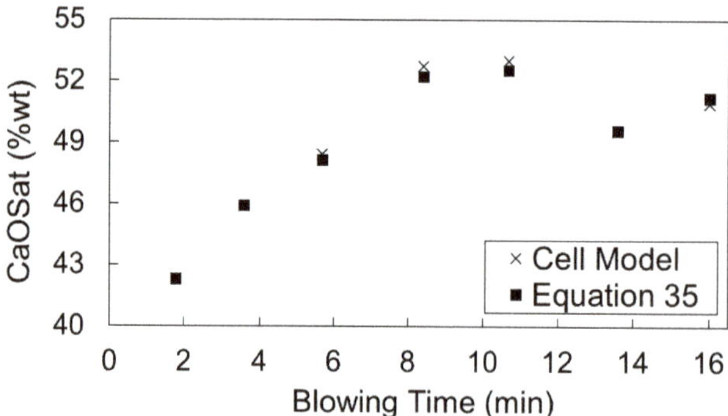

Figure 4. $(\%CaO_{sat})$ obtained using the Cell Model and Equation (43) for the Cicutti et al. [13] slag data.

The rate at which heat is absorbed by the flux particles is given by:

$$Q_{flux} = 4\pi \sum_i r_i^2 h_i n_i (T_s - T_i) \quad i \in \{L, D\} \tag{44}$$

where T_s is the temperature of the slag, T_i is the flux temperature, h_i is the heat transfer coefficient determined using data and n_i is the number of flux particles. Assuming a uniform temperature for the flux particle, an energy balance gives Equation (45) for the temperature of an iron ore particle:

$$W_i C_{P,i} \rho_i \frac{dT_i}{dt} = 4\pi r_i^2 h_i (T_s - T_i) \quad i \in \{L, D\} \tag{45}$$

where $C_{P,i}$ is the heat capacity and $W_i = 4/3\rho_i \pi r^3$ is the mass of one single flux particle.

2.6. Decarburization in the Emulsion Zone

In Figure 5, the initial carbon composition and diameter of a metal droplet are represented by $C_{i,0}$ and D_0, respectively. The metal droplet is formed from the interaction of the oxygen jet and the metal bath at the impact zone at time t_0, and ejected to the slag–metal–gas emulsion. The droplet stays in the emulsion zone for R_t seconds, then it returns to the metal bath at time $t_0 + R_t$ having a composition $C_{i,f}$ and diameter D_f. While in the emulsion zone, carbon in the metal droplet can reduce FeO in slag according to reaction 3.

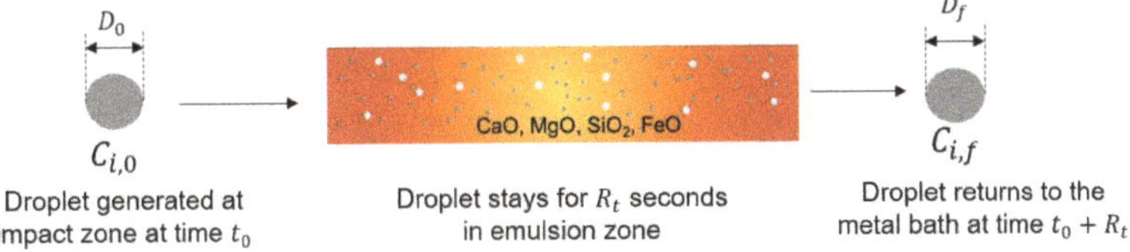

Figure 5. Droplet decarburization.

The droplet generation rate R_B, given by Equation (46), is calculated using the empirical expression proposed by Subagyo et al. [31] modified by a parameter p_{DG} to account for possible differences between the conditions at which the correlation was derived and a given BOF operation:

$$R_B = p_{DG} \frac{\dot{V}_{O_2} N_B^{3.2}}{(2.6 \times 10^6 + 2 \times 10^{-4} N_B^{12})^{0.2}} \tag{46}$$

where $\dot{V}_{O_2}$ is the volumetric flow rate of oxygen and N_B is the blowing number [31].

Since millions of droplets are generated at every point, it can become computationally expensive to track the composition of the individual metal droplets. Therefore, an algebraic equation to calculate the final carbon content of the metal droplets was developed and the following assumptions were made in order to make the problem tractable:

- All droplets decarburize immediately after they are ejected from the impact zone
- The carbon content of the metal droplets in the emulsion zone is approximated as the average carbon content of the individual metal droplets
- Except for carbon, the mass of other elements in the metal droplets stays constant whilst the droplet travels through the emulsion zone

The first-principles model developed by Kadrolkar and Dogan [10] for the droplet decarburization was used to generate data for the final carbon content ($C_{C,f}$) of individual droplets with respect to its initial carbon content $C_{C,0}$, the slag temperature T_s and composition. Data were generated for the range of initial carbon content between 0.3% and 5% and slag temperature of 1623–2153 K, and for the slag composition provided by Cicutti et al. [13]. Equations (47) and (48) give a good description of the generated data with an R^2 of 0.96 and 0.92, respectively.

$$\Delta C_C = C_{C,0} - C_{C,f} = 0.9514 \left(\frac{0.001 T_s}{C_{C,0}}\right)^{-1.4345} \quad (\%FeO) > 10 \tag{47}$$

$$C_{C,f} = 0.67492 C_{C,0}^{1.2261 C_{C,0}/(\%FeO)} \quad (\%FeO) \leq 10 \tag{48}$$

In Figure 6, the average final carbon content of the droplets reported by Cicutti et al. [13], the values predicted by Kadrolkar and Dogan [10]'s model and what was obtained using Equation (47) are shown. Similar to the first-principles model proposed by Kadrolkar and Dogan [10], the final carbon content of the droplets predicted by Equation (47) is largely within the range of values reported by Cicutti [13] for a real BOF operation. These results indicate that Equation (47) approximates the kinetics of droplet decarburization reasonably well and can therefore be used within the BOF model to describe the kinetics of decarburization in the emulsion zone.

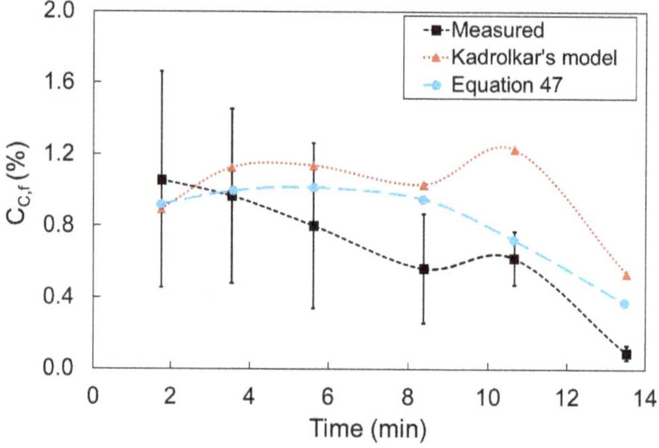

Figure 6. Comparison of final carbon content of metal droplets in the emulsion reported by Cicutti et al. [13], and the values predicted using Kadrolkar and Dogan [10]'s first-principles model and Equation (47) as a function of blow time.

Using Equations (47) and (48) and the previously stated assumptions, and introducing a term $\alpha_{\mu,D}^{p}$ to account for changes in the slag viscosity, it is possible to obtain Equation (49) for the rate of carbon removal:

$$\dot{W}_{C,s} = R_B \left(1 - \frac{1 - 0.01 C_{C,0}}{1 - 0.01 C_{C,f}} \right) \alpha_{\mu,D}^{p} \tag{49}$$

with $\alpha_{\mu,D}^{p}$ given by:

$$\alpha_{\mu,D}^{p} = \exp(-\alpha_{\mu}^{p} \mu_s) \tag{50}$$

where $\dot{W}_{C,s}$ is the rate of carbon removal, α_{μ}^{p} is a parameter and μ_s is the slag viscosity. The initial carbon content of the droplet $C_{C,0}$ is equal to the carbon content of the metal bath C_b at the time of ejection. The term $\alpha_{\mu,D}^{p}$ is necessary because the effect of slag viscosity is not taken into account in Equations (47) and (48). However, it is likely that the high slag viscosity at the beginning of the blow significantly decreases the decarburization rate. At high viscosities the slag becomes less fluid, negatively impacting the rate of FeO mass transfer to the droplet surface and potentially leading to FeO depletion in the neighborhood of the droplet, decreasing the decarburization rate. The value of α_{μ}^{p} can be determined using dynamic data.

The mass of carbon $W_{C,s}$ in the emulsion is then given by:

$$\frac{dW_{C,s}}{dt} = (0.01 R_B C_{C,0} - \dot{W}_{C,s}) - 0.01 \overline{C}_{C,s} \frac{W_{D,s}}{R_t} \tag{51}$$

where the first term on the right hand side is the amount of carbon entering the emulsion zone minus the carbon removed via decarburization, and the second term is the flow rate of carbon returning to the metal bath. $\overline{C}_{C,s}$ is the average carbon content of the droplets in the emulsion calculated by:

$$\overline{C}_{C,s} = 100\frac{W_{C,s}}{W_{D,s}} \tag{52}$$

and $W_{D,s}$ is the total mass of droplets in the emulsion determined as:

$$\frac{dW_{D,s}}{dt} = (R_B - \dot{W}_{C,s}) - \frac{W_{D,s}}{R_t} \tag{53}$$

The droplet residence time R_t is calculated using the following empirical expression:

$$R_t = 5 + 20e^{-0.3\frac{(\%FeO)}{C_{C,0}}} \tag{54}$$

As long as the residence time R_t of the droplets in the emulsion is small, $\overline{C}_{C,s}$ is approximately equal to $C_{C,f}$ and the simplifying assumption of uniform carbon content for the droplets in the emulsion does not significantly impact the final result, while avoiding the need to track the behavior of individual droplets.

2.7. Implementation

The mathematical model presented in the previous sections was implemented as a DAE system using CasADi [12] with Python 3.7. An index-1 DAE in semi-explicit form can be written in the form given by Equations (55) and (56):

$$\dot{x}(t) = f(x(t), z(t), u(t), p) \tag{55}$$

$$0 = h(x(t), z(t), u(t), p) \tag{56}$$

where f and h are the differential and algebraic functions, respectively, x and z are the differential and algebraic states, u and p represent the control variables and time-independent parameters, and t is time. Within CasADi, the DaeBuilder class was selected since it is capable of performing model reduction by eliminating algebraic variables that can be explicitly calculated. The resulting DAE system was solved using the variable step size DAE solver IDAS [32]. For implementation of the mathematical model, the following modifications were made where necessary:

- Equations of the form:

$$y(a) = \frac{1}{a} \tag{57}$$

where $a \to 0$, were rewritten as:

$$y(a) = \frac{1}{a + \epsilon} \tag{58}$$

where ϵ is a positive small number. This strategy was used on the submodels for flux dissolution, iron ore and scrap melting to prevent division by zero since either the radius or thickness of the particles continuously decreases with time and can eventually equal zero.
- Piecewise functions of the form:

$$y(c) = \begin{cases} y_1(c) & a > b \\ y_2(c) & a \leq b \end{cases} \tag{59}$$

were rewritten using hyperbolic tangent functions:

$$\tilde{y}(c) = y_1(0.5\tanh(\gamma(a - b)) + 0.5) + y_2(0.5\tanh(\gamma(b - a)) + 0.5) \tag{60}$$

where γ is an adjustable parameter that controls the steepness of the continuous switching function approximation. This was used for flux dissolution (Equation (42)), scrap melting (Equation (29)), decarburization in the emulsion zone (Equations (47) and (48)), among others for a smooth transition and to ensure differentiability.

- Flux additions: Flux and iron ore can be added at anytime during a blow, and each individual addition is modeled as shown in Section 2.5. To model the indiviudal flux additons a new variable t_{ij}, where i is the flux type (lime, dolomite, iron ore) and j is the addition number (first, second, third), is defined for the flux addition time. Given the radius r_{ij} of the flux added at time t_{ij}, Equation (41) for lime dissolution rate can be reformulated as:

$$\frac{dr_{ij}}{dt} = \begin{cases} 0 & t < t_{ij} \\ k_L \frac{\rho_s}{100\rho_L}(\%CaO_s - \%CaO_{sat}) & t \geq t_{ij} \end{cases} \tag{61}$$

which was implemented using a hyperbolic tangent function.

3. Results and Discussion

3.1. System Parameters and Input Data

The parameters α^p introduced during the model development to account for distinct BOF operations and differences between the conditions at which the respective equations were derived and the operating conditions were manually adjusted for a data set available in the literature for a 200 ton furnace [13], as well as for the data provided by Plant A for 70 heats for a 250 ton furnace. The final values of the parameters α^p are given in Table 1.

Table 1. The values of model parameters for simulation and optimization. The heat transfer coefficients h are given in $Wm^{-2}K^{-1}$, and α^p_{Fe} is given in $kgm^{-2}Pa^{-1}s^{-1}$. All the other parameters are dimensionless.

Parameters	Cicutti	Plant A
α^p_l	0.2	0.07
$\alpha^p_{O_2}$	1	1.16
$\alpha^p_{Si,C}$	10	2
$\alpha^p_{CO_2}$	1	1
$\alpha^p_{C_c}$	3	3.07
α^p_{DG}	1	1
α^p_{Fe}	7×10^{-6}	2.0×10^{-5}
α^p_{Si}	7	7
α^p_L	30	30
α^p_D	20	20
α^p_μ	0.5	0.5
h_L, h_D	1000	1000
h_{ore}	2500	2500
h_{s-b}	80,000	80,000

In Cicutti et al. [13]'s study, lime is added before the blow starts and at every minute up to 7 min, whereas dolomite is added before the blow starts and again at 7 min. Ar/N$_2$ gas is continuously injected from the bottom of the furnace to aid with stirring. In Plant A operations, lime and dolomite are added only before the blow starts, scrap selection varies for each heat, and the furnace design does not allow for bottom stirring.

3.2. Simulation Results for Cicutti's Operations

Cicutti et al. [13]'s data have been used to validate several dynamic models developed for the BOF [6,8,9,33]. Information regarding the hot metal and scrap compositions is shown in Table 2. For the mass of flux, gravel and iron ore refer to Cicutti et al. [13].

Table 2. Mass and composition of hot metal and scrap types added to the BOF, and average scrap thickness [22].

Input	Hot Metal	Heavy Scrap	Light Scrap	Pig Iron	External Scrap
Mass (kg)	170,000	3570	10,000	12,140	4284
Thickness (mm)	-	100	25	200	500
C (%)	4	0.08	0.08	4.5	0.05
Si (%)	0.33	0.05	0.05	0.5	0.001
Mn (%)	0.52	0.3	0.3	0.5	0.2
P (%)	0.066	-	-	-	-
S (%)	0.015	-	-	-	-

Figure 7 shows the metal bath temperature and carbon content predicted by the model described in this article, as well as the data published by Cicutti et al. [13].

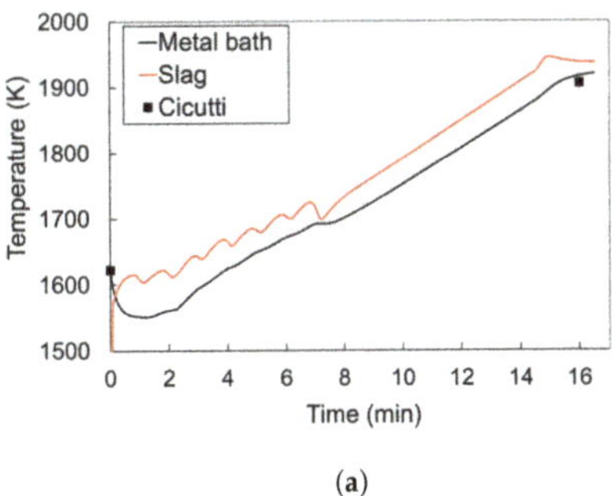

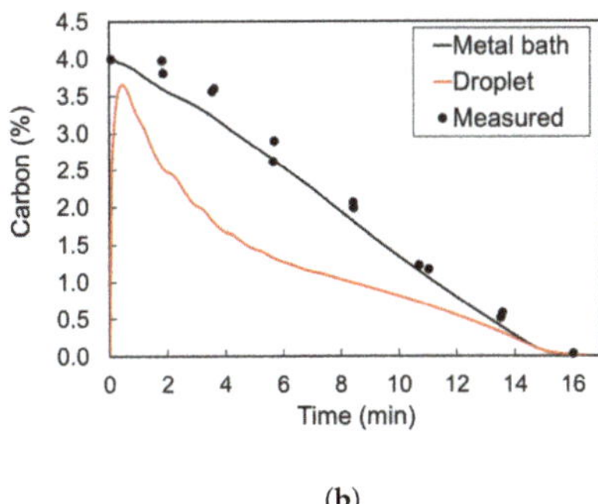

Figure 7. (a) Comparison between measured [13] and predicted values for the temperature of liquid metal and slag. (b) Comparison between measured [13] and predicted values for the carbon content of liquid metal and the returning metal droplets.

In the first few minutes, the bath temperature decreases due to the heat absorbed by the scrap. Thereafter, the temperature increases approximately linearly as heat is released by the oxidation reactions. The oscillations in the slag temperature during the first half of the blow are due to the flux additions, done at every minute. Figure 7b shows that the carbon content predicted by the model for the bulk metal agrees well with the measured data. The predicted final carbon content of the returning droplets is also shown in Figure 7b. Due to the high interfacial area, the rate of carbon refining at the droplet level is significantly higher than that for the bulk metal, thus the lower carbon content.

The decarburization rate in the emulsion zone is shown in Figure 8a. The extent of droplet decarburization is primarily dependent on the initial carbon content of the liquid metal droplet when it is ejected from the impact zone [10]. At low initial carbon contents the liquid metal droplets do not bloat and their residence time in the emulsion zone decreases significantly [10,34]. Due to that the contribution of the emulsion zone to the total decarburization rate decreases significantly towards the end of the blow. It follows that at high carbon contents it is possible to increase the decarburization rate in the emulsion zone by increasing the droplet generation rate.

It is possible to identify the three decarturization periods characteristic of the BOF operation on the total decarburization rate graph shown in Figure 8a:

- Period I: As the silicon content of the metal bath decreases, the decarburization rate increases
- Period II: Decarburization rate stays approximately constant
- Period III: Decarburization rate is controlled by mass transfer of carbon in the metal bath

The change from Period II to III is shown by the dashed line in Figure 8a, but it can also be seen in the compostion of the gases exiting the furnace in Figure 8b. As the decarburization at the impact zone decreases, more oxygen becomes available for the post-combustion reaction explaining the increase in the percentage of CO_2 at approximately 14 min. The percentage of CO_2 in Figure 8b is slightly higher than would normally be observed in practice. This can be due to the ideal assumption that all the oxygen not used in the oxidation reactions is consumed in the post-combustion of CO.

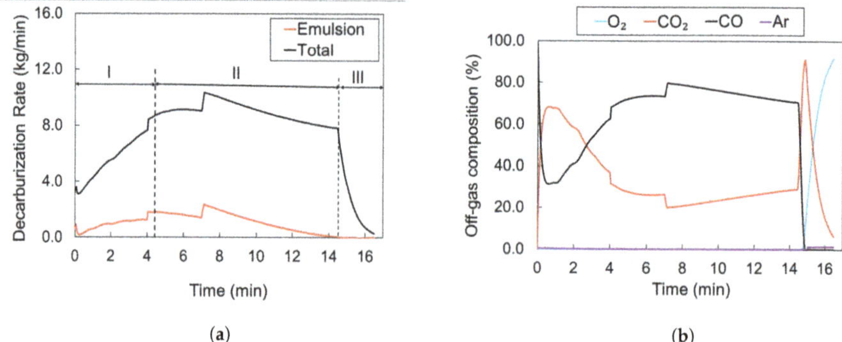

(a)   (b)

Figure 8. (**a**) Total decarburization rate and decarburization rate at the emulsion zone and (**b**) composition of the off-gas stream exiting the BOF.

The profile for the lance height and oxygen flow rate is shown in Figure 9a. The effect of lance height changes on the decarburization rate is clear at 4 min and 7 min in Figure 8a: Lowering the lance height increases the droplet generation rate, as well as the rate constants for the decarburization reactions taking place at the impact zone, which leads to a higher decarburization rate.

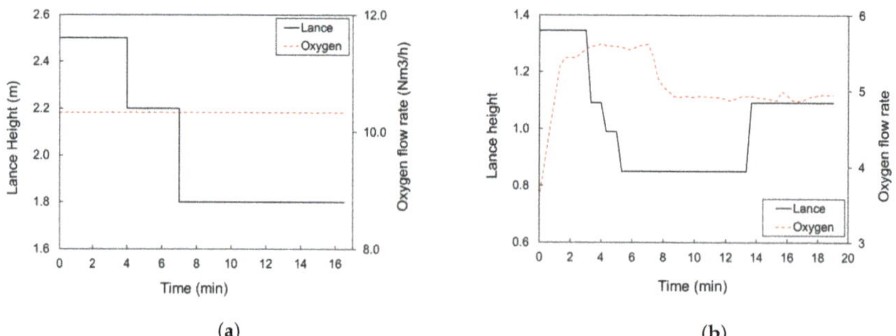

(a)   (b)

Figure 9. (**a**) Control profile for Cicutti et al. [13]'s data and (**b**) scaled control profiles for a heat from Plant A.

The contribution of the decarburization in the emulsion to the total decarburization is significantly lower for the present work than previously suggested [8,17]. Rout et al. [17] suggested that 76% of the total decarburization happens in the emulsion, while in the modeling approach adopted by Sarkar et al. [8] decarburization only takes place in the emulsion zone. For the current paper, the emulsion zone was responsible for 15% of the total carbon removed. The reason why the contribution of the decarburization in the emulsion is lower for the present study is because of the significantly lower droplet generation rate R_B. Sarkar et al. [8] modified the droplet generation rate (Equation (46)) by a factor of 15. Using a modified correlation for R_B, Rout et al. [35] obtained a droplet generation rate similar in magnitude to Sarkar et al. [8]. It is not currently viable to measure how much decarburization occurs at the impact and emulsion zones individually, and it may be the case that a different set of parameter values yields approximately the same total decarburization rate. However, taking into account the gradual decrease in the decarburization rate in the emulsion zone in Figure 8a, it can be inferred that for a very high contribution of the emulsion to the total decarburizaton rate, Period II would no longer be characterized by an approximately constant decarburization rate.

The slag composition throughout the blow is presented in Figure 10 for SiO_2, CaO, FeO and MgO. There is a good agreement between the values predicted by the model and the data. The large content of silicon dioxide in the slag during the first minute is due to 800 kg of gravel addition. A large SiO_2 content increases the slag viscosity, reducing the decarburization rate at the emulsion and allowing FeO to build up. Flux dissolution slows down significantly at high slag viscosities, but as the blow proceeds SiO_2 gets diluted by FeO and the flux dissolution rate increases. The error between the model prediction and measured data is, most likely, due to the treatment of slag as a homogeneous phase for the density and viscosity calculations [36].

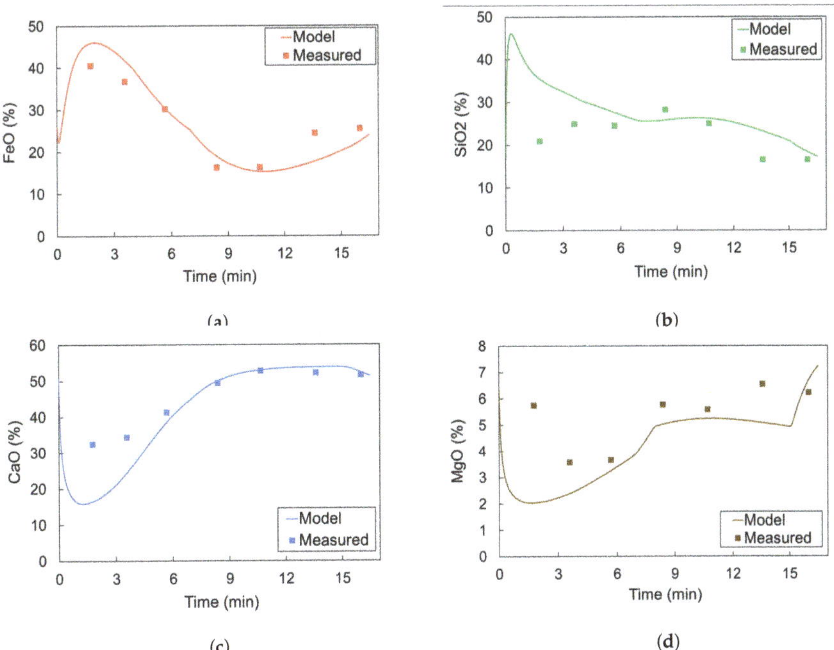

Figure 10. Evolution of slag composition for the Cicutti data [13] and model prediction: (**a**) FeO, (**b**) SiO_2, (**c**) CaO, (**d**) MgO.

A comparison between the carbon content prediction by the present and previous [6,8,9] studies is shown in Figure 11. Only for the current model, the carbon content prediction starts from time zero, and enregy balances are included; moreover, the quality of the prediction itself is quite good compared with previous works. This is also the only study for which the mathematical model was transposed as a DAE system and integrated using a variable step size solver, whilst in the aforementioned works integration was carried out using a fixed step size. The first main benefit stemming from the current implementation is the reduced computational time as shown in Table 3. Moreover, the convergence is taken care of by the integrator, and convergence studies based on step size are not required. Secondly, the dynamic model can be easily built within an optimization framework to determine the optimal input trajectories.

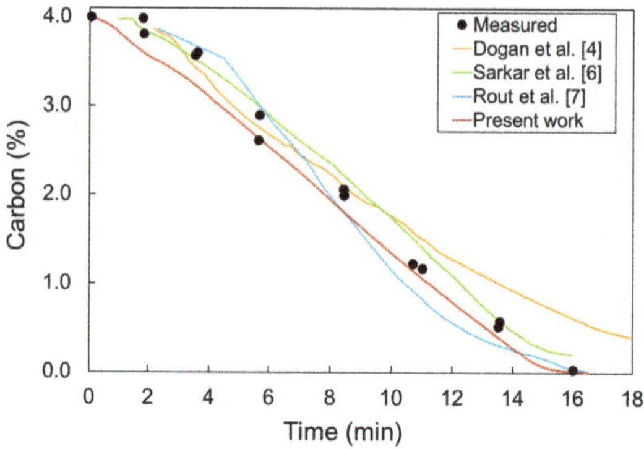

Figure 11. Comparison of the carbon content prediction by different models [6,8,9] and the measured values [13] for a 200-ton furnace.

Table 3. Simulation time for Cicutti et al. [13]'s data required in different studies.

Model	Computer	Software	Solution Time
Dogan et al. [6]	Pentium (R) 4 CPU 3.00 GHz and 3GB of RAM	Scilab	240 min
Sarkar et al. [8]	Not given	Matlab	27 min
Rout et al. [9]	Intel(R) Core(TM) i5-4570 CPU @3.20 GHz and 8 GB RAM	Matlab	20 min
Present study	Intel(R) Core(TM) i7-7700 CPU @3.16GHz and 16.0 GB RAM	Python 3.7	0.036 min

3.3. Simulation Results for Plant A Operations

The developed framework model was also used to simulate 71 heats using the data provided by Plant A for its BOF operation. The model parameters, shown in Table 1, were manually adjusted based on data collected at the end of the blow for the end-point carbon content, slag composition and temperature. The extreme conditions in the BOF often prevent samples from being collected during the blow, therefore there is no data regarding the state of the system during the blow.

The goodness of the prediction was evaluated in terms of the average of the model predictions and process data for the 71 heats. Figure 12 shows the quality of the prediction for the end-point slag composition together with the standard deviation. The results indicate that the model is able to predict the slag composition at the end of the blow reasonably well.

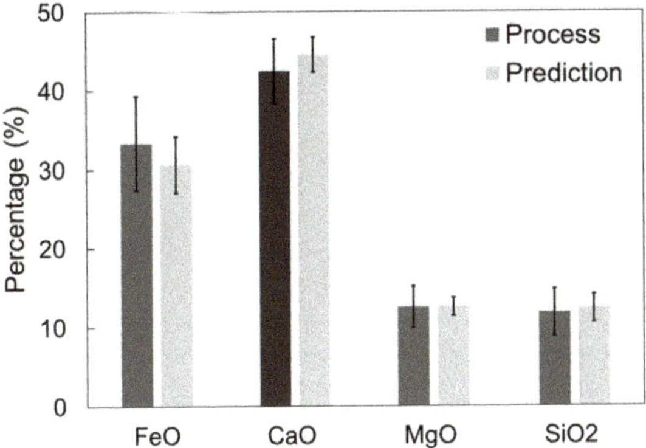

Figure 12. Average process and predicted values for the end-point slag composition in weight percentage.

The carbon content of the metal bath can be measured using a carbon probe before tapping, or a sample is collected and the measurement is done in a laboratory setting, with the second giving the more accurate results. However, laboratory results are not available for all the heats for the current study. The average end-point carbon content and temperature measured using the probe and the model's prediction for the 71 batches are shown in Figure 13, based on which it is possible to conclude that the model performs fairly well. The model was able to predict the end-point carbon content of 80% of the heats with a precision of ±0.03%, and the end-point temperature of 61% of the heats with a precision of ±30 K.

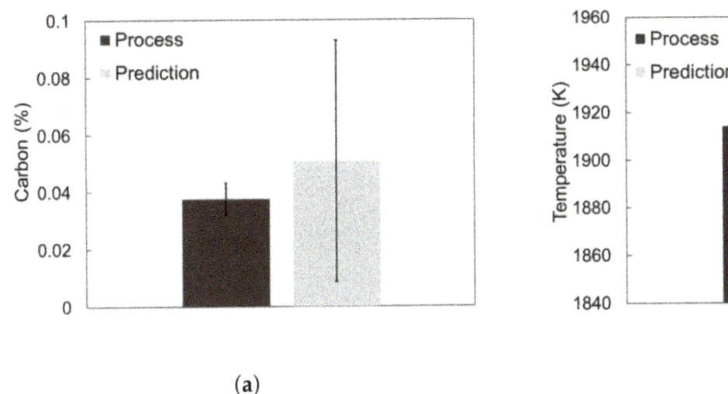

(a)

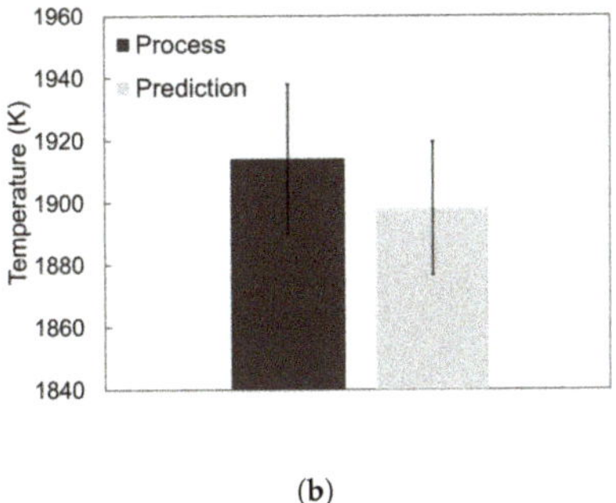

(b)

Figure 13. Average and standard deviation of the model predictions and process data for the end-point (**a**) carbon content of the liquid metal and (**b**) temperature of the liquid metal.

The prediction error can be attributed to several factors. If the furnace is used back-to-back, there is always some slag left over from one heat to another, however its temperature and mass are not measured and the initial mass of slag present in the system is not accurately known. Moreover, the initial temperature of the hot metal charged to the furnace is also not precisely known since it is measured, on average, 40 minutes before the blow starts. Furthermore, due to the lack of data, it is assumed that all scrap types have similar properties, which can lead to inaccurate prediction of the heat needed for complete melting. Another potential reason could be the assumption that the off-gas and slag have the same temperature; however, the data available is not sufficient to estimate the actual off-gas temperature during the process. This, with the absence of sub-models to account for phosphorus oxidation, manganese oxidation and the formation of higher iron oxides, can explain the poorer performance of the model in predicting the end-point temperature.

For each heat, the mass of iron ore, flux, scrap and hot metal charged to the furnace is different, explaining the variation in the predicted and measured values. Since the inputs are always changing, it makes it challenging to identify the primary source for larger deviations between the simulated and actual data. Plant A has three different input profiles for the oxygen flow rate and lance height, however the overall quality of the predictions is not significantly affected by the different control profiles indicating that the model can potentially be used to simulate a wide range of operating conditions.

The trajectories for the temperature, carbon content and slag composition are shown in Figure 14 for one of the heats. We observe that the trajectories are similar to the Cicutti case. The content of SiO_2 is lower for Plant A operation at the beginning of the blow because no gravel is added. Moreover, the rate of silicon oxidation is slower due to the lack of bottom stirring. As mentioned before, at low SiO_2 contents, decarburization in the emulsion is enhanced due to the reduced slag viscosity. This is evident in Figure 14a by the larger difference between the carbon content of the droplets and the carbon content of the metal bath at high FeO contents at the beginning of the blow. As expected, due to the lack of bottom stirring, the FeO content of the slag increases much faster than that for Cicutti's case toward the end of the blow. The rapid increase can also be explained by the higher lance height towards the end of the blow compared to the mid-blow. The scaled control profiles for the oxygen flow rate and lance height for the given heat are shown in Figure 9b.

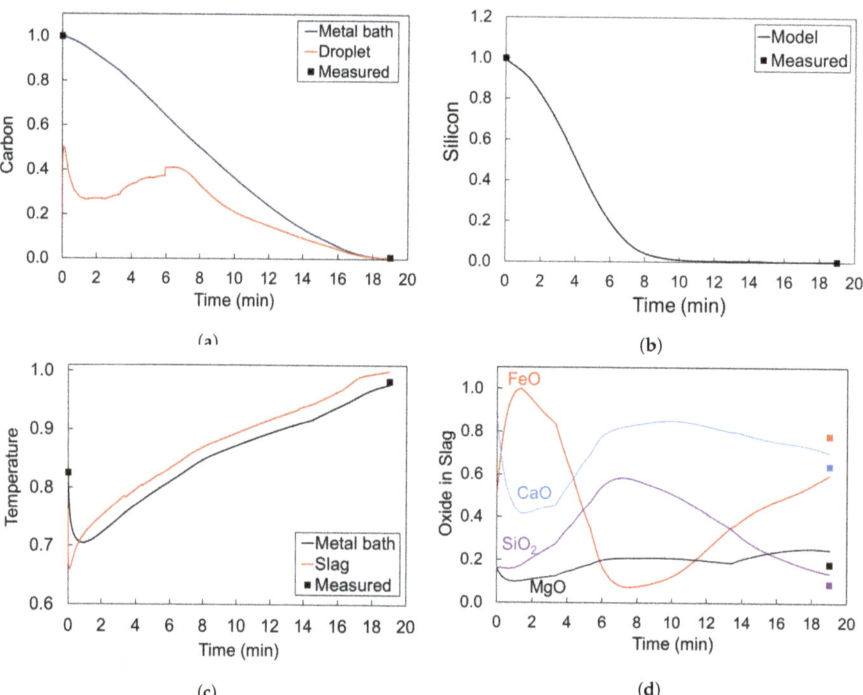

Figure 14. Evolution of: (**a**) Carbon content of liquid metal and final carbon content $C_{C,f}$ of the metal droplets, (**b**) silicon content of liquid metal, (**c**) slag and metal bath temperature and (**d**) slag composition. The values have been scaled for proprietary reasons.

Figure 14 indicates that the model is able to predict the end-point carbon content, silicon content and temperature of liquid metal reasonably well. It also provides insights about the trajectory followed by the variables and how the change in one or more of the inputs affects the system. The developed framework could potentially aid in finding more profitable operation modes as well as in understanding the more complex relationships between process variables.

4. Conclusions

The dynamic mathematical model presented in this study extends the work of Dogan et al. [6] by incorporating slag formation and energy balances. The models for the decarburization in the emulsion zone and scrap melting were updated with more recent findings, and an empirical relationship to calculate the calcium oxide saturation was obtained. Therefore, the main phenomena taking place in the BOF were accounted for, making the model suitable for further studies.

The mathematical model was translated as a DAE system to an open source environment (Python with CasADi [12]). The current implementation allowed for a significantly shorter simulation time compared with previous studies [6,8,9]. Moreover, as more complex and detailed models for the phenomena taking place in the BOF are developed, they can be incorporated relatively easily using the current framework.

It was shown that the updated model gives a better prediction of the carbon content trajectory for Cicutti et al. [13]'s data compared with the previous version, and more recently developed dynamic models for the BOF [8,9]. The dynamic model was also used to simulate 71 heats for a real industrial BOF operation. The model predictions for the end-point carbon content, slag composition and temperature agree reasonably well with the data for a wide range of operating conditions.

Author Contributions: Conceptualization, D.D., C.S and N.D.; Methodology, D.D., C.S and N.D.; Software, D.D.; Validation, D.D., C.S. and N.D.; Formal Analysis, D.D., C.S. and N.D.; Investigation, D.D., C.S. and N.D.; Resources, D.D., C.S. and N.D.; Data Curation, D.D.; Writing – Original Draft Preparation, D.D.; Writing – Review & Editing, D.D., C.S. and N.D.; Visualization, D.D.; Supervision, C.S. and N.D.; Project Administration, C.S. and N.D.; Funding Acquisition, C.S. and N.D. All authors have read and agreed to the published version of the manuscript.

Funding: This research was funded by OCE (Ontario Centres of Excellence), Praxair, McMaster Steel Research Center (SRC) and McMaster Advanced Control Consortium (MACC).

Acknowledgments: We thank Ameya Kadrolkar for making the model for droplet decarburization available for this work. We are also grateful to Eugene Pretorious and John Thomson for helpful discussions and valuable suggestions.

Conflicts of Interest: The authors declare no conflict of interest.

References

1. World Steel Association. Fact Sheet: Steel and Raw Materials. 2016. Available online: https://www.worldsteel.org/publications/fact-sheets.html (accessed on 27 March 2019).

2. Molloseau, C.; Fruehan, R. The reaction behavior of Fe-CS droplets in CaO-SiO 2-MgO-FeO slags. *Metall. Mater. Trans. B* **2002**, *33*, 335–344. [CrossRef]

3. Chen, E.; Coley, K.S. Kinetic study of droplet swelling in BOF steelmaking. *Ironmaking Steelmaking* **2010**, *37*, 541–545. [CrossRef]

4. Jalkanen, H. Experiences in physicochemical modelling of oxygen converter process (BOF). Adv. Process. Met. Mater. **2006**, *2*, 541–554.

5. Kattenbelt, C.; Roffel, B. Dynamic modeling of the main blow in basic oxygen steelmaking using measured step responses. *Metall. Mater. Trans. B* **2008**, *39*, 764–769. [CrossRef]

6. Dogan, N.; Brooks, G.A.; Rhamdhani, M.A. Comprehensive model of oxygen steelmaking part 1: Model development and validation. *ISIJ Int.* **2011**, *51*, 1086–1092. [CrossRef]

7. Lytvynyuk, Y.; Schenk, J.; Hiebler, M.; Sormann, A. Thermodynamic and Kinetic Model of the Converter Steelmaking Process. Part 1: The Description of the BOF Model. *Steel Res. Int.* **2014**, *85*, 537–543. [CrossRef]

8. Sarkar, R.; Gupta, P.; Basu, S.; Ballal, N.B. Dynamic modeling of LD converter steelmaking: Reaction modeling using Gibbs' free energy minimization. *Metall. Mater. Trans. B* **2015**, *46*, 961–976. [CrossRef]

9. Rout, B.K.; Brooks, G.; Rhamdhani, M.A.; Li, Z.; Schrama, F.N.; Sun, J. Dynamic Model of Basic Oxygen Steelmaking Process Based on Multi-zone Reaction Kinetics: Model Derivation and Validation. *Metall. Mater. Trans. B* **2018**, *49*, 537–557. [CrossRef]

10. Kadrolkar, A.; Dogan, N. The Decarburization Kinetics of Metal Droplets in Emulsion Zone. *Metall. Mater. Trans. B* **2019**, *50*, 2912–2929. [CrossRef]

11. Shukla, A.K.; Deo, B.; Robertson, D.G.C. Scrap Dissolution in Molten Iron Containing Carbon for the Case of Coupled Heat and Mass Transfer Control. *Metall. Mater. Trans. B* **2013**, *44*, 1407–1427. [CrossRef]

12. Andersson, J.A.E.; Gillis, J.; Horn, G.; Rawlings, J.B.; Diehl, M. CasADi—A software framework for nonlinear optimization and optimal control. *Math. Program. Comput.* **2019**, *11*, 1–36. [CrossRef]

13. Cicutti, C.; Valdez, M.; Pérez, T.; Petroni, J.; Gomez, A.; Donayo, R.; Ferro, L. Study of slag-metal reactions in an LD-LBE converter. In Proceedings of the 6th International Conference on Molten Slags, Fluxes and Salts, Helsinki, Finland, 12-17 June 2000.

14. Deo, B.; Boom, R. *Fundamentals of Steelmaking Metallurgy*; Prentice Hall International: Upper Saddle River, NJ, USA, 1993.

15. Koria, S.C.; Lange, K.W. Penetrability of impinging gas jets in molten steel bath. *Steel Res.* **1987**, *58*, 421–426. [CrossRef]

16. Kitamura, S.y.; Kitamura, T.; Shibata, K.; Mizukami, Y.; Mukawa, S.; Nakagawa, J. Effect of stirring energy, temperature and flux composition on hot metal dephosphorization kinetics. *ISIJ Int.* **1991**, *31*, 1322–1328. [CrossRef]

17. Rout, B.K.; Brooks, G.; Rhamdhani, M.A.; Li, Z.; Schrama, F.N.; Overbosch, A. Dynamic Model of Basic Oxygen Steelmaking Process Based on Multizone Reaction Kinetics: Modeling of Decarburization. *Metall. Mater. Trans. B* **2018**, *49*, 1022–1033. [CrossRef]

18. O'Malley, R.J. The Heating and Melting of Metallic DRI Particles in Steelmaking Slags. Ph.D. Dissertation, Massachusetts Institute of Technology, Cambridge, CA, USA, 1983.

19. Zhang, L. Modeling on melting of sponge iron particles in iron-bath. *Steel Res.* **1996**, *67*, 466–474. [CrossRef]

20. Pineda-Martínez, E.; Hernández-Bocanegra, C.A.; Conejo, A.N.; Ramirez-Argaez, M.A. Mathematical Modeling of the Melting of Sponge Iron in a Bath of Non-reactive Molten Slag. *ISIJ Int.* **2015**, *55*, 1906–1915. [CrossRef]

21. Shukla, A.K. Dissolution of Steel Scrap in Molten Metal During Steelmaking. Ph.D. Dissertation, Indian Institute of Technology, Kanpur, India, 2011.

22. Dogan, N. Mathematical Modelling of Oxygen Steelmaking. Ph.D. Dissertation, Swinburne University of Technology, Melbourne, Australia, 2011.

23. Caffery, G.; Rafiei, P.; Honeyands, T.; Trotter, D.; Marketing, B.B. Understanding the Melting Characteristics of HBI in Iron and Steel Melts. In Asssociation for Iron & Steel Technology, Proceedings of the AISTech Conference, Nashville, TN, USA, 15–17 September 2004, Volume 1, p. 503.

24. Gaye, H.; Wanin, M.; Gugliermina, P.; Schittly, P. Kinetics of scrap dissolution in the converter: Theoretical model and plant experimentation. In Proceedings of the 68th Steelmaking Conference, Detroit, USA, 14–17 April 1985; pp. 91–103

25. MacRosty, R.D.; Swartz, C.L. Dynamic optimization of electric arc furnace operation. *AIChE J.* **2007**, *53*, 640–653. [CrossRef]

26. Bekker, J.G.; Craig, I.K.; Pistorius, P.C. Modeling and Simulation of an Electric Arc Furnace Process. *ISIJ Int.* **1999**, *39*, 23–32. [CrossRef]

27. Dogan, N.; Brooks, G.A.; Rhamdhani, M.A. Kinetics of flux dissolution in oxygen steelmaking. *ISIJ Int.* **2009**, *49*, 1474–1482. [CrossRef]

28. Hamano, T.; Horibe, M.; Ito, K. The dissolution rate of solid lime into molten slag used for hot-metal dephosphorization. *ISIJ Int.* **2004**, *44*, 263–267. [CrossRef]

29. Umakoshi, M.; Mort, K.; Kawai, Y. Dissolution rate of burnt dolomite in molten FetO-CaO-SiO2 slags. *Trans. Iron Steel Inst. Jpn.* **1984**, *24*, 532–539. [CrossRef]

30. Kadrolkar, A.; Andersson, N.Å.; Dogan, N. A Dynamic Flux Dissolution Model for Oxygen Steelmaking. *Metall. Mater. Trans. B* **2017**, *48*, 99–112. [CrossRef]

31. Subagyo.; Brooks, G.A.; Coley, K.S.; Irons, G.A. Generation of Droplets in Slag-Metal Emulsions through Top Gas Blowing. *ISIJ Int.* **2003**, *43*, 983–989. [CrossRef]

32. Hindmarsh, A.C.; Brown, P.N.; Grant, K.E.; Lee, S.L.; Serban, R.; Shumaker, D.E.; Woodward, C.S. SUNDIALS: Suite of nonlinear and differential/algebraic equation solvers. *ACM Trans. Math. Softw.* **2005**, *31*, 363–396. [CrossRef]

33. Lytvynyuk, Y.; Schenk, J.; Hiebler, M.; Sormann, A. Thermodynamic and kinetic model of the converter steelmaking process. Part 2: The model validation. *Steel Res. Int.* **2014**, *85*, 544–563. [CrossRef]

34. Brooks, G.; Pan, Y.; Coley, K.; others. Modeling of trajectory and residence time of metal droplets in slag–metal–gas emulsions in oxygen steelmaking. *Metall. Mater. Trans. B* **2005**, *36*, 525–535. [CrossRef]

35. Rout, B.K.; Brooks, G.; Rhamdhani, M.A.; Li, Z.; others. Modeling of droplet generation in a top blowing steelmaking process. *Metall. Mater. Trans. B* **2016**, *47*, 3350–3361. [CrossRef]

36. Mills, K.C. Viscosities of Molten Slags In *Slag Atlas*; 2nd ed.; Verlag Stahleisen: Düsseldorf, Germany, 1995; p. 353.

Article

Modeling the Effect of Scrap on the Electrical Energy Consumption of an Electric Arc Furnace

Leo S. Carlsson *, Peter B. Samuelsson and Pär G. Jönsson

Royal Institute of Technology , Brinellvägen 23, 114 28 Stockholm, Sweden; petersam@kth.se (P.B.S.); parj@kth.se (P.G.J.)
* Correspondence: leoc@kth.se

Received: 29 July 2020; Accepted: 24 August 2020; Published: 26 August 2020

Abstract: The melting time of scrap is a factor that affects the Electrical Energy (EE) consumption of the Electric Arc Furnace (EAF) process. The EE consumption itself stands for most of the total energy consumption during the process. Three distinct representations of scrap, based partly on the apparent density and shape of scrap, were created to investigate the effect of scrap on the accuracy of a statistical model predicting the EE consumption of an EAF. Shapley Additive Explanations (SHAP) was used as a tool to investigate the effects by each scrap category on each prediction of a selected model. The scrap representation based on the shape of scrap consistently resulted in the best performing models while all models using any of the scrap representations performed better than the ones without any scrap representation. These results were consistent for all four distinct and separately used cleaning strategies on the data set governing the models. In addition, some of the main scrap categories contributed to the model prediction of EE in accordance with the expectations and experience of the plant engineers. The results provide significant evidence that a well-chosen scrap categorization is important to improve a statistical model predicting the EE and that experience on the specific EAF under study is essential to evaluate the practical usefulness of the model.

Keywords: electrical energy consumption; Electric Arc Furnace; scrap melting; statistical modeling

---

1. Introduction

Electrical Energy (EE) can account for between 40–66% of the total energy usage during the Electric Arc Furnace (EAF) process, which is a number that highlights the importance of further improvements in modeling of the EE consumption [1]. The energy losses, which partly governs the EAF process energy dynamics, are mostly related to off-gases, slag, dust, furnace cooling, electrical and radiative losses. Most of these energy losses are closely linked to the total process time of any given heat. The process time itself is influenced by numerous impositions. One of these is the melting time of the charged raw materials, which in the scope of this study is primarily steel scrap.

Many articles have studied and proposed statistical models predicting the EE consumption of the EAF [1,2]. However, only a handful of studies have used scrap types as input variables to a statistical model predicting the EE of an EAF [2–7]. Neither of these studies analyze the contributions of each scrap type on specific predictions by the statistical model. Verifying the effects of the input variables on the complete prediction space is paramount to evaluate the practical usefulness of any statistical model let alone to make the users, i.e., process engineers, trust the model. Only when these two conditions are met can the model possibly be used to solve practical problems. An example of a practical problem where an EE prediction model can be used is when determining the EE requirement for the EAF in a Demand Side Management (DSM) system, which optimizes the processes in a steel plant with respect to the available power in the transmission line [8].

Shapley Additive Explanations (SHAP) is a recent development in the field of interpretable machine learning [9], and has previously been used to analyze a statistical model predicting the EE of an EAF producing stainless steel [10]. Tap-to-Tap time (TTT), delays, and total charged weight were found to be the three most influencing variables. TTT and total charged weight were correctly interpreted by the model with respect to what is known from process metallurgical experience. The delay variable was incorrectly interpreted by the model, which was concluded to be due to the high correlation between delays with TTT.

The aim of the current study is to investigate the effect of scrap types on a statistical model predicting the EE of an EAF. Although verifying the effects of other input variables on the model output is important, it is not the main focus of this article. For such an analysis we refer to previous studies [2,10]. To investigate the effects of scrap, three distinct representations of scrap types based on the plant scrap codes, scrap physical shape, and scrap apparent density, respectively, will be used in the models. The reasons are two-fold. First, to provide the steel plant engineers with an intuitive and simple method to categorize scrap for modeling purposes. Second, to find the optimal scrap representation for the prediction problem with respect to the accuracy and precision of the statistical model.

In addition to its scrap-oriented focus, this study further builds on the modeling methodology presented in two previous studies [2,10]. Four different data cleaning strategies will be used to investigate the effect of data cleaning on the accuracy of the statistical model and an additional non-linear statistical model framework will be employed; Random Forests (RF). Furthermore, this study uses data from an EAF producing steel for tubes, rods, and ball-bearing rings and not from an EAF producing stainless steel, thus broadening the application of the modeling methodology.

The results demonstrated that the three subsets of input variables provided by the scrap representations all increase the performance of the models. Using SHAP, it was found that heavy scrap, i.e., scrap with low surface-area-to-volume ratio, contributed to an increased EE consumption while steel sheets, a scrap type with high surface-area-to-volume ratio, contributed to a decreased EE consumption. These findings were confirmed by the steel plant engineers to agree well with previous experiences using these scrap types as raw material.

2. Background

2.1. Melting of Steel Scrap in Liquid Steel

2.1.1. Driving Forces

The driving forces in scrap melting are present in any process in which scrap melting occurs. However, the driving factors vary significantly between the EAF and Basic Oxygen Furnace (BOF). This section highlights the driving factors in scrap melting with respect to the EAF in the steel plant of study. The goal is not to create an extensive review over the various research topics governing the melting of scrap, but rather to motivate the scrap representations used in the experimental part of this study. A comprehensive review in the field was compiled by Freidrich [11]. Prominent later developments in the field of scrap melting have been summarized in a recent review [12].

The melting of solid scrap in liquid steel is dominated by several factors. These are temperature gradients between solid scrap and liquid steel, concentration gradients between solid scrap and liquid steel, the freezing effect, the rate of stirring of the steel melt. These phenomena are explained further below.

Temperature gradients between the solid scrap and the steel melt is one of the most important driving factors in the melting of steel scrap. The melting rate, in m/s, can be determined by the following equation:

$$\frac{dx}{dt} = h \cdot \frac{T_{HM} - T_{liq}}{\rho_{scr} \cdot (H_s + (T_{HM} - T_{liq}) \cdot c_p)} \tag{1}$$

where T_{HM} is the temperature of the molten steel and T_{liq} is the scrap melting temperature. Furthermore, H_s is the heat of melting of scrap, c_p is the specific heat of scrap, ρ_{scr} is the density of the scrap metal, and h is the heat transfer coefficient in the interface of the molten steel and scrap [12]. The higher the temperature gradient, $T_{HM} - T_{liq}$, the faster the steel scrap will melt.

Alloying element gradients also contribute to the melting of scrap in a process known as dissolution. In this case, alloying elements migrate to the solid-liquid metal interface. The most dominant alloying element in this process is carbon. However, in the BOF, the carbon concentration difference can exceed 4 wt-% while in the EAF the carbon concentration difference seldom exceeds 1 wt-%.

Assuming that the dissolution rate by carbon can be determined by the shortest length of the scrap, the following equation can be used:

$$\frac{dx}{dt} = \frac{\beta \cdot (C_l - C_i)}{\rho_s (C_l - C_0)} \tag{2}$$

where β is the mass transfer coefficient of carbon, C_l is the carbon content in the liquid steel, C_0 is the initial carbon content in the steel scrap, and C_i is the carbon content in the solid-liquid interface [13].

A similar equation can be determined for silicon, which can also be an alloying element to account for should the difference in silicon concentration between steel scrap and the molten steel be large.

The **freezing effect** occurs in the solid-liquid interface when the scrap first comes in contact with the liquid steel. A solidified shell is formed due to the large temperature difference between the two. This means that the volume of the scrap increases initially. The solidified shell is proportional to the surface area of the scrap that is submerged in the hot metal or molten steel. Hence, the reduction in the steel scrap size does not occur instantly, rather it decreases after the solidified shell has melted.

The **stirring** velocity is the velocity of the melt in the boundary layer between the melt and the scrap surface area. Numerous studies have related the stirring velocity to the mass transfer coefficient on scrap in liquid steel. However, there exists a wide range of reported mass transfer coefficient values for scrap in liquid steel under forced convection [12]. Nevertheless, a commonly deduced relationship between the mass transfer coefficient and the stirring velocity may be written as follows:

$$h_{scr} = c \cdot u^p \tag{3}$$

where h_{scr} is the mass transfer coefficient under forced convection, c and p are constants that are determined experimentally. u is the average stirring power, which is related to the average stirring velocity due to the physical relationship between energy, momentum and velocity. The stirring power is governed by, for example, oxygen blowing and carbon boil.

Furthermore, the effect of stirring on the melting rate of scrap in the EAF is low compared to the BOF since the stirring is more intense in the latter, i.e., higher stirring velocity governed by the stirring power per unit volume. Furthermore, one should not expect the stirring to be very intense in the EAF since the liquid steel depth is low and the solid-to-liquid ratio is high prior to the final stages of the process, i.e., superheating, which hampers the flow velocity of the liquid steel. The device that primarily facilitates stirring in the EAF is oxygen lancing, but other devices such as porous plugs and induction stirring enhance the stirring.

2.1.2. Scrap Surface-Area-to-Volume Ratio

It is evident that the aforementioned factors are dependent on the surface-area-to-volume ratio of the scrap pieces. On the one hand, the effect from temperature and alloying element gradients influence on the complete surface area exposed to the steel melt. On the other hand, the mass of the steel scrap piece determines the melting time since more mass needs to be heated (Equation (1)) and more mass of the alloying elements have to be transported (Equation (2)). The mass is proportional to the volume of the scrap piece. Thus, the surface-area-to-volume ratio can be expressed either as a function of the surface area and volume or as a function of the surface area, apparent density of scrap, and mass of the scrap piece:

$$R_{SV} = \frac{A}{V} = \frac{A \cdot \rho_s}{m} \tag{4}$$

Hence, to facilitate lower melting times one should use scrap that has a high surface-area-to-volume ratio. The surface-area-to-volume ratios of some elementary geometrical shapes are presented below to illuminate the effects of the surface-area-to-volume ratio in scrap melting. However, real scrap pieces are often of more complex geometric shapes.

Sphere: $\frac{A}{V} = \frac{4\pi r^2}{\frac{4\pi r^3}{3}} = \frac{3}{r}$

Cylinder: $\frac{A}{V} = \frac{2\pi r l + 2\pi r^2}{\pi r^2 l} = 2(\frac{1}{r} + \frac{1}{l})$

Cube: $\frac{A}{V} = \frac{6l^2}{l^3} = \frac{6}{l}$

Square plate: $\frac{A}{V} = \frac{2l^2 + 4lt}{l^2 t} = 2(\frac{1}{t} + \frac{2}{l})$

The thickness is defined as the thinnest dimension of the scrap piece. For the cylinder, one ought to keep $r << l$. For the square plate, one ought to keep $t << l$. The cube and sphere are equidistant from the center of the scrap piece, which means that one should keep the length and radius as small as possible, respectively.

2.1.3. The Steel Plant of Study

The dissolution of steel scrap in molten steel due to carbon content gradients between the steel scrap and molten metal are not significant in the EAF of study. The steel plant does not produce high Si steels nor does the carbon content vary significantly between the hot heel and the charged scrap. The temperature gradients between the steel scrap and molten steel will be similar for all heats which have the 5–10-ton hot heel remaining in the furnace at the start of the heat. However, some heats will be produced without this initial 5–10 ton of hot heel. This will affect the melting through temperature gradients mainly for the scrap charged by the first basket. The main source of stirring in the EAF of study is by oxygen lancing. The stirring is mainly facilitated by CO from carbon boil.

2.2. The Electric Arc Furnace

2.2.1. Process

The steel plant of study uses an EAF with a nominal charging capacity of 110 ton and a transformer system of 80 MVA. The steel products produced are rods, tubes, and ball-bearing rings. The EAF does not use a pre-heater but uses a hot heel, which is molten steel left over from each previous heat. The amount of hot heel is 5–10 ton. During operation oxyfuel burners are used to remove the cold spots between the furnace electrodes. The burners can also function as oxygen lances to inject oxygen to the molten metal.

The process begins with a default mode where 5–10 ton of hot heel and a full basket of scrap are present. The scrap basket is layered with different scrap types according to a pre-specified recipe. When the basket of scrap has been charged into the furnace, the lowermost scrap mixes with the hot heel and creates a mixture of scrap and partially molten steel. The melting phase starts when the transformer is powered on and the electrode arcs are bored down into the upper layer of the scrap. This process proceeds until enough space is available for the second basket of scrap to be charged. In this instance, the amount of molten steel has increased but partially molten steel scrap is still present in the steel bath. Furthermore, heaps of scrap are also present around the electrode arcs. The scrap from the second basket gets piled on top of these heaps of scrap. The second melting phase starts in a similar manner as the first melting phase. When a clear visible molten steel bath is present, the refining phase starts. Oxygen lancing and injected fine carbon facilitate a foaming slag which increases the energy yield from the arcs to the steel bath. Some partially molten steel scraps are present in molten bath during the refining phase. These are usually large and bulky pieces of scrap that require longer exposure time to the molten steel to completely melt. Oxygen lancing also provides stirring of the steel bath, which increases the melting rate of the remaining steel scraps. The contribution of the stirring is expected to be minor compared to stirring using, for example, an electromagnetic field. Finally, the steel is tapped into a ladle and the steel is further treated in downstream processes.

The described EAF process is illustrated in Figure 1.

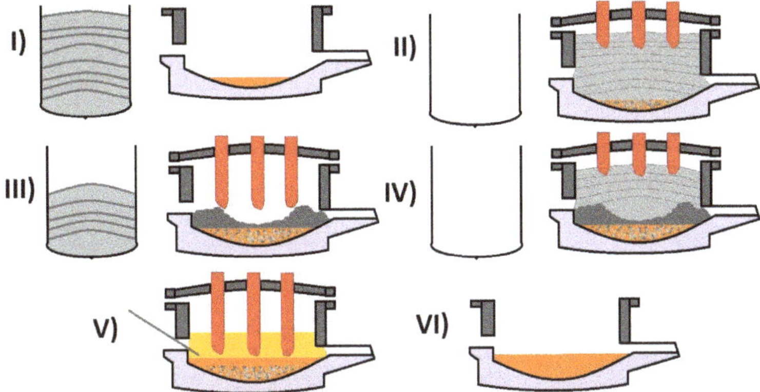

Figure 1. The Electric Arc Furnace (EAF) process in the steel plant considered in the current article. It shows the process divided into six parts. (**I**) The first scrap basket is fully charged with up to 10 layers of scrap. The furnace shell contains hot heel. (**II**) The scrap from the first basket has been charged into the furnace whereupon the lowermost scrap layers intermix with the hot heel. The furnace arcs are bored down into the scrap and the transformer is powered on. (**III**) After the first melting phase the furnace contains a steel scrap-melt mixture and heaps of partially solid scrap. The second scrap basket is charged in as much to fully fill the furnace in the next step. (**IV**) Charging of the second basket. (**V**) Refining of the molten steel where injected carbon and oxygen lancing facilitate an even layer of foaming slag. (**VI**) The heat is ready for tapping.

As a last step, any preparations to the furnace are made before the next heat. This can, for example, be replacement of electrodes or fettling of the furnace refractories. The steel plant of study also replaces the hot heel periodically to allow for inspection of the furnace bottom refractories. This means that some heats do not have a hot heel present at the start of the heat.

2.2.2. Charging of Scrap Baskets

Several aspects that affect the scrap charging either directly or indirectly must be highlighted. The various steel grades have pre-specified recipes, i.e., charge types, which are used to determine how much of each scrap type that will be charged into the baskets. This, in its turn, is determined by the selection of primary scrap and alternative scrap. Primary scrap is always used if it is available in the scrap yard. Alternative scrap is used if the primary scrap is not available. Each primary scrap type has a matching alternative scrap type. Hence, the recipe is a representation of an ideal charged basket, but not necessarily representative of the actual charged basket.

Available scrap types on the market also affect what charging strategies are possible. The available scrap types in the market set the limit to what performance can be expected with regards to the EE consumption. If a major scrap type that requires less EE to melt is not available, then the EE of the next heats could be expected to increase, if a scrap type with higher apparent density is used instead. Since the scrap turn-over rate can be as low as one week for several scrap types, varying available scrap types are thus to be expected. Customer demand also influences the scrap requirements since scrap with higher amount of tramp elements cannot be used for steel types with high performance requirements. The market factors and customer demand variations will not be accounted for in the numerical experiments. However, it is important to keep these factors in mind since they indirectly influence the EE consumption.

Some charging strategies are subsequently described. The first basket is usually filled with scrap to 75–100% of the total volume of the basket. The total volume of a scrap basket is 65 m^3. The second scrap basket is filled to reach the pre-specified amount of molten steel, which is approximately 100 tonnes. A third scrap basket is used if the amount of required scrap was not satisfied by the first two baskets.

Each scrap basket can be made up of a total of 10 layers of different scrap types. The scrap types in each layer are pre-specified by the scrap recipe. Combined, the layers will consist of a blend of various scrap types. Bulky scrap with high apparent density (1.4 $\geq$ tons/m^3) such as internal casting residuals are put in the bottom layers of the basket. This ensures that the bulkier scrap gets a longer exposure to the steel melt and thus reduces the risk of solid scrap pieces in the latter stages of the EAF process. Consequently, scrap with lower apparent density (0.4–1.0 tons/m^3) such as sheet and turnings are put in the intermediate and upper layers. These scrap types do not require as long exposure to the steel melt compared to other scrap types. Furthermore, the main source of heat for melting these scrap types, most likely, comes from the radiative heat transfer from the electrode arcs.

Some effect of compaction of scrap types with lower apparent densities is expected due to the weight imposed by the upper layers of scrap. Hence, the apparent density of these scrap types will be higher in a fully charged basket compared to when they are solely present in the basket or in the scrap yard. The apparent density of the scrap in the charged furnace can be assumed to have a reasonable agreement with the apparent density of the scrap in the scrap bucket post-charging.

2.2.3. Parameters Governing the EE Consumption

Several previous articles have presented estimates of the energy sources and energy sinks during the EAF process [14–18]. These reported values have been compiled in Table 1 and presents a guidance regarding which input variables that need to be considered when predicting the EE consumption. The reason behind the large percentage differences in energy sources and energy sinks is because one of the studies compiled information from 16 EAF [14], some of which used up to 90% Direct Reduced Iron (DRI) and 10% scrap as raw materials. Due to the gangue content of the DRI, mainly SiO_2, more slagformers must be added which increases the energy consumption. Furthermore, DRI contains some remaining iron oxides that will be reduced by carbon in the melt. This also requires additional energy. In addition, the amount of injected oxygen per ton charged raw material ranged from 5 m^3/t to 40 m^3/t, which contributes to the large difference in oxidation of alloying elements [14]. The other articles provided data on one EAF each, three of which used 100% scrap [15,17,18] and one that used an unspecified mix of scrap and DRI [16].

A further discussion about the choice of input variables related to the energy balance equation, but not related to scrap charging, has been published previously [1].

Table 1. Synthesized values of energy sources and energy sinks reported in [14–18].

	Energy Factor	**% of Total Energy Sources or Energy Sinks**
In	Electric	40–66%
	Oxidation of alloying elements	20–50%
	Burner fuel	2–11%
Out	Liquid steel	45–60%
	Slag and dust	4–10%
	Off-gas	11–35%
	Cooling	8–29%
	Radiation and electrical losses	2–6%

In addition, the EE consumption is partly governed by the selected scrap mix in the steel plant. This has been schematically illustrated in Figure 2. The available scrap in the scrap yard sets the limit to the charged scrap mixture in each basket. The charging strategy, comprising of both the scrap types and the layering of the basket, determines the exposed surface area of each scrap piece to the hot heel and the furnace arcs during the process. The burners and oxygen lancing facilitate the melting of the scrap. These factors in combination determine the aggregated melting time of the baskets combined. An increased melting time contributes to a longer TTT, which increases the total heat losses during the process. The increased heat loss has to be counteracted by increased amount of energy sources, which is a combination of EE consumption, burner fuel, and oxidation by alloying elements in the scrap (Table 1).

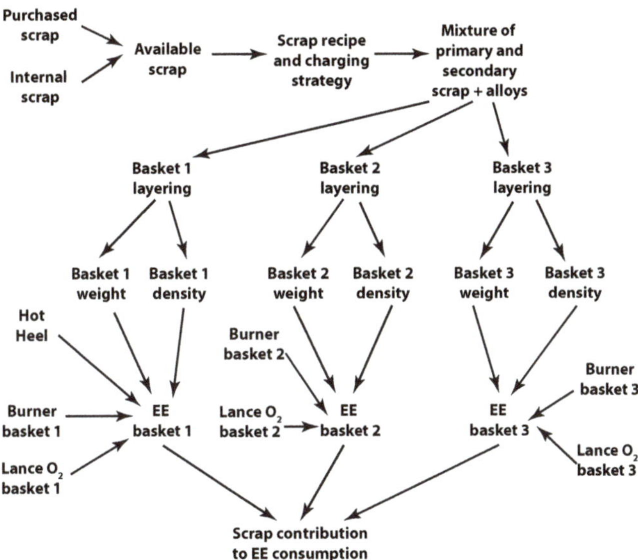

Figure 2. The relationship between the available scrap and the scrap contribution to the Electrical Energy (EE) consumption.

2.2.4. Non-Linearity

The non-linearity of the EAF process with respect to EE is caused by numerous factors. One of these factors is related to the various sub-processes of the EAF as well as the resulting total process time. For example, the heat loss through radiation and cooling losses are more prevalent during the refining stage, when the steel is molten, compared to the first charging and melting phases. Furthermore, the radiative heat transfer is proportional to T^4, which adds to the non-linearity. By varying the refining times and melting times, the resulting contribution to the EE consumption can vary significantly between heats.

Delays imposed by up- and downstream processes as well as by the process itself add to the non-linearity of the process with respect to the EE consumption. The reason is because the process time is no longer a result of the total charged scrap weight or produced steel grade. Furthermore, the delays are very hard to predict and should therefore be considered to be an external non-linear factor imposed on the EAF at all times. The effects of delays have been extensively discussed in [1].

The charged scrap types also add to the non-linearity of the process since combining varying amounts of each scrap type will affect the melting behavior. For example, large and bulky pieces of steel scrap must be exposed to the steel melt for longer times than thinner pieces of steel scrap.

Larger fractions of heavy scrap increase the melting time, which in turn increases the EE consumption. The general strategy is to charge bulkier scrap types in the lower layers of the scrap bucket. This is regardless of the amount of bulky scrap in the heat. Failing to expose bulkier scrap to the hot heel and molten steel means that partially molten pieces of scrap will remain during the latter stages of the refining process. Hence, the melting time becomes the factor that determines the TTT of the heat and hence also the EE consumption.

2.3. Statistical Modeling

2.3.1. Inherent Traits

There is one key difference that separates statistical models from physico-chemical models and that is the connection between the input and output values. Physico-chemical models present their prediction, i.e., output value, based on pre-determined equations that use the input values. These equations are related to established physical and chemical laws. Statistical models, on the other hand, interpret the values of the input variables in the context of previously observed input values and output values. Hence, the output of a statistical model is purely based on probability and does not necessarily adhere to established physical and chemical laws. The connection between the two is purely dependent on the data that is used to adapt the statistical model to the prediction problem of interest. This leads to three distinct traits unique to statistical models. These are data quality, data variability, and correlation.

Data quality is related to how close the registered value is to the true value that is intended to be measured. Uncertainties are imposed and the performance of a statistical model is reduced if the data quality is low. There are numerous sources that can affect the data quality. Two examples are the manual logging of data, which is prone to human error, and the definition of a variable in the logging system, which may differ from what is measured in reality. One common data quality issue in the steel industry is the precision of measurement equipment. For example, scales weighting raw materials can have a precision of ± 100 kg and temperature sensors can have a precision of $\pm 5\,^{\circ}$C. In this study, data quality refers to the extent to which the data is affected by the aforementioned examples.

Data variability is a requirement because variations of the values in previously observed data are what the statistical model learn. An input variable that is constant is useless to a statistical model. A constant variable in a physico-chemical model is not useless. A straightforward example is the latent heat of melting of steel, which is an important component of a physico-chemical model predicting the temperature of steel in, for example, the EAF.

Correlation is a metric that indicates the relation between random variables. Strongly correlated input variables are similar and therefore redundant as input variables to a statistical model. In this case, only one of the input variables should be selected. Weakly correlated variables, on the other hand, may be redundant. The degree of redundancy for weakly correlated variables depends on the intra-correlation between the input variable and the other input variables. For example, scrap type A may be redundant if the total scrap weight and scrap types B and C are included in a potential model. The sum of scrap type B and C implies that scrap type A must be the difference between the total scrap weight and scrap types B and C. In this case, adding scrap type A as input variable would likely not increase the accuracy of the statistical model since there is no new information gained from this input variable. It is important to note that correlation *does not* imply causation. However, statistical models lack the ability to distinguish between the two. Even though a correlation can shed light on areas where causation may exist, it is the task of the practitioner with domain-expertise to separate the causative relations from the non-causative relations. This stresses the importance of possessing knowledge about the domain in which the statistical model is used.

2.3.2. The Abstract Case

Supervised statistical models will be used in this paper which means that each row of input data, i.e., variables, has a corresponding output data point. This framework can be explained by the following steps:

1. Select statistical model framework. The available hyper-parameters are unique to the statistical model framework and are selected by the modeler.
2. Train the model using a set of matching input and output data. Continue the training phase until the accuracy of the model converges.
3. Test the model with a set of previously unseen data.
4. Evaluate the models practical usefulness using the accuracy on the test data.
5. If the accuracy is good enough, deploy the model in a production environment.

In a practical context, the model hyper-parameters (step 1) are chosen based on a comprehensive hyper-parameter search, also known as grid-search. During the parameter search, several models are trained for each combination of hyper-parameters. The combination with the highest and most stable accuracy is the most optimal hyper-parameter selection.

Non-linear supervised statistical models should be used when predicting the EE consumption of an EAF. This is because some important input variables governing the EAF process are non-linearly related to the EE consumption. The statistical model framework should always be chosen based on the nature of the prediction problem.

Although non-linear statistical models are excellent at learning complex relations between variables, these types of models are susceptible to overfitting. Overfitting means that the statistical model has adapted itself too well to a particular set of data, i.e., the training data, in such a way that it cannot predict well on future data. Combating this phenomenon is important since the relations between the variables are expected to change from the training data to test data. This is the natural course of any steel plant process. The strategies to reduce overfitting will be explained further in Section 3.4.1 where the specific model frameworks used in the numerical experiments are presented.

2.3.3. Previous Studies

Statistical models have previously been used as a tool to predict the EE consumption of the EAF. A comprehensive review of the subject has recently been published [1]. However, only four of the previous studies have used some representation of scrap types as part of the input variables in the models and as part of the model analysis [3–6].

The first study used the weight of shredded scrap as the only scrap type variable for a Multivariate Linear Regression (MLR) model [6]. The coefficient for this variable is negative, which indicates that less EE is needed than what is normally required when more shredded scrap is added. The model was then used on data from 5 different EAF, all of which used various amounts of other scrap types which were not taken into consideration by the model.

The second study used response graphs to investigate the total EE prediction response by each scrap type in the first and second baskets [4]. However, a response graph only displays the total EE prediction when varying one single input variable and does not reveal the specific contribution by each input variable.

The third study used Partial Least Squares (PLS) regression to model the EE consumption of two different EAF [3], one of which is the steel plant governing the data in the current study. However, the significance of each scrap type representation was only given by an ad-hoc subjective measure, as indicated by the descriptive words *low* and *high*.

The last study used the statistical modeling frameworks ML, RF, and Artificial Neural Networks (ANN) to predict the EE consumption [5]. The effect of the scrap types on the EE was only reported for the MLR model since the model coefficients reveal the impact of each scrap type on the EE consumption. The values of the coefficients were then compared with experience-based values.

However, some assumptions were made regarding the comparison since the MLR model used kWh/t charged scrap and the experience-based values were reported in kWh/t tapped steel.

A recent study, published after the review, used Kolmogorov–Smirnov (KS) tests and correlation metrics to highlight the change of all input variables between the training and test data [2]. In addition, permutation feature importance was used to investigate the importance of each input variable to the model prediction for both data sets. In combination, the KS tests and permutation feature importance produced evidence that some of the input variables had the main influence in the performance reduction of the model. However, the focus was never to investigate the specific effects of the scrap types on the EE consumption of the selected models.

3. Method

3.1. Representing Scrap Types

The scrap types will be represented in three distinct ways. The first representation will use the scrap codes from the steel plant of study, i.e., scrap type. The second representation is based on a visual categorization of each scrap code with the aim to provide an intuitive categorization for the steel plant engineers as well as to minimize the number of distinct scrap categories. The third representation is based on the estimated apparent density of each scrap code. The estimated apparent densities are both from established technical specifications as well as from estimations conducted by the plant engineers. The relations between the three scrap representations can be seen in Figure 3. Henceforth, the term *scrap representation* will refer to either of the three distinct scrap representations and the term *scrap category* will be used to specify a specific category in either of the visual or apparent density categorizations. The term *scrap type* will be used to refer to a specific scrap type as defined by the steel plant coding system.

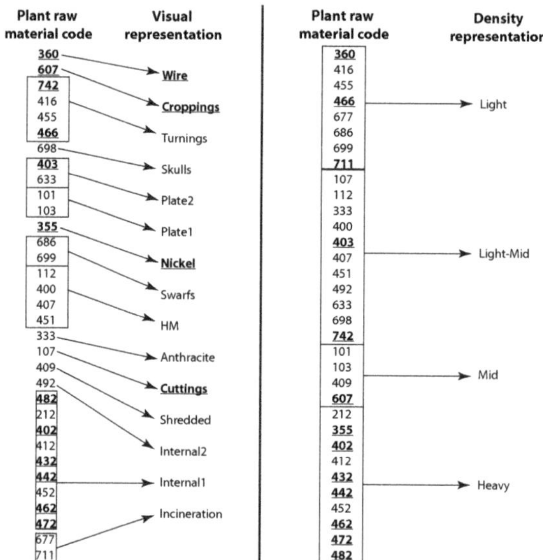

Figure 3. The relationship between the steel plant scrap types and the two scrap representations based on visual and estimated apparent density properties, respectively. The bold underlined scrap types and categories occur in less than 10% of the heats and are therefore bundled together into two aggregate variables, SCR_{Aggr} and *Aggregate*, for the plant scrap representation and visual scrap representation, respectively.

Each of the three representations are further described in detail.

3.1.1. Steel Plant Scrap Yard System

The steel plant uses an internal coding system for the raw material types, which follow a logic that is based on where the raw material is sourced from, its quality, composition, and shape. The majority of the raw material types represent steel scrap. Although nickel granulates and anthracite, i.e., coal, have distinct coding, these raw materials will be referred to as scrap types well knowing that they are not scrap. The main reason is that the focus of this study is on the effect of steel scrap on the EE consumption of the EAF. The effect of alloying elements is not of interest. Furthermore, nickel granulates and anthracite represent a small fraction of the total amount of charged raw materials for all steel grades.

The composition of the scrap can vary from low-to high alloying elements such as carbon, chromium, nickel, and molybdenum. Furthermore, quality-hampering impurities such as copper and tin are prevalent in some scrap types that are considered of low quality. The scrap quality is also dependent on the confidence in the actual content of the scrap. In general, more reliability is put on quality from internal sourced scrap and scrap from other steel plants than on purchased scrap from municipal waste incineration plants. The shapes of scrap can vary from thin plates to bulky residuals from foundries and own arising scrap.

14 of the 33 scrap types were charged in less than 10% of the heats represented by the data. To reduce the number of variables in the numerical experiments, a variable was created as the sum of the charged weights of these scrap types. This variable is called SCR_{Aggr}, see also Figure 3.

The goal of including this representation in the numerical experiments is to investigate if the most granular representation of the available scrap, in the steel plant of study, is the most optimal with respect to the accuracy of the EE consumption prediction models.

3.1.2. Visual Categorization

The melting rate of scrap in the EAF is predominantly dependent on the surface area available to heat transferring media such as the hot heel and radiation from the arc plasma. By categorizing the scrap types according to their distinct physical shape, two benefits can be achieved. First, scrap types with similar shapes are expected to have closely related melting performance in the EAF due to their similar surface-area-to-volume ratios. Second, the number of variables will be reduced which is beneficial from a statistical modeling point of view.

The difference between the categories Internal1 and Internal2 is based on the apparent density and the difference between the categories Plate1 and Plate2 is based on the thickness of the plates. They were separated because a difference in apparent density and plate thickness will affect the melting time of the scrap pieces.

4 of 14 visual categories were charged in less than 10% of the heats represented by the data. As with the steel plant scrap representation, a variable was created as the sum of the charged weight of these scrap types to reduce the number of variables. This variable is called *Aggregate*.

3.1.3. Density Categorization

All scrap types have an estimated apparent density conducted by either established standards or by the steel plants engineers. The main goal of this scrap representation is to categorize each scrap type into four categories, which are based on the estimated apparent density, i.e., density interval. All scrap types have an estimated apparent density range or a lower bound apparent density value.

The following categories were created, based on ranges of reported apparent densities:

- **Light:** 0.3–1.0 ton/m^3
- **Light-Mid:** 0.56–1.0 ton/m^3
- **Mid:** 0.7–1.4 ton/m^3
- **Heavy:** >1.4 ton/m^3

Intuitively, scrap types with lower apparent densities should melt faster than scrap types with higher apparent densities. Since the density of the metal pieces can be assumed to be the same, the variation in apparent density is a measure of "porosity" of the scrap types. Higher "porosity" leads to more surface area available for heat transfer, which in its turn leads to higher melting rates and consequently shorted melting times.

3.2. Data Governing the EAF

3.2.1. Variable Selection

The selected variables are based on the discussion regarding the non-linearity of the EAF process, reported energy sources and energy sinks, the correlative relationships of the variables to the EE consumption, and the scrap variables governed by the three scrap representations (see Section 2 and Section 3.1). The selected variables from each type are shown in Table 2.

Table 2. The variables used in the models.

Variables	Unit	Definition
Electrical Energy (EE)	kWh	The electrical energy consumption for the heat.
Total Weight	kg	The sum of all charged scrap types.
Tap-to-Tap time (TTT)	min	The time between the end of the tapping from the previous heat to the end of tapping of the current heat.
TCB2	min	Time between start of heat until charging of the second basket.
Burner oil	kg	Total amount of oil added by burner.
Burner O_2	Nm^3	Total amount of oxygen added by burner.
O_2-lance	Nm^3	Total amount of oxygen added by lance.
Injected carbon	kg	Total amount of carbon injection.
Lime and dolomite	kg	Total lime and dolomite added.
No. Charges	–	Total number of scrap baskets added.
SCR_{101}	kg	Above 3 mm thick plate. Apparent density above 1.0 ton/m^3.
SCR_{103}	kg	Thin plate, cuttings of rolled thin plate. Apparent density above 1.0 ton/m^3.
SCR_{212}	kg	Heavy cuttings of internal scrap. Apparent density above 1.4 ton/m^3.
SCR_{333}	kg	Anthracite (carbon). Apparent density 0.72 ton/m^3.
SCR_{400}	kg	Purchased rebar and plates. Apparent density 0.56–1.0 ton/m^3.
SCR_{407}	kg	Heavy melting mix (HM1). Apparent density above 0.7 ton/m^3.
SCR_{409}	kg	Shredded scrap. Apparent density above 1.0 ton/m^3.
SCR_{412}	kg	Internal scrap. Apparent density above 1.4 ton/m^3.
SCR_{416}	kg	Turnings. Apparent density above 0.4 ton/m^3.
SCR_{451}	kg	Purchased scrap (HM1 and HM2). Apparent density above 0.7 ton/m^3.
SCR_{452}	kg	Internal scrap. Apparent density above 1.4 ton/m^3.
SCR_{455}	kg	Turnings. Apparent density above 0.4 ton/m^3.
SCR_{492}	kg	Mixed internal scrap. Apparent density above 0.7 ton/m^3.
SCR_{633}	kg	Si-rich plate. Apparent density 0.7–1.0 ton/m^3.
SCR_{677}	kg	Incineration scrap. Apparent density above 0.5 ton/m^3.
SCR_{686}	kg	Grinding swarfs and grinding swarf briquettes. Apparent density 0.4–1.0 ton/m^3.
SCR_{698}	kg	Skulls. Apparent density above 0.7 ton/m^3.
SCR_{699}	kg	Grinding swarfs and grinding swarf briquettes. Apparent density 0.3–1.0 ton/m^3.
SCR_{711}	kg	Incineration scrap. Apparent density above 0.5 ton/m^3.
SCR_{Aggr}	kg	Sum of scrap types charged in less than 10% of the heats.
Incineration	kg	
Heavy melting scrap (HM)	kg	
Plate1	kg	
Plate2	kg	
Internal1	kg	
Internal2	kg	Scrap representation based on the shape of the scrap type. See Section 3.1.2 and Figure 3.
Shredded	kg	
Swarfs	kg	
Turnings	kg	
Carbon	kg	
Skulls	kg	
Aggregate	kg	
Heavy	kg	
Mid	kg	Scrap representation based on the reported and estimated apparent density ranges or values of each respective
Light-Mid	kg	scrap type in the steel plant. See Section 3.1.3 and Figure 3.
Light	kg	
Hot heel	–	1 if hot heel is present at the start of the heat, else 0. To account for the heat transfer by the hot heel.
Furnace shell number	–	An ordinary variable counting the number of heats since the last furnace barrel maintenance.

3.2.2. Variable Batches

The variable batches are based on the variables motivated in Section 3.2.1 as a starting point. All 8 variable batches from this group can be seen in Table 3 and the variables in each variable group are shown in Table 4.

Table 3. The domain-specific variable batches. The variables present in each variable group are shown in Table 4.

Variable Group	Variable Batch	1	2	3	4	5	6	7	8
	Base	x	x	x	x	x	x	x	x
	Plant scrap category			x	x				
	Visual scrap category					x	x		
	Density scrap category							x	x
	Furnace related	x		x		x		x	

The reason the scrap representations will not be used together in any of the variable batches is due to physical consistency. Besides being redundant, there is a physical logic tied to each scrap representation. Using a mixture of scrap representations will not indicate which scrap representation is the most optimal to use. One of the aims of this study is to investigate the best scrap representation with respect to the performance of the models on test data.

Table 4. Input variables for each variable group. There is a total of 48 input variables.

Variable Group	Variables	No. Variables	Variable Group	Variables	No. Variables
Base	Total Weight TTT TCB2 Burner oil Burner O_2	9		SCR_{686} SCR_{698} SCR_{699} SCR_{711} SCR_{Aggr}	
	O_2-lance Injected carbon Lime and dolomite No. Charges		**Visual scrap representation**	Incineration HM Plate1 Plate2	12
Plant scrap representation	SCR_{101} SCR_{103} SCR_{112} SCR_{333} SCR_{400} SCR_{407} SCR_{409} SCR_{412}	20		Internal1 Internal2 Shredded Swarfs Turnings Carbon Skulls Aggregate	
	SCR_{416} SCR_{451} SCR_{452} SCR_{455}		**Density scrap representation**	Heavy Mid Light-Mid Light	4
	SCR_{492} SCR_{633} SCR_{677}		**Furnace related**	Hot heel Furnace shell number	2

3.3. Data Treatment

3.3.1. Purpose

To ensure the reliability and validity of a statistical model, the data which is used to create the model must be treated. The reason is because statistical models adapt their coefficients solely based on data, as opposed to physical models, which have pre-determined coefficients. By including data that is of low quality, the statistical model will inherit that quality when making predictions on new, previously unseen, data. In general, data treatment is a double-edged sword. On the one hand, a model should be able to predict well on any future data. On the other hand, all data sets contain data points that represent extreme cases. Any statistical regression model will predict these extreme cases with a low accuracy, since the coefficient adaptation algorithm is based on minimizing the error on the entire data set included in the training phase. Any extreme case receives a lower priority due to its rarity. Hence, a successful modeling effort strikes a balance between these two opposing effects.

Data treatment methods can be divided into two categories, domain-specific methods and statistical methods. The two disparate data treatment methods will be described further.

3.3.2. Domain-Specific Methods

This method uses the knowledge and experience in the domain from which the data originates. Domain-specialization and manual treatment of the data are required. This can be both expensive and time consuming but the end-result, i.e., the cleaned data, is expected to be of higher quality than using pure statistical outlier detection algorithms, which do not adhere to the domain-specific considerations. In the scope of the EAF process, an example of a domain-specific treatment is the removal of instances of data that are improbable. If the registered charged content of scrap is 190 t while the maximum capacity of the furnace is 120 t, then that instance should be removed. Another example is the removal of heats that are not part of regular production. For example, testing the effects of new scrap types, or delivery batches.

3.3.3. Statistical Methods

Statistical data treatment methods refer to algorithms that identify anomalies in the distributions in data sets. The algorithms are purely mathematical and do not take into account any domain-specific reasons behind anomalies or domain-specific relations between the variables. The advantage of this method is that it is easy to apply and does require very little, if any, domain-specific knowledge. However, applying outlier detection algorithms to multiple variables will potentially remove most of the data. A simple example is the various types of scrap types used in production, which are prone to extreme (very high or very low) values since some scrap types are not available at all times. Applying a statistical cleaning algorithm to all scrap types will omit most of the data. Hence, statistical treatment methods must be used with caution and only on well-selected variables that are expected to impact the predictions dramatically. For example, total charged scrap weight and TTT.

Tukey's fences [19], which is based on the interquartile range of a distribution, will be used in the numerical experiments. The method removes data points that are present outside the range defined as:

$$q_1 - \epsilon(q_1 - q_3) \leq x_j \leq q_3 + \epsilon(q_1 - q_3) \tag{5}$$

where q_1 and q_3 are the first and third quartiles of variable j, respectively. ϵ is a pre-specified constant indicating how far out the outlier must be before being cleaned. $\epsilon = 3.0$ will be used in the numerical experiments which removes extreme outliers [19].

Since this method is based on the quartiles of the distribution under consideration, it is less sensitive to skewed distributions compared to cleaning by omitting data points outside of $\pm 3\sigma$ from the mean. This inherent characteristic of Tukey's fences is advantageous, since most variables governing the EAF process are non-Gaussian. See Figure 4 for examples.

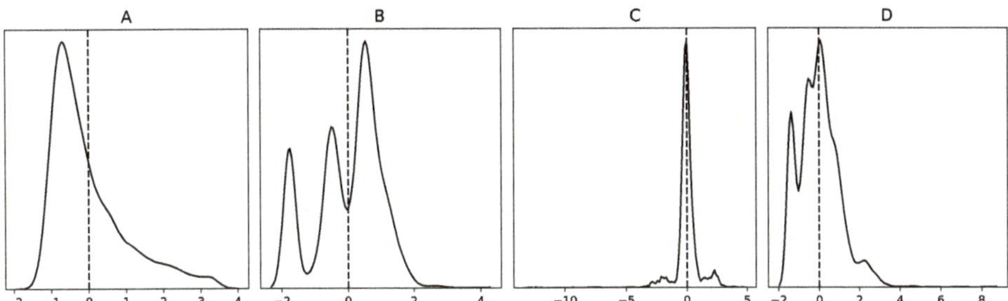

Figure 4. The distributions for four variables governing the EAF under study highlighting the absence of the Gaussian distribution. All values are normalized, and the dashed lines indicate the mean values. (**A**): TTT. (**B**): Charged weight of internal scrap. (**C**): Total charged weight of raw materials. (**D**): Charged weight of shredded scrap.

3.3.4. Applied Data Treatments

Due to the opposing effects of data treatment as mentioned earlier, it is impossible to know which approach strikes a good balance between model generalizability and model accuracy. Therefore, four different data treatment approaches will be used in the modeling to investigate the influence of the different approaches.

The first data treatment approach was conducted by a senior engineer at the steel plant. This data treatment was done by manually inspecting each row of the data set and flagging rows which contained values that were not consistent with the data instance as a whole. This data treatment is a combination of domain-expertise and some subjectivity of the senior engineer. However, because the data is inherently coupled with the steel plant it originates from, using on-plant experts is critical to a successful data treatment operation. This data treatment is referred to as *Expert*.

For the following three data treatment approaches, two filter steps were applied to remove unrealistic heats with respect to events in time. The first filter removed heats where the timestamp of charging the second heat was negative respect to the start of the heat. The second filter removed heats where the charging of the second basket occurred after the heat ended. Applying these filters were enough to remove all unrealistic heats.

The second data cleaning approach was conducted by the authors of this study which was based only on domain-specific knowledge. Two cleaning steps were applied. The first cleaning step removed heats with TTT at, or above, 180 min since these heats are likely experiencing a longer delay in the process or a scheduled stop. Usually, the TTT is aimed at 60–70 min. The second cleaning step removed heats with a Total charged weight at, or above, 141 ton. This is a limit set by the steel plant for abnormally large charge weights. This data treatment approach is referred to as *Domain-specific*.

The third data treatment used Tukey's fences to remove clear outliers, see Section 3.3.3. Tukey's fences were calculated and applied to each of the following input variables using the training data: Total Weight, TTT, Time until Charging of Basket 2 (TCB2), Burner O_2, Burner oil, O_2-lance, and Injection carbon. Each 'fence' was then applied to the training and test data, respectively. This data treatment approach is referred to as *Tukey*.

The fourth data treatment approach used the second and third data treatment approaches, in order. This data treatment approach is referred to as *Domain-specific Tukey*.

Each of the four described data treatment approaches will be used in the modeling to investigate which one of the data treatments is the most optimal for a model applied in practice.

3.4. Modeling the EE Consumption

3.4.1. Statistical Modeling Frameworks

To consider the non-linearity of the EAF process, a non-linear statistical modeling framework must be used. The numerical experiments in this study will use two different non-linear statistical modeling frameworks; ANN and RF. ANN has been used in a previous article using the same methodology as described in this paper [2]. RF will be used to broaden the scope of models used to predict the EE consumption using the same methodology. Furthermore, SHAP interactions, an interpretable machine learning algorithm that calculates the interactions between the input variables with respect to the output variable can be used for RF models [20]. This is not the case for ANN models where only regular SHAP values can be used. SHAP will be further explained in Section 3.5.1.

Artificial Neural Networks: This model framework uses a fully connected network of nodes to make predictions [21]. The first layer, which is known as the input layer, receives the values from the input variables. The values are then propagated through the intermediate layers, which are known as hidden layers, to the last layer. The last layer is the output layer where the prediction is made. See Figure 5 for an illustration of an arbitrary ANN model.

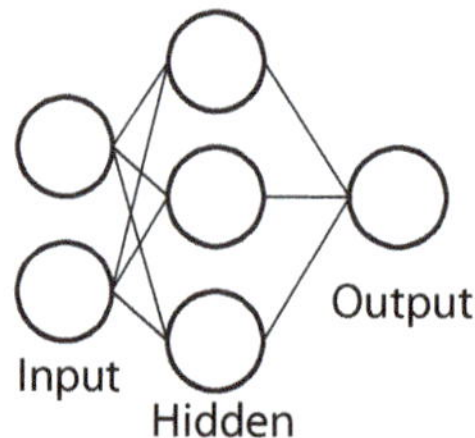

Figure 5. An Artificial Neural Network (ANN) for predicting an output value based on two input values [2]. It has one hidden layer with three nodes. The lines between the nodes illustrate that the ANN is fully connected and the forward flow of calculations in the network.

By changing the number of nodes in the hidden layers and the number of hidden layers, one can alter the complexity of the model. Increasing the number of hidden layers and nodes increase the complexity and enable the model to learn more complicated relations between the input variables and the output variable.

In the output layer, and in each of the hidden layers, each node multiplies a weight value with each of the value propagated by the previous layer. The resulting weight-value factors are summed together. Mathematically, this process can be expressed as:

$$s_j = \sum_{i=1}^{P} w_i \cdot x_i \tag{6}$$

where j is the j:th node in the current layer and P is the number of nodes in the preceding layer.

Each value, s_j, is then fed into an activation function, and the resulting value is propagated to the next layer in the network. Two commonly used activation functions are the hyperbolic tangent (tanh) and the logistic sigmoid.

During the training phase, the weights are updated in the direction of minimizing the overall loss of the predictions on the training data. Since the output variable is connected to the input variables by the network weights, it is possible to mathematically express the loss as a function of the network weights. Finding an optimal local minimum, with respect to the overall loss, in the weight space requires a sophisticated algorithm. These algorithms are known as gradient-descent algorithms as their function is to descent to the most optimal local minima in loss space [21].

Given enough hidden layers and nodes, an ANN can learn any complex relationship between variables even though the relationships are not valuable for prediction purposes. This overfitting phenomenon can be reduced by splitting the training data into two sets. The first set of data is used to adapt the weights while the other set is used to calculate the loss after each weight update.

Random Forest: This statistical modeling framework is a model made of two, or more, decision trees. See Figure 6 for an illustration of a simple decision tree for prediction purposes. The RF model framework was first reported by L. Breiman [22]. RF belongs to the statistical model group known as ensemble models, which is a group of statistical models that is made up of two or more models that when combined, aim to increase the prediction accuracy.

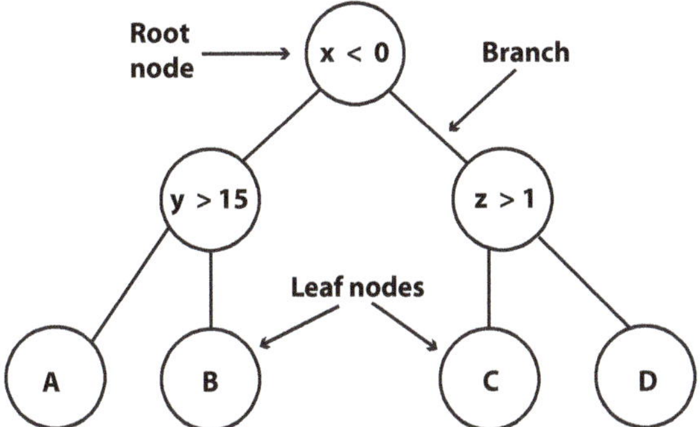

Figure 6. A simple decision tree sorting points on the $\{x, y, z\}$ coordinate system. Points satisfying the conditions on each node proceed on the left branch, the rest proceed to the right branch. The points $A = \{0, 42, 5\}$, $B = \{-10, -10, -10\}$, $C = \{31, 4, 0\}$, and $D = \{2, 4, 61\}$, are sorted in the decision tree.

In an RF model, each decision tree is trained on a sub-sample of the complete training data set. This sub-sample is drawn *with replacement* from the complete training data set, a process known as bootstrapping. By training each decision tree on a sub-sample of the training data set the overfitting of the RF model is reduced, since each decision tree becomes specialized on one segment of sample space. Furthermore, the optimal split for the next branch from each node in the decision tree is selected using a random selection of a pre-specified number of the total available input variables. This procedure also reduces overfitting since each decision tree now has a higher probability of being diverse with respect to the other trees in the model. Using many trees created by a random selection of features and data points, RF type models have proven to converge such that overfitting does not become a problem [22]. In the prediction phase of an RF model, each decision tree predicts the value of the output variable. In a prediction of regression type, i.e., when predicting continuous values, the prediction by the RF model is the average of the predictions from all decision trees. The prediction of one data instance, x_k, in an arbitrary RF model is illustrated in Figure 7.

To optimize an RF model for the task at hand, i.e., improve the accuracy, one must search for an optimal combination of hyper-parameters. The most important hyper-parameters are the maximum tree depth, i.e., number of splits from the root node, the number of decision trees, and the maximum number of features used to find the optimal condition when splitting a node [23].

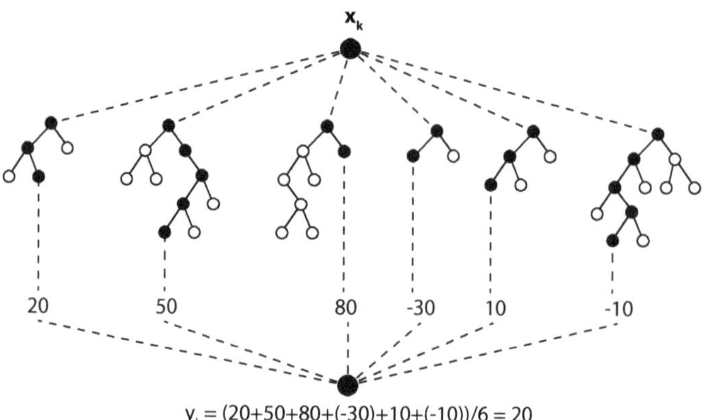

Figure 7. An arbitrary RF model consisting of 6 decision trees. The prediction, y_k, of the data instance, x_k, is determined by averaging the sum of the outputs from the decision trees. The filled nodes show the path x_k has taken in each decision tree.

3.4.2. Parameter Optimization

The variations of parameters for each statistical model framework, RF and ANN, can be seen in Tables 5 and 6, respectively. Each combination of parameters represents one model type. The data cleaning strategies and variable batches were included as additional, model framework independent, parameters. By including the cleaning strategies as a parameter, it is possible to find out which cleaning strategy achieves the most optimal trade-off between accuracy and data omission. By including the variable batches as a parameter, it is possible to find the model with a relative high accuracy but with the least amount of input variables. One should always abide to the concept of model parsimony, which is to select the simplest model among a set of models with next to the same performance.

Table 5. Parameter combinations used for the RF models. Each value is separated by a comma. m is the number of input variables. Each combination of parameters represents one model type.

Parameter	Variations	#Combinations
Number of trees	10, 30, 50, 70, 90, 110, 130, 150, 170, 190, 210, 230, 250	13
Max tree depth	2, 3, 4, 5, 6, 7, 8, 9, 10, Unlimited	10
Max features in split	$v, \sqrt{v}$	2
Cleaning strategies	See Section 3.3	4
Variable batches (domain-specific)	See Section 3.2.1	8
	Total:	8320

Table 6. Parameter combinations used for the ANN models. Each value is separated by a comma. The topology (z) and (z,z), indicate one and two layers with z nodes in each layer, respectively. Each combination of parameters represents one model type.

Parameter	Variations	#Combinations
Activation function	Hyperbolic tangent function, Logistic sigmoid	2
Learning rate	0.1, 0.01, 0.001	3
Topology	(z) and (z,z) in z ∈ 1, 3, ..., 29	30
Cleaning strategies	See Section 3.3	4
Variable batches (domain-specific)	See Section 3.2.1	8
	Total:	5760

3.4.3. Selection of Training and Test Data

Selecting the test data from a random sub-sample of the complete data makes the training and test data become chronologically intertwined. This is a shortcoming because it does not reflect the practical purpose of a statistical model predicting the EE of an EAF [1]. From a process perspective, a statistical model will predict on heats that are from a future point in time with respect to the heats whose data have been used to adapt the parameters of the model, i.e., training data. To account for this shortcoming, the test data will be selected in chronological order from the training data. The test data will be all heats produced from 1st of February 2020 to the 28th of February 2020 and the training data will the heats produced from the 10th of November 2018 through January 2020. The start date of the training data was selected based on a furnace upgrade that was completed on the 9th of November 2018. This amounted to 4032 training data points and 263 test data points before data treatment. The heats in the training data and test data will be referred to as *training heats* and *test heats*, respectively.

3.4.4. Model Performance Metrics

The performance of the models will be compared using two fundamental metrics. These are the coefficient of determination, R^2, and the regular error metric.

The adjusted-R^2 value should be used instead of the regular R^2 value when comparing models that use different number of input variables. The reason is because each additional input variable increases the R^2-value when the number of data points is fixed [24]. The adjusted-R^2 value can be calculated as follows:

$$\bar{R}^2 = 1 - (1 - R^2)\frac{n-1}{n-v-1} \tag{7}$$

where n is the number of data points, is the number of input variables, and R^2 is the regular R-square value.

The regular error metric was chosen in favor of the absolute error metric, because in a practical context, an overestimated prediction of EE is vastly different from an underestimated prediction of EE. The regular error metric can be defined as follows:

$$E_i = y_i - \hat{y}_i \tag{8}$$

where y_i is the true value, $\hat{y}_i$ as the predicted value, $i \in 1, 2, \ldots, n$, and n is the number of data points. Using all the data points under consideration, the standard deviation, mean, minimum, and maximum, error values are defined the ordinary way.

The EE consumption, and therefore the unit of error, will be expressed in kWh/heat rather than in kWh/t tapped steel. The main reason for this choice is that the former varies only by consumed EE while the latter varies both by consumed EE and the yield. This is a more challenging problem as it requires the statistical model to adapt to both the consumed EE as well as the yield, which is defined as the weight of tapped steel divided by weight of charged scrap. Furthermore, the tap weight is dependent on factors such as the slag created by oxidation in the process and on the total amount of dust generated. These two examples are, in turn, affected by process times, amount of oxygen injected, additives, and the charge mix.

3.4.5. Model Selection

Each model type, which is defined by one of the parameter combinations shown in Tables 5 and 6, will be instantiated 10 times. The reason for this is to investigate the stability of each model type and to reduce the impact of randomness, which is prevalent in each of the statistical model frameworks. In RF, the randomness is introduced by random selection of data points for each tree and the random selection of input variables to determine each split. In ANN, the randomness is partly governed by the random selection of the initial values of the network weights.

The aggregate statistical metrics based on the 10 model instances of each model type are presented in Table 7.

Table 7. The adjusted-R^2 and error metric variants that are used to evaluate the performance of the aggregated model instances.

Symbol	Definition
$\bar{R}^2_\mu$	Mean adjusted R-square of the 10 model instances on the test data
$\bar{R}^2_\sigma$	Standard deviation of adjusted R-square of the 10 model instances on the test data
$\bar{R}^2_{min}$	Minimum adjusted R-square of the 10 model instances on the test data
$\bar{R}^2_{max}$	Maximum adjusted R-square of the 10 model instances on the test data
Δ_μ	Mean error of the mean error of the 10 model instances on the test data
Δ_σ	Standard deviation of the mean error of the 10 model instances on the test data.
Δ_{min}	Minimum error of the mean error of the 10 model instances on the test data.
Δ_{max}	Maximum error of the mean error of the 10 model instances on the test data.

To determine the stability of the models, which is influenced by the underlying randomness, the idea behind the model selection criteria was to keep the difference between $\bar{R}^2_{max}$ and $\bar{R}^2_{min}$ as low as possible. Hence, the algorithm for selection the best model type of each variable batch can be expressed as follows:

1. Select all models that pass the following condition: $\bar{R}^2_{max} - \bar{R}^2_{min} \leq 0.05$.
2. Select the model(s) with the highest $\bar{R}^2_\mu$. If the number of models exceeds 1, then proceed to the next step, otherwise select the one model.
3. Select the model with $min(\bar{R}^2_{max} - \bar{R}^2_{min})$.

The adjusted-R^2 was chosen to determine the stability of the models because it indicates the goodness of fit. The other error metrics, based on the model error, do not relate to goodness of fit.

Using the above algorithm, the number of models will be reduced to 8, which is equal to the number of variable batches. From this model subset, the model with the fewest number of input variables will be selected, given that more than one model have the same $\bar{R}^2_\mu$-value. This ensures that the model selection adheres to the concept of model parsimony, i.e., that the simplest model, which is the model with least number of input variables, is selected when more than one model have the same performance.

3.5. Model Evaluation and Analysis

3.5.1. Shapley Additive Explanations (SHAP)

Any prediction by the statistical model, f, can be explained by a linear combination of the contributions by all input variables:

$$f(x) = \phi_0 + \sum_{i=1}^{M} \phi_i \tag{9}$$

where ϕ_0 is the contribution when information about the variables are not present. In practice, this will always be the mean value of all prediction of the statistical model. This is sensible since the mean value is always the most optimal choice if complementary information is lacking.

Each SHAP value, ϕ_i, can be calculated using the following expression [9]:

$$\phi_i(f, x) = \sum_{Z \subseteq \hat{Z} \setminus \{i\}} \frac{|Z|!(M - |Z| - 1)!}{M!} [f_x(Z \cup \{i\}) - f_x(Z)] \tag{10}$$

where Z is a subset of the set of all input variables, $\hat{Z}$, and M is the total number of input variables. $\hat{Z} \setminus \{i\}$ is the set of all input variables excluding variable i and $Z \cup \{i\}$ is a subset of the set of all input variables including variable i. Hence, each SHAP value, ϕ_i, represents the average contribution by an input variable on the output variable of all combinations in which that input variable is presented to the statistical model, f.

The number of calculations required to calculate Equation (10) scales drastically when the number of input variables and data points increases. To reduce the number of calculations, one can use an approximative method known as the Kernel SHAP method, which assumes variable independence and model linearity. Assuming variable independence means that Kernel SHAP will produce value combinations that are unrealistic for variables that are dependent. For example, high amount of burner oil will be matched with low oxygen through the burners. The linearity assumption is analogous to linear approximation of functions in mathematics where the space in the vicinity of a data point is assumed to be linear. Furthermore, Kernel SHAP is model independent which means that it can be applied to any supervised statistical model. It has previously been used to analyze an ANN predicting the EE of an EAF [10].

RF enables the use of the Tree SHAP method, which is adapted to tree-based statistical models. Tree SHAP can calculate the exact SHAP values because it does not assume variable independence. Furthermore, the algorithm is computationally efficient compared to the regular SHAP method [20], which means that the SHAP values can be calculated within a reasonable timeframe. By not assuming variable independence, Tree SHAP adheres to the true behavior of the model as opposed to Kernel SHAP, which uses a simplified representation of the original model. This is important since the aim of SHAP is to explain the behavior of the statistical model as accurately as possible.

Tree SHAP also provides the ability to use SHAP interaction values, which calculates the interaction effects between the input variables as contributions to the prediction [25]. Instead of receiving one SHAP value per input variable and prediction, the interaction value calculation provides an $M x M$ matrix per prediction, where M is the number of input variables. The diagonal of the matrix contains the main interaction values, $\phi_{i,i}$, which are the contributions by each input variable without the influence of the other input variables. The upper and lower triangles of the matrix contain the one-to-one interaction values, $\phi_{i,j}$, which show the contributions to the prediction by each input variable pair. The SHAP interaction value is equally split such that $\phi_{i,j} = \phi_{j,i}$, which means that the total contribution of the input variable pair (i, j) is $\phi_{i,j} + \phi_{j,i}$.

The SHAP interaction values can be calculated by [25]:

$$\phi_{i,j}(f, x) = \sum_{Z \subseteq \hat{Z} \setminus \{i,j\}} \frac{|Z|!(M - |Z| - 2)!}{2(M - 1)!} \nabla_{i,j}(Z) \tag{11}$$

where

$$\nabla_{i,j}(Z) = f_x(Z \cup \{i, j\}) - f_x(Z \cup \{i\}) - f_x(Z \cup \{j\}) + f_x(Z) \tag{12}$$

The SHAP interaction values can be expressed as the regular SHAP values as follows:

$$\phi_i = \phi_{i,i} + \sum_{i \neq j}^{M} \phi_{i,j} \tag{13}$$

Henceforth, the SHAP interaction values relate to the model prediction as:

$$\sum_{i=0}^{M} \sum_{j=0}^{M} \phi_{i,j}(f,x) = f(x) \tag{14}$$

This means that the SHAP interaction values are a more granular representation of the contribution by each input variable on the prediction as opposed to regular SHAP values. On the one hand, the main interaction values provide us with the interaction of each input variable unaffected by the influences of the other input variables. On the other hand, the one-to-one interaction values between the input variables provide us with information on how the interaction between the input variables adds to the prediction contribution.

The aim of showing the main interaction effects on EE by the input variables is to investigate if the models adhere to the underlying relationships between the input variables and the output variables. One must bear in mind that this analysis is univariate with respect to the output variable. Hence, it is not possible to draw any conclusions regarding the intra-relationships between the input variables that together affect the output variable.

There is also one concept known as SHAP feature importance. It is defined as the mean absolute value of the regular SHAP values. See Equation (15).

$$FI_j = \frac{1}{n} \sum_{i=1}^{n} \phi_j^i \tag{15}$$

where FI_j is the SHAP feature importance for variable j and n is the number of data points.

SHAP feature importance measures the global importance of each input variable and is a more trustworthy measure of feature importance than the traditional permutation-based feature importance. The main reason is that SHAP feature importance is rooted in solid mathematical theory while permutation-based feature importance is based on the empirical evidence provided by random permutations.

In the numerical experiments, the package *shap*, with the method Tree Explainer, in Python will be used to calculate SHAP. The method 'tree path dependent' will be used since it adheres to the variable dependence among the input variables. The software and hardware used in the numerical experiments can be seen in Appendix A, Tables A1 and A2.

3.5.2. Correlation Metrics

Two different correlation metrics will be used to investigate the intra-correlation between input variables as well as the correlation between the input variables and EE, i.e., the output variable. By studying the resulting correlation values with domain-specific knowledge, it is possible to verify the connection between the model prediction and the input variables. After all, the intra-correlation between input variables and their respective correlations with the output variable is what the model learns during the training phase. The two correlation metrics are explained further.

Pearson correlation: The Pearson correlation metric that can only detect linear relationships between two random variables [26]. It assumes values between -1 and 1, where the former is a perfect negative relationship between the variables and the latter is a perfect positive relationship, i.e., the variables are identical. A value of 0 indicates that the variables have no relation.

The Pearson correlation coefficient is defined as follows:

$$\rho_{X,Y} = \frac{covariance(X,Y)}{\sigma_X \sigma_Y} \tag{16}$$

where σ_X and σ_Y are the standard deviations of the two random variables X and Y, respectively.

Distance correlation (dCor): Although dCor cannot distinguish between positive and negative correlative relationships between variables, it is able to detect non-linear relationships between variables [27]. This is important since some variables governing the EAF process have a non-linear relationship to EE. By using dCor and Pearson in tandem, it is possible to get a clearer picture of the relationships between the variables governing the statistical models.

The mathematical expression for dCor has the same form as the Pearson correlation coefficient:

$$dCor(V_1, V_2) = \frac{dCov(V_1, V_2)}{\sqrt{dVar(V_1)dVar(V_2)}} \tag{17}$$

where $dVar(V_1)$ and $dVar(V_2)$ are the distance variance of the random variables V_1 and V_2, respectively. $dCov(V_1, V_2)$ is the corresponding distance covariance. The square root of $dVar(V_1)$ and $dVar(V_2)$ gives the distance standard deviations.

dCor assumes values between 0 and 1, where the former indicates that the variables are independent, and the latter indicates that the variables are identical. dCor has been used previously when evaluating the variables governing statistical models predicting the EE of the EAF [2,10].

3.5.3. Charge Types

Three distinctly different charge types, i.e., scrap recipes, with different tramp element contents will be used in the analysis to exemplify the contribution to EE by the scrap type or scrap category for each of the chosen charge types. The first charge type, denoted A, has the lowest level of tramp elements of the charge types used in the steel plant. It therefore requires higher amounts of scrap types where the amount of tramp elements such as Cu and Sn are well-known. Examples are residual scrap from the forging mill and purchased scrap from other steel mills. The second charge type, denoted B, does not have as strict requirements as charge type A. Hence, a higher amount of scrap types with lower qualities can be used. A typical lower quality scrap type is heavy melting scrap (HM) which can have relatively high amounts of tramp elements. The last charge type, denoted C, can have higher contents of tramp elements. Hence, purchased scrap from other steel plants and residual scrap are used to a lesser extent.

It is important to note that alloying elements promoting the desired steel properties such as Ni, Mo, and Cr come either from own arising scrap or from purchased scrap with low level of impurities. In the cases the internal scrap is not used, pure alloying elements such as Ni-granulated must be used instead. This is a more expensive route with respect to the total cost of raw materials.

4. Results and Discussion

4.1. EE consumption Models

$\bar{R}_{\mu}^2$-values for all variable batches and data cleaning strategies are shown in Figure 8. Consistency throughout the variable batches can be observed for all four cleaning strategies when the RF model framework is used. The consistency among the variable batches means, for the RF model framework that the deciding factor for the performance is the variable batch and not the hyper-parameters of the statistical model framework. This is a wanted outcome since the performance of a model should mainly be dependent on factors stemming from the application domain and not based on an optimization of parameters in an abstract framework.

Variable batches (VB) 1 and 2 produce models with the lowest $\bar{R}_{\mu}^2$-values. These are also the VB that *do not* use scrap representation variables. This provides evidence that scrap types are relevant factors when determining the EE consumption of the EAF and that scrap should not be treated collectively using only the total weight of all charged scrap.

The best performing models are those from VB 5 and 6 which use categories from the visual scrap representation. This provides evidence that a categorization based on scrap shapes is an optimal approach when creating a statistical model predicting the EE in the steel plant under study.

The performance of the models using the visual scrap representation are, performance-wise, followed by the models using the plant scrap representation (VB 3 and 4) and the density scrap representation (VB 7 and 8), respectively. Essentially, this indicates that a too fine or a too coarse representation of the charged scrap are sub-optimal for a statistical model predicting the EE. The steel plant scrap representation has numerous scrap types that contain the same scrap with respect to shape and dimension. The only difference is varying alloying content from Ni, Cr, and Mo, which does not significantly affect the melting time.

The coarse representation is based on the apparent density of the scrap, which does not take into account the shape of the scrap. Scrap shapes are closely related to the area-to-volume ratio, which is the strongest factor determining the melting time of scrap in the EAF since the stirring in the EAF is low during a large part of the melting phase. For example, HM has the same apparent density as skulls, which consist of bulky mixtures of solid slag and metal that takes long time to melt. Likewise, thin and thick plate have similar densities but different area-to-volume ratios.

The consistency among the four sets of VB between the four cleaning strategies for both statistical model frameworks further strengthens the evidence regarding the effects of the chosen scrap representations on the predictive performance of the models.

The performance and meta data of the best models and meta data from each cleaning strategy and model framework are shown in Table 8. In general, the ANN models perform similarly or better than the RF models with regards to the $\bar{R}_\mu^2$-values. However, the ANN models always have a smaller mean error and standard deviation of error, i.e., Δ_μ and Δ_σ.

With regards to the modeling meta data, the ANN and RF models had the same VB for the best models on the data from each cleaning strategy. This is also a wanted outcome based on the same reasoning as before regarding the importance priority between the domain-specific factors and the abstract model-based factors.

The total amount of cleaned data points was 25.8 percentages higher for the *Expert* cleaning strategy compared to the *Domain-specific* cleaning strategy. The $\bar{R}_\mu^2$-values only increased slightly using the *Expert* cleaning strategy; 0.035 and 0.039 for the RF and ANN models, respectively. Using the statistical cleaning method Tukey's fences, the amount of data cleaned were 11 and 13.3 percentages higher than the *Domain-specific* cleaning strategy. As opposed to the *Expert* cleaning strategy, the $\bar{R}_\mu^2$-values were worse for the models involving Tukey's fences. *Tukey* reported reduced $\bar{R}_\mu^2$-values of 0.049 and 0.077 for the RF and ANN models, respectively. For the *Tukey-Domain-specific* the reduction in $\bar{R}_\mu^2$ were 0.053 and 0.078, respectively. These results lead to two important findings. First, the usage of statistical cleaning heuristics results in a model performance that is sub-par to models using data cleaned by the usage of domain-specific knowledge; the *Expert* and *Domain-specific* cleaning strategies. Second, using data cleaned by an *Expert* yields models with the best performance, which illuminates the importance of knowledge about the specific EAF operations one intends to model. However, the large relative percentage of data loss using the *Expert* cleaning strategy (34.2%) as opposed to the *Domain-specific* cleaning strategy (10%) tilts the chosen cleaning strategy in favor of the latter since the data loss percentage directly relates to the percentage of future heats the model can predict on. This finding is closely tied to the practical usefulness of the model.

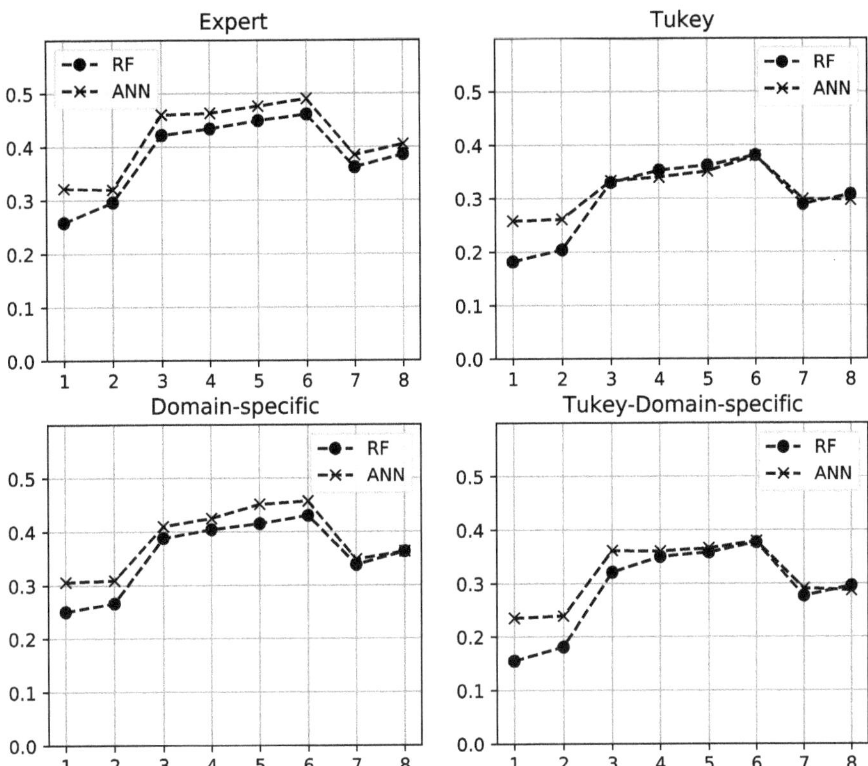

Figure 8. $\bar{R}_{\mu}^2$-values for each variable batch (VB 1-8 on the abscissa) and cleaning strategy. **VB 1–2:**Without scrap representation. **VB 3–4:** Steel plant scrap representation. **VB 5-6:** Visual scrap representation. **VB 7–8:** Density scrap representation.

Table 8. The performance and meta data of the best models, and the meta data for the four cleaning strategies. The values in parentheses show the values from the ANN models.

Cleaning Type	Expert	Domain-Specific	Tukey	Tukey-Domain-Specific
Train/test split	2571/187	3625/241	3179/212	3089/207
% cleaned train/test	36.2%/28.9%	10.1%/8.4%	21.2%/19.4%	23.4%/21.3%
% total cleaned data	35.8%	10.0%	21.0%	23.3%
Variable batches	6 (6)	6 (6)	6 (6)	6 (6)
No. Variables	21 (21)	21 (21)	21 (21)	21 (21)
$\bar{R}_{\mu}^2$	0.461 (0.490)	0.430 (0.457)	0.381 (0.380)	0.377 (0.379)
Δ_{μ} (kWh/heat)	408 (238)	369 (209)	252 (214)	239 (126)
Δ_{σ} (kWh/heat)	1938 (1863)	2173 (2063)	2062 (1995)	2042 (1976)
Δ_{min} (kWh/heat)	-4496 (-5316)	-7064 $-(7756)$	-5061 (-5464)	-5079 (-4262)
Δ_{max} (kWh/heat)	7113 (5699)	7833 (7735)	7810 (7608)	7942 (7162)

4.2. Analysis of the Selected Model

The selected model for the analysis is the RF model with a variable batch 6 using domain-specific cleaning. This model was selected based on the relatively high $\bar{R}_\mu^2$-value (0.457) while still keeping 90% of the data compared to the model using the expert cleaning type, which provided the highest $\bar{R}_\mu^2$-value (0.490) but only keeping 64.2% of the data. Hence, the chosen model can be used in 90% of future heats given that the test data and training data come from the same distribution.

The main interactions of the variables Plate1, Internal1, HM, and Shredded on the EE consumption and the charged weight distributions of these scrap categories for each of the charge types A, B, and C, can be seen in Figure 9.

The EE contribution by Plate1 on charge type A is lower than for B and C. However, charge type B can have both a positive and negative EE contribution since the densest part of the distribution is present in the steepest EE interaction change for Plate1. Hence, it is hard to conclude whether charge type B gets a similar contribution as does charge type A or charge type C. Charge type C is commonly using zero, or next to zero, amount of Plate1. Hence there is a large contribution to EE by Plate1 for this charge type. The EE contribution by Internal1 is slightly lower for charge type A than charge types B and C. The EE contribution by HM on steel type A is higher than for charge types B and C, the latter two of which are charged similarly across all heats. For the Shredded scrap category, all charge types receive similar contributions to EE.

Based on these highlights, it is expected that charge type A has the lowest EE consumption, that charge type C has the highest EE consumption, and that the EE consumption of charge type B is in-between those of type A and type C. The following EE consumption, relative to the EE consumption by charge type A, was obtained based on the data from the heats used to create Figure 9: charge type B $= 0.99$ and charge type C $= 1.02$.

On the other hand, charge type B requires slightly less EE consumption on average than charge type A (1.00). This is likely because Plate1 being closer to a negative contribution to EE for charge type B than was previously anticipated. Furthermore, the analysis only focused on 4 of the 12 scrap categories, out of which 3 are of particular interest to the selected charge types. In addition, the model also uses a total of 21 input variables to predict the EE consumption of any given heat. However, given these caveats, the analysis is in line with what could be expected from the model analyzed as well as from expert domain knowledge.

The main utility of SHAP main interaction values is that it provides clarity on how specific values governed by the input variable distribution contribute to the EE consumption prediction. The SHAP main interaction values only show the univariate relationship between the input variable and the EE consumption prediction. It is possible to use the SHAP interaction effects between the input variables as explained in Section 3.5.1. However, the number of SHAP plots to analyze will be equal to an additional 20 for each input variable; giving a total of 441 SHAP interaction plots for a complete analysis of the selected model. Although experience and knowledge about the specific EAF can guide the selection of SHAP interaction plots, a further analysis of the SHAP plots is best left as a future point of study. The above analysis shows what can be done with the available tools. Although interesting, an exhaustive analysis is for obvious reasons out of scope of the present paper.

The SHAP main interaction effects on EE by each scrap category of the selected model can be observed in Figure 10. Thin plate, i.e., Plate1, has been confirmed by the steel plant engineers to contribute to less EE. This is evident since a steep drop can be observed. The reduced EE by Plate1 is eventually flattened out and increases when the amount of Plate1 approaches the upper limit of the furnace capacity. Internal1 contains heavy scrap with an apparent density of over 1.4 ton/m^3. Heavy scrap takes longer time to melt which contributes to a steadily rising EE with an increased amount of Internal1 scrap. This has also been confirmed by the steel plant engineers, which refer to the use of Internal1 as the reciprocal of Plate1 in the steel plant charging strategies. One could observe a slight decrease in the EE contribution by increasing the amount of Plate2 and a slight increase in EE contribution for Internal2. However, these scrap categories consist of less than 1% of the total charged weight in the studied heats. Hence, it is difficult to draw any clear conclusion on their contribution to the EE. Shredded scrap is charged based on operating practices rather than for specific charging strategies which results in steel with low amount of tramp elements. The EE contribution is decreasing with increased amount of Shredded scrap from the nominal amount used. This was not confirmed by the process engineers. The decreasing EE contribution could be a model artifact or because the shredded scrap does contribute to a decrease EE consumption in the process. The latter could be the case since shredded scrap is easily melted due to its high surface-area-to-volume ratio.

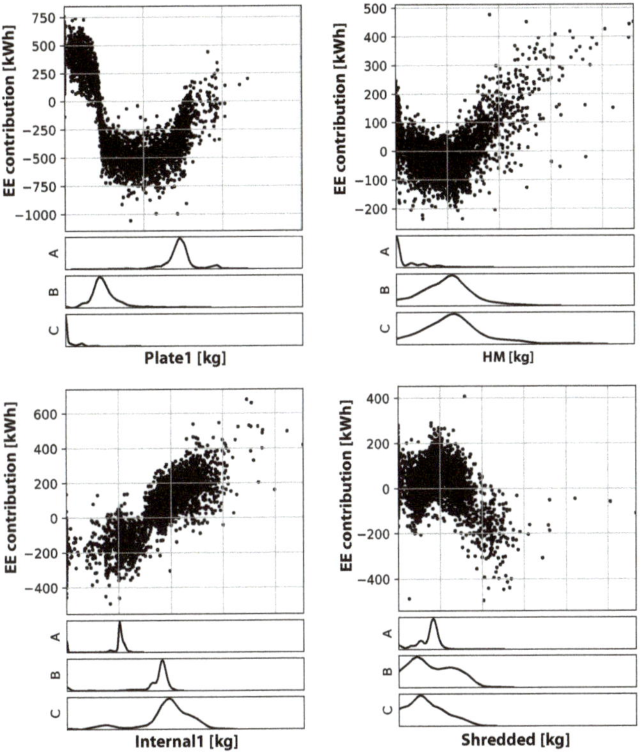

Figure 9. SHAP main interactions on EE by Plate1, Internal1, HM, and Shredded scrap categories. The three probability density plots below each plot show the distribution of the scrap category for each of the charge types A, B, and C, respectively. See Section 3.5.3 for an explanation of the different steel types. The y-axis of the scatter plot shows the SHAP main interactions on EE for each variable and the y-axes of the probability density plots show the frequency of each charge weight. The x-axes of both the SHAP plot and the probability density plot for each charge type are the amount charged of the scrap category. The values on the x-axes has been omitted due to proprietary reasons. The values governing the plots are from both the training and test data.

Incineration scrap is charged in low amounts, corresponding to only approximately 1.5% of the total charging weight on average for all heats and approximately 5% for the heats using charge types with lower requirements on impurity tramp elements. The increase in EE consumption is likely due to the melting requirements of the scrap weight rather than due to the surface-area-to-volume ratio. Also, skulls require higher EE according to the SHAP main interaction effect. This was confirmed by the steel plant engineers which reported that skulls are difficult to melt. Skulls are large concrete-like pieces of slag and metal mixture.

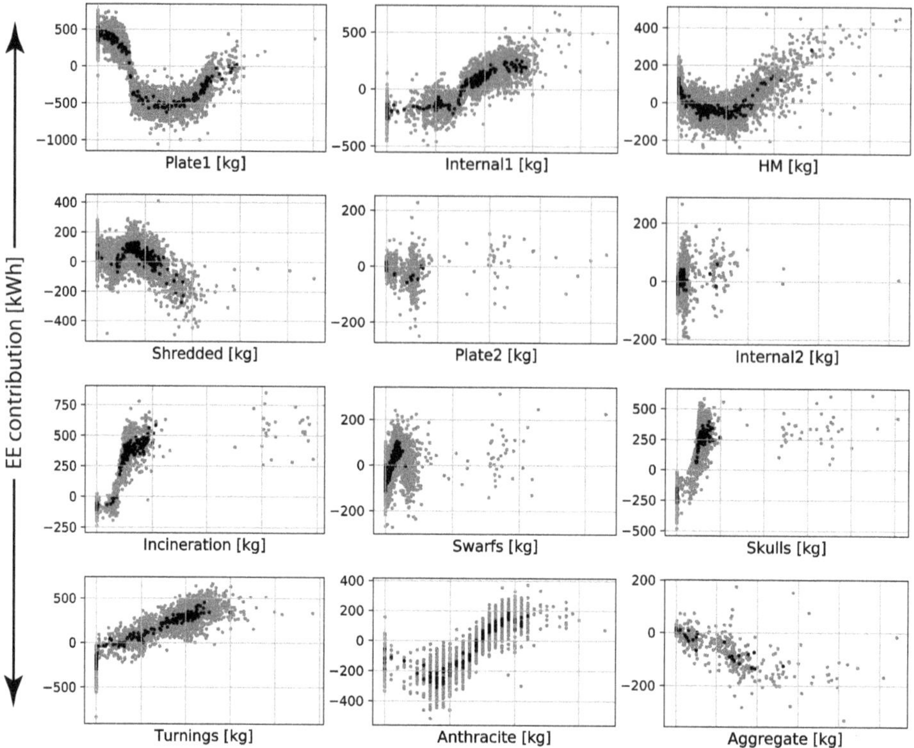

Figure 10. SHAP main interaction effects for the scrap categories for the selected model. The y-axes show the main interaction effect on EE while the x-axes show the amount of each charged scrap category. The values on the x-axes has been omitted due to proprietary reasons. The grey dots represent values from the training data and the black dots represent values from the test data.

The SHAP main interaction values for the base variables can be seen in Figure 11. Here, the focus will be on the variables whose EE contribution is counter-intuitive from the standpoint of a practitioner in physico-chemical modeling. Specifically, these are Burner O_2, Carbon Injection, Burner oil, and Lance O_2. According to the steel plant engineers, Burner oil is only effective up to a certain amount of *kg*, which is when the burner oxygen is used in tandem with burner oil. This agrees well with the observation from Figure 11. The burners are used in their maximum capacity when melting scrap. However, the burners still need to be active for the remainder of the heat to prevent the burners from getting clogged by slag and scrap. Thus, Burner oil is closely related to TTT, which is the reason Burner oil in higher amounts contributes positively to EE. Carbon injection, which should contribute to more heat generated by carbon boil, contributes positively to EE. This was confirmed by the steel plant engineers to be related to the continuous injection of carbon fines throughout the heat. As soon as liquid steel is present, carbon fines are injected to facilitate foaming slag. Similar to Burner oil, Carbon Injection is also closely related to TTT. In the steel plant of study, Lance O_2 is only used to clear the slag door to enable sampling of the steel melt temperature and composition. Therefore Lance O_2 does not have a consistent contribution to the EE.

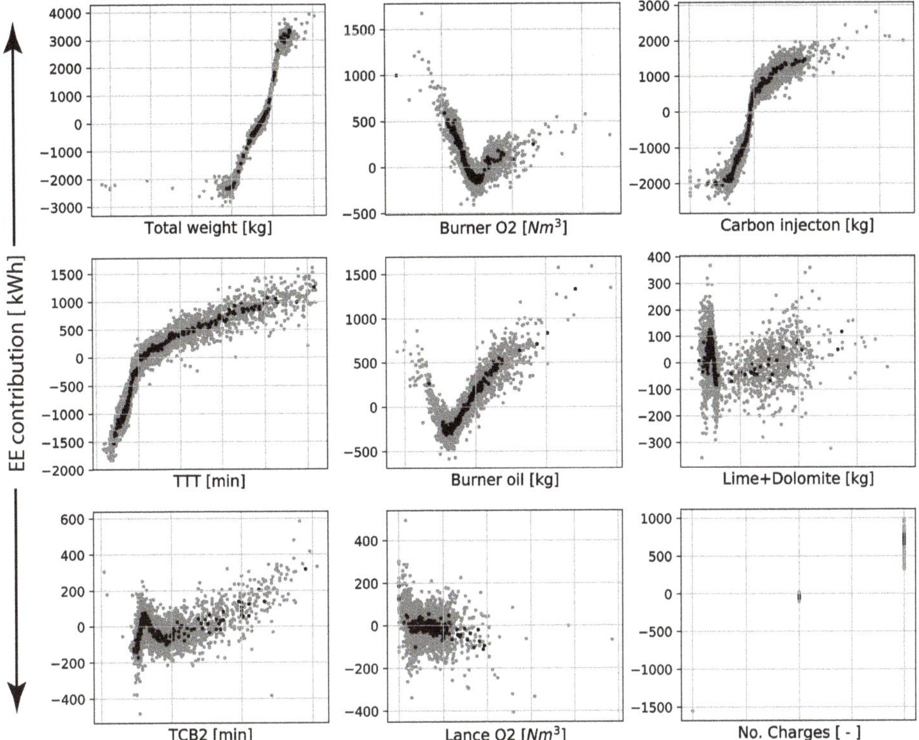

Figure 11. SHAP Main interaction effects for the base variables for the selected model. The y-axes show the main interaction effect on EE while the x-axes show the increasing amount of each variable. See Table 2 for details about the definition of each variable. The values on the x-axes has been omitted due to proprietary reasons. The grey dots represent values from the training data and the black dots represent values from the test data.

The positive relationship Injection Carbon to EE can also be observed by its Pearson correlation coefficient in Table 9. For both the Pearson correlation and dCor, Injection Carbon has the highest value with respect to the EE. In addition, the SHAP feature importance (Figure 12) regards the Carbon Injection as almost twice as important as the Total weight when the model predicts the EE consumption.

The sometimes counter-intuitive relations between the input variables to the EE consumption prediction emphasize the importance of not only having a firm understanding of the physico-chemical and process experience on the relations governing the EAF. It is also important to understand the relationship between the governing factors to the EE consumption of the specific EAF that one intends to model.

It is important to observe that the total weight and TTT are among the three most important variables for the model when predicting on EE. Both variables are also correctly considered by the model with respect to what is known from process metallurgical experience. This agrees well with the results from the previously reported SHAP analysis of a statistical model, created by the authors of the present study, predicting the EE of an EAF [10].

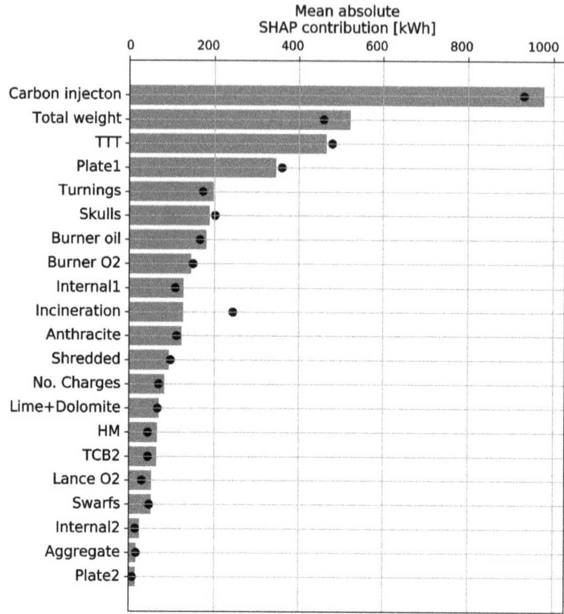

Figure 12. SHAP feature importance for the selected model as defined in Equation (15). The bars show the feature importance on the training data and the dark dots show the feature importance on the test data.

Table 9. Correlation between the input variables and the EE consumption as defined in the transformer system. The input variables are ordered by dCor in the training data set. The values in parentheses are the Pearson correlation coefficients.

Input Variable	Training	Test	Input Variable	Training	Test
Carbon injection	0.39(0.40)	0.33(0.31)	TCB2	0.12(0.13)	0.14(0.07)
Total weight	0.38(0.36)	0.35(0.37)	HM	0.10(0.10)	0.22(0.21)
TTT	0.28(0.31)	0.33(0.31)	Lance O2	0.09(0.07)	0.16(0.13)
Burner oil	0.24(0.24)	0.33(0.29)	Anthracite	0.09(0.05)	0.23(0.17)
Plate1	0.20(-0.17)	0.34(-0.31)	Internal2	0.08(0.05)	0.34(0.06)
Turnings	0.20(0.20)	0.28(0.24)	Swarfs	0.07(0.04)	0.28(0.20)
Skulls	0.18(0.15)	0.35(0.34)	Shredded	0.05(-0.03)	0.22(0.22)
Charges	0.17(0.19)	0.20(0.21)	Lime+Dolomite	0.04(0.01)	0.13(-0.13)
Internal1	0.16(0.16)	0.22(0.13)	Plate2	0.04(-0.02)	0.07(-0.01)
Burner O2	0.15(0.13)	0.14(0.0)	Aggregate	0.04(-0.04)	0.10(-0.07)
Incineration	0.15(0.10)	0.33(0.30)			

5. Conclusions

The main aim of this study was to investigate the effects of scrap in the performance of a statistical model predicting the EE consumption of an EAF. This was done using three distinct representations of the scrap types as well as SHAP main interaction effects which reveals the contributions from the scrap variables on each specific EE consumption prediction. In addition, the study extended a previously reported methodology used to create statistical models predicting the EE consumption of an EAF [2]. This was achieved by investigating the effects on the model performance using four different data cleaning strategies, an additional statistical model framework, RF, and data from an EAF producing steel for tubes, rods, and ball-bearing rings as opposed to stainless steel.

The main conclusions of the study may be summarized as follows:

5.1. Analysis of the Models

- All models using variables from any of the three scrap representations performed better than the models using only the total charged scrap weight as a scrap representation. This consistency provides evidence that more granular scrap representations are important when modeling the EE consumption of an EAF using steel scrap as the main raw material.
- The models using the visual scrap representation had the highest $\bar{R}^2_\mu$-values regardless of cleaning strategy. This provides evidence that a scrap representation based on the shape of the scrap, i.e., surface-area-to-volume ratio, is the most optimal to use. The result agrees well with the surface-area-to-volume ratio and the physico-chemical relationships on the scrap melting rate governed by temperature gradients, alloying gradients, and the freezing effect. In addition, this representation is also intuitive and easy to apply from the perspective of steel plant engineers and scrap yard operators.
- Data cleaning strategies using domain-specific knowledge from either an expert from the steel plant of study or a domain expert provide models with the highest $\bar{R}^2_\mu$-values compared to pure statistical cleaning methods. This further emphasizes the importance of domain-specific knowledge when modeling the EAF.

5.2. Analysis of the Selected Model

- The effects of the most important scrap categories on the EE consumption for the selected charge types A, B, and C are transparently presented by SHAP main interaction values.
- SHAP main interaction values revealed the contribution by each scrap category on the complete prediction space for the selected model. Plate1 and Internal1 were confirmed by the steel plant engineers to agree well with their expectations. However, a more thorough investigation must be conducted to evaluate the reasons behind the effects of the other scrap categories. SHAP interaction values between the input variables is advised as a starting point.
- The experience and knowledge provided by the steel plant engineers were essential to determine the reasons behind the counter-intuitive responses by Injected Carbon, Lance O_2, Burner oil, and Burner O_2 on the EE consumption. This indicates that the participation by steel plant engineers is of outmost importance to evaluate the trustworthiness of a practical model.
- The authors question whether one should use the Carbon Injection as part of EE prediction models in the steel plant of study, given the specific operational practice. It is likely that the Carbon Injection variable creates a model artifact for the selected model.
- The test data coincides with the training data in the SHAP plots. This raises the trust in the model since the model response on previously unseen heats is within the range of SHAP values from the heats used to adapt the model parameters.

6. Further Work

A continuation of the present study is to systematically evaluate the EE consumption of each scrap category by performing controlled experiments. This would require a similar approach taken to achieve the experiments in this study. Specifically, modeling according to the methodology presented in this study, categorizing the scrap according to one of the provided scrap representations, and cooperating with the steel plant engineers. As a final step, the resulting EE prediction model should be implemented in the existing EAF process control system. Only then can the true effect by the model on the resulting, actual, EE consumption be investigated.

Author Contributions: Conceptualization, L.S.C. and P.B.S.; methodology, L.S.C. and P.B.S.; software, L.S.C.; validation, L.S.C. and P.B.S.; formal analysis, L.S.C. and P.B.S.; investigation, L.S.C. and P.B.S.; resources, P.B.S. and P.G.J.; data curation, L.S.C.; writing–original draft preparation, L.S.C.; writing–review and editing, L.S.C. and P.B.S.; visualization, L.S.C.; supervision, P.B.S. and P.G.J.; project administration, P.B.S. and P.G.J.; funding acquisition, P.G.J. All authors have read and agreed to the published version of the manuscript.

Funding: This research was funded by "Hugo Carlssons Stiftelse för vetenskaplig forskning" in the form of a scholarship granted to the corresponding author.

Acknowledgments: We want to thank the plant engineers Patrik Undvall and Jan Pettersson at the Ovako Steel Hofors mill for their support and data provisioning during this project.

Conflicts of Interest: The authors declare no conflict of interest.

Nomenclature

T_{HM}	Temperature of the molten steel
T_{liq}	Scrap melting temperature
ρ_{scr}	Density of the scrap metal
H_s	Heat of melting of scrap
c_p	Specific heat of scrap
h	Heat transfer coefficient in the interface of the molten steel and scrap
β	Mass transfer coefficient of carbon
C_l	Carbon content in liquid steel
C_i	Carbon content in the solid-liquid interface
ρ_s	Apparent density of scrap
C_0	Initial carbon content in the steel scrap.
h_{scr}	Mass transfer coefficient under forced convection
c	Experimentally determined constant
p	Experimentally determined constant
u	Average stirring power in the boundary between the melt and scrap surface area
R_{SV}	Surface-area-to-volume ratio
A	The total area of an arbitrary scrap piece
V	The total volume of an arbitrary scrap piece
l	The length of an arbitrary scrap piece
r	The radius of a cylindrical or spherical scrap piece
t	The thickness of a square plate
m	The mass of a scrap piece
T	Temperature of the surface of a body emitting thermal radiation
q_1	First quartile
q_3	Third quartile
ϵ	Pre-specified constant for outlier extremity
x_j	An instance of variable j
σ	Standard deviation
v	Number of input variables
P	Number of nodes in the previous layer
s_j	Summation of the input values for j:th node in the current layer
w_i	Weight of node i in the previous layer
x_i	Value of node i in the previous layer
x_k	A data instance to be predicted by a RF
y_k	The predicted value of data instance x_k by a RF
R^2	Coefficient of determination
$\bar{R}^2$	Coefficient of determination adjusted for number of data points and variables
n	Number of data points
E_i	Regular error for data point i
y_i	True value of the output variable for data point i
$\hat{y}_i$	Predicted value of the output variable for data point i

$\overline{R}^2_\mu$	Mean adjusted R-square of the 10 model instances on test data
$\overline{R}^2_\sigma$	Standard deviation of adjusted R-square of the 10 model instances on test data
$\overline{R}^2_{min}$	Minimum of adjusted R-square of the 10 model instances on test data
$\overline{R}^2_{max}$	Maximum of adjusted R-square of the 10 model instances on test data
Δ_μ	Mean error of the mean error of the 10 model instances on the test data
Δ_σ	Standard deviation of the mean error of the 10 model instances on the test data
Δ_{min}	Minimum of the mean error of the 10 model instances on the test data
Δ_{max}	Maximum of the mean error of the 10 model instances on the test data
f	An arbitrary statistical model
f_x	A simplified representation of f
ϕ_0	The contribution to the prediction by f when all information is absent
ϕ_i	The contribution to the prediction by f by variable i
Z	A subset of a set of all input variables
$\hat{Z}$	The set of all input variables
$|Z|$	The number of variables in the subset Z
$\hat{Z} \setminus \{i\}$	The set of all input variables excluding variable i
$Z \cup \{i\}$	The subset of a set of all input variables including variable i
$Z \cup \{j\}$	The subset of a set of all input variables including variable j
$Z \cup \{(i,j)\}$	The subset of a set of all input variables including variable i and j
M	The total number of input variables
$\phi_{i,i}$	The main interaction value for variable i
$\phi_{j,i}$	The interaction effect of input variable i imposed on variable j
$\phi_{i,j}$	The interaction effect of input variable j imposed on variable i
FI_j	SHAP feature importance for variable j
ϕ_j^i	Regular SHAP value for variable j and data instance i
$\rho_{X,Y}$	Pearson correlation coefficient between random variables X and Y
σ_X	Standard deviation of random variable X
σ_Y	Standard deviation of random variable Y
V_1	Random variable
V_2	Random variable
$dCor\,(V_1, V_2)$	Distance correlation between V_1 and V_2
$dCov\,(V_1, V_2)$	Distance covariance between V_1 and V_2
$\sqrt{dVar\,(V_1)}$	Distance standard deviation for V_1
$\sqrt{dVar\,(V_2)}$	Distance standard deviation for V_2

Abbreviations

The following abbreviations are used in this manuscript:

EE	Electrical Energy
EAF	Electric Arc Furnace
BOF	Basic Oxygen Furnace
DRI	Direct Reduced Iron
DMS	Demand Side Management
MLR	Multivariate Linear Regression
PLS	Partial Least Squares regression
ANN	Artificial Neural Network
RF	Random Forest
KS	Kolmogorov–Smirnov
dCor	Distance Correlation
SHAP	SHapley Additive Explanations
TTT	Tap-to-Tap Time
TCB2	Time to Charging of Basket 2
HM	Heavy Melting Scrap
VB	Variable Batch

Appendix A

Table A1. Hardware specifications for the numerical experiments.

Computer model	Dell Latitude E5570
CPU	Intel Core i7 2376 MHz
RAM	16,203 MB

Table A2. Software specifications for the numerical experiments.

Purpose	Software/Package	Version
Operating system	Microsoft Windows 7 Professional	6.1.7601 Service Pack 1 Build 7601
Programming language	Python 3	3.7.1
Python distribution	Anaconda 3	4.6.7
Data handling	pandas	0.23.4
	numpy	1.17.4
Statistical modeling	scikit-learn	0.20.1
SHAP values	shap	0.8.1
Distance correlation	dcor	0.3
Visualization	matplotlib	3.0.2
	seaborn	0.9.0

References

1. Carlsson, L.S.; Samuelsson, P.B.; Jönsson, P.G. Predicting the Electrical Energy Consumption of Electric Arc Furnaces Using Statistical Modeling. *Metals* **2019**, *9*, 959. [CrossRef]
2. Carlsson, L.S.; Samuelsson, P.B.; Jönsson, P.G. Using Statistical Modeling to Predict the Electrical Energy Consumption of an Electric Arc Furnace Producing Stainless Steel. *Metals* **2020**, *10*, 36. [CrossRef]
3. Sandberg, E.; Lennox, B.; Undvall, P. Scrap management by statistical evaluation of EAF process data. *Control. Eng. Pract.* **2007**, *15*, 1063–1075. [CrossRef]
4. Baumert, J.C.; Vigil, J.; Nyssen, P.; Schaefers, J.; Schutz, G.; Gillé, S. *Improved Control of Electric arc Furnace Operations by Process Modelling*; European Commission: Luxembourg, 2005; ISBN 92-894-9789-0.
5. Haupt, M.; Vadenbo, C.; Zeltner, C.; Hellweg, S. Influence of Input-Scrap Quality on the Environmental Impact of Secondary Steel Production. *J. Ind. Ecol.* **2016**, *21*, 391–401. [CrossRef]
6. Köhle, S.; Hoffmann, J.; Baumert, J.; Picco, M.; Nyssen, P.; Filippini, E. *Improving the Productivity of Electric arc Furnaces*; European Commission: Luxembourg, 2003; ISBN 92-894-6136-5.
7. Chen, C.; Liu, Y.; Kumar, M.; Qin, J. Energy Consumption Modelling Using Deep Learning Technique—A Case Study of EAF. In Proceedings of the 51st CIRP Conference on Manufacturing Systems, Stockholm, Sweden, 16–18 May 2018.
8. Manana, M.; Zobaa, A.; Vaccaro, A.; Arroyo, A.; Martinez, R.; Castro, P.; Laso, A.; Bustamante, S. Increase of capacity in electric arc-furnace steel mill factories by means of a demand-side management strategy and ampacity techniques. *Int. J. Electr. Power Energy Syst.* **2021**, *124*, 106337. [CrossRef]
9. Lundberg, S.M.; Lee, S.I. A Unified Approach to Interpreting Model Predictions. In *Advances in Neural Information Processing Systems 30*; Guyon, I., Luxburg, U.V., Bengio, S., Wallach, H., Fergus, R., Vishwanathan, S., Garnett, R., Eds.; Curran Associates, Inc.: New York, NY, USA, 2017; pp. 4765–4774.
10. Carlsson, L.S.; Samuelsson, P.B.; Jönsson, P.G. Interpretable Machine Learning—Tools to Interpret the Predictions of a Machine Learning Model Predicting the Electrical Energy Consumption of an Electric Arc Furnace. *Steel Res. Int.* **2020**, 2000053. [CrossRef]
11. Friedrichs, H.A. *Schmelzen und Lösen, Stahleisen-Sonderberichte*, 1st ed.; Verlag Stahleisen GmbH: Düsseldorf, Germany, 1984; Volume 12.

12. Penz, F.M.; Schenk, J. A Review of Steel Scrap Melting in Molten Iron-Carbon Melts. *Steel Res. Int.* **2019**, *90*, 1900124. [CrossRef]

13. Deng, S.; Xu, A.; Yang, G.; Wang, H. Analyses and Calculation of Steel Scrap Melting in a Multifunctional Hot Metal Ladle. *Steel Res. Int.* **2019**, *90*, 1800435. [CrossRef]

14. Kirschen, M.; Badr, K.; Pfeifer, H. Influence of Direct Reduced Iron on the Energy Balance of the Electric Arc Furnace in Steel Industry. *Energy* **2011**, *36*, 6146–6155. [CrossRef]

15. Sandberg, E. Energy and Scrap Optimisation of Electric arc Furnaces by Statistical Analysis of Process Data. Ph.D. Thesis, Luleå University of Technology, Luleå, Sweden, 2005.

16. Pfeifer, H.; Kirschen, M. Thermodynamic analysis of EAF electrical energy demand. In Proceedings of the 7th European Electric Steelmaking Conference, Venice, Italy, 26–19 May 2002.

17. Steinparzer, T.; Haider, M.; Zauner, F.; Enickl, G.; Michele-Naussed, M.; Horn, A. Electric Arc Furnace Off-Gas Heat Recovery and Experience with a Testing Plant. *Steel Res. Int.* **2014**, *85*, 519–526. [CrossRef]

18. Keplinger, T.; Haider, M.; Steinparzer, T.; Trunner, P.; Patrejko, A.; Haselgrübler, M. Modeling, Simulation, and Validation with Measurements of a Heat Recovery Hot Gas Cooling Line for Electric Arc Furnaces. *Steel Res. Int.* **2018**, *89*, 1800009. [CrossRef]

19. Tukey, J.W. *Exploratory Data Analysis*, 1st ed.; Pearson: London, UK, 1977; ISBN 978-0201076165.

20. Lundberg, S.M.; Erion, G.; Chen, H.; DeGrave, A.; Prutkin, J.M.; Nair, B.; Katz, R.; Himmelfarb, J.; Bansal, N.; Lee, S.I. From local explanations to global understanding with explainable AI for trees. *Nat. Mach. Intell.* **2020**, *2*, 56–67. [CrossRef] [PubMed]

21. Goodfellow, I.; Bengio, Y.; Courville, A. *Deep Learning*; MIT Press: Cambridge, MA, USA, 2016; Available online: http://www.deeplearningbook.org (accessed on 18 August 2020).

22. Breiman, L. Random Forests. *Mach. Learn.* **2001**, *45*, 5–32. [CrossRef]

23. Cutler, A.; Cutler, D.; Stevens, J. *Ensemble Machine Learning: Methods and Applications*; Springer: Berlin/Heidelberg, Germany, 2011; Volume 45, Chapter 5, pp. 157–175.

24. Ajossou, A.; Palm, R. Impact of Data Structure on the Estimators R-square and Adjusted R-square in Linear Regression. *Int. J. Comput. Math.* **2013**, *20*, 84–93.

25. Lundberg, S.M.; Erion, G.G.; Lee, S.I. Consistent Individualized Feature Attribution for Tree Ensembles. *arXiv* **2018**, arXiv:1802.03888.

26. De Siqueira Santos, S.; Yasumasa Takahashi, D.; Nakata, A.; Fujita, A. A comparative study of statistical methods used to identify dependencies between gene expression signals. *Brief. Bioinform.* **2014**, *15*, 906–918. [CrossRef] [PubMed]

27. Székely, G.; Rizzo, M. Brownian distance covariance. *Ann. Appl. Stat.* **2009**, *3*, 1236–1265. [CrossRef] [PubMed]

 *processes*

Article

Development of an Electric Arc Furnace Simulator Based on a Comprehensive Dynamic Process Model

Thomas Hay [1,*], **Thomas Echterhof** [1] **and Ville-Valtteri Visuri** [2]

[1]   Department for Industrial Furnaces and Heat Engineering, RWTH Aachen University, Kopernikustr. 10, 52074 Aachen, Germany; echterhof@iob.rwth-aachen.de

[2]   Process Metallurgy Research Unit, University of Oulu, P.O. Box 4300, University of Oulu, 90014 Oulu, Finland; ville-valtteri.visuri@oulu.fi

*   Correspondence: hay@iob.rwth-aachen.de; Tel.: +49-241-8026-074

Received: 22 October 2019; Accepted: 11 November 2019; Published: 14 November 2019

Abstract: A simulator and an algorithm for the automatic creation of operation charts based on process conditions were developed on the basis of an existing comprehensive electric arc furnace process model. The simulator allows direct user input and real-time display of results during the simulation, making it usable for training and teaching of electric arc furnace operators. The automatic control feature offers a quick and automated evaluation of a large number of scenarios or changes in process conditions, raw materials, or equipment used. The operation chart is adjusted automatically to give comparable conditions at tapping and allows the assessment of the necessary changes in the operating strategy as well as their effect on productivity, energy, and resource consumption, along with process emissions.

Keywords: electric arc furnace; simulation; process model; steelmaking

---

1. Introduction

The electric arc furnace (EAF) process is the main process for recycling of ferrous scrap [1] and the second most important steelmaking process route in terms of global steel production [2]. EAF process models have proven to be useful for improving process understanding and control as well as resource and energy efficiency by providing information that cannot be measured directly during the process due to the extreme conditions inside the furnace. Numerous models have been developed using different approaches both for the complete process as well as local phenomena or single process phases. However, few simulators allow for real-time manipulation of simulations by the user. Logar et al. [3] describe a simulator based on their process model and the World Steel Association provides an online EAF simulator on their website [4]. The simulator of Logar et al. [3] is based on a previously developed process model considering detailed heat transfer [5,6] and thermochemical equations [7] as well as an electrical model [8], whereas the World Steel Association gives only limited information about the workings of their model [9]. Other published models run simulations based on predetermined input data and although they can be used to study different scenarios, they do not adapt to current simulation results or user inputs.

In order simulate the process independent of plant data and cover extreme cases that may occur with unusual user input, a comprehensive mechanistic modelling approach is necessary. Several such models are available in the literature such as those of Bekker et al. [10–12] and MacRosty et al. [13–15] who introduced some of the simplifications and assumptions that Logar et al. based their model on, as well as several others that have been published with a varying degree of detail [16–31]. Both Meier et al. [32–37] and Fathi et al. [38] have published models developed on the basis of the work of Logar et al. [3,5–8]. With the exception of Logar et al., all these models rely on predetermined input from measured data and, as far as can be determined from the published information, no other simulators that provide direct feedback and control for the user are available. Meier implemented the automatic control of the model on the basis of current simulation results [35], but his model does not allow direct user input during the simulation. The model and simulator presented here are based on a further development of the model proposed by Meier [39]. Although the general model structure and most of the assumptions and simplifications from Logar et al. [3,5–8] remain, the model has been adjusted for a different solver algorithm [36] and substantially modified [32–34,36,37,39] making it one of the most comprehensive and up-to-date EAF process models currently available. The differences compared to the model published by Logar et al. include a more efficient numerical method for solving the model equations [36], increased detail in the gas phase with additional species and reactions [34,37], a more comprehensive model of the radiative heat transfer considering the influence of the gas phase [32,37], the treatment of different carbon carriers [33], and the refinement of the thermochemical model for the interaction of slag and melt with each other as well as the injected oxygen and carbon [39].

2. Process Model

The model presented here as a basis for the simulator is essentially the state of development documented in a previous publication [39]. The model is executed in the MATLAB environment. Within the model there are eight different zones (i.e., control volumes), each of which is homogeneous in terms of composition and temperature. Separate energy and mass balances are defined for each zone. The zones are the (1) solid scrap, (2) liquid metal, (3) solid slag formers, (4) liquid slag, (5) gas phase, (6) water cooled wall and roof, and (8) the electrode(s). The electric arc is treated simply as a heat source whereas the air-cooled bottom vessel and lower wall section are represented as heat sinks. Figure 1 shows a schematic illustration of all zones and heat sinks/sources within the model.

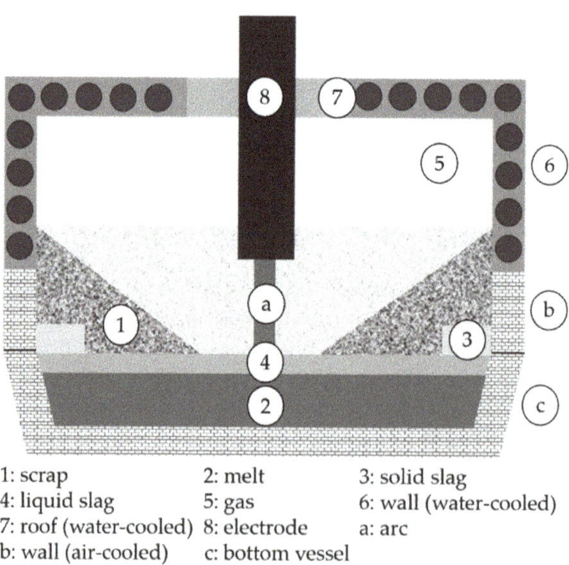

1: scrap   2: melt   3: solid slag
4: liquid slag   5: gas   6: wall (water-cooled)
7: roof (water-cooled) 8: electrode a: arc
b: wall (air-cooled)   c: bottom vessel

Figure 1. Model zones.

Relevant phenomena such as heat exchange through radiation and direct contact, phase changes and mass transfer between zones, chemical reactions, as well as the change in geometry during meltdown, are considered in the model. Simulation results have been validated using extensive data from industrial furnaces [32–34,37,39].

The operation of the furnace is characterized by continuous as well as discontinuous mass and energy flows. For the charging of scrap baskets, the amount and composition of scrap, slag formers, and coal charged with each basket can be considered with an unlimited number of baskets per heat. The charging in bulk of coal, slag formers, or alloys without scrap can be accounted for as well by defining the desired amount and composition and charging it independently. Due to the discontinuous change of conditions inside the furnace resulting from charging of material in bulk, the integration is stopped and restarted with new initial values adjusted according to the added material for each charging event.

The time-dependent model input or operation chart consists of the electric energy input characterized by voltage and current as well as the mass flows of oxygen for lances, burners and post-combustion, natural gas for burners, coal for carbon lancing, slag former injection, and off-gas extraction. The cooling of the wall and roof is determined on the basis of the mass flows and inlet temperature of the respective cooling water flows. For the simulation based on plant data, these inputs are determined from measured values or estimates and stored for each heat at the beginning of the simulation to be evaluated at every time step of the simulation. Additional parameters describing furnace-specific properties such as the geometry and empirical factors adjusted for each individual plant are stored in a separate Excel file. Different parameter sets can be defined and automatically simulated in succession. By comparing the results for different parameter settings, empirical parameters can be adjusted for different furnaces or process conditions and the impact of parameters such as the composition of slag formers or coal and gases can be evaluated. For the automatic control of the simulation as implemented by Meier [35], the masses charged in bulk are predetermined while the time of charging, mass flows, and electric energy input are determined on the basis of the progress of the simulation and rules derived from the operation of a real-life furnace. For the simulator, the operation chart and charging of baskets can be determined by the user during the simulation.

Figure 2 shows the structure of the model with its different modules and the exchange of information between them. Essentially, the main model runs independent from the type of data use and the change between the three running modes (measurement-based simulations, automatic control, and user-input-controlled simulations) affects only the data module and its interaction with the main model.

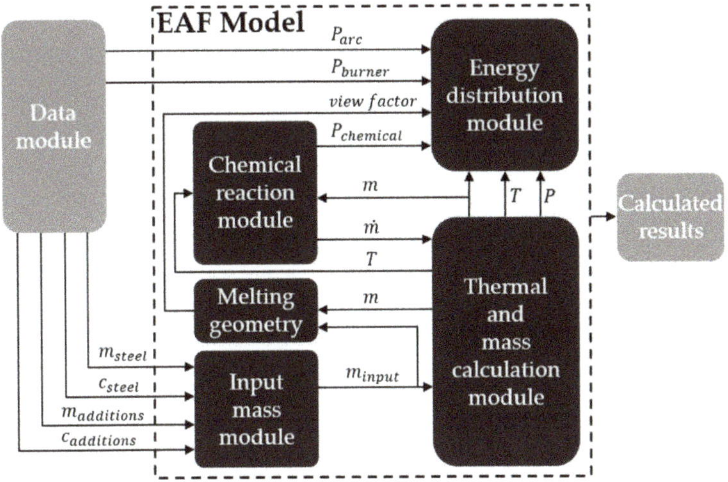

Figure 2. Model structure (translated from [37]). EAF: electric arc furnace.

The data module determines both the initial conditions for each basket that are calculated in the input mass module and the time-dependent input during the simulation that is transferred to the energy distribution module for each time step. Figure 3 shows how the structure of the data module and the data that is exchanged with the EAF model when measured data is used. The input is predetermined for each point of time and together with the geometry and other furnace-specific parameters determines the input for the model. The validation data is not used directly as model input but can be compared to model results and to adjust empirical parameters.

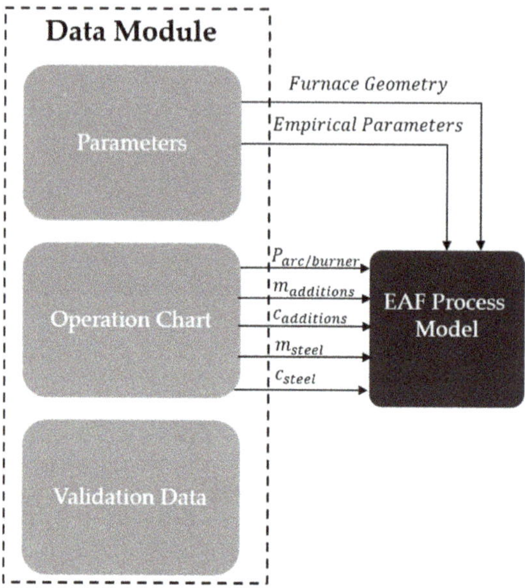

Figure 3. Data module with measured data.

Figure 4 shows the structure of the data model for the automatic control (left) and simulator (right) modes. Because the input is no longer predetermined for the full simulation, the time and current state have to be passed back to the data module for each step. In the case of the automatic control mode, the current input is then determined on the basis of the current set of parameters, rules, and minimum/maximum values and passed back to the EAF model. In the simulator mode, the user interface is updated with the current model results and the input for the model is updated with those set in the interface so that any changes made by the user since the last update are carried over to the EAF model. Furthermore, for the simulator mode, the selected speed is passed to the model as an additional input parameter, whereas in all other cases the model always runs with the highest attainable speed by default.

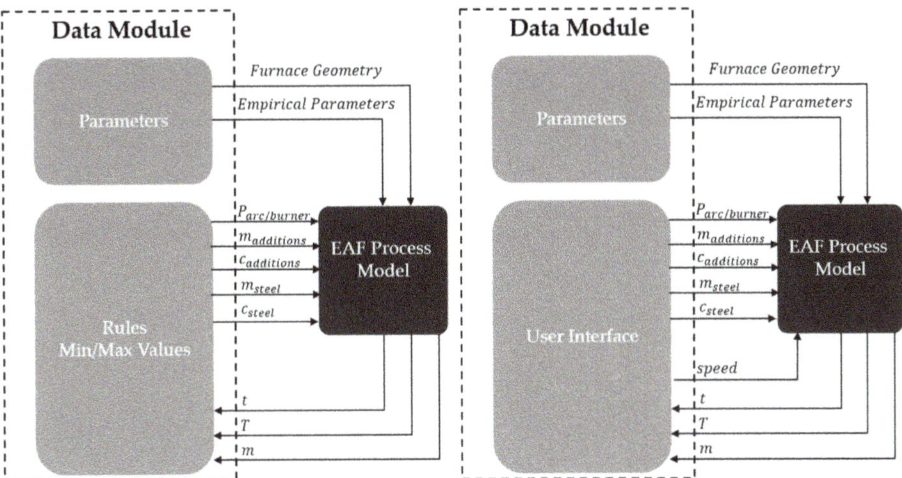

Figure 4. Data module for automatic control and simulator modes.

3. Automatic Control Mode

The automatic control mode of the simulation implemented by Meier [35] was developed to evaluate different scenarios such as the use of alternative materials of varying quantity and composition, for instance, the replacement of coal with alternative carbon carriers and the use of oxygen from different sources with varying oxygen and nitrogen content, or different operating strategies for the control of post-combustion, burners, and carbon and oxygen lances. The simulations are automatically adjusted to reach the predetermined conditions at tapping for all scenarios so that their impact on energy consumption, operating cost, and carbon emissions can be evaluated.

3.1. Input

Different scenarios for which automatic operating strategies are to be generated and compared can be defined by the user in an additional worksheet of the Excel file used for the furnace parameters. First, the masses and compositions of scrap, two coal types, and two slag formers (usually lime and dolomite) charged with three scrap baskets are defined. Although the input file is currently limited to this, the model allows the addition of both more baskets and additional types of coal or slag formers. The scrap composition can be defined separately for each basket. Furthermore, the initial conditions before charging of the first basket, namely, the mass, temperature, and composition of the hot heel present in the furnace have to be set. In addition to the values defining the initial conditions and charged material, the minimum and maximum values have to be set for the electric power and voltage, oxygen input for lances, post-combustion and burners, carbon injection, off-gas mass flow, slag former injection, natural gas for burners, and the voltage. The variation of these parameters between the selected minimum and maximum values over the course of each simulation is determined according to a set of rules that in turn can be adjusted through the parameters shown and described in Table 1.

One minimum and several maximum values can be given for each parameter. The model automatically selects all possible combinations and runs a simulation for each case. This can be combined with the variation of parameters, such as the composition of coal and slag formers, allowing a large number of scenarios to be simulated and compared automatically.

Table 1. Parameters for automatic control.

Parameter	Unit	Description
$t_{stop\text{-}delay}$	s	Time to raise electrodes and open roof
$t_{start\text{-}delay}$	s	Time to close roof and lower electrodes
$V_{scrap\text{-}max}$	%	Maximum fraction of furnace volume that can be filled with scrap
$P_{start\text{-}reduction}$	%	Reduction of electric power during bore down (until arc is covered by scrap)
$P_{refine\text{-}reduction}$	%	Reduction of electric power during refining (flat bath)
$P_{wall\text{-}reduction}$	%	Reduction of electric power when water-cooled wall overheats
$T_{wall\text{-}crit}$	K	Critical wall temperature for power reduction
$O_{lance\text{-}min}$	%	Fraction of maximum value during reduced lancing
$C_{lance\text{-}burner}$	%	Oxygen lancing increased when burner power below this fraction
$C_{lance\text{-}bath}$	%	Influence of free bath surface on oxygen lancing
$C_{post\text{-}scrap}$	%	Post-combustion reduced when remaining scrap below this fraction
$t_{post\text{-}delay}$	s	Post-combustion starting with delay after power on
$C_{burner\text{-}scrap}$	%	Burner power reduced when remaining scrap below this fraction
$C_{carbon\text{-}batth}$	%	Carbon lancing initiated when free bath surface above this fraction
$C_{carbon\text{-}scrap}$	%	Carbon lancing reduced when remaining scrap below this fraction

3.2. Control of Operation Chart Parameters

Using predefined rules, the model determines the desired value for each parameter from the selected minimum and maximum values during the simulation, replacing the fixed operation chart determined from measured data. The rules are derived from the current operating strategies and operation charts from the regular operation of different EAF. The simulation starts with the charging of the first basket and is terminated after the scrap charged with the final basket has melted and a predefined carbon content and temperature of the melt have been reached. After and before the charging of each basket all values of the operation chart are set to their minimum (power off) values for the delays defined by $t_{start\text{-}delay}$ and $t_{stop\text{-}delay}$ to account for the power-off time associated with the raising and lowering of the electrodes, as well the opening and closing of the roof as necessary, when scrap is charged into the furnace. The condition for charging the next basket is that the sum of the currently remaining solid scrap volume and the volume of the scrap charged with the next basket are below the maximum allowed scrap volume $V_{scrap\text{-}max}$. If no additional basket is defined and the conditions for tapping are met, all operation chart parameters are reduced to their minimum values and the simulation is terminated. The charging of the second and additional baskets is triggered by an event function that constantly checks if the necessary conditions have been met. Initially, once the conditions are met for charging of the next basket, all operation chart parameters are reduced to their respective minimum values and the simulation continues until $t_{stop\text{-}delay}$ has passed. The simulation is then stopped and reinitiated with new starting values and continues with a power-off period determined by $t_{start\text{-}delay}$ and the following power-on period just as with the first basket.

A hyperbolic tangent function is used to allow rapid but continuous changes between states where necessary. It can assume values between one and zero and is based on a controlling variable and a threshold. $\varphi(a,b)$ indicates a hyperbolic tangent function that has the value zero as long as the variable a is smaller than the threshold b, and rapidly changes to one if a increases to values higher than b. For example, the factor indicating the charging of the next basket would be based on the maximum allowed scrap volume $V_{scap\text{-}max}$, the scrap volume in the next basket $V_{scrap\text{-}next}$, and the actual scrap volume V_{scrap}, as shown in Equation (1):

$$\Phi_{basket} = \varphi\left(V_{scrap}, V_{scrap\text{-}max} - V_{scrap\text{-}next}\right). \tag{1}$$

It would become zero once the free volume is large enough to charge the next basket. Such factors are used for all conditions and delays defining the operation chart.

The electric power is reduced for the initial period after charging until bore down has progressed far enough to allow the arc to burn inside the scrap, at which point significantly less radiation is received by the wall and roof, and maximum power can be used. Later during the process, power is reduced once the full bath surface is uncovered and the flat bath phase begins. During any stage of the process, the power will be reduced if the wall temperature reaches the critical temperature $T_{wall\text{-}crit}$. The voltage is assumed to be a linear function of the power and the current is determined accordingly.

The oxygen lance mass flow is initially set to the percentage defined by $O_{lance\text{-}min}$ and increases once the burner power conditions set by $C_{lance\text{-}burner}$ is fulfilled with another increase to its maximum value once the complete bath surface is free of solid scrap. The ratio of the first and second increase is determined by the parameter $C_{lance\text{-}bath}$. Post-combustion oxygen flow is started with the maximum value once the additional delay $t_{post\text{-}delay}$ has passed and a minimum of 5% of the scrap has melted. It is stopped once the mass of solid scrap remaining divided by the initial amount of scrap has reached $C_{post\text{-}scrap}$. The burners are started with full power and both oxygen and natural gas input are stopped once the remaining scrap has reached $C_{burner\text{-}scrap}$. Carbon injection is started after meltdown of the last scrap charge has progressed far enough to create a free bath surface as defined by $C_{carbon\text{-}bath}$. Once the melt temperature approaches the desired tapping temperature or the amount of remaining scrap reaches $C_{carbon\text{-}scrap}$, carbon injection is phased out. Oxygen lancing is reduced proportionally with the reduction in carbon injection.

Off-gas extraction is started at the minimum mass flow and increased with the injection of gases and carbon. Because the injection of slag formers using lances is not practiced at the furnaces this automatic control was initially designed for, no rules have been defined for this purpose and the mass flow is permanently set to zero. The practice could, however, be included by simply defining the necessary factors and rules, as the process model does include the necessary equations and simulations based on measured data have been run successfully for furnaces where the injection of lime is practiced. Cooling water flows and inlet temperatures are assumed to be constant for these simulations.

4. Simulator Mode

Direct real-time feedback and control are necessary for the simulator function. Therefore, a user interface has to be implemented to control the simulator and display current results while the simulation is running. Figure 5 shows the currently used interface which displays the simulation time, current melt temperature, remaining solid scrap mass, and the cooling water outlet temperatures of wall and roof to indicate the progress of the process. The user can enter values for electric power, oxygen, carbon, and lime injection, and select burners 'on' or 'off' to control the simulation. The button 'charge basket' allows for the charging of the masses selected for scrap, coal, lime, and dolomite in the corresponding fields.

Power [MW]	Oxygen Lance [kg/s]	Carbon Lance [kg/s]	Speed	Time [s]	Melt Temperature [K]	Solid Mass [kg]
0	0	0	1	0	1950	0

Burner	Postcombustion Oxygen [kg/s]	Lime Lance [kg/s]			Wall Cooling Temperature [K]	Roof Cooling Temperature [K]
ON	0	0			0	0

Scrap [kg]	Coal [kg]	Lime [kg]	Dolomite [kg]			
0	0	0	0	Charge Basket		
				Tap Steel		

Figure 5. User interface.

The model allows the display of any value calculated during the simulation, parameters such as the total energy consumption or off-gas composition and temperature that can be used for control of industrial furnaces that can easily be added to the display. The same is true for information that is not attainable for a real-world furnace such as temperatures and energy or mass flows inside the furnace. Furthermore, every operation chart parameter can be added and manipulated by the user. Values for off-gas mas flow, arc voltage, and cooling water flows are currently set automatically but could be added to the interface instead. Simulations are started by charging the first basket and ended using the button 'tap steel'. Pressing these buttons during the simulation triggers an event and the sequence of determining new starting values and reinitiating the integration, which is described for the simulation using measured data or automatic control.

4.1. Simulation Speed

The EAF model runs simulations significantly faster than the real process times—the process taking roughly one hour per heat in reality can be simulated in less than a minute using a tabletop computer (3.4 GHz, 4 core central processing unit, 32 GB random-access-memory). In order to allow the user to make inputs and evaluate the current results, the simulation has to be slowed down. Therefore, the simulation is paused after each second of simulated process time, the user interface is updated, and values changed by the user after the previous update are returned to the process model. The duration of the pause depends on the actual speed of the simulation and the desired speed set by the user. If the simulation is progressing slower than the desired simulation rate, the pause is set to the smallest possible value that allows the user interface to be updated. Otherwise, the pause is held long enough to allow the simulation to synchronize with the desired simulation rate. The simulation speed can be adjusted by the user during the simulation.

4.2. Model Adjustments

Due to the different nature of the simulation input, several changes to the model and solver are required to insure fast and stable operation when using the simulator. When measured data is used as input, the operation chart values can be interpolated between time steps and the full operation chart is available at all times. For the automatic control, smooth changes are insured through the use of a hyperbolic tangent function and the rules remain the same throughout the simulation, allowing for stable simulation. Furthermore, both the automatic control and measured data from industrial operation usually insure smooth and consistent progression of the process, whereas with the simulator erratic user inputs and far more extreme and unusual states can be encountered, posing a significant challenge for the process model.

4.2.1. Continuity of Operation Chart

When using the simulator, the model has to be adjusted for unpredictable and discontinuous changes in operation chart values due to user input. Whenever such sudden changes are detected by the ODE15s solver used for the EAF model, negative time steps can occur, meaning that the solver will jump back to an earlier time and recalculate with a smaller step-size. Therefore, it is not possible to always use the value currently set in the interface. Instead, the operation chart is extended with the current time and values every time the interface is updated, and if negative steps occur then the previous values are used accordingly.

4.2.2. Pressure Oscillations

In rare cases, usually when reinitiating electric energy input after long power-off periods, the calculated pressure can start oscillating, causing the simulation to become very slow and eventually crash. This is caused by a feedback loop from the coupling of leak air ingress with the pressure and temperature of the gas phase. The pressure is calculated using the ideal gas law according to Equations (2) and (3):

$$P = \frac{RT\sum \frac{m_i}{M_i}}{V},$$ (2)

$$\frac{dP}{dt} = \frac{RT\sum \frac{dm_i}{dt}\frac{1}{M_i}}{V} + \frac{\frac{dT}{dt}\sum \frac{m_i}{M_i}}{V},$$ (3)

where P is the pressure, t is the time, T is the gas phase temperature, R is the gas constant, and m_i and M_i are the masses and molar masses of species in the gas phase, respectively. V is the volume of the gas phase and is assumed to remain constant, whereas the mass of the gas phase depends on the density and composition of the gas.

The leak air intake is assumed to be a linear function of the pressure inside the furnace, as shown in Equation (4):

$$\dot{m}_{leak-air} = a + b(P - c)$$ (4)

where a, b, and c are empirical parameters fitted to match measured data and optimize stability. In some cases, an increased intake of leak air will cause the temperature to drop, with a resulting reduction in pressure that in turn will increase the leak air ingress and vice versa. This can lead to unstable behavior and pressure oscillations with large amplitudes. This problem was addressed by defining the leak air ingress as a differential variable according to Equation (5).

$$\frac{d\dot{m}_{leak-air}}{dt} = 50(a + b(P - c) - \dot{m}_{leak-air}).$$ (5)

This allows the solver to detect steep changes in the leak air intake and adjust the time-step accordingly so that oscillations can be avoided. The difference in simulation results between Equations (4) and (5) is negligible.

4.2.3. Additional Stability Improvements and Model Acceleration

In addition to the pressure oscillations, certain unusual cases exist where masses would reach small negative values causing the model to crash. This was rectified by using the 'NonNegative' option of the ODE15s solver to insure the necessary adjustments of time-steps when masses approach zero with a steep gradient and the adjustment of some empirical model parameters, especially in several hyperbolic tangent functions that reduce chemical reaction rates when the mass of a reactant is small. Furthermore, the model was accelerated by making the integration more efficient. The ODE15s solver numerically calculates a Jacobian matrix that leads to a high number of function evaluations with a single variable changing. By detecting this behavior and using previous results for expensive analytical calculations, such as the determination of chemical activities and radiative heat transfer, if none of the relevant inputs have changed, the simulation speed is increased significantly. Furthermore, the number of function evaluations could be reduced by using the 'JPattern' option. Overall this made the simulation more stable and significantly faster [39]. Crashes or non-physical results such as negative masses have not been observed with the new model, even with unrealistic inputs and extreme conditions.

5. Results and Discussion

Validation of the EAF process model using extensive measured process data from a 140 t direct current (DC) furnace can be found in previous publications where the thermochemistry of the gas phase [34] and liquid phases [39], the behavior of different carbon carriers [33], the heat transfer [32], as well as the performance of the complete model [37], were evaluated thoroughly, which is not reproduced here. The model has also been tested with data from additional furnaces including both alternating current (AC) and DC technology and has given satisfactory results after adjustment of the empirical parameters during preliminary research. Instead, various simulations where both were run in automatic control and simulator mode were set up to evaluate the model stability and speed as well as its capability to reproduce existing operation strategies and adjust them for an altered set of boundary conditions.

5.1. A Case Study for Different Operating Modes

The automatic control was initially adjusted to reproduce the results obtained from simulations with measured data. Inputs with the scrap baskets as well as maximum and minimum mass flows were set to match those documented for an industrial 140 t DC furnace. After adjusting the automatic control to match the simulation on the basis of measured data for the operation chart, a scenario was tested where the oxygen used for the furnace operation was taken from a different source with an oxygen content of 40% instead of the 99% used for the initial case, with the remaining fraction consisting of nitrogen in both cases. The mass flows for oxygen lancing, burners, and post-combustion were adjusted to have the same mass flow of pure oxygen. Due to the decreased oxygen content, this led to an increased total mass flow and more nitrogen being injected together with the oxygen. The operation chart was automatically adjusted to reach the same tapping temperature and maximum carbon content, resulting in an increased energy consumption (both electrical and chemical) and an increased tap-to-tap time. The following cases were studied:

- Case 1, indicating the results obtained by adjusting the automatic control to reproduce the measured operation chart;
- Case 2, indicating the results from the same control settings with the decreased oxygen content.

The oxygen shown in the following discussion is the actual mass of pure oxygen; therefore, between Case 1 and Case 2 the total oxygen consumption increased by 6.5%, whereas, due to the increased nitrogen fraction, the nitrogen carried into the furnace with the injected oxygen for Case 2 was 64 times that of Case 1. Figure 6 shows the measured electrical power (real) during the heat compared to the automatic control for Case 1 and Case 2. The time was normalized using tap-to-tap time of the measured heat, the power was normalized using the maximum measured value.

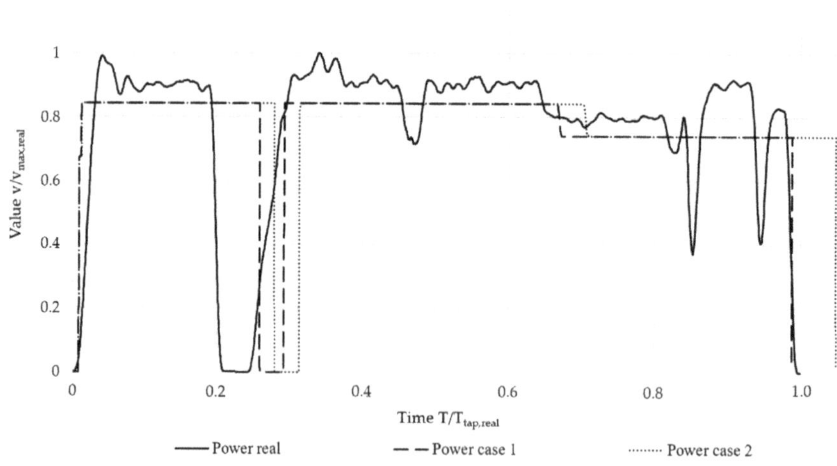

Figure 6. Electric power comparison.

Although the measured value fluctuated, the automatic control gave constant values. This had little impact on the overall consumption if the mean measured power was selected for automatic control. As can be seen from the reduction in power to zero, for the measured heat, the second basket was charged at about 0.23 process time, whereas automatic control charging occurred at roughly 0.3 for Case 1 and slightly later for Case 2. The time when the second basket was charged depended on the initial density of the scrap as well as other parameters, which could vary between baskets, and the progression of the process was not reproduced exactly here. Therefore, the reduction of power and mass flows associated with the charging of the second basket occurred slightly later in the simulations when compared to the measured operation chart. At about 0.65 and 0.7, the power was reduced for the flat bath phase for Case 1 and Case 2, respectively.

Figure 7 shows the consumption of natural gas denoted by CH_4 and injected coal denoted by C for the measured heat and Cases 1 and 2. Mass flows and time were normalized the same way as in Figure 6. For the first basket, the burners both in Case 1 and Case 2 showed almost identical behavior as was measured. Due to the delayed charging of the second basket, burner operation was delayed as well for the second basket, with a larger delay for Case 2 as meltdown of the initial basket was slower when compared to Case 1. Carbon lancing started slightly later than for the measured operation chart and the mass flow was reduced at around 80% of the process time.

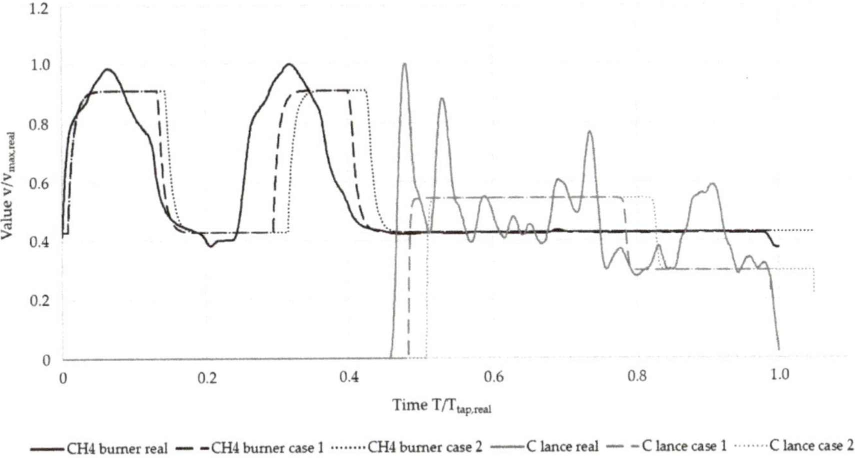

Figure 7. Natural gas and injected coal comparison.

Figure 8 shows the normalized mass flows of post-combustion and lanced oxygen. The oxygen mass flow for natural gas combustion is not shown as its profile was similar to that of the natural gas shown in Figure 7. The post-combustion oxygen followed a profile comparable to that of the burners and showed good agreement with the measured values. With the automatic control, oxygen lancing was increased for a short period before the second basket was charged. At roughly 60% process time, the mass flow was increased to its maximum value and reduced again at around 80% for the flat bath phase until tapping. The measured value showed a smoother progression during meltdown and fluctuated more during the flat bath phase; however, the general profile and the total oxygen consumption could be reproduced during automatic control.

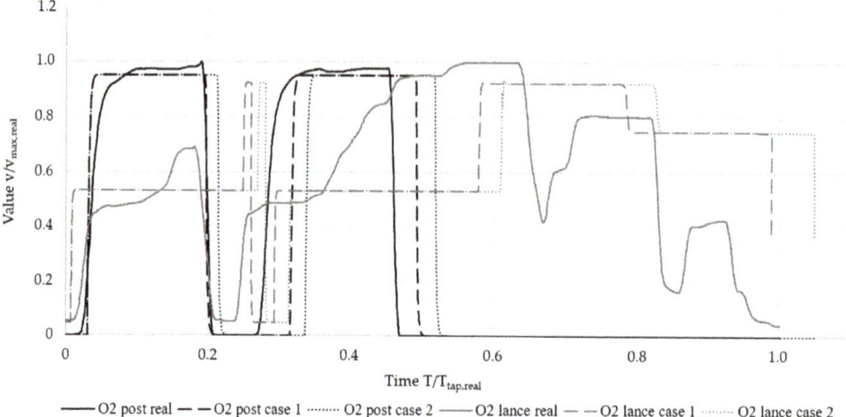

Figure 8. Oxygen comparison.

Figure 9 shows the progression of the simulated melt temperature for the measured operation chart and Cases 1 and 2. The delayed charging of the first basket was visible between 0.2 and 0.4 process time, as the temperature drop associated with the charging of cold scrap occurred later under automatic control. Although the temperature profile of Case 1 closely followed that of the measured case and reached tapping temperature almost simultaneously, the increased energy demand for Case 2 was visible in the lower temperature during the process and the delayed achievement of the desired tapping temperature. Overall, the automatic control implemented was able to reproduce the progression of the heat compared to the measured operation chart and indicate what impact a different oxygen source would have under otherwise similar conditions, showing the increased tap-to-tap time and consumption of electrical and chemical energy when the same operating strategies were applied in both cases.

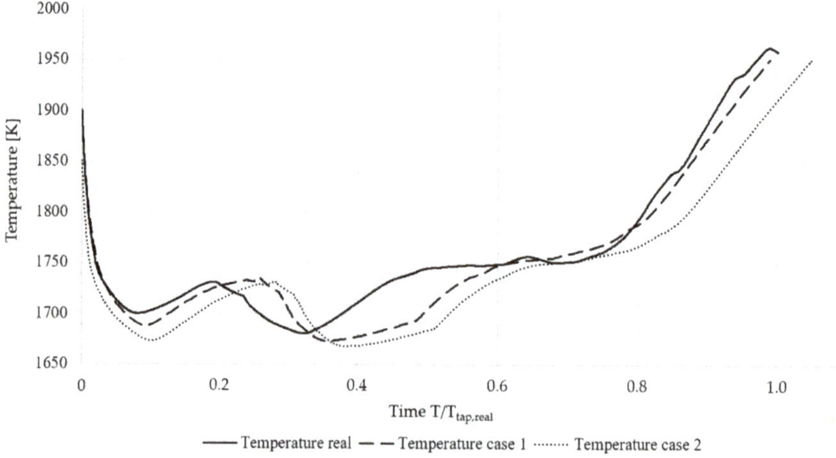

Figure 9. Melt temperature comparison.

Table 2 shows the resulting consumption of electric power, oxygen, natural gas, and coal, as well as the extracted off-gas and tap-to-tap time for the two cases relative to the values measured for the complete heat. Again, the increases in electrical and chemical energy used for Case 2 were visible.

Table 2. Calculated performance indicators for the study cases.

Parameter	Case 1	Case 2
Electric energy	1	1.06
Oxygen through lance	1.1	1.17
Oxygen for post-combustion	1	1.06
Injected carbon	0.9	0.96
Off-gas	1.06	1.13
Natural gas	0.99	1.04
Oxygen for natural gas burners	1	1.07
Total oxygen	1.07	1.14

5.2. Speed and Stability

For the simulator mode, the model speed and stability are most important. After charging a basket, the simulation is initiated and the solver takes 3–5 s to run the initial steps. After this initial delay, the simulator can be run at higher speed, allowing up to 20 seconds of the process to be simulated per second of real-time. Therefore, periods during meltdown or refining where no user input is necessary can be simulated at high speeds, whereas lower speeds can be selected during phases where frequent user inputs are necessary.

Figure 10 shows the progression of the selected simulation speed and the actual speed attained by the simulator during an example heat where two baskets were charged and a total process time of 2646 s was simulated in 272 s of real-time. The speed actually attained was calculated for each second of simulated process time. For most of the time, the simulation matched the selected speed; however, there were numerous instances where the speed dropped below the selected speed which was then compensated by an increased speed until the simulation and the target speed synchronized again. This never took more than a few seconds and the actual delay between the target time and the simulation stayed within less than 20 s. The short periods of re-synchronization were barely noticeable for the user, and inputs could be made precisely at the desired time and state of the simulated process.

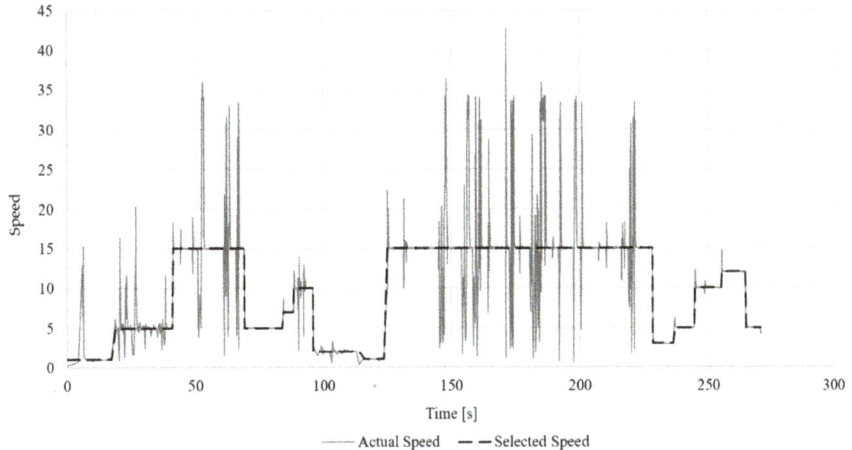

Figure 10. Selected and attained simulation speed.

Figure 11 shows the simulated and the target time during the first 55 s of the same simulation. The delay during initialization is visible at 0–6 s, after which the simulation synchronized with the desired speed and closely matched the target time. After the initiation of the basket, deviations from the target time remained small and lasted for no more than 5 seconds.

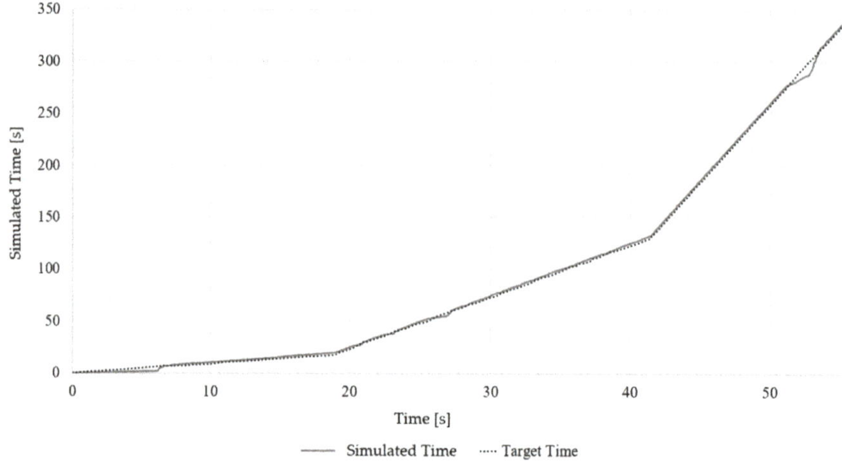

Figure 11. Simulated and target time.

Although results obtained with the automatic control and simulator cannot be validated using measured data, the internal consistency of the model can be evaluated using energy and mass balances. An energy balance using all heat flows calculated during the simulation yielded and error in the order of magnitude of 10^{-10} kWh, which was assumed to be a numerical error and was irrelevant for the simulation results. The mass balance prepared for each element gave a maximum error of 0.1% of the respective total mass, which was of no relevance for the simulation results.

The above examples illustrate the potential of the simulator developed. The simulator can be run in different modes depending on the aim of the simulation and, consequently, the simulator is applicable for different purposes such as process development, online predictions, and training of electric arc furnace operators. Simulators of this type are also applicable for dynamic optimization problems [40] and as soft sensors for evaluating parameters that are crucial for the EAF operators [3]. Another notion highlighting the importance of developing fundamental process models further is the fact that the computational load of comprehensive computational-fluid-dynamics-based approaches [41] remains—at least for the time being—too high for online applications.

5.3. Further Research

The Simulator mode allows direct user input and feedback during the simulation through a simple graphical interface. Although the current demonstrator is limited, all operation chart values can easily be added and manipulated during the simulation and additional settings such as compositions of charged materials for each basket could be added. All intermediate results of the model, such as temperatures, heat flows, and masses, are available during the simulation.

A target value is selected (for example 100 MW of electrical power) for both the automatic control and the simulator modes, and that exact value was used during the simulation, whereas in reality many operation chart values, especially the electrical power, current, and voltage, will fluctuate due to random influences such as the behavior of the electric arcs. In further work, a stochastic element could be added to introduce such fluctuations during simulations and produce more realistic behavior. Furthermore, it may be necessary to adjust the density of the scrap for each basket separately to better reproduce the behavior seen in real-life operation of the furnace.

In addition, further improvement of the underlying process model itself is of course possible. For example, a better representation of the melting behavior of scrap and adjustments for the use of other input materials, such as direct reduced iron, hot briquetted iron, or hot metal; improved description of the foaming of the slag; as well as the automatic adjustment of model parameters for different furnaces are possible areas of further research.

6. Conclusions

An automatic control and a simulator were developed on the basis of a comprehensive dynamic EAF process model that has previously been validated exhaustively using data from industrial furnaces. The automatic control is capable of reproducing real-live operation charts and adjusting them for different operating conditions. It can be used to evaluate various scenarios such as new control strategies, different materials for injection and charging, or the installation of new equipment. The rules currently used are partially based on parameters that cannot be measured or observed directly in the real-live process, and adjusted rules that are based more closely on the actual parameters used in process control such as total electric energy input and off-gas composition may be more useful for future work. The rules and parameters can be adjusted easily to allow different furnaces and operating strategies to be replicated and evaluated.

The simulator was shown to be stable and fast enough to run simulations based on user input in real-time as well as with higher speeds selected by the user with no noticeable delay apart from a few seconds during initialization after charging. Additional inputs and outputs can be added to make the simulator more versatile. For use in teaching and training, a more refined user interface and documentation would also be useful.

Due to the modular structure of the EAF process model simulations based on measured data, automatic control and simulator input can be run using the same basic model and structure with changes only necessary to the input module and the evaluation and output of the results, with the flexible solver algorithm allowing for stable and fast simulations for all cases. The model can therefore be used to evaluate the current state of the process and identify potential for improvement, to create and assess different scenarios and operating strategies, and for training and teaching using the simulator.

Author Contributions: Conceptualization, methodology, software, and writing—original draft preparation by T.H.; supervision, project administration, and writing—review and editing by T.E.; writing—review and editing by V.-V.V.

Funding: This research received no external funding

Conflicts of Interest: The authors declare no conflict of interest.

References

1. Madias, J. *Electric Furnace Steelmaking: Treatise in Process Metallurgy—Volume 3: Industrial Processes*; Elsevier: Oxford, UK, 2014; pp. 271–300.
2. *Steel Statistical Yearbook 2018*; World Steel Association: Brussels, Belgium, 2018.
3. Logar, V.; Škrjanc, I. Development of an Electric Arc Furnace Simulator Considering Thermal, Chemical and Electrical Aspects. *ISIJ Int.* **2012**, *52*, 1924. [CrossRef]
4. World Steel Association. Electric Arc Furnace Simulation. Available online: https://steeluniversity.org/product/electric-arc-furnace-simulation/ (accessed on 12 July 2019).
5. Logar, V.; Škrjanc, I. Modeling and Validation of the Radiative Heat Tranfer in an Electric Arc Furnace. *ISIJ Int.* **2012**, *52*, 1225–1232. [CrossRef]
6. Logar, V.; Dovžan, D.; Škrjanc, I. Modeling and Validation of an Electric Arc Furnace Part 1, Heat and Mass Transfer. *ISIJ Int.* **2012**, *52*, 402–412. [CrossRef]
7. Logar, V.; Dovžan, D.; Škrjanc, I. Modeling and Validation of an Electric Arc Furnace Part 2, Thermo-chemistry. *ISIJ Int.* **2012**, *52*, 413–423. [CrossRef]
8. Logar, V.; Dovžan, D.; Škrjanc, I. Mathematical modeling and experimental validation of an electric arc furnace. *ISIJ Int.* **2011**, *51*, 382–391. [CrossRef]
9. *Electric Arc Furnace Simulation User Guide*; University of Liverpool: Liverpool, UK, 2006.
10. Bekker, J.G.; Craig, I.K.; Pistorius, P.C. Model predictive control of an electric arc furnace off-gas process. *Control Eng. Pract.* **2000**, *8*, 445–455. [CrossRef]

11. Bekker, J.G.; Craig, I.K.; Pistorius, P.C. Modeling and Simulation of an Electric Arc Furnace Process. *ISIJ Int.* **1999**, *39*, 23–32. [CrossRef]

12. Bekker, J.G. Modelling and Control of an Electric Arc Furnace Off-Gas Process. Master's Thesis, University of Pretoria, Pretoria, South Africa, 1998.

13. MacRosty, R.D.M.; Swartz, C.L.E. Dynamic optimization of electric arc furnace operation. *AIChE J.* **2007**, *53*, 640–653. [CrossRef]

14. MacRosty, R.D.M.; Swartz, C.L.E. Dynamic modeling of an industrial electric arc furnace. *Ind. Eng. Chem. Res.* **2005**, *44*, 8067–8083. [CrossRef]

15. MacRosty, R.D.M.; Swartz, C.L.E. Optimization as a tool for process improvement in EAF operations. In Proceedings of the Iron and Steel Technology Conference AISTech, Charlotte, NC, USA, 9–12 May 2005.

16. Matson, S.; Ramirez, W.F. Optimal Operation of an Electric Arc Furnace. In Proceedings of the 57th Electric Furnace Conference, Pittsburgh, PA, USA, 14–16 November 1999.

17. Matson, S.A.; Ramirez, W.F. The dynamic modeling of the electric arc furnace. In Proceedings of the 55th Electric Furnace Conference, Chicago, IL, USA, 9–12 November 1997.

18. Matson, S.A.; Ramirez, W.F.; Safe, P. Modeling an EAF using Dynamic Material and Energy Balances. In Proceedings of the 56th Electric Furnace Conference, New Orleans, LA, USA, 15–18 November 1998.

19. Deneys, A.C.; Peaslee, K.D. Post-Combustion in the EAF—A Steady State Simulation Model. In Proceedings of the 55th Electric Furnace Conference, Chicago, IL, USA, 9–12 November 1997.

20. Shah, D.H.; Peaslee, K.D. Post Combustion in EAF-Dynamic Simulation Model. In Proceedings of the 56th Electric Furnace Conference, New Orleans, LA, USA, 15–18 November 1998.

21. Cameron, A.; Saxena, N.; Broome, K. Optimizing EAF Operations by Dynamic Process Simulation. In Proceedings of the 56th Electric Furnace Conference, New Orleans, LA, USA, 15–18 November 1998.

22. Cameron, A. Optimising electric arc furnace operations. *Steel Times* **1999**, *227*, 7–10.

23. Modigell, M.; Traebert, A.; Monheim, P. A modelling technique for metallurgical processes and its applications. *AISE Steel Technol.* **2001**, *78*, 45–47.

24. Traebert, A.; Modigell, M.; Monheim, P.; Hack, K. Development of a modelling technique for non-equilibrium metallurgical processes. *Scand. J. Metall.* **1999**, *28*, 285–290.

25. Morales, R.D.; Rodríguez-Hernández, H.; Conejo, A.N. A mathematical simulator for the EAF steelmaking process using direct reduced iron. *ISIJ Int.* **2001**, *41*, 426–436. [CrossRef]

26. Nyssen, P.; Colin, R.; Knoops, S.; Junque, J. On-line EAF control with a dynamic metallurgical model. In Proceedings of the 7th European Electric Steelmaking Conference (EEC), Venice, Italy, 26–29 May 2002.

27. Nyssen, P.; Marique, C.; Prüm, C.; Bintner, P.; Savini, L. A New Metallurgical Model for the Control of EAF Operations. In Proceedings of the 6th European Electric Steelmaking Conference, Düsseldorf, Germany, 13–15 June 1999.

28. Nyssen, P.; Monfort, G.; Junque, J.L.; Brimmeyer, M.; Hubsch, P.; Baumert, J.C. Use of a dynamic metallurgical model for the on-line control and optimization of the electric arc furnace. In Proceedings of the STEELSIM, 2nd International Conference of Simulation and Modelling of Metallurgical Processes in Steelmaking, Graz/Seggau, Austria, 12–14 September 2007.

29. Nyssen, P.; Ojeda, C.; Baumert, J.C.; Picco, M.; Thibaut, J.C.; Sun, S.; Waterfall, S.; Ranger, M.; Lowry, M. Implementation and on-line use of a dynamic process model at the ArcelorMittal-Dofasco electric arc furnace. In Proceedings of the STEELSIM, 4th International Conference of Simulation and Modelling of Metallurgical Processes in Steelmaking, METEC InSteelCon, Düsseldorf, Germany, 27 June–1 July 2011.

30. Arnout, S.; Verhaeghe, F.; Blanpain, B.; Wollants, P.; Hendrickx, R.; Heylen, G. A Thermodynamic Model of the EAF Process for Stainless Steel. *Steel Res. Int.* **2006**, *77*, 317–323. [CrossRef]

31. Kho, T.S.; Swinbourne, D.R.; Blanpain, B.; Arnout, S.; Langberg, D. Understanding stainless steelmaking through computational thermodynamics Part 1: Electric arc furnace melting. *Trans. Inst. Min. Metall. Sect. C* **2010**, *119*, 1–8. [CrossRef]

32. Meier, T.; Gandt, K.; Hay, T.; Echterhof, T. Process Modeling and Simulation of the Radiation in the Electric Arc Furnace. *Steel Res. Int.* **2018**, *89*, 1700487. [CrossRef]

33. Meier, T.; Hay, T.; Echterhof, T.; Pfeifer, H.; Rekersdrees, T.; Schlinge, L.; Elsabagh, S.; Schliephake, H. Process Modeling and Simulation of Biochar Usage in an Electric Arc Furnace as a Substitute for Fossil Coal. *Steel Res. Int.* **2017**, *88*, 1600458. [CrossRef]

34. Meier, T.; Gandt, K.; Echterhof, T.; Pfeifer, H. Modeling and Simulation of the Off-gas in an Electric Arc Furnace. *Metall. Mater. Trans. B* **2017**, *48*, 3329–3344. [CrossRef]

35. Meier, T. Energetische Analyse des Sauerstoffeinsatzes im Elektrolichtbogenofen Mithilfe Eines Selbstregelnden Dynamischen Prozessmodells. Master's Thesis, RWTH Aachen University, Aachen, Germany, 2017.

36. Meier, T.; Logar, V.; Echterhof, T.; Igor, Š.; Pfeifer, H. Modelling and Simulation of the Melting Process in Electric Arc Furnaces—Influence of Numerical Solution Methods. *Steel Res. Int.* **2016**, *87*, 581–588. [CrossRef]

37. Meier, T. Modellierung und Simulation des Elektrolichtbogenofens. Ph.D. Thesis, RWTH Aachen University, Aachen, Germany, 2016.

38. Fathi, A.; Saboohi, Y.; Škrjanc, I.; Logar, V. Comprehensive Electric Arc Furnace Model for Simulation Purposes and Model-Based Control. *Steel Res. Int.* **2017**, *83*, 1600083. [CrossRef]

39. Hay, T.; Reimann, A.; Echterhof, T. Improving the Modeling of Slag and Steel Bath Chemistry in an Electric Arc Furnace Process Model. *Metall. Mater. Trans. B* **2019**, *50*, 2377–2388. [CrossRef]

40. Saboohi, Y.; Fathi, A.; Škrjanc, I.; Logar, V. Optimization of the Electric Arc Furnace Process. *IEEE Trans. Ind. Electron.* **2019**, *66*, 8030–8039. [CrossRef]

41. Odenthal, H.-J.; Kemminger, A.; Krause, F.; Vogl, N. A Holistic CFD Approach for Standard and Shaft-Type Electric Arc Furnaces. In Proceedings of the AISTech 2017, Nashville, TN, USA, 8–11 May 2017.

Article

3D Integrated Modeling of Supersonic Coherent Jet Penetration and Decarburization in EAF Refining Process

Yuchao Chen, Armin K. Silaen and Chenn Q. Zhou *

Center for Innovation through Visualization and Simulation (CIVS), Purdue University Northwest, 2200 169th Street, Hammond, IN 46323, USA; chen2058@pnw.edu (Y.C.); asilaen@pnw.edu (A.K.S.)
* Correspondence: czhou@pnw.edu; Tel.: +1-219-989-2665

Received: 25 May 2020; Accepted: 13 June 2020; Published: 17 June 2020

Abstract: The present study proposes a complete 3D integrated model to simulate the top-blown supersonic coherent jet decarburization in the electric arc furnace (EAF) refining process. The 3D integrated model avoids the direct simulation of the supersonic coherent jet interacting with the liquid steel bath and provides a feasible way to simulate the decarburization in the liquid steel-oxygen two-phase reacting flow system with acceptable computational time. The model can be used to dynamically predict the details of the molten bath, including 3D distribution of in-bath substances, flow characteristics and bath temperature and provide a basis for optimizing the decarburization rate or other required parameters during the refining process.

Keywords: supersonic coherent jet; decarburization; steel refining; EAF; CFD

---

1. Introduction

Electric arc furnace (EAF) steelmaking featuring high efficiency and energy saving has become one of the major steelmaking methods in the world. The steel refining stage is seen as a final period to improve the thermal homogenization of the bath and adjust the metallurgical parameters of the steel grade. During this period, oxygen injection is desired to help to stir the molten bath, remove impurities and further improve the quality of the liquid steel. In the 1990s, Praxair, Inc. introduced the supersonic coherent jet technology (Cojet®) for oxygen injection in EAF [1], which significantly improves the jet performance and makes the jet to have deeper penetration depth, less splashing and, most importantly, more oxygen delivered to the molten bath. Therefore, this technology has been widely-used in EAF operation. Generally, the oxygen carried by jet will dissolve and generate numerous in-bath bubbles and the turbulence created by those bubbles will result in the intensive stirring effect. Meanwhile, the bubbles containing the oxygen will effectively react with the carbon in the bath to form the CO bubbles, which achieves the purpose of the decarburization and the further bath stirring. Except for the carbon, other impurities including phosphorus, sulfur, aluminum, silicon and manganese may also be dissolved in the molten bath. Therefore, oxygen is also responsible for removing those impurities during the refining stage. The carbon and impurity removal process involves different exothermic oxidation reactions leading to the consistent rising of the in-bath temperature. Once the carbon content is reduced to a critical value and the bath temperature reaches the desired temperature, the liquid steel is ready for tapping.

Among the research of the EAF refining process, most researchers choose to conduct the study separately for either the gas phase or the liquid phase, namely the supersonic coherent jet part or the molten bath part. For the supersonic coherent jet part, Anderson et al. [2] designed an experiment aimed to generate the supersonic coherent jet and measured the corresponding characteristics of the jet. Alam et al. [3] developed a numerical model with a revised $k - \varepsilon$ turbulence model to simulate the supersonic coherent jet and validated the model by comparing the measurement from Anderson's experiment. Liu et al. [4] experimentally and numerically studied the effect of different ambient temperatures and different oxygen preheating temperatures on jet performance. Li et al. [5] investigated the jet performance using different shrouding flame created by different types of fuel. As for the molten bath part, previous works can be further subdivided into the study of molten bath dynamics and the study of in-bath chemical reactions. Cafeery et al. [6] and Li [7] simulated the stirring process for both top-blown and bottom-blown EAF to investigate the mixing efficiency and the bath homogeneity. Ramirez et al. [8] compared different operation conditions of the furnace in the refining process in order to eliminate the flow dead zones in the bath and improve the heat dissipation of the arc. On the other hand, Szekely et al. [9] developed a mathematical model to predict the stainless steel decarburization process. Zhu [10] further analyzed the bath reaction mechanism to study the effect of different carbon contents on deoxidizer consumption.

The research mentioned above separately provide useful information for modeling the EAF refining from two perspectives but still lack a complete mathematical description of this process. Theoretically, the numerical simulation based on the finite element or finite difference method could be a feasible way to achieve the goals but considering that the simulation of such a complex process (high-speed jet interacting with the liquid phase, post-combustion, in-bath multiphase chemical reactions, etc.) may lead to either unaffordable computational time or numerical convergence issues, no related research using this method have been reported so far. In fact, some researchers, like Memoli et al. [11], attempted to adopt a jet cavity to connect both the top-blown jet part and the molten bath part for modeling the EAF refining process, which makes it possible to establish a mathematical model that contains the complete phenomena in the process. However, their models only considered the refining using the top-blown supersonic conventional jet (without the shrouding flame), which obviously makes a difference in the case using the top-blown supersonic coherent jet. In addition, the calculated 3D jet cavity was only used to provide the effective contact area with the molten bath for decarburization simulation. Therefore, the proposed decarburization simulation is still based on a zero-dimensional mathematical model in essence, which cannot provide a detailed 3D information of the molten bath.

Based on the previous statement, the study of the EAF refining process presented in this paper focuses on using the computational fluid dynamics (CFD) technique to simulate the top-blown supersonic coherent jet decarburization process. A complete 3D integrated model with good accuracy has been proposed, which can be used to predict the complete phenomena in the refining process and provide detailed 3D distribution of in-bath substances as well as the flow characteristics for further investigation. More information of the methodology for this 3D integrated model will be illustrated below.

2. Methodology

The EAF refining process is a complex, high-temperature physicochemical process. Theoretically it may be feasible to directly conduct the CFD simulation of the entire EAF refining process, however, obvious limitations can be found including the numerical instability of simulation and the extremely high computational costs. Therefore, the attempt of using a fully-coupled CFD model to concurrently capture the multi-physical phenomena, such as the combustion flame, the supersonic compressible flow, the jet penetration and the multiphase reacting flow, is difficult as expected.

To avoid the aforementioned difficulties, the complex phenomena during the refining stage can be classified into three major categories based on their major physical principles, including (1) the supersonic coherent jet above the liquid steel bath; (2) the interaction between the coherent jet and the liquid steel; (3) the stirring and the chemical reactions inside the liquid steel bath. The present study modeled each part separately and then made the overall integration to predict the entire refining process. This methodology can greatly ensure the simulation accuracy of each part with acceptable computational time, meanwhile facilitate the targeted control and analysis of various key parameters at the same time. The methodology of developing the proposed 3D integrated model is given in Figure 1 and the details are described below:

- The supersonic coherent jet with the revised $k - \varepsilon$ turbulence model will be firstly simulated in an open space under the actual high ambient temperature conditions inside the furnace to obtain the jet characteristics, which will be used for the subsequent estimation and simulation.
- A theoretical interface will then be calculated to represent the jet penetration cavity inside the liquid steel bath based on the supersonic coherent jet characteristics at the bath surface. This method is based on the energy balance between the injected jet and penetrated bath and enables us to avoid the direct simulation of the supersonic coherent jet interacting with the liquid steel bath.
- The geometry of the bottom section of the EAF with the above-estimated jet cavity can be established and reasonable boundary conditions need to be defined on the cavity surface according to the results from part 1, so that the thermodynamic and kinetic coupled multiphase reacting flow simulation can be performed to predict in-bath stirring and decarburization process for the refining stage.

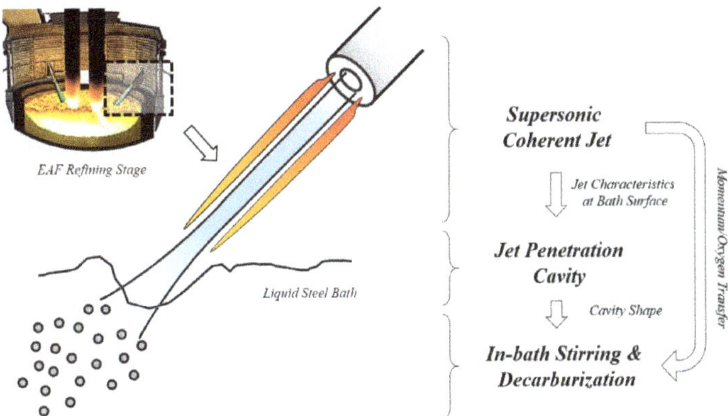

Figure 1. Methodology of developing the 3D integrated model for electric arc furnace (EAF) steel refining simulation.

The CFD simulation of part 1 and part 3 mentioned above is based on solving the appropriate Navier-Stokes equations with the required source terms of the specific phenomenon and incorporating them into the finite volume method (FVM) of Patankar [12]. A commercial software ANSYS® FLUENT 19.1 (ANSYS Inc., Washington County, PA, USA, 2019) [13] was adopted for the simulation.

This 3D integrated model can eliminate the compatibility issue of different CFD models during multi-physics simulations, especially for the consideration of the interaction between multi phases. Besides, the adoption of the current model makes the three-dimensional modeling of the decarburization process possible, which provides the way for a more comprehensive analysis of the chemical reaction rate and the species distribution inside the liquid steel bath. The complete model will be detailed in the next section.

3. Numerical Model

3.1. Supersonic Coherent Jet Modeling

The supersonic coherent jet is modeled with the assumption that the jet flow is conducted in a steady, compressible, non-isothermal process. The corresponding Navier-Stokes equations were solved with the modification of the $k - \varepsilon$ turbulence model, which aims to improve the prediction accuracy of the jet potential core length and oxygen delivery rate and provide correct numerical conditions to estimate the subsequent jet penetration cavity shape and simulate the decarburization process. The governing equations solved are listed below.

The continuity conservation equation can be expressed by:

$$\nabla\cdot\left(\rho\vec{v}\right) = 0,\tag{1}$$

The momentum conservation equation is represented as:

$$\nabla\cdot\left(\rho\vec{v}\vec{v}\right) = -\nabla p + \nabla\cdot\left(\overline{\overline{\tau}}\right) + \rho\vec{g} + \vec{F},\tag{2}$$

where ρ is the fluid density; $\vec{v}$ is the velocity vector; p is the static pressure; $\overline{\overline{\tau}}$ is the stress tensor; $\vec{g}$ is the acceleration of gravity and $\vec{F}$ is the external body force.

The energy conservation equation can be written as:

$$\nabla\cdot\left[\vec{v}(\rho E + p)\right] = \nabla\cdot\left[\left(k + \frac{c_p\mu_t}{Pr_t}\right)\nabla T - \sum_j h_j\vec{J}_j + \left(\overline{\overline{\tau}}_{eff}\cdot\vec{v}\right)\right] + S_h,\tag{3}$$

where E is the total energy related to the sensible enthalpy h; k is the thermal conductivity; c_p is the specific heat; μ_t is the turbulent viscosity; Pr_t is the turbulent Prandtl number whose default value is 0.85 for the $k - \varepsilon$ turbulence model. For the free shear flow with high heat transfer simulation, the appropriate turbulent Prandtl number should be set as 0.5 according to the suggestions by Wilcox [14] and Alam [3]. However, the shrouding combustion flame around the primary supersonic oxygen jet prevents the entrainment of ambient gas into the center jet, which further impacts the generation of the free shear layer. Thus, 0.85 was still adopted for the turbulent Prandtl number to estimate the turbulent thermal conductivity in the current model. $\vec{J}_j$ is the diffusion flux of substance j and S_h is the volumetric heat sources including the heat of chemical reaction during the simulation.

Considering the supersonic state of the primary oxygen jet, the flow turbulence can be resolved through a time-averaged velocity scalar. In the present study, the averaged Reynolds stresses term was determined using the modified $k - \varepsilon$ turbulence model originally proposed by Launder and Spalding [15]. The governing equations of the turbulent kinetic energy k and turbulence dissipation rate ε can be expressed by:

$$\frac{\partial}{\partial x_i}(\rho u_i k) = -\rho\overline{u_i u_j}\frac{\partial u_j}{\partial x_i} + \frac{\partial}{\partial x_i}\left(\mu + \frac{\mu_t}{\sigma_k}\frac{\partial k}{\partial x_i}\right) - \rho\varepsilon - \rho\varepsilon M_\tau^2,\tag{4}$$

$$\frac{\partial}{\partial x_i}(\rho u_i \varepsilon) = -C_{\varepsilon 1}\rho\overline{u_i u_j}\frac{\varepsilon}{k}\frac{\partial u_j}{\partial x_i} + \frac{\partial}{\partial x_j}\left(\mu + \frac{\mu_t}{\sigma_\varepsilon}\frac{\partial k}{\partial x_j}\right) - C_{\varepsilon 2}\rho\frac{k^2}{\varepsilon},\tag{5}$$

where μ is the molecular viscosity; σ_k and σ_ε are the turbulent Prandtl number for k and ε, whose values are 1.0 and 1.3, respectively; M_τ is the turbulent Mach number that can be defined as:

$$M_\tau = \frac{\sqrt{2k}}{a},\tag{6}$$

where a is the acoustic velocity; $C_{\varepsilon 1}$ and $C_{\varepsilon 2}$ are constants whose values are 1.44 and 1.92, respectively.

The turbulent viscosity μ_t used in Equations (4) and (5) is defined by:

$$\mu_t = C_\mu\rho\frac{k^2}{\varepsilon},\tag{7}$$

where C_μ is a constant value originally equal to 0.09 for the standard $k - \varepsilon$ turbulence model. In order to consider the influence of the entrained ambient gas that is reduced by the shrouding combustion flame, C_μ was modified according to the formula proposed by Alam et al. [3]. The original value of C_μ was divided by a variable C_T to include the effects of the local total temperature gradient in estimating the turbulent viscosity, thereby further reducing the mixed growth rate of the shear layer to accurately simulate the jet potential core length [16]. The modified C_T can be expressed as:

$$C_\mu = \frac{0.09}{C_T} \tag{8}$$

and

$$C_T = 1 + \frac{C_1 T_g{}^m}{1 + C_2 f(M_\tau)}, \tag{9}$$

where C_1, C_2 and m is constantly equal to 1.2, 1.0 and 0.6, respectively; T_g is the normalized local total temperature gradient, which can be calculated by:

$$T_g = \frac{k^{\frac{3}{2}} |\nabla T_t|}{\varepsilon\, T_t}, \tag{10}$$

where T_t is the local total temperature of the flow field; $f(M_\tau)$ is a function that further considers the influence of turbulent Mach number, which can be estimated by:

$$f(M_\tau) = \left(M_\tau{}^2 - M_{\tau 0}{}^2\right) H(M_\tau - M_{\tau 0}), \tag{11}$$

where $H(x)$ is the Heaviside function; $M_{\tau 0}$ is a constant equal to 0.1 [17]. All aforementioned modifications of the standard $k - \varepsilon$ turbulence model are incorporated into the CFD-solver Fluent through the user-defined function (UDF) code based on C language and compiled in the CFD solver for the simulation.

In order to capture the shrouding combustion flame, the species transport model with the eddy dissipation concept (EDC) [18] was employed to simulate the 28-step natural gas-oxygen combustion reactions. The Discrete Ordinates (DO) radiation model with Weighted-Sum-of-Gray-Gases Model (WSGGM) [19] was adopted to model the radiation heat transfer phenomenon for the combustion.

The numerical simulation domain of the supersonic coherent jet is shown in Figure 2, which contains 3 million computational cells totally. Total computational time is around 15 h if using 80 cores in the High Performance Computing (HPC) cluster to obtain the converged results. The simulation domain is a cylindrical-shaped vessel originating from the exit of the converging-diverging nozzle where the nozzle structure is ignored. The dimension of the vessel is much larger than the burner, which can be used to simulate the supersonic coherent jet behavior in the open space. Therefore, except for the wall where the nozzle exit is located, the other walls of the vessel are set as outlets. More detail on burner operating conditions and other information are mentioned in another published paper [20].

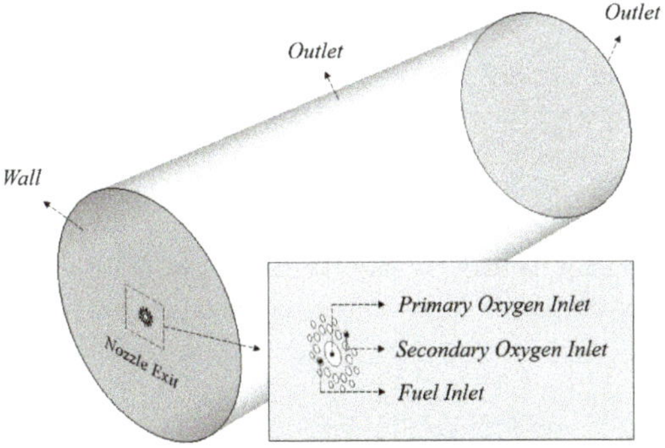

Figure 2. 3D computational domain of the supersonic coherent jet modeling.

3.2. Jet Penetration Cavity Estimation

The present study utilized a novel method to consider the jet penetration in the liquid steel bath so that the direct simulations of interaction between multi phases can be avoided. The basic idea is to calculate a theoretical interface to represent the jet penetration cavity inside the liquid steel bath and this cavity will be estimated based on the characteristics of the coherent jet reaching the bath surface and used as the physical boundary of the computational domain for subsequent decarburization simulations. The shape of the cavity interface is assumed to be a revolution paraboloid according to Memoli et al. [11], which is more precise for the coherent jet with high momentum, as its penetration depth is greater than the radius of its cross-section. The three-dimensional mathematical expression of a revolution paraboloid in Cartesian coordinate can be written as:

$$z = \frac{x^2 + y^2}{c}, \tag{12}$$

where c is the constant need to be defined by a given volume of the jet penetration cavity and the penetration depth.

The volume of the jet penetration cavity V can be determined by calculating liquid steel replaced by the gas flow based on the impulsive balance at the cavity interface if ignoring the impact of the liquid steel surface tension [21]. The expression of the jet cavity volume can be written as:

$$V = \frac{\pi \rho_j v_j^2 d_j^2}{4 g \rho_s}, \tag{13}$$

where ρ_j and ρ_s are the density of primary oxygen jet and liquid steel, respectively; v_j and d_j are the primary oxygen jet velocity and diameter when at bath surface, which can be determined through the supersonic coherent jet modeling of a given distance from the nozzle exit to the bath.

Jet penetration depth D refers to an empirical formula derived by Ishikawa et al. [22], which describes the penetration depth created by the turbulent jet. For the supersonic coherent jet, the constants in the formula need to be modified accordingly. The empirical formula shows the relationship between the jet penetration depth of a single-hole or multi-holes nozzle and the burner operating conditions, which can be expressed as:

$$D = \gamma_{h_0} e^{-\frac{\sigma_1 L}{\gamma_{h_0} \cos \theta}} \tag{14}$$

$$\gamma_{h_0} = \sigma_2 \left(\frac{\dot{V}}{\sqrt{3} n d} \right), \tag{15}$$

where L is the axial distance between the nozzle exit to the bath surface; θ is the angle of the jet inclination; $\dot{V}$ is the volume flow rate of primary oxygen jet; n is the number of the nozzle and equal to 1 for the current study; d is the nozzle exit diameter for primary oxygen jet; σ_1 and σ_2 are two constants originally equal to 1.77 and 1.67, respectively and those two parameters are determined through experiments for a specific type of coherent jet used in the present study.

The actual refining process has the slag layer covering the liquid steel bath to protect the arc and reduce heat radiation loss. The coherent jet needs to pass through the slag layer before reaching the liquid steel bath. During this period, the jet will lose some of its momentum. Therefore, the jet penetration depth should be shorter than the one without the slag layer. In the current model, the slag layer is assumed to be converted equivalently to a corresponding liquid steel layer to include its effect on the jet penetration depth. The equivalent slag layer height h_s can be estimated by:

$$h_s = \frac{\rho_{sl}}{\rho_s} h_{sl}, \tag{16}$$

where ρ_{sl} and h_{sl} are the values for slag layer density and slag layer height, respectively. The actual jet penetration depth D_{act} reads as:

$$D_{act} = D - h_s. \tag{17}$$

Once the constant c is determined by solving Equations (13) to (17), the theoretical parabolic jet cavity interface can be defined and included as the physical boundary for the computational domain of the bottom section of the EAF for the decarburization simulation. This eliminates the need to include the consideration of supersonic jets and its interaction with the liquid surface in the decarburization simulation. The estimation of the three-dimensional jet penetration cavity based on actual burner operating conditions is illustrated in Figure 3 and the computational domain with five jet penetration cavities established according to the actual burner arrangement provided by industry is given in Figure 4. This computational domain is going to be used in subsequent decarburization simulations.

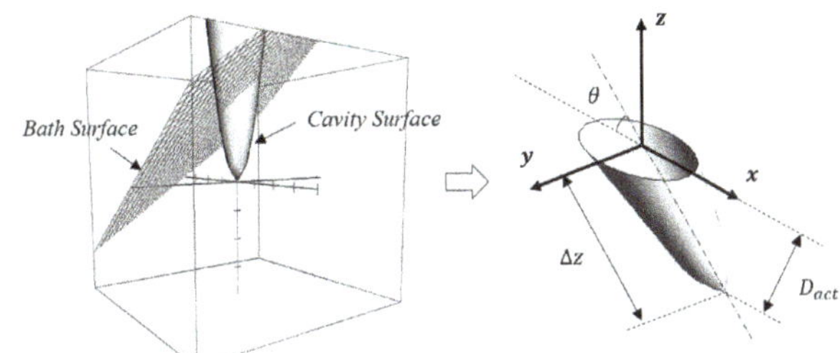

Figure 3. Sketch of 3D jet penetration cavity estimation.

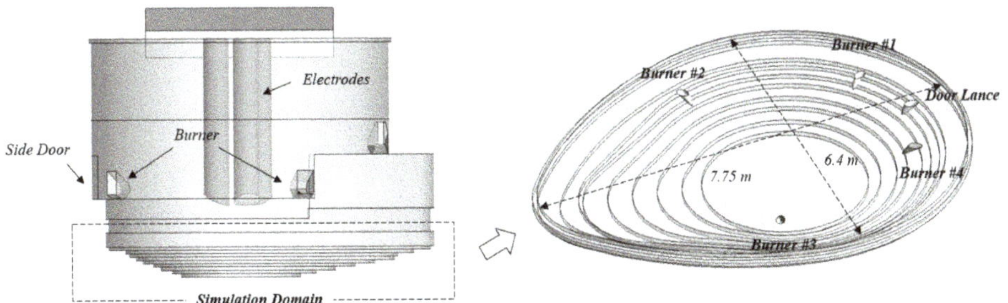

Figure 4. 3D computational domain with jet penetration cavities for decarburization modeling.

Notice that when the supersonic coherent jet impinges on the liquid steel bath forming the jet penetration cavity, the exchange of energy and substance occurs intensively between the gas phase and liquid phase. Therefore, the jet penetration cavity surface, as the physical boundary of the computational domain, needs to establish appropriate boundary conditions to consider the energy and substance transfer during the jet impingement. In the present study, both jet momentum transfer and delivery of the oxygen were considered. Based on the energy balance on the cavity surface, the jet momentum transferred to the liquid steel bath $P_{s,avg}$ can be expressed as:

$$P_{s,avg} = \alpha \rho_{O_2} v_{O_2}{}^2 A = \frac{\alpha \rho_{O_2}{}^2 A}{\rho_s}\left[\frac{1}{\Delta z}\int_{z_2}^{z_1} v_{O_2}(z)dz\right]^2, \tag{18}$$

where α is the transferable percentage of the jet total momentum at liquid steel bath, which is 0.06 according to the reference [23]; v_{O_2} is average jet velocity along cavity centerline; A is the cavity surface area; Δz is the length of the cavity centerline, which is equal to $z_1 - z_2$.

The amount of oxygen delivered to the liquid steel $m_{O_2,\,avg}$ through the jet cavity can be estimated by calculating the average oxygen distribution along the cavity centerline:

$$m_{O_2,\,avg} = \frac{1}{\Delta z} \int_{z_2}^{z_1} m_{O_2}(z)dz. \tag{19}$$

3.3. Decarburization Modeling

The current decarburization modeling focuses on the refining process after the solid scrap is completely melted down into a flat bath. Therefore, the scrap melting phenomenon is not included in the present study for the sake of simplicity. The model proposed here considers a liquid steel-oxygen two-phase reacting flow system inside the flat bath and the simulation domain, as mentioned above, only includes the bottom section of EAF with estimated jet penetration cavities, through which the oxygen enters the domain to react with carbon and other impurities. The injected oxygen also results in two main effects on the system including the stirring of the liquid steel bath and the bath temperature rise due to the heat released by the oxidation reactions. In the present study, the oxidations of the carbon, iron and manganese as a mixture of liquid steel by the injected oxygen are listed in Table 1:

Table 1. Oxidation reactions (A) to (C) considered in present study.

Reaction (A)	$C + \frac{1}{2}O_2(g) = CO(g)$
Reaction (B)	$Fe + \frac{1}{2}O_2(g) = FeO$
Reaction (C)	$Mn + \frac{1}{2}O_2(g) = MnO$

Oxidation reactions take place in cells of the simulation domain that contain the oxygen. The oxidation rates of carbon, iron and manganese at high carbon content are mainly limited by the amount of oxygen contained in the same cell. If the oxygen is sufficient, the rate equations can be written as suggested by Wei and Zhu [24]:

$$-\frac{W_s}{100\,M_C}\frac{d[\%C]}{dt} = \frac{2\eta_C Q_{O_2}}{22,400}x_C \tag{20}$$

$$-\frac{W_s}{100\,M_{Mn}}\frac{d[\%Mn]}{dt} = \frac{2\eta_{Mn} Q_{O_2}}{22,400}x_{Mn}, \tag{21}$$

where W_s and Q_{O_2} is the mass of liquid steel and the volume of oxygen in the corresponding cell, respectively; M_i is the mole mass of each substance; η_i is the efficiency factor of each substance, which is a function of total mixing of the system and can be estimated based on the work done by Shukla et al. [25]; x_i is the oxygen distribution ratios of each substance and is assumed to be proportional to the Gibbs free energies of corresponding oxidation reactions:

$$x_C = \frac{\Delta G_C}{\Delta G_C + \Delta G_{Fe} + \Delta G_{Mn}} \tag{22}$$

$$x_{Mn} = \frac{\Delta G_{Mn}}{\Delta G_C + \Delta G_{Fe} + \Delta G_{Mn}}, \tag{23}$$

where the Gibbs free energies ΔG_i of respective substance can be defined as:

$$\Delta G_C = \Delta G_C^0 + RT\ln\left[\frac{P_{CO}}{a_C \cdot a_{O_2}^{0.5}}\right] \tag{24}$$

$$\Delta G_{Fe} = \Delta G_{Fe}^0 + RT\ln\left[\frac{a_{FeO}}{a_{Fe} \cdot a_{O_2}^{0.5}}\right] \tag{25}$$

$$\Delta G_{Mn} = \Delta G_{Mn}^0 + RTln\left[\frac{a_{MnO}}{a_{Mn}\cdot a_{O_2}^{0.5}}\right],\tag{26}$$

where ΔG_i^0 and a_i is the standard Gibbs free energy and the activity of each substance in the bath respectively; R is gas constant; P_{CO} is the partial pressure of carbon monoxide.

At low carbon content, the oxidation rate of carbon is no longer controlled by the oxygen contained in the cell. Instead, the mass carbon transfer rate to liquid steel will directly impact the decarburization rate, which can be expressed as:

$$-W_s\frac{d[\%C]}{dt} = -\rho_s k_C A_{inter}\left([\%C] - [\%C]_e\right),\tag{27}$$

where A_{inter} is the bubble inter-surface area; $[\%C]_e$ is carbon equilibrium concentration in the molten bath; k_C is the carbon mass transfer coefficient through the oxygen bubble surface which can be calculated by [26]:

$$k_c = 0.59\cdot[D_C\cdot(u_{rel}/d_B)]^{0.5},\tag{28}$$

where D_C is the diffusion coefficient of carbon; u_{rel} is relative velocity of liquid steel; d_B is the bubble diameter.

The oxides formed through Reaction (A) to Reaction (C) may gradually float upwards to the top surface, which is the lower surface of the slag layer that is not included in the simulation domain. Practically, the oxides will accumulate at the slag layer and have further reactions there. The absorption of the oxides by slag layer can be achieved computationally by removing the corresponding oxides that are in contact with the domain top surface. The process described above is illustrated in Figure 5.

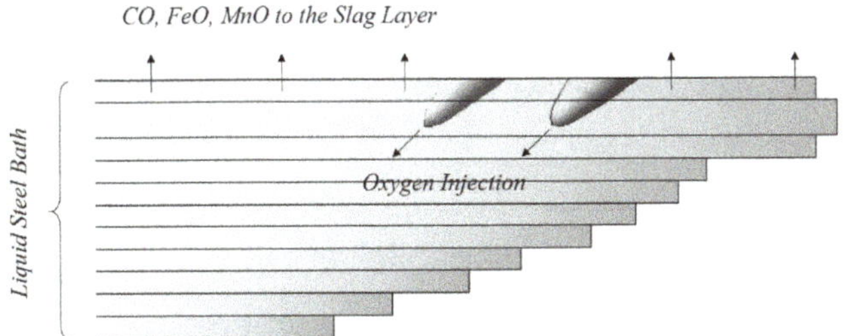

Figure 5. Oxygen injection through jet penetration cavities and absorption of oxides.

During the refining stage, the temperature of the liquid steel bath increases due to the energy released by the oxidation reactions. The amount of energy released to the bath can be estimated by the oxidation rates and the oxidation enthalpies ΔH_i of each reaction, where ΔH_i is a function of bath temperature and taken from reference [27]. Thus the rate of energy-generating in a cell due to the oxidations can be expressed as:

$$\frac{dE_{reac}}{dt} = \sum \Delta H_i W_s\frac{d[\%i]}{dt},\tag{29}$$

where i represents the carbon, iron and manganese considered in the liquid steel bath.

The current liquid steel-oxygen two-phase reacting flow system was solved in the numerical simulation domain given in Figure 4, which has 2.5 million computational cells totally. By adopting the Eulerian model with the appropriate source terms compiled through user-defined function (UDF) code, the model is able to achieve the above-described simulation of in-bath oxidation reactions and heat release. The total computational time needed to simulate 1000 s decarburization process is around 50 h if using 0.05 s time step size and 80 cores in the HPC cluster.

4. Simulation Results and Discussions

4.1. Model Validaiton

The validation of the coherent jet modeling in ambient air with and without shrouding flame was conducted based on the research works done by Anderson et al. [2]. The simulation was set up according to the reported experiment. The comparison of the jet axial velocity distribution between the present study and the measurement data is given in Figure 6a. The jet with the shrouding flame can maintain its initial velocity for a longer distance compared with the jet without the shrouding flame. The coherent jet with Mach 2.1 at nozzle exit has the potential core length around 48 De (where De is the diameter of the converging-diverging nozzle exit), which is 2.5 times longer compared with the conventional jet in this validation. On the contours given in Figure 6b, the entrainment of the ambient air is blocked by the shrouding combustion flame, which significantly reduces the turbulence effect around the primary oxygen jet leading to the great increase of the potential core length. The average difference in this validation is 5.9% compared with experimental data, which shows good accuracy of current supersonic coherent jet modeling.

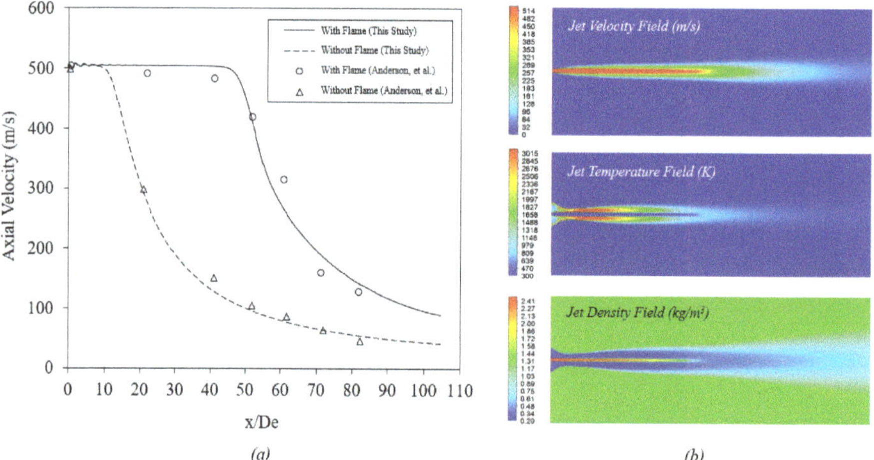

Figure 6. (**a**) Jet axial velocity distribution with & without shrouding flame; (**b**) Velocity, temperature, density field distribution of the supersonic coherent jet.

The validation of the jet penetration cavity estimation was done by comparing the predicted jet penetration depth with the measurement data provided by Praxair under the same burner operating conditions. It should be noted that the burner configuration of the coherent jet used for current validation differs from the above-mentioned coherent jet validation. The key parameters of bath for jet penetration cavity estimation are given in Table 2. The jet penetration depth is measured under the fixed burner operating conditions. The distance between the jet nozzle exit and the liquid steel bath surface is successively increased to reach different penetration depths, thereby obtaining six sets of available data for the validation as shown in Figure 7a. The parity plot of the jet penetration depth inside the liquid steel bath is given in Figure 7b showing the comparison of the estimation in the present study and the experimental data by Anderson, et al. The jet penetration reduces exponentially as the nozzle moves away from the liquid surface. This is because when the nozzle-liquid surface distance is greater than the jet potential core length, no doubt increasing the distance (cases #4 to #6 in Figure 7a) will decrease the jet penetration depth since smaller momentum the jet will have when reaching the liquid bath surface. On the other hand, if the nozzle-liquid surface distance is less than the jet potential core length, the jet maintains the same axial velocity but the area of the jet mixing zone in the radial direction increases with the increase of the nozzle-liquid surface distance (cases #1 to #3 in Figure 7a). This results in a reduction of the jet maximum pressure gradient in the mixing zone, thereby reducing the jet penetration ability inside the liquid steel bath. The tendency of the estimation in the present study meets the above description and the error compared with measurement data is less than 5% showing a good accuracy of the estimation.

Table 2. Key parameters of bath for jet penetration cavity estimation.

Key Parameters	Values
Liquid steel density	7700 kg/m^3
Slag layer density	4350 kg/m^3
Slag layer height	0.381 m
Angle of jet inclination	40° from horizontal

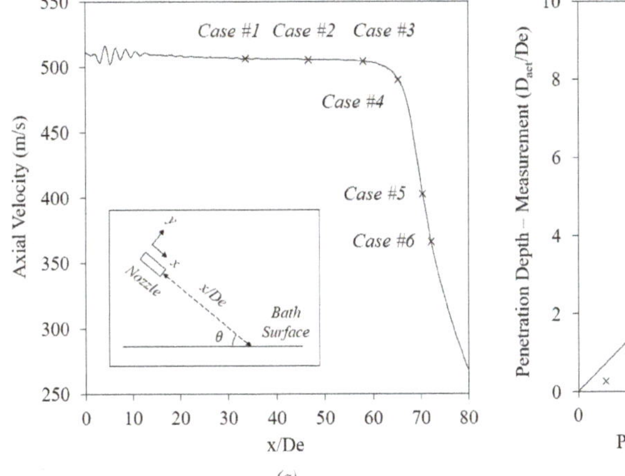

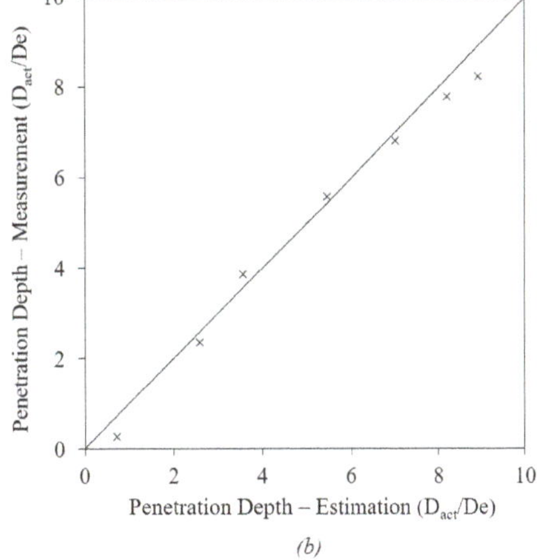

(a)

(b)

Figure 7. (**a**) Different nozzle-liquid surface distance for specific burner configuration; (**b**) Parity plot of penetration depth inside liquid steel bath.

The current 3D integrated model was adjusted accordingly to predict the decarburization process of the industrial electric arc furnace for model validation. Under the same furnace operating conditions, the percentage error of the decarburization rate prediction given by the model is less than 7% compared with industrial data.

4.2. Stirring Mechanism

Generally, there are different stirring mechanisms that affect the flow characteristics of the molten bath, including the momentum stirring, the bubble stirring and the electromagnetic stirring. In addition, the buoyancy force arising from the bath temperature gradient and the bath concentration gradient will also make contributions to stir the bath. In the present study, the electromagnetic stirring is assumed to be neglected in this alternating current (AC) EAF since the electromagnetic field induced is limited to the small region around the arcs and has a minor effect on the molten bath flow [28–30]. The stirring caused by the buoyancy-driven flow is always included in the model by applying the Boussinesq hypothesis for the liquid steel phase since it is a natural phenomenon built in any thermal system. As for the rest two stirring mechanisms, that is, the momentum stirring and the bubble stirring, they are the main stirring power provided by the supersonic coherent jet to the refining process, which are valuable and feasible to be investigated using the 3D integrated model proposed above.

In the actual steel refining process, the distance between the burner nozzle exit and the liquid steel bath surface is usually controlled to be within the jet potential core length to guarantee the coherent jet can maintain a relatively high kinetic energy when reaching the liquid steel bath. In this case, the coherent jet is able to push the liquid steel aside to form a cavity and transfer the momentum to the bath simultaneously. Generally, only a portion of the jet momentum can be transferred to a liquid steel bath and further used to generate the stirring. Sano et al. [23] reported that only 6% of the total jet momentum is transferable to the liquid steel and this value has also been proved to be suitable for the current simulation under the given burner operating condition. The stirring generated by this direct momentum transfer is called the momentum stirring. Another critical stirring mechanism during the refining process is the bubble stirring. The decarburization reaction will generate the CO bubbles that float upward together with oxygen bubbles quickly, creating a strong stirring power inside the liquid steel bath due to the bubble-liquid drag force. The CO bubbles will eventually be absorbed by the slag layer and subsequent chemical reactions will occur there. It has been reported that bubble stirring plays a key role in the homogenization of the flow field, therefore the corresponding investigation conducted below is aimed to reveal the impact of those two stirring mechanisms on the development of the flow field.

The comparison of the flow filed with and without the bubble stirring is given in Figure 8. Figure 8a shows the case only considering the momentum stirring. It can be seen that the high momentum transferred from the surface of the jet penetration cavity pushes the liquid steel downward, resulting in a high-speed liquid steel flow in the axial direction of the cavity together with vortexes generated on both sides. The volume-averaged liquid steel velocity in the bath fluctuates most frequently around a value of 0.01425 m/s. Figure 8b shows the case considering both the momentum stirring and the bubble stirring. Because of the bubble stirring, the high-velocity region occurs surrounding the jet penetration cavity instead and the direction of the vortex changes significantly. This is mainly due to the intensive oxidation reactions in that region generating a large number of oxides and the stirring intensity of the momentum transfer became much weaker compared to that of the floating CO and oxygen bubbles. In addition, significant turbulence is generated through bubble stirring in the liquid steel bath as well, which can lead to a better mixing. The volume-averaged velocity of the liquid steel bath is approximately 0.1485 m/s, which is about 10 times as much as the case only considering the momentum stirring. From this comparison, it can be seen that bubble stirring greatly promotes the homogenization of the liquid steel bath and is one of the most important stirring mechanism in the EAF refining stage.

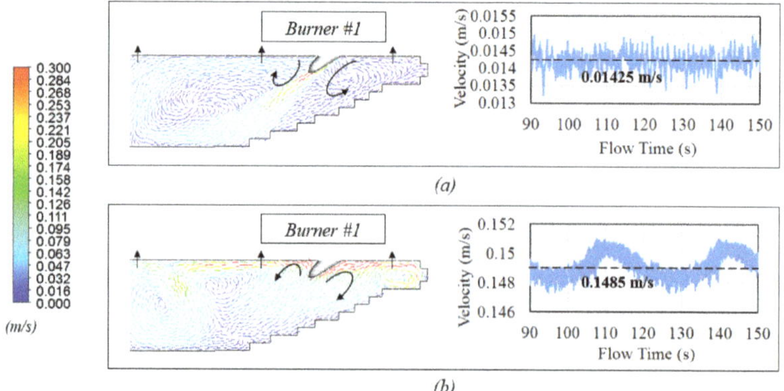

Figure 8. (a) Momentum stirring without bubble stirring; (b) Momentum stirring together with bubble stirring.

The bath mixing efficiency evaluation for both cases was conducted as well and the results are given in Figure 9. A tracer is introduced in the center of the liquid steel bath and the simulations stop once the tracer is fully diffused inside the bath. During the simulation, the area-averaged concentration of the tracer is monitored at three different horizontal planes in the bath to evaluate the mixing time. The vertical distance between those three horizontal planes and the bottom of the furnace is 0.07 m (plane monitor 1), 0.47 m (plane monitor 2) and 0.57 m (plane monitor 3), respectively. Noticed that, plane monitor 1 is close to the furnace bottom surface, which is aimed to avoid the dead zone and guarantee the full diffusion of the tracer in the domain. Figure 9a shows the contours of the in-bath tracer mass fraction variation over time. Figure 9b,c show the molar concentration variation of the tracer over time. When the variation in the molar concentration of the tracer is negligible, the mixing time required for the liquid steel bath to reach full diffusion can be obtained. It can be seen that without the bubble stirring, the mixing time is estimated to be 1665.3 s, which is almost 9.5 times longer than the time needed for the case with the bubble stirring. Therefore, the bubble stirring needs to be considered in the stirring mechanism for the future simulation of the liquid steel flow field.

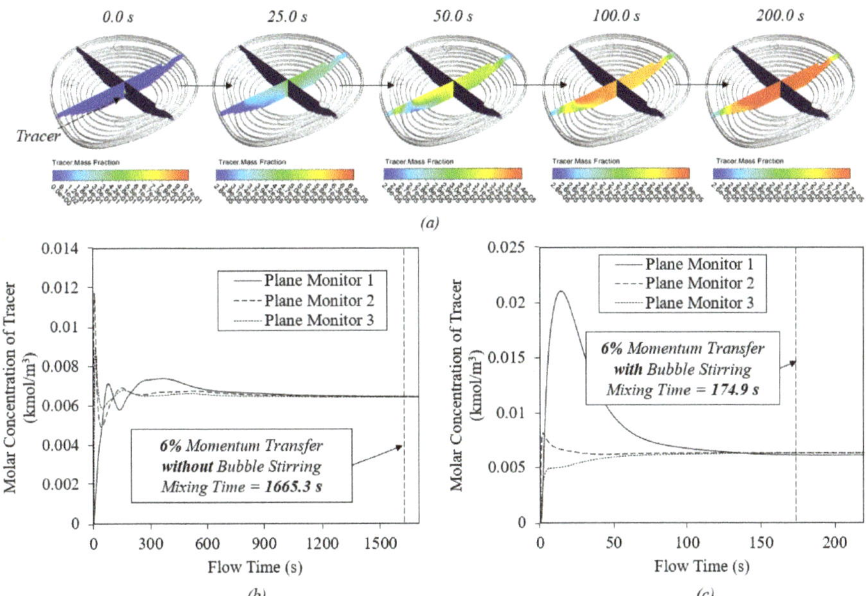

Figure 9. (**a**) In-bath tracer mass fraction over time (with momentum transfer and bubble stirring); (**b**) Mixing time of flow field without bubble stirring; (**c**) Mixing time of flow field with bubble stirring.

4.3. Decarburization Rate and Bath Temperature Rising Rate

The thermodynamic and kinetic coupled two-phase reacting flow simulation can be performed by the proposed 3D integrated model to reveal the details of species concentrations and temperature distribution inside the liquid steel bath and to predict the overall decarburization rate and bath temperature rising rate in the refining stage.

Generally, the average carbon content and the bath temperature are two important indicators for the operators to decide whether liquid steel meets the requirement of tapping. The volume-averaged carbon mass fraction and bath temperature predicted by this 3D integrated model are plotted in Figures 10 and 11. From the simulation results, after 800 s (around 13 min), the bath temperature will be increased to 1918 K with carbon content reduced to 0.056%. The percentage of carbon reduction and bath temperature increasing is 86.3% and 5.7% respectively. It is reported that the time required to reach the same carbon content and bath temperature during the actual refining process is about 12 to 15 min and current results give a correct prediction by comparing with the data published in the reference [31].

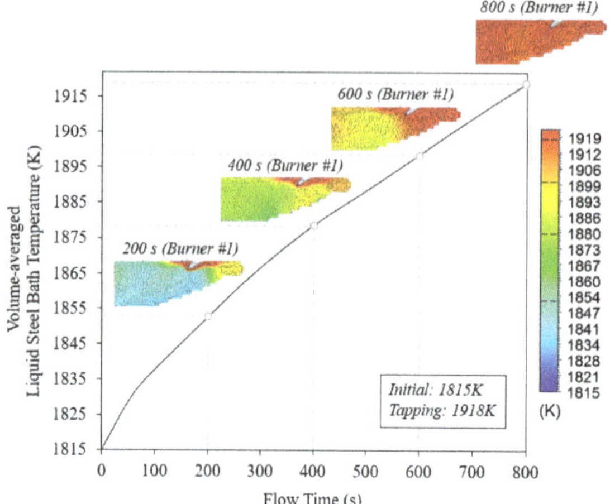

Figure 10. Volume-averaged liquid steel bath temperature variation over time.

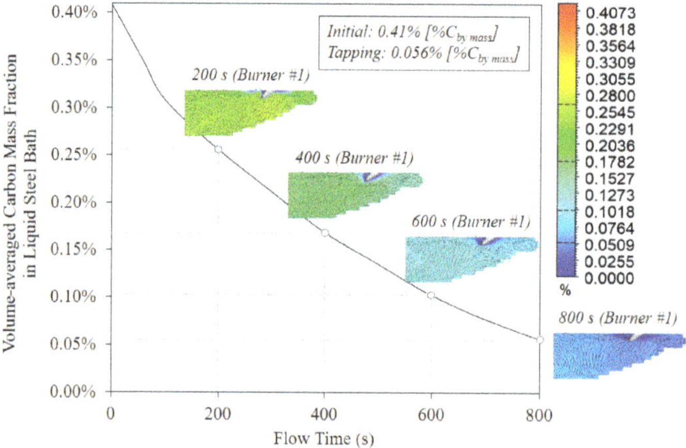

Figure 11. Volume-averaged carbon mass fraction variation in liquid steel bath over time.

The contours given in Figures 10 and 11 also show the detailed distribution for both in-bath carbon content and temperature field, which are plotted on the cross-section plane through burner #1 at 200 s, 400 s, 600 s and 800 s. The initial temperature and carbon content are assumed to be 1815 K and 0.41% by mass respectively. It can be seen that the carbon content maintains relatively low around the jet penetration cavity region due to a large amount of the oxygen bubbles injected from the cavity surface. The oxygen will first oxidize the reactive substances in this area and the excess oxygen will either travel with the flow to other areas away from the cavity or float upward by the buoyancy and be absorbed at the slag layer bottom surface (domain top surface). Generally, due to the concentration difference and stirring effect, the carbon will continue to move to the vicinity of the jet penetration cavity and react with the remaining or newly injected oxygen to generate the CO bubbles. The CO bubbles can float up with the flow causing the aforementioned in-bath bubble stirring and be absorbed once reaching the domain top surface. Moreover, since the oxygen reactions mainly occur around the cavity, a large amount of chemical energy will be released there to heat up the liquid steel.

Therefore, a red region representing high temperature can be seen around the jet penetration cavity in the contours. This thermal effect will spread to the entire liquid steel bath over time, resulting in significant bath temperature rise.

4.4. Carbon Distribution in Liquid Steel Bath

The current burner arrangement of the furnace is based on the arrangement commonly used in the industry, that is, the four burners are pointed 45-degree downward. The current simulation also takes the decarburization effect of door lance into account, which is used to make an immerged oxygen injection. Thus a total of five jet penetration cavities was established in the simulation domain.

Compared with other models, the current proposed 3D integrated model can analyze the detailed variation of the carbon distribution inside the liquid steel bath using the CFD technique. The simulation results are plotted in Figure 12. It can be seen that the current burner arrangement will result in an uneven carbon distribution in the liquid steel bath and the decarburization rate in front part of the furnace is much slower than elsewhere. When the refining progresses reaches around 200 s (about 3 min), the average carbon content in front part of the furnace is about twice higher than that of other places. It is not difficult to tell from the arrangement of the burners that most of the burners are located in the middle or rear of the furnace, which leads to the issues including the weak stirring and less oxygen injection in front of the furnace. The decarburization in this front area mainly depends on the overall flow pattern in the bath, that is, the liquid steel carrying high carbon content flows from the front of the furnace to the middle and rear of the furnace under specific rotating pattern so that the carbon content can be reacted with rich oxygen injected from the burner in that area. Obviously, this way of decarburization highly depends on the overall bath flow pattern and has a relatively low decarburization rate in the front of the furnace, which may result in a potential issue.

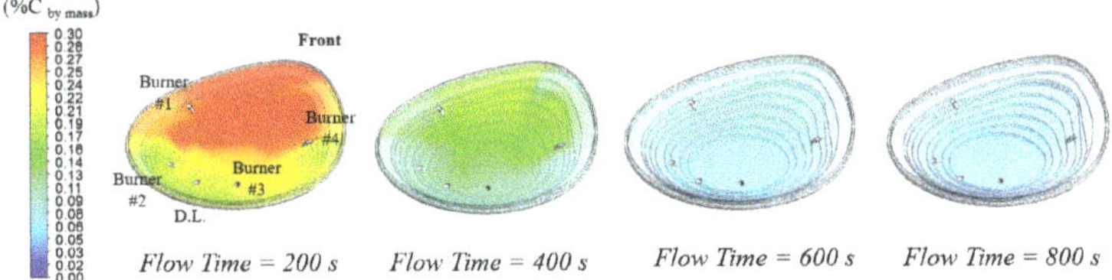

Figure 12. 3D carbon content distribution in the bath over time.

Figure 13 plots the detailed carbon mass fraction distribution on a plane (0.5 m from furnace bottom) close to the end of the refining process. The fraction of in-bath carbon content required for the liquid steel tapping is usually to be 0.03% to 0.05% by mass. It can be seen from the figure that the aforementioned uneven distribution of carbon content still exists at the end of the refining. In the actual operation of EAF refining stage, the operator typically inserts the test rod into the liquid steel bath through the side door to measure the carbon content, whose value is used to represent the average carbon content of the entire bath being tested. Once the temperature and carbon content meets the requirement, the liquid steel will be tapped. However, the measurement at this time only reflects the actual carbon content at the rear of the furnace. According to the previous analysis, the front of the furnace was not well stirred in the entire refining process. Thus, the actual carbon content was much higher there compared with the carbon content at the rear of the furnace. If the measured value is used to represent the average in-bath carbon content under this situation, it can potentially under predict the actual carbon content and further affect the quality of tapped liquid steel. By using this 3D integrated model, multiple cases can be simulated with acceptable computational time to optimize the burner arrangement to achieve higher stirring and decarburization rate.

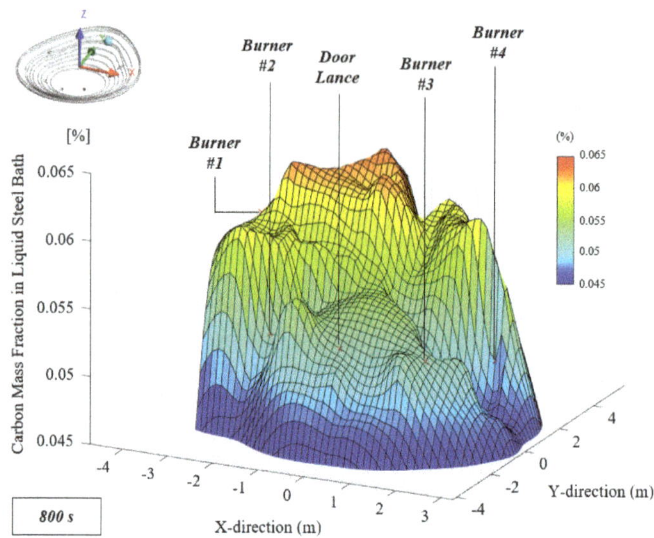

Figure 13. Carbon mass fraction on plane (0.5 m from furnace bottom) close to the end of refining stage.

5. Conclusions

A complete 3D integrated model including the coherent jet modeling, the jet penetration cavity estimation and the decarburization modeling was proposed in this study and the validations for three parts were conducted respectively. The 3D integrated model can avoid the direct simulation of the supersonic coherent jet interacting with the liquid steel bath and provide a possible method to simulate the liquid steel-oxygen two-phase reacting flow system for the decarburization prediction with acceptable computational time. The conclusions made by adopting this 3D integrated model in EAF refining simulation are listed below:

- The stirring mechanism was analyzed using the model and the results indicated that the bubble stirring greatly promotes the homogenization of the liquid steel bath and is one of the most important stirring mechanism need to be considered in the EAF refining simulation.
- The decarburization rate and bath temperature distribution were investigated as well. The 3D integrated model indicated that decarburization mainly occurs around the jet penetrating cavity and due to the oxidation reaction, a large amount of chemical energy will be released there to increase the bath temperature. The overall bath decarburization rate and temperature rising rate predicted by the model has good agreement with reference data.
- 3D carbon distribution in the liquid steel bath was also investigated by the model. The results illustrated that the burner arrangement considered in present study results in an uneven bath decarburization rate, which is mainly due to the less oxygen-blowing and weak bath-stirring in front of the furnace. The uneven carbon concentration may lead to the under-prediction of the actual carbon content inside the furnace and further affect the quality of tapped liquid steel.

Author Contributions: Conceptualization, Y.C.; methodology, Y.C.; software, Y.C.; validation, Y.C.; formal analysis, Y.C.; investigation, Y.C.; resources, Y.C.; data curation, Y.C.; writing—original draft preparation, Y.C.; writing—review and editing, Y.C. and A.K.S.; supervision, C.Q.Z.; project administration, C.Q.Z.; funding acquisition, C.Q.Z. All authors have read and agreed to the published version of the manuscript.

Funding: This research was funded by Steel Manufacturing Simulation and Visualization Consortium (SMSVC).

Acknowledgments: The authors would like to thank the Steel Manufacturing Simulation and Visualization Consortium (SMSVC) members for funding this project. The Center for Innovation through Visualization and Simulation (CIVS) at Purdue University Northwest is also gratefully acknowledged for providing all the resources required for this work. The authors also appreciate the great help from Yury Krotov (Steel Dynamics, Inc.), Joseph Maiolo (Praxair) and Hamzah Alshawarghi (Praxair).

Conflicts of Interest: The authors declare no conflict of interest. The funders had no role in the design of the study; in the collection, analyses or interpretation of data; in the writing of the manuscript or in the decision to publish the results.

References

1. Mathur, P.; Messina, C. Praxair CoJet™ Technology–Principles and Actual Results from Recent Installations. *AISE Steel Technol.* **2001**, *78*, 21–25.
2. Anderson, J.E.; Farrenkopf, D.R. Coherent gas jet. U.S. Patent 5,823,762, 20 October 1998.
3. Alam, M.; Naser, J.; Brooks, G.; Fontana, A. Computational fluid dynamics modeling of supersonic coherent jets for electric arc furnace steelmaking process. *Metall. Mater. Trans. B* **2010**, *41*, 1354–1367.
4. Liu, F.; Zhu, R.; Dong, K.; Hu, S. Flow field characteristics of coherent jet with preheating oxygen under various ambient temperatures. *ISIJ Int.* **2016**, *56*, 1519–1528. [CrossRef]
5. Li, Z.L.; Zhang, L.L.; Cang, D.Q. Temperature corrected turbulence model for supersonic oxygen jet at high ambient temperature. *ISIJ Int.* **2017**, *57*, 602–608. [CrossRef]
6. Caffery, G.; Warnica, D.; Molloy, N.; Lee, M. Temperature homogenisation in an electric arc furnace steelmaking bath. In *Proceedings of the International Conference on CFD in Mineral and Metal Processing and Power Generation*; CSIRO: Canberra, Australia, 1997; pp. 87–99.
7. Li, B. Fluid flow and mixing process in a bottom stirring electrical arc furnace with multi-plug. *ISIJ Int.* **2000**, *40*, 863–869. [CrossRef]
8. Ramírez, M.; Alexis, J.; Trapaga, G.; Jönsson, P.; Mckelliget, J. Modeling of a DC electric arc furnace—Mixing in the bath. *ISIJ Int.* **2001**, *41*, 1146–1155. [CrossRef]
9. Szekely, J.; Asai, S. Decarburization of stainless steel: Part II. A mathematical model and a process optimization for industrial scale systems. *Metall. Trans.* **1974**, *5*, 1573–1580. [CrossRef]
10. Zhu, F.Y.; Chi, H.B.; Jiang, X.Y.; Mao, W.D. Process and Practice of EAF with De-P Hot Metal Charging for Melting Stainless Steel. In *Materials Science Forum*; Trans Tech Publications Ltd.: Zurich, Switzerland, 2007; Volume 561, pp. 1063–1066.
11. Memoli, F.; Mapelli, C.; Ravanelli, P.; Corbella, M. Simulation of oxygen penetration and decarburisation in EAF using supersonic injection system. *ISIJ Int.* **2004**, *44*, 1342–1349. [CrossRef]
12. Patankar, S. *Numerical Heat Transfer and Fluid Flow*; Taylor & Francis: Abingdon, UK, 2018.
13. *ANSYS FLUENT Release 19.1*; ANSYS Inc., ANSYS Academic Research: Canonsburg, PA, USA, 2013.
14. Wilcox, D.C. *Turbulence Modeling for CFD*; DCW Industries: La Canada, CA, USA, 1998; Volume 2.
15. Wolfshtein, M.; Lin, A.; Naot, D. *Mathematical Models of Turbulence*; No. BOOK; Academic Press: Cambridge, MA, USA, 1972.
16. Alam, M.; Naser, J.; Brooks, G. Computational fluid dynamics simulation of supersonic oxygen jet behavior at steelmaking temperature. *Metall. Mater. Trans. B* **2010**, *41*, 636–645. [CrossRef]
17. Abdol-Hamid, K.S.; Pao, S.P.; Massey, S.J.; Elmiligui, A. Temperature corrected turbulence model for high temperature jet flow. *J. Fluids Eng.* **2004**, *126*, 844–850. [CrossRef]
18. Magnussen, B. On the structure of turbulence and a generalized eddy dissipation concept for chemical reaction in turbulent flow. In Proceedings of the 19th Aerospace Sciences Meeting, St. Louis, MO, USA, 12–15 January 1981; p. 42.
19. Yin, C. Prediction of air-fuel and oxy-fuel combustion through a generic gas radiation property model. *Appl. Energy* **2017**, *189*, 449–459. [CrossRef]
20. Tang, G.; Chen, Y.; Silaen, A.K.; Krotov, Y.; Riley, M.F.; Zhou, C.Q. Effects of fuel input on coherent jet length at various ambient temperatures. *Appl. Therm. Eng.* **2019**, *153*, 513–523. [CrossRef]
21. Banks, R.B.; Chandrasekhara, D.V. Experimental investigation of the penetration of a high-velocity gas jet through a liquid surface. *J. Fluid Mech.* **1963**, *15*, 13–34. [CrossRef]
22. Ishikawa, H.; Mizoguchi, S.; Segawa, K. A model study on jet penetration and slopping in the LD converter. *Tetsu-to-Hagané* **1972**, *58*, 76–84. [CrossRef]

23. Sano, M.; Mori, K. Fluid flow and mixing characteristics in a gas-stirred molten metal bath. *Trans. Iron Steel Inst. Jpn.* **1983**, *23*, 169–175. [CrossRef]
24. Wei, J.H.; Zhu, D.P. Mathematical modeling of the argon-oxygen decarburization refining process of stainless steel: Part I. Mathematical model of the process. *Metall. Mater. Trans. B* **2002**, *33*, 111–119. [CrossRef]
25. Shukla, A.K.; Deo, B.; Millman, S.; Snoeijer, B.; Overbosch, A.; Kapilashrami, A. An insight into the mechanism and kinetics of reactions in BOF steelmaking: Theory vs. practice. *Steel Res. Int.* **2010**, *81*, 940–948. [CrossRef]
26. Saint-Raymond, H.; Huin, D.; Stouvenot, F. Mechanisms and modeling of liquid steel decarburization below 10 ppm carbon. *Mater. Trans. JIM* **2000**, *41*, 17–21. [CrossRef]
27. Chen, J.X. *Handbook on Common Using Data, Graphs and Tables in Steelmaking*; Metallurgical Industry Press: Beijing, China, 1984.
28. David, F.; Tudorache, T.; Firteanu, V. Numerical evaluation of electromagnetic field effects in electric arc furnaces. In *The International Journal for Computation and Mathematics in Electrical and Electronic Engineering*; COMPEL: Melbourne, Australia, 2001.
29. OJP, G.; RamíRez-Argáez, M.A.; AN, C. Effect of arc length on fluid flow and mixing phenomena in AC electric arc furnaces. *ISIJ Int.* **2010**, *50*, 1–8.
30. Widlund, O.; Sand, U.; Hjortstam, O.; Zhang, X. Modelling of electric arc furnaces (EAF) with electromagnetic stirring. In Proceedings of the 4th International Conference on Modelling and Simulation of Metallurgical Processes in Steelmaking (SteelSim), Metec InSteelCon 2011, Stahlinstitut VDEh, Düsseldorf, Germany, 27 June–1 July 2011.
31. Pretorius, E.; Oltmann, H.; Jones, J. *EAF Fundamentals*; LWB Refractories: New York, NY, USA, 2010.

 *processes*

Article

Numerical Modeling of Equal and Differentiated Gas Injection in Ladles: Effect on Mixing Time and Slag Eye

Luis E. Jardón-Pérez [1], Carlos González-Rivera [1], Marco A. Ramirez-Argaez [1] and Abhishek Dutta [2,*]

[1] School of Chemistry, National Autonomous University of Mexico (UNAM), C.P. Mexico City 04510, Mexico; dregwar@gmail.com (L.E.J.-P.); carlosgr@unam.mx (C.G.-R.); marco.ramirez@unam.mx (M.A.R.-A.)
[2] Department of Materials (MTM), KU Leuven, Kasteelpark Arenberg 44 bus 2450, B-3001 Heverlee-Leuven, Belgium
*   Correspondence: abhishek.dutta@kuleuven.be

Received: 10 July 2020; Accepted: 30 July 2020; Published: 2 August 2020

Abstract: Ladle refining plays a crucial role in the steelmaking process, in which a gas stream is bubbled through molten steel to improve the rate of removal of impurities and enhance the transport phenomena that occur in a metallurgical reactor. In this study, the effect of dual gas injection using equal (50%:50%) and differentiated (75%:25%) flows was studied through numerical modeling, using computational fluid dynamics (CFD). The effect of gas flow rate and slag thickness on mixing time and slag eye area were studied numerically and compared with the physical model. The numerical model agrees with the physical model, showing that for optimal performance the ladle must be operated using differentiated flows. Although the numerical model can predict well the hydrodynamic behavior (velocity and turbulent kinetic energy) of the ladle, there is a deviation from the experimental mixing time when using both equal and differentiated gas injection at a high gas flow rate and a high slag thickness. This is probably due to the insufficient capture of the velocity field near the water–oil (steel–slag) interface and slag emulsification by the numerical model, as well as the complicated nature of correctly simulating the interaction between both gas plumes.

Keywords: secondary refining; numerical model; dual gas injection; slag eye; mixing time

1. Introduction

Secondary refining in steelmaking consists of removing impurities from liquid steel through desulfurization, deoxidation and inclusion removal [1]. All aspects of refinement require agitation of the steel in its molten liquid form to accelerate steel–slag exchanges and mixing phenomena. Liquid steel is agitated by injecting gas through porous plugs located in the bottom of the ladle, which produces a movement of recirculation in the liquid steel, causing homogenization of thermal and chemical gradients, acceleration of metal–slag reactions, removal of gaseous species and flotation/precipitation of non-metallic inclusions present in the molten metal towards the slag to be removed [2]. A ladle usually has a cylindrical or truncated cone shape with a metal casing, covered inside with refractory brick. At the bottom is the porous plug, where inert gas (Ar) is injected. It usually also has graphite electrodes on top to maintain the temperature of the liquid mixture. It also has a hopper for the addition of alloys, mainly ferroalloys, and a powder injection system for the processes that require it. The slag layer plays a decisive role in refining the steel in the ladle. It is used to perform the desulfurization reaction, as well as to prevent oxidation of the metal and reduce heat dissipation. As the key to obtaining low sulfur containing steel, the efficiency and productivity of the desulfurization process depend largely on: (a) the consecutive kinetic steps, which consist of two main processes, namely, the chemical reactions at the interface and interphase mass transfer of sulfur from metal to slag phase, and (b) mixing within steel–slag phases. The efficiency of these physicochemical processes depends largely on the mixing degree of the molten steel by gas injection; thus, mixing time has been used extensively as a measure of the efficiency of the process. As the ladle is agitated by gas injection, the argon bubbles break up the slag layer, exposing a certain area of liquid metal to the atmosphere, called the slag eye or open eye. This phenomenon is harmful, because it is a site for reoxidation and nitrogen pickup in the bath. During ladle operation, this can be envisaged by the reaching of low mixing times using violent gas stirring. However, large gas flow rates lead to big slag eye areas as well. The behavior of the slag layer and mixing phenomena in the ladle are highly influenced by the argon stirring rates, the number of nozzles and their configuration. Thus, there is a balance needed for the ladle refining process which requires high mixing times but low slag eye areas [3,4].

Mixing time is a parameter that measures the mixing efficiency of the primary phase (liquid steel) in a bath. Mathematically, it is defined as the time at which there is a chemical homogeneity of 95% in the steel [5]. Mixing time helps to quantify the degree of agitation needed to homogenize the liquid contents after a step change in the composition inside a ladle. Researchers [6,7] have studied the mixing time of a two-phase system using central gas injection. Joo and Guthrie [8] concluded that the more a nozzle is moved from the center towards the mid-radius position, the more the mixing time decreases, which was confirmed by Krishnapisharody et al. [9]. It is important to note that both these research groups injected a tracer above the plume area to measure mixing time. Khajavi and Barati et al. [10] stated that an overlying slag layer has a significant effect on mixing time. In the case of dual gas injection, Chattopadhyay et al. [11] identified all possible dual purging locations that produced the least mixing time and compared their result with single purging experiments. They concluded that dual purging can reduce mixing time to a great extent even in the lower flowrate range. The promising results obtained initially by Liu et al. [12] and Haiyan et al. [13], showing the possibility of improving the mixing in a ladle by performing differentiated gas blowing, raised interest in the study of the effects of different gas blown modes for dual injection (equal and differentiated) on the ladle performance. Tang et al. [14] found both the ladle mixing and its exposed slag eye area are remarkably affected by the bottom gas blowing modes. In most cases, the differentiated mode can decrease the mixing time and slag eye area under given gas flowrates, compared with the equal gas blown mode. They also found that, generally, a relatively small angle between the porous plugs and a small radial position is beneficial to a decrease in the mixing time, whereas a relatively far plug radial position leads to a smaller slag eye. Jardón-Pérez et al. [15] used a physical model of a gas-stirred ladle with dual plugs to study the effect of gas flow ratio (equal or differentiated), gas flow rate, and slag thickness on mixing time and open eye area, using particle image velocimetry (PIV) to reveal the flow structure and using planar laser-induced fluorescence (PLIF) to determine the mixing time. They also performed a multi-objective optimization using a genetic algorithm, similar to Mazumdar et al. [16]. Their results revealed that the differentiated injection ratio significantly changes the flow structure and greatly influences the behavior of the system regarding mixing time and open eye area. Their results suggest that for optimal performance the ladle must be operated using a differentiated flow ratio. Liu et al. [17] recently made a comprehensive review of the variables considered in the study of mixing time in ladle metallurgy in the last three decades. Among all the variables, they found that the gas flow rate is by far the most important variable affecting the mixing time. In general terms, the effect of increasing the gas flow rate increases the slag eye area and an increase in the slag thickness decreases the slag eye area.

The dynamics of gas–liquid interaction in a metallurgical reactor such as a ladle furnace is similar to a bubble column reactor [17], although it is more complex, since the process is a triphasic system. Gathering experimental data in such high temperature environment is quite complex; thus, researchers use scaled down water (physical) models and numerical models using computational fluid dynamics (CFD) to understand the hydrodynamics and mixing processes [5]. Li et al. [18] developed a CFD model to analyze the transient three phase flow in an argon-stirred ladle with one and two off-centered porous plugs. The effect of the argon gas flow rate on the spout height and slag eye area was discussed.

The slag layer behavior and the interface phenomena were also analyzed. Haiyan et al. [13] found that the use of different flowrates can significantly reduce the mixing time, compared with the mixing time reached when using the same flow rate. They performed a validation of their CFD model, which suggests that the difference in mixing time arises due to the different flow behavior and the associated changes in stirring energy dissipation, which in turn could explain the observed decrease in mixing time. In the case of differentiated flows, the simulation showed that the eyes of the loops of the two plumes are not located at the same height. This is due to the difference in the gas flow rates injected at each plug (i.e., the strong plume forms a larger circulation loop stirring most of the ladle, whereas the weak plume forms a smaller circulation loop). Using a coupled Eulerian–Lagrangian model, Conejo et al. [19] studied mixing time and slag eye area fitted with dual plugs as a function of operating variables, namely, gas flow rate, radial nozzle position, separation angle between nozzles and partitioning flow rate. They suggested that if mixing time is the parameter of primary interest then nozzle configuration with equal flow partitioning (50%:50%) between the nozzles should suffice for both low and high gas flow rates, whereas if slag eye is the parameter of primary interest then a configuration with non-equal flow partitioning (25%:75%) between the nozzles should be preferred. Villela-Aguilar et al. [20] performed a multiphase numerical simulation to analyze the effects of the gas flow rate, radial position and angle between plugs and differentiated flows in two plugs on the mixing time in a secondary refining ladle. They found that the angle of separation between the plugs is the most relevant variable to reduce mixing time. They also found that it is possible to reduce the mixing time by using a good differentiated configuration in both gas flow and location of the porous plugs. According to a review by Liu et al. [17], there is still room for further improvement of the numerical model regarding the representation of the turbulence and the slag–steel interactions. One of the least studied variables, in the case of dual gas injection points, is the effect of the use of different gas flow rates in each plug (i.e., a gas blowing ratio different from 50%:50%).

In the present study, the effect of gas injection in equal (50%:50%) and differentiated (75%:25%) flows, along with the gas flow rate and slag thickness, on mixing time and slag eye were studied using a numerical model of a ladle and were compared with previously-obtained experimental data based on PIV measurements of the hydrodynamics and PLIF measurements of mixing time for model validation. This study aims to improve the mixing time in a secondary refining ladle and to identify a balanced compromise between mixing time and slag eye, while improving numerical modeling practices using experimental data on differentiated dual gas injection modes in ladles.

2. Methodology: Numerical Model Development

A Eulerian multi-phase mathematical model simulating the physical model described in this study was developed under the following assumptions: (i) steady state, (ii) a symmetry plane is considered for the dual gas injection system and thus only half of the ladle is solved; (iii) Newtonian and incompressible fluids for both liquids and gas phases; (iv) isothermal flow and (v) constant bubble diameter of 0.01 m. The latter is certainly an oversimplification, by neglecting bubble disintegration and coalescence phenomena under real dynamic bubble size distributions. However, using the Eulerian model most regions of the ladle are fairly well predicted, as presented in Section 3.1, both in turbulence and velocity magnitude, except for the water-oil-bubble open eye regions. Governing equations for the 3D Eulerian-Eulerian multi-phase algorithm include mass conservation equation, Navier–Stokes equations and the k-ε realizable turbulence model for the water phase.

The volume of the q-phase, V_q, is given by the volume integral:

$$V_q = \int_V \alpha_q \, dV \tag{1}$$

where α_q is the volume fraction of phase q, and the sum of volume fractions must be equal to one according to:

$$\sum_{q=1}^{n} \alpha_q = 1 \tag{2}$$

The continuity equation for each q-phase is:

$$\nabla \cdot \left(\alpha_q \, \rho_q \vec{v}_q \right) = 0 \tag{3}$$

where ρ_q and $\vec{v}_q$ are the density and velocity vector of the q-phase, respectively.

The momentum conservation equation for the q-phase is:

$$\nabla \cdot \left(\alpha_q \, \rho_q \vec{v}_q \vec{v}_q \right) = - \alpha_q \, \nabla P + \nabla \cdot \left(\alpha_q \, \mu_{ef,q} \left(\nabla \vec{v}_q + \left(\nabla \vec{v}_q \right)^T \right) \right) + \alpha_q \, \rho_q \vec{g} + \vec{F}_T \tag{4}$$

For the water phase:

$$\mu_{ef,l} = \mu_{lam,l} + \mu_{t,l} \tag{5}$$

For other phases:

$$\mu_{ef,q} = \mu_{lam,q} \tag{6}$$

where P, $\mu_{ef,q}$, $\vec{g}$ are the pressure, effective viscosity of the q-phase and the gravity acceleration, respectively. The effective viscosity for the water phase is the sum of the molecular viscosity ($\mu_{lam,l}$) and the turbulent viscosity ($\mu_{t,l}$) defined by the turbulence model employed. For the other phases $\mu_{lam,q}$ is only the laminar viscosity of every fluid, and the subscripts l and *lam* stand for water and laminar, respectively. The term $\vec{F}_T$ is the contribution of all interphase forces. The only force considered is the drag force, and the virtual mass, lift and turbulent dispersion forces are neglected.

In this study, a sensitivity analysis was performed to choose the best turbulence model that predicts correctly the turbulence measured in the ladle. From all the models tested, the best option is the realizable k-ε turbulence model [21]. This model solves two additional conservation equations applicable only to the water liquid phase, one of these equations being the conservation of turbulent kinetic energy, k:

$$\nabla \cdot \left(\alpha_l \rho_l k \vec{v}_l \right) = \nabla \cdot \left(\alpha_l \left(\mu_{lam,l} + \frac{\mu_{t,l}}{\sigma_k} \right) \nabla k \right) + \alpha_l G_{k,l} + \alpha_l G_b - \alpha_l \rho_l \varepsilon + \alpha_l \rho_l \pi_{k,l} \tag{7}$$

The conservation equation for the dissipation of the turbulent kinetic energy, ε:

$$\nabla \cdot \left(\alpha_l \rho_l \varepsilon \vec{v}_l \right) = \nabla \cdot \left(\alpha_l \left(\mu_{lam,l} + \frac{\mu_{t,l}}{\sigma_\varepsilon} \right) \nabla \varepsilon \right) + \alpha_l C_1 \left(\frac{\varepsilon}{k} G_b C_3 \right) - \alpha_l \rho_l C_2 \frac{\varepsilon^2}{k + \sqrt{\frac{\varepsilon \mu_{lam,l}}{\rho_l}}} + \alpha_l \rho_l \pi_{\varepsilon,l} \tag{8}$$

C_1, C_2, C_3, σ_k and σ_ε are constants of the model, with values of 1.44, 1.92, 1.3, 1.0 and 1.2 respectively. G_k is the production of turbulent kinetic energy due to the mean velocity gradients of the water phase and G_b is the additional turbulent kinetic energy produced by the bubbles. $\pi_{k,l}$ and $\pi_{\varepsilon,l}$ are the turbulent interaction terms defined by Troshko-Hassan [22] as:

$$\pi_{k,l} = C_{ke} \sum_{p=1}^{m} K_{gl} \left| \vec{v}_g - \vec{v}_l \right|^2 \tag{9}$$

$$\pi_{\varepsilon,l} = C_{td} \frac{3 C_A \left| \vec{v}_g - \vec{v}_l \right|}{2 d_g} \pi_{k,l} \tag{10}$$

where C_{ke} and C_{td} are model constants with values of 0.75 and 0.45, respectively. K_{gl} is the covariance of the velocities of the continuous phase l and the disperse phase g. d_g is the bubble diameter; the velocity difference between the gas and liquid phase, $\left(\vec{v}_g - \vec{v}_l\right)$, is defined as the relative velocity, $\vec{v}_{rel}$; whereas C_A is the drag coefficient determined through a sensitivity analysis (not mentioned in this study) performed that revealed the best option was the well-known symmetric model:

$$C_A = \begin{cases} \dfrac{24\left(1+0.15Re^{0.687}\right)}{Re} & Re \leq 1000 \\ 0.44 & Re > 1000 \end{cases} \tag{11}$$

where Re is the Reynolds number.

The interphase forces, $\vec{F}_T$ are limited to the drag force, $\vec{F}_A$ as follows:

$$\vec{F}_T = \vec{F}_A = \frac{3\alpha_g \alpha_l \rho_l C_A}{4d_g}\left(\vec{v}_g - \vec{v}_l\right) \tag{12}$$

Finally, to compute mixing time in the water phase l, a single conservation equation for a tracer chemical species i is solved in transient mode with initial conditions of zero concentration of solute everywhere except for the pulsed tracer injected at the free surface above the plume of high flow, as follows:

$$\frac{\partial}{\partial t}\left(\rho_l w_{i,l}\right) + \nabla \cdot \left(\rho_l \vec{u}_l w_{i,l}\right) = \nabla \cdot \left(\rho_l D_{i,l} + \frac{\mu_{t,l}}{Sc_t}\right)\nabla w_{i,l} \tag{13}$$

The geometry of the water (physical) model that was used in PIV experiments is made in 3D using ANSYS Design Modeler 19.0 (Ansys Inc., Canonsburg, PA, USA) for the numerical model. The diameter and height of the cylindrical ladle are 0.185 m and 0.214 m (0.17 m liquid level), respectively, similar to that of the physical model [15] that corresponds to a 1:17 scale ratio of an industrial-size ladle of 200-ton capacity. Figure 1 shows an illustration of the mesh built with approximately 350,000 elements with average orthogonality, skewness and aspect ratios of 0.98, 0.1 and 1.99, respectively. The choice of the mesh elements was based on a sensitivity analysis using approximately 200,000 and 500,000 mesh elements.

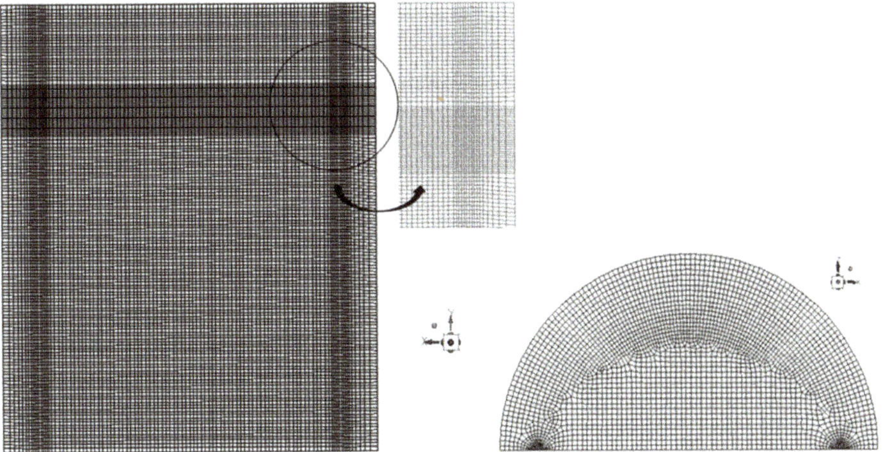

Figure 1. Computational domain of the ladle used for the numerical simulations presented in this study. The two vertical sections indicating the gas injection inlets and the top horizontal section indicating the slag layer are comparatively denser than the remaining mesh domain. The inset shows the mesh density for the slag zone which is different from the melt zone. The top portion of the mesh (half along the symmetry plane) is shown separately.

Non-slip boundary conditions at the bottom and lateral walls have been used, whereas the standard wall functions have been used to connect the turbulent core of the fluid with the laminar flow near the walls. The gas injection inlets at the nozzle positions and an open boundary to the atmosphere at the top of the system allow the outflow of the gas phase. The complete set of boundary conditions can be found in Table 1.

Table 1. Boundary conditions used in the numerical model presented in this study.

Boundary	Mass Transport Condition	Momentum Transport Condition
Inlets	velocity inlet of air with turbulent intensity	velocity inlet of air with turbulent intensity
Outlet	pressure outlet with air backflow	pressure outlet with air backflow
bottom wall	impermeable boundary	no slip with standard wall functions
lateral wall	impermeable boundary	no slip with standard wall functions

The system of partial differential equations was numerically solved and the solution in pseudo-transient mode was considered to be converged when the residuals of all conservation equations were below 1×10^{-3}. Approximately 1200 iterations were required to get the convergence in around 36 h of CPU time in a computer with 8 MB in RAM with an Intel Core$^\circledR$ i7-3770 processor of 3.4 GHZ. Table 2 lists all the numerical simulations performed in this study which is based on a full factorial experimental design at two levels with the three variables, namely, gas flow rate, dual gas injection ratio and (slag) oil thickness, as mentioned in Jardón-Pérez et al. [15].

Table 2. Design of experiment with high and low values of the three variables, namely, gas flow rate, dual gas injection ratio and (slag) oil thickness for the eight case studies presented in this study.

Cases	Experiment Number	(Slag) Oil Thickness (hs) (%)	Gas Flow Rate (Q) (L/min)	Dual Gas Injection Ratio (P) (%/%)
1	a	3	1.54	50/50
2	b	3	2.22	50/50
3	c	3	1.54	25/75
4	d	3	2.22	25/75
5	e	5	1.54	50/50
6	f	5	2.22	50/50
7	g	5	1.54	25/75
8	h	5	2.22	25/75

3. Results and Discussion

3.1. Model Validation

Figure 2 shows the flow patterns obtained through the PIV technique reported in Jardón-Pérez et al. [15] at different operating conditions listed in Table 2, whereas Figure 3 shows the flow patterns obtained with the numerical model presented in this study. As seen in these figures, a reasonable agreement is observed between experimental and numerical results for all the cases. For equal (50%:50%) dual gas injection, two symmetric toroids are formed at each side of the plume, whereas for unequal (25%:75%) injection, symmetry vanishes and the high flow rate circulation expands at the expense of the low flow rate circulation loop. An increase in the gas flow rate increases the expansion of the plume in general, and for unequal injection further expands the high-flow rate circulation loop. The increment in (slag) oil thickness mitigates the inertia of the plumes, reducing the velocities of the liquid, and diminishes the loop expansions. An increase in (slag) oil thickness lowers the velocity of the liquid in the vicinity of the oil layer at the top of the ladle. In general, the experimental results were predicted successfully with the model, but some differences were also perceived. The numerical predictions using CFD do not account for the expansion of the bigger loop for unequal injection as occurs in PIV-measured flow patterns. As an illustration, Figure 4 shows measured

(a) and calculated (b) streamlines obtained for experiment b (see Table 2) under equal dual gas injection, showing good agreement between both predicted and measured streamlines; however, under unequal dual gas injection for experiment g (see Table 2), calculated (d) streamlines do not capture the expansion of the strong plume and the contraction of the small loop as seen in the measured (c) streamlines. In general, the magnitudes of the liquid velocity were slightly underestimated. Mean value was overestimated (see Table 3), but the distribution is quite different, especially in differentiated injection, since the model does not accurately predict the interaction of the plumes. In the case of differentiated dual gas injection, the interaction between the recirculation loops was not observed. This is probably due to the drag effect, which was not successfully implemented, and represents a challenging issue in numerical modeling, from the hydrodynamic point of view, of the more complex multiple differentiated dual gas injections, compared to the traditional equal dual gas injection. Despite the differences, the numerical model of the ladle furnace in steady state showed good results, with a reasonable agreement with the experimentally measured liquid flow patterns reported in Jardón-Pérez et al. [15]. A more quantitative validation is shown in Figure 5 by comparing experimental (continuous line) and numerical (dotted line) mean velocity radial profiles at h/H = 0.8 for experiments b (a) and g (b); axial profiles at r/R = −0.75 for experiments b (c) and g (d); and axial profiles at r/R = 0.75, for experiments b (e) and g (f). In all these cases, a good agreement between prediction and measurement is achieved, with, in general, the numerical velocity profiles slightly over-predicting the measured results, but with the simulations capturing the measured liquid motion in the plumes.

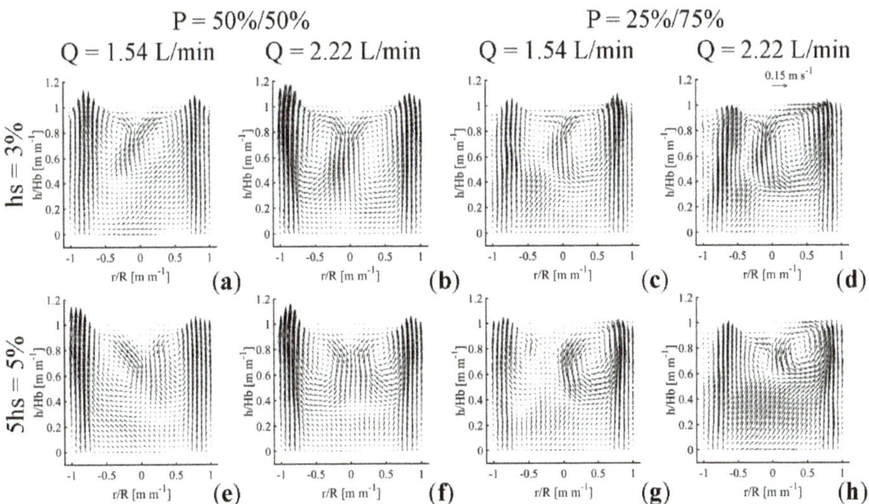

Figure 2. Flow patterns of the eight case studies obtained with the experimental model (particle image velocimetry (PIV) technique) in the longitudinal plane. (**a**) through (**h**) are the experiments described in Table 2.

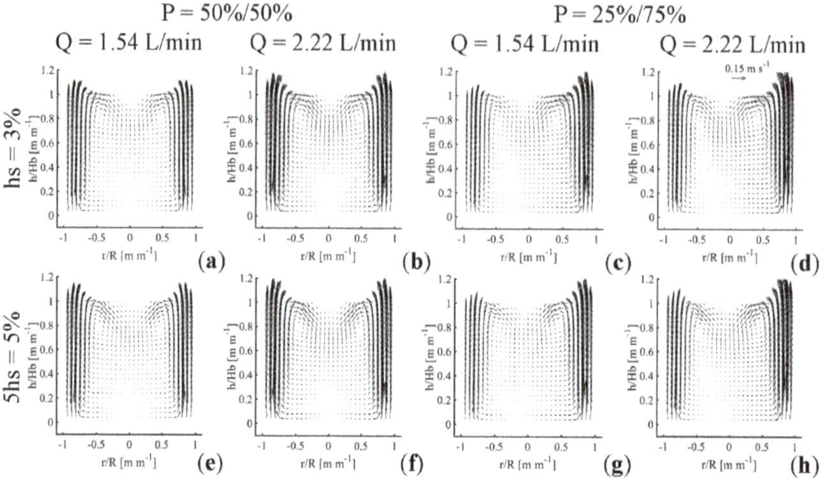

Figure 3. Flow patterns of the eight case studies obtained with the numerical model and shown along the same longitudinal plane. (**a**) through (**h**) are the experiments described in Table 2. The cases presented in this study are in the same order as in the experimental study of Jardón-Pérez et al. [15].

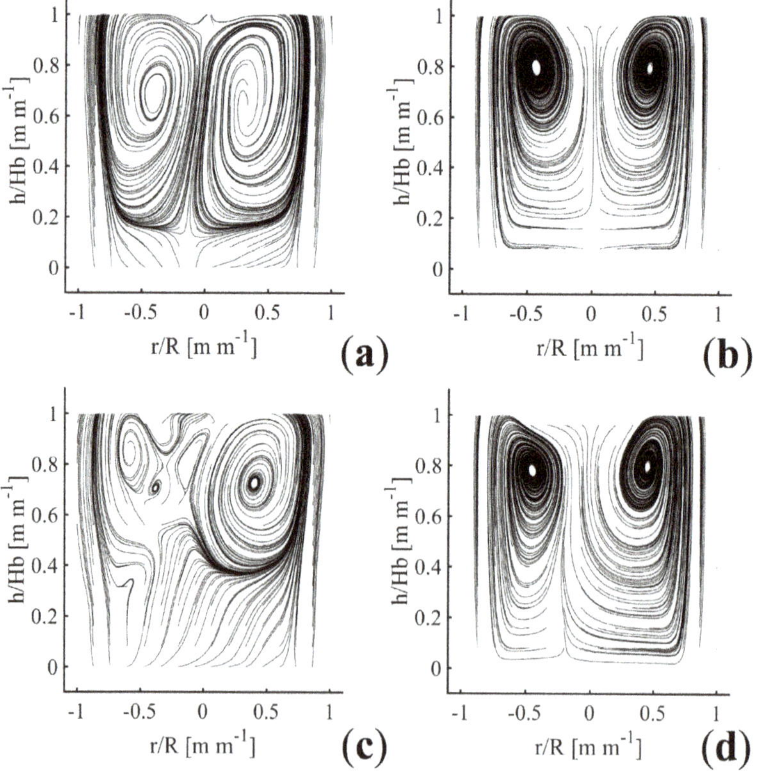

Figure 4. Streamlines obtained for experiment b (Table 2), measured (**a**) and calculated (**b**); and for experiment g (Table 2), measured (**c**) and calculated (**d**).

Table 3. Mean values of velocity v ($\times 10^{-2}$ m/s) for the experimental and numerical model along the longitudinal plane (symmetry plane) for the eight cases presented in this study.

(Slag) Oil Thickness (hs)	Low 1.54 L/min Gas Flow Rate (Q)		High 2.22 L/min Gas Flow Rate (Q)	
	50%:50% Dual Gas Injection Ratio	25%:75% Dual Gas Injection Ratio	50%:50% Dual Gas Injection Ratio	25%:75% Dual Gas Injection Ratio
3% oil thickness				
experimental	4.36 ± 2.61	5.33 ± 3.17	4.18 ± 2.72	4.53 ± 3.16
numerical	4.91 ± 4.94	5.98 ± 6.01	5.03 ± 5.24	5.91 ± 6.14
difference (%)	12.81	12.13	20.46	30.37
5% oil thickness				
experimental	4.62 ± 2.86	4.74 ± 3.09	3.60 ± 2.27	4.53 ± 2.77
numerical	5.07 ± 5.15	5.30 ± 5.38	4.68 ± 5.00	5.52 ± 5.87
difference (%)	9.85	11.87	30.07	22.01

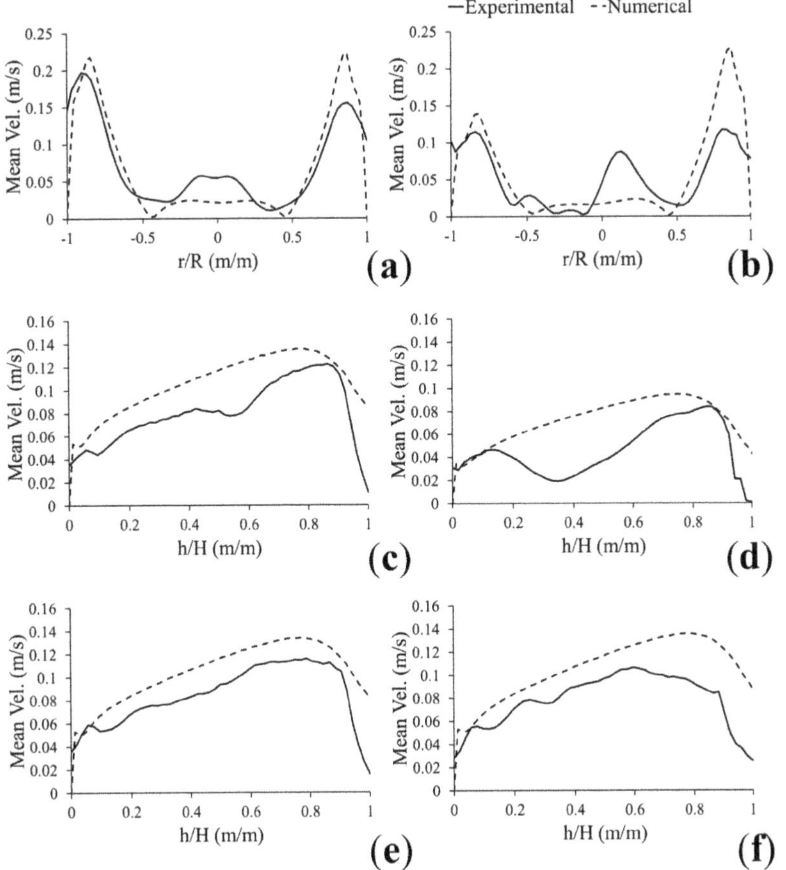

Figure 5. Comparison of experimental (continuous line) and numerical (dotted line) mean velocity radial profiles at h/H = 0.8 for experiments b (**a**) and g (**b**); axial profiles at r/R = −0.75 for experiments b (**c**) and g (**d**); and axial profiles at r/R = 0.75 for experiments b (**e**) and g (**f**).

3.2. Turbulence Modeling

Figure 6 shows the simulated profiles of turbulent kinetic energy (k) at the same operating conditions of Figure 2 (see Table 2 for the sequence of the experiments). A reasonable agreement can be seen in Table 4 between the experimental and numerical results of turbulence, showing that the k-ε realizable turbulence model was an appropriate choice to represent turbulence in the three-phase fluid flow system implemented in this study. Due to the turbulence promoted by the bubbles, high turbulence zones corresponding to the two plume zones were observed. By increasing the gas flow rate (Figure 6b,d) the values of turbulent kinetic energy k in the circulation loops increase. A thicker slag reduces turbulence and the interaction between the loops. Although the magnitudes of the turbulent kinetic energy k in the plane overestimated the experimental results in almost all cases (in comparison with Figure 4 of Jardón-Pérez et al. [15]), the above-mentioned main effects of the three variables on k were successfully predicted with the numerical model. However, some features were not captured by the model, such as the drag effect (comparing Figure 2h with Figure 3h), where the smaller plume zone was not attracted to the center due to the influence of the loop.

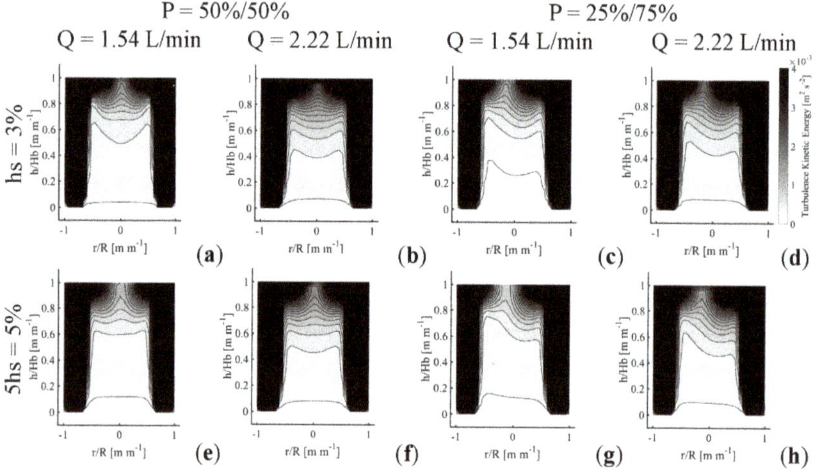

Figure 6. Contours of turbulent kinetic energy (k) of the eight case studies obtained with the numerical model and shown along the same longitudinal plane. (**a**) through (**h**) are the experiments described in Table 2. The cases presented in this study are in the same order as in the experimental study of Jardón-Pérez et al. [15].

Table 4. Mean values of turbulent kinetic energy k ($\times 10^{-3}$ m^2/s^2) for the experimental and numerical model along the longitudinal plane (symmetry plane) for the eight cases presented in this study.

	Low 1.54 L/min Gas Flow Rate (Q)		High 2.22 L/min Gas Flow Rate (Q)	
(Slag) Oil Thickness (hs)	50%:50% Dual Gas Injection Ratio	25%:75% Dual Gas Injection Ratio	50%:50% Dual Gas Injection Ratio	25%:75% Dual Gas Injection Ratio
3% oil thickness				
experimental	0.74 ± 0.51	1.18 ± 0.75	0.83 ± 0.55	1.05 ± 0.75
numerical	0.78 ± 1.02	1.15 ± 1.44	1.00 ± 1.37	1.24 ± 1.66
difference (%)	5.86	2.56	19.88	18.73
5% oil thickness				
experimental	0.97 ± 0.61	1.11 ± 0.72	0.60 ± 0.39	0.78 ± 0.57
numerical	0.91 ± 1.17	1.14 ± 1.43	0.89 ± 1.23	1.11 ± 1.49
difference (%)	6.64	1.98	46.87	41.88

3.3. Slag Eye Modeling

Tang et al. [14] compared the change in slag eye area between equal and differentiated flows at various gas injection parameters, including the angle between the dual plugs from 45° to 180° and its radial position from r/R = 0.5 to r/R = 0.7 and the gas flow rate. They found that in most cases the proportion of slag eye in the differentiated flow system is smaller than that in the equal flow system. Their results show that the proportion of slag eye in differentiated flow first decreases and then increases with the increasing relative angle of plugs at the same flowrate, and the exposed areas of the slag eyes of the two modes generally decrease with plug position from 0.55R to 0.70R, apparently due to the obstacle of the ladle wall and its absorption of the stirring energy. The slag eye is the smallest at 0.7R–90° and 0.7R–135°. Conejo et al. [19] measured the change in the slag eye area between equal and differentiated flows for two nozzle radial positions (0.7R/0.7R and 0.7R/0.5R) and two angles (45° and 90°). Their results show that the use of differentiated flow increases the area of slag eye for an angle of 45° and equal plug positions, whereas there is a decrease in the eye when using unequal plug positions and an angle between plugs of 90°. From these results, it can be seen that the change in slag eye becomes rather complex in the presence of multiple variables.

Figure 7 presents the time-averaged predicted slag eye of the eight experiments (see Table 2) using the developed numerical model. The effect of the three variables, namely, gas flow rate, dual gas injection ratio and (slag) oil thickness on the size of the slag eye was predicted. As seen in the experiment, an increase in gas flow rate increases the extent of slag eye area in all cases, whereas the opposite effect is obtained with an increase of (slag) oil thickness. A good agreement between the experimental results (see Figure 5 of Jardón-Pérez et al. [15]) and the numerical results can be seen in Table 5, with slag eye area as a percentage of the total surface. Again, the drag effect generated with a differentiated gas injection was not captured completely by the model, i.e., the smaller slag eye is not attracted to the center as observed in the experimental results, but the shapes of the slag eye were correctly predicted. For example, for the case with 5% (slag) oil thickness and differentiated 2.22 L/min gas flow rate, presented in Figure 8, the numerical size of the big eye is slightly under-predicted, whereas the small eye is over-predicted, when compared against the measured eyes. As mentioned earlier, the velocity field near the water–oil (steel–slag) interface and the slag eye size features are not properly captured by the numerical model, which opens an opportunity for improvement of the physical description of the interaction between the two phases through the modification of forces such as drag and surface tension.

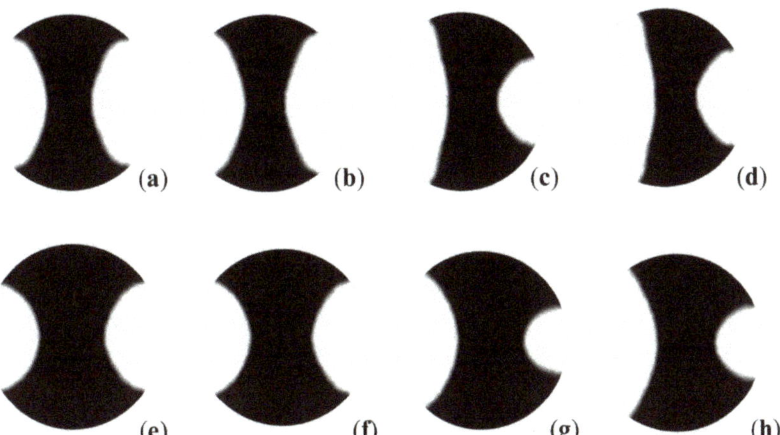

Figure 7. Time-averaged prediction of slag eyes at different operating conditions obtained with the numerical model. (**a**) through (**h**) are the experiments described in Table 2. The cases presented in this study are in the same order as in the experimental study of Jardón-Pérez et al. [15].

Table 5. Slag eye area as a percentage of the total surface area for different operating conditions experimentally obtained through image analysis from both experimental and numerical models.

(Slag) Oil Thickness (hs)	Low 1.54 L/min Gas Flow Rate (Q)		High 2.22 L/min Gas Flow Rate (Q)	
	50%:50% Dual Gas Injection Ratio	25%:75% Dual Gas Injection Ratio	50%:50% Dual Gas Injection Ratio	25%:75% Dual Gas Injection Ratio
3% oil thickness				
experimental	39.40 ± 2.27	45.40 ± 3.53	51.47 ± 1.49	58.35 ± 1.97
numerical	45.35	49.75	49.84	55.21
difference (%)	15.10	9.59	3.17	5.38
5% oil thickness				
experimental	34.13 ± 1.79	34.21 ± 2.96	43.99 ± 2.06	49.88 ± 3.03
numerical	30.07	38.31	33.89	38.93
difference (%)	11.90	11.99	22.95	21.94

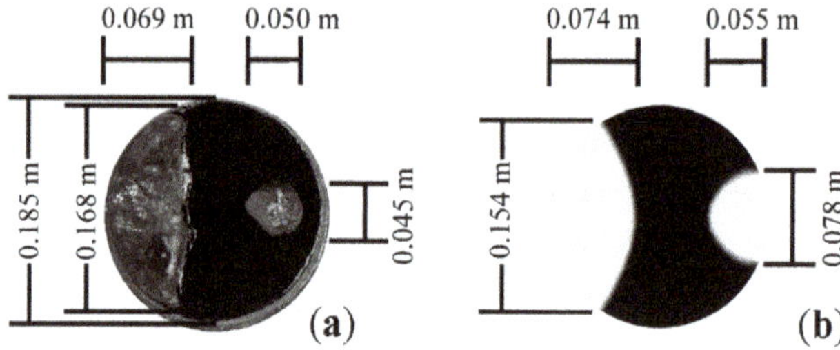

Figure 8. Comparison of the time-averaged photograph of the slag eye obtained with the (**a**) experimental model (operating condition of differentiated dual gas injection with 5% (slag) oil thickness and 2.22 L/min gas flow rate) and the corresponding time-averaged (**b**) numerical model prediction. Dimensions of measured and predicted open slag eyes are also indicated.

3.4. Mixing Time Modeling

Mixing phenomena in metallurgical steel ladles by bottom gas injection involve three phases, namely, liquid molten steel, liquid slag and gaseous argon. Mixing time is the time needed for an established constant flow field to disperse an injected non-reactive tracer until the required mixing criterion is achieved. This criterion is generally set to 95%, although 99% and 99.5% [23,24] can be used as well. Typically, the mixing efficiency of a ladle is qualified as the mixing time (τ_m) and is an important indicator for the hydrodynamic performance to quantify the degree of agitation, which influences secondary refining treatment in a ladle through desulphurization. Lou and Zhu [25] modeled desulfurization and inclusion removal in a ladle with multiple nozzle configurations. They concluded that dual nozzle configurations have a higher desulfurization rate than single nozzle configurations for the same flow rate. Zhu et al. [26] investigated the mixing phenomena in argon-stirred ladles with six types of tuyere arrangement. They concluded that mixing time is greatly influenced by the position of the tracer, and that mixing time decreases with increasing gas flow rate, although the effect is not so remarkable. Haiyan et al. [13] studied the effect of gas flow rate on the mixing phenomenon in a bottom-stirring ladle with dual plugs. They found that compared with the same flowrate for the two plugs, the mixing time is lower with different flow rates when the plug positions are located at 0.64R. This is mainly due to the strong gas plume, which produces a larger circulation flow to stir the ladle. A weaker plume forms a smaller one, which weakens the interference and collision from the two plumes, thus reducing the mixing time. Recently, Ramasetti et al. [27] calculated the mixing time of tracer addition into the metal bath of a ladle, which decreased when the argon flow rate was increased.

Nunes et al. [28] were among the earliest to use a dye tracer to determine the mixing time in a ladle, thus eliminating the dependence on the monitoring position. For the measurement of mixing time, techniques such as using a pH meter or conductivity meter are the most common, although more recently dye tracers and laser-induced fluorescence (LIF) have been used. To measure the time needed to reach a stable state, usually a criterion of ± 5% of variation of property is used, which can be measured by pH in a pH meter, conductivity in a conductivity meter, and tracer concentration and luminescence in LIF. In the experimental study of Jardón-Pérez et al. [15], mixing time in a gas-stirred ladle was measured by means of the novel technique of planar laser-induced fluorescence (PLIF) that uses Rhodamine 6G as a tracer. PLIF determinations were performed at two different planes and pH probe determinations were performed at two different locations. The results were then compared, which showed not only the accuracy of the PLIF method, but also that it is less sensitive to the location of the measurement than the pH probe method. A detailed explanation of the technique can be found in Jardón-Pérez et al. [29]. Ascanio [30] does not recommend the technique for highly aerated flows, because the presence of bubbles complicates the calibration and the measurement of the obtained concentration contour.

From the experimental study of Jardón-Pérez et al. [15], it was found that all cases gave smaller mixing times with the same conditions of (slag) oil thickness and dual gas injection ratio. The increase in the thickness gave slightly higher mixing times. The numerical model of the mixing time presented in this study gave very similar results to the experimental mixing time for both 50%:50% equal flow and 25%:75% dual differentiated flow. As with the experimental study, the effect of a 25%:75% dual injection ratio is slightly less significant with conditions of high gas flow rate (2.22 L/min) as compared with conditions of low gas flow rate (1.54 L/min) for 3% (slag) oil thickness than with 5% (slag) oil thickness (see Table 6). The deviation from the experimental mixing time was found to be the greatest for 25%:75% dual differentiated flow at a high gas flow rate (2.22 L/min), whereasthe deviation was found to be the lowest for 25%:75% dual differentiated flow at low gas flow rate (1.54 L/min) both at 5% oil thickness. The difference between experimental and predicted mixing times can be explained by comparing the experimental (Figure 2) and numerical (Figure 3) flow patterns. As can be seen, the area of the low-velocity zone with dual injection is higher for the numerical model than the physical model, mainly due to the difficulty of simulating the interaction between both the gas plumes, especially when a differentiated injection is used. This is because the interaction between the low-injection and the high-injection plumes (see Figure 3c,d,g,h) shows a large deformation in both plume structures due to the drag force that the high-injection plume exerts on the low-injection plume.

Table 6. Mixing time in seconds for different operating conditions experimentally (obtained through the planar laser-induced fluorescence (PLIF) method) and compared with the numerical model.

(Slag) Oil Thickness (hs)	Low 1.54 L/min Gas Flow Rate (Q)		High 2.22 L/min Gas Flow Rate (Q)	
	50%:50% Dual Gas Injection Ratio	25%:75% Dual Gas Injection Ratio	50%:50% Dual Gas Injection Ratio	25%:75% Dual Gas Injection Ratio
3% oil thickness				
experimental	8.04 ± 0.57	7.24 ± 0.80	6.84 ± 0.26	6.57 ± 0.40
numerical	9.67	8.71	8.18	7.07
difference (%)	20.28	20.31	19.56	7.67
5% oil thickness				
experimental	10.09 ± 1.04	9.35 ± 1.13	7.18 ± 0.62	5.92 ± 0.45
numerical	12.53	9.49	8.82	8.16
difference (%)	24.15	1.47	22.79	37.85

Figure 9 shows the trend in the model predictions and the experimental results reported in Jardón-Pérez et al. [15] based on all the cases implemented in this study. In this Figure, it is seen that the model is able to capture the more important issues regarding the behavior of the system under different experimental conditions. Considering that differences between predicted and experimental measurements increase for differentiated gas injection and high slag thickness (Cases 7 and 8 in Figure 9 and Table 2), the results suggest that these are places in which to seek improvements in the numerical model, especially in high slag thickness. This is probably because with a higher slag thickness the influence of slag emulsification on mixing time becomes prominent, since the numerical model does not consider the emulsified droplets. This is also shown clearly in Figure 10, which depicts the difference in the model prediction compared with the experimental results reported in Jardón-Pérez et al. [15] for the cases of low gas flow rate (Figure 10a) and high gas flow rate (Figure 10b), thus exploring the ability of the numerical model to capture the two limit cases analyzed in this study.

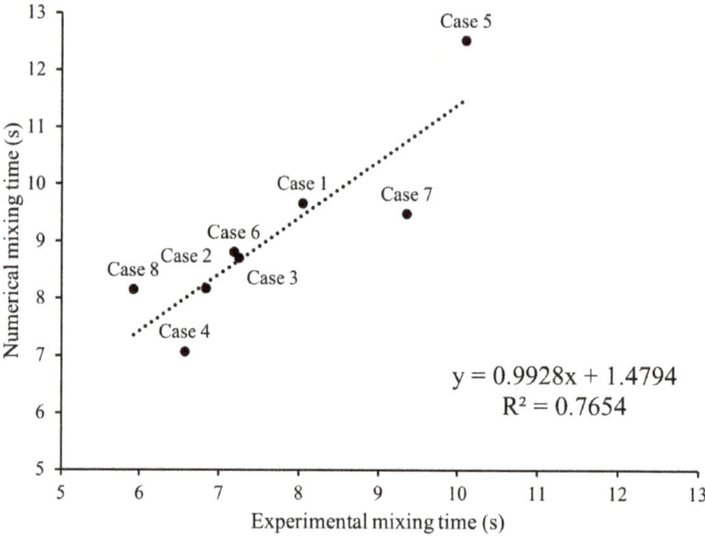

Figure 9. Comparison between the model prediction (dotted line) and experimental mixing time (black dots) obtained by Jardón-Pérez et al. [15].

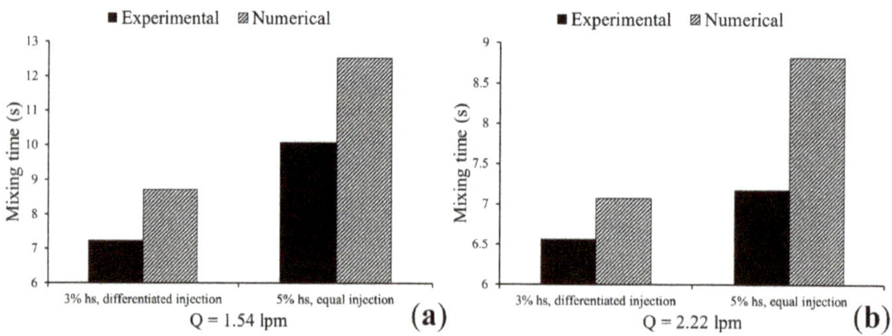

Figure 10. Effect of the gas flow rate on the mixing time predicted by the model and experimental measurements for (**a**) low gas flow and (**b**) differentiated gas flow.

In both figures, the best match is obtained for the case of low gas flow rate and low slag thickness. At a high flow rate, the differences between experimental and predicted results increase for all cases (see Table 6) and these differences are bigger for differentiated gas injection, independently of the gas flow rates. The results suggest that modeling efforts must be focused on the improvement of the turbulence model, as well as finding approaches to simulate the effect of the slag layer (mainly the slag emulsification) on the flow dynamics of the stirred melt and to simulate changes in the flow plume linked to the change in the mode of gas injection.

However, even if numerical values of both slag eye area and mixing time show differences between numerical and experimental values (see Tables 5 and 6), the effect of the studied variables is correctly predicted by the mathematical model, i.e., an increase in gas flow rate causes an increase in the exposed area and a decrease in the mixing time, an increase in slag height causes a decrease in the slag eye area and an increment in the mixing time, and finally, the use of differentiated gas injection causes a slight increment in the exposed area but also a decrement in the mixing time compared with equal injection. Therefore, the CFD numerical model could be used to study the main effects of operation variables on both mixing time and slag eye area.

4. Conclusions

In the numerical study of the effect of dual gas injection using equal (50%:50%) and differentiated (75%:25%) flows and variable (slag) oil thickness compared with the experimental results obtained previously, the following conclusions can be made:

1. The numerical model using CFD predicts the hydrodynamic behavior of the ladle well, in comparison with the physical model. Turbulent kinetic energy is adequately and qualitatively predicted, although it is somewhat overestimated. It can be said that the model qualitatively predicts the influence of the gas flow, the distribution of the flows and the level of slag on the distribution of velocities and turbulence.
2. The predicted slag eye shows a good agreement with the experimental results with slag eye area as a percentage of the total surface. However, due to the interphase interaction, the slag eye from differentiated gas injection is not captured completely by the model.
3. The numerical model does not fully predict the effect of differentiated gas injection, since the drag model used does not exactly simulate the interaction between both recirculation zones, hence predicting a smaller area of low-velocity zones.
4. There is a deviation in predicted mixing time from experimental mixing time for both equal and differentiated gas injection, which becomes significant at a high gas flow rate and a high slag thickness.

The anomaly in the observations is attributed to the insufficient capture of the hydrodynamic behavior near the water–oil (steel–slag) interface, probably caused by the emulsification of slag, which might also influence changes in the gas plumes, linked to the change in the mode of gas injection. This is coupled with the difficulty of accurately predicting the complex interaction between the gas plumes, especially in differentiated injection, which causes a variation in the predicted values of mixing time and slag eye area, although the effect of the variables (gas flow rate, slag height and injection mode) can be studied with an in-depth understanding of the mathematical model presented.

Author Contributions: Conceptualization, M.A.R.-A. and A.D.; methodology, L.E.J.-P.; software, L.E.J.-P. and A.D.; validation, L.E.J.-P. and C.G.-R.; formal analysis, C.G.-R.; investigation, L.E.J.-P. and C.G.-R.; resources, L.E.J.-P.; writing—original draft preparation, L.E.J.-P.; writing—review and editing, A.D. and C.G.-R.; supervision, A.D.; funding acquisition, M.A.R.-A. All authors have read and agreed to the published version of the manuscript.

Funding: This research was funded by [DGAPA UNAM] grant number [PAPIIT IN115619] and L.E.J.-P. received PhD scholarship by [CONACYT] grant number [CVU 624968].

Conflicts of Interest: The authors declare no conflict of interest.

References

1. Ghosh, A. *Secondary Steelmaking: Principles and Applications*; CRC Press LLC: Boca Raton, FL, USA, 2001.
2. Du, S. Improving process design in steelmaking. In *Fundamentals of Metallurgy*; Elsevier Inc.: Amsterdam, The Netherlands, 2005; pp. 369–398.
3. Mazumdar, D.; Evans, J.W. Macroscopic Models for Gas Stirred Ladles. *ISIJ Int.* **2004**, *44*, 447–461. [CrossRef]
4. Amaro-Villeda, A.M.; Ramírez-Argáez, M.A.; Conejo, A.N. Effect of Slag Properties on Mixing Phenomena in Gas-stirred Ladles by Physical Modeling. *ISIJ Int.* **2014**, *54*, 1–8. [CrossRef]

5.   Hoang, Q.N.; Ramírez-Argáez, M.A.; Conejo, A.N.; Blanpain, B.; Dutta, A. Numerical Modeling of Liquid–Liquid Mass Transfer and the Influence of Mixing in Gas-Stirred Ladles. *JOM* **2018**, *70*, 2109–2118. [CrossRef]

6.   Asai, S.; Okamoto, T.; He, J.-C.; Muchi, I. Mixing Time of Refining Vessels Stirred by Gas Injection. *Trans. Iron Steel Inst. Jpn.* **1983**, *23*, 43–50. [CrossRef]

7.   Sano, M.; Mori, K. Fluid flow and mixing characteristics in a gas-stirred molten metal bath. *Trans. Iron Steel Inst. Jpn.* **1983**, *23*, 169–175. [CrossRef]

8.   Joo, S.; Guthrie, R.I.L. Modeling flows and mixing in steelmaking ladles designed for single- and dual-plug bubbling operations. *MTB* **1992**, *23*, 765–778. [CrossRef]

9.   Krishnapisharody, K.; Irons, G.A. An Analysis of Recirculatory Flow in Gas-Stirred Ladles. *Steel Res. Int.* **2010**, *81*, 880–885. [CrossRef]

10. Khajavi, L.T.; Barati, M. Liquid Mixing in Thick-Slag-Covered Metallurgical Baths—Blending of Bath. *Met. Mater. Trans. B* **2010**, *41*, 86–93. [CrossRef]

11. Chattopadhyay, K.; Sengupta, A.; Ajmani, S.K.; Lenka, S.N.; Singh, V. Optimisation of dual purging location for better mixing in ladle: A water model study. *Ironmak. Steelmak.* **2009**, *36*, 537–542. [CrossRef]

12. Liu, H.; Qi, Z.; Xu, M. Numerical Simulation of Fluid Flow and Interfacial Behavior in Three-phase Argon-Stirred Ladles with One Plug and Dual Plugs. *Steel Res. Int.* **2011**, *82*, 440–458. [CrossRef]

13. Haiyan, T.; Xiaochen, G.; Guanghui, W.; Yong, W. Effect of Gas Blown Modes on Mixing Phenomena in a Bottom Stirring Ladle with Dual Plugs. *ISIJ Int.* **2016**, *56*, 2161–2170. [CrossRef]

14. Tang, H.; Liu, J.; Zhang, S.; Guo, X.; Zhang, J. A novel dual plugs gas blowing mode for efficient ladle metallurgy. *Ironmak. Steelmak.* **2019**, *46*, 405–415. [CrossRef]

15. Jardón-Pérez, L.E.; González-Morales, D.R.; Trápaga, G.; González-Rivera, C.; Ramírez-Argáez, M.A. Effect of Differentiated Injection Ratio, Gas Flow Rate, and Slag Thickness on Mixing Time and Open Eye Area in Gas-Stirred Ladle Assisted by Physical Modeling. *Metals* **2019**, *9*, 555. [CrossRef]

16. Mazumdar, D.; Dhandapani, P.; Sarvanakumar, R. Modeling and Optimisation of Gas Stirred Ladle Systems. *ISIJ Int.* **2017**, *57*, 286–295. [CrossRef]

17. Liu, Y.; Ersson, M.; Liu, H.; Jönsson, P.G.; Gan, Y. A Review of Physical and Numerical Approaches for the Study of Gas Stirring in Ladle Metallurgy. *Met. Mater. Trans. B* **2018**, *50*, 555–577. [CrossRef]

18. Li, B.; Yin, H.; Zhou, C.Q.; Tsukihashi, F. Modeling of Three-phase Flows and Behavior of Slag/Steel Interface in an Argon Gas Stirred Ladle. *ISIJ Int.* **2008**, *48*, 1704–1711. [CrossRef]

19. Conejo, A.N.; Mishra, R.; Mazumdar, D. Effects of Nozzle Radial Position, Separation Angle, and Gas Flow Partitioning on the Mixing, Eye Area, and Wall Shear Stress in Ladles Fitted with Dual Plugs. *Met. Mater. Trans. B* **2019**, *50*, 1490–1502. [CrossRef]

20. Villela-Aguilar, J.D.J.; Ramos-Banderas, J.Á.; Hernández-Bocanegra, C.A.; Urióstegui-Hernández, A.; Solorio-Díaz, G. Optimization of the Mixing Time Using Asymmetrical Arrays in Both Gas Flow and Injection Positions in a Dual-plug Ladle. *ISIJ Int.* **2020**, *60*, 1172–1178. [CrossRef]

21. Shih, T.-H.; Liou, W.W.; Shabbir, A.; Yang, Z.; Zhu, J. A new k-ε eddy viscosity model for high reynolds number turbulent flows. *Comput. Fluids* **1995**, *24*, 227–238. [CrossRef]

22. Troshko, A.A.; Hassan, Y.A. A two-equation turbulence model of turbulent bubbly flows. *Int. J. Multiph. Flow* **2001**, *27*, 1965–2000. [CrossRef]

23. Krishnakumar, K.; Ballal, N.B.; Sinha, P.K.; Sardar, M.K.; Jha, K.N. Water Model Experiments on Mixing Phenomena in a VOD Ladle. *ISIJ Int.* **1999**, *39*, 419–425. [CrossRef]

24. González-Bernal, R.; Solorio-Diaz, G.; Ramos-Banderas, A.; Torres-Alonso, E.; Hernández-Bocanegra, C.A.; Zenit, R. Effect of the Fluid-Dynamic Structure on the Mixing Time of a Ladle Furnace. *Steel Res. Int.* **2018**, *89*, 1700281. [CrossRef]

25. Lou, W.; Zhu, M. Numerical Simulation of Slag-metal Reactions and Desulfurization Efficiency in Gas-stirred Ladles with Different Thermodynamics and Kinetics. *ISIJ Int.* **2015**, *55*, 961–969. [CrossRef]

26. Zhu, M.-Y.; Inomoto, T.; Sawada, I.; Hsiao, T.-C. Fluid Flow and Mixing Phenomena in the Ladle Stirred by Argon through Multi-Tuyere. *ISIJ Int.* **1995**, *35*, 472–479. [CrossRef]

27. Ramasetti, E.; Visuri, V.-V.; Sulasalmi, P.; Fabritius, T.; Saatio, T.; Li, M.; Shao, L. Numerical Modeling of Open-Eye Formation and Mixing Time in Argon Stirred Industrial Ladle. *Metals* **2019**, *9*, 829. [CrossRef]

28. Nunes, R.P.; Pereira, J.A.M.; Vilela, A.C.F.; Laan, F.T.V. Visualisation and analysis of the fluid flow structure inside an elliptical steelmaking ladle through image processing techniques. *J. Eng. Sci. Technol.* **2007**, *2*, 139–150.

29. Jardón-Pérez, L.E.; Amaro-Villeda, A.; González-Rivera, C.; Trápaga, G.; Conejo, A.N.; Ramírez-Argáez, M.A. Introducing the Planar Laser-Induced Fluorescence Technique (PLIF) to Measure Mixing Time in Gas-Stirred Ladles. *Met. Mater. Trans. B* **2019**, *50*, 2121–2133. [CrossRef]

30. Ascanio, G. Mixing time in stirred vessels: A review of experimental techniques. *Chin. J. Chem. Eng.* **2015**, *23*, 1065–1076. [CrossRef]

Article

Physical Simulation of Molten Steel Homogenization and Slag Entrapment in Argon Blown Ladle

Fu Yang, Yan Jin *, Chengyi Zhu, Xiaosen Dong, Peng Lin, Changgui Cheng, Yang Li, Lin Sun, Jianhui Pan and Qiang Cai

The State Key Laboratory of Refractories and Metallurgy, Wuhan University of Science and Technology, 947 Heping Avenue, Qingshan District, Wuhan 430081, China
* Correspondence: jinyan@wust.edu.cn; Tel.: +86-156-9718-0966

Received: 20 May 2019; Accepted: 27 June 2019; Published: 24 July 2019

Abstract: Argon stirring is one of the most widely used metallurgical methods in the secondary refining process as it is economical and easy, and also an important refining method in clean steel production. Aiming at the issue of poor homogeneity of composition and temperature of a bottom argon blowing ladle molten steel in a Chinese steel mill, a 1:5 water model for 110 t ladle was established, and the mixing time and interface slag entrainment under the different conditions of injection modes, flow rates and top slag thicknesses were investigated. The flow dynamics of argon plume in steel ladle was also discussed. The results show that, as the bottom blowing argon flow rate increases, the mixing time of ladle decreases; the depth of slag entrapment increases with the argon flow rate and slag thickness; the area of slag eyes decreases with the decrease of the argon flow rate and increase of slag thickness. The optimum argon flow rate is between 36–42 m^3/h, and the double porous plugs injection mode should be adopted at this time.

Keywords: secondary refining; water model; mixing time; slag entrapment

1. Introduction

The major tasks in the production of clean quality steel include the removing of inclusions and unwanted impurities, the secondary refining process is the important step of cleansing molten steel, and argon bottom blowing is one of the most popular ladle metallurgical methods in all types of ladle metallurgy processes as it is economical and easy [1]. During the argon bottom blowing, the molten steel homogenization and entrapment of the top slag have a great effect on the result of the ladle metallurgy. The water model [2–6] and mathematical models [7–9] are often used to simulate metal bath in the ladle during the argon bottom blowing, as the flow field in the ladle can be observed in the laboratory without the interfering of high temperature molten steel, splashing and dust in the actual teeming ladle. Some researchers found that eccentric bottom blowing is in favor of bath mixing in the ladle [7,10]. Some investigators also found that the position of porous plugs, gas flow rate and the size of ladle have an abundant influence on bath mixing in the ladle [4,5,11–17]. It was discovered that the top slag layer could consume some part of kinetic energy of flow and enlarge the mixing time of the ladle [18–22]. Some researchers studied the flow field and inclusion removal in the ladle during argon blowing by studying the bubble motion in the process of argon stirring [23,24]. The results of Luis E. Jardón-Pérez's research show that the ladle must be operated using a differentiated flow ratio for optimal performance [25]. With the help of the water model the argon bottom-blowing was improved dramatically. However, as the shortening of mixing time and decreasing top slag entrapment are a pair of contradictions during the argon bottom-blowing, the more reasonable and general compromise method should be studied based on the study of the detailed data of flow field in the metal bath of the ladle with the water model.

In this paper, based on the experiment of the water model, the bottom-blowing ladle of a Chinese steel mill was studied, and the effect of the layout of porous plug, argon flow rate and properties of top slag to ladle stirring efficiency, level fluctuation and slag entrapment. Based on the results, the operation of bottom-blowing of the ladle was improved.

2. Water Model

In order to uniform the composition and temperature of metal bath in the ladle, a 1:5 downscale water model was established, with the geometrical similarity of the actual ladle in a Chinese steel mill. Water and air were chosen to simulate molten steel and argon, the dimension data of the water model and prototype were shown in Table 1, and a schematic diagram of the prototype was in Figure 1.

Table 1. The dimension data of the water model and prototype.

Parameters	Prototype	Water Model
Top diameter of ladle/mm	3034	606.8
Bottom diameter of ladle/mm	3000	600
Ladle height/mm	3950	790
Blowing mode	Bottom blowing through porous plug	Bottom blowing through porous plug
Bottom blowing gas	Argon	Air
The density of blowing gas/kg/m^3	1.78	1.29
The density of liquid/kg/m^3	7020	1000
The temperature of liquid/K	1853	293

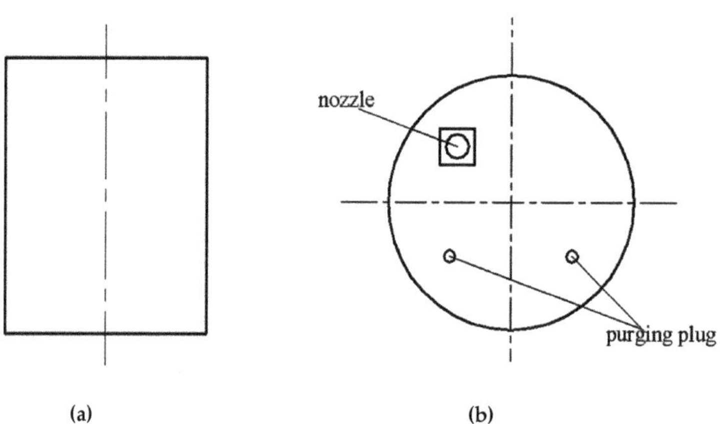

Figure 1. Schematic diagram of the ladle prototype; (**a**) vertical section; (**b**) porous plug arrangement.

The major forces affecting the flow of molten steel in the ladle include float force, viscous force and gravity. According to similarity rules, modified Freud number can characterize the kinetic similarity of the argon blowing system in the ladle with bottom blowing of argon. In our study work, the water model should have the same modified Freud number as the prototype, as Equation (1):

$$(Fr')_m = (Fr')_p \tag{1}$$

That is,

$$\frac{\rho_{air} \cdot u^2_{water}}{(\rho_{water} - \rho_{air}) \cdot g \cdot H_m} = \frac{\rho_{Ar} \cdot u^2_{steel}}{(\rho_{steel} - \rho_{Ar}) \cdot g \cdot H_p} \tag{2}$$

where, ρ_{air}, ρ_{water}, ρ_{Ar}, ρ_{steel} are the densities of air, water, argon and molten steel respectively, kg/m^3; g is the acceleration of gravity, m/s^2; u_{water}, u_{steel} are the characteristic velocities of air and argon respectively, m/s; H is the height of steel bath in the ladle, m.

The characteristic velocity u can be expressed with Equation (3):

$$u = \frac{4Q}{\pi \cdot d^2}$$

(3)

where, Q is the gas flow rate, m^3/h; d is the equivalent diameter of the porous plug, m.

Based on the data of Table 1, Equations (2) and (3), the relation between gas flow rates in the water model and in the actual teeming ladle, i.e., Q_m and Q_p, was derived as Equation (4).

$$Q_m = 0.00794Q_p$$

(4)

According to the range of flow rate of argon blown in the prototype ladle, that is 12–50 m^3/h, the air flow rates were calculated from Equation (4) and listed in Table 2.

Table 2. The flow rate of bottom gas in the water model and prototype ladle.

The flow rate in prototype/m^3/h	12	18	24	30	36	42	48
The flow rate in water model/m^3/h	0.095	0.143	0.191	0.238	0.286	0.333	0.381

As the flow behaviour of molten steel-slag was influenced by interfacial tension of molten steel and slag, the weber numbers of the model should be equivalent to that of the prototype to insure the kinetic similarity at the interface between the molten steel and slag, Equation (5).

$$We_m = We_p$$

(5)

That is,

$$\frac{\rho_{water} \cdot u_{water}^2}{\left[g \cdot \sigma_{water-oil} \cdot (\rho_{water} - \rho_{air})\right]^{1/2}} = \frac{\rho_{steel} \cdot u_{steel}^2}{\left[g \cdot \sigma_{steel-slag} \cdot (\rho_{steel} - \rho_{slag})\right]^{1/2}}$$

(6)

where, $\sigma_{water-oil}$ is the interfacial tension between the water and oil, N/m; $\sigma_{seel-slag}$ is the interfacial tension between the steel and slag, N/m.

In the water model experiment, aviation kerosene and vacuum pump oil were mixed in a certain proportion to obtain a mixture oil with the same kinematic viscosity of the top slag.

The slag layer thickness of the prototype ladle is 60–100 mm, and the oil layer thickness (OLT) in the water model experiment is 12–20 mm according to the similarity ratio of 1:5. Five oil layer thicknesses were selected in the experiment, as shown in Table 3, to study the effect of the slag layer thickness to level fluctuation and slag entrapment of steel bath in the ladle.

Table 3. The thicknesses of the top oil layer.

Item	1	2	3	4	5
Oil layer thickness/mm	12	14	16	18	20

There were two methods of eccentric bottom blowing in the prototype ladle: One is a single porous plug with eccentric distance of 0.6 R, in which the eccentric distance is the distance between the centers of the porous plug and ladle; the other is a double porous plug with an intersection angle of 100° and eccentric distance of 0.6 R.

The schematic diagram of the water model experiment setup was shown in Figure 2. In the experiments, the mixing time of the model was measured through the stimulus-response method with a tracer of KCl solution, flow field, level fluctuation and slag entrapment of steel bath in the model ladle was recorded by high speed digital camera.

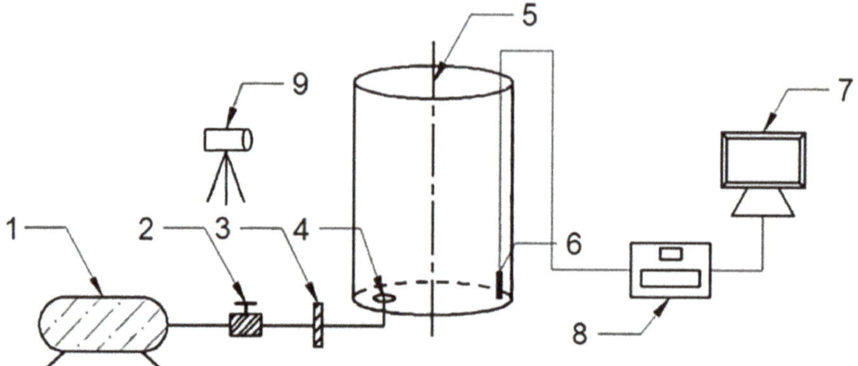

Figure 2. Water model experiment setup: (**1**) Air compressor; (**2**) pressure gauge; (**3**) air flow rate controller; (**4**) porous plug; (**5**) model ladle; (**6**) conductivity probe; (**7**) computer; (**8**) conductivity meter; (**9**) high speed digital camera.

3. Results and Discussion

3.1. Mixing Time

In the water model experiment, seven air flow rates, that is 0.095, 0.143, 0.191, 0.238, 0.286, 0.333, 0.381 m^3/h, were used to bottom-blow into the model ladle with/without an oil layer covered through single or double porous plugs.

The influences of the bottom air flow rate, slag layer and the number of porous plugs to the mixing time were shown in Figure 3.

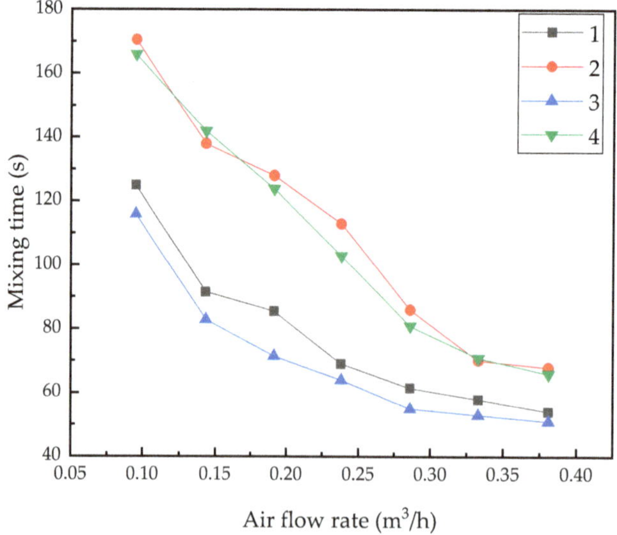

Figure 3. The relation between mixing time and air flow rate in the ladle with single porous plug and double porous plugs: (**1**) Single porous plug without oil layer; (**2**) single porous plug with 16 mm OLT; (**3**) double porous plugs without oil layer; (**4**) double porous plugs with 16 mm OLT.

From Figure 3, it was found that as the increase of the bottom air flow rate, the mixing time of the metal bath in ladle decreased. When the bottom blowing air flow rate was equal to 0.095 m^3/h (corresponding 12 m^3/h in prototype), the mixing time of the steel bath was relatively long, as stirring power produced by the dispersing small bubbles from porous plugs was too small and the circulating flow rate in steel bath was weak.

As the bottom blowing air flow rate was increased from 0.143 m^3/h to 0.286 m^3/h (corresponding from 18 m^3/h to 36 m^3/h in the prototype), the mixing time was reduced abundantly. When the bottom air flow rate was above 0.143 m^3/h, the bubble group in water model was transferred from dispersing small bubbles to spherical bubbles or coronal bubbles group. The stirring energy by the bubble group increased abundantly, so was the circulating flow in the steel bath of the ladle, which decreased the mixing time markedly.

When the bottom blowing flow rate increased above 0.286 m^3/h, the mixing time of the steel bath increased slowly with the bottom blowing flow rate. The reason was that when the flow rate exceeds 0.286 m^3/h, the diameter of the plume caused by bubble groups did not increase further, more bubbles were blown into the plume, bubble coalescence and breaking were more frequent, and more energy was exhausted in bubble coalescence and breaking, instead of driving circulate flow in the bath. At the same time, there were more energy consumed by the surface rise and splashing, which were caused by the escaping of a large number of bubbles. Therefore, when the bottom blowing flow rate increased above 0.286 m^3/h (i.e., 36 m^3/h in prototype), the increase of the bottom blowing gas flow rate could not improve the mixing of the bath in the ladle, and there was an obvious inflection point on the mixing time curve.

The slag layer could influence the mixing time, Figure 3. It was shown that at the same bottom blowing flow rate, the mixing time without the oil layer was obviously shorter than the mixing time with OLT of 16 mm. The reason was that the horizontal flow at the surface was obstructed by the slag layer at the bath top, so was the circulating flow in the bath. It can also be seen from the figure that the air flow rate at mixing time inflection without the oil layer in the ladle was 0.286 m^3/h, and the air flow rate at mixing time inflection point with the oil layer in the ladle was 0.333 m^3/h.

As there were two plumes in the ladle with double porous plugs, the intersection area of bubble columns doubled, and the mixing of the bath improved obviously. From Figure 3, it was found that the mixing time of the bath in the ladle with double porous plugs shortened abundantly.

The relation between the mixing time and bottom blowing flow rate in the ladle with double porous plugs was shown in Figure 4. It was found that with the increase of the bottom gas flow rate, the mixing time of the bath in the ladle was decreased. When the flow rate is above 0.333 m^3/h (i.e., 42 m^3/h in prototype), the mixing time was reduced slowly as the increase of the flow rate, and the trend was similar to the ladle with a single porous plug. The slag layer obstructed the mixing in the ladle with double porous plugs, as the work by viscous force at the surface of the bath in the ladle consumed the kinetic energy of plumes driven by bubbles blown from double porous plugs. The thickness of the slag layer had a large influence to the mixing time of the bath in the ladle, as the mixing time of the bath in the ladle decreased with the increase of thickness of the slag layer.

In summary, for the actual operation of the prototype ladle, the best option of the bottom gas flow rate was 36–42 m^3/h for the ladle. When the bottom flow rate in the actual ladle increased above that range, the mixing time could not be reduced effectively.

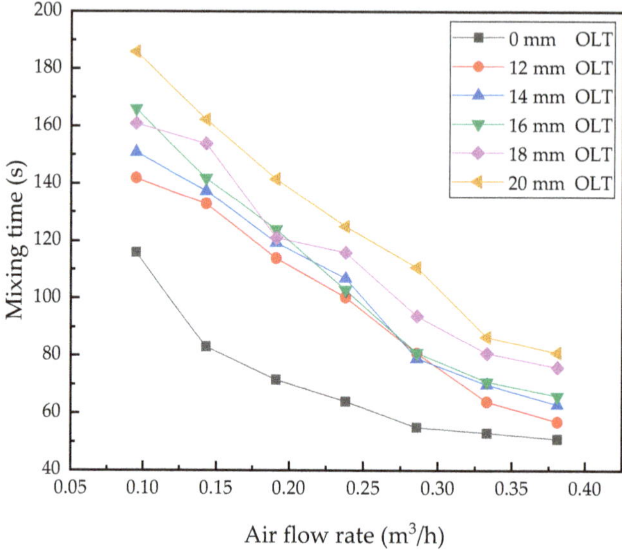

Figure 4. The relation of mixing time and air flow rate in the ladle with double porous plugs.

3.2. The Entrapment of Slag in Ladle

For the entrapment of the top slag in the ladle, the influence of the bottom gas flow rate, number of porous plugs and thickness of the slag layer to the top slag entrapment in the ladle was studied with an image processing method in the water model.

3.2.1. The Entrapment of Slag in Ladle with Single Porous Plug

In the water model, the entrapment of the top slag during bottom blowing was videoed to analyse the influence of the bottom gas flow rate to entrapment, Figure 5, and in the experiment the thickness of oil (simulating the slag layer) was 14 mm. It was found that as the bottom blowing gas flow rate was less than 0.143 m^3/h, the fluctuation at the interface between the oil and water was gentle (Figure 5a,b), the escape of bubbles caused the little disturb at the interface between the oil and water and the horizontal flow at the interface driven by the upper flow around the bubble column caused the oil layer thickening at the area around the bubble escaping region. As the bottom flow rate was increased above 0.143 m^3/h, the escape of bubbles caused the obvious disturb at the interface between the oil and water, the horizontal flow at the interface caused the oil layer obviously thickening at the area around the bubble escaping region, as shown in Figure 5c–f, the oil bump was formed at that area, the shear of horizontal flow resulted in the small droplet divided from the oil bump and the entrapment was formed, most of the oil droplets were soon floated up to the top oil layer. When the bottom gas flow rate was above 0.333 m^3/h (Figure 5g,h), the entrapment worsened, most of the oil droplets were not floated up to the top oil layer, instead, they were dragged into the deep region in the water model by the downward flow.

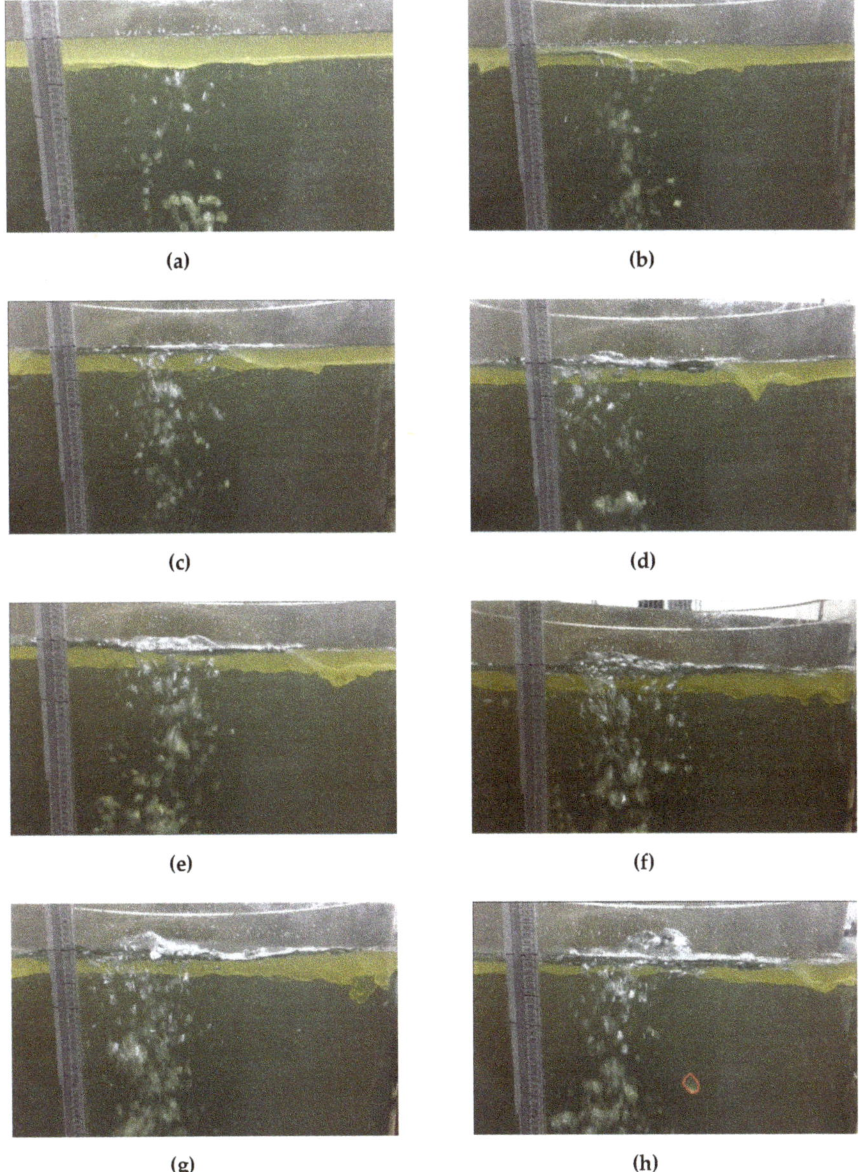

Figure 5. The influence of the air flow rate to the entrapment in the water model simulating the ladle with a single porous plug and OLT of 14 mm: (**a**) Q = 0.0475 m³/h; (**b**) Q = 0.095 m³/h; (**c**) Q = 0.143 m³/h; (**d**) Q = 0.191 m³/h; (**e**) Q = 0.238 m³/h; (**f**) Q = 0.286 m³/h; (**g**) Q = 0.333 m³/h; (**h**) Q = 0.381 m³/h.

The area without the slag covering, which was caused by the bubbles-escaping at the top of the bath, was called the slag eye. The area of the slag eye in the water model with a single porous plug at different bottom gas flow rates and slag thicknesses was summarized in Table 4, which was measured from the digital image of the top of the water model during bottom blowing. From Table 4, it was found that the area of the slag eye increased as the increase of the bottom gas flow rate. The reason was that as the bottom gas flow rate increased, the rise velocity of the plume increased and then the horizontal velocity at the liquid-oil interface was also raised, the increase of the horizontal velocity raised the repulsive force at the interface between the water and oil and the area of the slag eye increased. On the other hand, as the raise of the top oil thickness, the work to repulsive the oil at the top of the water model, which overcame the interfacial tension and viscous force, increased and the area of the slag eye decreased at the same bottom gas flow rate.

Table 4. The areas of the slag eye at different air flow rate and oil thickness in the water model with a single porous plug.

Bottom Gas Flow Rate, m^3/h	Area (%) of Slag Eye to Top Area of Water Model				
	12 mm OLT	**14 mm OLT**	**16 mm OLT**	**18 mm OLT**	**20 mm OLT**
0.0475	6.79	6.30	6.58	2.83	2.88
0.095	15.06	10.40	11.24	8.60	5.62
0.143	17.05	14.87	12.80	11.06	8.46
0.191	19.63	18.16	16.92	13.43	12.77
0.238	20.22	19.90	17.51	14.84	13.87
0.286	25.37	20.46	18.39	14.32	15.38
0.333	26.44	22.96	19.92	17.60	17.67
0.381	29.91	23.14	24.60	20.25	18.26

Figure 6 showed the relation between entrapping the depth of the top oil in the water model with a single porous plug, bottom flow rate and oil thickness. It was shown that at the same bottom flow rate the entrapping depth increased as the increase of oil thickness, and at the same oil thickness the entrapping depth increased as the bottom flow rate, but when the bottom flow rate was above a certain level, the entrapping depth would not increase as the increase of the flow rate. The reason was that as the bottom flow rate increased, more of the oil layer at the top of the water model was repulsed from the bubble escaping region of the plume, the horizontal velocity of water increased, the velocity of the downward flow around the plume also increased, the bump of the oil layer increased, Figure 5, the shear forced on the bump of the oil layer increased, and all these factors caused oil entrapping in the water model raising. From Figure 6, the critical bottom flow rates for oil entrapment in the water model were in the range of 0.095–0.143 m^3/h (18–24 m^3/h for prototype ladle).

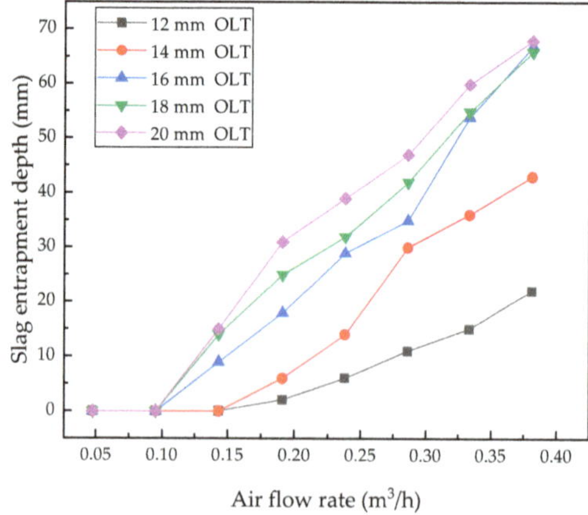

Figure 6. The influence of air flow rate to entrapping depth in the water model with a single porous plug.

3.2.2. The Entrapment of Slag in Ladle with Double Porous Plugs

The entrapment of the slag in the ladle with double porous plugs was shown in Figure 7. For the water model with double porous plugs, the region between the porous plugs and the adjoining wall of the water model was signed as Region A, and the region between the two porous plugs was signed as Region B, as shown in Figure 8a. At Region A, the flow of water was similar to that of the water model with a single porous plug. However at Region B, the oil layer was pushed by the flows of counter directions from the two porous plugs, the bump of the oil layer was higher than that in the water model with a single porous plug, and the shear from the horizontal flow was stronger intensively, as shown in Figure 7. The plumes from the two porous plugs caused the vortex at Region B. So, there was oil entrapped into the vortex, which was called as vortex entrapment.

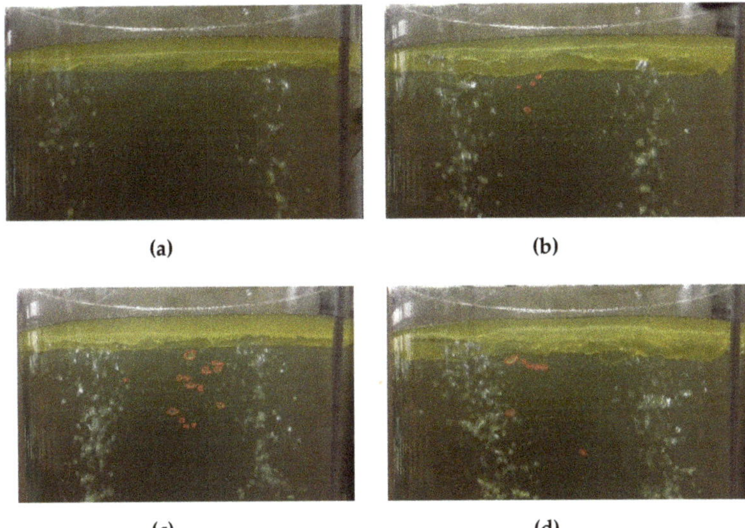

Figure 7. Entrapment in the water model at a different bottom flow rate (16 mm OLT): (**a**) Q = 0.095 m^3/h; (**b**) Q = 0.191 m^3/h; (**c**) Q = 0.286 m^3/h; (**d**) Q = 0.381 m^3/h.

Figure 8. The slag eyes in the water model with double porous plugs (16 mm OLT): (**a**) Q = 0.286 m^3/h, (**b**) Q = 0.381 m^3/h.

For the slag eye, when the bottom flow rate was small, there were two slag eyes formed at the top of the water model. When the bottom flow rate was large, the two slag eyes were merged and formed the goggle type slag eye, Figure 8. When the goggle type slag eye was formed, there was no slag floated at the Region B, Region A was the only region with slag entrapment.

Table 5 showed the areas of slag eyes at different flow rate and oil thickness. With the increase of the bottom flow rate the area of slag eyes increased for the same oil layer thickness, and with the oil layer thickness increasing the area of slag eyes decreased in the water model with double porous plugs, the trend of which was just like that in the water model with a single porous plug. Compared with Table 4, the slag eye is easier to form in the double porous plugs ladle, and the area of the slag eye is usually larger than that in a single porous plug ladle.

Table 5. The areas of the slag eye at a different bottom flow rate and top oil thickness in the water model with a double porous plug.

Bottom Gas Flow Rate, m³/h	Area (%) of Slag Eye to Top Area of Water Model				
	12 mm OLT	14 mm OLT	16 mm OLT	18 mm OLT	20 mm OLT
0.095	13.38	11.55	9.27	7.73	4.67
0.191	20.22	17.95	18.07	13.75	10.11
0.286	26.97	20.67	23.21	19.98	12.59
0.381	30.95	26.28	28.22	20.37	15.67

As the increase of the bottom flow rate, the entrapping depth increased in the water model with double porous plugs, shown in Figure 9, the trend of which was similar to that in the water model with a single porous plug. In the water model with double porous plugs, the critical bottom flow rate causing entrapment was 0.095 m³/h (12 m³/h for prototype ladle). Compared with Figure 6, it was found that the entrapping depth was higher obviously in the water model with double porous plugs, because there was vortex entrapment formed in Region B.

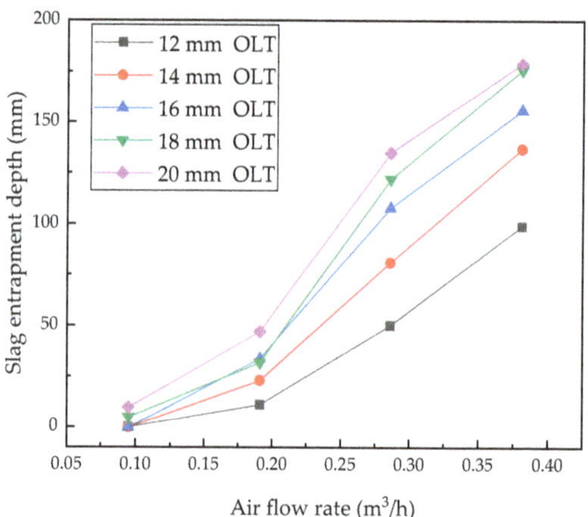

Figure 9. The influence of a double porous plug bottom flow rate to entrapping depth in the water model.

3.3. Comprehensive Analysis of Mixing and Slag Entrapment

The results of mixing time show that the optimum gas flow rate for the prototype ladle is 36–42 m³/h, the ladle with double porous plugs should be selected at the same time. Slag thickness has a significant influence on the entrapping depth of the slag and slag eye area. In actual production, slag entrapment is beneficial to improve the refining effect, therefore, more consideration should be given to the area of the slag eye and bath mixing.

The bubble motion in the water model experiment was shown in Figure 10. Bubble motion is accompanied by the process of coalescence, collapse and re-coalescence. In the process of bubble floating up, the size and shape of the bubble have changed. When the diameter of the bubble exceeds 1 cm, the shape of the bubble changes into a spherical corona. In low viscosity liquids, the rising velocity of the spherical coronal bubble is independent of the properties of liquids and can be calculated by Equation (7) [26].

$$u_b = 1.02\left(\frac{gd_b}{2}\right)^{\frac{1}{2}}$$

(7)

where, u_b is the bubble velocity, m/s; g is the acceleration of gravity, m/s²; d_b is the bubble diameter, m.

(a)   (b)   (c)   (d)   (e)

Figure 10. The bubble motion in the water model: (**a**) Q = 0.0475 m³/h; (**b**) Q = 0.095 m³/h; (**c**) Q = 0.143 m³/h; (**d**) Q = 0.191 m³/h; (**e**) Q = 0.238 m³/h.

In order to apply the research results to more ladles, a dimensionless treatment was carried out for the slag layer thickness, gas flow rate and mixing time. Assuming that the diameter of the bubble is 1 cm, the product of the bubble velocity and area of the gas ports is taken as the characteristic flow rate.

$$Q_b = S \cdot u_b$$

(8)

$$S = \pi\left(r_1^2 + r_2^2 + \cdots + r_n^2\right)$$

(9)

where, Q_b is the characteristic flow rate, m³/s; $r_1, r_2, \ldots, r_n$ are the radius of gas ports, m; S is the area of gas ports of porous plugs, m². The radius of gas port in the water model experiment is 12 mm.

The dimensionless gas flow rate was treated according to Equation (10), and the dimensionless slag layer thickness was treated according to Equation (11).

$$Q^* = \frac{Q}{3600Q_b}$$

(10)

$$h^* = \frac{h}{h_l}$$

(11)

where, V^* is the dimensionless gas flow rate; V is the gas flow rate, m³/h; h^* is the dimensionless thickness of the slag layer; h is the thickness of the oil layer, m; h_l is the depth of the molten bath, m. In the water model, the water depth is 0.643 m.

Assuming that the shape of the bubble is a spherical corona and does not deform in the process of bubble flotation, the residence time t_b of the bubble in the water model can be calculated. Taking residence time t_b as the characteristic time, the dimensionless mixing time can be calculated by Equation (12).

$$t_b = \frac{h_l}{u_b} \tag{12}$$

$$t^* = \frac{t}{t_b}$$

where, t_b is the characteristic time, s; h_l is the depth of the water, m; t^* is the dimensionless time; t is the mixing time, s.

The relation between the dimensionless mixing time and dimensionless gas flow rate in the ladle with double porous plugs was shown in Figure 11, the 0, 0.0187, 0.0218, 0.0249, 0.028 and 0.0311 are dimensionless thickness in Figure 11 correspond to without oil and with oil layer of 12, 14, 16, 18, 20 mm. The trend of mixing time is consistent with Figure 4. In the water model, the optimum dimensionless flow rate is 0.157–0.183 for the ladle with double porous plugs.

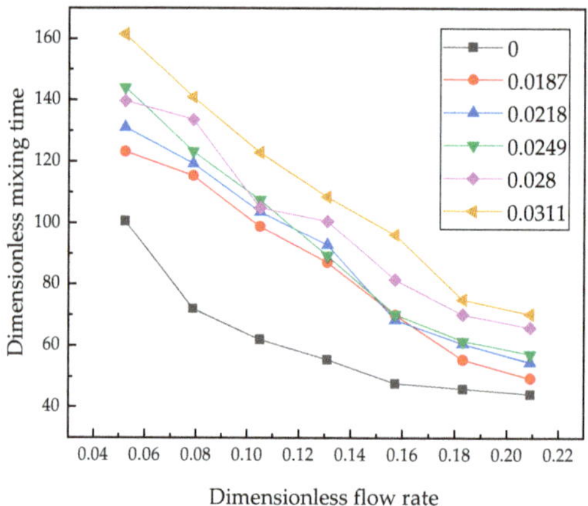

Figure 11. The relation of dimensionless mixing time and dimensionless flow rate in the ladle with double porous plugs.

Using multiple linear regression, the relationship between dimensionless mixing time, dimensionless flow rate and dimensionless oil layer thickness was fitted by Equation (13). The coefficient of determination (R square) of fitting Equation (13) is equal to 0.942, which indicates that the equation has a high fitting degree and may be used to calculate the mixing time of the ladle with double porous plugs.

$$t^* = 125.3601 + 1476.75327 \cdot h^* - 502.76857 \cdot Q^* \tag{13}$$

The area of the slag eye in the water model with a single porous plug at different dimensionless flow rates and dimensionless oil thicknesses was shown in Figure 12. It can be seen from the figure that when the dimensionless flow rate increases to 0.366 (corresponding to the prototype ladle flow rate of 42 m³/h), further increasing flow rate has little effect on the slag eye area, the dimensionless flow rate corresponding to the prototype ladle is 37.595. Therefore, considering the mixing time and slag entrainment, the optimized injection mode for the prototype ladle is the double porous plug with a flow rate of 36–42 m³/h.

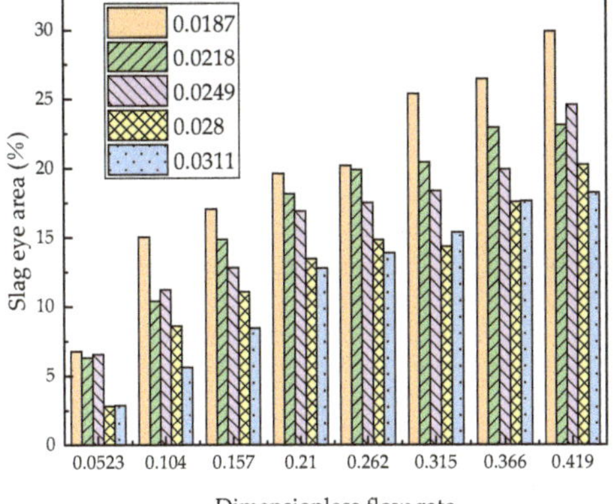

Figure 12. The areas of slag eye at different flow rates and oil thickness in water model with single porous plug.

Using multiple linear regression, the relationship between the area of the slag eye, dimensionless flow rate and dimensionless oil layer thickness was fitted by Equation (14). The coefficient of determination (R-square) of fitting Equation (14) is equal to 0.93188, which indicates that the equation has a high fitting degree and may be used to calculate the area of the slag eye of a single porous plug ladle.

$$S_\% = 21.30098 - 660.60213 \cdot h^* + 45.37523 \cdot Q^* \tag{14}$$

where, $S_\%$ is the percentage of the slag eye area to the molten bath surface area.

Equations (13) and (14) use dimensionless experimental data, and the determinant coefficient shows that equations have a high fitting degree, so for the general ladles, Equations (13) and (14) can be used to calculate the mixing time and bare steel area by flow rate and slag layer thickness. However, in the process of dimensionless data processing, the characteristic flow rate is a fixed value, geometric similarity ratio has an effect on dimensionless flow rate, so the equations can not be directly used in the ladle. Therefore, when using Equations (13) and (14) in generic ladles, the model flow rate calculated by the geometric similarity ratio of 1:5 can be used to calculate the mixing time and slag hole area.

4. Conclusions

Through the establishment of the water model, the influence of the bottom blowing flow rate on the mixing time and slag entrapment were studied. Through dimensionless treatment and multivariate linear regression, the equations which may be used to calculate the mixing time and slag eye area of the ladle are obtained and the conclusions are as follows:

(1) The bath mixing in the ladle is affected by the number of porous plugs, flow rate and slag layer. Under the same blowing flow rate, the mixing time of double porous plugs is shorter than that of a single porous plug. The mixing time of the two methods eccentric blowing is basically the same, and the mixing time decreases with the increase of the blowing flow rate, and increases with the increase of the slag layer thickness. There is an inflection point in the mixing time curve, the flow rate at the point is 0.333 m³/h (corresponding 42 m³/h in prototype), the mixing time before the inflection point changes significantly, but after the inflection point, the mixing time changes slowly. The mixing time of the ladle without the slag layer is significantly shorter than that with the slag layer.

(2) The critical bottom flow rates for oil entrapment in the water model were in the range of 0.095–0.143 m^3/h (18–24 m^3/h for the prototype ladle) for the ladle with a single porous plug, and the critical bottom flow rate causing entrapment was 0.095 m^3/h (12 m^3/h for the prototype ladle) for the ladle with a double porous plug, the double porous plug is easier to the entrapped slag.

(3) The entrapped slag depth increases with the argon flow rate and slag thickness; the area of slag eyes increases with the argon flow rate and decreases with slag thickness.

(4) Considering the mixing time and slag entrainment, the optimized injection mode for the prototype ladle is a double porous plug with a flow rate of 36–42 m^3/h.

Author Contributions: Conceptualization, Y.J., X.D.; Methodology, Y.J., F.Y.; Software, X.D., P.L.; Validation, P.L., F.Y. and X.D.; Formal analysis, Y.J., F.Y.; Investigation, P.L.; Resources, C.Z.; Data curation, L.S., J.P., and Q.C.; Writing—original draft preparation, F.Y.; Writing—review and editing, F.Y., Y.J.; Visualization, X.D.; Supervision, Y.L.; Project administration, Y.J.; Funding acquisition, C.Z., C.C.

Funding: This research was funded by the National Nature Science Foundation of China (NSFC), grant number 51674180 and grant number 51474163.

Acknowledgments: The technical assistance in the water model experiment provided by Yue Yu is greatly acknowledged.

Conflicts of Interest: The authors declare no conflict of interest.

References

1. Zhu, M.Y.; Lou, W.T.; Wang, W.L. Research Progress of Numerical Simulation in Steelmaking and Continuous Casting Processes. *Acta Metall. Sin.* **2018**, *54*, 131–150.

2. Parra, F.D.M.; Argáez, M.A.R.; Conejo, A.N.; Gonzalez, C. Effect of Both Radial Position and Number of Porous Plugs on Chemical and Thermal Mixing in an Industrial Ladle Involving Two Phase Flow. *ISIJ Int.* **2011**, *51*, 1110–1118. [CrossRef]

3. Villeda, A.M.A.; Argáez, M.A.R.; Conejo, A.N. Effect of Slag Properties on Mixing Phenomena in Gas-stirred Ladles by Physical Modeling. *ISIJ Int.* **2014**, *54*, 1–8. [CrossRef]

4. Zheng, W.; Tu, H.; Li, G.Q.; Shen, X.; Xu, Y.L.; Zhu, C.Y.; Lu, K. Physical simulation of refining process optimization for bottom argon blowing in a 250 t ladle. *J. Univ. Sci. Technol. Beijing* **2011**, *36*, 53–59.

5. Liu, Z.Q.; Li, L.M.; Cao, M.X.; Li, B.K. Water model of mixing time and critical flow rate in a gas-stirred ladle. *J. Mater. Metall.* **2016**, *15*, 176–180.

6. Conejo, A.N.; Kitamura, S.; Maruoka, N. Effects of Top Layer, Nozzle Arrangement, and Gas Flow Rate on Mixing Time in Agitated Ladles by Bottom Gas Injection. *Metall. Mater. Trans. B* **2013**, *44*, 914–923. [CrossRef]

7. Li, L.M.; Liu, Z.Q.; Li, B.K.; Matsuura, H.; Tsukihashi, F. Water Model and CFD-PBM Coupled Model of Gas-Liquid-Slag Three-Phase Flow in Ladle Metallurgy. *ISIJ Int.* **2015**, *55*, 1337–1346. [CrossRef]

8. Geng, D.Q.; Lei, H.; He, J.C. Optimization of mixing time in a ladle with dual plugs. *Int. J. Miner. Metall. Mater.* **2010**, *18*, 709–714. [CrossRef]

9. Zhu, M.Y.; Inomoto, T.; Sawada, I.; Hsiao, T.C. Fluid fow and mixing phenomena in the ladle stirred by argon through multi-tuyere. *ISIJ Int.* **1995**, *35*, 472–479. [CrossRef]

10. Van-Khang, N.; Bao, Y.P.; Wang, M.; Lin, L.; Li, X. Flow characteristics of argon bottom blowing in an ellipse ladle. *J. Univ. Sci. Technol. Beijing* **2014**, *36*, 1–5.

11. Liu, H.P.; Qi, Z.Y.; Xu, M.G. Numerical Simulation of Fluid Flow and Interfacial Behavior in Three-phase Argon-Stirred Ladles with One Plug and Dual Plugs. *Steel Res. Int.* **2011**, *82*, 440–458. [CrossRef]

12. Perez, L.E.J.; Amaro-Villeda, A.; Conejo, A.N.; Gonzalez-Rivera, C.; Ramirez-Argáez, M.A. Optimizing gas stirred ladles by physical modeling and PIV measurements. *Mater. Manuf. Process.* **2017**, *33*, 1–9.

13. Amaro-Villeda, A.M.; Bello, G.J.A.; Ramirez-Argáez, M.A. Experimental Study on Mixing in Gas-Stirred Ladles with and without the Slag Phase through a Water Physical Model. *MRS Online Proc. Libr. Arch.* **2012**, *1373*. [CrossRef]

14. Zhao, L.H.; Ma, W.J.; Wang, M. Physical modeling of argon bottom blowing refining in a 100 t ladle. *J. Univ. Sci. Technol. Beijing* **2014**, *36*, 140–144.

15. Chattopadhyay, K.; SenGupta, A.; Ajmani, S.K.; Lenka, S.N.; Singh, V. Optimisation of dual purging location for better mixing in ladle: A water model study. *Ironmak. Steelmak.* **2009**, *36*, 537–542. [CrossRef]

16. Cho, S.H.; Kim, C.W.; Han, J.W.; You, B.D.; Kim, D.S. Effect of Melt Depth and Nozzle Type on the Mixing Behavior in Bottom-Blown Steelmaking Ladle—A Water Model Approach. *Mater. Sci. Forum* **2006**, *510–511*, 494–497. [CrossRef]

17. Zhan, Z.H.; Li, Q.C.; Yin, S.B.; Zhang, B.Q. Physical simulation of mixing time and critical flow rate of bottom blowing argon in a 135 t LF ladle. *Contin. Cast.* **2018**, *43*, 29–33.

18. Amaro-Villeda, A.M.; Conejo, A.; Ramirez-Argáez, M.A. Effect of Slag on Mixing Time in Gas-Stirred Ladles Assisted with a Physical Model. *MRS Online Proc. Libr.* **2012**, *1485*, 101–106. [CrossRef]

19. Cho, S.H.; Hong, S.H.; Han, J.W.; You, B.D. Effect of Slag Layer on Flow Patterns in a Gas Stirred Ladle. *Mater. Sci. Forum* **2006**, *510–511*, 490–493. [CrossRef]

20. Patil, S.; Kumar, D.S.; Peranandhanathan, M.; Mazumdar, D. Mixing Models for Slag Covered, Argon Stirred Ladles. *ISIJ Int.* **2010**, *50*, 1117–1124. [CrossRef]

21. Ramasetti, E.K.; Visuri, V.V.; Sulasalmi, P.; Mattila, R.; Fabritius, T. Modeling of the Effect of the Gas Flow Rate on the Fluid Flow and Open-Eye Formation in a Water Model of a Steelmaking Ladle. *Steel Res. Int.* **2019**, *90*. [CrossRef]

22. Ramírez-Argáez, M.A.; Dutta, A.; Amaro-Villeda, A.; González-Rivera, C.; Conejo, A.N. A Novel Multiphase Methodology Simulating Three Phase Flows in a Steel Ladle. *Processes* **2019**, *7*, 175. [CrossRef]

23. Gou, D.Z.; Wang, W.X.; Geng, D.Q.; Lei, H. Bubble Coalescence/Breakage and Movement in the Ladle with Argon Blowing. *J. Northeast. Univ. Nat. Sci.* **2018**, *39*, 195–199.

24. Pan, S.M.; Chiang, J.D.; Hwang, W.S. Simulation of Large Bubble/Molten Steel Interaction for Gas-Injected Ladle. *J. Mater. Eng. Perform.* **1999**, *8*, 236–244. [CrossRef]

25. Jardón-Pérez, L.E.; González-Morales, D.R.; Trápaga, G.; González-Rivera, C.; Ramírez-Argáez, M.A. Effect of Differentiated Injection Ratio, Gas Flow Rate, and Slag Thickness on Mixing Time and Open Eye Area in Gas-Stirred Ladle Assisted by Physical Modeling. *Metals* **2019**, *9*. [CrossRef]

26. Xiao, X.G.; Xie, Y.G. *Yejin Fanying Gongchengxue Congshu*, 1st ed.; Metallurgical Industry Press: Beijing, China, 1997; p. 40.

 *processes* 

Article

Bubble Motion and Interfacial Phenomena during Bubbles Crossing Liquid–Liquid Interfaces

Hongliang Zhao [1,2], Jingqi Wang [1], Wanlong Zhang [1], Mingzhuang Xie [1], Fengqin Liu [1,2,*] and Xiaochang Cao [3]

[1] School of Metallurgical and Ecological Engineering, University of Science and Technology Beijing, Beijing 100083, China; zhaohl@ustb.edu.cn (H.Z.); w1719210242@163.com (J.W.); 18811343560@163.com (W.Z.); b20170143@xs.ustb.edu.cn (M.X.)

[2] Beijing Key Laboratory of Green Recycling and Extraction of Metal, Beijing 100083, China

[3] School of Mechanical Engineering, Dongguan University of Technology, Dongguan 523808, China; caoxc@dgut.edu.cn

* Correspondence: liufq@ustb.edu.cn

Received: 11 September 2019; Accepted: 3 October 2019; Published: 10 October 2019

Abstract: In metallurgical and chemical engineering processes, the gas–liquid–liquid multiphase flow phenomenon is often encountered. The movement of bubbles in the liquid, and the influence of bubbles on the liquid–liquid interface, have been the focus of extensive research. In the present work, an air–water–oil system was used to explore the movement of bubbles and the phenomenon that occurs when bubbles pass through an interface with various oil viscosities at various gas flow rates. The results show that bubble movement is greatly influenced by the viscosity of the oil at low gas flow rates. The type of phase entrainment and the jet height was changed when increasing the gas flow rate. The stability of the water–oil interface was enhanced with increasing viscosity of the oil phase.

Keywords: bubble motion; interfacial phenomena; entrainment; moving path

1. Introduction

The gas-injection technique has been widely adopted in the pyrometallurgy smelting processes of ferrous and nonferrous metals and in the recovery of secondary resources, as well as in chemical engineering processes, such as extraction processes. All the above processes generally involve a complex gas–liquid–liquid multiphase system in the vessel. The flow, mixing, transfer, and reaction among the components of the multicomponent fluid play important roles in increasing smelting efficiency and improving product quality. One phenomenon involves the gas bubbles crossing the liquid–liquid interface. Experimentally investigating bubble motion and interfacial phenomena during the blowing–smelting process at high temperatures is difficult [1–3], except in a few simple cases [4,5]. Most research on this subject has been carried out using cold-water model experiments, theoretical analysis, and numerical simulation techniques.

Reiter et al. [6,7] studied the interaction between single bubbles and a liquid–liquid interface system. The bubble motion at the interface (e.g., residence time and velocity), the interfacial phenomena (e.g., liquid "jet" and interfacial area), and the phase entrainment (e.g., number and size of droplets) were measured using high-speed photography. Dietrich et al. [8] used the Particle Image Velocimetry (PIV) technique to describe the flow fields around a bubble crossing the interface. Dayal [9] analyzed slag specimens collected from the slag–metal interface in an industrial 65-ton ladle furnace and explained the slag–metal interface phenomenon on the basis of cold-water experiment results. Kobyas [10] established a model to explain iron droplet formation and behavior in slag when gas bubbles pass through the molten iron–slag interface. Ueda [11] and Kochi [12] used a CFD model based on the finite volume method to predict the flow field and the penetration stage when a bubble rises through a water–oil interface. Some other simulation methods, including smoothed particle hydrodynamics (SPH) [13] and the multiphase particle method [14], have also been used to model gas bubbles passing through liquid–liquid interfaces.

In the current study, we simulated a slag–metal system using a water–oil system under cold experimental conditions and investigated the movement of bubbles and their behavior when they passed at various flow rates through a water–oil interface. The effects of different oil viscosities on bubble behavior were studied. We found that bubble movement is greatly influenced by the viscosity of the oil at low gas flow rates, whereas the movement of bubbles is more complex at high flow rates.

2. Experimental

A cold-water model was established for investigating both bubble motion (e.g., the path of movement, rising velocity, breakage, and coalescence) for a bubble crossing the liquid–liquid interface and the variation of the interfacial phenomena with bubble motion. Water and silicone oil were selected to investigate the liquid–liquid movement. Air was injected from a bottom nozzle to the lower phase (water). The nozzle diameter was 2 mm. The vessel was 500 mm in length (L) and 100 mm in width (W). The liquid height was 210 mm, and the height ratio between the water and silicone oil was 2:1. Silicone oils with different viscosities (shown in Table 1) were used [15,16], and the water was colored red to obtain a clear interface. The bubble and interface movements were recorded with a high-speed camera (HiSpec 5). The gas dispersion process was recorded from the moment of injection and 500 frames per second were captured until the flow field was stable for a duration of 30 s. The experimental system is shown in Figure 1.

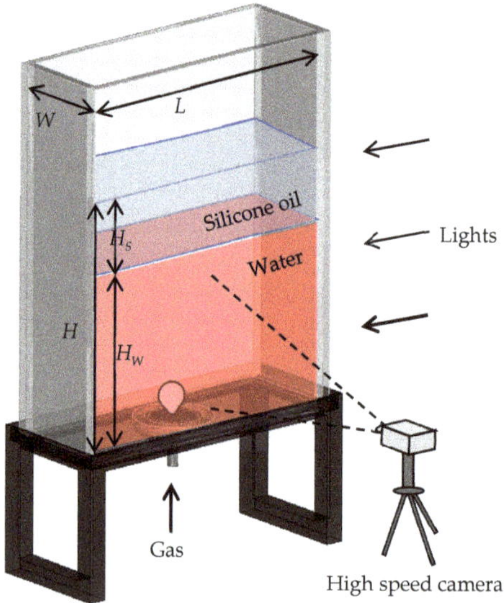

Figure 1. Experimental setup.

Table 1. Properties of the silicone oils used in the experiments.

No.	Kinematic Viscosity, cSt	Density, kg/m^3	Interface Tension, mN/m
1	50	956	30.6
2	100	957	30.7
3	200	961	36.9

3. Results and Discussions

3.1. Bubble Shape and Moving Path

Figure 2 shows the bubbles' shape changing with the gas flow rate. At a low gas flow rate of 20–40 mL/min, the bubbles with diameters of 2–3 mm were spherical in shape. At an increased gas flow rate of 60–500 mm, the shape of the injected bubbles changed to be ellipsoidal and the bubbles' diameters were 3–5 mm. With a further increase of gas flow (1000–1500 mL/min), the bubbles became mushroom-shaped and more irregular. The bubbles' sizes were about 10–20 mm.

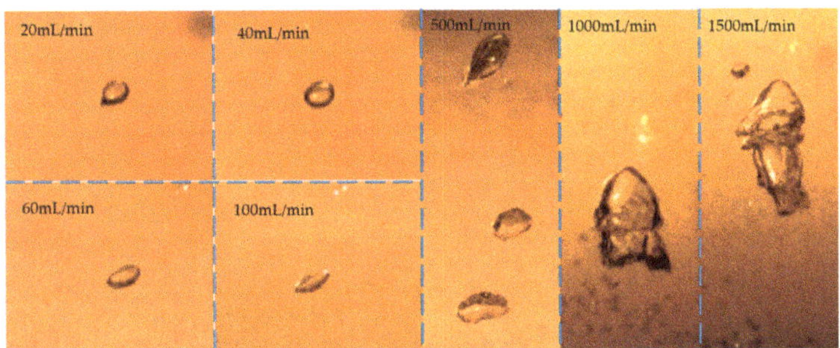

Figure 2. Bubble shape at different gas flow rate with oil phase viscosity of 50 cSt.

Figure 3 shows the bubble movement path and the interface fluctuation when it passed through the water–oil interface. The gas formed spherical bubbles after being ejected through the nozzle, and the shape of the bubbles subsequently changed from ellipsoidal or coronal during ascension because of the difference in internal and external pressures. When a bubble was close to the water–oil interface, its movement path deviated, resulting in a different ascending trajectory because of interface fluctuation caused by the impact of previous bubbles. The velocity of a bubble decreased rapidly, and the bubble surface pressure became uniform when it passed through the water–oil interface. It then ascended in a spherical shape while driving the interface upward. In addition, when the bubble entered the oil phase from the water, the surface of the bubble covered the water-phase liquid film, resulting in entrainment.

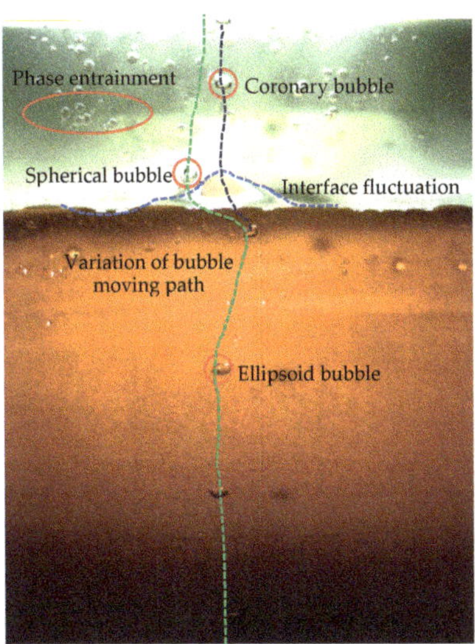

Figure 3. Bubble shape, movement path, and interfluctuation.

3.2. Bubble Residence Time at Liquid–Liquid Interface

Figure 4 shows that the heights to which the bubble rose varied with time when oil phases with different viscosities were used. The difference in density and interfacial tension between the oil phases is very small (Table 1). Parameters T_1, T_2, and T_3 are the times for a bubble to cross the water–oil interface. The interface is more easily broken when the oil has a low kinematic viscosity, thus, the bubble would cross using a shorter route and in less time. With an increase in the oil kinematic viscosity, the "elasticity" of the interface is enhanced, where bubbles should overcome the higher viscous resistance to rising. The bubbles require more time to cross an oil phase with higher viscosity.

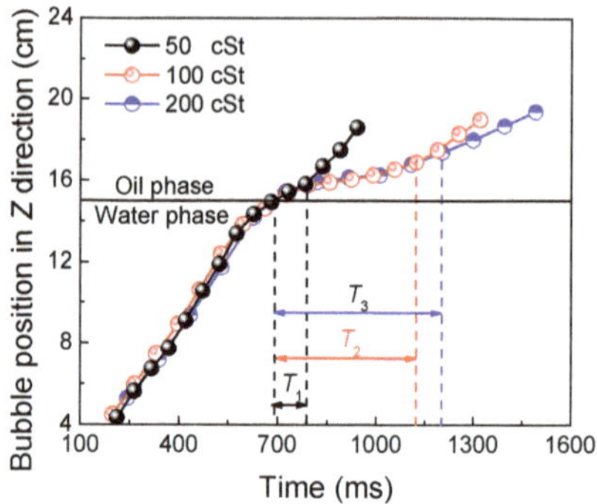

Figure 4. Bubble rising heights at different times (Q = 50 mL/min).

3.3. Bubble Rising Velocity

Figure 5 shows the distribution of bubble velocity at different heights with a low gas flow rate Q = 50 mL/min. As a result of the low gas flow rate, bubbles formed with lower frequency, lower velocity, and a smaller size after the gas was emitted from the nozzle. Thus, small bubbles interacted one by one with the water–oil interface. This figure reveals that the velocity of bubbles increased linearly in the water-phase (black points) and decreased when they began to pass through the interface (red points). When a bubble left the surface, its velocity was minimum. It then entered the oil phase. However, the velocity was lower in the oil phase than in the water-phase because of the high viscosity of the silicone oil (blue points). In addition, region 2 became larger with increasing oil viscosity.

As shown in Figure 6, about five bubbles were tracked to measure the bubbles' rising velocity. The blue points show the bubbles with a decreasing velocity caused by interface interaction, the red points show bubbles with a decreasing velocity caused by bubble breakage and the green points show the bubbles with an increasing velocity caused by bubble aggregation. When the gas flow rate was increased, the bubble flow was no longer one by one but rather continuous and even. The shape of the interface, therefore, changed dramatically. Because of the instability of the bubble shape, fragmentation, accumulation, and the influence of interface fluctuation, the change in large-bubble velocity can be divided into four stages during the rising process. The bubble velocity varied with height. In region 1, the bubbles' velocity increased during the rising process. When the bubble moved to region 2, the velocity of a bubble was reduced because the shape of the bubble changed during ascent. The shape and velocity of the bubble affected each other, so the velocity changed periodically. In region 3, the improvement in bubble shape caused the increasing of bubbles' velocity. The velocity was hampered by the fluctuation of the interface in region 4. As the viscosity increases, the difference in velocity between the bubbles at the top and bottom becomes smaller and the location of bubble aggregation becomes closer to the nozzle.

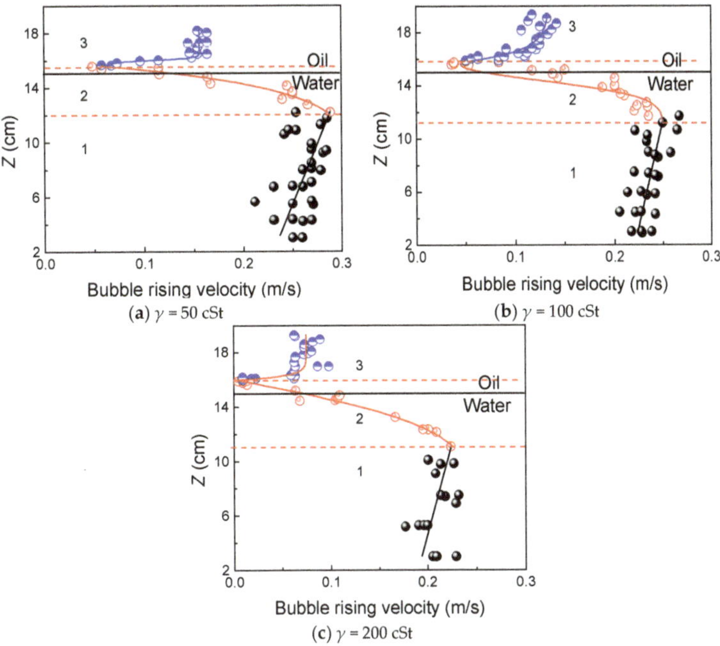

Figure 5. Rising-bubble velocity distribution (Q = 50 mL/min).

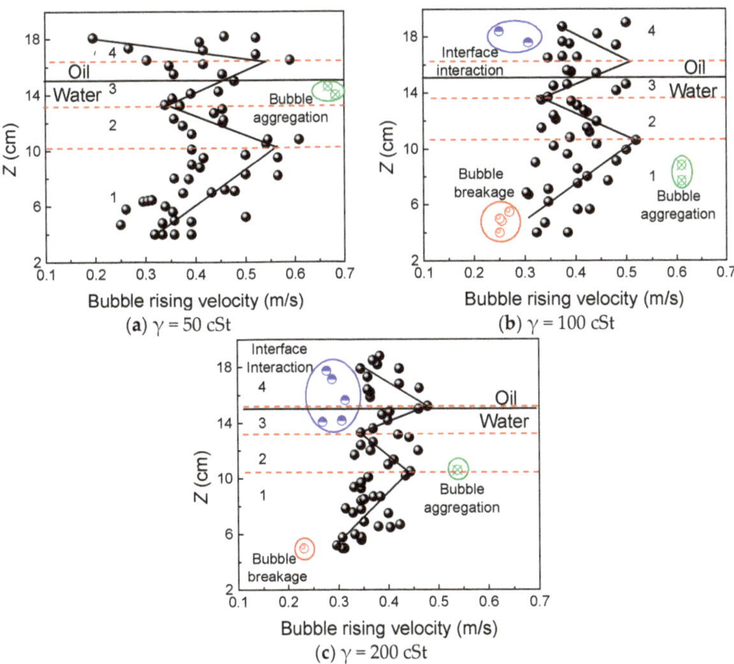

Figure 6. Rising-bubble velocity distribution (Q = 1000 mL/min).

3.4. Liquid–Liquid Interface Distribution and Phase Entrainment

Entrainment occurs when a bubble enters a new phase. Entrainment occurs via four main mechanisms, as shown in Figure 7:

(a)   Firstly, as shown in Figure 7a, the micro water-phase entrainment in the oil phase was generated by the interfacial fracture when a bubble crosses the interface at a low gas flow rate ($Q < 50$ mL/min). The micro water-phase entrainment generated in this case was sensitively disturbed by the fluid flow and dropped very slowly without gas injection. The micro water-phase entrainment is difficult to separate from the oil phase layer. When increasing oil viscosity, the micro water-phase entrainment will be intensified.

(b)   The second mechanism of water-phase entrainment at the mesoscale was an oil film escaping from the bubble surface, as shown in Figure 7b. This path was the main entrainment generation method at gas flow rates of <100 mL/min. The size of the mesoscale entrainment increases when increasing the gas flow rate and separates from the oil phase layer quickly without gas injection.

(c)   As shown in Figure 7c, at large gas flow rates (100–500 mL/min), the unstable water/oil interface generated would reach the top surface of the oil phase. Large-scale water-phase entrainment occurred via breakage of the column interface. The large-scale entrainment in this case would quickly drop to the water phase. Increasing the oil viscosity could strengthen the water/oil interface, which is more difficult to break.

(d)   The last mechanism was associated with higher gas flow rates (>1000 mL/min). The oil and water mixed and penetrated each other with strong stirring of the gas phase, as shown in Figure 7d. There was no clear water/oil interface. The mixed entrainment took a long time to separate.

Figure 7. Four mechanisms of liquid entrainment with an oil–oil viscosity of 50 cSt for (**a**) $Q = 50$ mL/min; (**b**) $Q = 100$ mL/min; (**c**) $Q = 250$ mL/min and (**d**) $Q = 4000$ mL/min.

In summary, entrainment in the oil phase was mainly produced via mechanisms (a) microscale entrainment, (b) mesoscale entrainment, (c) large-scale entrainment and (d) mixed entrainment with an increased gas flow rate.

3.5. Phase Distribution

Figure 8 showed the phase distribution of gas–water–oil at different viscosities and gas flow rates. With an increased gas flow rate, the frequency with which bubbles impacted the interface also increased. When the gas flow rate was increased to 500 mL/min, the bubble size increased, bubble breakage and aggregation increased, and a stable cylindrical interface formed between these two phases. When the gas flow rate was increased to 1500 mL/min, the cylindrical interface became unstable. When it was increased to 4000 mL/min, the interface disappeared, and the two phases were completely mixed. At the same gas flow rate, with an increased oil viscosity, the interface was more stable and fluctuated less; similarly, less water entrainment occurred (50–100 mL/min), the bubble column became more narrow and stable (500–1500 mL/min), and the two phases mixed less thoroughly (4000 mL/min).

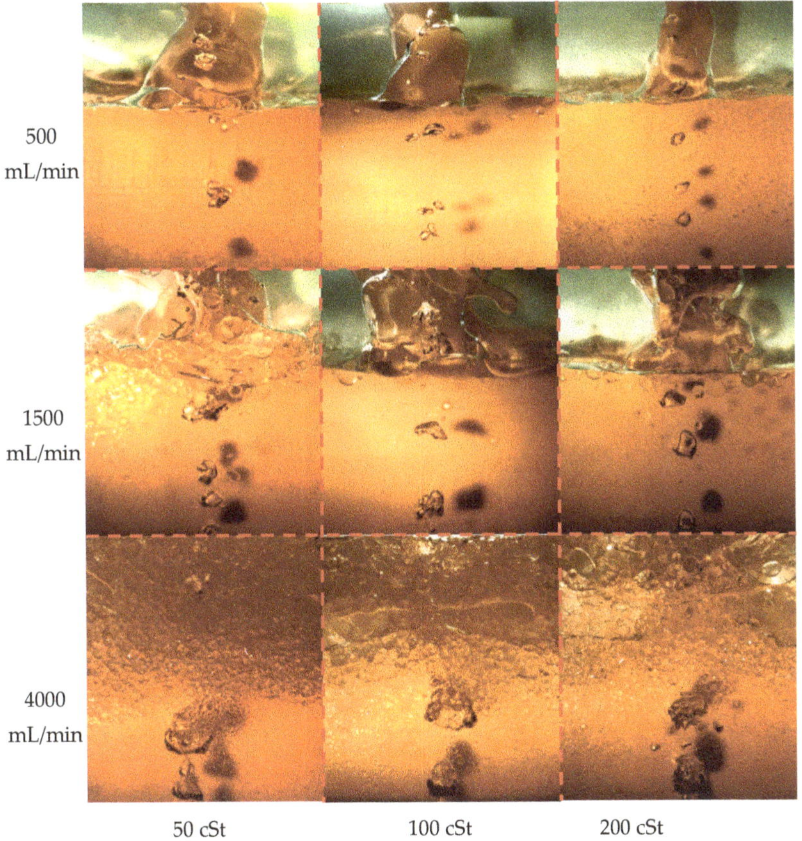

Figure 8. Gas–liquid–liquid distribution with different gas flow rates and different silicone oil viscosities.

The height of the water–oil interface changed as the bubble crossed the water–oil interface and was defined as the difference between the height of static water–oil interface and the interface height when bubbles had just separated from the water–oil interface. The jet height, h, could be measured from the high-speed images shown in Figure 9.

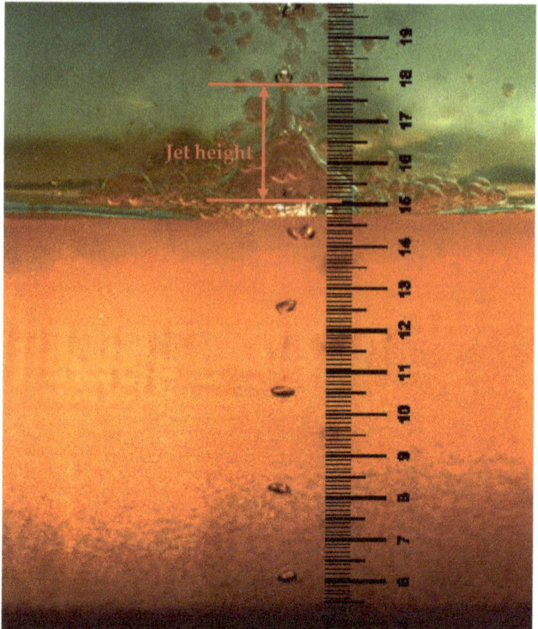

Figure 9. Measuring method of water–oil interface jet height.

Figure 10 shows the jet height of the water–oil interface caused by rising bubbles. The interface was impinged with continuous single bubbles when the gas flow rate was less than Q_1, Q_2, and Q_3 for different oils. When the gas flow rate was increased beyond Q_i, the aggregation between bubbles was improved, thus, the impingement was changed by multiple bubbles and the jet height was also substantially increased. The stability of the water–oil interface was enhanced, and the Q_i was also increased with increasing viscosity of the oil phase.

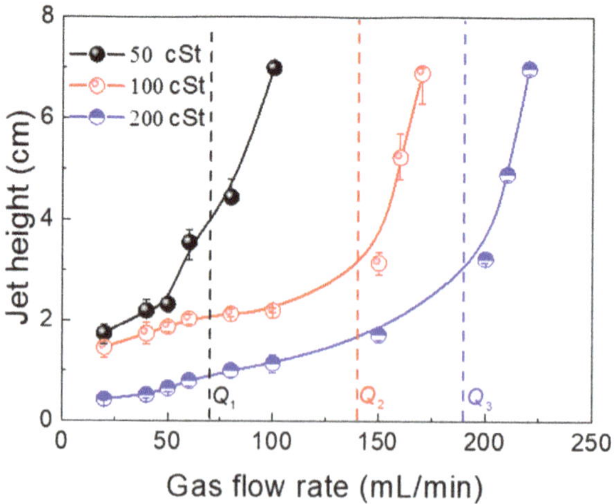

Figure 10. Jet height with different gas flow rates.

When the gas flow rate was less than Q_i, the jet height and gas flow rate showed a certain linear relationship and the slope approximated to 0.01. The intercept, b, was related to viscosity and decreased with an increase in oil phase viscosity. When the gas flow rate was greater than Q_i, the jet height increased exponentially with an increase in gas flow rate. The exponential factor approximated to 0.01 and the coefficient K changed slightly with the increase in oil phase viscosity. The relationship between h and Q could be approximately expressed as follows:

$$\begin{cases} h \approx 0.01Q + b & (Q < Q_i) \\ h \approx K(Q - Q_i)^{0.6} & (Q > Q_i) \end{cases}$$

4. Conclusions

Through observation of the phenomenon of bubbles crossing the water–air interface under cold-water model conditions, we deduced the following:

(1). The second phase (oil phase) can reduce the rise rate of a bubble as it passes through the liquid–liquid interface. The greater the viscosity, the more time required for a bubble to pass through the interface.

(2). The bubble rise rate decreases sharply as it crosses the interface because it must overcome interfacial tension and viscous drag during this process. Therefore, when the atmospheric amount is used, the bubble is mainly affected by the "liquid column" and deformation during ascension. The velocity distribution is M-shaped at this time.

(3). Phase entrainment can occur via four mechanisms, and increasing the viscosity of the second phase (oil phase) can suppress the generation of the fine entrained phase.

Author Contributions: Project Administration and Writing—Original Draft Preparation, H.Z.; Data Analysis, J.W.; Experiment and Investigation, W.Z.; Supervision, F.L.; Writing—Review & Editing, X.C., Formal Analysis and Methodology, M.X.

Funding: This research was funded by "The Fundamental Research Funds for the Central Universities, grant number FRF-TP-19-016A3", "The Guangxi Innovation-Driven Development Project, grant number AA18242042-1" and "The University in Guangdong Province Grant, grant number 2015KQNCX157".

Conflicts of Interest: The authors declare no conflicts of interest.

References

1. Su, C.J.; Chou, J.M.; Liu, S.H. Effect of gas bottom blowing condition on mixing molten iron and slag inside ironmaking smelter. *Mater. Trans.* **2009**, *50*, 1502–1509. [CrossRef]

2. Ma, J.; Sun, Y.S.; Li, B.W. Spectral collocation method for transient thermal analysis of coupled conductive, convective and radiative heat transfer in the moving plate with temperature dependent properties and heat generation. *Int. J. Heat Mass Transf.* **2017**, *114*, 469–482. [CrossRef]

3. Ma, J.; Sun, Y.S.; Li, B.W. Simulation of combined conductive, convective and radiative heat transfer in moving irregular porous fins by spectral element method. *Int. J. Therm. Sci.* **2017**, *118*, 475–487. [CrossRef]

4. Han, Z.; Holappa, L. Mechanisms of iron entrainment into slag due to rising gas bubbles. *ISIJ Int.* **2003**, *43*, 292–297. [CrossRef]

5. Beskow, K.; Dayal, P.; Björkvall, J.; Nzotta, M.; Sichen, D. A new approach for the study of Slag–Metal interface in steelmaking. *Ironmak. Steelmak.* **2006**, *33*, 74–80. [CrossRef]

6. Reiter, G.; Schwerdtfeger, K. Characteristics of entrainment at liquid/liquid interfaces due to rising bubbles. *ISIJ Int.* **1992**, *32*, 57–65. [CrossRef]

7. Reiter, G.; Schwerdtfeger, K. Observations of physical phenomena occurring during passage of bubbles through liquid/liquid Interfaces. *ISIJ Int.* **1992**, *32*, 50–56. [CrossRef]

8. Dietrich, N.; Poncin, S.; Pheulpin, S.; Li, H.Z. Passage of a bubble through a Liquid–Liquid interface. *AICHE J.* **2008**, *54*, 594–600. [CrossRef]

9. Delobelle, P.; Oytana, C. Study of Slag-Metal interface in ladle treatment. *Ironmak. Steelmak.* **2006**, *33*, 454–464.

10. Kobayashi, S. Iron droplet formation due to bubbles passing through molten iron/slag interface. *ISIJ Int.* **1993**, *33*, 577–582. [CrossRef]

11. Ueda, Y.; Kochi, N.; Uemura, T.; Ishii, T.; Iguchi, M. Numerical observation of flow field around the water column behind a rising bubble through an oil water interface. *ISIJ Int.* **2011**, *51*, 1940–1942. [CrossRef]

12. Kochi, N.; Ueda, Y.; Uemura, T.; Ishii, T.; Iguchi, M. Numerical simulation on penetration stage of a rising bubble through an Oil-Water Interface. *ISIJ Int.* **2011**, *51*, 1011–1013. [CrossRef]

13. Grenier, N.; Touzé, D.L.; Colagrossi, A.; Antuono, M.; Colicchio, G. Viscous bubbly flows simulation with an interface SPH model. *Ocean. Eng.* **2013**, *69*, 88–102. [CrossRef]

14. Natsui, S.; Takai, H.; Kumagai, T.; Kikuchi, T.; Suzuki, R.O. Multiphase particle simulation of gas bubble passing through liquid/liquid interfaces. *Mater. Trans.* **2014**, *55*, 1707–1715. [CrossRef]

15. Dehkordi, P.B.; Colombo, L.P.M.; Mohammadian, E.; Arnone, D.; Azdarpour, A.; Sotgia, G. Study of viscous Oil-Water-Gas slug flow in a horizontal pipe. *J. Pet. Sci. Eng.* **2019**, *178*, 1–13. [CrossRef]

16. Kajero, O.T.; Mukhtar, A.; Lokman, A.; James, A.B. Experimental study of viscous effects on flow pattern and bubble behavior in small diameter bubble column. *Phys. Fluids* **2018**, *30*, 093101. [CrossRef]

Article

Titanium Distribution Ratio Model of Ladle Furnace Slags for Tire Cord Steel Production Based on the Ion–Molecule Coexistence Theory at 1853 K

Jialiu Lei [1,*], Dongnan Zhao [1], Wei Feng [1] and Zhengliang Xue [2]

[1] School of Materials Science and Engineering, Hubei Polytechnic University, Huangshi 435000, China; zhaodongnan@hbpu.edu.cn (D.Z.); fengwei@hbpu.edu.cn (W.F.)

[2] The State Key Laboratory of Refractories and Metallurgy, Wuhan University of Science and Technology, Wuhan 430081, China; xuezhengliang@wust.edu.cn

* Correspondence: leijialiu@hbpu.edu.cn; Tel.: +86-0714-635-8328

Received: 25 September 2019; Accepted: 19 October 2019; Published: 1 November 2019

Abstract: High-strength tire cord steel is mainly used in radial ply tires, but the presence of brittle Ti inclusions can cause failure of the wires and jeopardize their performance in production. In order to control the titanium content during steel production, a thermodynamic model for predicting the titanium distribution ratio between CaO–SiO_2–Al_2O_3–MgO–FeO–MnO–TiO_2 slags during the ladle furnace (LF) refining process at 1853 K has been established based on the ion–molecule coexistence theory (IMCT), combined with industrial measurements, and the effect of basicity on the titanium distribution ratio was discussed. The results showed that the titanium distribution ratio predicted by the developed IMCT exhibited a dependable agreement with the measurements, and the optical basicity is suggested to reflect the correlation between basicity and the titanium distribution ratio. Furthermore, quantitative titanium distribution ratios of TiO_2, $CaO·TiO_2$, $MgO·TiO_2$, $FeO·TiO_2$, and $MnO·TiO_2$ were acquired by the IMCT model, respectively. Calculation results revealed that the structural unit CaO plays a pivotal role in the slags in the de-titanium process.

Keywords: titanium distribution ratio; thermodynamic model; ion–molecule coexistence theory; LF refining slags

---

1. Introduction

Titanium is a common microalloy addition to steels. It can be used to inhibit grain growth, reduce the incidence of transverse cracking in niobium-containing steels production, stabilize the alloy against sensitization to intergranular corrosion, and improve the service performance [1–4]. However, a low titanium content is demanded for special kinds of steel production, such as in high-strength tire cord steel, to enhance drawing and twisting performances.

As a product with superior quality to wire rods, tire cord steel is mainly used in radial ply tires. Before it is made, the steel wire is drawn from 5.5 mm to 0.15 mm in diameter and subjected to cyclic stress in the drawing and twisting process. Therefore, breakage of steel wire during fabrication is a crucial issue. This filament break is especially sensitive with the existence of angular and non-deformable Ti inclusions, such as titanium nitride (TiN) or titanium carbonitride (Ti(CN)) [5–9]. This causes a decrease in fatigue performance and can seriously affect traffic safety.

Therefore, the issue of the control of titanium content has received considerable critical attention. To control the titanium content during steel production, it is essential to study the titanium distribution ratio between steel and slag. To date, there has been limited theoretical and experimental studies implemented on the titanium distribution behavior in slags; acquiring relevant parameters at elevated temperatures between steel and slag is arduous and costly. It is quite essential to establish a thermodynamic model for calculating the titanium distribution ratio between steel and slag. The ion–molecule coexistence theory (IMCT) has been efficaciously applied to describe phosphate capacity, manganese distribution, sulfide capacity, and so on, as shown in Table 1 [10–23]. In the IMCT, the defined mass action–concentration (MAC) is consistent with the classical concept of activity in the slag.

Table 1. Applications of the ion–molecule coexistence theory (IMCT) model during ironmaking and steelmaking processes.

Slag Systems	Applications	Ref.
$CaO-SiO_2-FeO-MgO-MnO-Al_2O_3$	A thermodynamic model for predicting the manganese distribution ratio and manganese capacity of the slags was developed based on the IMCT. The established model was successfully applied to not only manganese equilibrium experiments but also industrial production.	[10]
$CaO-SiO_2-MgO-FeO-MnO-Al_2O_3-TiO_2-CaF_2$	A thermodynamic model for calculating the manganese distribution ratio between the slags and carbon saturated liquid iron was built based on the IMCT. The predicted manganese distribution ratio by IMCT had a good linear relationship with measurements expect individual points.	[11]
$CaO-SiO_2-MgO-FeO-Fe_2O_3-Al_2O_3-P_2O_5$	A thermodynamic model for predicting the phosphorus distribution ratio of the slags was developed based on the IMCT. The developed model was successfully applied to not only phosphorus equilibrium experiments, but also industrial production in Hismelt smelting reduction vessels.	[12]
CaO-based Slags	A thermodynamic model for predicting phosphorus partition between CaO-based slags during hot metal dephosphorization pretreatment was established based on the IMCT. The established model was verified as effective through comparing with measured results and predicted ones by other models.	[13]
$CaO-SiO_2-MgO-FeO-Fe_2O_3-MnO-Al_2O_3-P_2O_5$	A thermodynamic model for calculating the phosphorus distribution ratio between steelmaking slags and molten steel was built based on the IMCT. The built IMCT prediction model was verified with measured and some other reported models.	[14]
$CaO-FeO-Fe_2O_3-Al_2O_3-P_2O_5$	Thermodynamic models for predicting the phosphorus distribution ratio and phosphorus capacity of the slags during refining were developed based on the IMCT. The developed models were verified with experimental results and reported models.	[15]
$CaO-SiO_2-FeO-Fe_2O_3-P_2O_5$	Defined enrichment possibility and enrichment degree of solid solutions containing P_2O_5 from the developed IMCT model were verified from experimental results.	[16]
CaO-based Slags	Coupling relationships between dephosphorization and desulfurization abilities or potentials for CaO-based slags during the refining process of molten steel were proposed based on the IMCT. The proposed model was verified as effective and feasible through investigating the effect of slag composition.	[17]
$CaO-SiO_2-MgO-Al_2O_3$	A sulfide capacity prediction model of the slags was developed based on the IMCT. The developed model had a higher accuracy than other reported sulfide capacity prediction models.	[18]
$CaO-SiO_2-MgO-FeO-MnO-Al_2O_3$	A thermodynamic model for calculating the sulfur distribution ratio between ladle furnace (LF) refining slags and molten steel was established based on the IMCT. The model was verified with the measured and the calculated sulfur distribution ratio by Young's model and the KTH model in LF refining.	[19]
$CaO-SiO_2-MgO-FeO-MnO-Al_2O_3$	A sulfide capacity prediction model of the LF refining slags was built based on the IMCT. The built sulfide capacity prediction model was verified with the measured and calculated by Young's model and the KTH model in LF refining.	[20]
$CaO-FeO-Fe_2O_3-Al_2O_3-P_2O_5$	A thermodynamic model for predicting the sulfide capacity of the slags at various oxygen potentials was developed based on the IMCT. The built model was verified through comparing the determined sulfide capacity, and could be applied to precisely predict sulfide capacity.	[21]
$CaO-FeO-Fe_2O_3-Al_2O_3-P_2O_5$	A thermodynamic model for predicting the sulfur distribution ratio between the slags and liquid iron was built based on the IMCT. The developed model was verified with measured data of sulfur distribution equilibrium from the literatures.	[22]
$CaO-SiO_2-MgO-FeO-Fe_2O_3-MnO-Al_2O_3-P_2O_5$	The defined oxidation ability of metallurgical slags based on the IMCT was verified by comparisons with the reported activity in the selected Fe_tO-containing slag systems.	[23]

To improve the application domain of IMCT, in this paper, a titanium distribution ratio model of $CaO-SiO_2-Al_2O_3-MgO-FeO-MnO-TiO_2$ slags was built based on the IMCT, combined with industrial measurements. From the results, the titanium distribution behavior during ladle furnace (LF) refining in high-strength tire cord steel production is further revealed.

2. Materials and Methods

2.1. Production Procedure and Materials

The following process was adopted for the production of high-strength tire cord steel in Baosteel (Wuhan Branch): basic oxygen furnace (BOF) → tapping → ladle furnace (LF) → soft blowing → continuous casting (CC) → rolling. Figure 1 shows a schematic diagram of the production process.

Figure 1. Production process of high-strength tire cord steel.

Both the liquid steel and balanced slag were sampled at the end point of the LF refining process at about 1853 K. The orthonormal chemical components of molten steel and slags for 16 heats are given in Table 2.

Table 2. Chemical components of liquid steel and slags at the end point of LF refining (wt %).

Slag Composition							Metal Composition			
CaO	SiO$_2$	Al$_2$O$_3$	MgO	FeO	MnO	TiO$_2$	C	Si	Mn	Ti
39.19	40.50	7.97	8.47	1.36	2.17	0.34	0.81	0.20	0.47	0.0008
38.56	40.95	7.73	8.89	1.27	2.28	0.31	0.80	0.19	0.48	0.0008
35.19	43.10	6.82	10.79	1.41	2.52	0.18	0.83	0.18	0.46	0.0006
34.70	42.92	6.46	10.97	1.81	2.95	0.19	0.80	0.20	0.46	0.0007
39.47	41.52	7.19	8.17	1.61	1.78	0.25	0.80	0.20	0.45	0.0006
37.08	45.82	3.78	8.10	1.61	3.40	0.20	0.82	0.20	0.46	0.0007
34.28	45.19	6.07	11.09	1.36	1.69	0.33	0.82	0.19	0.47	0.0012
36.11	44.54	4.79	9.12	1.73	3.49	0.22	0.81	0.18	0.46	0.0008
40.04	39.73	8.87	8.38	1.28	1.38	0.32	0.82	0.18	0.48	0.0007
33.25	45.28	7.74	10.12	1.62	1.71	0.27	0.80	0.19	0.47	0.0012
34.52	47.74	3.52	10.72	1.11	2.14	0.25	0.81	0.19	0.46	0.0010
37.23	43.69	6.48	9.12	1.34	1.86	0.28	0.82	0.20	0.47	0.0008
42.90	43.04	5.46	5.98	1.00	1.29	0.33	0.80	0.18	0.45	0.0007
36.52	46.23	4.75	8.65	1.14	2.44	0.26	0.83	0.20	0.46	0.0009
34.51	46.01	5.54	10.61	0.94	2.14	0.24	0.80	0.19	0.46	0.0009
31.73	45.69	6.82	9.78	1.08	4.70	0.21	0.81	0.20	0.48	0.0010

2.2. Establishment of the IMCT Model

Based on the assumptions inherent in the IMCT, the dominant features of the IMCT model for the activities of the structural units in the slag can be summarized briefly as follows:

(1) The constitutional units in the slag consist of simple ions, ordinary molecules, and complicated molecules;

(2) Complex molecules are generated by the reactions of bonded ion couples and simple molecules under kinetic equilibrium;

(3) The activity of each constituent in the slag equals the MAC of the structural unit at the steelmaking temperature;

(4) The chemical reactions comply with the law of mass conservation.

The calculations were based on actual production involving CaO–SiO$_2$–Al$_2$O$_3$–MgO–FeO–MnO–TiO$_2$ slag systems. The initial numbers of moles for each composition in 100 g of CaO–SiO$_2$–Al$_2$O$_3$–MgO–FeO–MnO–TiO$_2$ slag were $a = n^0_{CaO}, b = n^0_{SiO_2}, c = n^0_{Al_2O_3}, d = n^0_{MgO}, e = n^0_{FeO}, f = n^0_{MnO},$

and $g = n^0_{TiO_2}$, respectively. The balanced mole number of each constituent unit in the slag was defined as n_i, and N_i denotes the MAC of each constitutional unit. The MAC is equivalent to the classical definition of activity based on the IMCT and can be acquired as

$$N_i = \frac{n_i}{\sum n_i} \qquad (1)$$

where $\sum n_i$ is the total mole number of each constitutional unit in equilibrium.

According to the IMCT, at 1853 K, the slag system contains five simple ions (Ca^{2+}, Fe^{2+}, Mg^{2+}, Mn^{2+}, and O^{2-}) and three ordinary molecules (Al_2O_3, SiO_2, and TiO_2). Based on the reported phase diagrams, 44 types of complex molecules can be generated at the steelmaking temperature [24,25]. The abovementioned structural units and their parameters are listed in Table 3.

Table 3. Parameters of structural units in the slag system at 1853 K.

Items	Constitutional Units	Balanced Mole Number	MACs
Simple cations and anions	$Ca^{2+} + O^{2-}$	$n_1 = n_{Ca^{2+}} = n_{O^{2-}} = n_{CaO}$	$N_1 = \frac{2n_1}{\sum n_i} = N_{CaO}$
	$Mg^{2+} + O^{2-}$	$n_4 = n_{Mg^{2+}} = n_{O^{2-}} = n_{MgO}$	$N_4 = \frac{2n_4}{\sum n_i} = N_{MgO}$
	$Fe^{2+} + O^{2-}$	$n_5 = n_{Fe^{2+}} = n_{O^{2-}} = n_{FeO}$	$N_5 = \frac{2n_5}{\sum n_i} = N_{FeO}$
	$Mn^{2+} + O^{2-}$	$n_6 = n_{Mn^{2+}} = n_{O^{2-}} = n_{MnO}$	$N_6 = \frac{2n_6}{\sum n_i} = N_{MnO}$
Simple molecules	SiO_2	$n_2 = n_{SiO_2}$	$N_2 = \frac{n_2}{\sum n_i} = N_{SiO_2}$
	Al_2O_3	$n_3 = n_{Al_2O_3}$	$N_3 = \frac{n_3}{\sum n_i} = N_{Al_2O_3}$
	TiO_2	$n_7 = n_{TiO_2}$	$N_7 = \frac{n_7}{\sum n_i} = N_{TiO_2}$
Complex molecules	$CaO{\cdot}SiO_2$	$n_8 = n_{CaO{\cdot}SiO_2}$	$N_8 = \frac{n_8}{\sum n_i} = N_{CaO{\cdot}SiO_2}$
	$3CaO{\cdot}2SiO_2$	$n_9 = n_{3CaO{\cdot}2SiO_2}$	$N_9 = \frac{n_9}{\sum n_i} = N_{3CaO{\cdot}2SiO_2}$
	$2CaO{\cdot}SiO_2$	$n_{10} = n_{2CaO{\cdot}SiO_2}$	$N_{10} = \frac{n_{10}}{\sum n_i} = N_{2CaO{\cdot}SiO_2}$
	$3CaO{\cdot}SiO_2$	$n_{11} = n_{3CaO{\cdot}SiO_2}$	$N_{11} = \frac{n_{11}}{\sum n_i} = N_{3CaO{\cdot}SiO_2}$
	$3CaO{\cdot}Al_2O_3$	$n_{12} = n_{3CaO{\cdot}Al_2O_3}$	$N_{12} = \frac{n_{12}}{\sum n_i} = N_{3CaO{\cdot}Al_2O_3}$
	$12CaO{\cdot}7Al_2O_3$	$n_{13} = n_{12CaO{\cdot}7Al_2O_3}$	$N_{13} = \frac{n_{13}}{\sum n_i} = N_{12CaO{\cdot}7Al_2O_3}$
	$CaO{\cdot}Al_2O_3$	$n_{14} = n_{CaO{\cdot}Al_2O_3}$	$N_{14} = \frac{n_{14}}{\sum n_i} = N_{CaO{\cdot}Al_2O_3}$
	$CaO{\cdot}2Al_2O_3$	$n_{15} = n_{CaO{\cdot}2Al_2O_3}$	$N_{15} = \frac{n_{15}}{\sum n_i} = N_{CaO{\cdot}2Al_2O_3}$
	$CaO{\cdot}6Al_2O_3$	$n_{16} = n_{CaO{\cdot}6Al_2O_3}$	$N_{16} = \frac{n_{16}}{\sum n_i} = N_{CaO{\cdot}6Al_2O_3}$
	$CaO{\cdot}TiO_2$	$n_{17} = n_{CaO{\cdot}TiO_2}$	$N_{17} = \frac{n_{17}}{\sum n_i} = N_{CaO{\cdot}TiO_2}$
	$3CaO{\cdot}2TiO_2$	$n_{18} = n_{3CaO{\cdot}2TiO_2}$	$N_{18} = \frac{n_{18}}{\sum n_i} = N_{3CaO{\cdot}2TiO_2}$
	$4CaO{\cdot}3TiO_2$	$n_{19} = n_{4CaO{\cdot}3TiO_2}$	$N_{19} = \frac{n_{19}}{\sum n_i} = N_{4CaO{\cdot}3TiO_2}$
	$3Al_2O_3{\cdot}2SiO_2$	$n_{20} = n_{3Al_2O_3{\cdot}2SiO_2}$	$N_{20} = \frac{n_{20}}{\sum n_i} = N_{3Al_2O_3{\cdot}2SiO_2}$
	$Al_2O_3{\cdot}TiO_2$	$n_{21} = n_{Al_2O_3{\cdot}TiO_2}$	$N_{21} = \frac{n_{21}}{\sum n_i} = N_{Al_2O_3{\cdot}TiO_2}$
	$2MgO{\cdot}SiO_2$	$n_{22} = n_{2MgO{\cdot}SiO_2}$	$N_{22} = \frac{n_{22}}{\sum n_i} = N_{2MgO{\cdot}SiO_2}$
	$MgO{\cdot}SiO_2$	$n_{23} = n_{MgO{\cdot}SiO_2}$	$N_{23} = \frac{n_{23}}{\sum n_i} = N_{MgO{\cdot}SiO_2}$
	$MgO{\cdot}Al_2O_3$	$n_{24} = n_{MgO{\cdot}Al_2O_3}$	$N_{24} = \frac{n_{24}}{\sum n_i} = N_{MgO{\cdot}Al_2O_3}$
	$MgO{\cdot}TiO_2$	$n_{25} = n_{MgO{\cdot}TiO_2}$	$N_{25} = \frac{n_{25}}{\sum n_i} = N_{MgO{\cdot}TiO_2}$
	$2MgO{\cdot}TiO_2$	$n_{26} = n_{2MgO{\cdot}TiO_2}$	$N_{26} = \frac{n_{26}}{\sum n_i} = N_{2MgO{\cdot}TiO_2}$
	$MgO{\cdot}2TiO_2$	$n_{27} = n_{MgO{\cdot}2TiO_2}$	$N_{27} = \frac{n_{27}}{\sum n_i} = N_{MgO{\cdot}2TiO_2}$
	$2FeO{\cdot}SiO_2$	$n_{28} = n_{2FeO{\cdot}SiO_2}$	$N_{28} = \frac{n_{28}}{\sum n_i} = N_{2FeO{\cdot}SiO_2}$
	$FeO{\cdot}Al_2O_3$	$n_{29} = n_{FeO{\cdot}Al_2O_3}$	$N_{29} = \frac{n_{29}}{\sum n_i} = N_{FeO{\cdot}Al_2O_3}$
	$FeO{\cdot}TiO_2$	$n_{30} = n_{FeO{\cdot}TiO_2}$	$N_{30} = \frac{n_{30}}{\sum n_i} = N_{FeO{\cdot}TiO_2}$
	$2FeO{\cdot}TiO_2$	$n_{31} = n_{2FeO{\cdot}TiO_2}$	$N_{31} = \frac{n_{31}}{\sum n_i} = N_{2FeO{\cdot}TiO_2}$
	$MnO{\cdot}SiO_2$	$n_{32} = n_{MnO{\cdot}SiO_2}$	$N_{32} = \frac{n_{32}}{\sum n_i} = N_{MnO{\cdot}SiO_2}$
	$2MnO{\cdot}SiO_2$	$n_{33} = n_{2MnO{\cdot}SiO_2}$	$N_{33} = \frac{n_{33}}{\sum n_i} = N_{2MnO{\cdot}SiO_2}$
	$MnO{\cdot}Al_2O_3$	$n_{34} = n_{MnO{\cdot}Al_2O_3}$	$N_{34} = \frac{n_{34}}{\sum n_i} = N_{MnO{\cdot}Al_2O_3}$
	$MnO{\cdot}TiO_2$	$n_{35} = n_{MnO{\cdot}TiO_2}$	$N_{35} = \frac{n_{35}}{\sum n_i} = N_{MnO{\cdot}TiO_2}$
	$2MnO{\cdot}TiO_2$	$n_{36} = n_{2MnO{\cdot}TiO_2}$	$N_{36} = \frac{n_{36}}{\sum n_i} = N_{2MnO{\cdot}TiO_2}$
	$2CaO{\cdot}Al_2O_3{\cdot}SiO_2$	$n_{37} = n_{2CaO{\cdot}Al_2O_3{\cdot}SiO_2}$	$N_{37} = \frac{n_{37}}{\sum n_i} = N_{2CaO{\cdot}Al_2O_3{\cdot}SiO_2}$
	$CaO{\cdot}Al_2O_3{\cdot}2SiO_2$	$n_{38} = n_{CaO{\cdot}Al_2O_3{\cdot}2SiO_2}$	$N_{38} = \frac{n_{38}}{\sum n_i} = N_{CaO{\cdot}Al_2O_3{\cdot}2SiO_2}$
	$2CaO{\cdot}MgO{\cdot}2SiO_2$	$n_{39} = n_{2CaO{\cdot}MgO{\cdot}2SiO_2}$	$N_{39} = \frac{n_{39}}{\sum n_i} = N_{2CaO{\cdot}MgO{\cdot}2SiO_2}$
	$3CaO{\cdot}MgO{\cdot}2SiO_2$	$n_{40} = n_{3CaO{\cdot}MgO{\cdot}2SiO_2}$	$N_{40} = \frac{n_{40}}{\sum n_i} = N_{3CaO{\cdot}MgO{\cdot}2SiO_2}$
	$CaO{\cdot}MgO{\cdot}SiO_2$	$n_{41} = n_{CaO{\cdot}MgO{\cdot}SiO_2}$	$N_{41} = \frac{n_{41}}{\sum n_i} = N_{CaO{\cdot}MgO{\cdot}SiO_2}$
	$CaO{\cdot}MgO{\cdot}2SiO_2$	$n_{42} = n_{CaO{\cdot}MgO{\cdot}2SiO_2}$	$N_{42} = \frac{n_{42}}{\sum n_i} = N_{CaO{\cdot}MgO{\cdot}2SiO_2}$
	$2MgO{\cdot}2Al_2O_3{\cdot}5SiO_2$	$n_{43} = n_{2MgO{\cdot}2Al_2O_3{\cdot}5SiO_2}$	$N_{43} = \frac{n_{43}}{\sum n_i} = N_{2MgO{\cdot}2Al_2O_3{\cdot}5SiO_2}$
	$CaO{\cdot}TiO_2{\cdot}SiO_2$	$n_{44} = n_{CaO{\cdot}TiO_2{\cdot}SiO_2}$	$N_{44} = \frac{n_{44}}{\sum n_i} = N_{CaO{\cdot}TiO_2{\cdot}SiO_2}$

The MACs for all the complex molecules can be determined using the reaction equilibrium constants K_i, $N_1(N_{CaO})$, $N_2(N_{SiO_2})$, $N_3(N_{Al_2O_3})$, $N_4(N_{MgO})$, $N_5(N_{FeO})$, $N_6(N_{MnO})$, and $N_7(N_{TiO_2})$, which are listed in Table 4.

Table 4. Reaction formulas, Gibbs free energies, and mass action–concentrations (MACs) [26–32].

Reaction Formulas	$\Delta G^{\theta}/(\mathrm{J\cdot mol^{-1}})$	MACs
$(Ca^{2+}+O^{2-})+(SiO_2)=(CaO\cdot SiO_2)$	$\Delta G^{\theta}=-21757-36.819T$	$N_8=K_1N_1N_2$
$3(Ca^{2+}+O^{2-})+2(SiO_2)=(3CaO\cdot2SiO_2)$	$\Delta G^{\theta}=-236972.9+9.6296T$	$N_9=K_2N_1^3N_2^2$
$2(Ca^{2+}+O^{2-})+(SiO_2)=(2CaO\cdot SiO_2)$	$\Delta G^{\theta}=-102090-24.267T$	$N_{10}=K_3N_1^2N_2$
$3(Ca^{2+}+O^{2-})+(SiO_2)=(3CaO\cdot SiO_2)$	$\Delta G^{\theta}=-118826-6.694T$	$N_{11}=K_4N_1^3N_2$
$3(Ca^{2+}+O^{2-})+(Al_2O_3)=(3CaO\cdot Al_2O_3)$	$\Delta G^{\theta}=-21757-29.288T$	$N_{12}=K_5N_1^3N_3$
$12(Ca^{2+}+O^{2-})+7(Al_2O_3)=(12CaO\cdot7Al_2O_3)$	$\Delta G^{\theta}=617977-612.119T$	$N_{13}=K_6N_1^{12}N_3^7$
$(Ca^{2+}+O^{2-})+(Al_2O_3)=(CaO\cdot Al_2O_3)$	$\Delta G^{\theta}=59413-59.413T$	$N_{14}=K_7N_1N_3$
$(Ca^{2+}+O^{2-})+2(Al_2O_3)=(CaO\cdot2Al_2O_3)$	$\Delta G^{\theta}=-16736-25.522T$	$N_{15}=K_8N_1N_3^2$
$(Ca^{2+}+O^{2-})+6(Al_2O_3)=(CaO\cdot6Al_2O_3)$	$\Delta G^{\theta}=-22594-31.798T$	$N_{16}=K_9N_1N_3^6$
$(Ca^{2+}+O^{2-})+(TiO_2)=(CaO\cdot TiO_2)$	$\Delta G^{\theta}=-79900-3.35T$	$N_{17}=K_{10}N_1N_7$
$3(Ca^{2+}+O^{2-})+2(TiO_2)=(3CaO\cdot2TiO_2)$	$\Delta G^{\theta}=-207100-11.51T$	$N_{18}=K_{11}N_1^3N_7^2$
$4(Ca^{2+}+O^{2-})+3(TiO_2)=(4CaO\cdot3TiO_2)$	$\Delta G^{\theta}=-293301-18.446T$	$N_{19}=K_{12}N_1^4N_7^3$
$3(Al_2O_3)+2(SiO_2)=(3Al_2O_3\cdot2SiO_2)$	$\Delta G^{\theta}=-4351-10.46T$	$N_{20}=K_{13}N_2^2N_3^3$
$(Al_2O_3)+(TiO_2)=(Al_2O_3\cdot TiO_2)$	$\Delta G^{\theta}=-5439-8.351T$	$N_{21}=K_{14}N_3N_7$
$2(Mg^{2+}+O^{2-})+(SiO_2)=(2MgO\cdot SiO_2)$	$\Delta G^{\theta}=-56902-3.347T$	$N_{22}=K_{15}N_2N_4^2$
$(Mg^{2+}+O^{2-})+(SiO_2)=(MgO\cdot SiO_2)$	$\Delta G^{\theta}=23849-29.706T$	$N_{23}=K_{16}N_2N_4$
$(Mg^{2+}+O^{2-})+(Al_2O_3)=(MgO\cdot Al_2O_3)$	$\Delta G^{\theta}=-18828-6.276T$	$N_{24}=K_{17}N_3N_4$
$(Mg^{2+}+O^{2-})+(TiO_2)=(MgO\cdot TiO_2)$	$\Delta G^{\theta}=-25104+2.804T$	$N_{25}=K_{18}N_4N_7$
$2(Mg^{2+}+O^{2-})+(TiO_2)=(2MgO\cdot TiO_2)$	$\Delta G^{\theta}=-17154-10.878T$	$N_{26}=K_{19}N_4^2N_7$
$(Mg^{2+}+O^{2-})+2(TiO_2)=(MgO\cdot2TiO_2)$	$\Delta G^{\theta}=-18619-7.99T$	$N_{27}=K_{20}N_4N_7^2$
$2(Fe^{2+}+O^{2-})+(SiO_2)=(2FeO\cdot SiO_2)$	$\Delta G^{\theta}=-9395-0.227T$	$N_{28}=K_{21}N_2N_5^2$
$(Fe^{2+}+O^{2-})+(Al_2O_3)=(FeO\cdot Al_2O_3)$	$\Delta G^{\theta}=-59204+22.343T$	$N_{29}=K_{22}N_3N_5$
$(Fe^{2+}+O^{2-})+(TiO_2)=(FeO\cdot TiO_2)$	$\Delta G^{\theta}=27293.76-26.25T$	$N_{30}=K_{23}N_5N_7$
$2(Fe^{2+}+O^{2-})+(TiO_2)=(2FeO\cdot TiO_2)$	$\Delta G^{\theta}=-33913.08+5.86T$	$N_{31}=K_{24}N_5^2N_7$
$(Mn^{2+}+O^{2-})+(SiO_2)=(MnO\cdot SiO_2)$	$\Delta G^{\theta}=38911-40.041T$	$N_{32}=K_{25}N_2N_6$
$2(Mn^{2+}+O^{2-})+(SiO_2)=(2MnO\cdot SiO_2)$	$\Delta G^{\theta}=36066-30.669T$	$N_{33}=K_{26}N_2N_6^2$
$(Mn^{2+}+O^{2-})+(Al_2O_3)=(MnO\cdot Al_2O_3)$	$\Delta G^{\theta}=-45116+11.81T$	$N_{34}=K_{27}N_3N_6$
$(Mn^{2+}+O^{2-})+(TiO_2)=(MnO\cdot TiO_2)$	$\Delta G^{\theta}=-24662+1.254T$	$N_{35}=K_{28}N_6N_7$
$2(Mn^{2+}+O^{2-})+(TiO_2)=(2MnO\cdot TiO_2)$	$\Delta G^{\theta}=-37620-1.672T$	$N_{36}=K_{29}N_6^2N_7$
$2(Ca^{2+}+O^{2-})+(Al_2O_3)+(SiO_2)=(2CaO\cdot Al_2O_3\cdot SiO_2)$	$\Delta G^{\theta}=-116315-38.911T$	$N_{37}=K_{30}N_1^2N_2N_3$
$(Ca^{2+}+O^{2-})+(Al_2O_3)+2(SiO_2)=(CaO\cdot Al_2O_3\cdot2SiO_2)$	$\Delta G^{\theta}=-4148-73.638T$	$N_{38}=K_{31}N_1N_2^2N_3$
$2(Ca^{2+}+O^{2-})+(Mg^{2+}+O^{2-})+2(SiO_2)=(2CaO\cdot MgO\cdot2SiO_2)$	$\Delta G^{\theta}=-73638-63.597T$	$N_{39}=K_{32}N_1^2N_2^2N_4$
$3(Ca^{2+}+O^{2-})+(Mg^{2+}+O^{2-})+2(SiO_2)=(3CaO\cdot MgO\cdot2SiO_2)$	$\Delta G^{\theta}=-205016-31.798T$	$N_{40}=K_{33}N_1^3N_2^2N_4$
$(Ca^{2+}+O^{2-})+(Mg^{2+}+O^{2-})+(SiO_2)=(CaO\cdot MgO\cdot SiO_2)$	$\Delta G^{\theta}=-124683+3.766T$	$N_{41}=K_{34}N_1N_2N_4$
$(Ca^{2+}+O^{2-})+(Mg^{2+}+O^{2-})+2(SiO_2)=(CaO\cdot MgO\cdot2SiO_2)$	$\Delta G^{\theta}=-80333-51.882T$	$N_{42}=K_{35}N_1N_2^2N_4$
$2(Mg^{2+}+O^{2-})+2(Al_2O_3)+5(SiO_2)=(2MgO\cdot2Al_2O_3\cdot5SiO_2)$	$\Delta G^{\theta}=-14422-14.808T$	$N_{43}=K_{36}N_2^5N_3^2N_4^2$
$(Ca^{2+}+O^{2-})+(TiO_2)+(SiO_2)=(CaO\cdot TiO_2\cdot SiO_2)$	$\Delta G^{\theta}=-122591.2+10.88T$	$N_{44}=K_{37}N_1N_2N_7$

The mass conservation equations for the CaO–SiO_2–Al_2O_3–MgO–FeO–MnO–TiO_2 slag balanced with bulk steel can be built based on the definitions of n_i and N_i for each structural unit as

$$a=\sum n_i\left(\begin{array}{l} 0.5N_1+N_8+3N_9+2N_{10}+3N_{11}+3N_{12}+12N_{13}+N_{14}+N_{15}+N_{16}+\\ N_{17}+3N_{18}+4N_{19}+2N_{37}+N_{38}+2N_{39}+3N_{40}+N_{41}+N_{42}+N_{44} \end{array}\right) \tag{2}$$

$$b=\sum n_i\left(\begin{array}{l} N_2+N_8+2N_9+N_{10}+N_{11}+2N_{20}+N_{22}+N_{23}+N_{28}+N_{32}+\\ N_{33}+N_{37}+2N_{38}+2N_{39}+2N_{40}+N_{41}+2N_{42}+5N_{43}+N_{44} \end{array}\right) \tag{3}$$

$$c=\sum n_i\left(\begin{array}{l} N_3+N_{12}+7N_{13}+N_{14}+2N_{15}+6N_{16}+3N_{20}+\\ N_{21}+N_{24}+N_{29}+N_{34}+N_{37}+N_{38}+2N_{43} \end{array}\right) \tag{4}$$

$$d = \sum n_i \left(\begin{array}{c} 0.5N_4 + 2N_{22} + N_{23} + N_{24} + N_{25} + 2N_{26} + \\ N_{27} + N_{39} + N_{40} + N_{41} + N_{42} + 2N_{43} \end{array} \right) \tag{5}$$

$$e = \sum n_i (0.5N_5 + 2N_{28} + N_{29} + N_{30} + 2N_{31}) \tag{6}$$

$$f = \sum n_i (0.5N_6 + N_{32} + 2N_{33} + N_{34} + N_{35} + 2N_{36}) \tag{7}$$

and

$$g = \sum n_i \left(\begin{array}{c} N_7 + N_{17} + 2N_{18} + 3N_{19} + N_{21} + N_{25} + N_{26} + \\ 2N_{27} + N_{30} + N_{31} + N_{35} + N_{36} + N_{44} \end{array} \right) \tag{8}$$

Based on the theory that the total MAC of each constitutional unit in CaO–SiO_2–Al_2O_3–MgO–FeO–MnO–TiO_2 slag with a fixed amount is equal to unity, Equation (9) can be derived as

$$\sum_{i=1}^{44} N_i = 1 \tag{9}$$

Equations (2)–(9) represent the MAC calculation model for each constitutional unit in CaO–SiO_2–Al_2O_3–MgO–FeO–MnO–TiO_2 slag systems. The activity of each constituent in the slag at the refining temperature can then be obtained.

Based on the IMCT, the simple molecule TiO_2 in the refining slags can be combined with ordinary molecules—such as CaO, Al_2O_3, MgO, FeO, MnO, and $CaO+SiO_2$—to form 13 stable de-titanium products as TiO_2, $CaO{\cdot}TiO_2$, $3CaO{\cdot}2TiO_2$, $4CaO{\cdot}3TiO_2$, $Al_2O_3{\cdot}TiO_2$, $MgO{\cdot}TiO_2$, $2MgO{\cdot}TiO_2$, $MgO{\cdot}2TiO_2$, $FeO{\cdot}TiO_2$, $2FeO{\cdot}TiO_2$, $MnO{\cdot}TiO_2$, $2MnO{\cdot}TiO_2$, and $CaO{\cdot}TiO_2{\cdot}SiO_2$, respectively. According to the reported expression of the manganese distribution ratio [10,11], the titanium distribution calculation model can be described as

$$\begin{aligned} L_{Ti} = \frac{(\%TiO_2)}{[\%Ti]} = \ & L_{Ti,TiO_2} + L_{Ti,CaO{\cdot}TiO_2} + L_{Ti,3CaO{\cdot}2TiO_2} + L_{Ti,4CaO{\cdot}3TiO_2} + L_{Ti,Al_2O_3{\cdot}TiO_2} \\ & + L_{Ti,MgO{\cdot}TiO_2} + L_{Ti,2MgO{\cdot}TiO_2} + L_{Ti,MgO{\cdot}2TiO_2} + L_{Ti,FeO{\cdot}TiO_2} \\ & + L_{Ti,2FeO{\cdot}TiO_2} + L_{Ti,MnO{\cdot}TiO_2} + L_{Ti,2MnO{\cdot}TiO_2} + L_{Ti,CaO{\cdot}TiO_2{\cdot}SiO_2} \\ = \ & M_{TiO_2} \cdot \sum n_i (N_{TiO_2} + N_{CaO{\cdot}TiO_2} + 2N_{3CaO{\cdot}2TiO_2} + 3N_{4CaO{\cdot}3TiO_2} \\ & + N_{Al_2O_3{\cdot}TiO_2} + N_{MgO{\cdot}TiO_2} + N_{2MgO{\cdot}TiO_2} + 2N_{MgO{\cdot}2TiO_2} + N_{FeO{\cdot}TiO_2} \\ & + N_{2FeO{\cdot}TiO_2} + N_{MnO{\cdot}TiO_2} + N_{2MnO{\cdot}TiO_2} + N_{CaO{\cdot}TiO_2{\cdot}SiO_2}) / [\%Ti] \end{aligned} \tag{10}$$

where L_{Ti} is the total titanium distribution ratio; $L_{Ti,i}$ represents the respective titanium distribution ratio of structure unit i containing TiO_2; N_i stands for the MAC of structure unit i; $\sum n_i$ denotes the sum of mole numbers for each structure unit in equilibrium (mol); and M_{TiO2} is the molar mass of TiO_2 (g/mol). Based on the IMCT model, the total titanium distribution ratio can then be acquired.

According to the calculation results of the IMCT model, the MACs of $3CaO{\cdot}2TiO_2$, $4CaO{\cdot}3TiO_2$, $Al_2O_3{\cdot}TiO_2$, $2MgO{\cdot}TiO_2$, $MgO{\cdot}2TiO_2$, $2FeO{\cdot}TiO_2$, $2MnO{\cdot}TiO_2$, and $CaO{\cdot}TiO_2{\cdot}SiO_2$ were lower than 10^{-5}. Therefore, their changes were ignored in the following discussion. In this case, Equation (10) can be simplified as

$$\begin{aligned} L_{Ti} = \frac{(\%TiO_2)}{[\%Ti]} &= L_{Ti,TiO_2} + L_{Ti,CaO{\cdot}TiO_2} + L_{Ti,MgO{\cdot}TiO_2} + L_{Ti,FeO{\cdot}TiO_2} + L_{Ti,MnO{\cdot}TiO_2} \\ &= M_{TiO_2} \cdot \frac{\sum n_i (N_{TiO_2} + N_{CaO{\cdot}TiO_2} + N_{MgO{\cdot}TiO_2} + N_{FeO{\cdot}TiO_2} + N_{MnO{\cdot}TiO_2})}{[\%Ti]} \end{aligned} \tag{11}$$

3. Results and Discussion

3.1. Comparison of Predicted and Measured Titanium Distribution Ratios

Comparisons between the calculated titanium distribution ratio $L_{Ti,cal}$ based on the IMCT and the measured $L_{Ti,mea}$ for $CaO–SiO_2–Al_2O_3–MgO–FeO–MnO–TiO_2$ slags balanced with liquid steel at 1853 K in the LF process are expounded in Figure 2. It reveals that the titanium distribution ratio predicted by the developed IMCT model exhibited a dependable agreement with the industrially measured results. Moreover, the predicted $L_{Ti,cal}$ values were all higher than the measured $L_{Ti,mea}$, which was due to the fact that the calculated values were acquired in an ideal equilibrium state, while during actual industrial production, the slag-metal reaction was in a local equilibrium or quasi-equilibrium state.

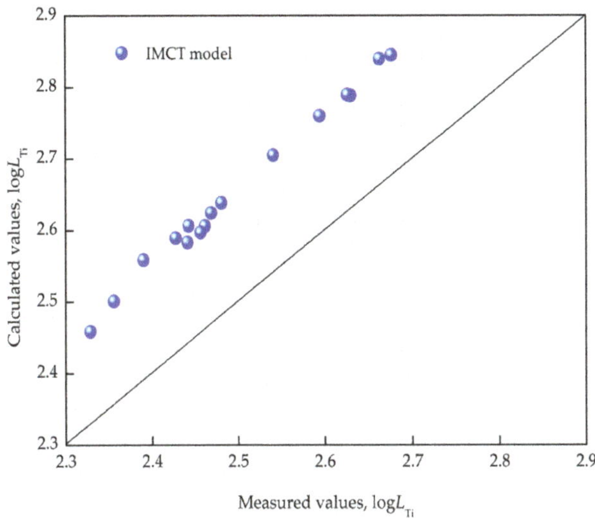

Figure 2. Comparisons between measured and calculated titanium distribution ratios for $CaO–SiO_2–Al_2O_3–MgO–FeO–MnO–TiO_2$ slags.

To verify the accuracy of the IMCT model, the mean deviations (Δ) of predictions by the IMCT model can be calculated as

$$\Delta = \frac{1}{Z}\sum_{n=1}^{Z}\left|\frac{\log L_{Ti,mea} - \log L_{Ti,cal}}{\log L_{Ti,mea}}\right| \times 100\% = 6.29\% < 6.3\% \tag{12}$$

where $L_{Ti,cal}$ and $L_{Ti,mea}$ are the calculated and measured titanium distribution ratios, respectively; and Z denotes the number of measured data. The mean deviation was lower than 6.3%, indicating that the titanium distribution ratio model can responsibly predict the maximum de-titanium potential of $CaO–SiO_2–Al_2O_3–MgO–FeO–MnO–TiO_2$ slags balanced with liquid steel at 1853 K in LF refining, and it can provide guidance for the design of a refining slag system.

3.2. Influence of Basicity on the Titanium Distribution Ratio

The relation between $L_{Ti,cal}$ or $L_{Ti,mea}$ and binary basicity (($\%CaO$)/($\%SiO_2$)), complex basicity (($\%CaO$) + 1.4($\%MgO$))/(($\%SiO_2$) + ($\%Al_2O_3$)), or optical basicity Λ ($\Lambda = \sum x_i \cdot \lambda_i$, where x_i is the mole fraction of a component, and λ_i is the optical basicity of a component in slag) obtained by using Pauling electronegativity [33] are depicted in Figures 3–5 (where R is the linear correlation coefficient), respectively. It is evident that (1) the relationship between $\log L_{Ti,cal}$ by the IMCT model or

$\log L_{Ti,mea}$ and optical basicity had a better dependence than that with the binary or complex basicity of $CaO–SiO_2–Al_2O_3–MgO–FeO–MnO–TiO_2$ slags and (2) raising the basicity can give rise to a distinctly increasing titanium distribution ratio. Linear fit results indicated that the optical basicity could better reflect the structure of the slags, and it is suggested to reflect the correlation between the titanium distribution ratio and basicity of the slags. The reasons for the different linear fits of these models is attributed to the different definitions of basicity. As for optical basicity, all the components in the slag are taken into account, which can reflect the whole features of the slag, while for binary basicity or complex basicity, the de-titanium contributions of other components were ignored.

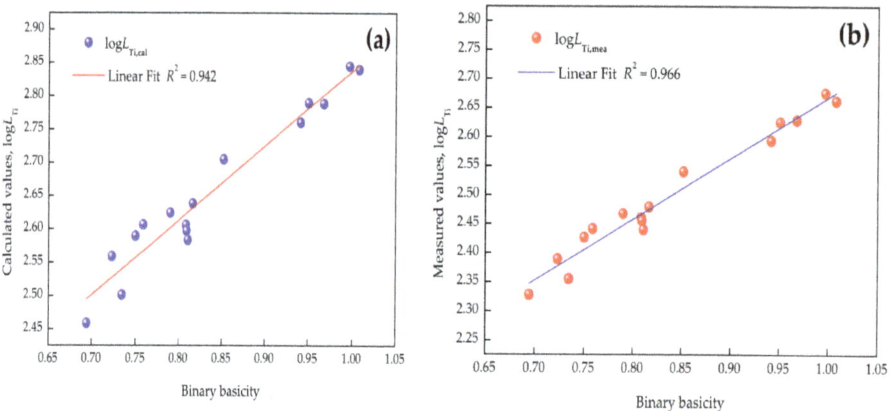

Figure 3. Correlation between binary basicity and titanium distribution ratio: (a) calculated by IMCT; (b) industrial measurements.

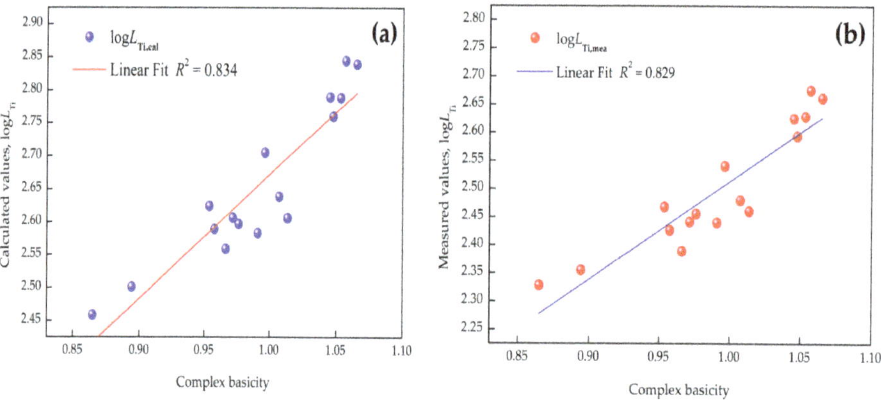

Figure 4. Correlation between complex basicity and titanium distribution ratio: (a) calculated by IMCT; (b) industrial measurements.

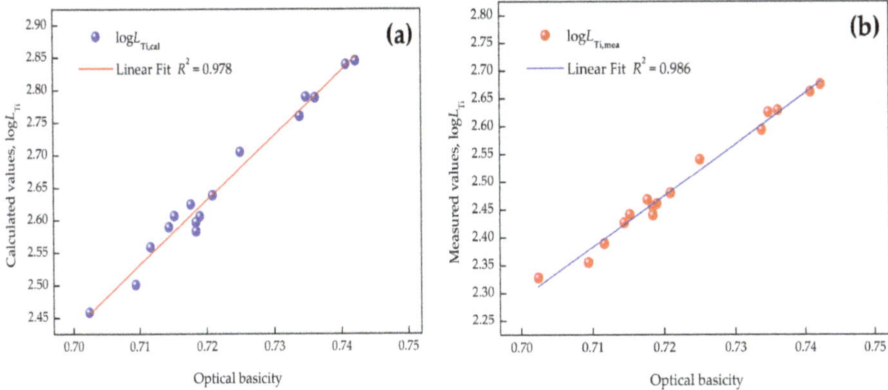

Figure 5. Correlation between optical basicity and titanium distribution ratio: (**a**) calculated by IMCT; (**b**) industrial measurements.

3.3. Contribution Ratio of the Respective Titanium Distribution Ratio Based on the IMCT

In addition to the total titanium distribution ratio of LF refining slags predicted by the established IMCT model, the respective titanium distribution ratio in the present slag system also can be determined. In order to expound the titanium distribution behavior in the LF process, the respective titanium distribution ratios of TiO_2, $CaO \cdot TiO_2$, $MgO \cdot TiO_2$, $FeO \cdot TiO_2$, and $MnO \cdot TiO_2$ in the slags can be acquired by the built IMCT model, respectively, and correlations between the quantitative titanium distribution ratio $L_{Ti,i,cal}$ of the five structural units and the $L_{Ti,cal}$ is described in Figure 6. It is evident that $L_{Ti,i,cal}$ had a nice linear relationship with $L_{Ti,cal}$, and the respective contribution rates of the structural units containing TiO_2 to total titanium distribution ratios in the slags can be defined by the fitted line gradient.

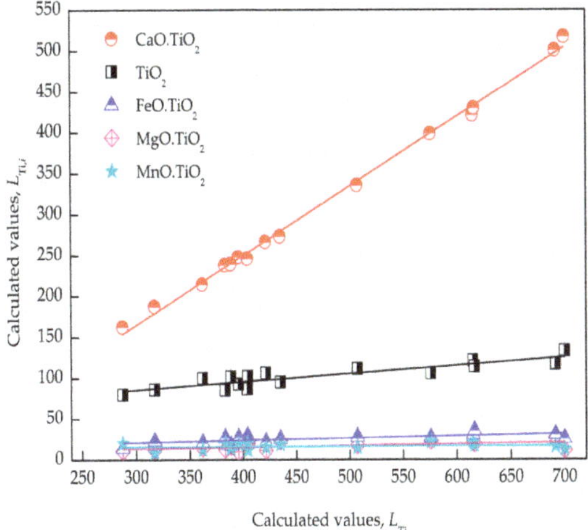

Figure 6. Correlation between the respective titanium distribution ratio and the total titanium distribution ratio of the slags, the solid line of color red, black, blue, magenta, and cyan represent the contribution ratio of $CaO \cdot TiO_2$, TiO_2, $FeO \cdot TiO_2$, $MgO \cdot TiO_2$, and $MnO \cdot TiO_2$, respectively.

The regression relationships between $L_{Ti,i,cal}$ and $L_{Ti,cal}$ for de-titanium products in the slags are listed in Table 5. Table 5 shows that the contribution rates of the five structural units to the predicted $L_{Ti,cal}$ by the IMCT model were approximately 9.97%, 84.96%, 2.03%, 2.65%, and 0.39%, respectively.

It can be concluded from the acquired industrial results that the constitutional unit CaO plays a pivotal role in $CaO–SiO_2–Al_2O_3–MgO–FeO–MnO–TiO_2$ slags in the de-titanium process.

Table 5. Linear regression expression of $L_{Ti,i,cal}$ against $L_{Ti,cal}$ for the structural units containing TiO_2 and their average contribution rates to $L_{Ti,cal}$ based on IMCT.

Constitutional Units	Expression of $L_{Ti,i,cal}$ against $L_{Ti,cal}$	Average Contribution Rate/%
TiO_2	$L_{Ti,TiO_2,cal} = 55.5334 + 0.0997L_{Ti,cal}$	9.97
$CaO·TiO_2$	$L_{Ti,CaO·TiO_2,cal} = -90.7440 + 0.8496L_{Ti,cal}$	84.96
$MgO·TiO_2$	$L_{Ti,MgO·TiO_2,cal} = 7.3030 + 0.0203L_{Ti,cal}$	2.03
$FeO·TiO_2$	$L_{Ti,FeO·TiO_2,cal} = 13.2733 + 0.0265L_{Ti,cal}$	2.65
$MnO·TiO_2$	$L_{Ti,MnO·TiO_2,cal} = 14.6340 + 0.0039L_{Ti,cal}$	0.39

4. Conclusions

A thermodynamic calculation model for the titanium distribution ratio of $CaO–SiO_2–Al_2O_3–MgO–FeO–MnO–TiO_2$ slags in the LF process at 1853 K has been established based on the IMCT. The built IMCT model has been tested against industrial measurements. The key findings can be drawn as follows:

(1) The established IMCT model for calculating the titanium distribution ratio exhibited a dependable agreement with the measurements, and the model can be responsibly applied to predict the maximum de-titanium potential in the LF process at metallurgical temperatures.

(2) The titanium distribution ratio will increase with the rise of basicity, and the optical basicity is suggested to describe the correlation between basicity and de-titanium ability of the slag. Higher optical basicity is in favor of the de-titanium process.

(3) The respective titanium distribution ratios of structural units containing TiO_2 can be acquired by the built IMCT model. The contribution rates of TiO_2, $CaO·TiO_2$, $MgO·TiO_2$, $FeO·TiO_2$, and $MnO·TiO_2$ to total de-titanium potential were approximately 9.97%, 84.96%, 2.03%, 2.65%, and 0.39%, respectively, revealing that the structural unit CaO plays a pivotal role in the slags in the de-titanium process.

Author Contributions: Methodology, W.F. and J.L.; Formal analysis, D.Z.; Investigation, Z.X.; Writing—review and editing, J.L.

Funding: This research was funded by the National Natural Science Foundation of China (grant nos. 51704105, 51874214, 21603070).

Acknowledgments: The authors gratefully acknowledge the resources partially provided by the State Key Laboratory of Refractories and Metallurgy, Wuhan University of Science and Technology.

Conflicts of Interest: The authors declare no conflicts of interest.

References

1. Abushosha, R.; Vipond, R.; Mintz, B. Influence of titanium on hot ductility of as cast steels. *Mater. Sci. Technol.* **1991**, *7*, 613–621. [CrossRef]

2. Chen, Z.; Li, M.; Wang, X.; He, S.; Wang, Q. Mechanism of floater formation in the mold during continuous casting of Ti-stabilized austenitic stainless steels. *Metals* **2019**, *9*, 635. [CrossRef]

3. Karmakar, A.; Kundu, S.; Roy, S.; Neogy, S.; Srivastava, D.; Chakrabarti, D. Effect of microalloying elements on austenite grain growth in Nb–Ti and Nb–V steels. *Mater. Sci. Technol. A* **2014**, *30*, 653–664. [CrossRef]

4. Reyes-Calderón, F.; Mejía, I.; Boulaajaj, A.; Cabrera, J.M. Effect of microalloying elements (Nb, V and Ti) on the hot flow behavior of high-Mn austenitic twinning induced plasticity (TWIP) steel. *Mater. Sci. Eng. A* **2013**, *560*, 552–560. [CrossRef]

5. Cui, H.Z.; Chen, W.Q. Effect of boron on morphology of inclusions in tire cord steel. *J. Iron Steel Res. Int.* **2012**, *19*, 22–27. [CrossRef]

6. Wu, S.; Liu, Z.; Zhou, X.; Yang, H.; Wang, G. Precipitation behavior of Ti in high strength steels. *J. Cent. South Univ.* **2017**, *24*, 2767–2772. [CrossRef]

7. Li, J.Y.; Zhang, W.Y. Effect of TiN inclusion on fracture toughness in ultrahigh strength steel. *ISIJ Int.* **1989**, *29*, 158–164. [CrossRef]

8. Petit, J.; Sarrazin-Baudoux, C.; Lorenzi, F. Fatigue crack propagation in thin wires of ultra high strength steels. *Procedia Eng.* **2010**, *2*, 2317–2326. [CrossRef]

9. Liu, H.Y.; Wang, H.L.; Li, L.; Zheng, J.Q.; Li, Y.H.; Zeng, X.Y. Investigation of Ti inclusions in wire cord steel. *Ironmak. Steelmak.* **2011**, *38*, 53–58. [CrossRef]

10. Duan, S.C.; Guo, X.L.; Guo, H.J.; Guo, J. A manganese distribution prediction model for $CaO–SiO_2–FeO–MgO–MnO–Al_2O_3$ slags based on IMCT. *Ironmak. Steelmak.* **2017**, *44*, 168–184. [CrossRef]

11. Duan, S.C.; Li, C.; Guo, X.L.; Guo, H.J.; Guo, J.; Yang, W.S. A thermodynamic model for calculating manganese distribution ratio between $CaO–SiO_2–MgO–FeO–MnO–Al_2O_3–TiO_2–CaF_2$ ironmaking slags and carbon saturated hot metal based on the IMCT. *Ironmak. Steelmak.* **2017**, *45*, 655–664. [CrossRef]

12. Li, B.; Li, L.; Guo, H.; Guo, J.; Duan, S.; Sun, W. A phosphorus distribution prediction model for $CaO–SiO_2–MgO–FeO–Fe_2O_3–Al_2O_3–P_2O_5$ slags based on the IMCT. *Ironmak. Steelmak.* **2019**. [CrossRef]

13. Yang, X.M.; Li, J.Y.; Chai, G.M.; Duan, D.P.; Zhang, J. A Thermodynamic model for predicting phosphorus partition between CaO–based slags and hot metal during hot metal dephosphorization pretreatment process based on the ion and molecule coexistence theory. *Metall. Mater. Trans. B* **2016**, *47*, 2279–2301. [CrossRef]

14. Yang, X.M.; Duan, J.P.; Shi, C.B.; Zhang, M.; Zhang, Y.L.; Wang, J.C. A thermodynamic model of phosphorus distribution ratio between $CaO–SiO_2–MgO–FeO–Fe_2O_3–MnO–Al_2O_3–P_2O_5$ slags and molten steel during a top–bottom combined blown converter steelmaking process based on the ion and molecule coexistence theory. *Metall. Mater. Trans. B* **2011**, *42*, 738–770. [CrossRef]

15. Yang, X.M.; Zhang, M.; Chai, G.M.; Li, J.Y.; Liang, Q.; Zhang, J. Thermodynamic models for predicting dephosphorisation ability and potential of $CaO–FeO–Fe_2O_3–Al_2O_3–P_2O_5$ slags during secondary refining process of molten steel based on ion and molecule coexistence theory. *Ironmak. Steelmak.* **2016**, *43*, 663–687. [CrossRef]

16. Li, J.Y.; Zhang, M.; Guo, M.; Yang, X.M. Enrichment mechanism of phosphate in $CaO–SiO_2–FeO–Fe_2O_3–P_2O_5$ steelmaking slags. *Metall. Mater. Trans. B* **2014**, *45*, 1666–1682. [CrossRef]

17. Yang, X.M.; Li, J.Y.; Zhang, M.; Yan, F.J.; Duan, D.P.; Zhang, J. A further evaluation of the coupling relationship between dephosphorization and desulfurization abilities or potentials for CaO–based slags: Influence of slag chemical composition. *Metals* **2018**, *8*, 1083. [CrossRef]

18. Shi, C.B.; Yang, X.M.; Jiao, J.S.; Li, C.; Guo, H.J. A sulphide capacity prediction model of $CaO–SiO_2–MgO–Al_2O_3$ ironmaking slags based on the ion and molecule coexistence theory. *ISIJ Int.* **2010**, *50*, 1362–1372. [CrossRef]

19. Yang, X.M.; Shi, C.B.; Zhang, M.; Chai, G.M.; Wang, F. A thermodynamic model of sulfur distribution ratio between $CaO–SiO_2–MgO–FeO–MnO–Al_2O_3$ slags and molten steel during LF refining process based on the ion and molecule coexistence theory. *Metall. Mater. Trans. B* **2011**, *42*, 1150–1180. [CrossRef]

20. Yang, X.M.; Zhang, M.; Shi, C.B.; Chai, G.M.; Zhang, J. A sulfide capacity prediction model of $CaO–SiO_2–MgO–FeO–MnO–Al_2O_3$ slags during the LF refining process based on the ion and molecule coexistence theory. *Metall. Mater. Trans. B* **2012**, *43*, 241–266. [CrossRef]

21. Yang, X.M.; Li, J.Y.; Zhang, M.; Chai, G.M.; Zhang, J. Prediction model of sulfide capacity for $CaO–FeO–Fe_2O_3–Al_2O_3–P_2O_5$ slags in a large variation range of oxygen potential based on the ion and molecule coexistence theory. *Metall. Mater. Trans. B* **2014**, *45*, 2118–2137. [CrossRef]

22. Yang, X.M.; Li, J.Y.; Zhang, M.; Zhang, J. Prediction model of sulphur distribution ratio between $CaO–FeO–Fe_2O_3–Al_2O_3–P_2O_5$ slags and liquid iron over large variation range of oxygen potential during secondary refining process of molten steel based on ion and molecule coexistence theory. *Ironmak. Steelmak.* **2016**, *43*, 39–55. [CrossRef]

23. Yang, X.M.; Zhang, M.; Zhang, J.L.; Li, P.C.; Li, J.Y.; Zhang, J. Representation of oxidation ability for metallurgical slags based on the ion and molecule coexistence theory. *Steel Res. Int.* **2014**, *85*, 347–375. [CrossRef]

24. Verein Deutscher Eisenhuttenleute (VDEh). *Slag Atlas*, 2nd ed.; Woodhead Publishing Limited: Cambridge, UK, 1995.

25. Chen, J.X. *Common Charts and Databook for Steelmaking*, 2nd ed.; Metallurgical Industry Press: Beijing, China, 2010.

26. Zhang, J. *Computational Thermodynamics of Metallurgical Melts and Solutions*; Metallurgical Industry Press: Beijing, China, 2007.

27. Rein, R.H.; Chipman, J. Activities in the liquid solution SiO$_2$–CaO–MgO–Al$_2$O$_3$ at 1600 °C. *Trans. Met. Soc. AIME* **1965**, *233*, 415–425.

28. Turkdogan, E.T. *Physical Chemistry of High Temperature Technology*; Academic Press: New York, NY, USA, 1980; pp. 8–12.

29. Gaye, H.; Welfringer, J. *Proceedings of the Second International Symposium on Metallurgical Slags and Fluxes*; Fine, H.A., Gaskell, D.R., Eds.; TMS–AIME: Lake Tahoe, NV, USA, 1984; pp. 357–375.

30. Ban-ya, S.; Chiba, A.; Hikosaka, A. Thermodynamics of Fe$_t$O–M$_x$O$_y$(M$_x$O$_y$=CaO, SiO$_2$, TiO$_2$, and Al$_2$O$_3$) binary melts in equilibrium with solid iron. *Tetsu-To-Hagane* **1980**, *66*, 1484–1493. [CrossRef]

31. Timucin, M.; Muan, A. Activity–composition relations in NiAl$_2$O$_4$–MnAl$_2$O$_4$ solid solutions and stabilities of NiAl$_2$O$_4$ and MnAl$_2$O$_4$ at 1300 °C and 1400 °C. *J. Am. Ceram. Soc.* **1992**, *75*, 1399–1406. [CrossRef]

32. Barin, I.; Knacke, O.; Kubaschewski, O. *Thermochemical Properties of Inorganic Substances (Supplement)*; Springer: New York, NY, USA, 1977; pp. 392–445.

33. Pauling, L. *The Nature of Chemical Bond*; Cornell University Press: Ithaca, NY, USA, 1960.

Review

Physical and Mathematical Modelling of Mass Transfer in Ladles due to Bottom Gas Stirring: A Review

Alberto N. Conejo

School of Metallurgical and Ecological Engineering, University of Science and Technology, 30 Xueyuan Road, Haidian District, Beijing 100083, China; aconejonava@hotmail.com or aconejo@ustb.edu.cn

Received: 9 April 2020; Accepted: 8 May 2020; Published: 27 June 2020

Abstract: Steelmaking involves high-temperature processing. At high temperatures mass transport is usually the rate limiting step. In steelmaking there are several mass transport phenomena occurring simultaneously such as melting and dissolution of additions, decarburization, refining (De-P and De-S), etc. In ladle metallurgy, refining is one of the most important operations. To improve the rate of mass transfer bottom gas injection is applied. In the past, most relationships between the mass transfer coefficient (*mtc*) and gas injection have been associated with stirring energy as the dominant variable. The current review analyzes a broad range of physical and mathematical modeling investigations to expose that a large number of variables contribute to define the final value of the *mtc*. Since bottom gas injection attempts to improve mixing phenomena in the whole slag/steel system, our current knowledge shows limitations to improve mixing conditions in both phases simultaneously. Nevertheless, some variables can be optimized to reach a better performance in metallurgical ladles. In addition to this, the review also provides a state of the art on liquid–liquid mass transfer and suggests the current challenges in this field.

Keywords: mass transfer coefficient; mixing time; physical modeling; mathematical modeling; kinetic models

---

1. Introduction

Many studies have been conducted to describe transport phenomena in ladles and the results are summarized in several reviews [1–4]. The previous reviews have included mass transfer to a limited extent. The reviews from Mazumdar et al. [1,2] were focused on solid–liquid systems. Sichen [3] discussed limitations of the two-film theory and Liu et al. [4] summarized mixing phenomena by physical and mathematical modeling due to gas stirring in ladles, indicating that a limited number of studies involved mass transfer. Ghotli et al. [5] reviewed liquid–liquid mass transfer in mechanically stirred vessels. Mechanical and gas stirring involve different operational parameters. In steelmaking, in particular in ladle metallurgy, its rate of production and final steel quality rely on mixing efficiency both liquid steel and liquid slag. The current review will primarily focus on the physical and mathematical modeling work that involves liquid–liquid mass transfer due to bottom gas injection.

The molar flux of species j (N_j) is proportional to the concentration gradient according with Fick's law for diffusion without convection and shown below. This form of the equation is valid for dilute systems, typical of steelmaking. The proportionality constant is the *mass transfer coefficient* (*mtc*). Depending on the experimental conditions the units for the *mtc* can change. To avoid confusion the units are briefly revised. If the concentration units for C_j are in mol/cm^3 then the units for k_j are cm/s [6].

$$N_j = -D_j\left(\frac{\partial C_j}{\partial y}\right) = -\frac{D_j}{\delta_m}\Delta C_j = k_j\left(C_j^b - C_j^{eq}\right) \tag{1}$$

where N_j is the molar flux in mol/cm^2·s, C_j is the concentration in mol/cm^3, D_j is the diffusion coefficient in cm^2/s, y is distance in cm, δ_m is the diffusion boundary layer thickness in cm, k_j is the mass transfer coefficient in cm/s. Superscripts *b* and *eq* represent bulk and equilibrium values, respectively. Subscript j represents a chemical species.

Since the molar flux (N_j) is the amount of material transferred per unit area and unit time, the experimental measurement of the *mtc* requires knowledge of the interfacial area (A). In most cases this value is unknown. In this condition, the mass transfer coefficient is reported as the product, (k_j·A) called *volumetric mass transfer coefficient (vmtc)*, with units cm^3/min. If the volume of the liquid remains constant, we denote this *vmtc* as $\left(k_j \cdot a\right)$, with units min^{-1}

$$N_j = \frac{1}{A}\frac{\partial n_j}{\partial t} = k_j\left(C_j^b - C_j^{eq}\right) \tag{2}$$

$$\frac{\partial n_j}{\partial t} = V\frac{\partial C_j}{\partial t} = \left(k_j \cdot A\right)\left(C_j^b - C_j^{eq}\right) \tag{3}$$

$$\frac{\partial C_j}{\partial t} = \left(k_j \cdot \frac{A}{V}\right)\left(C_j^b - C_j^{eq}\right) = \left(k_j \cdot a\right)\left(C_j^b - C_j^{eq}\right) \tag{4}$$

The effect of the stirring conditions on the mass transfer coefficient has been experimentally measured for different systems: (i) gas–solid–liquid system: a typical example is the melting of additions by mechanical stirring (solid-liquid system) or by gas stirring (gas-solid-liquid system), (ii) gas–liquid system: some examples are gas absorption from the atmosphere into liquid steel, absorption of elements when different types of gases are injected (nitrogen, carbon dioxide, etc.) or oxygen injection for decarburization, and (iii) gas–liquid–liquid system: the main examples are slag/metal interfacial reactions. To clarify terms used in this work, in the gas–liquid system the gas is a reacting species that dissolves in the liquid in contrast to the gas–liquid–liquid system where the gas is an inert species and where an impurity dissolved in the lower liquid phase diffuses to the upper liquid phase. This review will focus primarily on the study of the *mtc* in the gas–liquid–liquid system, however, as an introduction the other systems are briefly reviewed in the beginning in Sections 2 and 3, followed by a detailed review in Section 4 on liquid–liquid mass transfer involving both physical and mathematical modeling studies. Section 5 provides a final assessment of our current understanding on this subject and suggest guidelines for further research.

2. Mass Transfer during the Melting Rate of Additions (Solid–Liquid and Gas–Solid–Liquid Systems)

The effect of the stirring conditions on the mass transfer coefficient (*mtc*) during the melting rate of additions in ladles has been investigated in detail using a rotating cylinder electrode (RCE) immersed in a liquid and has resulted in many semi-empirical correlations involving dimensionless numbers, as shown in Table 1. Most of these correlations involve three dimensionless numbers; Sherwood (Sh), Reynolds (Re), and Schmidt (Sc). This is the result of a simple dimensional analysis, describing the mass transfer coefficient (k) as a function of the fluid's kinematic viscosity (ν), fluid's velocity (U), diffusivity (D) of transferred species, as well as a characteristic length of the reactor (*l*), under isothermal conditions.

$$k = f(U, D, \nu, l) \tag{5}$$

This system involves five variables and can be described with two dimensions (L,T), therefore, in accordance with the π-theorem, it can be defined with three π-dimensionless groups, as follows:

$$\pi_1 = k(l)^{a_1}(D)^{b_1} \tag{6}$$

$$\pi_2 = U(l)^{a_2}(D)^{b_2} \tag{7}$$

$$\pi_3 = \nu(l)^{a_3}(D)^{b_3} \tag{8}$$

Applying the principle of dimensional homogeneity, the resulting π-groups are:

$$\frac{kl}{D} = f\left(\frac{Ul}{D}\right)^a \left(\frac{\nu}{D}\right)^b \tag{9}$$

Alternatively:

$$Sh = f\left(Re^a Sc^b\right) \tag{10}$$

If the stirring conditions produce slag emulsification, the previous analysis should be extended to include surface tension [7]. In the mass transfer model developed by Oeters and Xie [8] this relationship holds for two cases under non-turbulent flow; a liquid in contact with a free surface and a liquid in contact with a solid wall. In the first case the velocity at the interface is the same as the velocity in the bulk and in the second case the velocity of the liquid at the interface is zero. Both are limiting cases for the liquid–liquid interface.

The first systematic correlation involving Sherwood (Sh), Reynolds (Re), and Schmidt (Sc) numbers, describing the *mtc* under turbulent flow without gas injection was reported by Eisenberg et al. in 1955. It has been confirmed to remain acceptable for the dissolution rate of iron into liquid steel by subsequent investigations [9–12]. An important parameter is the velocity of the fluid. If the experiments are carried out without gas injection, that velocity can be estimated from the peripheral velocity of the RCE. Under multi-phase flow conditions, the velocity components of the fluid are needed in order to compute the *mtc*. This information can be obtained with the development of mathematical models [13–16], by direct measurements, for example with particle image velocimetry (PIV) or laser doppler velocimetry (LDV) [17], by photographic analysis [18,19], and also with an energy balance [18].

Another group of correlations have been reported using the mass transfer Stanton number (St). It has been applied in the dissolution rate of solid lime into liquid slag [20–22].

The gas injection position is an important variable because it affects mixing phenomena. Wright [23] reported that the dissolution rate of a steel rod under natural convection was higher when placed in the center in comparison with an off-center position, on the contrary Koria [19] reported a higher dissolution rate when the rod was located off-center, under central bottom gas injection conditions. Alloy additions in the ladle should be made under conditions that enhance its melting rate.

Table 1. Mass transfer correlations for solid–liquid and gas–solid–liquid systems.

	Year	Authors	Solid Bar	r/R	Mass Transfer Correlations
1	1955	Eisenberg et al.	$(C_6H_5CO_2H)_{(s)}$	-	$Sh = 0.079Re^{0.7}Sc^{0.356}$
2	1967	Kosaka and Minowa	Steel bar	-	$Sh = 0.064Re^{0.75}Sc^{0.33}$
3	1974	Kim and Pehlke	Iron bar	-	$Sh = 0.112Re^{0.67}Sc^{0.356}$
4	1985	Shigeno et al.	Steel bar	-	$Sh = 0.051Re^{0.78}Sc^{0.33}$
5	1979	Szekely et al.	Graphite bar	0	$Sh = 2 + 0.72Re^{0.75}Ti^{0.25}Sc^{0.33}$
6	1989	Wright	Steel bar *	-	$Sh = 0.13(Gr{\cdot}Sc)^{0.75}$
7	1990	Mazumdar et al.	$(C_6H_5CO_2H)_{(s)}$	0	$Sh = 0.73Re^{0.57}Ti^{0.32}Sc^{0.33}$
8	1992	Mazumdar et al.	Steel rod	0	$k = 7.8 \times 10^{-3} Q^{0.19}$
9	2008	Kitamura et al.	Solid lime	-	$St\, Sc^{0.66} = 0.378Re^{-0.31}$

* natural convection, where: $Sh = k_m L/D$, $St = k_m/U$, $Re = \rho UL/\mu$, $Sc = \mu/\rho D$, $Ti = \sqrt{U_{rms}^2/\overline{U}_0}$. k_m is the convective *mtc*, D represents mass diffusivity, μ is the dynamic viscosity of the fluid, ρ is the density of the fluid, U the velocity of the fluid, $\overline{U}_0$ is the velocity at the center line of the rising two phase plume and U_{rms} is the rms or fluctuating velocity.

3. Mass Transfer due to Gas Absorption (Gas–Liquid System)

Gas absorption is an important phenomenon in steelmaking that covers the absorption of undesirable gases from the atmosphere. Maruoka et al. [24] investigated the removal of oxygen by water modeling with different layouts of gas injection. The whole experimental data was described by a relationship between the *vmtc* and the product of the ladle eye times the bubble velocity, subsequently, the *vmtc* was defined in terms of the gas flow rate and the number of nozzles $\left(\propto Q^{0.87} N^{0.13}\right)$. Kato et al. [25] measured the absorption of oxygen in a water model and found that the *mtc* is higher for bottom gas injection in comparison with top gas injection. Another group of studies have been carried out on the absorption of CO_2 in aqueous-NaOH solutions [26,27]. Inada et al. [26] reported that increasing the number of nozzles decreases the *vmtc* per one nozzle. They compared one, three, and five nozzles. These reports found an exponential relationship between the *mtc* and stirring energy, with an exponent in the range from 0.65 to 0.8. Rui et al. [28] also measured the rate of absorption of CO_2 and found that the *mtc* is higher for one oval snorkel in comparison with a circular snorkel, if the nozzle radial position is located between the center and half radius.

Mass transfer in solid–liquid, gas–liquid, and liquid–liquid systems, even if gas stirring is not involved, has many similarities. In all of these systems mass transfer is controlled by diffusion coefficients, velocities (solid, liquid or gas phases), physical properties for the phase involved, etc. At the same time there are important differences. If mass transfer from a solid is involved, the first step is melting and then in a second step is the dissolution process. Mass transfer from a gas phase is different to the case that involves mass transfer due to chemical reaction at the slag/metal interface, not only because the phases involved are different but the chemical reaction itself. The main point is to understand that different variables operate in those processes. In the following sections the review will focus on the previous work that has been developed to identify the variables that affect the rate of mass transfer in gas–liquid–liquid systems.

4. Slag/Metal Interfacial Mass Transfer due to Bottom Gas Injection (Gas–Liquid–Liquid Systems)

4.1. Physical Modeling

The *mtc* in bubble driven systems has been investigated by physical and mathematical modeling and is usually reported as a function of the stirring energy or gas flow rate since these terms are proportional to each other. Table 2 summarizes a large number of correlations between the *mtc* and gas stirring conditions. In addition to ladles, some of the correlations reported include bottom gas injection in other metallurgical reactors like the QBOP.

The first systematic physical modeling work to describe mass transfer as a function of gas injection was conducted by Richardson et al., starting in the late 1960s [29,30]. In one group of experiments [29], bubbles from 3–47 mL were passed through a column containing Hg and aqueous iron chloride. This system can be adjusted to get mass transfer control in either phase. It was found that mass transfer increased with the bubble size and height of the liquid. These authors [30] were the first ones to apply a turbulence theory to describe the *mtc* as a function of gas flow rate and reactor's dimensions. Li and Yin [31] also reported the effect of the bubble size, however instead of the *mtc* increasing with an increase in bubble size they reported a decrease, for nozzle diameters in the range from 1 to 3 mm. Inada and Watanabe [26] as well as Fruehan and Martonik [32] reported that the nozzle diameter has minimal or no effect on the *vmtc*, respectively. Riboud and Olette [33,34] investigated the desulphurization of liquid steel and reported the *mtc* as a function of the specific volumetric gas flow rate across the slag/metal interface, similar to the previous work by Richardson et al.

Up to 1975 mixing phenomena for bubble stirred systems were related with the gas flow rate. In this year Nakanishi et al. reported a correlation between mixing time and stirring energy [35] and eventually this idea was also used to report correlations between mass transfer and stirring energy. Stirring energy is directly proportional to the gas flow rate $(\varepsilon \propto Q^n)$. The exponent n is close to the unity [36]. This result can be theoretically derived from a relationship reported by Asai et al. [37] where $\varepsilon \propto U^3$ and the experimental data from Lehner et al. [38] where $U \propto Q^{1/3}$, then $\varepsilon \propto Q$. Engh and Oeters have also confirmed the linear relationship [39,40]. Many equations have been reported to compute stirring energy as a function of gas flow rate [41], one that is simple and accurate was reported by Mazumdar and Guthrie [42], as follows:

$$\varepsilon = \frac{\rho_l g H_l Q}{\rho_l \pi R^2 H_l} = \frac{gQ}{\pi R^2} \tag{11}$$

where ε is stirring energy in Watts/ton, ρ_l is the density of the liquid in ton/m^3, g is the acceleration due to gravity at the surface of Earth, 9.81 m/s^2, H_l is the height of the liquid in m, R is the radius of the reactor in m, Q is the gas flow rate in Nm^3/s.

Nakanishi et al. [43] reported results from a water model scaled from a Q-BOP indicating a change in the rate of mass transfer at about 80 Nl/min with bottom gas injection. Ishida et al. [44] and Berg et al. [45] also reported a change in the rate of mass transfer for a slag/metal system and in both cases a sharp transition for a stirring energy at 60 Watt/ton was reported. Umezawa et al. [46,47] employing also a steel/slag system and a much larger range in stirring energies did not report a change in the rate of mass transfer. Similar results were reported in a subsequent study with mechanical stirring [47]. In addition to correlations between the *mtc* and stirring energy, Clinton et al. [48] found a relationship between the *mtc* and the superficial velocity, and Minda et al. [49] found that increasing the volume of slag increases the *mtc*, conducting experiments at high stirring conditions.

In 1983 Asai et al. [37] published a review with 12 correlations including their own water modeling work. They analyzed the value of the exponent on the stirring energy by increasing the gas flow rate and suggested a drastic change from about 0.25 to more than unity, at about 450 W/m^3 (60 W/ton). They also analyzed that the low value can be predicted by the penetration theory under laminar flow. This work was updated to 15 correlations in 1988 [50]. This work clearly shows that gas flow rate is one of the main variables affecting the rate of mass transfer. Patel et al. [51] used an aqueous-NaOH solution covered by hexane and iodine dissolved in water as a tracer, which, when transferred into hexane it changes color to violet. The iodine equilibrium concentration was reported as 0.081% for an initial concentration of 0.16%. Their results were explained on the basis of Higbie's penetration theory.

The rate of mass transfer can change as a function of the gas flow rate. The change in slope is defined with a different correlation. Sawada et al. [52] reported one single correlation to describe the *mtc* including the reactor diameter to summarize their water modeling results. Ooga et al. [53] reported one change of slope that increased as the gas flow rate also increased. In this work slag emulsification was evident at a critical gas flow rate. The critical gas flow rate increased from 0.6 to 1.2 Nl/min by increasing the diameter of the ladle from 11 to 18 cm. Endo et al. [54] reported three changes in the slope, from low to high and then low again. Hirasawa et al. [55–59] carried out experiments at high temperatures analyzing the effect of reactor diameter, gas flow rate, height of liquid, and slag thickness. They reported three slopes. The slope decreased from the first to the second region at a critical gas flow rate and then increased again from the second to the third region. A strong slag emulsification was observed only in the third region. For the first region the critical gas flow rate changed with reactor diameter. The increase of the *mtc* in the third region was attributed to slag emulsification. They explained that in region two the motion of the slag increased but led to suppression of the motion of the lower phase due to a higher slag viscosity and a large interfacial tension. They reported that increasing the height of the liquid up to a H/D ratio of one, the *mtc* also increased and then remained constant, in regions one and two. Their work was the first attempt to unify all experimental data applying Davies theory of turbulence. Mukawa et al. [60] also reported a similar correlation for the *mtc*.

Fruehan et al. [61–63] investigated for the first time the relationship between mass transfer and mixing time in a ladle. They analyzed the following variables: gas flow rate, number of nozzles, nozzle radial position, slag volume, slag viscosity, and nozzle diameter. Mixing time is a parameter that measures the mixing efficiency of the primary phase (liquid steel), on the other hand, mass transfer measures the diffusion of species from the primary phase to the upper slag layer. Although both phenomena are affected by the mixing conditions, the final effect of the gas injection system can have dramatically different effects. Their conclusions are listed below:

i   At high gas flow rates mixing time and mass transfer cannot be improved simultaneously. Mass transfer is improved with central gas injection because the slag layer has better mixing conditions, however this position is the worst case for mixing liquid steel. At low gas flow rates, the rate of mass transfer is independent of the gas injection layout.

ii   Mass transfer shows three slopes that increase as the gas flow rate (Q) also increases. The change in slope is due to slag emulsification and occurs at about 5.3 W/ton. The nozzle diameter and slag viscosity do not have any effect at low Q but at high Q a decrease in slag viscosity and an increase in the nozzle diameter increase the *mtc*. When the slag viscosity increases it also increases the critical gas flow rate for slag emulsification. The authors also reported that increasing the slag thickness, at any Q, the *mtc* increases.

Figure 1 shows the results reported by Kim and Fruehan [61] on mixing time and mass transfer for central and off-central gas injection. It is clear that central gas injection has poor mixing conditions in the bath but on the other hand, the mass transfer coefficient is higher for central gas injection, an indication of better mixing conditions in the slag layer.

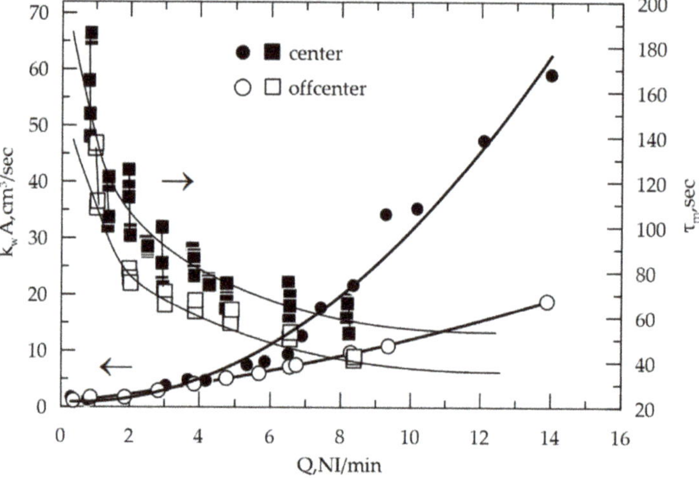

Figure 1. Effect of gas flow rate on both mixing time and the mass transfer coefficient, comparing central and off-central gas injection. Adapted from Kim and Fruehan [61].

Tata steel researchers reported that mixing time and mass transfer can be improved simultaneously applying a gas injection layout that involves differential gas flow rates in each nozzle, however this finding was not reported for a ladle but for a QBOP converter [64]. The stirring conditions in both reactors are quite different as will be explained below. They applied three configurations and the best case was defined when the gas flow rate was linearly increased from one side to the opposite side, using eight nozzles located at about 58% of the radius of the bottom.

Mietz et al. [65] reported results from three different geometric scale ladles using centric and eccentric gas injection. They confirmed that centric gas injection yields a higher rate of mass transfer in particular if a critical gas flow rate for emulsification is exceeded. The critical stirring energy in the three water models was about 15 W/ton. A further analysis also indicated that emulsification is much higher with centric gas injection in comparison with eccentric gas injection [66].

There are few studies regarding the effect of the number of nozzles in ladles on the *mtc*. Endo et al. [54] reported that the rate of mass transfer is higher with two porous plugs in comparison with one porous plug, in particular at high gas flow rates. The same report also cites a work from Aichi steel confirming similar results. A more recent work by Lou and Zhu [67] compared the desulphurization (DeS) ratio in a ladle with one and two nozzles and found that two nozzles (180° separation angle) yield a higher DeS ratio (48%), compared to 46% for central gas injection and 44% for off-center gas injection.

Koria et al. [68–70] studied mass transfer in a water model from a QBOP process. Under similar experimental conditions they reported two different correlations with exponents on the stirring energy of 0.4 and 1. These authors noticed the different time scales between mixing time and mass transfer. The time scale for mass transfer is higher than that for mixing time. Oeters et al. [71] conducted mass transfer experiments on desulphurization at high temperatures. They applied the boundary layer theory to predict the *mtc* with a very good agreement and also reported a change in slope due to slag emulsification at a critical stirring energy of about 4 W/ton. Kitamura et al. [72] reported data for high temperature experiments describing a correlation for the *mtc* in terms of the dimensions of the reactor and also reported a value of activation energy for the rate of desiliconization of hot metal. Koria [73] reported water modeling results on *mtc* for top and bottom gas injection in a QBOP model, defining a correlation that indicates the contribution from both top and bottom gas injection. The work reported by Lachmund et al. [74,75] on an industrial ladle furnace indicates a linear relationship between the *mtc* and stirring energy, one special feature in this work is the large values on stirring energies up to 200 W/ton at atmospheric pressure that led to a large slag emulsification. Most probably the authors have used the gas flow rate at (TP) and not (STP) conditions, which would explain such a high value. They reported the formation of approximately 5×10^7 slag droplets, which increased the surface area up to 173 m^2. Sulasalmi et al. [76] and Senguttuvan and Irons [77] reported the relationship between a liquid's velocity and the generation rate of the interfacial area and residence time of the slag droplets due to slag emulsification.

Gas injection in an industrial ladle furnace is carried out by porous plugs, however, most of the physical modeling work is carried out using nozzles. Few investigations have compared mixing phenomena with both injection elements. Stapurewicz and Themelis [78] compared mass transfer from a gas phase into a liquid with both injection systems and found a higher absorption rate with a porous plug due to the formation of fine bubbles in a water model that largely increase the reaction interface (up to 230%), this result however could be misleading because bubble formation from a porous plug in a water model is different to the argon/liquid steel system. The bubbles in a gas/metal system are bigger because of the non-wetting conditions and their final volume is dominated by the properties of the fluid. An example of the large differences in physical properties in both systems is the surface tension/density ratio; for the air/steel system is three times higher in comparison with the air/water system, 251 and 73, respectively [79]. Mori [80] has even questioned the results from water models to describe high-temperature slag/metal systems on the basis of a low interfacial tension for the water/oil system in comparison with slag/metal systems.

The effect of pressure on the *mtc* has not been considered in the previous correlations since the ladle furnace works under atmospheric conditions, however, the same ladle can operate under vacuum when transferred to the RH or tank degasser stations. The effects of gas stirring under vacuum are enhanced because there is a considerable increment in the bubble size and therefore for the same gas flow rate the *mtc* increases in comparison with atmospheric conditions. Lachmund et al. reported that gas stirring is five times more intense under vacuum (less than 4 mbar) compared with atmospheric conditions [74]. Under vacuum conditions, Sakaguchi and Ito [27] found a correlation of the form $ka \propto \varepsilon^{0.71}$.

From Table 2, it can be seen that using physical modeling to study mass transfer the results are generally expressed by two relationships, on the contrary, for a steel-slag system there is in general only one relationship. The reason for this behavior is the formation of an oil emulsion at a critical gas flow rate in the oil–water system. Understanding this phenomenon is extremely important to explain mass transfer. As emulsification increases the surface area between oil and water also increases and therefore the rate of mass transfer rate increases. Oil emulsification occurs when the difference in density between oil and water phases is small [50,81]. In a water model the density ratio (water/oil) is close to one and in a few cases can reach up to 1.5, on the contrary, for the steel/slag system the density ratio (steel/slag) is higher than 2.5. Therefore, while in a water model it is easy to form an emulsion in a real steel/slag system it is more difficult.

Asai et al. [50] developed a model to predict slag emulsification based on an energy balance. The critical metal velocity for slag emulsification was defined for the condition when the kinetic energy exceeds the sum of surface energy and energy due to buoyancy. The critical velocity was found to decrease from about 35 to less than 20 cm/s when the density ratio decreases. Since the density ratio in a water model is lower than in the real steel/slag system, it emulsifies with lower velocities. Later on, Wei and Oeters [82] in 1992 reported a robust theoretical model that predicts the rate of slag droplets formation and its size from the interfacial velocity. Savolainen et al. [83] studied in detail the effects of slag thickness (h_s), density differences ($\Delta\rho$), oil viscosity (μ), and interfacial tension (σ) of three oil/water systems on slag emulsification. They reported that both the critical velocity for emulsification and droplet size increases by increasing h_s, μ, and σ, on the other hand, increasing $\Delta\rho$ also increases the critical velocity but the droplet size decreases. It should be noted that these results were obtained employing oils with densities close to water. In addition to the physical properties of the liquids, the critical gas flow rate or liquid velocity also change depending on nozzle position [66].

In some metal/slag systems slag emulsification has been reported [44,45], however in most of these cases this is because they reproduce conditions of the BOP where the rates of gas injection are very high and lead to the formation of slag emulsions but this is not the case of ladle furnace conditions. In a ladle furnace, for example a ladle of 210 ton of nominal capacity with top diameter of 384 cm, bottom diameter 351 cm, and height of liquid steel of 283 cm, operating with two porous plugs and a maximum gas flow rate in each plug of 40 Nm^3/hr (approximately 12 Nl/min·ton), the maximum stirring energy is about 20 W/ton. This is an extremely high value that produces a large ladle eye and should not be used for a long time. This upper value on stirring energy can be taken as a reference to understand slag emulsification in the previous studies. Several of those studies indicated a critical stirring energy of 60 W/ton to promote slag emulsification, if this is the case then it should be more appropriate of a BOP but not for ladle furnace conditions. For ladle furnace conditions, the reported values from 4 to 6 W/ton would be more appropriate to define the onset for slag emulsification. As a general rule, in the relationships of the form $k \propto Q^n$, the value of n would be less than one if slag emulsification is not present.

Several tracers have been used to measure the *mtc*. In a ladle furnace the main impurity to be removed during steelmaking is sulphur. It is also known that the rate of desulphurization from liquid steel follows first order kinetics, therefore in a water model the tracer should behave similar to sulphur not only in terms of its partition ratio but also show a similar order of reaction to simplify the analysis of the *mtc*. The sulphur partition ratio (L_S) is in the range from 100 to 500 under equilibrium conditions [84–86]. In practice this value ranges from 20 to 100 depending on the stirring conditions and slag chemical composition [87]. Koria et al. [73] has reported that the partition ratio of benzoic acid between water and paraffin oil was 4.5. Kim and Fruehan [61] reported that thymol dissolved in water in contact with a mixture of paraffin oil–cotton seed oil (50/50) can reach a partition ratio higher than 350 and therefore it was suggested as an ideal tracer. The author has undertaken extensive work with a similar system and the maximum value was in the order of 50. Mietz et al. [66] used a mixture of iodium and potassium iodide dissolved in water and in contact with cyclohexane, reaching a partition ratio of 10. The magnitude of the *mtc* will depend on the type of tracer, its initial concentration, and the final equilibrium concentration.

Table 2. Correlations between mtc (or $vmtc$) and stirring energy ε (W/ton) or Q (Nl/min).

	Year	Authors	Process	System	Tracer	Gas	Correlation	ε
1	1968	Subramanian and Richardson	L-L column	In (Hg)–Fe^{2+} (H$_2$O)	Indium	Air	$k \propto V_b^{0.42}$	$3 < V_b < 47$ mL
2	1969	Patel et al.	L-L column	Water–hexane	Iodine	N$_2$	$k \propto Q^{0.72}$	$0.08 < Q < 0.4$
3	1974	Richardson et al.	L-L column	Molten salt–lead	Pb^{2+}	-	$k \propto \left(Q/d_c^2\right)^{0.5}$	-
4	1980	Lehner et al.	60 ton ladle	Steel–slag	Cu	Ar	$k \propto Q^{0.33}$	$8.3 < Q < 83.3$
5	1980	Nakanishi et al.	(Q-BOP)	Water–paraffin	Naphtol	Air	$kA = 3.7Q^{0.36}$	$30 < Q < 80$
							$kA = 1.5 \times 10^{-5} Q^{3.0}$	$80 < Q < 200$
6	1981	Ishida et al.	2.5 ton (LMF)	Steel–slag	Sulphur	Ar	$ka = 0.013\varepsilon^{0.25}$	$\varepsilon < 60$
							$ka = 8 \times 10^{-6} \varepsilon^{2.1}$	$\varepsilon > 60$
7	1981	Umezawa et al.	Mech and gas	Steel–slag	P	Ar	$ka = 1.56\varepsilon^{0.60}$	$40 < \varepsilon < 280$
8	1982	Clinton et al.	Contactors	Water–mercury	Quinone	N$_2$	$k \propto U_s^{0.6-0.8}$	$0.06 < Q < 0.4$
9	1982	Riboud- Olette	Ladle	Steel–slag	S	Ar	$k = 500\left(\frac{D_s Q}{A}\right)^{3.0}$	-
10	1983	Minda et al.	Ladle	Steel–slag	Cr$_2$O$_3$	Ar	$kA \propto \varepsilon^{0.9}$	$\varepsilon > 5000$
11	1983	Umezawa et al.	Mech. stirring	Steel–slag	P	-	$ka = 0.14\varepsilon^{0.58}$	$300 < \varepsilon < 2800$
12	1983	Asai et al.		Water–tetraline	Benzoic acid	Air	$ka \propto \varepsilon^{0.36}$	$Q < 150$
							$ka \propto \varepsilon^{1.0}$	$150 < Q < 650$
13	1984	Sawada et al.	Ladle	Water–paraffin	Naphtol	Air	$k \propto (\varepsilon/d_c)^{0.5}$	$1 < \varepsilon < 20$
14	1985	Berg et al.	6 ton ladle	Steel–slag	Sulphur	Ar	$k \propto \varepsilon^{0.3}$	$\varepsilon < 60$
							$k \propto \varepsilon^{1.3}$	$\varepsilon > 60$
15	1985	Schlarb-Frohberg	BOP	Water–white oil	Caprylic acid	Air	$kA = 3.7 \times 10^{-5} Ar^{0.9}$	$Q < 700$
16	1985	Ooga et al.	ladle	Water– benzene	Benzoic acid	N$_2$	$kA = \varepsilon^{0.66}$	$Q < 0.6$
							$kA = \varepsilon^{1.1}$	$0.6 < Q < 3$

Table 2. *Cont.*

	Year	Authors	Process	System	Tracer	Gas	Correlation	ε
17	1985	Endo et al.	VOD	Water–benzene	Cu^{2+}		$kA \propto \varepsilon^{0.4}$	$\varepsilon < 4$
							$kA \propto \varepsilon^{0.9}$	$4 < \varepsilon < 20$
18	1987	Hirasawa et al.	ladle	Cu (Si)–slag	Si	Ar	$k \propto \left(\varepsilon/d_c^2\right)^{0.5}$	$Q < 1$
19	1987	Kim and Fruehan	Ladle	Water(paraffin/cotton seed oils)	Thymol	Air	$kA = \varepsilon^{0.6}$	$Q < 5$
							$kA = \varepsilon^{2.51}$	$5 < Q < 12$
20	1988	Koria and George	BOP	Water–paraffin	Benzoic acid	Air	$ka \propto Q^{0.449}$	$1.4 < Q < 5$
21	1989	Matway et al.	BOP	Water–paraffin	Naphtol	Air	$kA \propto Q^{0.91}$	$10 < Q < 100$
22	1989	Wright	Ladle	Steel–slag	Carbon	N_2	$k \propto Q^{0.21}$	$Q < 6$
23	1990	Koria and Pal	BOP	Water–paraffin	Benzoic acid	Air	$ka = 2.8\times10^{-3}\varepsilon^{1.05}$	$1.1 < Q < 6.2$
24	1990	Dang and Oeters	Ind. furnace	Steel–slag	Sulphur	Ar	$k = 8.3\times10^{-5}Q^{0.168}$	$0.2 < \varepsilon < 35$
25	1991	Kitamura et al.	Ind. furnace	Steel–slag	P	Ar	$k \propto \left(\varepsilon H^2/D\right)^{0.5}$	$\varepsilon > 60$
26	1992	Koria	QBOP	Water–paraffin	Benzoic acid	Air	$k \propto (\varepsilon_b)^{0.76}(\varepsilon_t)^{0.78}$	$1 < Q < 15$
27	1995	Mukawa et al.	BOP	Steel–slag	P, Si	Ar	$k \propto \left(\varepsilon/d_c^2\right)^{0.7}$	$6 < Q < 1 \times 10^5$
28	1995	Xie and Oeters	ladle	Steel–slag	P, Si	Ar	$k_i \propto Q^{0.168}$	$0.4 < \varepsilon < 27$
29	1996	Li and Yin	Glass	Hg–ZnFe^{2+}(H$_2$O)	Fe^{2+}	Air	$k \propto \left(Q/d_n^{0.33}\right)^{0.05}$	$0.05 < Q < 0.02$
30	2003	Lachmund et al.	ladle	Steel–slag	S	Ar	$k \propto \varepsilon$	$5 < \varepsilon < 300$

where k is the mass transfer coefficient in units length/time, ka and kA are volumetric mass transfer coefficients with units; time^{-1} and lenght^3/time, respectively, ε is stirring energy in W/ton, ε_b represents stirring energy from bottom gas injection, ε_t represents stirring energy from top gas injection, Q is the gas flow rate in Nl/min, U$_s$ is the superficial velocity, d_c is the diameter of the crucible or reactor, d_n is the nozzle diameter in mm. Due to the broad number of studies, units should be checked in the original sources.

All equations in Table 2, except one, do not include the nozzle diameter which could be an indication that the inlet kinetic energy is not relevant. Lehrer reported that 96% of the inlet kinetic energy is consumed at the orifice region and only 6% is transferred to the liquid [88]. Taniguchi et al. [18] and Wright [23] also found similar results. On the other hand, the effect of the nozzle diameter on mass transfer has been reported by Kim and Fruehan [61] and Jiang et al. [89], for a liquid–liquid–gas and gas–liquid system, respectively. In the first case, the rate of mass transfer increased with an increase in nozzle diameter from 2 to 4.8 mm but only at high gas flow rates, and in the second case decreased when the nozzle diameter was increased from 1 to 3 mm and then remained almost constant with nozzle diameters up to 7 mm.

To define the activation energy for mass transfer requires information on the *mtc* as a function of temperature. Robison and Pehlke studied the reduction of chromium oxide from the slag [90]. They indicated that if the reaction rate is controlled by the interfacial chemical reaction and the area remains constant then a change in the stirring energy will not change the reaction rate, and also defined a range from 65 to 100 kcal/mole if the controlling mechanism is mass transfer from the slag phase and 20 kcal/mole if it is due to mass transport in the metal phase. Kitamura et al. [72] reported an activation energy of 30 kcal/mol for the rate of desiliconization of hot metal and Kang et al. [91] a value of 28 kcal/mol for the oxidation of aluminum by silica from the slag.

The *mtc* is an essential component in the development of a kinetic model, for example, when dealing with the desulphurization rate in the ladle furnace. The previous expressions from Table 2 can be used to simplify the calculation of the metal mtc (k_m). In general mass transfer is controlled by diffusion from the primary steel phase, therefore only k_m is needed. If the process is controlled by diffusion in both phases, the slag mass transfer coefficient is also required. A practical approach that has been used in many investigations [72,92–94] is to assume that the slag *mtc* (k_s) is 10 times slower compared with the steel

$$k_s = k_m/10. \qquad (12)$$

Mukawa et al. [60] confirmed this relationship by empirically adjusting parameters in a mathematical model.

Table 3 summarizes the main findings obtained by the previous physical modeling work. It is important to notice the large number of variables that affect the rate of mass transfer.

Table 3. Effect of process variables on the mass transfer coefficient (k) due to bottom gas injection in ladles.

Gas flow rate or stirring energy	Q or ε	k increases with gas flow rate (or stirring energy). Change in k at critical Q for slag emulsification.
Nozzle's radial position	r/R	k increases with central gas injection and decreases if the nozzle is off center
Bubble diameter	d_B	k increases with large bubbles (9–20 mm) and decreases with small bubbles (1–3 mm)
Nozzle diameter	d_n	Mixed results; it has no effect, k increases by increasing d_n at high Q, k decreases by increasing d_n
Superficial velocity	U_s	k increases with the superficial velocity of the liquid
Slag volume	W_s	k increases with the volume of slag
Reactor's diameter	d_c	k increases with the reactor diameter
Slag viscosity	μ_s	k increases by decreasing the slag viscosity, at high Q
Liquid's height	h_m	k increases by increasing the height of the liquid metal up to a critical value
Type of nozzle	-	k in one report was higher for porous plugs in comparison with nozzles for the case of gas–liquid mass transfer
Number of nozzles	N	k increases with number of porous plugs, from one to two

4.2. Mathematical Modelling of Mass Transfer in the Ladle

The *mtc* can be determined empirically through experimental work by physical modeling. This approach has several limitations: (i) lacks generality; is valid for specific cases, (ii) in general is a constant value, (iii) it is usually reported only in terms of one variable; stirring energy.

In order to overcome the previous limitations, the *mtc* has been explained and derived theoretically with different models. Application of this models to the entire gas/steel/slag system requires the development of numerical solutions. There are three main mass transfer theories:

(1) The film theory developed by Lewis and Whitman [95]. It is the simplest and most commonly used theory. Most of the experimental determination of the *mtc* is based on this theory [61,71]. It assumes that mass transfer occurs on both sides of the interface, flow is in steady state, and the equilibrium conditions are instantaneously reached at the interface. On these assumptions the following relationships are derived:

$$N_{i,m} = -D_{i,m}\frac{\partial C_{i,m}}{\partial y} = -D_m\frac{\left(C_{i,int}^m - C_{i,b}^m\right)}{\delta_m} = N_{i,s} = -D_s\frac{\left(C_{i,b}^s - C_{i,int}^s\right)}{\delta_s} \tag{13}$$

$$\text{if, } L = \frac{C_{i,int}^s}{C_{i,int}^m}$$

$$k = \frac{D}{\delta}, k_m = \frac{D_m}{\delta_m}, k_s = \frac{D_s}{\delta_s}, \text{ then}$$

$$N_i = \frac{k_m k_s}{k_s + k_m/L}\left(C_{i,b}^m - \frac{C_{i,b}^s}{L}\right) = k_o\left(C_{i,b}^m - \frac{C_{i,b}^s}{L}\right) \tag{14}$$

$$k_o = \frac{1}{\frac{1}{k_m} + \frac{1}{k_s L}} \tag{15}$$

where $C_{i,int}^j$ is the concentration at the interface on the side of the j-phase, L is the partition equilibrium ratio, δ is the thickness of the diffusion boundary layer, k_m is the *mtc* for species i in the metal phase, k_s is the *mtc* for species i in the slag phase and k_o the overall *mtc*. If the partition ratio (L) is large the second term can be neglected and the process is controlled by mass transfer in the metal phase. Kang et al. [96] reported that the partition ratio for sulphur should be higher than 100 to assume mass transfer control. Although metal mass transfer control is the most common case in steelmaking, slag mass transfer control or mixed control is also possible. Deo and Boom [97] suggested slag mass transfer control under the following conditions: (a) when the bulk slag phase is pre-saturated with the element to be transferred from the metal and the concentration of the element in the metal phase is high, (b) when the slag is highly viscous and has poor mixing conditions so that $k_s << k_m$, (c) when the reaction product layer that forms on the slag side is solid, so that it does not allow for proper homogenization of the bulk slag phase, (d) when the slag has a low sulfide capacity. Notice that based on the boundary layer theory, the *mtc* is proportional to D to the first power; $k_i \propto D$.

(2) The penetration theory was developed by Higbie [98] and assumes that liquid packets at the interface are periodically renewed by new fresh fluids coming from the bulk, each fluid packet is in contact with the interface for a given time. The boundary layer thickness is much larger than in the film theory. For the metal phase:

$$N_{i,m} = -D_{i,m}\frac{\partial C_{i,m}}{\partial y} = \sqrt{\frac{4D_{i,m}U}{\pi L}}\left(C_{i,int}^m - C_{i,b}^m\right) = k_m\left(C_{i,int}^m - C_{i,b}^m\right) \tag{16}$$

$$k_m = \sqrt{\frac{4D_{i,m}}{\pi\,t}} \qquad (17)$$

where t is the contact time of surface renewal. Its value is specific for a given system. Notice that based on the penetration theory, the *mtc* is proportional to the square root of D; $k_i \propto D^{1/2}$. Szekely [99] derived an alternate form of the *mtc* that gives a similar result to the penetration theory. He first derived an expression for heat transfer at the slag/metal interface due to bubble stirring under transient conditions and then applied the same treatment to mass transfer. The final result is expressed in terms of the diffusion coefficients, the equilibrium partition ratio, and the time interval between the arrival of two successive bubbles (t_e).

$$k_m = \left\{1 + \frac{1}{L}\left(\frac{D_{i,m}}{D_{i,s}}\right)^{\frac{1}{2}}\right\}^{-1} \sqrt{\frac{4D_{i,m}}{\pi\,t_e}} \qquad (18)$$

The time interval, t_e, was computed from the number of bubbles produced per unit time (N_b), the cross-sectional area of the bath (A_B), and the projected area of the bubble (A_b)

$$t_e = \frac{A_B}{A_b N_b} \qquad (19)$$

The specific value of the time interval depends on the phenomenon investigated, for example Bafghi et al. [100] defined its value as a function of the frequency of CO formation due to the reduction of FeO during slag foaming and the final expression for the *mtc* was defined in terms of the mass of slag and FeO.

(3)   Surface renewal theories: there is a large number of models that propose how to estimate the contact time of surface renewal. Danckwerts [101] suggested that this time is not constant and follows a normal distribution, t is replaced by a parameter s that defines the rate of replacement. Dong et al. [102] found this parameter to be the ratio between the normal fluctuating velocity U_o, at a depth l_o

$$k_m = 0.4\sqrt{\frac{D_{i,m}U_o}{l_o}}. \qquad (20)$$

The Large Eddy Model (LEM) suggested by Fortescue and Pearson [103] assumes that the larger eddies are dominant on mass transfer because they contain most of the turbulent energy,

$$k_m = 0.4\sqrt{\frac{D_{i,m}U_{rms}}{l}}. \qquad (21)$$

where U_{rms} is the rms velocity of the turbulence and l is the length scale of the large eddies.

Kolgomorov [104] in 1941 formulated the modern concepts of turbulence analyzing the interaction between large and small eddies "decaying" into one another, stating that the small eddies are statistically homogeneous and isotropic. Based on his derivation of length scale and velocity scale, the exposure time can be defined. Applying this concept Banerjee et al. [105] as well as Lamont and Scott [106] suggested that the eddies in the boundary layer are usually small in size and therefore more important. The Small Eddy Model (SEM) is defined by the following equation

$$k_m = 0.4\sqrt{D_{i,m}}\left(\frac{\varepsilon}{v}\right)^{0.25}. \qquad (22)$$

where $D_{i,m}$ is the molecular diffusivity, ε is the energy dissipation rate, and v is the kinematic viscosity. A similar equation was also reported by Miyauchi and Kataoka [107] and Ruckenstein [108], with different constants.

According with the surface renewal theories, k is proportional to the square root of the diffusion coefficient. Since the $D_{i,s}$ of the slag is approximately two orders of magnitude of that of the metal, $D_{i,m}$, we get $k_s = k_m/10$. This is another way to validate Equation (12).

Banerjee et al. [109] have proposed equations for both large and small eddies. The small eddies model for un-sheared interfaces when the far-field turbulence is homogeneous and isotropic, was called the Surface Divergent Model (SDM), defined as follows

$$k_m = 0.3U_L \sqrt{D_{i,m}} (ScRe_t)^{-0.5} \left[0.3\left(2.83(Re_t)^{\frac{3}{4}} - 2.14(Re_t)^{\frac{2}{3}}\right)\right]^{\frac{1}{4}}. \tag{23}$$

The previous mass transfer models have been compared. Theofanous et al. [110] found that LEM and SEM give good results at small and large turbulent Reynolds numbers, respectively. De Oliveira reported similar results for LEM and SDM [111].

The mathematical models developed to study mixing and mass transfer in ladles is extensive [4,112]. In a mathematical model the velocities of the fluids can be computed allowing the use of the previous mass transfer theories. The simplest numerical modeling approach to study mass transfer under isothermal conditions involves two main parts; development of a fluid dynamics model coupled with a mass transfer model. There are three well defined numerical algorithms; Quasi-single or pseudo-single-phase, Eulerian-Lagrangian (E-L) and Eulerian-Eulerian (E-E). Depending on the commercial code, they can assume different names, for example in ANSYS-Fluent the Lagrangian algorithm is called discrete particle model (DPM). The volume of fluid (VOF) model is a separate Eulerian algorithm that allows to track interfaces. In the most recent mathematical models [67,76,77,113], liquid steel and liquid slag are usually described by an Eulerian algorithm, its interface by the VOF algorithm, and the motion of the bubbles by a Lagrangian algorithm. The governing equations can be described by the following general transport equation for multiphase flow

$$\frac{\partial(\alpha_i \rho_i \phi_i)}{\partial t} + \nabla \cdot (\alpha_i u_i \phi_i) = \nabla \cdot \left(\alpha_i \Gamma_{\varnothing_{i,i}} \nabla \phi_i\right) + \alpha_i S_{\varnothing_i}. \tag{24}$$

where ϕ_i is the variable to be solved, α_i is the volume fraction of the phase, ρ_i is the density of the phase, u_i is the velocity, $\Gamma_{\varnothing_{i,i}}$ is the exchange coefficient, t is time, and S is the source term associated with the creation or destruction of ϕ_i. In the species transport equation, the source term is the rate of mass transport.

Once the flow reaches steady state the mass transfer model is solved. It can be solved only to compute the *mtc* or coupled with a kinetic model to predict, for example, the rate of desulphurization. The numerical calculation of *mtc's* can be made using empirical correlations or with the use of mass transfer theories.

The first numerical models involving mass transfer used a Quasi single-phase approach and axisymmetric gas injection, for example Mazumdar et al. [15] computed the fluid's velocity and then used a previously developed empirical correlation to compute the *mtc*. Costa and Tavares [114] used a similar approach. Ahmadi et al. [115–117] did measurements on the melting rate of a Si rod in liquid aluminum and also developed a 3D numerical model. They applied Mazumdar's correlation to predict the *mtc*. The results gave a similar order of magnitude but the correlation underpredicted the experimental data. They also modified the correlation proposed by Churchill and Bernstein [118] for forced convection, adding the turbulent Reynolds number, reporting some improvements with the following correlation

$$\overline{Sh} = 0.3 + \frac{0.62Re_d^{\frac{1}{2}}Sc^{\frac{1}{3}}}{\left[1 + (0.4/Sc)^{\frac{2}{3}}\right]^{\frac{1}{4}}} \left[1 + \left(\frac{Re_d}{282,000}\right)^{\frac{5}{8}}\right]^{\frac{4}{5}} Re_T^{0.066}. \tag{25}$$

Taniguchi et al. [119] calculated the *mtc* applying both the small-eddy model and the penetration theory for a gas–liquid system. The velocity predictions were validated with experimental data using LDV. The time of surface renewal was estimated from the ratio (diameter/velocity)$_{\text{bubble}}$. They found better agreement with the experimental data in the plume region using the small-eddy model. Lou and Zhu [120], Cao et al. [121], Hoang et al. [122], and Karouni et al. [123] have also reported a good agreement using the small eddy model. Cao et al. [121] employed a diffusivity value for the species in the metal side of 7.0×10^{-9} m^2/s and two orders of magnitude lower for the species in the slag side, 7.0×10^{-11} m^2/s.

De Oliveira et al. [111] have employed the following correlation suggested by Banerjee based on the large eddy model, to estimate the mass transfer coefficient in a continuous casting mold

$$k_m = 0.095 u^*(Sc)^{-0.5}.(25) \tag{26}$$

where u^* is the friction velocity at the interface.

Xie and Oeters [124] followed a different approach to study a multi-reaction system. They applied the boundary layer theory and estimated the fluid's velocities using a relationship with the gas flow rate. The mass transfer coefficient for each one of the chemical species involved was defined by an equation of the form $k_i \propto Q^{0.168}$. Their model gave satisfactory agreement below the stirring energy for slag emulsification, the critical value was found in the order of 4 W/ton. This authors also discussed the effect of sulfur on the *mtc*. In one group of experiments sulphur increased the *mtc* but in another case its effect was null and they attributed this behavior to different concentrations of oxygen. The rate of desulphurization (DeS) is higher in low oxygen melts and this explains the two cases comparing Al and Si deoxidation. In Al-deoxidation melts the oxygen content is lower and gives higher rates of DeS. Jun et al. [125] and Ying [126] indicate that oxygen, which is a surfactant, retards the absorption of nitrogen because the surface velocity decreases due to the Marangoni effect. On the contrary, Mendes indicates that increasing the concentration of surfactants increase the rate of mass transfer due to an increase in interfacial convection [127]. This subject requires further investigation.

One of the limitations in developing mathematical models to define the *mtc* and its subsequent use to predict mass transfer rates, is the large computational time involved. Cao et al. [121] suggested to decouple the simultaneous computation of the *mtc* with a fluid flow model and the multi component reaction kinetics model in order to save time. Van Ende and Jung [94] opted for the use of a semiempirical relationship and even a number that just fits model predictions with experimental data [128].

In most of the previous work on mass transfer in metallurgical reactors it is assumed a homogeneous bath, but this is not entirely true, especially at the beginning of the process. The extent of homogenization depends on several factors such as the gas flow rate and injection layout. Mietz and Bruhl [129] studied the effect of the volume of dead zones on the mass transfer rates and found a large discrepancy comparing ideal mixing and real mixing conditions. Eventually, depending on the gas flow rate, the concentrations become similar when mixing is complete and the dead zones are decreased. However, since the time-scale for mixing time in water models or prototypes is in the order of seconds, varying from 10 to 180 s [130,131] and the time-scale for mass transfer is at least one order of magnitude higher, it is valid to assume an homogeneous concentration in mass transfer studies. In regard to the time-scale for mass transfer experiments in water models it is important to consider that the values reported are relative. We have carried out extensive work [132] that shows that the rate of mass transfer depends on many variables; gas flow rate, injection layout, physicochemical properties of phases involved, etc., therefore, the actual ratio between the two scales is variable.

It has been pointed out that the main purpose in defining the value of the *mtc's* is to use them in a kinetic model. It can be slag-metal refining, lime dissolution, formation mechanisms of non-metallic inclusions, etc. In the previous paragraphs some of the reports that have used mass transfer theories to compute the *mtc* have been mentioned. Table 4 summarizes how the *mtc* has been defined in different kinetic models.

Table 4. Methods employed to compute k in reported kinetic models.

Method	References
Correlation $k \propto \varepsilon^n$	Singh et al. [133], Zhang et al. [93], Van Ende and Jung [94]
Experimental work	Choi et al. [134], Harada et al. [135], Kang et al. [91], Roy et al. [136]
Correlations from D. Analysis	Wei et al. [137], Sulasalmi et al. [138], Huang et al. [139]
Boundary layer theory	Xie and Oeters [124], Chen et al. [140]
Higbie's penetration theory	Taniguchi et al. [119]
Large Eddy Model (LEM)	De Oliveira et al. [111], Deo and Grieveson [141],
Small Eddy Model (SEM)	Taniguchi et al. [119], Lou and Zhu [120], Cao et al. [121], Hoang et al. [122] and Karouni et al. [123]

It has also been mentioned before that the interfacial area is a variable difficult to measure. Without information of the real value, the volumetric mass transfer coefficient is reported instead. There are few reports about modeling the interfacial area during bottom gas injection. Cao et al. [121] reported a mathematical model that predicts the interfacial area. Their results indicate that the interfacial area is lower than the static area due to formation of the slag eye (s). Its value decreases as soon as the slag eye forms and tends to stabilize, showing a dynamic fluctuation depending on the formation slag droplets entrapped in liquid steel. This is true if there no slag emulsification included in the model. Sulasalmi et al. [138] reported a kinetic model incorporating the growth in interfacial area due to formation of slag droplets. Calculating the kinetics of emulsions requires the knowledge of the droplets residence time, its size distribution, and its generation rate. Unfortunately, very few investigations are currently available for kinetic models of slag–metal systems including the emulsification rate.

When the gas flow rate is increased there are two opposing effects, on one side, the slag droplets increase the interfacial area but also the area of the slag eye increases, decreasing the interfacial area. Zhang et al. [93] reported that bottom gas injection promotes the removal of non-metallic inclusions, however at a critical gas flow rate, due to the enlargement of the slag eye and the corresponding reoxidation, the amount of total oxygen increases, increasing again the amount of non-metallic inclusions. Lou and Zhu [120] also reported the existence of a maximum gas flow rate to reach the maximum rate of desulphurization, this value was reported to be 200 Nl/min in a ladle with 80 ton of liquid steel, equivalent to 7 Watt/ton. These results are extremely important because they clearly indicate the need to define limits to the maximum gas flow rate to reach the higher rate of mass transfer from impurities from liquid steel to the slag but taking also into consideration liquid steel reoxidation due to slag eye formation. Another solution is increasing from one to two porous plugs because in this way it is possible to increase the gas flow rate, increasing the interfacial area but also decreasing the slag-eye area [133].

5. Final Remarks

All the physical and mathematical modeling research involving liquid–liquid mass transfer due to bottom gas injection that has been described in this review has resulted in an extensive number of correlations that together with turbulence models and theories provide a solid understanding of the computation of the mass transfer coefficients, nevertheless, in spite of this progress no single equation includes all the variables that affect the rate of mass transfer and therefore they cannot provide a unified approach. Wilson and Macleod [142] presented a similar result in a review on gas–liquid mass transfer.

Based on this review it has been found that the following variables affect the mass transfer coefficient: (1) mass transfer is affected primarily by the gas flow rate, however, is not only the amount of stirring energy and flow velocities resulting from bottom gas injection that defines the value of the *mtc* but also; (2) how this gas is injected (number of nozzles, its radial position, type of nozzles and nozzle diameter), (3) the concentration of surfactants in the liquid phases, (4) slag emulsification (which occurs above a critical gas flow rate) with its corresponding number of droplets, size and residence time, (5) interfacial area as a function of gas flow rate, (6) slag and steel physicochemical properties (diffusion coefficients, viscosity, relative densities, etc.), (7) bubble size and frequency, (8) reactor dimensions (diameter, height of the liquid, aspect ratio). From this list, gas flow rate and slag emulsification has been reported to have the largest influence. The author has argued that in the real steel/slag system slag emulsification in a ladle furnace can have a less significant role than is currently attributed because previous physical models have been developed using oils that do not represent the real slag phase. Slag emulsification in a ladle furnace is very important, however the upper limits of slag emulsification should be properly demonstrated by physical and mathematical modelling studies.

The correlations involving stirring energy and the *mtc* have been the most common but also the ones excluding most of the process variables affecting the rate of mass transfer. Dimensional analysis and mass transfer theories have included most of the relevant variables. Hirasawa et al. [59] reported the first attempt of a unified approach, however, their correlation does not include slag emulsification.

$$\frac{kd_c}{D} = C\left[\left(\frac{4Qd_c}{D\pi d_c^2}\right)\left(\frac{\rho g d_c^2}{\sigma}\right)\left(\frac{h}{d_c}\right)\right]^{\frac{1}{2}} \tag{27}$$

where C is a constant, h is the height of the liquid, σ is the interfacial tension.

In dimensionless form:

$$Sh = C'\left[(Pe)\left(\frac{\rho g d_c^2}{\sigma}\right)\left(\frac{d_B Re^{-n}}{d_c}\right)\right]^{\frac{1}{2}} \tag{28}$$

Reiter and Schwerdtfeger [143] also used dimensional analysis to describe mass transfer from bubbles in a thick upper phase, however this system is not representative of ladle furnace conditions.

Mathematical models involving CFD can estimate fluid flow patterns, with this information all the mass transfer theories can be applied to compute the *mtc*. It has been shown that in general all mass transfer theories can reproduce the experimental data, however, there is still some degree of empiricism with this approach. It is still necessary to adjust constants in those models.

Suggestions for future research: Future physical and mathematical modeling work to define the mass transfer coefficients in liquid–liquid mass transfer systems due to bottom gas injection should incorporate all the variables indicated previously, in particular those variables less studied; effect of nozzle radial position and separation angle, diameter and type of nozzles and the role of surfactants. In this effort dimensional analysis is most recommended. In physical modelling, proper similarity of slag emulsification with the real steel/slag system is needed in order to assess its real contribution to mass transfer. Mathematical modelling should incorporate a realistic approach as much as possible, which in addition to bubble size distribution should also evaluate the real contribution of slag emulsification and interfacial area under ladle furnace conditions. Once this work is completed, the final step will be to unify all experimental and numerical data.

The main purpose of bottom gas injection is to improve mixing phenomenon. The few experimental information available shows that a given injection layout can promote either mixing of liquid steel (by decreasing mixing time) or the rate of mass transfer (removal of impurities to the upper slag phase). One of the challenges ahead is to find conditions to improve both processes simultaneously.

Funding: This research received no external funding.

Acknowledgments: I want to acknowledge the support from the University of Science and Technology Beijing (USTB) to carry out this work during the pandemic of Coronavirus Covid-19.

Conflicts of Interest: The author declares no conflict of interest.

References

1. Mazumdar, D.; Guthrie, R.I.L. The Physical and Mathematical Modelling of Gas Stirred Ladle Systems. *ISIJ Int.* **1995**, *35*, 1–20. [CrossRef]
2. Mazumdar, D.; Evans, J.W. Macroscopic models for gas stirred ladles. *ISIJ Int.* **2004**, *44*, 447–461. [CrossRef]
3. Sichen, D. Modeling related to secondary steel making. *Steel Res. Int.* **2012**, *83*, 825–841. [CrossRef]
4. Liu, Y.; Ersson, M.; Liu, H.; Jönsson, P.G.; Gan, Y. A Review of Physical and Numerical Approaches for the Study of Gas Stirring in Ladle Metallurgy. *Metall. Mater. Trans. B Process Metall. Mater. Process. Sci.* **2019**, *50*, 555–577. [CrossRef]
5. Ghotli, R.A.; Abdul Aziz, A.R.; Ibrahim, S. Liquid-liquid mass transfer studies in various stirred vessel designs. *Rev. Chem. Eng.* **2015**, *31*, 329–343. [CrossRef]
6. Themelis, N.J. *Transport and Chemical Rate Phenomena*; Gordon and Breach: London, UK, 1995.
7. Veeraburus, M.; Philbrook, W.O. Observations on liquid-liquid mass transfer with bubble stirring. In *Physical Chemistry of Process Metallurgy*; Pierre, G.S., Ed.; Interscience: New York, NY, USA, 1959; p. 559.
8. Oeters, F.; Xie, H. Contribution to the theoretical description of metal-slag reaction kinetics. *Steel Res.* **1995**, *66*, 409–415. [CrossRef]
9. Gabe, D.R. The rotating cylinder electrode. *J. Appl. Electrochem.* **1974**, *4*, 91–108. [CrossRef]
10. Kosaka, M.; Minowa, S. Dissolution of Steel Cylinder into Liquid Fe-C Alloy. *Tetsu to Hagane* **1967**, *53*, 983–997. [CrossRef]
11. Kim, Y.U.; Pehlke, R.D. Mass Transfer During Dissolution of a Solid Into Liquid in the Iron-Carbon System. *Met. Trans* **1974**, *5*, 2527–2532. [CrossRef]
12. Shigeno, Y.; Tokuda, M.; Ohtani, M. The dissolution rate of graphite into Fe–C melt containing sulphur or phosphorus. *Trans. Japan. Inst. Met.* **1985**, *26*, 33–43. [CrossRef]
13. Szekely, J.; Lehner, T.; Chang, C.W. Flow Phenomena Mixing and Mass Transfer in Argon-Stirred Ladles. *Ironmak. Steelmak.* **1979**, *7*, 285–293.
14. Mazumdar, D.; Kajani, S.K.; Ghosh, A. Mass transfer between solid and liquid in vessels agitated by bubble plume. *Steel Res.* **1990**, *61*, 339–346. [CrossRef]
15. Mazumdar, D.; Narayan, T.; Bansal, P. Mathematical modelling of mass transfer rates between solid and liquid in high-temperature gas-stirred melts. *Appl. Math. Model.* **1992**, *16*, 255–262. [CrossRef]
16. Szekely, J.; Grevet, H.H.; El-Kaddah, N. Melting rates in turbulent recirculating flow systems. *Int. J. Heat Mass Transf.* **1984**, *27*, 1116–1121. [CrossRef]
17. Singh, A.K.; Mazumdar, D. Mass transfer between solid and liquid in a gas-stirred vessel. *Metall. Mater. Trans. B Process Metall. Mater. Process. Sci.* **1997**, *28*, 95–102. [CrossRef]
18. Taniguchi, S.; Ohmi, M.; Ishiura, S.; Yamauchi, S. Cold model study on the effect of gas injection upon the melting rate of a solid sphere in a liquid bath. *Trans. Iron Steel Inst. Japan* **1983**, *23*, 565–570. [CrossRef]
19. Koria, S.C. Model investigations on liquid velocity induced by submerged gas injection in steel bath. *Steel Res.* **1988**, *59*, 484–491. [CrossRef]
20. Matsushima, M.; Yadoomaru, S.; Mori, K.; Kawai, Y. A Fundamental Study on the Dissolution Rate of Solid Lime into Liquid Slag. *Trans. Iron Steel Inst. Japan* **1977**, *17*, 442–449. [CrossRef]
21. Taira, S.; Nakashima, K.; Mori, K. Kinetic Behavior of Dissolution of Sintered Alumina Into Cao-Si02&Al203 Slags. *ISIJ Int.* **1993**, *33*, 116–123.
22. Kitamura, S.; Shibata, H.; Maruoka, N. Kinetic Model of Hot Metal Dephosphorization by Liquid and Solid Coexisting Slags. *Steel Res. Int.* **2008**, *79*, 586–590. [CrossRef]
23. Wright, J.K. Steel dissolution in quiescent and gas stirred Fe/C melts. *Metall. Trans. B* **1989**, *20*, 363–374. [CrossRef]
24. Maruoka, N.; Lazuardi, F.; Maeyama, T.; Kim, S.-J.; Conejo, A.N.; Shibata, H.; Kitamura, S.-Y. Evaluation of bubble eye area to improve gas/liquid reaction rates at bath surfaces. *ISIJ Int.* **2011**, *51*, 236–241. [CrossRef]
25. Kato, Y.; Fujii, T.S.; Habu, Y. Gas-liquid mass transfer of bottom and top blowing in comverters by water modeling. *Testu to Hagane* **1983**, *69*, S1011.
26. Inada, S.; Watanabe, T. The Model Experiment on the Reaction between the Liquid and the Swarms of Gas Bubbles by NaOH-CO2 System. *Tetsu to Hagane* **1977**, *63*, 37–44. [CrossRef]
27. Sakaguchi, K.; Ito, K. Measurement of the Volumetric Mass Transfer Coefficient of Gas-stirred Vessel under Reduced Pressure. *ISIJ Int.* **1995**, *35*, 1348–1353. [CrossRef]

28. Rui, Q.; Jiang, F.; Ma, Z.; You, Z.; Cheng, G.; Zhang, J. Effect of elliptical snorkel on the decarburization rate in single snorkel refining furnace. *Steel Res. Int.* **2013**, *84*, 192–197. [CrossRef]

29. Subramanian, N.; Richardson, F.D. Mass transfer across interfaces agitated by large bubbles. *JISI* **1968**, *June*, 576–583.

30. Richardson, F.D.; Robertson, D.G.C.; Staples, B.B. Mass Transfer Across Metal/Slag Interfaces Stirred by Bubbles. In Proceedings of the The Darken Conference on Physical Chemistry in Metallurgy, Pittsburgh, PA, USA, 23–25 August 1976; pp. 25–48.

31. Li, X.; Yin, Z. Mass transfer to liquid-liquid interface. *Acta Metall. Sin.* **1996**, *9*, 151–156.

32. Fruehan, R.J.; Martonik, L.K. Physical Behavior and Liquid-Phase Mass Transfer of Submerged Gas Jets in Liquids. In Proceedings of the 3rd International Iron and Steel Congres, Chicago, IL, USA, 16–20 April 1978; pp. 229–238.

33. Riboud, O.V.; Olette, M. Mechanisms of some of the reactions involved in secondary refining. In Proceedings of the 7th International Conference on Vacuum Metallurgy (ICVM), Tokyo, Japan, 26–30 November 1982; pp. 879–889.

34. Gaye, H.; Gatellier, C.; Riboud, P.V. Physico-Chemical Aspects of the Ladle Desulphurization of Iron and Steel. *Foundry Process.* **1988**, 333–356.

35. Nakanishi, K.; Fujii, T.; Szekely, J. Possible relationship between energy dissipation and agitation in steel-processing operations. *Ironmak. Steelmak.* **1975**, *2*, 193–197.

36. Ilegbusi, O.J. Role of gas plumes in agitation and mass transfer in metallurgical systems. *Steel Res.* **1994**, *65*, 534–540. [CrossRef]

37. Asai, S.; Kawachi, M.; Muchi, I. Mass transfer rate in ladle refining processes. In *SCANINJECT III: Proceedings of the Refining of Iron and Steel by Powder Injection, Lulea Sweden, 15–17 June 1983*; MEFOS: Lulea, Sweden, 1983; pp. 12:1–12:29.

38. Lehner, T.; Carlsson, G.; Hsiao, T.C. On Fluid Flow and Metallurgical Reaction in Gas Stirred Melts. In *SCANINJECT II: Proceedings of the 2nd International Conference on Injection Metallurgy, Lulea, Sweden, 12–13 June 1980*; MEFOS: Lulea, Sweden, 1980; Article 22.

39. Engh, T.A. *Principles of Metal Refining*; Oxford University Press: Oxford, UK, 1992.

40. Oeters, F. *Metallurgy of Steelmaking*; Verlag Stahleisen mbH: Dusseldorf, Germany, 1994.

41. Conejo, A.N.; Lara, F.R.; Macias-Hernández, M.; Morales, R.D. Kinetic model of steel refining in a ladle furnace. *Steel Res. Int.* **2007**, *78*, 141–150. [CrossRef]

42. Mazumdar, D.; Guthrie, R.I.L. Mixing models for gas stirred metallurgical reactors. *Metall. Trans. B* **1986**, *17*, 725–733. [CrossRef]

43. Nakanishi, K.; Kato, Y.; Nozaki, T.; Emi, T. Cold Model Study on the Mixing Rates of Slag and Metal Bath in Q-Bop. *Tetsu-To-Hagane* **1980**, *66*, 1307–1316. [CrossRef]

44. Ishida, J.; Yamaguchi, K.; Sugiura, S.; Yamano, K.; Hayakawa, S.; Demukai, N. Effects of Stirring by Argon Gas Injection on Metallurgical Reactions in Secondary Steelmaking. *Denki-Seiko (Electric Furn. Steel)* **1981**, *52*, 2–8.

45. Berg, B.; Carlsson, G.; Bramming, M. Ladle Metallurgy-Influence of Different Stirring Methods. *Scand. J. Met.* **1985**, *14*, 299–305.

46. Umezawa, K.; Nishugi, S.; Arima, R.; Matsunaga, H. Development of De-P and De-S treatment methods for powder injection with CaO flux. *Testu to Hagane* **1981**, *67*, S182.

47. Umezawa, K.; Matsunaga, H.; Arima, R.; Tonomura, S.; Furugaki, I. The Influence of Operating Condition on Dephosphorization and Desulphurization Reactions of Hot Metal with Lime-based Flux. *Tetsu to Hagane* **1983**, *69*, 1810–1817. [CrossRef]

48. Clinton, S.D.; Perona, J.J. Mass Transfer in a Bubble-Agitated Liquid-Liquid System. *Ind. Eng. Chem. Fundamen.* **1982**, 269–271. [CrossRef]

49. Minda, A.Y.S.; Asaho, R.; Komamura, K.; Kato, Y. Study of stainless steel smelting by top blowing. *Testu to Hagane* **1983**, *69*, S1007.

50. Asai, S.; Muchi, I.; Kawachi, M. Fluid Flow and Mass Transfer in Gas Stirred Ladles. *Foundry Process.* **1988**, 261–292.

51. Patel, P.; Frohberg, M.G.; Papamantellos, D. Experimental studies of mass transfer between two immiscible liquids. *Trans. AIME* **1969**, *245*, 855–859.

52. Sawada, I.; Ohashi, T.; Kajioka, H. Mass transfer rate between slag and metal in a ladle due to bottom gas injection. *Tetsu to Hagane* **1984**, *70*, S161.

53. Ooga, Y.; Taniguchi, S.; Kikuchi, J. Fundamental research on the behavior of liquid-liquid substances under gas stirring. *Tetsu to Hagane* **1985**, *71*, S897.

54. Endo, S.; Hasegawa, M. Cold model study on effect of stirring on slag metal reaction. *Tetsu to Hagane* **1985**, *71*, S899.

55. Hirasawa, M.; Mori, K.; Sano, M.; Shimatani, Y.; Okazaki, Y. Effect of gas flow rate on slag/metal mass transfer. *Tetsu to Hagane* **1985**, *71*, S898.

56. Hirasawa, M.; Mori, K.; Sano, M.; Hatanaka, A.; Shimatani, Y.; Okazaki, Y. Kinetic Studies on the Rate of Reaction Between Molten Slag and Metal With Gas-Injection Stirring. *Tetsu to Hagane* **1987**, *73*, 1343–1349. [CrossRef]

57. Hirasawa, M.; Mori, K.; Sano, M.; Hatanaka, A.; Shimatani, Y.; Okazaki, Y. The Analysis System of Metal-side Stirring in a Slag-Metal Reaction with Gas-injection. *Testu to Hagane* **1987**, *73*, 1350–1357. [CrossRef]

58. Hirasawa, M.; Mori, K.; Sano, M.; Hatanaka, A.; Shimatani, Y.; Okazaki, Y. Rate of Mass Transfer Between Molten Slag and Metal Under Gas Injection Stirring. *Trans. ISIJ* **1987**, *27*, 277–282. [CrossRef]

59. Hirasawa, M.; Mori, K.; Sano, M.; Shimatani, Y.; Okazaki, Y. Correlation Equations for Metal-Side Mass Transfer in a Slag-Metal Reaction System With Gas Injection Stirring. *Trans. ISIJ* **1987**, *27*, 283–290. [CrossRef]

60. Mukawa, S.; Mizukami, Y. Effect of Stirring Energy and Rate of Oxygen Supply on the Rate of Hot Metal Dephosphorization. *ISIJ Int.* **1995**, *35*, 1374–1380. [CrossRef]

61. Kim, S.H.; Fruehan, R.J. Physical modeling of gas/liquid mass transfer in a gas stirred ladle. *Metall. Trans. B* **1987**, *18*, 673–680. [CrossRef]

62. Matway, R.J.; Fruehan, R.J.; Henein, H. Physical Modeling of Gas Injection in a Steelmaking Vessel-Mixing Times and Liquid/Liquid Mass Transfer Rates. *Iron Steelmak.* **1989**, *16*, 51–58.

63. Matway, R.J.; Fruehan, R.J.; Henein, H. Physical Modeling of Slag/Metal Reactions in Combined Blowing-Effect of Tuyere Location, Tuyere Size, and Gas Flowrate. *Iron Steelmak.* **1991**, *18*, 43–50.

64. Singh, V.; Lenka, S.N.; Ajmani, S.K.; Bhanu, C.; Pathak, S. A novel bottom stirring scheme to improve BOF performance through mixing and mass transfer modelling. *ISIJ Int.* **2009**, *49*, 1889–1894. [CrossRef]

65. Mietz, J.; Schneider, S.; Oeters, F. Model experiments on mass transfer in ladle metallurgy. *Steel Res.* **1991**, *62*, 1–9. [CrossRef]

66. Mietz, J.; Schneider, S.; Oeters, F. Emulsification and mass transfer in ladle metallurgy. *Steel Res.* **1991**, *62*, 10–15. [CrossRef]

67. Lou, W.; Zhu, M. Numerical simulation of slag-metal reactions and desulfurization efficiency in gas-stirred ladles with different thermodynamics and kinetics. *ISIJ Int.* **2015**, *55*, 961–969. [CrossRef]

68. Koria, S.C.; George, A. Selection of bottom injection parameters in combined blown steelmaking. *Ironmak. Steelmak.* **1988**, *15*, 127–133.

69. Koria, S.C.; Pal, S. Model study on mixing condition in combined blown steelmaking bath. *Ironmak. Steelmak.* **1990**, *17*, 325–332.

70. Koria, S.C.; Shamsi, M. Simulation of mass transfer from metal to slag in gas stirred ladles. *Ironmak. Steelmak.* **1990**, *17*, 401–409.

71. Deng, J.; Oeters, F. Mass transfer of sulfur from liquid iron into lime-saturated CaO-Al2O3-MgO-SiO2 slags. *Steel Res.* **1990**, *61*, 438–448. [CrossRef]

72. Kitamura, S.Y.; Kitamura, T.; Shibata, K.; Mizukami, Y.; Mukawa, S.; Nakagawa, J. Effect of Stirring Energy, Temperature and Flux Composition on Hot Metal Dephosphorization Kinetics. *ISIJ Int.* **1991**, *31*, 1322–1328. [CrossRef]

73. Koria, S.C. Studies of the bath mixing intensity in converter steelmaking processes. *Can. Metall. Q.* **1992**, *31*, 105–112. [CrossRef]

74. Lachmund, H.; Xie, Y.; Harste, K. Thermodynamic and kinetic aspects of the desulphurisation reaction in secondary metallurgy. *Steel Res.* **2001**, *72*, 452–459. [CrossRef]

75. Lachmund, H.; Xie, Y.; Buhles, T.; Pluschkell, W. Slag emulsification during liquid steel desulphurisation by gas injection into the ladle. *Steel Res.* **2003**, *74*, 77–85. [CrossRef]

76. Sulasalmi, P.; Visuri, V.V.; Kärnä, A.; Fabritius, T. Simulation of the effect of steel flow velocity on slag droplet distribution and interfacial area between steel and slag. *Steel Res. Int.* **2015**, *86*, 212–222. [CrossRef]

77. Senguttuvan, A.; Irons, G.A. Modeling of slag entrainment and interfacial mass transfer in gas stirred ladles. *ISIJ Int.* **2017**, *57*, 1962–1970. [CrossRef]

78. Stapurewicz, T.; Themelis, N.J. Mixing and mass transfer phenomena in bottom-injected gas-liquid reactors. *Can. Metall. Q.* **1987**, *26*, 123–128. [CrossRef]

79. Terrazas, M.S.C.; Conejo, A.N. Effect of nozzle diameter, nozzle radial position and a top slag layer on mixing time during bottom gas injection in metallurgical ladles. In Proceedings of the 6th International Congress on the Science and Technology of Steelmaking (ICS 2015), Beijing, China, 12–14 May 2015.

80. Mori, K. Kinetics of Fundamental Reactions Pertinent To Steelmaking Process. *Trans. Iron Steel Inst. Japan* **1988**, *28*, 246–261. [CrossRef]

81. Joo, S.; Guthrie, R.I.L. Modeling flows and mixing in steelmaking ladles designed for single- and dual-plug bubbling operations. *Metall. Trans. B* **1992**, *23*, 765–778. [CrossRef]

82. Wei, T.; Oeters, F. Model test for emulsion in gas-stirred ladles. *Steel Res.* **1992**, *63*, 60–68. [CrossRef]

83. Savolainen, J.; Fabritius, T.; Mattila, O. Effect of fluid physical properties on the emulsification. *ISIJ Int.* **2009**, *49*, 29–36. [CrossRef]

84. Hatch, G.G.; Chipman, J. Sulphur Equilibria between Iron Blast Furnace Slags and Metal. *JOM* **1949**, *1*, 274–284. [CrossRef]

85. Simeonov, S.; Ivanchev, I.; Hainadjiev, A. Sulphur equilibrium distribution between CaO-CaF2-Si02-Al203 slags and carbon-saturated iron. *ISIJ Int.* **1991**, *31*, 1396–1399. [CrossRef]

86. Andersson, M.; Hallberg, M.; Jonsson, L.; Jönsson, P. Slag-metal reactions during ladle treatment with focus on desulphurisation. *Ironmak. Steelmak.* **2002**, *29*, 224–232. [CrossRef]

87. Sardar, M.K.; Mukhopadhyay, S.; Majumder, S.; Mallick, S.; Singh, R.K. Effect of plug location on desulfurization characteristics of slag during ladle furnace operation. *Can. Metall. Q.* **2006**, *45*, 175–180. [CrossRef]

88. Lehrer, L.H. Gas agitation of liquids. *Ind. Eng. Chem. Process Des. Dev.* **1968**, *7*, 226–239. [CrossRef]

89. Jiang, F.; Song, Y.X.; Cheng, G.G. Effect of orifice configuration on the volumetric mass transfer coefficient during ladle purging. *Rev. Metall.* **2011**, *108*, 465–472. [CrossRef]

90. Robison, J.W.; Pehlke, R.D. Kinetics of Chromium Oxide Reduction From a Basic Steelmaking Slag By Silicon Dissolved in Liquid Iron. *Met. Trans* **1974**, *5*, 1041–1051. [CrossRef]

91. Kang, Y.B.; Kim, M.S.; Lee, S.W.; Cho, J.W.; Park, M.S.; Lee, H.G. A reaction between high Mn-High Al Steel and CaO-SiO2-Type molten mold flux: Part II. reaction mechanism, interface morphology, and Al 2O3 accumulation in molten mold flux. *Metall. Mater. Trans. B Process Metall. Mater. Process. Sci.* **2013**, *44*, 309–316. [CrossRef]

92. Conejo, A.N.; Kitamura, S.; Maruoka, N.; Kim, S.-J. Effects of top layer, nozzle arrangement, and gas flow rate on mixing time in agitated ladles by bottom gas injection. *Metall. Mater. Trans. B Process Metall. Mater. Process. Sci.* **2013**, *44*, 914–923. [CrossRef]

93. Zhang, Y.; Ren, Y.; Zhang, L. Kinetic study on compositional variations of inclusions, steel and slag during refining process. *Metall. Res. Technol.* **2018**, *115*, 415. [CrossRef]

94. Van Ende, M.A.; Jung, I.H. A Kinetic Ladle Furnace Process Simulation Model: Effective Equilibrium Reaction Zone Model Using FactSage Macro Processing. *Metall. Mater. Trans. B Process Metall. Mater. Process. Sci.* **2017**, *48*, 28–36. [CrossRef]

95. Lewis, W.K.; Whitman, W.G. Principles of Gas Absorption. *Ind. Eng. Chem.* **1924**, *16*, 1215–1220. [CrossRef]

96. Kang, J.G.; Shin, J.H.; Chung, Y.; Park, J.H. Effect of Slag Chemistry on the Desulfurization Kinetics in Secondary Refining Processes. *Metall. Mater. Trans. B Process Metall. Mater. Process. Sci.* **2017**, *48*, 2123–2135. [CrossRef]

97. Deo, B.; Boom, R. *Fundamentals of Steelmaking Metallurgy*, 1st ed.; Prentice-Hall: New York, NY, USA, 1993.

98. Higbie, R. The rate of absorption of a pure gas into a still liquid. *Trans. Am. Inst. Chem. Eng.* **1935**, *35*, 36–60.

99. Szekely, J. Mathematical model for heat or mass transfer at the bubble-stirred interface of two immiscible liquids. *Int. J. Heat Mass Transf.* **1963**, *6*, 417–422. [CrossRef]

100. Bafghi, M.S.; Kurimoto, H.; Sano, M. Effect of Slag Foaming on the Reduction of Iron Oxide in Molten Slag by Graphite. *ISIJ Int.* **1992**, *32*, 1084–1090. [CrossRef]

101. Danckwerts, P.V. Significance of liquid-film coefficients in gas absorption. *Eng. Process Dev.* **1951**, *43*, 1460–1467. [CrossRef]

102. Dong, L.; Johansen, S.T.; Engh, T.A. Mass transfer at gas-liquid interfaces in stirred vessels. *Can. Metall. Q.* **1992**, *31*, 299–307. [CrossRef]

103. Fortescue, G.E.; Pearson, J.R.A. Gas absorption into a turbulent liquid. *Chem. Eng. Sci.* **1967**, *22*, 1163–1176. [CrossRef]

104. Kolmogorov, A.N. The local structure of turbulence in incompressible viscous fluid for very large Reynolds number. *Dokl. Akad. Nauk SSSR* **1941**, *30*, 538–540.

105. Banerjee, S.; Rhodes, E.; Scott, D.S. Mass transfer to falling wavy liquid films in turbulent flow. *Eng. Chem. Fundam.* **1968**, *7*, 22–27. [CrossRef]

106. Lamont, J.C.; Scott, D.S. An eddy cell model of mass transfer into the surface of a turbulent liquid. *AIChE J.* **1970**, *16*, 513–519. [CrossRef]

107. Miyauchi, T.; Kataoka, H. Liquid film coefficient of mass transfer on free liquid surface. *J. Chem. Eng. Japan* **1970**, *3*, 257–258. [CrossRef]

108. Ruckenstein, E. Physical Models for Mass or Heat Transfer Processes. *Int Chem. Eng.* **1967**, *7*, 490.

109. Banerjee, S.; Lakehal, D.; Fulgosi, M. Surface divergence models for scalar exchange between turbulent streams. *Int. J. Multiph. Flow* **2004**, *30*, 963–977. [CrossRef]

110. Theofanous, T.G.; Houze, R.N.; Brumfield, L.K. Turbulent mass transfer at free, gas-liquid interfaces, with applications to open-channel, bubble and jet flows. *Int. J. Heat Mass Transf.* **1976**, *19*, 613–624. [CrossRef]

111. De Oliveira Campos, L.D. *Mass Transfer Coefficients across Dynamic Liquid Steel/Slag Interface*; L'Université de Bordeaux: Bordeaux, France, 2017.

112. Duan, H.; Zhang, L.; Thomas, B.G.; Conejo, A.N. Fluid Flow, Dissolution, and Mixing Phenomena in Argon-Stirred Steel Ladles. *Metall. Mater. Trans. B Process Metall. Mater. Process. Sci.* **2018**, *49*, 2722–2743. [CrossRef]

113. Cloete, S.W.P.; Eksteen, J.J.; Bradshaw, S.M. A mathematical modelling study of fluid flow and mixing in full-scale gas-stirred ladles. *Prog. Comput. Fluid Dyna mics* **2009**, *9*, 345–356. [CrossRef]

114. Costa, L.T.; Tavares, R.P. Multiphase Mass Mass Transfer in Iron and Steel Refining Processes. In *Mass Transfer—Advancement in Process Modelling*; Solecki, M., Ed.; InTech Open: London, UK, 2015; pp. 149–187. ISBN 9789535121923.

115. Seyed Ahmadi, M.; Argyropoulos, S.A.; Bussmann, M.; Doutre, D. Comparative Studies of Silicon Dissolution in Molten Aluminum Under Different Flow Conditions, Part I: Single-Phase Flow. *Metall. Mater. Trans. B Process Metall. Mater. Process. Sci.* **2015**, *46*, 1275–1289. [CrossRef]

116. Seyed Ahmadi, M.; Argyropoulos, S.A.; Bussmann, M.; Doutre, D. Comparative Studies of Silicon Dissolution in Molten Aluminum Under Different Flow Conditions Part II: Two-Phase Flow. *Metall. Mater. Trans. B* **2015**, *46*, 1290–1301. [CrossRef]

117. Seyed Ahmadi, M.; Bussmann, M.; Argyropoulos, S.A. Mass transfer correlations for dissolution of cylindrical additions in liquid metals with gas agitation. *Int. J. Heat Mass Transf.* **2016**, *97*, 767–778. [CrossRef]

118. Churchill, S.W.; Bernstein, M. A correlating equation for forced convection from gases and liquids to a circular cylinder in crossflow. *J. Heat Transfer* **1977**, *99*, 300–306. [CrossRef]

119. Taniguchi, S.; Kawaguchi, S.; Kikuchi, A. Fluid flow and gas-liquid mass transfer in gas-injected vessels. *Appl. Math. Model.* **2002**, *26*, 249–262. [CrossRef]

120. Lou, W.; Zhu, M. Numerical Simulation of Desulfurization Behavior in Gas-Stirred Systems Based on Computation Fluid Dynamics–Simultaneous Reaction Model (CFD–SRM) Coupled Model. *Metall. Mater. Trans. B Process Metall. Mater. Process. Sci.* **2014**, *45*, 1706–1722. [CrossRef]

121. Cao, Q.; Nastac, L.; Pitts-Baggett, A.; Yu, Q. Numerical Investigation of Desulfurization Kinetics in Gas-Stirred Ladles by a Quick Modeling Analysis Approach. *Metall. Mater. Trans. B Process Metall. Mater. Process. Sci.* **2018**, *49*, 988–1002. [CrossRef]

122. Hoang, Q.N.; Ramírez-Argáez, M.A.; Conejo, A.N.; Blanpain, B.; Dutta, A. Numerical Modeling of Liquid–Liquid Mass Transfer and the Influence of Mixing in Gas-Stirred Ladles. *JOM* **2018**, *70*, 2109–2118. [CrossRef]

123. Karouni, F.; Wynne, B.P.; Talamantes-Silva, J.; Phillips, S. Hydrogen Degassing in a Vacuum Arc Degasser Using a Three-Phase Eulerian Method and Discrete Population Balance Model. *Steel Res. Int.* **2018**, *89*, 1–11. [CrossRef]

124. Xie, H.; Oeters, F. Kinetics of mass transfer of manganese and silicon between liquid iron and slags. *Steel Res.* **1995**, *66*, 501–508. [CrossRef]

125. Jun, Z.; Shi, F.; Mukai, K.; Tsukamoto, H. Numerical analysis of nitrogen absorption rate accompanied with Marangoni convection in the molten iron under non-inductive stirring condition. *ISIJ Int.* **1999**, *39*, 409–418. [CrossRef]

126. Ying, Q. Mass Transfer Coefficients in Metallurgical Reactors. *J. Univ. Sci. Technol. 10* **2010**, *2*, 1–9.

127. Mendes, M.A. Surfactant Effects on Mass Transfer in Liquid-Liquid Systems. *Fluid Mech. Surfactant Polym. Solut.* **2004**, 39–56.

128. Pirker, S.; Gittler, P.; Pirker, H.; Lehner, J. CFD, a design tool for a new hot metal desulfurization technology. *Appl. Math. Model.* **2002**, *26*, 337–350. [CrossRef]

129. Mietz, J.; Bruhl, M. Model calculations for mass transfer with mixing in ladle metallurgy. *Steel Res.* **1990**, *61*, 105–112. [CrossRef]

130. Terrazas, M.S.C.; Conejo, A.N. Effect of Nozzle Diameter on Mixing Time During Bottom-Gas Injection in Metallurgical Ladles. *Metall. Mater. Trans. B Process Metall. Mater. Process. Sci.* **2015**, *46*, 711–718. [CrossRef]

131. Nuñez, D.A.; Ramirez-Argaez, M.A.; Conejo, A.N. Mathematical modeling of bottom gas injection in industrial metallurgical ladles in the presence of a top layer of slag. In Proceedings of the 8th Pacific Rim International Congress on Advanced Materials and Processing 2013 (PRICM 8), Waikoloa, HI, USA, 4–9 August 2013; Volume 4.

132. Likhachou, P.; Conejo, A. *Mass Transfer in Ladles Due to Bottom Gas Injection. Research in progress*, Beijing, China. 2020.

133. Singh, U.; Anapagaddi, R.; Mangal, S.; Padmanabhan, K.A.; Singh, A.K. Multiphase Modeling of Bottom-Stirred Ladle for Prediction of Slag–Steel Interface and Estimation of Desulfurization Behavior. *Metall. Mater. Trans. B Process Metall. Mater. Process. Sci.* **2016**, *47*, 1804–1816. [CrossRef]

134. Choi, J.Y.; Kim, D.J.; Lee, H.G. Reaction kinetics of desulfurization of molten pig iron using CaO-SiO2-Al2O3-Na2O slag systems. *ISIJ Int.* **2001**, *41*, 216–224. [CrossRef]

135. Harada, A.; Maruoka, N.; Shibata, H.; Kitamura, S.Y. A kinetic model to predict the compositions of metal, slag and inclusions during ladle refining: Part1. Basic Concept and Application. *ISIJ Int.* **2013**, *53*, 2110–2117. [CrossRef]

136. Roy, D.; Pistorius, P.C.; Fruehan, R.J. Effect of silicon on the desulfurization of Al-killed steels: Part II. Experimental results and plant trials. *Metall. Mater. Trans. B Process Metall. Mater. Process. Sci.* **2013**, *44*, 1095–1104. [CrossRef]

137. Wei, J.H.; Zhu, S.J.; Yu, N.W. Kinetic model of desulphurization by powder injection and blowing in RH refining of molten steel. *Ironmak. Steelmak.* **2000**, *27*, 129–137. [CrossRef]

138. Sulasalmi, P.; Visuri, V.V.; Kärnä, A.; Järvinen, M.; Ollila, S.; Fabritius, T. A Mathematical Model for the Reduction Stage of the CAS-OB Process. *Metall. Mater. Trans. B Process Metall. Mater. Process. Sci.* **2016**, *47*, 3544–3556. [CrossRef]

139. Huang, F.; Zhang, L.; Zhang, Y.; Ren, Y. Kinetic Modeling for the Dissolution of MgO Lining Refractory in Al-Killed Steels. *Metall. Mater. Trans. B Process Metall. Mater. Process. Sci.* **2017**, *48*, 2195–2206. [CrossRef]

140. Chen, Y.N.C.; Bao, Y.P.; Wang, M.; Zhao, L.H.; Peng, Z. A mathematical model for the dynamic desulfurization process of ultra-low-sulfur steel in the LF refining process. *Metall. Res. Technol.* **2014**, *111*, 37–43. [CrossRef]

141. Deo, B.; Grieveson, P. Kinetics of desulphurisation of molten pig iron. *Steel Res.* **1986**, *57*, 514–519. [CrossRef]

142. Wilson, G.T.; MacLeod, N. A critical appraisal of empirical equations and models for the prediction of the coefficient of reaeration of deoxygenated water. *Water Res.* **1974**, *8*, 341–366. [CrossRef]

143. Reiter, G.; Schwerdtfeger, K. Characteristics of Entrainment at Liquid/Liquid Interfaces due to Rising Bubbles. *ISIJ Int.* **1992**, *32*, 57–65. [CrossRef]

Article

Numerical Modeling of Transport Phenomena in the Horizontal Single Belt Casting (HSBC) Process for the Production of AA6111 Aluminum Alloy Strip

Usman Niaz [1], Mihaiela Minea Isac [1] and Roderick I. L. Guthrie [2,*]

[1] Mining & Materials Engineering Department, McGill University, Montreal, QC H3A 0C5, Canada;
    usman.niaz@mail.mcgill.ca (U.N.); mihaiela.isac@mcgill.ca (M.M.I.)

[2] McGill Metals Processing Centre, McGill University, Montreal, QC H3A 0C5, Canada

* Correspondence: roderick.guthrie@mcgill.ca

Received: 31 March 2020; Accepted: 27 April 2020; Published: 30 April 2020

Abstract: In this research study, numerical modelling and experimental casting of AA6111 strips, 250 mm wide, 6 mm thick, was conducted. The velocity of the molten AA6111 alloy at the nozzle slot outlet was raised to 2 m/s, whilst the belt speed was kept at 0.3 m/s. The numerical model demonstrates considerable turbulence/fluctuations in the flow of the molten AA6111 alloy in the HSBC process, rendering its free surface highly non-uniform and uneven. These discontinuites in the flow resulted from the sudden impact of molten metal onto the inclined refractory plane, and then onto the slowly moving belt. However, it has been determined that these surface variations are rapidly damped, and as such are not detrimental to final strip surface quality. Any surface perturbations remaining can be eliminated via hot plastic deformation. The experimental findings are in accordance with the model predictions. Furthermore, at high metal heads inside the delivery launder, the molten metal was observed to be flowing inwards towards the center of the strip, thereby filling the centre depression region, formed otherwise. The model predictions were validated against experimental findings. A surface roughness and microstructural analysis was also conducted to determine the surface and bulk quality of the as-cast strip.

Keywords: horizontal single belt casting process (HSBC); computational fluid dynamics (CFD); double impingement feeding system

---

1. Introduction

Near net shape casting (NNSC) processes can be regarded as an ideal method for metal sheet production. Apart from their low energy requirements, lower capital and operating costs, and smaller plant footprints than those associated with slab casters (Fe), and DC casters (Al), they also have promising metallurgical characteristics associated with much higher cooling rates possible [1–5]. However, they can also have drawbacks, regarding surface quality, which conventional plants can overcome, by scarfing and multi-pass rolling, to the final quality strip [6]. An advantage of NNSC processes is that a homogenous microstructure with almost zero macrosegregation and fine grain sized products can be attained, provided cooling rate conditions are met, as the cast product's dimensions approach the desired sheet specifications [4].

Therefore, fewer hot/cold reduction passes are required as opposed to conventional methods i.e., slab, or thin slab, casting (TSC) of steels, or direct chill (DC) casting for aluminum alloys. This undoubtedly brings down the overall cost of the operation. On the other hand, the high surface area-to-cast thickness is much higher (~100×), making surface oxidation more problematic, as well as surface quality and surface dimensions [3]. The three most commonly used NNSC processes, commercialized to date, are the twin roll casting (TRC), the earlier twin belt caster, and, most recently, the single belt caster [6]. Historically, the commercialization of the Hazelett twin belt caster in the 1930s by Clarence Hazelett, was an early breakthrough for NNSC. It has since been used around the world, ever since, for casting non-ferrous alloys, mainly aluminum, copper, and zinc, down to cast thicknesses of 13–21 mm [7]. All attempts with steel have failed.

As the name suggests, TRC utilizes two contra-rotating water-cooled rolls, onto which the molten metal is fed and cast as a thin strip. Invented by Henry Bessemer back in 1857 [8], and despite extensive research and development activities that followed, the world's first strip caster for aluminum was only commercialized in 1954, approximately 100 years after Bessemer's initial idea. Currently, FATA Hunter and Novelis PAE (previously SCAL-Pechiney) are leading companies manufacturing TRC equipment for aluminum strip production [9]. The cast strip thickness is almost 6 mm. The first TRC strip caster commercialized for steel was by Nippon Steel and Mitsubishi Heavy Industries Corporation in 2000. The caster could produce stainless steel strips, 2–5 mm thick, 0.76–1.5 m wide, in NSC's Hikari Works in Southern Japan, but was not a commercial success and was abandoned. Later, in 2002, NUCOR started the production of a low carbon steel strip, 1.7–1.9 mm thick via a TRC process, CASTRIP, and this continues to operate, commercially [6]. However, TRC, though a proven aluminum/steel strip manufacturing process, suffers from low annual production of steel vs. that of a conventional slab casting process, i.e., 400,000 tpy and 2,000,000 tpy, respectively [10].

An Introduction to Horizontal Single Belt Casting (HSBC) Process

The horizontal single belt casting (HSBC) process, independently conceived by Herbertson, Guthrie, Reichelt, and Schwerdtfeger, et al., emerged from a joint effort of BHP (Australia), McGill University (Canada), and the Hazelett Strip Casting Corporation (USA) as discussed above. This process is like the Pilkington float glass technology, in which the molten glass is continuously poured over a bath of molten tin where it solidifies into a continuous glass sheet [5].

The HSBC process avoids the multi-hot rolling-deformation steps required for direct chill (DC) cast material, together with its intermediate annealing steps, while producing the same desired final thicknesses of sheet products. As such, a large amount of energy can be saved. This is essentially true for all NNSC processes discussed above. The HSBC process features a compact design and provides for the better economic production of both ferrous and non-ferrous metallic products. The production of advanced high-strength steel (AHSS) strips, 16 mm thick, via the HSBC process at Salzgitter Group Steelworks, Germany, is an example that makes use of the advantages that the HSBC process offers, vs. conventional strip manufacturing processes. To date various ferrous/nonferrous alloys strips, up to a thickness of 16 mm, were successfully produced. For further reading, please refer to the available literature [11,12].

Simplistically, the HSBC process involves feeding molten metal on to an intensively cooled, moving belt, which acts as the mold. Depending on the metal head in the launder, the velocity of the molten metal issuing from the slot nozzle can be easily adjusted. However, the force of gravity is also equally responsible in further accelerating the molten metal before it contacts the moving belt, on which it solidifies [13–15]. The material produced via the HSBC process can then be processed downstream, by hot rolling, followed by cold rolling, as shown in Figure 1.

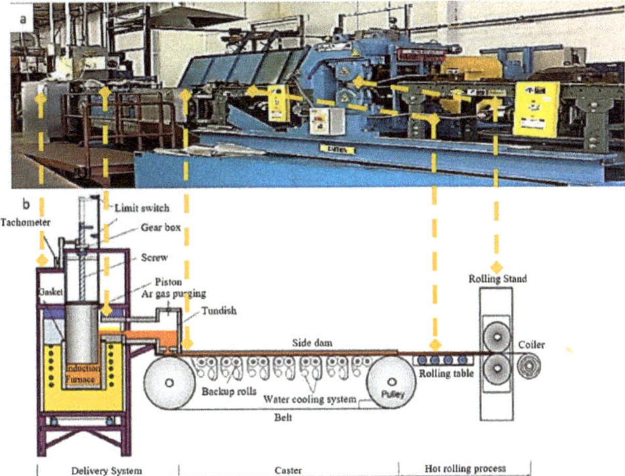

Figure 1. (**a**) A photograph of the HSBC pilot-scale system, and (**b**) A schematic of the HSBC pilot-scale machine located at MetSim Inc,'s High Temperature, Melting and Casting Laboratory, Quebec, Canada.

2. An Overview of Previous Studies Conducted on HSBC Metal Feeding Systems

Many variants of the feeding system have been investigated by researchers at McGill Metals Processing Centre, to date. They can be classified into two types of the metal delivery systems; either single-impingement or multi-impingement, based on how many times the molten metal encounters obstacles before reaching the cooling substrate, as shown in Figure 2 [2,3]. In a single impingement feeding system, the molten metal is abruptly stopped by the horizontally moving belt, since there is no intermediate obstacle in its way, which could decrease its kinetic energy. As a result, the molten metal tends to penetrate back into the quadruple region, i.e., the region where melt, refractory, air, and belt coexist, as determined by Sa Ge et al. [3], for the casting of plain carbon steel, employing a single impingement feeding system (Figure 3a) [3]. If the backflow is too excessive, it may lead to skull formation, thereby curtailing further casting as determined experimentally.

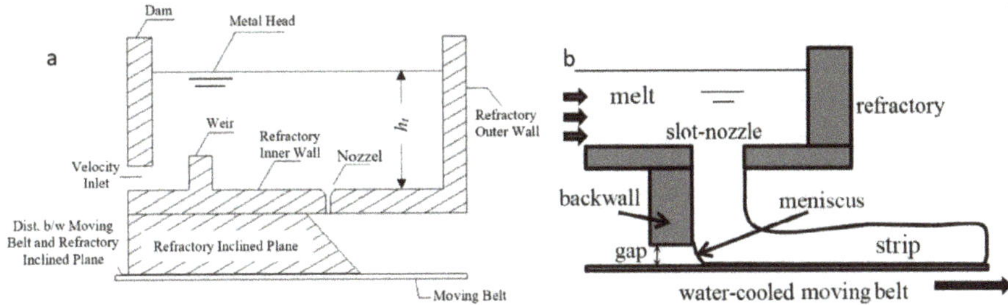

Figure 2. Different feeding systems for HSBC process, (**a**) Double impingement, (**b**) single impingement [1–4].

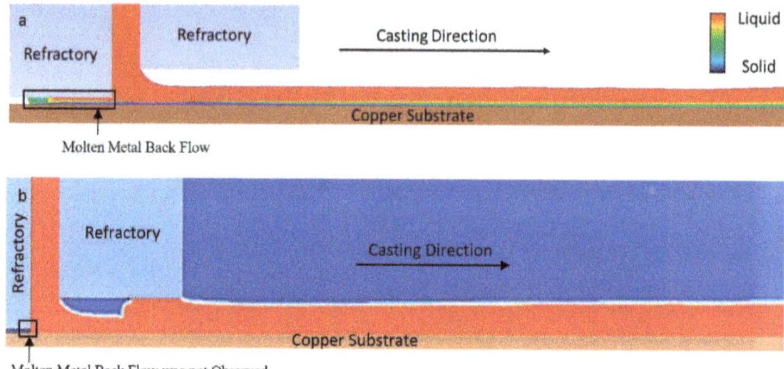

Figure 3. (**a**) Simulated flow of the molten plain carbon steel in single impingement feeding system, showing back-flow into the quadruple region; molten phase is colored red, fully solidified shell is in blue, and the partially solidified zone is in between [3], (**b**) Numerical simulations did not predict any back flow of molten Al-Mg-Sc-Zr into the quadruple region; the molten phase is coloured red, air is represented in blue. The numerical simulations were supported by experiments in these studies [13].

Additionally, it has been shown by Sa Ge [13], that a single impingement feeding system can be used to produce Al-Mg-Sc-Zr alloy strips, without any backflow into the quadruple region (see Figure 3b), as was found for steel casting. This could be due to the low density of Al-Mg-Sc-Zr alloy and a higher contact angle, i.e., a lower wettability between liquid Al-Mg-Sc-Zr and the alumina refractory [13].

In both these studies, the molten metal flowing over the moving belt was considered as being nearly iso-kinetic, a condition in which the velocities of the molten metal and belt approach each other. Furthermore, the as-cast thickness of the produced strips was ~3 mm [3,5].

As explained above, excessive backflow of molten metal into the quadruple region is not desired. This can be conveniently prevented by employing a double-impingement feeding system in the HSBC process [14]. In a double impingement feeding system the molten metal dispensing from the refractory nozzle slot first interacts with a 45° inclined refractory plane, followed by its second interaction with the moving belt, on to which it begins to solidify [14]. In this way, the final impact of the molten metal with the moving belt is not as rapid and abrupt, as it would be for a single-impingement feeding system [4]. Furthermore, in a double-impingement feeding system, the flow of the molten metal over the inclined refractory plane and the moving belt is entirely gravity-driven, unlike the variant of the single-impingement feeding system reported by Sa Ge for the casting of Al-Mg-Sc-Zr. For that, the flow of the molten metal is impeded by the refractory front wall, as shown in Figure 3b [13]. During continuous operation, the refractory material may abrade and embed small particles into the pool of molten metal. This could significantly decrease the bulk quality of the cast strip.

In this research study, non-isokinetic feeding of molten metal over the moving belt has been considered. This is significantly different from our group's previous research studies, for which only near iso-kinetic feeding was considered, as discussed above. Under non-isokinetic feeding conditions, as in the present case (the belt/side dam speed is considerably slower, i.e., 0.3 m/s, as compared to the molten metal velocity at the nozzle slot outlet, i.e., 2 m/s). The strip produced under this condition has ~6 mm thickness, which is thicker as compared to the strips obtained under iso-kinetic feeding (~3 mm). This allowed us to perform substantial hot deformation in order to produce a 1 mm, or lower, thickness of strip. Hot deformation is necessary, as it transforms the cast dendritic structure into fine equiaxed grains, leading to a far more uniform distribution of alloying elements throughout the sheet material. Another purpose of hot reduction is to squash/weld any pores, if present, in the cast strip, so as to improve its mechanical properties [15].

3. Objectives of the Present Research

The objective of this research study is to produce high-quality AA6111 aluminum alloy strips 250 mm wide, ~6 mm thick. Additionally, molten metal flow in a double-impingement feeding system has been analyzed under non-isokinetic feeding. Under these experimental conditions, the molten metal was observed to be flowing inwards, i.e., towards the center of the strip. This can usefully eradicate center shrinkage cavity defects formed otherwise. In order to investigate this phenomenon, a three-dimensional mathematical model was developed using Fluent software (14.5, Ansys, Inc., Canonsburg, PA, USA, 1970), and its accuracy was evaluated against experimental data. Thanks to these numerical simulations, we now understand the complex interaction of the molten metal with the inclined refractory plane and the moving belt that leads to the phenomenon of the molten metal's inward flow.

A horizontal single belt pilot caster installed at MetSim Inc., Montreal, QC, Canada, was used for the casting experiments. However, several modifications were applied to the existing caster, as it was not capable of producing 250 mm wide strips. These included the design of a new alumina refractory nozzle slot (250 mm wide and 3 mm thick), increasing the cooling capability of the moving steel belt, needed to completely solidify molten AA6111 strip before it exits the moving belt. Additionally, the caster modifications included enlarging the strip guidance system, and, lastly, the extension of the length and width of the run-out table, so as to accommodate the wider strip exiting the caster.

4. Aluminum Alloy Grade, AA6111, Used in the Present Research

Keeping in mind the suitability of the HSBC process to cast both ferrous and non-ferrous alloys, and knowing the usefulness of AA6111 aluminum alloy in the production of the lightweight body in white (BIW) automotive structures, AA6111 was selected for HSBC strip production. AA6111 is an alloy of Al-Mg-Si (Cu). It possesses higher mechanical strengths (i.e., 400 MPa), high formability, and good corrosion resistance. The main mechanism behind AA6111's increased strength is precipitation hardening, along with the solid solution and work strengthening [16,17]. The alloying elements present in AA6111 are as shown in Table 1.

Table 1. Chemical composition of the AA6111 strip produced by the HSBC process, determined using the spark OES technique.

			Alloying Elements, wt%					
Cu	Fe	Mg	Mn	Cr	Si	Ti	Zn	Al
0.5–0.9	>0.4	0.5–1.0	0.1–0.45	<0.1	0.6–1.1	<0.1	<0.15	Remaining

The casting of AA6111 strips via the HSBC process is comparatively new, and is presently in its development stages. It is, therefore, hoped that this paper will add fundamental knowledge to the information already existing on this subject, and help industries realize the versatility of the HSBC process to cast AA6111 aluminum alloy strips.

5. Details of the Experimental Procedure

AA6111 alloy was produced by first melting pure aluminum in a pre-heated induction furnace under a protective argon atmosphere, followed by the addition of Al-Mg, and Al-Mn alloys, etc. Good melt stirring was used, to ensure completely dissolved/mixing of the alloy additives into the pure aluminum. The melt was then de-gassed and Ti-B grain refiner was added in a conventional way. The AA6111 melt was then cast into the strips using the HSBC system. The step-by-step operation of the HSBC pilot caster is presented below.

The process started with the production of AA6111 alloy using the 600 lb induction melting furnace. Afterwards, the furnace was moved on rails to the casting station, where it was locked with the liquid metal delivery system. This consisted of a refractory cylinder (regulated by a servo motor) and a launder, as shown in Figure 1. Once the tight seal between the induction furnace and delivery system was ensured, the motorised refractory cylinder was allowed to enter into the induction furnace at a pre-selected speed, thereby displacing the molten metal into the launder. Once the molten metal reached the desired level within the launder, the stopper bar blocking the nozzle outlet was rapidly withdrawn, and liquid metal began to pour onto the belt. A weir and a dam were used to help prevent Al_2O_3 oxide skin from entering the nozzle slot, as well as to help in minimizing turbulence present within the flowing molten metal. The moving belt could also be equipped with two rotating side dams. Their purpose was to contain the molten metal once it leaves the nozzle slot, and to give a straight/smooth edge before it enters the minimill for hot reduction. To avoid any premature freezing, the entire delivery system and the refractory piston were preheated to approximately 500–550 °C, using electrical resistive heating systems.

To evaluate the bulk, as well as the surface, quality of the cast strips, samples were sectioned from the strip. All samples were polished and prepared for metallographic observations and analyzed under Leica DM IRM optical and Hitachi TM3030 scanning electron microscope. The surface roughness was measured using a 3D Nanovea profilometer. Results will be presented in later paragraphs.

6. Details of the Model Setup

For CFD studies, the three-dimensional, transient state, turbulent fluid flow was modeled using Fluent software (14.5, Ansys, Inc., Canonsburg, PA, USA, 1970). The code is based on the finite volume method (FVM) [18]. The simulation domain chosen to carry out this research study had the following dimensions, length (0.195 m), height (0.016 m), and width (0.05 m), as shown in Figure 4. The semi-implicit method for pressure linked equations (SIMPLE) was used for coupling pressure and velocity in the governing equations. More details can be found in the literature [19]. To improve on the accuracy, the advection terms were discretized using a 2nd-order upwind scheme over the entire simulation domain, whereas the diffusion term was approximated by the central differencing scheme. To stabilize the interactive process, an under-relaxation factor of 0.7 for the velocity and 0.3 for the pressure, were used. The solution process was iterated until the residuals of governing equations reduced to 1×10^{-7}. Different grids were tested until mesh-independent results were achieved. Finally, 2,867,541 hexahedral cells were identified as being an accurate, but less computationally intensive, exercise for obtaining the desired results. The molten metal was treated as a Newtonian, incompressible fluid, and all the physical properties were assumed to be constant (Tables 2 and 3).

Table 2. Physical properties of the Phases used in the model.

Operating Parameters/Assumptions	Value
Slot Nozzle Dimension	3×250 mm
Inlet Velocity	2 m/s
Surface Tension of the Melt in Air	0.914 N/m
Copper Substrate Longitudinal speed	0.3 m/s
Turbulence Model	SST k-ω
Contact Angle Between Melt and Alumina Refractory	135° [13]
Contact Angle Between Melt and Copper Substrate	105° [13]
Distance Between Stationary Inclined Refractory Plane and Moving Belt	0.4 mm

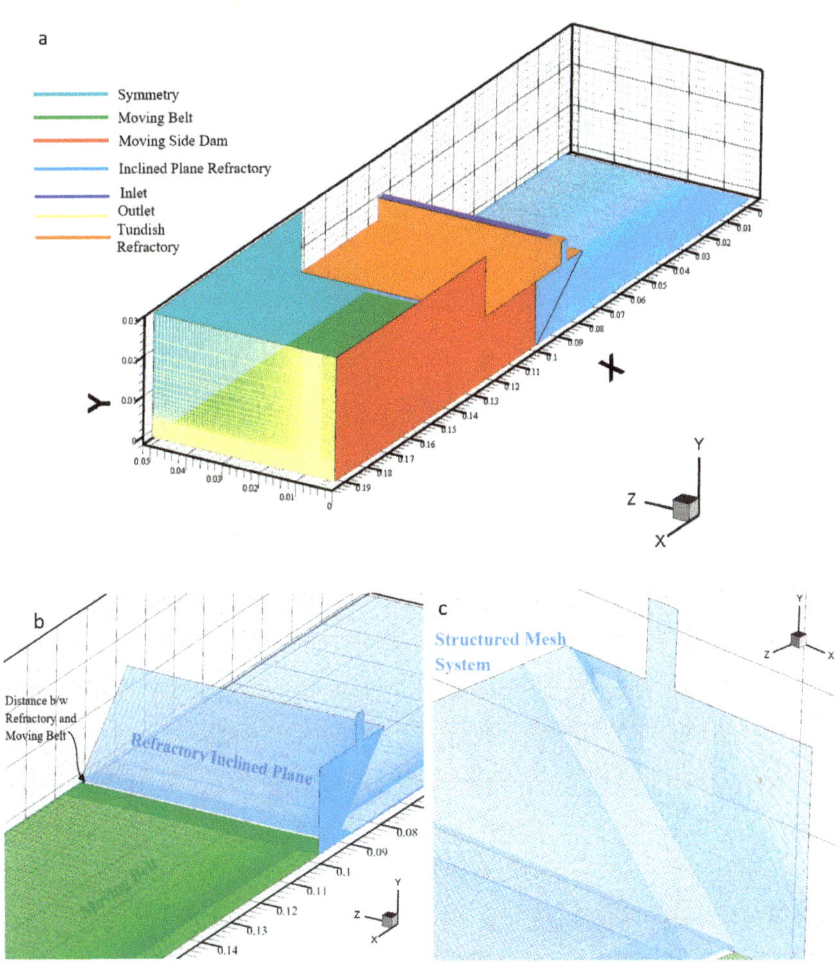

Figure 4. (**a**) Simulation domain containing hexahedral meshes (3D), (**b**) Mesh refinement at the nozzle outlet, edges and the quadruple region, (**c**) A closer look on the hexahedral mesh system at the inclined refractory plane. The dimensions are in meters

Table 3. Operating parameters and assumptions made in the model [2,13].

Property	AA6111	Air
Density, $\rho\left(\frac{Kg}{m^3}\right)$	2300	1.225
Specific Heat Capacity, $C_P\left(\frac{KJ}{KgK}\right)$	1.177	1.006
Thermal Conductivity, $K\left(\frac{W}{mK}\right)$	104	0.0242
Viscosity, $\mu\left(\frac{Kg}{ms}\right)$	0.001338	1.75×10^{-5}
Molecular Weight, $\left(\frac{Kg}{kmol}\right)$	26.98	28.97
Standard State Enthalpy $\left(\frac{J}{kgmol}\right)$	$1.100493e^7$	-
Reference Temperature (K)	298.15	298.15
Initial Temperature (K)	1000	300

7. Results and Discussion

It was observed through numerical simulation studies that in the HSBC process, the molten metal tends to laterally contract/shrink while exiting through the nozzle slot outlet, as shown in Figure 5. Due to this lateral contraction, the weight of the molten metal around the edges increases considerably as compared to the center. The heavier section accelerates downwards under the influence of gravity, eventually reaching a high terminal speed before it strikes the moving belt. This concept is further explained by plotting molten metal velocity, adjacent to the free molten metal/air interface and the moving belt, against distance in the positive z-direction. As expected, the magnitude of the velocity around the edges, for both cases, is high as compared to the center. These velocities are computed 5 mm away from the quadruple region (down the ramp), as shown in Figure 6.

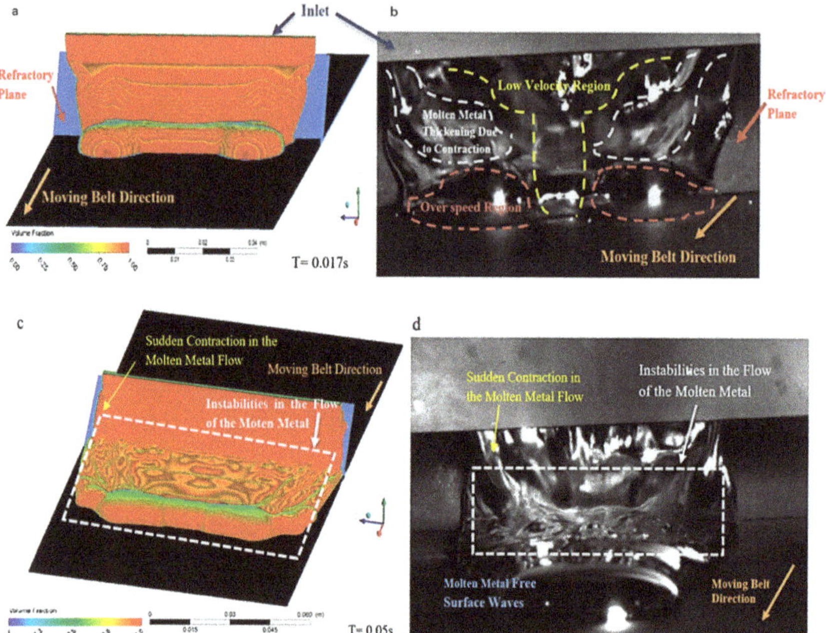

Figure 5. Molten AA6111 alloy flow in the HSBC process. (**a,c**) Simulated AA6111 flow showing metal's contraction after leaving the nozzle slot. (**b,d**) Actual Molten AA6111 flow in the HSBC process.

Additionally, it has been observed that the velocity of the molten metal adjacent to the moving belt is lower than the velocity at the free molten metal/air interface, as shown in Figure 6. This is due to the friction offered by the moving belt, which tends to slow down the velocity of the molten metal adjacent to it.

Based on the above discussion, it can be concluded that the amount of molten metal, delivered towards the edges, is considerably greater in comparison to the center, as shown in Figure 5a,b, owing to the initial contraction/shrinkage of the molten metal while exiting through the slot nozzle outlet. The net effect is an inward flow of the molten metal towards the center, as shown in Figure 7. This inward flow can be very beneficial, as it eradicates the center shrinkage cavity defect, formed otherwise, at low metal heads in the launder. This topic is further explained in the following paragraphs.

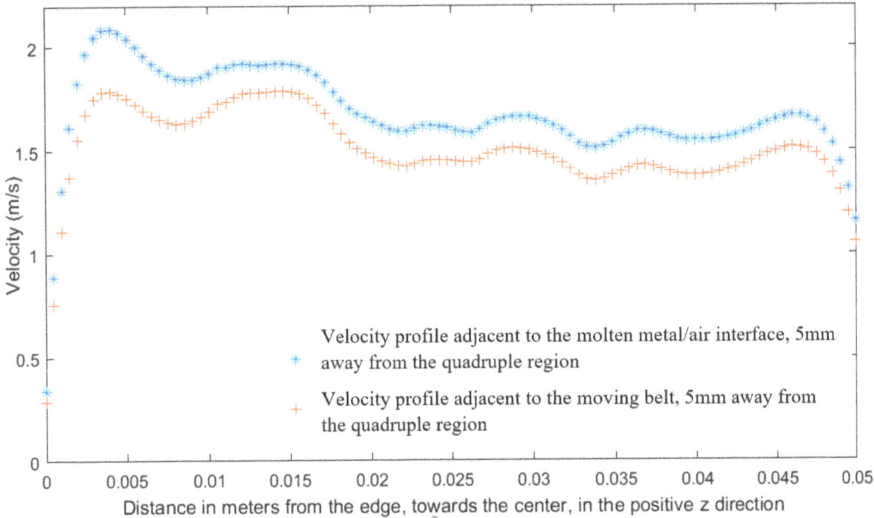

Figure 6. A plot of the velocity vs. distance (m) from the edge dam moving at 0.3 m/s.

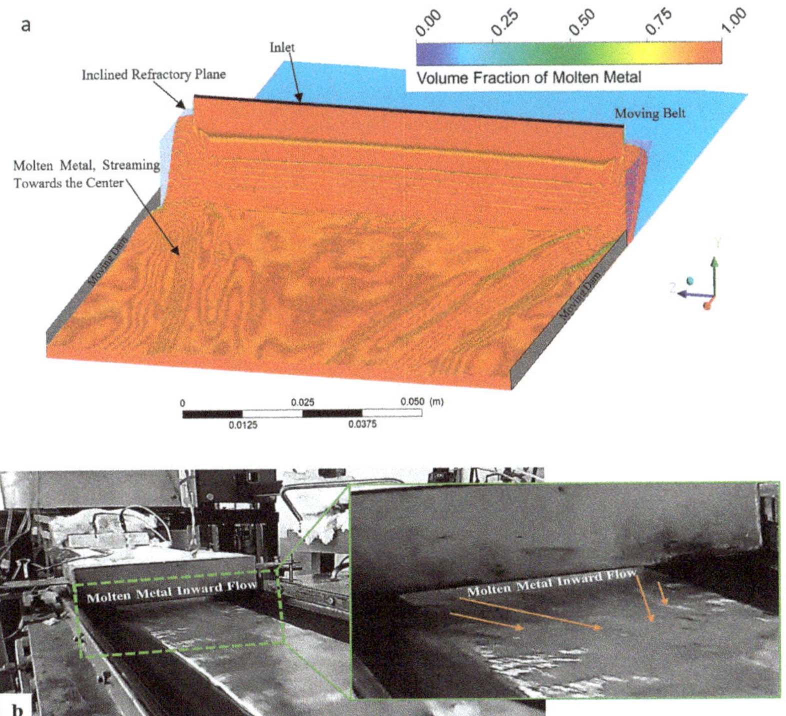

Figure 7. The inward flow of the molten metal over the moving belt. (**a**) Simulated; (**b**) actual.

7.1. Method to Eradicate Center Cavity Defects

As discussed above, in the HSBC process, employing a double impingement feeding system, the molten metal streams towards the center. Keeping in mind that at lower metal heads (<50 mm), the quantity of the molten metal delivered onto the moving belt is also low, as per Equation (1). As observed experimentally, under the rapid heat extraction rate to the moving belt (i.e., 500 K/sec) [3], the molten metal passing over the moving belt, tends to solidify almost instantaneously. Since the molten metal does not have enough time to level off before the completion of solidification, this results in a strip with a thicker edge, and a comparatively thinner center.

The opposite is true for high metal heads inside the tundish (>50 mm). In this case, the velocity of the molten metal exiting the refractory nozzle slot outlet is high enough to trigger a strong net inward flow. Under these conditions, the molten metal will have enough time to fill the center empty region and to evenly spread throughout the thickness of the strip before the completion of solidification. This helps to eliminate any center cavity defect and to achieve a uniform thickness of the strip across its width.

$$V = C_D \sqrt{2gh}$$
(1)

where V is the velocity, h is the molten metal head inside the tundish, and C_D is the coeffecient of discharge.

7.2. Iso-Surfaces of Z-Component of Velocities

The velocity vector can be resolved into three components, i.e., x, y, and z in which the z-velocity component represents a net inward flow of the molten metal. For these reasons, the isosurfaces of the z-component of the velocities were evaluated and are represented in Figure 8.

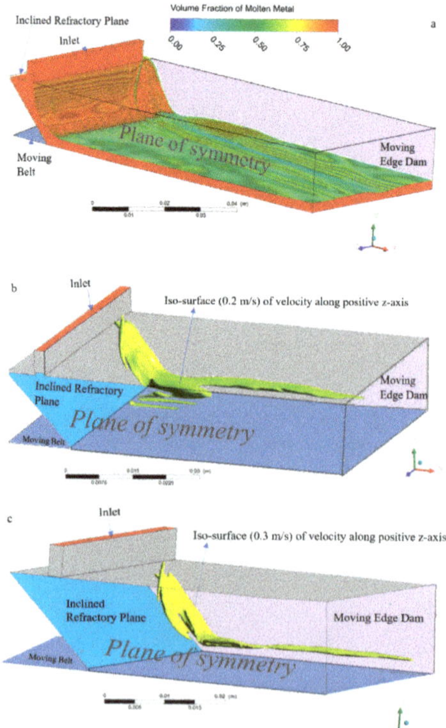

Figure 8. *Cont.*

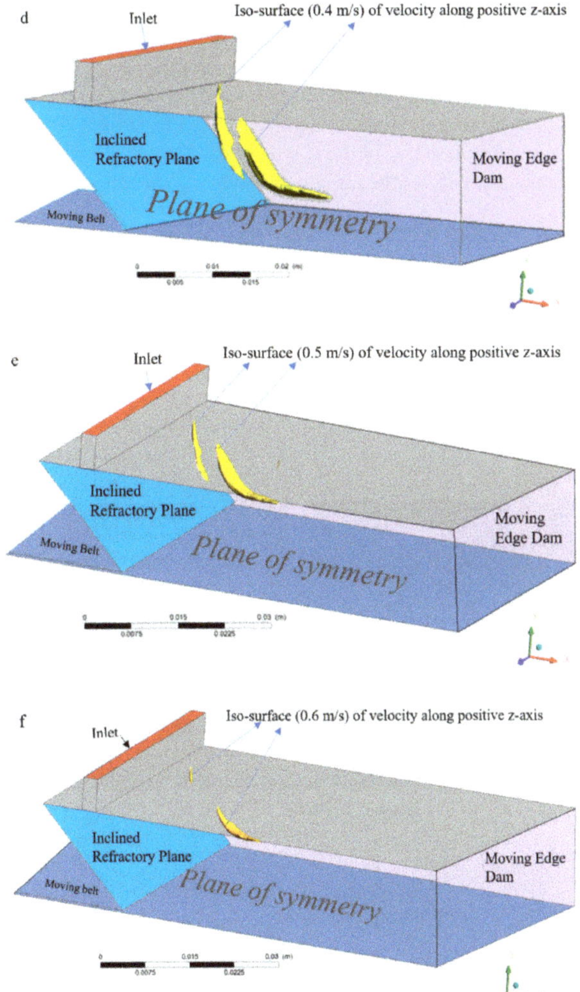

Figure 8. (**a**) Molten metal flow in the HSBC process. (**b**) Iso-velocity (0.2 m/s) along + z-direction. (**c**) Iso-velocity (0.3 m/s) along + z-direction. (**d**) Iso-velocity (0.4 m/s) along + z-direction. (**e**) Iso-velocity (0.5 m/s) along + z-direction. (**f**) Iso-velocity (0.6 m/s) along + z-direction.

As expected, the z-component of the velocity vector is maximal during the first instants of the molten metal contacting the moving belt, owing to the fact that molten metal, while flowing over an inclined plane, continuously accelerates under the force of gravity. Furthermore, the z-component of velocity was observed to be high adjacent to the molten metal/air interface, and almost zero near the moving belt. However, further downstream, over the belt, the z-component of velocity was observed to be decreasing with distance. This is essentially true, as there is no driving force that could help the molten metal to further accelerate over a horizontal moving belt.

7.3. Casting of AA6111 without the Use of Side Dams

The friction imposed by a slowly-moving belt reduces the velocity of the molten metal over the moving belt. As a result, the molten metal tends to spread outwards, as shown in Figure 5c,d. However, adjacent to the free interface, the flow of the molten metal is directed towards the center, as is evident from the z-component of iso-velocities presented above. Depending on the relative magnitude of these two opposing effects, the molten metal can either flow towards the center, or outwards.

However, by looking at Figure 7, it can be clearly seen that the molten metal is flowing towards the center. This is very beneficial, as it eliminates the need for side dams to control the outward flow of the molten metal. The successful casting of the AA6111 strip (Figure 7b), without the aid of side dams (See Figure 7), experimentally, verifies the numerical modeling predictions.

7.4. Pressure Distribution of Molten Metal and the Generation of a Vortex Near the Triple Point

It is observed via the numerical simulations, that the inclined refractory plane has the tendency to lessen, or moderate, the final impact of the molten metal on to the moving belt, by converting a part of the molten metal's kinetic energy into static pressure, as presented in Figure 9b.

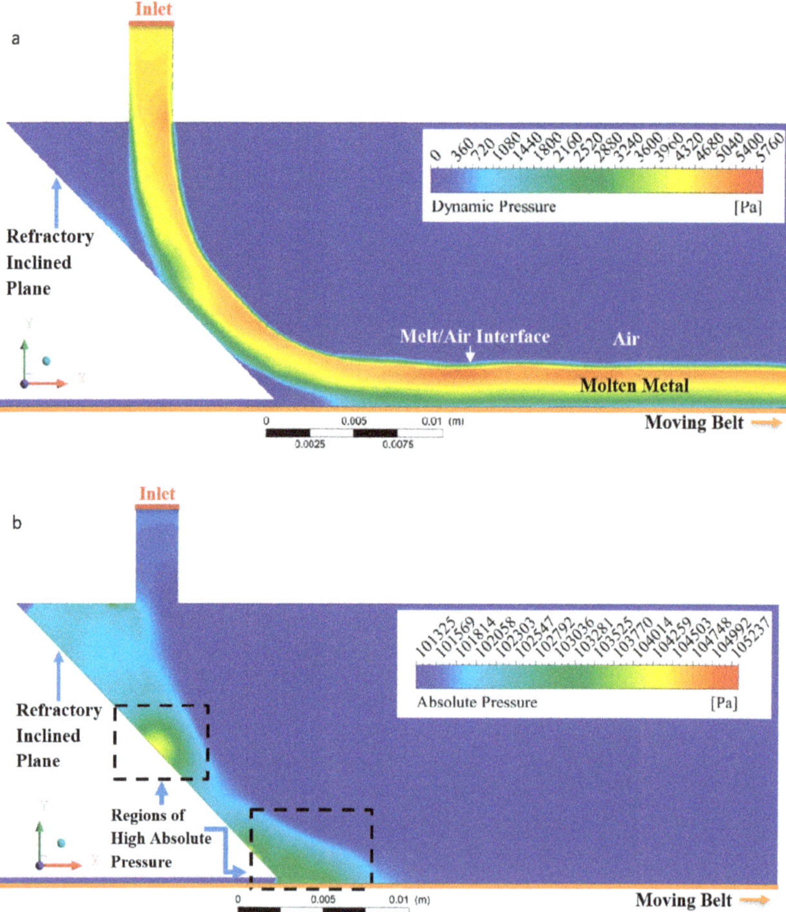

Figure 9. (a) Dynamic Pressure, $P_{dyn} = 0.5\rho \sum u_i^2$, where ρ is the density, u is the velocity. (b) Absolute pressure (103,281 Pa).

Additionally, the numerical simulations predict a considerably higher absolute pressure (low velocity) near the quadruple region (Figure 9b), due to sudden decrease in velocity of the molten metal by a slow-moving belt. This results in a part of the impinging molten metal climbing upwards, forming a vortex, as shown in Figure 10. Furthermore, the dynamic pressure, i.e., $\frac{1}{2}\rho u^2$ is observed to be highly adjacent to the molten melt/air interface. This result is due to the high velocity of the molten metal near the free surface, as shown in Figure 9a.

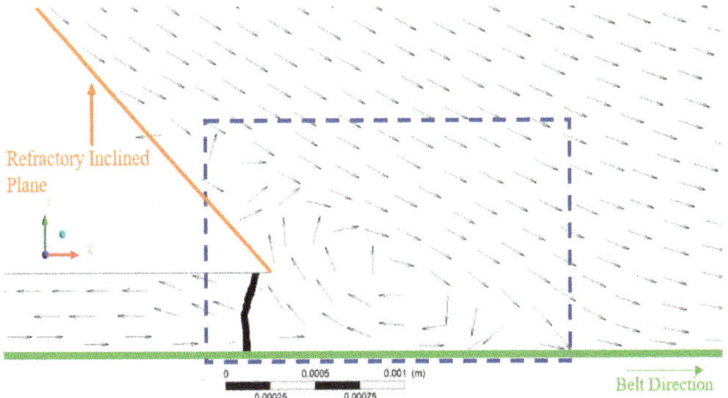

Figure 10. The swirling motion of the molten metal at the triple point, after the second impingement.

7.5. The Temperature of the Molten Metal at the Melt/Air Interface

The hydraulic jump on the inclined refractory plane could substantially degrade the surface quality of the cast product, owing to the generation of free surface waves/discontinuities [14]. As per the numerical simulations, the temperature over of the melt/air interface is above the liquidus for a considerable distance, as shown in Figures 11 and 12, even when considering perfect contact of molten metal with the moving belt, which is held constant at 300 K, by the cooling water under the belt. The experimental casting of the AA6111 alloy strip was in accord with numerical simulation predictions, in which the melt/air interface with the belt was observed to be in a liquid state for approximately the first meter along the moving belt. In this way, any molten metal surface discontinuities had enough time to settle down by the damping forces generated, before final solidification.

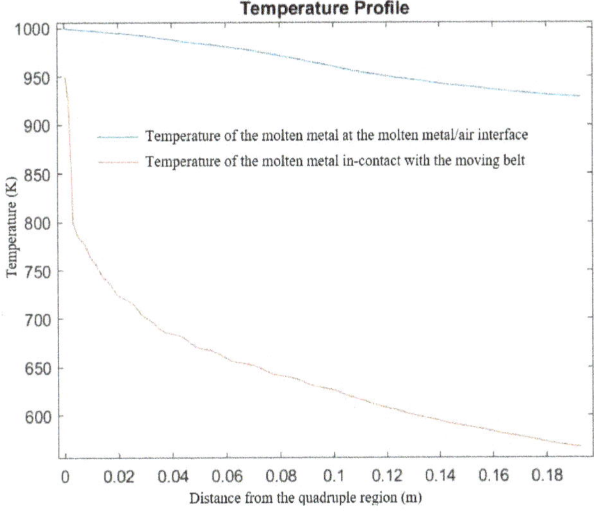

Figure 11. The predicted temperature distribution along the top and bottom faces of the strip along the casting direction.

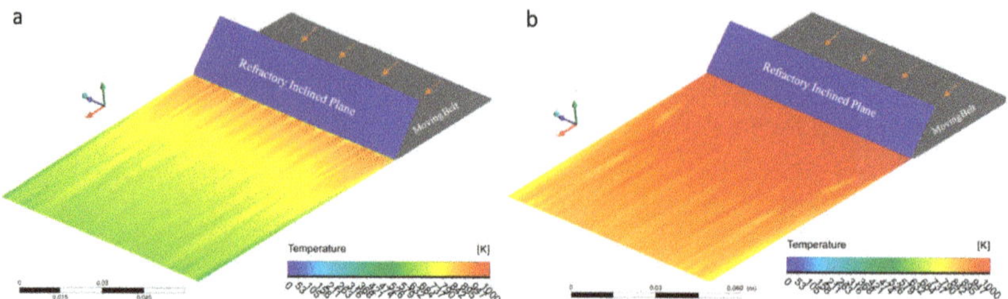

Figure 12. Contour of molten metal temperature in contact with (**a**) the moving belt and (**b**) the molten metal/air interface.

7.6. Characterization of the AA611 Alloy Cast Strip

A microstructural study has been carried out to analyze the quality of the cast strips. The samples for optical microscopy were ground down to 1200 grid, and then electropolished/etched using 2% perchloric acid ($HClO_4$) in alcohol. Micro images were taken using a Leica DM IRM optical microscope (Leica Microsystems, Concord, ON, Canada) and a TM3030 Scanning Electron Microscope (Hitachi, Pleasanton, CA, US). The microstructure consisted of fine equiaxed grains throughout the thickness of the strips (see Figure 13). The microstructure also contains porosities at various locations within the cast strip, very similar to DC cast product. The average grain size of the strip was found to be 85 μm, as shown in Figure 13.

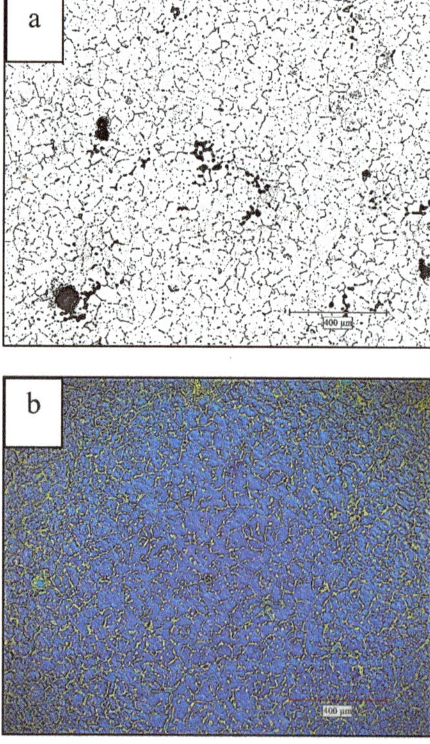

Figure 13. *Cont.*

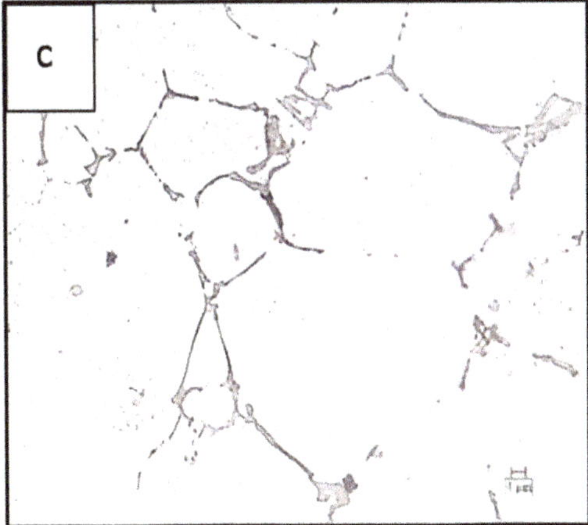

Figure 13. (**a**) Microstructure at the center of the strip (50×); (**b**) polarized micrograph (50×); (**c**) intermetallics observed at the grain boundaries as well as inside the grains (500×).

The inter-metallics are also observed to be distributed inside the grains, as well as at the grain boundaries of the cast microstructure, as shown in Figure 13c. X-ray micro-analyses revealed that inter-metallics dispersed throughout the cast structure have the following stoichiometry: $Al_{17}Cu_2Mg_3Si_3$, $Al_{20}Cu_2Mg_{2.5}Si_5$, whereas the elongated inter-metallics distributed at the grain boundaries are in the category of $Al_{17}(CuMg)_2(FeMn)Si_2$ or $Al_{25}(CuMg)_{4.5}(FeMn)Si_5$ [16,17]. These phases are clearly observed in the Figure 13c, at a higher magnification of 500×.

7.7. Energy Dispersive Spectroscopy (EDX) Analysis of the Cast AA6111 Alloy

Energy dispersive spectroscopy (EDX) analysis confirmed the presence of Al, Cu, Mg, and Si, in AA6111 alloy, as shown in Figure 14. Furthermore, there is a negligible macro-segregation of alloying elements in the cast strip, as shown in the chemical element's maps, obtained at 15 KV excitation voltage. This is caused by the rapid solidification of molten AA6111 alloy in the HSBC process, which resulted in a homogenous microstructure with fine equiaxed grains. Additionally, the elemental maps provide us with details of the chemical nature of the secondary phases. EDX analyses revealed that inter-metallics dispersed throughout the cast structure are rich in Cu and Mg, whereas the elongated inter-metallics found at the grain boundaries are concentrated in Si, Cu, and Mg.

7.8. Surface Roughness Measurement

The surface waviness of the top/bottom sides of the strip was determined using a Nanovea 3D profilometer (Nanovea, Irvine, California, United States). This technique works on the principle of measuring the physical wavelength of light and directly relating it to a specific height. This ensures accurate measurement of surface roughness/finish [20]. The scan length for all the measurements was 25 mm, whereas the scan speed was 0.1 mm/s. Ten random locations were selected for surface roughness measurements. These locations were randomly selected from all over the strip. The surface profiles are almost identical to one another. The upper surface roughness lies within the 125 μm (0.125 mm) range, as shown in Figure 15a, which is considerably smaller than the DC cast product roughness, i.e., 0.45 mm for AA5182 aluminium alloy (Figure 16a,b) [21]. Additionally, pinholes/blowholes were not detected on the surface of the cast strip (Figure 16c,d), unlike continuously cast products, which possessed defects on their surfaces, and require surface grinding prior to hot rolling [21].

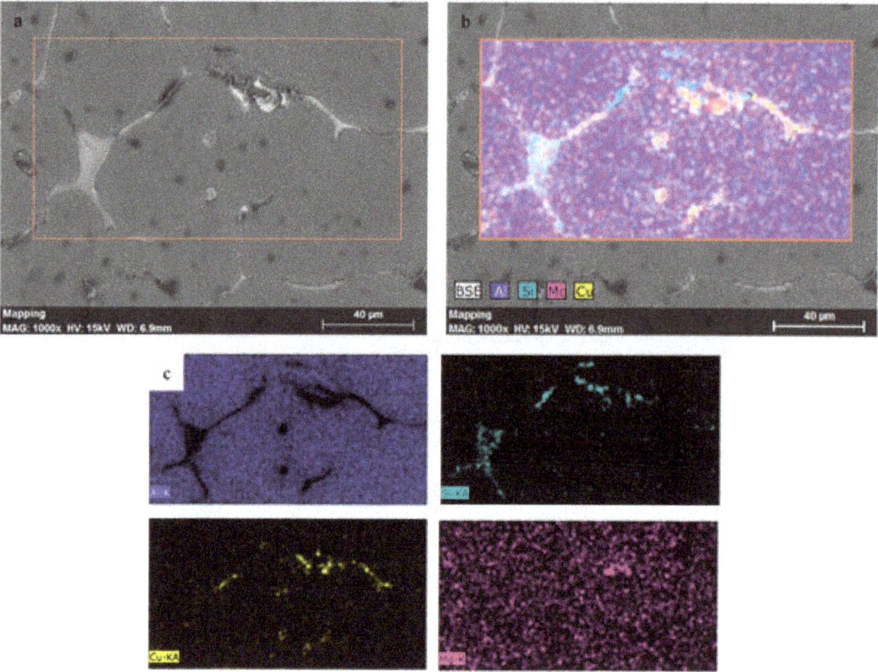

Figure 14. (**a**) Intermetallics distributed within grains and at grain boundaries (1000×). (**b**,**c**) Elemental maps showing the chemical content of the intermetallics.

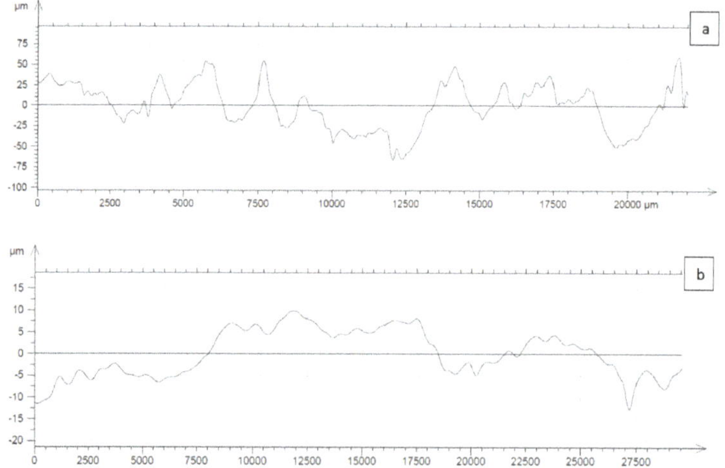

Figure 15. 3D profilometry results: (**a**) strip top surface roughness and (**b**) strip bottom surface roughness.

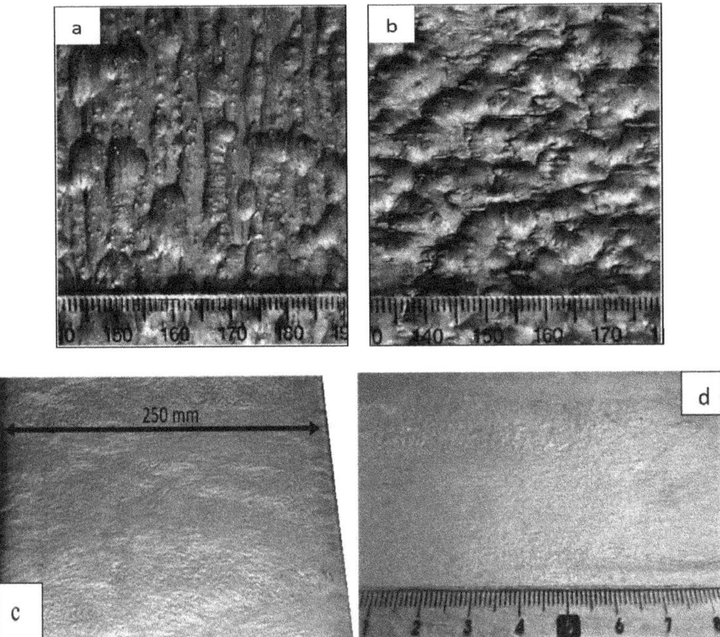

Figure 16. Strand surface morphologies of direct chill of (**a**) AA3004 (Al-1%Mn-1%Mg), (**b**) AA5182 (Al-4.5%Mg) [21], (**c**) surface morphology of AA6111 (Al-1.1%Si-1%Mg-0.45%Mn), 250 mm wide strip produced via the HSBC process. (**d**) Surface morphology of AA6111 (Al-1.1%Si-1%Mg-0.45%Mn), covering the central 80 mm length of the strip produced via the HSBC process.

The strip bottom surface roughness was also measured, and lay in the 20 μm range, as shown in Figure 15b. As evidenced by the results of the line scans, the bottom surface quality is much superior to the top surface. This fact is credited to the fact that the molten metal is in direct contact with the moving belt, and conforms to its shape, during the solidification process. On the other hand, the top surface of the cast strip is exposed to the atmosphere and is affected by disturbances in the flows of the molten metal.

8. Conclusions

This present paper discusses the casting conditions and analyses results on AA6111 alloy strips, 250 mm wide and 6 mm thick, produced via the HSBC process. Computational fluid dynamics (CFD) analyses were performed to examine molten AA6111 flows in the HSBC process so as to achieve uniform thickness and a good surface finish of the cast strip.

The following conclusions were drawn, using the double impingement with no side dams condition, for the liquid metal feeding system.

1. Under non-isokinetic feeding, the surface quality of the cast AA6111 alloy strip is not compromised by the generation of surface disturbances.
2. The AA6111 alloy molten stream shrinks from its two edges. This build-up of the mass around the corners eventually reaches a high terminal velocity. The net inward flow of the molten metal resulted, which filled the center shrinkage depressions.
3. It was also determined that the inclined refractory plane of the double-impingement metal feeding system has the tendency to lessen, or moderate, the final impact of molten metal with

the moving belt, as compared to a single-impingement metal feeding system, where the molten metal encounters an abrupt change in direction by the moving belt.

4. The swirling flow of the molten metal in an immediate vicinity of the triple point is due to the sudden vertical deceleration of the molten metal by the moving belt. However, the meniscus at the triple point was still observed to be stable and non-fluctuating.

5. The temperature of the molten metal within the immediate vicinity of the free surface, along the entire length of the simulation domain, remains above the liquidus temperature. Thus, any molten metal discontinuities formed at the free molten metal/air interface had enough time to be damped, prior to solidification.

Author Contributions: Methodology and writing: U.N.; supervision, technical advice, and access to experimental and analytical equipment: M.M.I. and R.I.L.G. All authors have read and agreed to the published version of the manuscript.

Funding: The authors would like to acknowledge the financial support received from the Natural Sciences and Engineering Research Council of Canada (NSERC), and the International Advisory Board of supporting companies of the McGill Metals Processing Centre (MMPC), as well as the technical support of MetSim Inc., in carrying out this research. The authors would also like to acknowledge the support received from ANSYS Inc. for software licensing to facilitate this research.

Conflicts of Interest: The authors declare no conflict of interest.

References

1. Celikin, M.; Li, D.; Calzado, L.; Isac, M.; Guthrie, R.I.L. Horizontal single belt strip casting (HSBC) of Al-Mg-Sc-Zr alloys. *Light Metals* **2013**, *2016*, 1037–1040.

2. Xu, M.; Isac, M.; Guthrie, R.I.L. A numerical simulation of transport phenomena during the horizontal single belt casting process using an inclined feeding system. *Metall. Mater. Tran. B* **2018**, *49*, 1003–1013. [CrossRef]

3. Ge, S.; Chang, S.; Wang, T.; Calzado, L.E.; Isac, M.; Kozinski, J.; Guthrie, R.I.L. Mathematical modeling and microstructure analysis of low carbon steel strips produced by horizontal single belt casting (HSBC). *Metall. Mater. Tran. B* **2016**, *47*, 1893–1904. [CrossRef]

4. Guthrie, R.I.L.; Isac, M. Horizontal single belt casting of aluminum and steel. *Steel Res. Int.* **2014**, *85*, 1291–1302. [CrossRef]

5. Ge, S.; Isac, M.; Guthrie, R.I.L. Progress in strip casting technologies for steel; technical developments. *ISIJ Int.* **2013**, *53*, 729–742. [CrossRef]

6. Sanders, R.E. Continuous casting for aluminum sheet: A product perspective. *JOM* **2012**, *64*, 291–301. [CrossRef]

7. Hazelett Process vs. Conventional Casting Processes. Available online: https://www.hazelett.com/process (accessed on 27 April 2020).

8. Matsushita, T.; Nakayama, K.; Fukase, H.; Osada, S. Development and commercialization of twin roll strip caster. *Facilities* **2009**, *5*, 6.

9. Hamer, S.; Taraglio, B.; Romanowski, C. Continuous casting and rolling of aluminum: Analysis of capacities, product ranges, and technology. *Light Metal Age-Chic.* **2002**, *60*, 6–17.

10. Isac, M.; Guthrie, R.I.L. The design of a new casting process: From fundamentals to practice. In *Treatise on Process Metallurgy*; Elsevier: Montreal, QC, Canada, 2012; pp. 555–583.

11. Spitzer, K.H.; Rüppel, F.; Viščorová, R.; Scholz, R.; Kroos, J.; Flaxa, V. Direct Strip Casting (DSC)-an option for the production of new steel grades. *Steel Res. Int.* **2003**, *74*, 724–731. [CrossRef]

12. Wans, J.; Geerkens, C.; Cremers, H.; Grethe, U.; Juchmann, P.; Schmidt-Jurgensen, R. Belt casting technology experiences based on the worldwide first BCT caster. *METEC 2nd ESTAD* **2017**, 111–117.

13. Ge, S.; Chattopadhyay, K.; Isac, M.; Guthrie, R.I.L. Mathematical modeling of transport phenomena in Horizontal Single Belt Casting (HSBC). In Proceedings of the 5th International Congress on the Science and Technology of Steelmaking, Dresden, Germany, 1–3 October 2012.

14. Niaz, U.; Isac, M.; Guthrie, R.I.L. Numerical modelling and experimental casting of 17% Mn–4% Al–3% Si–0.45% C wt-% TWIP steel via the horizontal single belt casting (HSBC) process. *Ironmak. Steelmak.* **2020**, *10*, 1–14. [CrossRef]

15. Sychkov, A.B.; Zhigarev, M.A.; Perchatkin, A.V.; Mazanov, S.N.; Zenin, V.S. The transformation of defects in continuous-cast semi-finished products into surface defects on rolled products. *Metallurgist* **2006**, *50*, 83–90. [CrossRef]

16. Li, D.; Shabestari, S.G.; Isac, M.; Guthrie, R.I.L. Studies in the Casting AA6111 Strip on a Horizontal, Single Belt, Strip Casting Simulator. *TMS* **2006**, *135*, 851–856.

17. Mukhopadhyay, P. Alloy designation, processing, and use of AA6XXX series aluminium alloys. *ISRN Met.* **2012**, *2012*, 165082. [CrossRef]

18. Hirt, C.W.; Nichols, B.D. Volume of fluid (VOF) method for the dynamics of free boundaries. *J. Comput. Phys.* **1981**, *39*, 201–225. [CrossRef]

19. *Manual, A Fluid Dynamics Analysis Program*; FLUENT Inc., 10 Cavendish Court. Centerra Resource Park: Lebanon, NH, USA, 1998.

20. Surface Roughness Determination Using 3D Nanovea Profilometer. Available online: https://nanovea.com/profilometry-roughness-finish (accessed on 27 April 2020).

21. Sengupta, J.; Thomas, B.G.; Wells, M.A. The use of water cooling during the continuous casting of steel and aluminum alloys. *Met. Mater. Trans. A* **2005**, *36*, 187–204. [CrossRef]

processes

Article

Understanding TiN Precipitation Behavior during Solidification of SWRH 92A Tire Cord Steel by Selected Thermodynamic Models

Lu Wang [1,2,3], Zheng-Liang Xue [1,2,*], Yi-Liang Chen [1] and Xue-Gong Bi [1,2]

[1] The State Key Laboratory of Refractories and Metallurgy, Wuhan University of Science and Technology, Wuhan 430081, Hubei, China; wanglu@wust.edu.cn (L.W.); chenyiliang456258@163.com (Y.-L.C.); bixuegong@wust.edu.cn (X.-G.B.)

[2] Key Laboratory for Ferrous Metallurgy and Resources Utilization of Ministry of Education, Wuhan University of Science and Technology, Wuhan 430081, Hubei, China

[3] Hubei Provincial Engineering Technology Research Center of Metallurgical Secondary Resources, Wuhan University of Science and Technology, Wuhan 430081, Hubei, China

* Correspondence: xuezhengliang@wust.edu.cn

Received: 15 November 2019; Accepted: 17 December 2019; Published: 19 December 2019

Abstract: Tire cord steel is widely used in the tire production process of the vehicle manufacturing industry due to its excellent strength and toughness. Titanium nitride (TiN) inclusion, existing in tire rod, has a seriously detrimental effect on the fatigue and drawing performances of the tire steel. In order to control its amount and morphology, the precipitation behavior of TiN during solidification in SWRH 92A tire cord steel was analyzed by selected thermodynamic models. The calculated results showed that TiN cannot precipitate in the liquid phase region regardless of the selected models. However, the precipitation of TiN in the mushy zone would occur at the final stage during the solidification process (at solid fractions greater than 0.98) if the LRSM (Lever-rule model was applied for the N and Scheil model for Ti) or Ohnaka models (without considering the effect of carbon on secondary dendrite arm spacing (SDAS)) were adopted. For the Ohnaka model, in the case when the effect of carbon on SDAS was considered, TiN would probably precipitate in the solid phase zone rather than precipitate in the liquid phase region or mushy zone.

Keywords: tire cord steel; TiN inclusion; solidification; segregation models

---

1. Introduction

Tire cord steel is a kind of high-carbon steel and thus possesses high strength and toughness. Therefore, it is widely used in the production of tires for cars and airplanes [1]. With the development of lightweight materials, the strength level of tire cord steel has become a significant factor to be considered. The main types of tire cord steel were SWRH 62A and SWRH 67A (about 1750 MPa) before the 1990s, and then after that SWRH 72A (about 1870 MPa). In the 21st century, the hypereutectoid tire cord steel (SWRH 82A) dominated the market due to its higher strength level [2,3]. Nowadays, the research on ultra-high strength level steel, such as SWRH 92A tire cord steel, has been drawing the increasing attention of manufacturing engineers.

Non-metallic inclusions, such as oxide-or Ti-bearing inclusions, existing in tire cord steel have serious detrimental effects on the drawing performance and fatigue properties. Furthermore, the properties of inclusions, such as size, composition, amount, and morphology, play a key role on the quality of steel [4,5]. At present, the damage problems of brittle oxide inclusions for steel can be better controlled by morphology-controlling technology. Nevertheless, for Ti-bearing inclusions, due to their non-deformable characteristic, they could cause a serious detrimental effect on the drawing performance, which would reduce the life of high-carbon steels, especially for the SWRH 82A and SWRH 92A tire cord steels. Titanium nitride (TiN), with high hardness and melting point, would cause filament breaks during wire drawing and rope stranding or deteriorate the fatigue properties of steels. Furthermore, TiN inclusion has more harmful effects on the material processing than those of oxide inclusions. For example, a TiN inclusion of 6 μm would cause a similar fatigue performance to an oxide inclusion of 25 μm [6]. Titanium carbonitride (TiC$_x$N$_{1-x}$, x represents the molar ratio of TiC in TiC$_x$N$_{1-x}$), a continuous solid solution formed via replacing partial moles of N in TiN crystal with C has similar properties to those of TiN. It also has a detrimental effect on the fatigue performance and, as a result, leads to wire breaking during the drawing and stranding processes [7]. It has been reported [8] that the molar ratio of TiC increases with increasing strength of tire core steel, which would cause a more seriously destructive effect. However, the value of x in TiC$_x$N$_{1-x}$ is still very small [8]. In other words, the main composition of Ti-bearing inclusion precipitated in tire cord steel is still TiN. Thus, it is important to control TiN inclusion to improve the performance of SWRH 92A tire cord steel.

Many researchers [2–4,9–14] have reported the precipitation behaviors of TiN inclusion in different types of steels over the decades. Jiang et al. [4] found that the solidification segregation ratio of Ti was far greater than that of N, and reported that TiN inclusion would not precipitate until the solid fraction reached 0.9 when using SWRH 82A tire cord steel. Cai et al. [9] showed that the precipitation of TiN could only occur in the solid–liquid two-phase region where the solid fraction was greater than 0.95, and the particle size of TiN decreased with increasing cooling rate. Similar results were also reported in other references [11–13]. However, Liu et al. [14] demonstrated that TiN would not precipitate in the liquid phase or mushy zone, but in austenite (γ-Fe). The precipitation temperature of 1598 K (below the solidus temperature in this reference) was calculated. Nowadays, in the industrial production of SWRH 92A tire cord steel, TiN inclusion always appears in samples even though the concentration of N and Ti are controlled at extremely low levels (0.0043 mass% and 0.0005 mass% for N and Ti, respectively). Therefore, the precipitation behavior of TiN in SWRH 92A tire cord steel remains a hard problem to be solved. In order to make the mechanism of TiN precipitation clearer, a series of relevant studies were initiated in this paper, to show guidance for the development of ultra-high strength grade steels.

2. Material and Equilibrium Solubility Product

The chemical composition of SWRH 92A tire cord steel studied (from a Chinese steel mill) is shown in Table 1. Elements O and N were analyzed by the ONH analyzer (TC500C, LECO Corporation, St. Joseph, MI, American), elements C and S were analyzed by the CS Analyzer (Model EMIA-820V), and the contents of other elements were analyzed by ICP technology (IRIS Advantage ER/S, Thermo Elemental Corporation, Waltham, MA, American). Content of C was about 0.9203 mass%.

Table 1. Chemical composition of studied SWRH 92A tire cord steel (mass%).

Elements	Si	P	S	O	Mn	N	Ti
Content/mass%	0.18	0.018	0.0064	0.0018	0.51	0.0043	0.0005

To evaluate the stage (liquid phase, mushy zone, or solid phase) at which the TiN inclusion would precipitate in the steel, the liquidus temperature (T_L) and solidus temperature (T_S) as well as the equilibrium solubility product of N and Ti were first calculated.

Equations (1) and (2) were employed to estimate T_L and T_S [15], respectively,

$$T_L = T_{Fe} - \sum \Delta t_L \cdot w_{[i]} \tag{1}$$

$$T_S = T_{Fe} - \sum \Delta t_S \cdot w_{[i]} \tag{2}$$

where T_{Fe} was the melting point of pure Fe, 1811 K; Δt_L and Δt_S were the reduced temperature values for element i when the mass fraction was 1 mass%, K, the corresponding values can be acquired from Table 2 [15]; $w_{[i]}$ represented the mass fraction of element i, 1 mass% was considered as the unit. Combining Table 1, Table 2, Equations (1) and (2), the values of T_L and T_S can be calculated, i.e., $T_L = 1748$ K, $T_S = 1636$ K.

Table 2. Values of Δt_L and Δt_S in Equations (1) and (2) [15], respectively.

Elements	C	Si	P	S	O	Mn	N	Ti
Δt_L	65	8	30	25	80	5	90	20
Δt_S	175	20	280	575	160	30	-	40

The chemical reaction for the formation of TiN in molten steel can be expressed by Equation (3),

$$[Ti] + [N] = TiN_{(s)} \tag{3}$$

Standard Gibbs free energy change ΔG_3^θ for Equation (3) can be derived from Equations (4)–(7) [8,16],

$$Ti(s) = Ti(l) \quad \Delta G_4^\theta = 15500 - 8T \ (J/mol) \tag{4}$$

$$Ti(l) = [Ti] \quad \Delta G_5^\theta = -69500 - 27.28T \ (J/mol) \tag{5}$$

$$\frac{1}{2}N_2(g) = [N] \quad \Delta G_6^\theta = 10500 + 20.37T \ (J/mol) \tag{6}$$

$$Ti(s) + \frac{1}{2}N_2(g) = TiN(s) \quad \Delta G_7^\theta = -334500 + 93T \ (J/mol) \tag{7}$$

Therefore, the expression of ΔG_3^θ can be obtained,

$$\begin{aligned}\Delta G_3^\theta &= -\Delta G_4^\theta - \Delta G_5^\theta - \Delta G_6^\theta + \Delta G_7^\theta \\ &= -291000 + 107.91T \ (J/mol)\end{aligned} \tag{8}$$

The reaction equilibrium constant K_3^θ for Equation (3) is shown as follows:

$$K_3^\theta = \frac{a_{TiN}}{a_{[Ti]} \cdot a_{[N]}} = \frac{1}{w_{[Ti]} \cdot w_{[N]} \cdot f_{[Ti]} \cdot f_{[N]}} \tag{9}$$

where a_{TiN}, $a_{[Ti]}$, and $a_{[N]}$ denote the activities of TiN, Ti, and N in molten steel, respectively, herein, $a_{TiN} = 1$; $w_{[Ti]}$ and $w_{[N]}$ denote the mass fractions of Ti and N in molten steel, respectively; $f_{[Ti]}$ and $f_{[N]}$ denote the activity coefficients of Ti and N, which can be estimated by Equations (10) and (11) [13], respectively.

$$\lg f_{[Ti]} = \lg f_{[Ti]}^{1873\ K} \cdot \left(\frac{2557}{T} - 0.365\right) \tag{10}$$

$$\lg f_{[N]} = \lg f_{[N]}^{1873\ K} \cdot \left(\frac{3280}{T} - 0.75\right) \tag{11}$$

where $\lg f_{[Ti]}^{1873\ K}$ and $\lg f_{[N]}^{1873\ K}$ are the interaction coefficients of Ti and N at 1873 K, which can be calculated by Equations (12) and (13) [7–9], respectively:

$$\lg f_{[Ti]}^{1873\ K} = \sum e_{Ti}^i \cdot w_{[i]} \tag{12}$$

$$\lg f_{[N]}^{1873\ K} = \sum e_N^i \cdot w_{[i]} \tag{13}$$

Due to the fact that the mass fraction of Fe is more than 90 mass% in molten steel, then the impact of second-order interaction coefficients can be ignored. Thus, the first-order interaction coefficients (as shown in Table 3) are used only during the calculation process [15,17].

Table 3. First-order interaction coefficients e_j^i of solute elements in molten steel at 1873 K [15,17].

e_j^i ($i\rightarrow$)	C	Si	P	S	O	N	Mn	Ti
e_{Ti}^i	−0.165	0.05	−0.0064	−0.11	−1.8	−1.8	0.0043	0.013
e_N^i	0.13	0.047	0.045	0.007	0.05	0	−0.021	−0.53

According to Equations (10)–(13), one can obtain,

$$\lg f_{[Ti]} + \lg f_{[N]} = \frac{24.3832}{T} - 0.0395 \tag{14}$$

For Equation (9), take the logarithm of 10 on both sides at the same time,

$$\lg K_3^\theta = -(\lg f_{[Ti]} + \lg f_{[N]}) - (\lg w_{[Ti]} + \lg w_{[N]}) \tag{15}$$

$$\lg K_3^\theta = \frac{\ln K_3^\theta}{2.303} = -\frac{\Delta G_3^\theta}{2.303RT} = \frac{15204.6}{T} - 5.6383 \tag{16}$$

Taking the equilibrium solubility product of Ti and N as K_3^{equ} ($K_3^{equ} = w_{[Ti]} \cdot w_{[N]}$), and combining Equations (14)–(16), one can obtain,

$$\lg K_3^{equ} = -\frac{15229}{T} + 5.6778 \tag{17}$$

By substituting T = 1636 K and T = 1748 K into Equation (17), respectively, the relationship between $w_{[Ti]}$ and $w_{[N]}$ can be obtained, as shown in Figure 1. Figure 1 shows that concentrations of N and Ti in the sample, see Point A in this figure, are much lower than those at the liquidus phase and solidus phase temperatures, which indicates that TiN will not precipitate in the liquid phase or mushy zone.

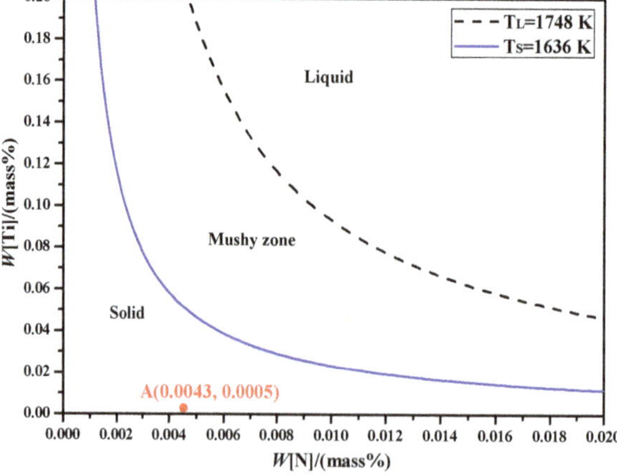

Figure 1. Required solubility product of N and Ti for the precipitation of TiN inclusion in SWRH 92A tire cord steel.

3. Thermodynamic Analysis

3.1. Segregation Models

The above results shown in Figure 1 are based on the uniform distributions for elements N and Ti in molten steel without considering micro-segregation. However, due to the decrease in solubility of solute elements N and Ti during the solidification process, micro-segregation will occur inevitability, which will further lead to the precipitation of TiN. With respect to this issue, several micro-segregation models were proposed to describe the concentration changes of solute elements as a function of the solid fraction, as listed in Table 4 [14].

Table 4. Micro segregation models for solute elements during the solidification process [14].

Models	Equation	Conditions	No
Lever-rule	$w_{[i]}/w_{[i]}^0 = [1 - (1-k_i)g]^{-1}$	Complete diffusion both in liquid and γ-Fe phase	(1)
Scheil model	$w_{[i]}/w_{[i]}^0 = (1-g)^{k_i-1}$	Complete diffusion in liquid and no diffusion in γ-Fe phase	(2)
Basic equations	$w_{[i]}/w_{[i]}^0 = [1 - (1-\phi k_i)g]^{(k_i-1)/(1-\phi k_i)}$		(3)
Brody–Fleming model	$\phi = 2\alpha$	Complete diffusion in liquid and finite diffusion in γ-Fe phase	(4)
Ohnaka model	$\phi = 4\alpha/(1+4\alpha)$		(5)
Clyne–Kurz model	$\phi = 2\alpha(1 - e^{-\frac{1}{\alpha}}) - e^{-\frac{1}{2\alpha}}$		(6)

where $w_{[i]}$ and $w_{[i]}^0$ denote the instantaneous and initial concentration of solute elements (N and Ti) in the liquid phase zone during the solidification process, respectively; k_i is the equilibrium distribution coefficient between liquid and γ-Fe phase, herein $k_C = 0.34$, $k_N = 0.48$, and $k_{Ti} = 0.30$ [18–20]; g represents the solid fraction; ϕ (in the range of 0–1) denotes the inverse diffusion coefficient and α is the Fourier parameter.

3.2. Usage of the LRSM Model

As shown in Table 4, it can be seen that the Lever-rule model is obtained based on the assumption that solute elements are completely diffused in both liquid and γ-Fe phases; however, the Scheil model neglects such diffusion in the γ-Fe phase, which means the solute elements are completely diffused in liquid and have no diffusion in the γ-Fe phase. Due to the fact that the diffusion coefficient of N is much larger than that of Ti in the γ-Fe phase, as comparison of Equations (18) and (19) indicates [18,21], and more obviously supported by Figure 2, so, it is reasonable to assume that solute element N is completely diffused in the γ-Fe phase and the diffusion in the γ-Fe phase for Ti is neglected. That is to say, the Lever-rule model is applied for the N and Scheil model for Ti. In the current paper, this model combination was named as the LRSM model. Then the corresponding concentration expressions of solute elements N and Ti can be described by Equations (20) and (21), respectively.

$$D_N^\gamma = 0.91 \exp(-168600/RT) \tag{18}$$

$$D_{Ti}^\gamma = 0.15 \exp(-250000/RT) \tag{19}$$

$$w_{[N]}^{act} = \frac{w_{[N]}^0}{1 - (1-k_N)g} \tag{20}$$

$$w_{[Ti]}^{act} = w_{[Ti]}^0 (1-g)^{k_{Ti}-1} \tag{21}$$

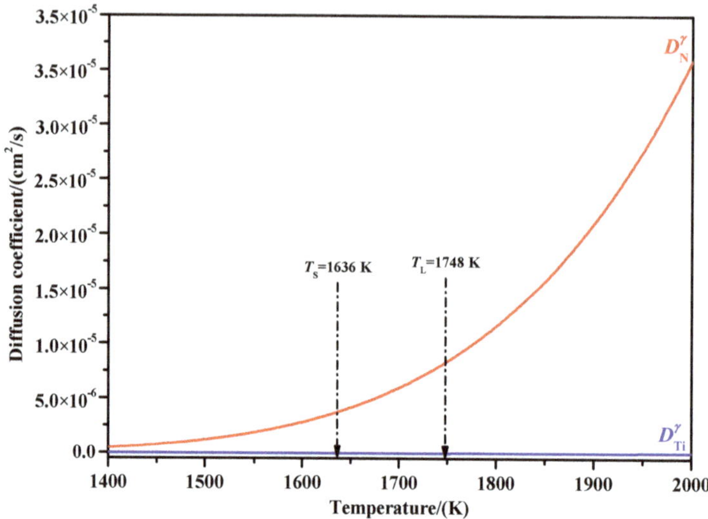

Figure 2. Diffusion coefficients of solute elements N and Ti in the γ-Fe phase at different temperatures.

Prior to analyzing the solidification process of molten steel, the actual solubility products of N and Ti should be calculated, which can be considered as Q_3^{act} ($Q_3^{\text{act}} = w_{[\text{N}]}^{\text{act}} \cdot w_{[\text{Ti}]}^{\text{act}}$), as shown by Equation (22),

$$Q_3^{\text{act}} = \frac{(1-g)^{k_{\text{Ti}}-1}}{1-(1-k_{\text{N}})g} \cdot w_{[\text{N}]}^0 \cdot w_{[\text{Ti}]}^0 \tag{22}$$

In addition, the relationship between solidification front temperature ($T_{\text{L-S}}$) and solid fraction (g) can be expressed by Equation (23) [22],

$$T_{\text{L-S}} = T_{\text{Fe}} - \frac{T_{\text{Fe}} - T_{\text{L}}}{1 - g\frac{T_{\text{L}}-T_{\text{S}}}{T_{\text{Fe}}-T_{\text{S}}}} \tag{23}$$

At the same time, by substituting Equation (23) into Equation (17), the relationship between $\lg K_3^{\text{equ}}$ and solid fraction (g) can also be obtained. As is well known, if the actual solubility product reaches the equilibrium value (or $\lg Q_3^{\text{act}} \geq \lg K_3^{\text{equ}}$), TiN will precipitate. The values of $\lg K_3^{\text{equ}}$ and $\lg Q_3^{\text{act}}$ (calculated by LRSM model) are depicted in Figure 3, from which it can be easily seen that TiN will only precipitate at the very late stage of the solidification process, with a solid fraction bigger than 0.9966. When substituting 0.9966 into Equation (23), the solidification front temperature ($T_{\text{L-S}} = 1637$ K) can be easily deduced, which is almost the same as the theoretical solidus temperature ($T_{\text{S}} = 1636$ K) of the studied tire cord steel. This result suggests that TiN will not precipitate in the mushy zone until nearly close to complete solidification.

3.3. Usage of Ohnaka Model on Considering the Effect of Carbon on SDAS L

The above analytical results (Figure 3) were obtained based on the assumption that N is completely diffused and Ti has no diffusion in the γ-Fe phase. In fact, both N and Ti would diffuse to some extent in the γ-Fe phase, as shown in Equations (18) and (19), respectively.

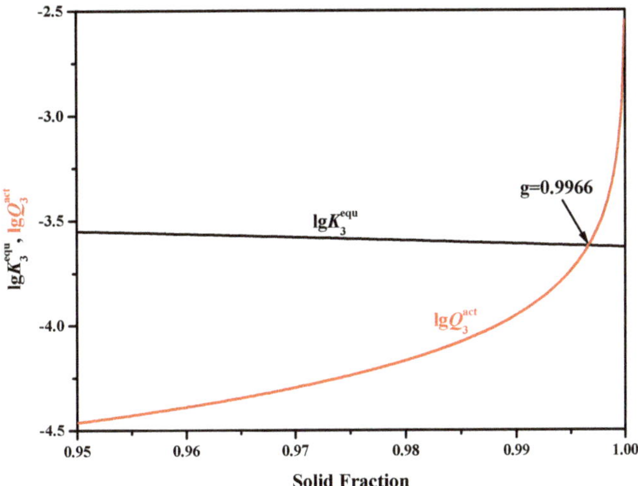

Figure 3. Comparison of equilibrium solubility product with the calculated value obtained by the LRSM (Lever-rule model was applied for the N and Scheil model for Ti) model.

In order to achieve a more realistic representation on the precipitation of TiN in SWRH 92A tire cord steel, the finite diffusion of the solute elements in the γ-Fe phase are now considered. The basic concentration expression is shown in Equation (24)). Herein, it can be easily found that if ϕ equals to zero, Equation (24) will change into the Scheil model; while if ϕ equals to one, Equation (24) will change into the Lever-rule model, instead. Furthermore, if the Brody–Fleming model [23] is adopted (as seen in Equation (25)), ϕ will be not physically reasonable when the Fourier parameter α is bigger than 0.5; the Clyne–Kurz model [24] lacks the actual physical meaning for ϕ if Equation (26) is used. In order to solve those problems, Ohnaka [25] presented a simple modification of ϕ based on comparison with the approximate solution of the diffusion equation, as shown in Equation (27) [25], which showed better agreement with the experimental data of Matsumiya et al. [26] than did predictions using Equation (25). Therefore, the Ohnaka model was used in this paper. The Fourier parameter α involves the diffusion coefficient D_i^{γ} (cm²/s), SDAS L [cm, the unit of L was converted from μm (calculated by Equations (29) and (31)) to cm for the calculation in Equation (28), corresponding to the unit of D_i^{γ}], and the local solidification time τ (s), as seen in Equation (28) [27],

$$w_{[i]}/w_{[i]}^0 = [1 - (1 - \phi k_i)g]^{(k_i-1)/(1-\phi k_i)} \tag{24}$$

$$\phi = 2\alpha \tag{25}$$

$$\phi = 2\alpha\left(1 - e^{-\frac{1}{\alpha}}\right) - e^{-\frac{1}{2\alpha}} \tag{26}$$

$$\phi = 4\alpha/(1 + 4\alpha) \tag{27}$$

$$\alpha = \frac{4D_i^{\gamma}\tau}{L^2} \tag{28}$$

The expressions of D_i^{γ} for solute elements N and Ti are shown in Equations (18) and (19), respectively; SDAS L (herein, the unit of L calculated by Equation (29) was μm) is related with the cooling rate R_C (K/s) and the carbon concentration $w_{[C]}$, as expressed by Equation (29) [28]; the local solidification time τ (s) was calculated by Equation (30) [21,29].

$$L = 143.9 \cdot R_C^{-0.3616} \cdot w_{[C]}^{(0.5501-1.996\cdot w_{[C]})}\ \left(w_{[C]}^0 > 0.15\right) \tag{29}$$

$$\tau = \frac{T_L - T_S}{R_C} \tag{30}$$

By substituting Equations (27)–(30) and the equilibrium distribution coefficients and diffusion coefficients of different solute elements into Equation (24), the corresponding segregation ratio of different solute elements during the solidification process can be calculated, as shown in Figure 4. From Figure 4 it can be seen that the segregation ratio of both N and Ti increased with the increasing solid fraction, and the final segregation ratios almost equaled each other although the cooling rates were different, which suggests that the effect of the cooling rate can be ignored. However, it can also be seen that the cooling rate has a certain effect on the intermediate segregation process for Ti, the faster the cooling rate, the larger the segregation ratio would be. As for N, however, the effect can be ignored.

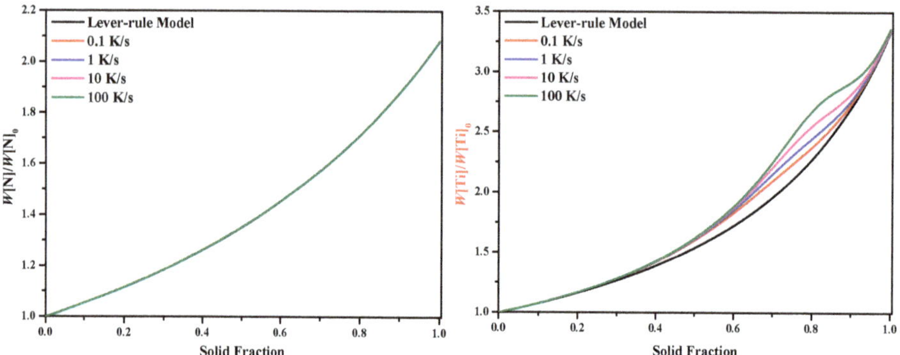

Figure 4. Effects of cooling rates on the segregation ratios of solute elements N and Ti during the solidification process (the results obtained by the Lever-rule model are given as a reference).

What is more, when comparing with the results obtained by the Lever-rule model, it is surprising to find that the final segregation ratio almost kept the same for both N and Ti. The possible reasons may be explained as follows. When the segregation of solute element carbon was considered, the concentration of carbon would increase during the solidification process (the plot was not given in this paper), then the value of SDAS L would decrease (according to Equation (29) in the current condition), which may accelerate the diffusion velocity of solute elements between the liquid and the γ-Fe phases and even arrive complete. Besides, when the cooling rate became slower, the diffusion of solute element between liquid and γ-Fe phases would be more complete due to the adequate time, as shown in Figure 5, which makes the inverse diffusion coefficient closer to one and the results are similar to the case obtained by the Lever-rule model. In addition, it can also be seen from Figure 5 that the inverse diffusion coefficients of N almost equaled one during the total solidification process at the different cooling rates, so the results were almost the same as those obtained by the Lever-rule model. As for Ti, the inverse diffusion coefficients gradually increased with increasing solid fraction. The slower the cooling rate, the larger the inverse diffusion coefficients would be. However, the final values almost equaled one for the four different cooling rates (0.1, 1, 10, and 100 K/s). That is to say, the Ohnaka model applied will change into the Lever-rule model at the end of the solidification process.

In addition, when the current Ohnaka model (considering the effect of carbon on SDAS L) was applied, precipitation of TiN during the solidification process would not happen because the value of the actual solubility product $\lg Q_3'^{act}$ was much smaller than that of the equilibrium value $\lg K_3^{equ}$, as can be seen in Figure 6. The current results clearly indicate that TiN cannot precipitate in the solid–liquid two-phase region (mushy zone).

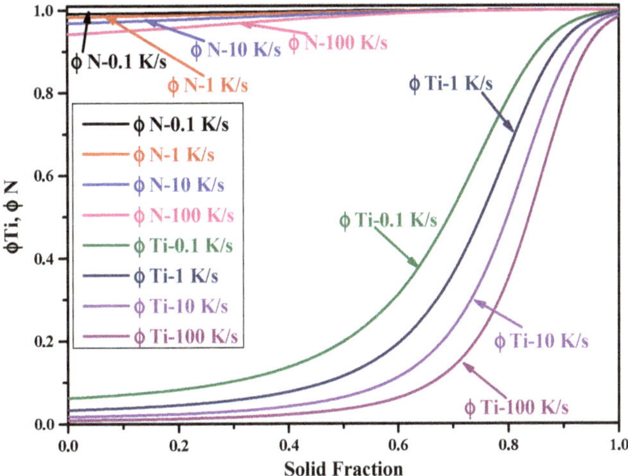

Figure 5. Inverse diffusion coefficients of solute elements N and Ti with different cooling rates.

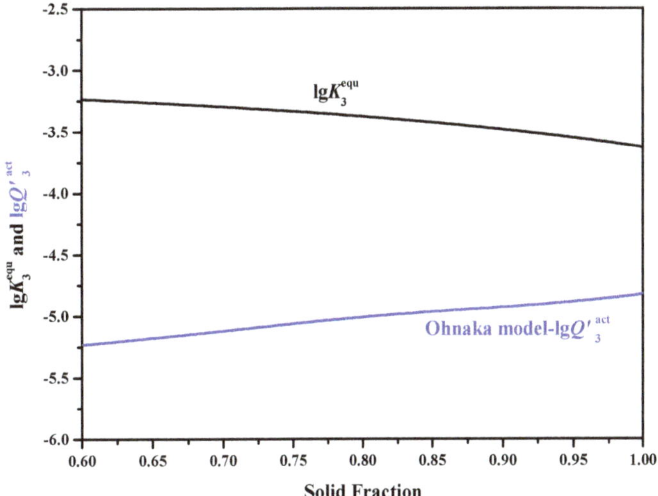

Figure 6. Comparisons of equilibrium solubility product with the calculated value obtained by the Ohnaka model (considering the effect of carbon on SDAS *L*).

3.4. Use of the Ohnaka Model without Considering the Effect of Carbon on SDAS L

As mentioned above, when considering the effect of carbon on SDAS *L*, TiN will not precipitate during the solidification process. However, in other references [13,22,30] on analyzing the segregation ratio of different solute elements, the SDAS *L* (herein, the unit of *L* calculated by Equation (31) is μm) was calculated as a function of the cooling rate R_C (K/s) only, see Equation (31),

$$L = 688 \cdot RC^{-0.36} \tag{31}$$

Similarly, by substituting Equations (27), (28), (30), and (31) and the equilibrium distribution coefficients and diffusion coefficients of different solute elements into Equation (24), the corresponding segregation ratios of the solute elements N and Ti during the solidification process can be obtained, as shown in Figure 7. From Figure 7 it can be seen that both N and Ti show a strong segregation tendency especially in the latter period and in comparison, the segregation ratio of Ti is much bigger than that of N. In addition, the effect of cooling rate on the segregation ratio can be nearly ignored (the plot is not given in the current paper). Therefore, it is possible for TiN to precipitate even though the initial concentrations of N and Ti are very low, as shown in Figure 8, and in which the critical solid fraction is a little smaller than that obtained by the LRSM model (0.98 vs. 0.9966). That is to say, TiN can precipitate a little earlier. Anyhow, TiN is only generated at the very late stage closing to the complete solidification of the molten steel.

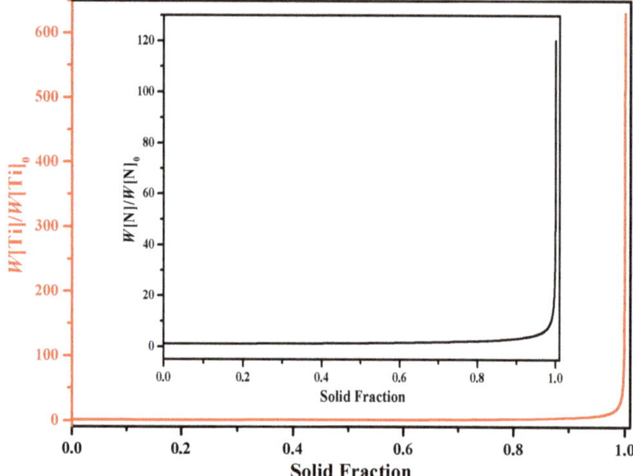

Figure 7. Segregation ratio of solute elements N and Ti during the solidification process when Equation (31) was used with a cooling rate of 10 K/s.

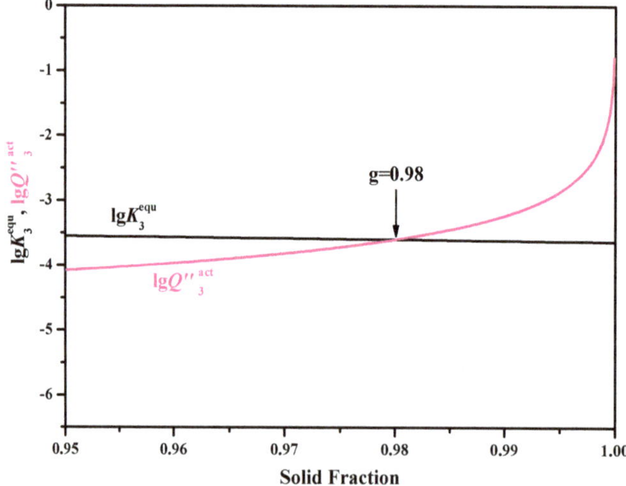

Figure 8. Comparison of equilibrium solubility product with the calculated values obtained by the Ohnaka model (without considering the effect of carbon on SDAS L).

Similar to the previous calculation, another surprising phenomenon is also found in this case. The segregation ratios of solute elements N and Ti are nearly the same as those obtained by the Scheil model, as shown in Figure 9. As mentioned above, when the inverse diffusion coefficient ϕ equals zero, Equation (24) will turn into the Scheil model. In the current discussion, ϕ nearly equals zero for both N and Ti, as seen in Figure 10. That is to say, both N and Ti almost completely diffuse in the liquid and have no diffusion in the γ-Fe phase in this situation. The possible reason for this result may be due to characteristics of the microstructure (mostly likely due to the much larger SDAS L) between the liquid and solid phases, which makes the inverse diffusion of solute elements insufficient.

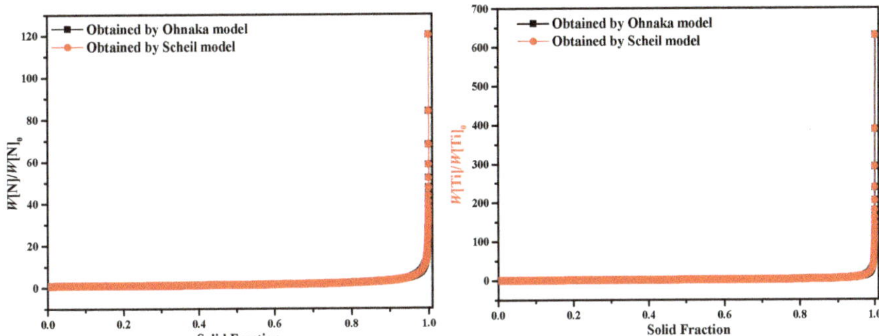

Figure 9. Comparison of segregation ratio of solute elements N and Ti when adopting the Ohnaka (without considering the effect of carbon on SDAS L) and Scheil models.

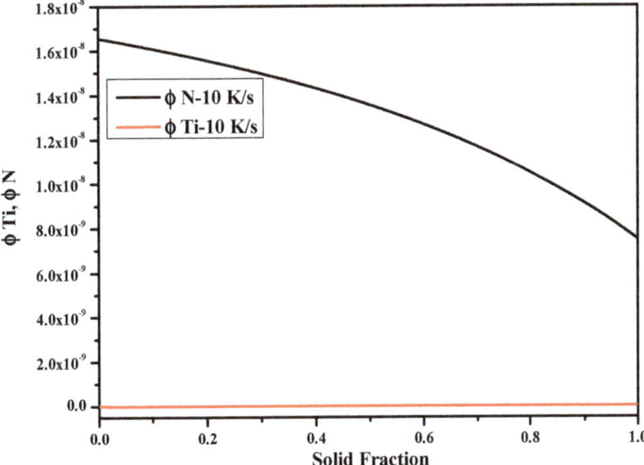

Figure 10. Inverse diffusion coefficients of solute elements N and Ti when adopting the Ohnaka model (without considering the effect of carbon on SDAS L).

4. Conclusions

According to the calculated results, it is possible to summarize the precipitation behavior of TiN inclusion in SWRH 92A tire cord steel during the whole solidification process, as follows:

(1) Precipitation of TiN will not occur in the liquid phase region regardless of the selected micro-segregation models.

(2) When adopting the LRSM and Ohnaka (without considering the effect of carbon on secondary dendrite arm spacing (SDAS L)) models, TiN will precipitate in the mushy zone at the very late

stage of the solidification process, with solid fractions larger than 0.9966 and 0.98, respectively. When considering the effect of carbon on SDAS L for the Ohnaka model, TiN will not precipitate in both the liquid phase and mushy zone.

(3) Results of different segregation models show that the Ohnaka model (considering the effect of carbon on SDAS L) is similar to the Lever-rule model at the very late stage during the solidification process; however, for the case without considering the effect of carbon on SDAS L, the result is similar to the Scheil model.

(4) Due to the fact that different segregation models may lead to different results, more attention should be paid to selecting the appropriate model or developing new models to analyze the actual segregation phenomena. Besides, further experimental work and theoretical analysis to understand in depth the precipitation behavior of TiN in SWRH 92A tire cord steel will be required in the future.

Author Contributions: Writing-original draft, methodology and review, L.W.; Methodology, visualisation and investigation, Z.-L.X.; Formal analysis, Y.-L.C.; Writing-review and editing, X.-G.B. All authors have read and agreed to the published version of the manuscript.

Funding: This research received no external funding.

Acknowledgments: The authors gratefully acknowledge the financial support from the National Natural Science Foundation of China, and the Special Project of Central Government for Local Science and Technology Development of Hubei Province (51874214 and 2019ZYYD076, respectively).

Conflicts of Interest: The authors declare no conflict of interest.

References

1. Yu, C.F.; Xue, Z.L.; Jin, W.T. Precipitation and Solid Solution of Titanium Carbonitride Inclusions in Hypereutectoid Tire Cord Steel. *J. Iron Steel Res. Int.* **2016**, *23*, 338–343. [CrossRef]

2. Jiang, Y.D.; Lei, J.L.; Zhang, J.; Xiong, R.; Zou, F.; Xue, Z.L. Effect of carbon content on Ti inclusion precipitated in tire cord steel. *J. Surf. Eng. Mater. Adv. Technol.* **2013**, *3*, 283–286. [CrossRef]

3. Lei, J.L.; Xue, Z.L.; Jiang, Y.D.; Zhang, J.; Zhu, T. Study on TiN precipitation during solidification for hypereutectoid tire crod steel. *Metal. Int.* **2012**, *17*, 10–16.

4. Jiang, Y.D.; Xue, Z.L.; Zhang, J. Genetic analysis for large TiN inclusions in wire rod for tire cord steel of SWRH82A. *J. Iron Steel Res. Int.* **2014**, *21*, 91–94. [CrossRef]

5. Zou, X.; Zhao, D.; Sun, J.; Wang, C.; Matsuura, H. An integrated study on the evolution of inclusions in EH36 shipbuilding steel with Mg addition: From casting to welding. *Metall. Mater. Trans. B* **2018**, *49*, 481–489. [CrossRef]

6. Yin, X.; Sun, Y.; Yang, Y.; Deng, X.; Barati, M.; McLean, A. Effect of alloy addition on inclusion evolution in stainless steels. *Ironmak. Steelmak.* **2017**, *44*, 152–158. [CrossRef]

7. Ma, W.J.; Bao, Y.P.; Zhao, L.H.; Wang, M. Control of the precipitation of TiN inclusions in gear steels. *Int. J. Miner. Metal. Mater.* **2014**, *21*, 234–239. [CrossRef]

8. Wang, L.; Xue, Z.L.; Zhu, H.Y.; Lei, J.L. Thermodynamic analysis of precipitation behavior of Ti-bearing inclusions in SWRH 92A tire cord steel. *Results Phys.* **2019**. [CrossRef]

9. Cai, X.F.; Bao, Y.P.; Wang, M.; Lin, L.; Dai, D.C.; Gu, C. Investigation of precipitation and growth behavior of Ti inclusions in tire cord steel. *Metall. Res. Technol.* **2015**, *112*, 407–418. [CrossRef]

10. Zhang, J.; Xue, Z.L.; Jiang, Y.D.; Lei, J.L.; Zou, F.; Xiong, R. Research on precipitation of Ti inclusions during solidification of tire cord steel SWRH82A. *Metal. Int.* **2014**, *19*, 34–38.

11. Jin, Y.; Du, S. Precipitation behaviour and control of TiN inclusions in rail steels. *Ironmak. Steelmak.* **2018**, *45*, 224–229. [CrossRef]

12. Liu, T.; Chen, D.; He, W.; Long, M.; Gui, L.; Duan, H.; Cao, J. Study on the Formation and Control of TiN Inclusion in Mushy Zone for High Ti Microalloyed Steel. In *TMS 2018 147th Annual Meeting & Exhibition Supplemetal Proceedings*; Springer: Cham, Switzerland, 2018; pp. 605–615.

13. Liu, H.; Wang, H.; Li, L.; Zheng, J.; Li, Y.; Zeng, X. Investigation of Ti inclusions in wire cord steel. *Ironmak. Steelmak.* **2011**, *38*, 53–58. [CrossRef]

14. Liu, Y.; Zhang, L.F.; Duan, H.; Zhang, Y.; Luo, Y.; Conejo, A.N. Extraction, thermodynamic analysis, and precipitation mechanism of MnS-TiN complex inclusions in low-sulfur steels. *Metall. Mater. Trans. A* **2016**, *47*, 3015–3025. [CrossRef]

15. Chen, J. *Manual of Chart and Data in Common Use of Steel Making*; Metallurgical Industry Press: Beijing, China, 2010.

16. Fu, J.; Zhu, J.; Di, L.; Tong, F.S.; Liu, D.L.; Wang, L.Y. Study on the precipitation behavior of TiN in the microalloyed steels. *Acta Metall. Sin.* **2000**, *36*, 801–804.

17. Tian, Q.; Wang, G.C.; Zhao, Y.; Li, J.; Wang, Q. Precipitation Behaviors of TiN Inclusion in GCr15 Bearing Steel Billet. *Metall. Mater. Trans. B* **2018**, *49*, 1149–1164. [CrossRef]

18. Pa, M.; Dp, D.; Chandra, T.; Cr, K. Grain growth predictions in microalloyed steels. *ISIJ. Int.* **1996**, *36*, 194–200.

19. Luo, S.; Wang, B.; Wang, Z.; Jiang, D.; Wang, W.; Zhu, M. Morphology of Solidification Structure and MnS Inclusion in High Carbon Steel Continuously Cast Bloom. *ISIJ Int.* **2017**, *57*, 2000–2009. [CrossRef]

20. Suzuki, S.; Weatherly, G.; Houghton, D. The response of carbo-nitride particles in hsla steels to weld thermal cycles. *Acta Metall.* **1987**, *35*, 341–352. [CrossRef]

21. Tian, Q.; Wang, G.C.; Shang, D.; Lei, H.; Yuan, X.; Wang, Q.; Li, J. In Situ Observation of the Precipitation, Aggregation, and Dissolution Behaviors of TiN Inclusion on the Surface of Liquid GCr15 Bearing Steel. *Metall. Mater. Trans. B* **2018**, *49*, 3137–3150. [CrossRef]

22. Ma, Z.; Janke, D. Characteristics of oxide precipitation and growth during solidification of deoxidized steel. *ISIJ Int.* **1998**, *38*, 46–52. [CrossRef]

23. Bower, T.F.; Brody, H.; Flemings, M.C. Measurements of solute redistribution in dendritic solidification. *Trans. Metall. Soc. AIME* **1966**, *236*, 624–634.

24. Clyne, T.; Kurz, W. Solute redistribution during solidification with rapid solid state diffusion. *Metall. Trans. A* **1981**, *12*, 965–971. [CrossRef]

25. Ohnaka, I. Mathematical analysis of solute redistribution during solidification with diffusion in solid phase. *Trans. Iron Steel Inst. Jpn.* **1986**, *26*, 1045–1051. [CrossRef]

26. Matsumiya, T.; Kajioka, H.; Mizoguchi, S.; Ueshima, Y.; Esaka, H. Mathematical analysis of segregations in continuously-cast slabs. *Trans. Iron Steel Inst. Jpn.* **1984**, *24*, 873–882. [CrossRef]

27. Lee, M.H.; Park, J.H. Synergistic effect of nitrogen and refractory material on TiN formation and equiaxed grain structure of ferritic stainless steel. *Metall. Mater. Trans. B* **2018**, *49*, 877–893. [CrossRef]

28. Won, Y.M.; Thomas, B.G. Simple model of microsegregation during solidification of steels. *Metall. Mater. Trans. A* **2001**, *32*, 1755–1767. [CrossRef]

29. Zhang, X.; Ma, G.J.; Liu, M.K. Micro-segregation model calculation of residual tin in boiler and pressure vessel steel. *Philos. Mag.* **2019**, *99*, 1041–1056. [CrossRef]

30. Goto, H.; Miyazawa, K.I.; Yamada, W.; Tanaka, K. Effect of cooling rate on composition of oxides precipitated during solidification of steels. *ISIJ Int.* **1995**, *35*, 708–714. [CrossRef]

Article

Experimental Study on Precipitation Behavior of Spinels in Stainless Steel-Making Slag under Heating Treatment

Jianli Li [1,2,3,*], **Qiqiang Mou** [1], **Qiang Zeng** [1] and **Yue Yu** [2,3]

1   The State Key Laboratory of Refractories and Metallurgy, Wuhan University of Science and Technology, Wuhan 430081, China
2   Hubei Provincial Key Laboratory for New Processes of Ironmaking and Steelmaking, Wuhan University of Science and Technology, Wuhan 430081, China
3   Key Laboratory for Ferrous Metallurgy and Resources Utilization of Ministry of Education, Wuhan University of Science and Technology, Wuhan 430081, China
*   Correspondence: jli@wust.edu.cn; Tel.: +86-134-1956-9956

Received: 3 July 2019; Accepted: 29 July 2019; Published: 1 August 2019

Abstract: The stability of chromium in stainless steel slag has a positive correlation with spinel particle size and a negative correlation with the calcium content of the spinel. The effect of heating time on the precipitation of spinel crystals in the CaO-SiO_2-MgO-Al_2O_3-Cr_2O_3-FeO system was investigated in the laboratory. Scanning electron microscopy with energy-dispersive and X-ray diffraction were adopted to observe the microstructure, test the chemical composition, and determine the mineral phases of synthetic slags, and FactSage7.1 was applied to calculate the crystallization process of the molten slag. The results showed that the particle size of the spinel crystals increased from 9.42 to 10.73 µm, the calcium content in the spinel crystals decreased from 1.38 at% to 0.78 at%, and the content of chromium in the spinel crystal increased from 16.55 at% to 22.78 at% with an increase in the heating time from 0 min to 120 min at 1450 °C. Furthermore, the species of spinel minerals remained constant. Therefore, an extension in the heating time is beneficial for improving the stability of chromium in stainless steel slag.

Keywords: stainless steel slag; heating time; Cr_2O_3; spinel; crystal size

1. Introduction

Stainless steel slag (SS slag) is a solid waste discharged from the production of stainless steel, and includes electric arc furnace (EAF) slag and argon-oxygen decarbonization furnace (AOD) slag [1]. The amount of SS slag is growing with the increase in demand for stainless steel. However, the toxicity of SS slag is extremely high due to its Cr^{6+} content [2,3], and untreated SS slag threatens the ecological environment and human health [4–6]. Thus, it improving the chromium stability in SS slag is urgent. Chromium spinels not only prevent the leaching of chromium, but can also inhibit the oxidation of Cr^{3+} [7], which is regarded as an ideal mineral phase for stabilizing chromium.

Thus far, much work has been focused on the relationship between spinel precipitation and chromium stability in order to prove that the stability of chromium can be enhanced by increasing spinel precipitation. Zeng et al. [8] found that when the size of the spinel crystals in the slag was increased from 5.77 µm to 8.40 µm, the concentration of the Cr (VI) leaching decreased from 0.1434 mg/L to 0.0021 mg/L. Zhao et al. [9] reported the same phenomenon. When the size of the was spinel increased from 6.0 µm to 17.3 µm, the concentration of the Cr (VI) leaching decreased from 1.24 mg/L to <0.01 mg/L. Therefore, the size of the spinel crystal plays an important role in the stability of chromium in steelmaking slags containing chromium. Regarding the size of the spinel crystal, Wang et al. [10]

adopted the addition of B_2O_3 to promote an increase in the spinel size from 7.87 μm to 12.72 μm and the concentration of chromium from 29.69% to 81.90%. Zhang et al. [11] reported that when the content of Al_2O_3 was increased from 0 to 15%, the spinel size increased from <20 μm to 23.50 μm. However, most of these investigations were mainly focused on the effect of additions to the spinel characteristics, and there have been few studies on the influence of heat treatment on the precipitation behavior of spinels in SS slag.

In this paper, the effect of heating time on the precipitation of spinels in Ca-SiO_2-MgO-Al_2O_3-Cr_2O_3-FeO was investigated based on characterizations of synthetic samples through scanning electron microscopy equipped with an energy dispersive spectrometer (SEM-EDS, NanoSEM400, FEI, Hillsboro, OR, USA), x-ray diffraction (XRD, X Pert Pro MPD, PANalytica, Almelo, The Netherlands), Image-Pro Plus 6.0 (IPP, Media Cybernetics, MD, USA), and FactSage 7.1 (GTT-Technologies, Aachen, Germany).

2. Experimental Process

The slag system was CaO-SiO_2-MgO-Al_2O_3-Cr_2O_3-FeO and the component of the slag sample is shown in Table 1. The raw materials included CaO, SiO_2, MgO, Al_2O_3, Cr_2O_3, and $FeC_2O_4 \cdot 2H_2O$ (Sinopharm Chemical Reagent Co., Ltd., Shanghai, China). First, the raw materials were weighed based on Table 1 and then mixed. 0.10 wt% H_3BO_3 (Sinopharm Chemical Reagent Co., Ltd., Shanghai, China) was added into the mixtures to prevent the pulverization of the synthetic samples. Second, mixtures weighing 200 g were put into a molybdenum crucible in the carbon-tube furnace. The furnace was warmed to 1600 °C at a heating rate of 10 °C/min under a nitrogen atmosphere and held for 30 min. Then, the temperature was decreased to 1450 °C at a rate of 20 °C/min, and the slag samples were taken out from the furnace and quenched with water to obtain synthetic slag samples at different sampling times, as shown in Figure 1. The sampling times were 0, 5, 10, 20, 30, 40, 60, 80, 100, and 120 min. Finally, the slag samples were ground, polished, and sprayed with gold powder.

Table 1. The chemical composition of stainless-steel slag, g.

CaO	SiO$_2$	MgO	Al$_2$O$_3$	Cr$_2$O$_3$	FeO	CaO%/SiO$_2$%
46.67	33.33	8.00	6.00	6.00	8.00	1.40

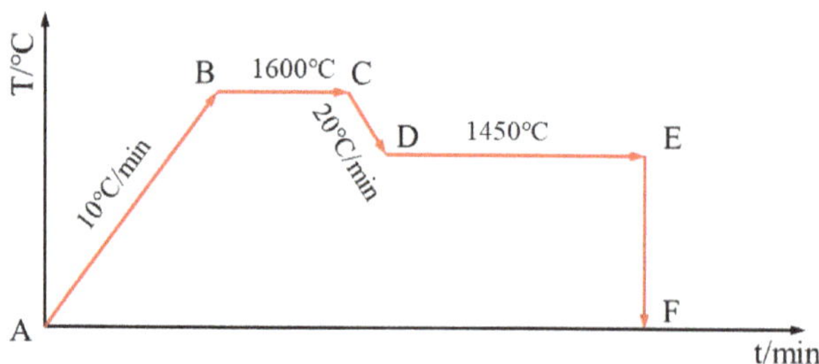

Figure 1. The heating procedure adopted in the experiments.

The characterization of the samples was conducted with SEM-EDS and XRD. The size of the spinel crystals was measured with Image-Pro Plus 6.0 (IPP). According to the Scheil-Gulliver equation, FactSage7.1 was applied to simulate the phase transformation and precipitation of the spinel phase during solidification of the molten slag. The specific setting conditions are shown as follows.

Database: FactPS, FToxide, FSstel;

Compound setting: idea gas, pure solid;

Solution phase setting: FToxid-SLAGA, FToxid-SPINA, FToxid-MeO_A, FToxid-bC2SA, FToxid-aC2SA, FToxid-Mel_A.

The FToxid-SLAGA was set as the target phase of Scheil-Gulliver cooling. The setting temperature was 2000 °C and the solidification step was 10 °C. The calculation process was terminated automatically when the target phase completely disappeared. The results were exported as pictures and edited with FactSage7.1.

3. Experimental Results

3.1. Mineral Phase of CaO-SiO$_2$-MgO-Al$_2$O$_3$-Cr$_2$O$_3$-FeO System

The XRD diffraction spectrum of the slag sample at different sampling times is shown in Figure 2. The intensity and position of the diffraction peaks were identical, indicating that the four slag samples contained the same mineral phase, namely, spinel and dicalcium silicate. In other words, the mineral phase structures remained constant as the heating time increased. Thus, prolonging the heating time at 1450 °C had no adverse effect on the mineral phase structures of the CaO-SiO$_2$-MgO-Al$_2$O$_3$-Cr$_2$O$_3$-FeO system.

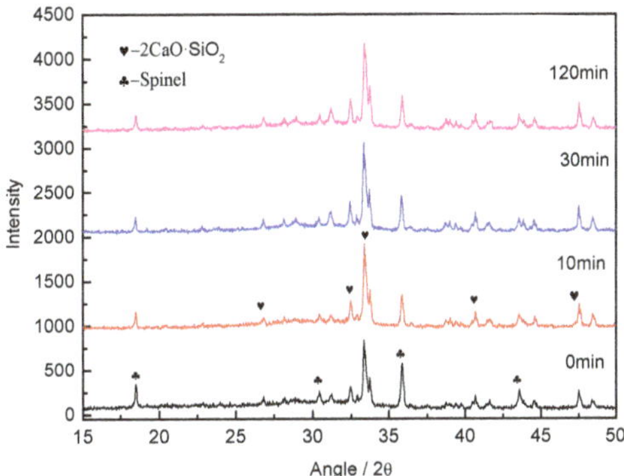

Figure 2. Effect of heat treatment time on the XRD diffraction pattern of the slag samples at 1450 °C.

The microstructures of the slag samples observed by SEM-EDS are shown in Figure 3. There were three kinds of mineral phases in all samples: (1) the white and regularly shaped phase is the spinel crystal, (2) the black striped phase is the α-C$_2$S phase, and (3) the hoar phase is the glassy matrix. The proportion of spinel phase with respect to the other phases remained nearly stable for all of the executed treatments.

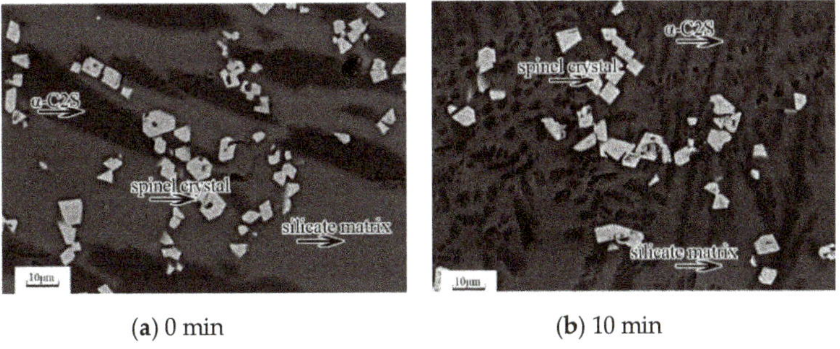

(**a**) 0 min   (**b**) 10 min

Figure 3. *Cont.*

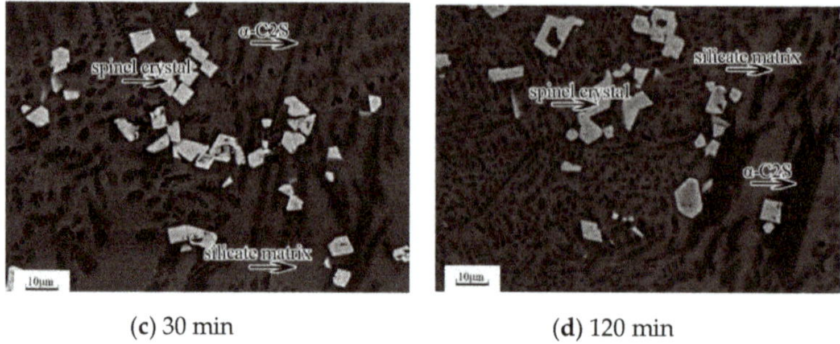

(**c**) 30 min (**d**) 120 min

Figure 3. Effect of heating time on the microstructure of the slag samples at 1450 °C.

3.2. Behavior of Spinel Crystal in CaO-SiO$_2$-MgO-Al$_2$O$_3$-Cr$_2$O$_3$-FeO System

The chemical compositions of spinel are shown in Table 2. When the heating time was prolonged from 0 min to 120 min, the calcium content decreased gradually from 1.38 at% to 0.68 at% and the chromium content increased from 16.55 at% to 22.78 at%. The content of calcium and iron showed a downward trend in spinel crystals. Nevertheless, the magnesium content showed an upward trend. As shown in Figure 4, the size of the spinel crystals increased from 9.42 μm to 10.73 μm when increasing the heating time from 0 min to 120 min, and the increase in the spinel size reached 13.91%.

Table 2. Chemical composition of the spinel crystals in Figure 4. Unit = atom%.

Heating Time	Cr	O	Fe	Mg	Al	Ca
0 min	16.55	67.09	4.84	6.96	3.17	1.38
10 min	20.56	60.91	5.18	8.28	3.76	1.32
30 min	20.94	61.27	4.65	8.31	3.55	1.22
120 min	22.78	60.22	4.25	8.44	3.61	0.68

Cao et al. [12] showed that the size of the spinel crystals increased from 10.8 μm to 24.8 μm during the cooling process from 1350 °C to 1250 °C and the increase reached 129% with respect to the initial size of the spinel crystals. The increase reached 108% (4.25 μm to 8.88 μm), with the silicon content increasing from 5 wt% to 15 wt% [13]. Moreover, with the addition of FeO increasing from 2 wt% to 20 wt%, the increase in size reached 46% [8]. Compared with the literature data, the increase in the spinel size was moderate in the case studied here.

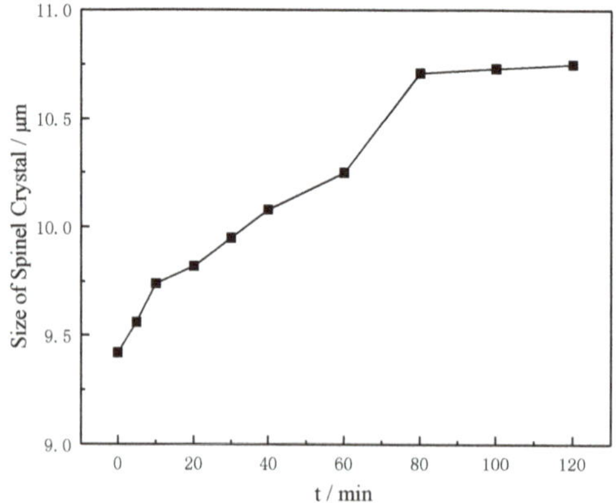

Figure 4. Effect of heat treatment time on the size of the spinel crystals at 1450 °C.

4. Discussion

4.1. Thermodynamic Considerations

The crystallization process of the molten slag is composed of nucleus formation and crystal growth, and is also determined by thermodynamic and kinetic factors. As shown in Figure 5, according to the calculation results of the solidification process through FactSage7.1, there are three phases including the liquid slag, spinel crystal, and α-C_2S in the CaO-SiO_2-MgO-Al_2O_3-Cr_2O_3-FeO system at 1450 °C. Combined with the preparation of synthetic slag, the nucleus could not form in the liquid phase, which transforms directly into the glassy matrix, whereas spinel crystal and α-C_2S were high-temperature precipitated phases in the CaO-SiO_2-MgO-Al_2O_3-Cr_2O_3-FeO system at 1450 °C. The heating time only affected the diffusion of the particles, and had little effect on mineral composition. These observations can explain why the mineral phases remained constant with the extension of heating time at 1450 °C.

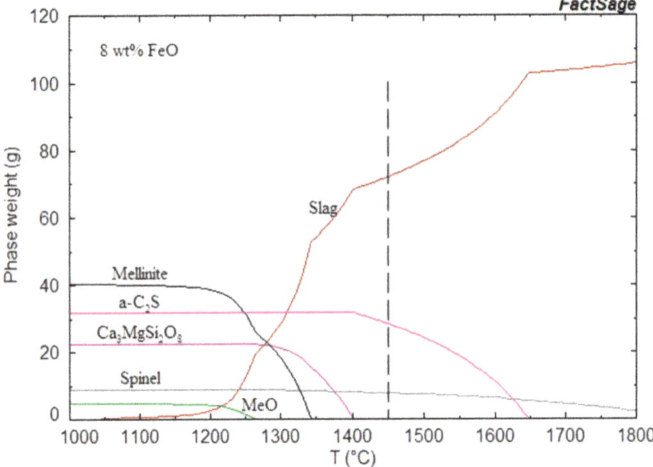

Figure 5. Theoretical analysis of the slag solidification process and selection of temperature.

Spinel crystal is the main mineral phase for improving chromium stability in the CaO-SiO_2-MgO-Al_2O_3-Cr_2O_3-FeO system [14–16]. Additionally, the reactions that form spinel crystals can occur between the MgO, Al_2O_3, Cr_2O_3, CaO, and FeO in the CaO-SiO_2-MgO-Al_2O_3-Cr_2O_3-FeO system. The reactions are shown in Equations (1)–(5).

$$(MgO) + (Cr_2O_3) = MgO{\cdot}Cr_2O_3(s) \tag{1}$$

$$(FeO) + (Cr_2O_3) = FeO{\cdot}Cr_2O_3(s) \tag{2}$$

$$(MgO) + (Al_2O_3) = MgO{\cdot}Al_2O_3(s) \tag{3}$$

$$(FeO) + (Al_2O_3) = FeO{\cdot}Al_2O_3(s) \tag{4}$$

$$(CaO) + (Cr_2O_3) = CaO{\cdot}Cr_2O_3(s) \tag{5}$$

$MgCr_2O_4$ formed by MgO and Cr_2O_3 is the most stable phase at high temperature [15]. During the heating period, iron and calcium elements that evolve into spinels through the isomorphic substitution are gradually replaced by magnesium. Consequently, the magnesium content increases, while the contents of iron and calcium decrease. Furthermore, chromium stability is affected by calcium content to a large extent [15]. The silicate microcrystals containing calcium are adsorbed into the spinel lattice during spinel growth due to the formation of a finite solid solution with

MgO·Cr$_2$O$_3$ [17]. Compared with other silicate mineral phases, dicalcium silicate is more soluble in water. The dissolution of chromium increases significantly as the content of dicalcium silicate in chromium spinels increases [17,18]. Thus, prolonging the heating time can reduce the content of dicalcium silicate in chromium spinels and improve their stability.

4.2. Precipitation and Growth of Spinel Crystals

The precipitation of spinel crystals includes nucleation and crystal growth. The slag system satisfies the basic requirements of supersaturation and supercooling for crystal nucleation and the essence of crystal nucleus growth is the transfer of atoms and other particles in liquid to the surface of the crystal nucleus. The composition of spinel crystals is obviously different from the chemical composition of slag and its growth rate depends on the long-range diffusion of solute atoms, which is essential to ensure the continuous growth of crystal [19,20]. For a slag system under constant temperature, the chemical reaction merely occurs between the free oxides such as (FeO), (Cr$_2$O$_3$), (MgO), (Al$_2$O$_3$), (CaO) based on slag molecular theory. The ability of molecules in slag to carry out chemical reactions is closely related to their activity, and the concentration of free oxides is the activity. Therefore, the reaction rate of Equations (1)–(5) are related to the concentrations of (FeO), (Cr$_2$O$_3$), (MgO), (Al$_2$O$_3$) and (CaO). As shown in Figure 6, various spinel crystals are precipitated when the slag with the concentration of solute atom C_0 is cooled to the temperature T. Furthermore, the concentration of solute atoms in the parent phase and precipitated phase at the phase interface is C_M and C_N, respectively. During the dt time, the phase boundary pushes the distance of dr toward the parent phase, and the quantity of solute atoms needed for the volume increase of the precipitated phase is $(C_N - C_M)\, dr$. Moreover, this part of the solute is provided by the diffusion of solute atoms in the parent phase.

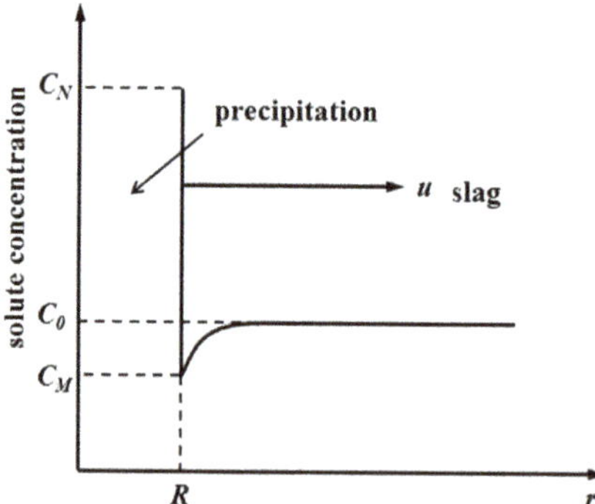

Figure 6. Concentration distribution of the solute atoms in slag during spinel crystal growth. C_0 represents the average concentration of solute atoms in the matrix, C_M represents the concentration of solute atoms in the matrix side at the crystal interface, C_N represents the concentration of solute atoms in the crystal side at the crystal interface, and u represents the growth rate of the crystal.

The growth rate of the spinel crystals can be reflected by Equations (6) and (7) [21]:

$$u = \frac{dr}{dt} = \frac{D}{c_N - c_M}\left(\frac{\partial c}{\partial r}\right)_{r=R} \tag{6}$$

$$D = \frac{k_B T}{3\pi d \mu} \tag{7}$$

where d, μ, T, and k_B refer to the particle diameter, melt viscosity, absolute temperature, and Boltzmann constant, respectively. $\left(\frac{\partial c}{\partial r}\right)_{r=R}$ refers to the concentration gradient of the particle near the phase interface.

It can be seen from the above equation that the growth rate of the precipitated phase crystal nucleus is directly proportional to the diffusion coefficient of the solute in the parent phase and the concentration gradient of the solute atom near the phase interface, while being inversely proportional to the difference in the equilibrium concentration between the two phases at the interface. For a system with a constant temperature, the diffusion coefficient D (cm^2/s, the order of magnitude is 10^{-5}–10^{-7} generally) is constant and is closely related to the viscosity in slag. In this way, the growth rate of spinel crystals mainly depends on the particle concentration difference ($C_N - C_M$) and concentration gradient ($(\frac{\partial c}{\partial r})_{r=R}$) at the interface of the spinel crystal. After the particle reaches the interface, it is rapidly consumed, meaning that the concentration difference of components near the phase interface is basically unchanged. For convenience of discussion, the ($C_N - C_M$) of components near the phase interface was considered approximately as a constant. Therefore, the effect of heating time on the growth rate of the spinel crystals mainly depends on the concentration gradient of the particles in the melt.

As shown in Figure 7, the theory of the spinel transition fraction X (X = process/final precipitation amount × 100) reached 88.37%, 92.25%, and 96.29% at 1450 °C, 1400 °C and 1350 °C, respectively. This reveals that the process of spinel crystallization is close to the terminal at the experimental temperature. In addition, during the solidification of the CaO-SiO_2-MgO-Al_2O_3-Cr_2O_3-FeO system, the contents of chromium in the residual liquid-slag and spinel crystals are as shown in Figure 8. At 1450 °C, the chromium content in the liquid phase was only 0.49%, which was far less than that in the spinel crystals. Obviously, the concentration gradient of particles was small due to the completion of the precipitation region of the spinel crystals and the limited content of chromium in the liquid phase. Moreover, the content of the residual liquid phase decreased with the completion of spinel crystal crystallization. The diffusion condition of the solute atoms becomes worse, which causes the diffusion coefficient D to become evidently low. Slow particle diffusion results in the prolongation of the heating time having little effect on the increase in the size of the spinel crystals. If the holding point is carried out at a higher temperature, the higher content of chromium in the liquid phase and the higher concentration gradient of the solute will increase, leading to increases in the growth rate of the spinel crystals and the size of spinels.

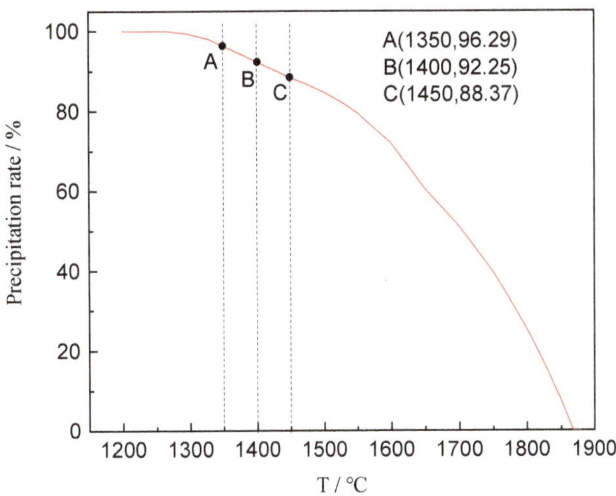

Figure 7. Theoretical transition fraction of a spinel crystal.

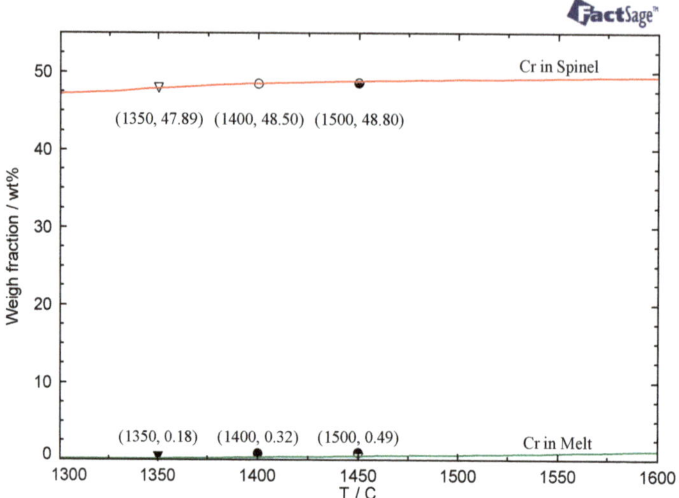

Figure 8. Content of chromium in the spinel crystal and liquid phase during slag solidification.

5. Conclusions

The precipitation behavior of spinel crystals in the CaO-SiO$_2$-MgO-Al$_2$O$_3$-Cr$_2$O$_3$-FeO system was studied by means of SEM-EDS, XRD, and IPP in the laboratory. Based on this study, the following conclusions were made:

(1) When the heating time increased from 0 min to 120 min at 1450 °C, the species of mineral phases precipitated from the CaO-SiO$_2$-MgO-Al$_2$O$_3$-Cr$_2$O$_3$-FeO system remained constant, and consisted of spinel and dicalcium silicate.

(2) The size of the spinel crystals increased from 9.42 to 10.73 μm with an increase in the heating time from 0 to 120 min. The increase in the spinel size reached 13.91%, which is considered moderate. The theoretical transformation fraction of the spinel reached 88.37% at 1450 °C. The crystallization process of the spinel occurred as a result of the low content of chromium and magnesium, forming the spinel in residual liquid, and the high viscosity of the solid-liquid mixture is a critical factor.

(3) The calcium content in the spinel crystal decreased gradually from 1.38 to 0.78 at%, while the chromium content increased from 16.55 to 22.78 at%. During the long-term heating, MgCr$_2$O$_4$ consisting of MgO and Cr$_2$O$_3$ was the most stable phase. The iron and calcium elements involved in the isomorphic substitution could be gradually replaced by magnesium in the spinel crystals.

Author Contributions: Y.Y. and Q.Z. helped to carry out the experimental work; Q.M. drafted the manuscript and conducted the experiments; J.L. analyzed the experimental data and modified and polished the draft.

Funding: The research was supported by the National Natural Science Foundation of China (No. 51404173), the Hubei Provincial Natural Science Foundation (No. 2016CFB579), the China Postdoctoral Science Foundation (No. 2014M562073), and the State Key Laboratory of Refractories and Metallurgy.

Conflicts of Interest: The authors declare no conflicts of interest.

References

1. Shen, H.; Forssberg, E.; Nordström, U. Physicochemical and mineralogical properties of stainless steel slags oriented to metal recovery. *Resour. Conserv. Recycl.* **2004**, *40*, 245–271. [CrossRef]
2. Kim, E.; Spooren, J.; Broos, K.; Nielsen, P.; Horckmans, L.; Geurts, R.; Vrancken, K.C.; Quaghebeur, M. Valorization of stainless steel slag by selective chromium recovery and subsequent carbonation of the matrix material. *J. Clean. Prod.* **2016**, *117*, 221–228. [CrossRef]

3. Lei, P.C.; Shen, X.J.; Li, Y.; Guo, M.; Zhang, M. An improved implementable process for the synthesis of zeolite 4A from bauxite tailings and its Cr^{3+} removal capacity. *Int. J. Miner. Metall. Mater.* **2016**, *23*, 850–857. [CrossRef]

4. Shen, H.; Forssberg, E. An overview of recovery of metals from slags. *Waste Manag.* **2003**, *23*, 933–949. [CrossRef]

5. Kamerud, K.L.; Hobbie, K.A.; Anderson, K.A. Stainless steel leaches nickel and chromium into foods during cooking. *J. Agric. Food Chem.* **2013**, *61*, 9495–9501. [CrossRef] [PubMed]

6. Wen, J.; Jiang, T.; Zhou, M.; Gao, H.Y.; Liu, J.Y.; Xue, X.X. Roasting and leaching behaviors of vanadium and chromium in calcification roasting-acid leaching of high-chromium vanadium slag. *Int. J. Miner. Metall. Mater.* **2018**, *25*, 515–526. [CrossRef]

7. Cao, L.H.; Liu, C.J.; Zhao, Q.; Jiang, M.F. Analysis on the stability of chromium in mineral phases in stainless steel slag. *Metall. Res. Technol.* **2017**, *115*. [CrossRef]

8. Zeng, Q.; Li, J.L.; Mou, Q.Q.; Zhu, H.Y.; Xue, Z.L. Effect of FeO on spinel crystallization and chromium stability in stainless steel-making slag. *JOM* **2019**, *71*, 2331–2337. [CrossRef]

9. Zhao, Q.; Liu, C.J.; Cao, L.H.; Zheng, X.; Jiang, M.F. Stability of chromium in stainless steel slag during cooling. *Minerals* **2018**, *8*, 445. [CrossRef]

10. Wang, W.; Liao, W.; Wu, X.R.; Li, L.S. Study on occurrence and concentrating behavior of chromium in stainless steel slag. *Multipurp. Util. Miner. Resour.* **2012**, *3*, 42–45.

11. Zhang, W.C.; Wu, X.R.; Wang, W.; Liao, W.; Li, L.S. Enrichment of chromium in stainless steel slag by modification. *J. Anhui Univ. Technol.* **2012**, *1*, 12–16.

12. Cao, L.H.; Liu, C.J.; Zhao, Q.; Jiang, M.F. Growth behavior of spinel in stainless steel slag during cooling process. *J. Iron Steel Res. Int.* **2018**, *25*, 1131–1139. [CrossRef]

13. Li, W.L.; Xue, X.X. Effects of silica addition on chromium distribution in stainless-steel slag. *Ironmak. Steelmak.* **2017**, *45*, 929–936. [CrossRef]

14. Pan, C.T.; Guo, P.M.; Pang, J.M.; Zhao, P. Research on six components slag system of stainless steel slag. *J. Iron Steel Res. Int.* **2013**, *29*, 305–311.

15. Li, J.L.; Xu, A.J.; He, D.F.; Yang, Q.X.; Tian, N.Y. Effect of FeO on the formation of spinel phases and chromium distribution in the $CaO\text{-}SiO_2\text{-}MgO\text{-}Al_2O_3\text{-}Cr_2O_3$, system. *Int. J. Miner. Metall. Mater.* **2013**, *20*, 253–258. [CrossRef]

16. Li, W.L.; Xue, X.X. Effect of cooling regime on phase transformation and chromium enrichment in stainless-steel slag. *Ironmak. Steelmak.* **2018**, *45*. [CrossRef]

17. Samada, Y.; Miki, T.; Hino, M. Prevention of chromium elution from stainless steel slag into seawater. *ISIJ Int.* **2011**, *51*, 728–732. [CrossRef]

18. Gelfi, M.; Cornacchia, G.; Roberti, R. Investigations on leaching behavior of EAF steel slags. In Proceedings of the 6th European Slag Conference, Madrid, Spain, 20–22 October 2010.

19. Parkinson, G.; Freij, S. Surface morphology and crystal growth mechanism of gibbsite in industrial Bayer liquors. *Hydrometallurgy* **2005**, *78*, 246–255.

20. Zhang, L.J.; Chen, Z.Y.; Hu, Q.M.; Yang, R. On the abnormal fast diffusion of solute atoms in, α-Ti: A first-principles investigation. *J. Alloy Compd.* **2018**, *740*, 156–166. [CrossRef]

21. Yong, Q.L. *Second Phase in Structural Steels*; Metallurgical Industry Press: Beijing, China, 2006; pp. 295–307.

Article

Thermal Behavior of a Rod during Hot Shape Rolling and Its Comparison with a Plate during Flat Rolling

Joong-Ki Hwang

School of Mechanical Engineering, Tongmyong University, Busan 48520, Korea; jkhwang@tu.ac.kr;
Tel.: +82-51629-1567

Received: 12 February 2020; Accepted: 6 March 2020; Published: 10 March 2020

Abstract: The thermal behavior of a rod during the hot shape rolling process was investigated using the off-line hot rolling simulator and numerical simulation. Additionally, it was compared with a plate during the flat rolling process to understand the thermal behavior of the rod during the hot rolling process in more detail. The temperature of the rod and plate during the hot rolling process was measured at several points with thermocouples using the rolling simulator, and then the measured temperature of each region of a workpiece was analyzed with numerical simulation. During hot rolling process, the temperature distribution of the rod was very different from the plate. The temperature deviation of the rod with area was much higher than that of the plate. The variation in effective stress of the rod along the circumferential direction can induce the temperature difference with area of the rod, whereas the plate had a relatively lower temperature deviation with area due to the uniform effective stress on the surface area. The heat generation by plastic deformation during the forming process also increased the temperature deviation of the rod with area, whereas strain distribution of the plate during flat rolling contributed to the uniformity of temperature of the plate with area. The higher temperature deviation of the rod along the circumferential and radial directions during the shape rolling process can increase the possibility of occurrence in surface defects compared to the plate during flat rolling.

Keywords: shape rolling; flat rolling; wire rod; temperature distribution

1. Introduction

Most steel products are manufactured by the hot rolling process because the hot rolling process is one of the most efficient plastic forming processes in metal forming industries. In this process, the initial large and thick material, which is called slab, bloom, and billet, is changed to the desired shape by passing through two counter-rotating rolls in the temperature range of 900–1200 °C. Many researchers have been devoted to measuring and predicting the temperature of steels during the hot rolling process to improve the quality of hot-rolled products because the thermal history of steel has a strong influence on the quality of final products [1–6]. Typically, the hot rolling process is categorized into two groups: flat rolling and shape rolling. Plate and strip are made via the flat rolling process, whereas wire, rod, and bar are manufactured via the shape rolling process. The deformation behavior and thermal history of a rod during the shape rolling process are more complex than those of a plate during the flat rolling process because the rod experiences three-dimensional (3D) deformation during the shape rolling process, as shown in Figure 1. In contrast, the plate is deformed under the plain strain condition, i.e., two-dimensional (2D) deformation. This different deformation behavior between the two processes may induce different thermal history between plate and rod, leading to the different microstructures and mechanical properties between the two processes.

During the last three decades, there have been considerable studies on the temperature of the hot rolling process. Most of the researchers have analyzed the rolling temperature of a workpiece on the basis of the numerical simulations such as finite difference method (FDM) and finite element method (FEM), and then the numerical results were validated by experimental measurement with optical pyrometers. Meanwhile, in the 1990s, several researchers investigated the temperature distribution of a plate during flat rolling using thermocouples in pilot rolling mills [7–13]. They investigated the effects of rolling speed, reduction ratio of height, oxide scale on the surface, and lubrication conditions on temperature distribution of the plate, and reported that the cooling rate of the plate along the height direction is different. It should be noted that most of the studies aforementioned were concentrated on the strip and plate during the flat rolling process.

In contrast, a little attention has been paid to the thermal behavior of shaped products during the hot rolling process because the prime concern of industrial shape rolling mills is to produce many products with an appropriate cross-sectional shape as fast as possible. In addition, considerable research on the thermal history of the rod was developed on the basis of the approach of plate rolling. Komori and Kato [14] analyzed the temperature distribution of a rod during the hot rolling process with roll shape using the combination of FEM and FDM. They showed that the roll velocity and friction coefficient greatly influences the temperature distribution at the roll–rod interface. They also reported that temperature distribution of the rod is dependent on the roll shape—square-diamond pass and oval-round pass have a different pattern of temperature distribution. Using the four-pass hot rod rolling experiments with low carbon steel, Said et al. [15] reported that 3D modeling is highly necessary to accurately predict the roll force, torque, and temperature of a rod during the shape rolling process. Yuan et al. [16] simulated the temperature and deformation of the rod during multi-pass rod rolling by 3D FEM, and the predicted temperature was compared with the measured temperature with a pyrometer. Xue and Liu [17] developed the 2D numerical model for calculating the temperature distribution of the rod during the multi-pass hot rolling process. Despite the 2D simulation, the calculation time was very short because the constitutive equation was not solved by iterative scheme. The results were validated by FEM simulation and industrial measurement. Serajzadeh et al. [18] developed a mathematical model based on FEM and phase transformation to predict temperature and microstructure distributions of a steel rod. They showed that the rolling speed affects the temperature of the rod. The increase in the rolling speed leads to a lower temperature drop at the surface area of the rod due to the reduced heat transfer between roll and rod stemming from the shorter contact time. Serajzadeh [19] also developed a mathematical 2D model that considered the various process parameters such as caliber shape and rolling speed. Kown et al. [1] studied the temperature distribution of the rod during hot rolling using 3D FEM analysis and experimental measurement. They insisted that the temperature in the surface area drastically decreases during rolling and recovers right after rolling due to the heat redistribution within the rod.

Meanwhile, steel wire, rod, and bar manufacturing industries have suffered from the several surface defect problems during hot rolling process: surface flaw, decarburization, and abnormal grain growth [20,21]. On the basis of the author's experiences, these surface defects are strongly related to the uneven temperature distribution of the rod during the shape rolling process in comparison with the plate during the flat rolling process [1,22]. Consequently, it is necessary to investigate the detailed temperature distribution of the rod during the hot rod rolling process. In addition, comparative study on the thermal behavior between rod and plate is necessary in order to completely understand the thermal behaviors and the formation of surface defects of the rod during the shape rolling process. On the basis of literature reviews, numerical simulation was mainly used to understand the thermal behaviors of the rod because it is difficult to measure the temperature of the rod during the hot rolling process. Additionally, most of the experimental studies were conducted using the radiation-type pyrometers and thermal cameras to measure the temperature of the rod. However, there are the following five limitations or errors in measuring the temperature of the rod during the hot rolling process:

(i) Pyrometer cannot measure the rod temperature at the rolling time due to the existence of facilities around rolls; thus, only equalized temperature along the radial direction of workpiece is measured, stemming from the rapid rebound of surface temperature. That is, the sharp temperature drop at the surface area by the roll contact cannot be detected.

(ii) During hot rod rolling, measurement errors are generated due to the severe rolling conditions such as the high rolling speed, the vibration of the rod, and the evaporation of coolant for the rolls.

(iii) The measured radiation energy of the rod by pyrometers tends to be lower than that of the actual value, owing to the shape of the round rod, indicating that pyrometers should be carefully validated to measure the shaped materials compared to the flat materials [23].

(iv) Radiation-type pyrometers measure the only surface temperature of a workpiece. Therefore, we cannot understand the temperature distribution of a workpiece along the radial direction.

(v) The radiation energy is highly dependent on the surface conditions of the rod. The rod has various oxide scales according to the process conditions and chemical compositions. In other words, measuring the surface temperature is not easy due to the thick oxide scales as shown in Figure 2.

Accordingly, measuring the accurate temperature of the rod with area is not easy. In addition, many researchers measured temperature of the rod at only one point, such as measuring the temperature of the plate. As shown in Figure 1b, the surface of the rod experiences different conditions of heat transfer with area, which is very different from the plate rolling. For instance, one location of the surface strongly makes contact with roll and other location of the surface softly makes contact with roll, which can produce different thermal behaviors of the rod along the circumferential direction, leading to different microstructures and mechanical properties along the circumferential direction of the rod. This different thermal history of the rod with area can induce the surface flaw during hot rod rolling [1]. Consequently, it is necessary to understand the thermal behavior of the rod during the hot rolling process to solve the quality problems in wire, rod, and bar industries and to produce high quality products. Overall, it is insufficient to measure the temperature by pyrometers and to measure the temperature at only the center and surface areas in order to understand the thermal behaviors of the rod during the shape rolling process due to the complex roll contact of the rod during the process. This is totally different from the plate during the flat rolling process, as shown in Figure 1. To the best of the author's knowledge, no study has been attempted to measure the rod temperature during shape rolling in detail, although several works exist that focus upon plate rolling.

Therefore, temperature of the rod was measured using thermocouples at several points via off-line rolling simulator in order to understand the thermal behavior of the rod during hot shape rolling process in more detail, which can help to improve the surface properties of the rod, for instance through surface flaw, decarburization, and abnormal grain growth, during the shape rolling process. For comparison purposes, temperature of the plate was also measured during the flat rolling process. Then, the measured temperature of the rod and plate with thermocouples was analyzed using a numerical simulation.

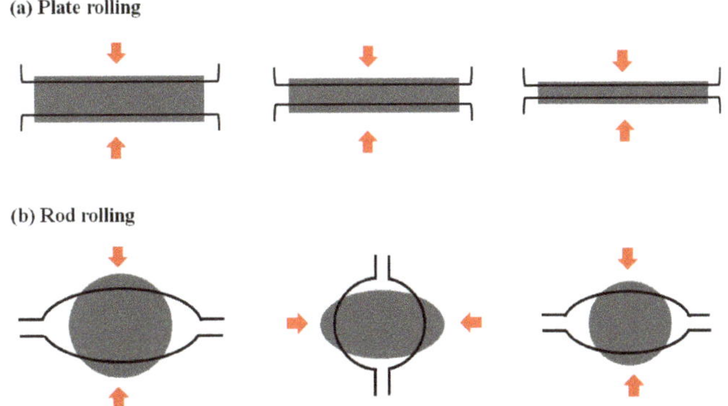

Figure 1. Schematic description of (**a**) plate and (**b**) rod during the rolling processes.

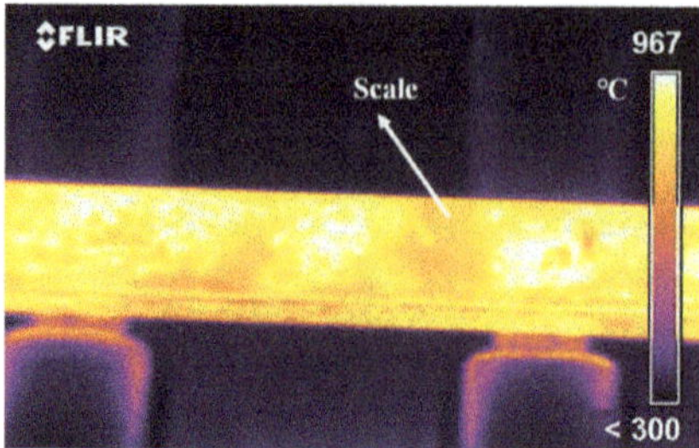

Figure 2. Thermal image of billet during hot rod rolling using an infrared thermo-camera.

2. Experimental Procedure and Numerical Simulation

2.1. Experiment Using Off-line Rolling Simulator

In order to understand the thermal histories of the rod and plate with area, a simulator for the hot rolling process was used. The rolling simulator mainly consisted of a couple of rolls, guide for workpiece, roller conveyor, and reheating furnace, as shown in Figure 3. Prior to the hot rolling test, the workpiece with four thermocouples was placed in a box-type reheating furnace. An oxide scale formation on the surface of a workpiece was suppressed using nitrogen gas in the reheating furnace. When the workpiece was heated to the final temperature of 1150 °C, it stayed for an additional 20 min for a homogenization. Then, the workpiece was withdrawn from the reheating furnace and rolled without lubrication.

For the flat rolling test, a rectangular plate of 150 × 30 mm was selected for an initial workpiece and it was hot-rolled using the flat rolls, as shown in Figure 4a. The round specimen with a diameter of 50 mm was also hot-rolled using the oval groove, as shown in Figure 4b. The ductile casting iron (DCI) rolls of 400 mm in diameter were rotated at a speed of 10 RPM. Plain low carbon steel, AISI 1020, was used and the analyzed chemical composition in weight percent was Fe-0.2C-0.4Mn. The specific operating conditions are summarized in Table 1. The caliber roll was designed to have a reduction of area (RA) per pass of 20%, which is a general rolling condition in hot rod rolling industries. RA per pass was calculated using the following equation:

$$RA = \frac{A_0 - A_f}{A_0} \times 100(\%)$$

(1)

where A_o and A_f are the areas of initial and final cross section, respectively. The reduction of height of a plate during flat rolling was selected to have a same average effective strain with the rod during shape rolling. The average effective strain of a rod during shape rolling was calculated using the model proposed by Lee et al. [24] that is based on the equivalent rectangle approximation method, which transforms a non-rectangular cross section shape into a rectangular shape, and the average effective strain (ε_p) is calculated as follows:

$$\varepsilon_p = \left[\frac{2}{3}\left(\varepsilon_1^2 + \varepsilon_2^2 + \varepsilon_3^2\right)\right]^{1/2} = \frac{2}{\sqrt{3}}\varepsilon_2\left[1 + \left(\frac{\varepsilon_1}{\varepsilon_2}\right)^2 + \left(\frac{\varepsilon_1}{\varepsilon_2}\right)\right]^{1/2}$$

(2)

where ε_1 and ε_2 are simply obtained by calculating the reduction ratio of width and height in equivalent rectangle approximation, respectively. In case of plate rolling, the ratio of ε_1 and ε_2 is very small in nature, and thus the average effective strain is represented as follows:

$$\varepsilon_p = \frac{2}{\sqrt{3}}\varepsilon_2$$

(3)

where ε_2 is calculated by the reduction ratio of height as follows:

$$\varepsilon_2 = \ln\left(\frac{H_i}{H_f}\right) \tag{4}$$

where H_i and H_f are the height of initial and final plate, respectively. The final height of a plate was chosen using Equation (3), as shown in Figure 4a.

Figure 4a,b shows the schematic description of the measurement points of temperature using thermocouples in workpieces. Four points were measured at the each process using chromel–alumel (K-type) thermocouples with an Inconel sheath. The response time of the thermocouple was improved by decreasing its diameter; therefore, 1.0 mm diameter thermocouple was selected, although it was easily breakable during the experiment under the severe working conditions such as the hot rolling test. Thermocouples were embedded in the 70 mm deep hole drilled from the tail end of a workpiece, as shown in Figure 4c, to easily handle the workpiece during the rolling test and to minimize the thermal disturbance on the surface of a workpiece [23]. One thermocouple was located at the center, and three of them were placed as close the surface as possible, i.e., the thermocouples of the surface were located at 1.5 mm from the outer surface. A multi-channel data recorder gathered temperature data during the test with a sampling time of 0.2 s. Because the thermocouples that are embedded in the surface region were easy to burn out during the hot rolling test, each test was repeated three times to ensure reliability and repeatability.

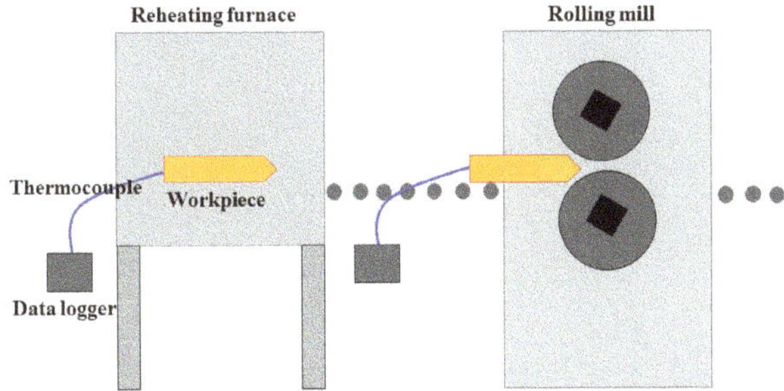

Figure 3. Schematic description of the off-line hot rolling simulator in this study.

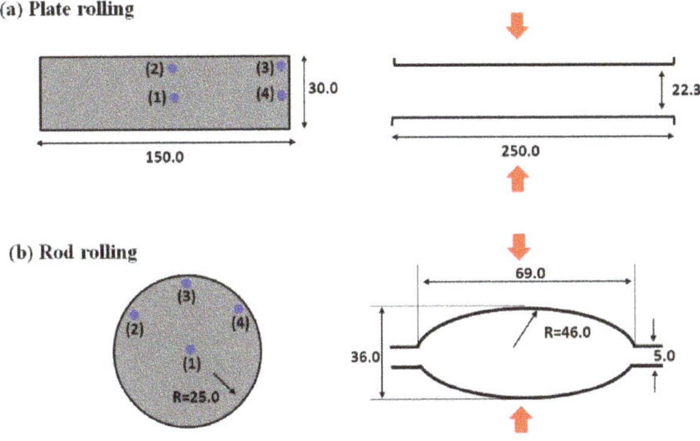

Figure 4. *Cont.*

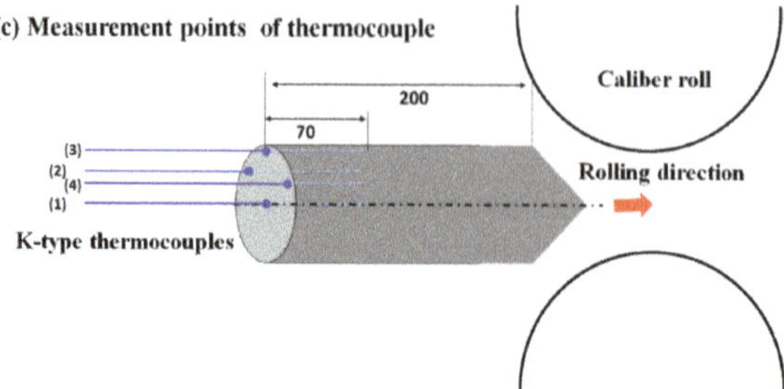

Figure 4. Schematic description showing the roll design and the measurement points of temperature using thermocouples: (**a**) cross section of flat rolling, (**b**) cross section of rod rolling, and (**c**) cross section of the longitudinal direction of the rod.

Table 1. Process parameters of hot rolling test using the off-line simulator in this experiment.

	Parameters	Value
Workpiece	Material	AISI 1020
	Initial temperature of workpiece	1150 °C
Rolling mill	Rolling speed	10 RPM
	Roll diameter	400 mm
	Temperature of roll	21 °C
Process conditions	Reduction ratio of height of plate	25.5%
	RA per pass of rod	20.2%
	Surrounding temperature	21 °C

2.2. Numerical Simulation

DEFORM FE commercial software was used in simulating the 3D complex distributions of temperature and strain in both rod and plate. The roll for rolling process was considered as a rigid body, and the workpiece was assumed to be isotropic material. The basic equation governing the temperature distribution of a workpiece during hot rolling is represented as follows:

$$\rho C_p \frac{\partial T}{\partial t} = k\left(\frac{\partial^2 T}{\partial x^2} + \frac{\partial^2 T}{\partial y^2} + \frac{\partial^2 T}{\partial z^2}\right) + Q \tag{5}$$

where ρ, C_p, k, and Q are the density, specific heat capacity, thermal conductivity of a workpiece, and volumetric rate of heat generation arising from the plastic deformation, respectively. Thermal properties of a workpiece such as thermal conductivity and specific heat were chosen from the library data provided by DEFORM FE software. That is, k and ρC_p values were approximately 31 W·m^{-1}K^{-1} and 4.3 N·mm^{-2}K^{-1}, respectively. To solve the above governing equation, the boundary conditions for a workpiece are expressed as follows:

$$k\left(\frac{\partial T}{\partial t}\right)_{surface} = \varepsilon\sigma(T^4 - T_a^4) + h_{conv}(T - T_a) + h_{cond}(T - T_R) \tag{6}$$

where ε, σ, T_a, and T_R are emissivity, Stefan–Boltzmann constant, ambient temperature, and roll temperature, respectively. Ambient and roll temperature is 25 °C, and ε is assumed to be 0.7. h_{conv} and h_{cond} are the convective heat transfer coefficient and conductive heat transfer coefficient, respectively, which is discussed in the next section. The shear friction coefficient of 0.6 was selected in the roll-workpiece interface [10,25,26], and other process parameters were identical to the experimental conditions, as given in Figure 4 and Table 1.

To shorten the calculation time, the quarter of full geometry was modeled stemming from the symmetrical condition of the rod and plate rolling processes, and a section of 500 mm in length was simulated. The caliber rolls had 400 mm in diameter, and the rolling speed was set as 10 RPM. The brick-type mesh was used, and the total number of mesh elements in the workpiece was approximately 21,600.

3. Results

3.1. Measured Temperature of a Workpiece with Area

Figure 5a represents the measured temperature profiles of the plate with area during flat rolling. Temperature of all the surface areas decreased when the plate was extracted from the reheating furnace; in particular, the temperature of corner area decreased fast due to the higher radiation heat transfer. In contrast, the surface areas were cooled slowly compared to the corner area, stemming from their geometric condition of the radiation heat transfer, i.e., flat plane. The center area of the plate was cooled slowly because the size of the initial workpiece was relatively large. When the workpiece was contacted with the cold roll, the upper surface and corner areas of the plate cooled very sharply due to the higher conduction heat transfer between the roll and plate. On the other hand, the non-contact surface, i.e., side, was cooled slowly because no conduction heat transfer occurred between the roll and plate in this area. Meanwhile, the temperature of the center slightly increased due to the heat generation by the plastic deformation. After passing through the rolling section, the temperatures of all the areas were equalized due to the higher thermal conductivity of the workpiece. That is, the center area with a higher temperature acted as a heat source for the surface areas with a lower temperature.

Figure 5b shows the temperature profiles of the rod with area during shape rolling, i.e., round-oval pass. The upper surface area, i.e., the hard contact surface, was cooled very fast. In other words, the lowest temperature of the rod was lower than 600 °C. This was a very surprising result considering the location of thermocouple in the surface—the surface indicated 1.5 mm from the outer surface. It is reasonable to induce that the temperature of the outer hard contact surface could be dropped below 500 °C. This low temperature at the rod surface was very different from the result of the plate during flat rolling, which is discussed in the next section. Ideally, the surface area at the right side had a similar temperature profile to the surface area at the left side, but they had somewhat different temperature profiles because the rod may not have rolled symmetrically in this study. That is, the temperature discrepancy between the right side surface and left side surface can be attributed to the error in alignment of a workpiece during the shape rolling process, which is frequently observed in an industrial rod rolling process. All the surface temperatures were recovered after rolling, owing to the redistribution of heat within the rod. Meanwhile, the center temperature of the rod increased when the rod passed through the roll bite, which was the result of the heat generation by plastic work. It is worth noting that the rod exhibited a higher temperature rise compared to the plate at the center area during the rolling process.

On the basis of the comparison study of temperature profiles of the plate and rod with area, it was found that the temperature profiles were very different between the two rolling processes. The rod had a higher temperature deviation with area in comparison with the plate. In particular, the surface temperature of the rod was very different along the circumferential direction during shape rolling. For instance, the temperature of the outer surface could be dropped below 500 °C at the hard contact area of the rod with rolls, which can make different microstructures and mechanical properties among surface areas [2,27]. Furthermore, it could induce surface defects of a rod during shape rolling, i.e., surface flaws [1,28], abnormal grain growths [29], decarburizations, and martensite transformations in the surface area. Accordingly, it is necessary to analyze the rod temperature for at least the following four regions in order to fully understand the thermal behavior of the rod during hot shape rolling: center, hard contact surface, soft contact surface, and non-contact surface. This means that the 1D or 2D approaches, which are generally used in plate rolling, were insufficient to describe the thermal behaviors of a rod during shape rolling. Overall, the center area had the maximum temperature in the both processes, and the minimum temperature appeared in the corner area during plate rolling and in the hard roll contact area during shape rolling.

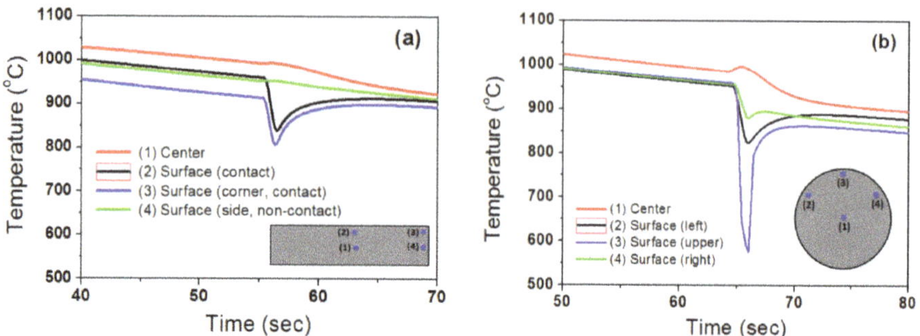

Figure 5. Measured temperature profiles of (**a**) the plate during flat rolling and (**b**) the rod during shape rolling with area.

3.2. Temperature Distribution of Workpiece by Numerical Simulation

To have more useful information on the thermal behavior of the rod during shape rolling, a numerical simulation was conducted. During the hot rolling process, heat is transferred with several mechanisms as shown in Figure 6, which is summarized as follows:

(i) The heat of a workpiece is dissipated by radiation because the temperature of the workpiece is relatively high during the hot rolling process, for instance, 800–1200 °C.

(ii) Forced convection heat transfer takes the heat from a workpiece because the workpiece moves during the hot rolling process. Additionally, the workpiece loses the heat by natural convection heat transfer via the derived air flow originated from the density difference of air between the surface of a workpiece and ambient.

(iii) The contact between the hot workpiece and cold rolls makes a strong conduction heat transfer; therefore, the surface area of a workpiece is chilled in a very short time as the workpiece contacts the rolls.

(iv) Heat is generated in the deformation zone, i.e., roll bite zone, due to the plastic deformation.

(v) Frictional stress between workpiece and roll produces heat.

All the heat transfer mechanisms need to be carefully considered to understand the thermal behavior of a workpiece and to improve the prediction accuracy of temperature in a workpiece. It is well known that heat transfer coefficient is dependent on several process parameters such as rolling speed, reduction ratio, lubrication condition, surface roughness of workpiece and roll, size and shape of a workpiece, and roll shape, which is the main reason why several researchers have used different values of parameter to simulate the hot rolling process, as summarized in Table 2. In this study, because the author was interested in the temperature distribution of plate and rod in the roll bite, the temperature near the roll bite was simulated during plate rolling and rod rolling. The contact heat transfer coefficient of 24 kW·m^{-2}K^{-1} was chosen to simulate the temperature distribution of both rod and plate, owing to the similar roll shape and process conditions in reference [2]. It should be noted that the rod and plate had a different heat transfer coefficient because of the different roll shape. However, it is difficult to find the optimum heat transfer coefficients of the two processes due to the limited experiments in this study. Accordingly, the same heat transfer coefficient was chosen for the two processes on the basis of the literature review (Table 2), and then the thermal behavior of the two processes was qualitatively compared.

Figure 7 shows the temperature distribution of the rolled workpieces calculated by the FE numerical analysis. Temperature was varied with rolling process and area. The center area had the highest temperature and the surface area tended to have the lowest temperature during hot rolling, which is consistent with the measured temperature profiles using thermocouples (Figure 5). In both rolling processes, the hard contact surface area with roll had the lowest temperature due to the high conduction heat transfer by roll contact, whereas the center area had the highest temperature originating from the relatively small heat loss during hot rolling process. The surface temperature was different between the plate and rod. The upper surface and lower surface of the plate had similar temperatures, whereas the rod had a temperature variation along the circumferential direction of the surface area. For more detailed information, the temperature distribution in the cross section of the rolled workpiece near the roll bite is presented in Figure 8. It is clear that the surface of the rod had a different temperature along the circumferential direction, which meant that the hard contact surface and soft contact surface with rolls experienced different thermal histories. The maximum temperature was similar between the two processes, but the minimum temperature was different in plate and rod—the minimum temperature of the plate was 668 °C, and that of the rod was 598 °C. Therefore, the temperature difference of the rod with area was higher than that of the plate, which is consistent with the result of measured temperature profiles using thermocouples. However, the deviation in temperature between the results of simulation and experiment was in existence. For instance, the surface temperature of the rod was estimated at 500 °C on the basis of the experiment, but the surface temperature of the rod was 598 °C from the numerical simulation, which was closely related to the selected heat transfer coefficient for the numerical simulation. That is, the heat transfer coefficient in this rolling condition was higher than the selected value of 24 $kW \cdot m^{-2}K^{-1}$.

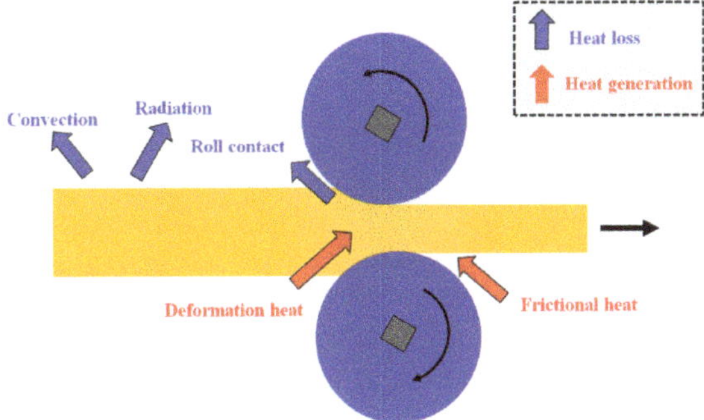

Figure 6. Schematic description showing the heat transfer mechanisms in a workpiece during the hot rolling process.

Table 2. Process parameters and values for heat transfer during the hot rolling process.

Contact Conduction ($kWm^{-2}K^{-1}$)	Convection ($Wm^{-2}K^{-1}$)	Ratio of Mechanical Work to Heat	Rolling Type	Reference
5	10	-	Shape rolling	[30]
10	10	-	Shape rolling	[31]
24	2.33	0.9	Round-oval	[2]
72	2.33	0.9	Square-diamond	[2]
4.8	30	-	Shape rolling	[32]
40	10	-	Flat rolling	[33]
45–85	-	0.85–0.95	Flat rolling	[10]
54–71	-	-	Flat rolling	[34]
7.6–17.6	-	-	Flat rolling	[35]
20–45			Flat rolling	[36]

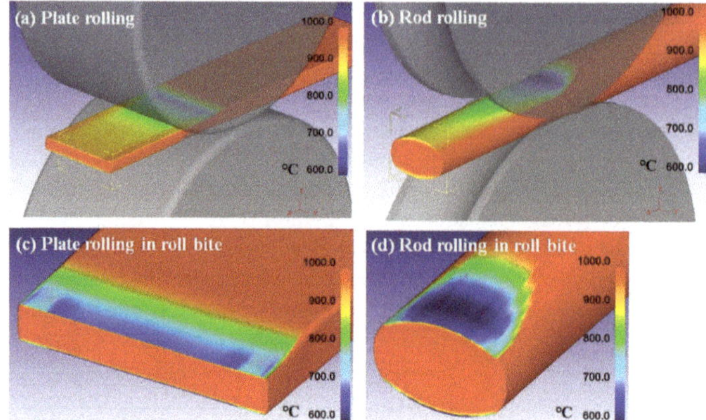

Figure 7. Comparison of contour maps of temperature during plate rolling and rod rolling.

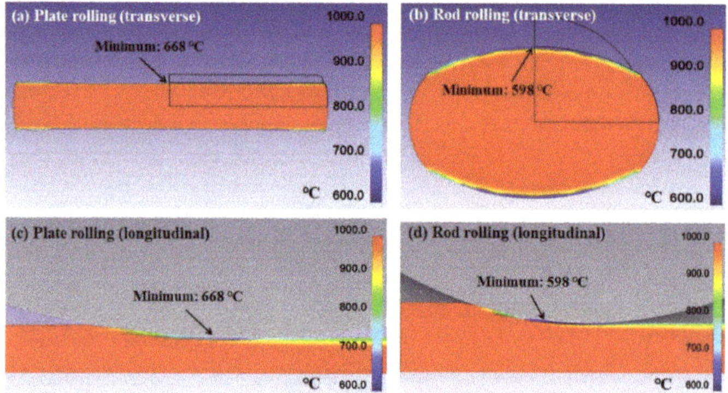

Figure 8. Comparison of contour maps of temperature in the roll bite during plate rolling and rod rolling.

4. Discussion

The most interesting point of the present study is the fact that the rod rolling process caused higher temperature deviation of the rod with area in comparison with the plate rolling. In particular, the rolled rod experienced very low temperature at the surface area.

4.1. Effect of Conduction Heat Transfer in the Workpiece-roll Interface

It is well known that the conduction heat transfer between the cold roll and hot workpiece is proportional to the contact pressure [8,37]. That is, the heat transfer coefficient by the roll contact increased along the arc of roll contact and reached the steady-state maximum values, and then decreased until the separation of the workpiece and roll [7]. Figure 9 shows the distribution of effective stress during plate and rod rolling processes. The plate had a uniform distribution of effective stress in the surface area. In contrast, the rod exhibited an inhomogeneous distribution of effective stress on the surface area—the upper surface area had the maximum value, and the effective stress gradually decreased along the circumferential direction from the upper surface. This different behavior of effective stress on the surface area between flat rolling and shape rolling made a different contact heat transfer coefficient between the two processes, leading to the different thermal history. In other words, the higher contact pressure in the upper surface of the rod during shape rolling led to the higher temperature drop in the upper surface area and the inhomogeneous temperature distribution along the circumferential direction of the rod. It is worth mentioning that the contact heat transfer coefficient also depended on rolling speed [7], reduction ratio [9], scale formation on the surface [35], and lubrication condition [7,36]. However, the effect of roll shape was only considered in this study because the other process parameters were set as identical in the two processes.

4.2. Effect of Heat Generation due to Plastic Deformation

The heat generation by the plastic deformation during rolling process affects the thermal behavior of a workpiece, which means that the temperature distribution of workpiece depends on the strain distribution of a workpiece with area. The temperature rise by plastic deformation was typically described as the following equation [38]:

$$\Delta T = \frac{\Delta Q}{\rho C_p} = \frac{\beta}{\rho C_p} \int_{\varepsilon_1}^{\varepsilon_2} \sigma d\varepsilon \tag{7}$$

where β and ΔT are the fraction coefficient between deformation work and heat energy and temperature rise, respectively. As given in Table 2, the fraction of deformation energy converted to heat energy is generally assumed to be 0.9 because only a small amount of deformation energy is stored within the workpiece as elastic energy. Because the temperature rise of a workpiece is highly related to the amount of plastic deformation according to Equation (7), the effective strain distribution of both plate and rod was evaluated, as shown in Figure 10a,b. It was clearly shown that the strain distribution of the plate was more homogeneous than that of the rod. Additionally, the pattern of effective strain distribution was totally different between the two processes. In the plate, the surface area had the maximum strain value and the center area had the minimum [39]. On the other hand, the center area had the maximum strain and the surface area had the minimum during shape rolling, which was well described in the profiles of effective strain along the horizontal and vertical directions of the two processes (Figure 10c). The higher effective strain in the center area of the rod during shape rolling can make a higher temperature rise in the center area, as shown in Figure 5b. Meanwhile, the relatively lower effective strain in the center area of the plate during flat rolling induced the lower temperature rise in the center area (Figure 5a). Comparing the distribution of effective strain in the plate, the rod had a higher temperature deviation with area because the temperature rise in the center area of the rod was higher than that of the plate. Overall, the center temperature of the rod became higher by the temperature rise due to the higher plastic deformation, whereas the surface temperature of the rod became lower by the hard contact with roll. This can make a higher temperature deviation of the rod with area during the shape rolling process. In contrast, strain distribution of the plate during flat rolling contributed to the uniformity of temperature distribution of the plate with area because the surface area had a higher effective strain during the flat rolling process.

The frictional stress in the roll–workpiece interface also increased temperature of a workpiece during the forming process, and the amount of temperature rise can be different between the two processes. Further experiments are necessary to reveal the frictional effect on temperature rise with rolling process, which is one of the valuable research topics in the field of forming industries. Overall, the higher temperature deviation of the rod along the circumferential and radial directions can induce the surface defects such as surface flaw, abnormal grain growth, and decarburization during the shape rolling process. In contrast, the influence of temperature deviation of the plate on surface defects is relatively weak during the flat rolling process.

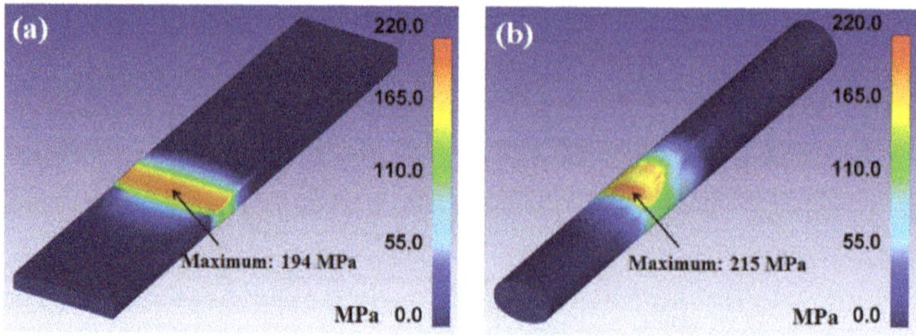

Figure 9. Contour maps of effective stress of (**a**) plate and (**b**) rod during the hot rolling process.

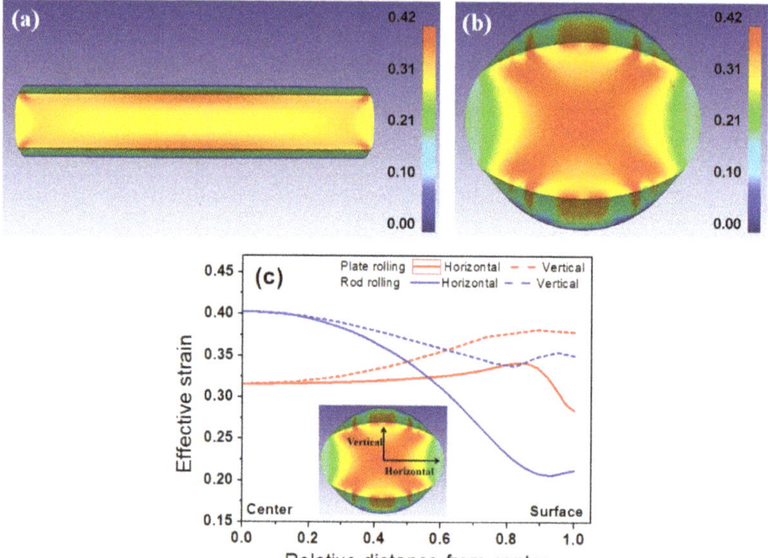

Figure 10. Contour maps of effective strain of (**a**) plate, (**b**) rod, and (**c**) comparison of profiles of effective strain along the horizontal and vertical directions of a workpiece.

5. Conclusions

On the basis of the comparative studies on the temperature distribution of a workpiece during hot plate rolling and rod rolling via an off-line simulator and numerical simulation, the following major conclusions were drawn:

1. The temperature distribution of the rod during shape rolling is different from that of the plate during flat rolling. The temperature deviation of the rod with area during shape rolling is much higher in comparison with the plate during flat rolling.

2. The measured temperature at the hard contact surface of the rod with roll is lower than 600 °C, whereas that of the plate is approximately 800 °C.

3. The higher variation in effective stress of the rod along the circumferential direction can induce the higher temperature difference with area of the rod compared to the plate during flat rolling.

4. The heat generation by the plastic deformation during forming process also increases the temperature deviation of the rod with area compared to the plate—higher effective strain at the center area of rod raises the temperature at center area of the rod. In contrast, strain distribution of the plate during flat rolling contributes to the uniformity of temperature distribution of the plate with area.

Funding: This work was supported by the National Research Foundation of Korea (NRF) grant, funded by the Korean government (MSIT) (NRF-2018R1D1A1B07050103).

Conflicts of Interest: The author declares no conflict of interest.

References

1. Kwon, H.C.; Lee, H.W.; Kim, H.Y.; Im, Y.T.; Park, H.D.; Lee, D.L. Surface wrinkle defect of carbon steel in the hot bar rolling process. *J. Mater. Process. Technol.* **2009**, *209*, 4476–4483. [CrossRef]
2. Lee, H.W.; Kwon, H.C.; Im, Y.T.; Hodgson, P.D.; Zahiri, S.H. Local austenite grain size distribution in hot bar rolling of AISI 4135 steel. *ISIJ Int.* **2005**, *45*, 706–712. [CrossRef]
3. Laasraoui, A.; Jonas, J.J. Prediction of temperature distribution, flow stress and microstructure during the multipass hot rolling of steel plate and strip. *ISIJ Int.* **1991**, *31*, 95–105. [CrossRef]
4. Nioi, M.; Pinna, C.; Celotto, S.; Swart, E.; Farrugia, D.; Husain, Z.; Ghadbeigi, H. Finite element modelling of surface defect evolution during hot rolling of Silicon steel. *J. Mater. Process. Technol.* **2019**, *268*, 181–191. [CrossRef]
5. Li, Y.; Cao, J.; Qiu, L.; Kong, N.; He, A.; Zhou, Y. Effect of strip edge temperature drop of electrical steel on profile and flatness during hot rolling process. *Adv. Mech. Eng.* **2019**, *11*, 1–11. [CrossRef]
6. Pesin, A.; Pustovoytov, D. Research of edge defect formation in plate rolling by finite element method. *J. Mater. Process. Technol.* **2015**, *220*, 96–106. [CrossRef]
7. Devadas, C.; Samarasekera, I.V.; Hawbolt, E.B. The thermal and metallurgical state of steel strip during hot rolling: Part 1. Characterization of heat transfer. *Metall. Trans. A* **1991**, *22A*, 307–319. [CrossRef]
8. Chen, W.C.; Samarasekera, I.V.; Hawbolt, E.B. Fundamental phenomena governing heat transfer during rolling. *Metall. Trans. A* **1993**, *24A*, 1307–1320. [CrossRef]
9. Chen, B.K.; Thomson, P.F.; Choi, S.K. Temperature distribution in the roll-gap during hot flat rolling. *J. Mater. Process. Technol.* **1992**, *30*, 115–130. [CrossRef]
10. Serajzadeh, S.; Taheri, A.K.; Nejati, M.; Izadi, J.; Fattahi, M. An investigation on strain inhomogeneity in hot strip rolling process. *J. Mater. Process. Technol.* **2002**, *128*, 88–99. [CrossRef]
11. Karagiozis, A.N.; Lenard, J.G. Temperature distribution in a slab during hot rolling. *J. Eng. Mater. Technol.* **1988**, *110*, 17–21. [CrossRef]
12. Hlady, C.O.; Brimacombe, J.K.; Samarasekera, I.V.; Hawbolt, E.B. Heat transfer in the hot rolling of metals. *Metall. Mater. Trans. B* **1995**, *26B*, 1019–1027. [CrossRef]
13. Shirizly, A.; Lenard, J.G. The effect of scaling and emulsion delivery on heat transfer during the hot rolling of steel strips. *J. Mater. Process. Technol.* **2000**, *101*, 250–259. [CrossRef]
14. Komori, K.; Kato, K. Analysis of temperature distribution in caliber rolling of a bar. *JSME Int. J.* **1989**, *32*, 208–216. [CrossRef]
15. Said, A.; Lenard, J.G.; Ragab, A.R.; Abo Elkhier, M. The temperature, roll force and roll torque during hot bar rolling. *J. Mater. Process. Technol.* **1999**, *88*, 147–153. [CrossRef]
16. Yuan, S.Y.; Zhang, L.W.; Liao, S.L.; Jiang, G.D.; Yu, Y.S.; Qi, M. Simulation of deformation and temperature in multi-pass continuous rolling by three-dimensional FEM. *J. Mater. Process. Technol.* **2009**, *209*, 2760–2766. [CrossRef]
17. Xue, J.; Liu, M. A new control volume-based 2D method for calculating the temperature distribution of rod during multi-pass hot rolling. *ISIJ Int.* **2013**, *53*, 1836–1840. [CrossRef]
18. Serajzadeh, S.; Mirbagheri, H.; Taheri, A.K. Modelling the temperature distribution and microstructural changes during hot rod rolling of a low carbon steel. *J. Mater. Process. Technol.* **2002**, *125*, 89–96. [CrossRef]
19. Serajzadeh, S. Prediction of microstructural changes during hot rod rolling. *Int. J. Mach. Tools Manuf.* **2003**, *43*, 1487–1495. [CrossRef]
20. Choi, S.; Lee, Y. An approach to predict the depth of the decarburized ferrite layer of spring steel based on measured temperature history of material during cooling. *ISIJ Int.* **2014**, *54*, 1682–1689. [CrossRef]
21. Hwang, J.K. Effects of process conditions, material properties, and initial shape of flaw on the deformation behavior of surface flaw during wire drawing. *ISIJ Int.* **2019**, *59*, 2052–2061. [CrossRef]
22. Kim, H.Y.; Kwon, H.C.; Lee, H.W.; Im, Y.T.; Byon, S.M.; Park, H.D. Processing map approach for surface defect prediction in the hot bar rolling. *J. Mater. Process. Technol.* **2008**, *205*, 70–80. [CrossRef]

23. Hwang, J.K. The temperature distribution and underlying cooling mechanism of steel wire rod in the Stelmor type cooling process. *Appl. Therm. Eng.* **2018**, *142*, 311–320. [CrossRef]
24. Lee, Y.; Kim, H.J.; Hwnag, S.M. Analytic model for the prediction of mean effective strain in rod rolling process. *J. Mater. Process. Technol.* **2001**, *114*, 129–138. [CrossRef]
25. Kwon, H.C.; Lee, Y.; Kim, S.Y.; Woo, J.S.; Im, Y.T. Numerical prediction of austenite grain size in round-oval-round bar rolling. *ISIJ Int.* **2003**, *43*, 676–683. [CrossRef]
26. Awais, M.; Lee, H.W.; Im, Y.T.; Kwon, H.C.; Byon, S.M.; Park, H.D. Plastic work approach for surface defect prediction in the hot bar rolling process. *J. Mater. Process. Technol.* **2008**, *201*, 73–78. [CrossRef]
27. Bontcheva, N.; Petzov, G. Total simulation model of the thermo-mechanical process in shape rolling of steel rods. *Comput. Mater. Sci.* **2005**, *34*, 377–388. [CrossRef]
28. Son, I.H.; Lee, J.D.; Choi, S.; Lee, D.L.; Im, Y.T. Deformation behavior of the surface defects of low carbon steel in wire rod rolling. *J. Mater. Process. Technol.* **2008**, *201*, 91–96. [CrossRef]
29. Ochiai, I.; Ohba, H.; Hida, Y.; Nagumo, M. Effect of precipitation behavior of AlN on abnormal growth of ferrite grains in low-carbon steel wires. *Tetsu-to-Hagane* **1984**, *70*, 2001–2008. [CrossRef]
30. Na, D.H.; Lee, Y. A study to predict the creation of surface defects on material and suppress them in caliber rolling process. *Int. J. Precis. Eng. Manuf.* **2013**, *14*, 1727–1734. [CrossRef]
31. Nalawade, R.S.; Puranik, A.J.; Balachandran, G.; Mahadik, K.N.; Balasubramanian, V. Simulation of hot rolling deformation at intermediate passed and its industrial validity. *Int. J. Mech. Sci.* **2013**, *77*, 8–16. [CrossRef]
32. Kim, S.Y.; Im, Y.T. Three-dimensional finite element analysis of non-isothermal shape rolling. *J. Mater. Process. Technol.* **2002**, *127*, 57–63. [CrossRef]
33. Bagheripoor, M.; Bisadi, H. Effects of rolling parameters on temperature distribution in the hot rolling of alumimum strips. *Appl. Therm. Eng.* **2011**, *31*, 1556–1565. [CrossRef]
34. Zhou, S.X. An integrated model for hot rolling of steel strips. *J. Mater. Process. Technol.* **2003**, *134*, 338–351. [CrossRef]
35. Li, Y.H.; Sellars, C.M. Comparative investigations of interfacial heat transfer behaviour during hot forging and rolling of steel with oxide scale formation. *J. Mater. Process. Technol.* **1998**, *80*, 282–286. [CrossRef]
36. Wu, J.; Liang, X.; Tang, A.; Pan, F. Interfacial heat transfer under mixed lubrication condition for metal rolling: A theoretical calculation study. *Appl. Therm. Eng.* **2016**, *106*, 1002–1009. [CrossRef]
37. Malinowski, Z.; Lenard, J.G.; Davies, M.E. A study of the heat-transfer coefficient as a function of temperature and pressure. *J. Mater. Process. Technol.* **1994**, *41*, 125–142. [CrossRef]
38. Curtze, S.; Kuokkala, V.T. Dependence of tensile deformation behavior of TWIP steels on stacking fault energy, temperature and strain rate. *Acta Mater.* **2010**, *58*, 5129–5141. [CrossRef]
39. Ding, J.; Zhao, Z.; Jiao, Z.; Wang, J. Central infiltrated performance of deformation in ultra-heavy plate rolling with large deformation resistance gradient. *Appl. Therm. Eng.* **2016**, *98*, 29–38. [CrossRef]

 *processes*

Article

Effect of Cambered and Oval-Grooved Roll on the Strain Distribution During the Flat Rolling Process of a Wire

Joong-Ki Hwang

School of Mechanical Engineering, Tongmyong University, Busan 48520, Korea; jkhwang@tu.ac.kr;
Tel.: +82-51-629-1567

Received: 20 April 2020; Accepted: 15 July 2020; Published: 20 July 2020

Abstract: The effect of the roll design on the strain distribution of the flat surface, lateral spreading, and the strain inhomogeneity of a flat-rolled wire were investigated during the flat rolling process. Oval-grooved and cambered rolls with various radii were applied to the flat rolling process based on a numerical simulation. The effective strain on the flat surface of the wire increased when using a cambered roll due to the highly intensified contact pressure on the flat surface, while the effective strain on the flat surface of the wire decreased when using an oval-grooved roll. Lateral spreading decreased when using an oval-grooved roll because the spread in the free surface area of the wire was highly restricted by the oval-grooved roll shape. In contrast, the spread in the surface area increased when using a cambered roll due to the less-restricted metal flow at the free surface. Accordingly, a cambered roll with a small radius is highly recommended in order to improve the surface quality of flat-rolled wires. This is beneficial for industrial plants because the cambered roll can be easily applied in flat rolling plants.

Keywords: roll design; flat-rolled wire; strain inhomogeneity; normal pressure; macroscopic shear bands

---

1. Introduction

Flat-rolled wires are widely used in windshield wipers, springs, guide rails, and saw blades [1]. There are two main issues in the flat rolling of wire. The first issue is shape control of the flat-rolled wires, because direct shape control is impossible during the flat rolling of wire; that is, lateral spreading occurs in the free surface of the wire [2]. Consequently, several studies have been conducted regarding the influence of process conditions, such as reductions in the height, initial wire size, friction, and rolling speed on the lateral spreading of a wire using empirical or numerical methods [1–7]. The studies reported that the lateral spreading of a wire increased with reductions in the height and friction, which was independent of the rolling velocity. The second issue is the inhomogeneity of the mechanical properties of flat-rolled wires. In particular, the low hardness on the flat surfaces of a wire is a crucial issue in the industry. Therefore, the strain distribution of flat-rolled wire has been investigated using finite element (FE) analysis and hardness tests [8–14]. Kazeminezhad and Karimi Taheri [9] reported that the strain inhomogeneity of a wire increased with decreasing the reduction in height and increasing the friction coefficient. Vallellano et al. [12] reported that the maximum contact pressure occurred in the roll entry zone due to the local inhomogeneity of deformation. Hwang [14] reported that the difference of the effective strain along the horizontal direction was much higher than that of the vertical direction, and the maximum difference of the effective strain occurred in between the center area and the free surface area of the flat-rolled wire. It is well known that the occurrence of macroscopic shear bands (MSBs) is highly related to the strain inhomogeneity of flat-rolled wire [8], because the occurrence of MSBs indicates that the deformation is highly inhomogeneous during the forming process. The occurrence of MSBs has been reported in several compression-type metal-forming processes, such as plain strain compression [15], uniaxial compression [16], flat rolling of wire [8], and flat roll drawing [17]. The restricted metal flow at the interface between the specimen and tool is the main reason for the occurrence of MSBs in specimens during the compression-type forming process. Therefore, the strain inhomogeneity can be improved by controlling the behavior of MSBs during the rolling process.

Over the past three decades, although several studies have reported the influence of process conditions on the strain inhomogeneity of flat-rolled wires [8–11], most of the studies have been conducted based on external process conditions, such as the roll diameter, reduction in thickness, rolling speed, and friction coefficient. Meanwhile, it may be inferred from experience that the roll shape is an important design parameter in shape rolling processes, such as caliber rolling [18,19] and flat rolling of wire. However, no studies have reported on the effect of the roll design on the strain distribution and shape control of a flat-rolled wire.

Therefore, the present study focuses on the effect of the roll design on the strain distribution in a flat-rolled wire in order to improve the homogeneity of mechanical properties and to increase the hardness on the flat surface of a wire. Oval-grooved roll and cambered roll techniques with various radii were applied to the flat rolling process based on numerical simulation, and then a general strategy for fabricating high-quality flat-rolled wire products was deduced, considering the process and working conditions in industries.

2. Numerical Procedures

The DEFORM FE commercial software developed by Scientific Forming Technologies Corporation in Ohio, USA, version 11.0 with a three-dimensional (3D) module was used to analyze the flat wire rolling process, because a flat-rolled wire experiences 3D inhomogeneous deformation during the process [14]. The workpiece was assumed to be an isotropic and rigid plastic material, while the effect of the strain rate was not considered in this study. In this case, the constitutive behavior was generally described by Hollomon's law [20,21], as follows:

$$\sigma = K\varepsilon^n \tag{1}$$

where K means the strength coefficient and n refers to the strain hardening exponent. The K and n values of the present material are chosen as 1980 and 0.54, respectively, based on the curve fitting of the tensile test in twinning-induced plasticity (TWIP) steel in [22]. Each value was inserted into the DEFORM commercial software model as follows:

$$\sigma = 1980\varepsilon^{0.54} \tag{2}$$

All rolls with 400 mm diameter were assumed to be rigid bodies. The rolling speed was set to 5 revolutions per minute (RPM) to ignore the effect of temperature rise. Oval-grooved rolls with radii of 10, 20, and 30 mm were applied to the flat rolling process to tailor the strain distribution, as shown in Figure 1b. A cambered roll was also applied with various radii to understand the effect of the roll shape on the strain distribution of a flat-rolled wire, as shown in Figure 1c. The oval-grooved roll and cambered roll were only used for the first rolling pass, while the same flat roll was applied to the second pass, as shown in Figure 1, to fabricate a flat-rolled wire with the desired shape for customers. The reduction in height (R_h) was calculated using the following equation:

$$R_h = \frac{h_0 - h_1}{h_0} \times 100(\%) \tag{3}$$

where h_0 and h_1 are the initial and final heights of the wire, respectively. R_h was about 22% at the first pass, regardless of the roll design. At the second pass, the R_h was approximately 18%, and thus the total R_h was 35%. The shear friction coefficient between the wire and the roll interface was selected as 0.3 [14].

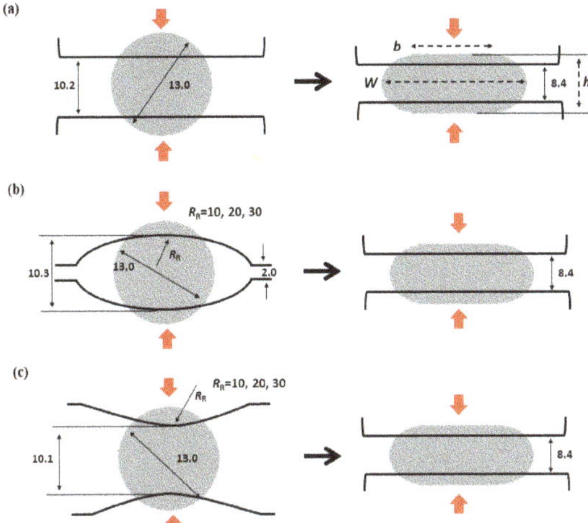

Figure 1. Schematic of the cross-sectional shape of the wire and the roll design during the flat rolling process with a (**a**) flat roll, (**b**) oval-grooved roll, and (**c**) cambered roll.

To reduce the computational cost, only one-quarter of the full geometry was calculated, owing to the symmetric condition of the flat-rolled wire. Approximately 15,000 brick-type elements were used, while 300 elements were used in the cross-section of the wire, as shown in Figure 2.

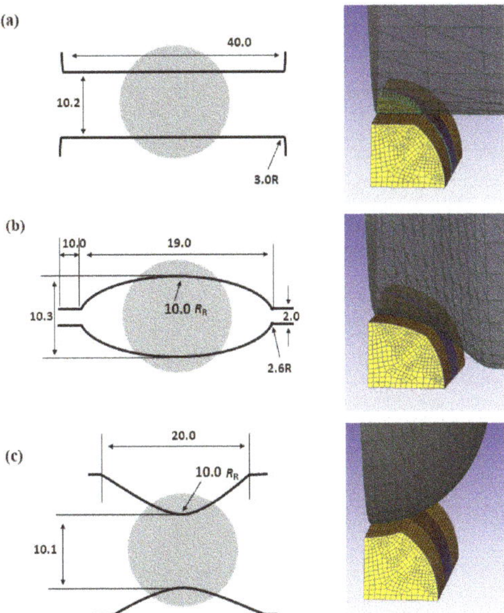

Figure 2. Detailed geometry and elements of the numerical modeling with a (**a**) flat roll, (**b**) oval-grooved roll with a radius of 10 mm, and (**c**) cambered roll with a radius of 10 mm.

3. Validation of the Numerical Model

The accuracy of the numerical model was validated prior to evaluating the results of the numerical simulation by comparing the numerically simulated geometries with the measured geometries of the flat-rolled TWIP steel wire. The analyzed chemical composition of TWIP steel is Fe-19.94Mn-0.60C-1.03Al (wt.%), which was fabricated by vacuum induction method. Prior to hot rolling, the ingot measuring 125 mm in thickness was homogenized at 1200 °C for 3 h. Then, the ingot was directly rolled onto a plate measuring 20 mm in thickness using multipass rolling at temperatures above 950 °C, followed by air cooling to simulate the hot rod rolling process. The hot-rolled plate was machined into several round bars with a diameter of 13 mm for the flat wire rolling test. The bar was rolled into the flat-rolled wire using flat rolls with a diameter of 400 mm at a rolling speed of 5 RPM. The other process conditions were kept the same as those of the numerical simulation.

Figure 3a shows the comparison of the cross-sectional shapes of flat-rolled wires based on the experimental and numerical simulation, as well as the used terminologies in this study, while Figure 3b compares the measured and numerically simulated width of the contact area (*b*) and the lateral spread (*W*) values with the total R_h. As expected, the *W* and *b* values increased when increasing the total R_h. Overall, the *W* and *b* values obtained by the numerical simulation were in good agreement with the experimental values. However, it was found that the deviation of the *W* value between the two results slightly increased with the total R_h. This inconsistency was related to the hardening model applied in this study [5] and the selected friction coefficient of 0.3.

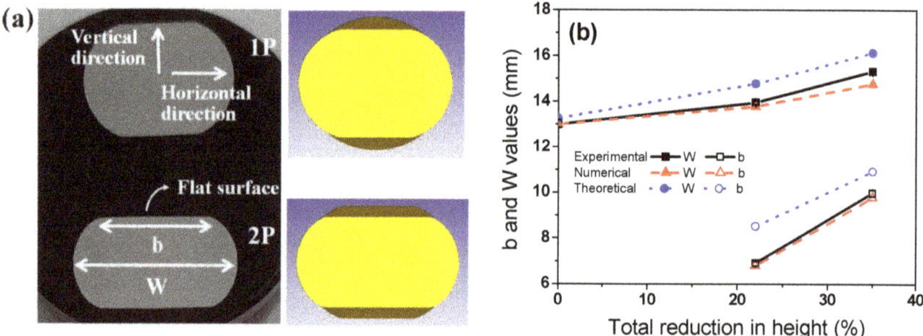

Figure 3. (**a**) Photograph of the caliber-rolled TWIP steel wire in this experiment and the shape of the deformed wire based on the numerical simulation. (**b**) Comparison of the measured, simulated, theoretically derived *W* and *b* values as a function of the total reduction in height.

Meanwhile, Kazeminezhad and Karimi Taheri [1,3] suggested the prediction of *b* and *W* values as a function of the total reduction in the height (Δh), h_0, and h_1 from the plain carbon steels as the following equations:

$$b = \sqrt{2\Delta h h_0} \tag{4}$$

$$\frac{W_1}{W_0} = 1.02\left(\frac{h_0}{h_1}\right)^{0.45} \tag{5}$$

where the subscripts of 0 and 1 indicate the initial and final values during the flat rolling process, respectively. As shown in Figure 3b, the *W* and *b* values of TWIP steel were lower compared to those of the plain carbon steel theoretically suggested in the relationships. Hwang [14] suggested that this result is highly related to the different strain hardening behavior between plain carbon steels and TWIP steel. Namely, the strain hardening exponents of TWIP steel are much higher than those of plain carbon steels [23]. Overall, it can be concluded from an engineering application point of view that the proposed FE analysis for the flat wire rolling process can be used to evaluate the characteristics of the shape and strain distribution with the process conditions.

4. Results and Discussion

4.1. Strain Distribution with Roll Design

Figure 4 shows a comparison of the contours of the von Mises strain (effective strain) and normal pressure of the flat-rolled wire with the representative roll designs (i.e., flat roll design, oval-grooved and cambered rolls with a radius of 10 mm). Clearly, the strain distribution in the flat-rolled wire was complex—it had two MSBs with high effective strain [8]. The center area tended to have a maximum strain, while the free surface area tended to have a minimum strain. However, the distribution of the effective strain of the wire was different with the roll design. In particular, the shapes of the MSBs changed with the roll design. During the first pass, the total width of the MSBs increased with the oval-grooved roll, while that of the MSBs decreased with the cambered roll. The different behavior of MSBs with the roll design was related to the normal pressure on the wire surface [24,25], as shown in Figure 4, indicating that the MSBs can be controlled by tailoring the contact pressure on the specimen during the rolling process. It should be noted that it is necessary to increase the strength on the flat surface of a wire, because the external stress was mainly imposed on the flat surface of the wires under service. For a better understanding of the strain distribution of the flat-rolled wire, the effective strain was extracted from the contour maps in both the cross-section and the flat surface, as shown in Figure 5 for the final product (i.e., after the second pass). The maximum effective strain in the center area decreased with the cambered roll, whereas it increased with the oval-grooved roll. Meanwhile, the minimum effective strain in the free surface area was similar regardless of the roll design, indicating that the cambered roll reduced the overall strain inhomogeneity of the flat-rolled wire.

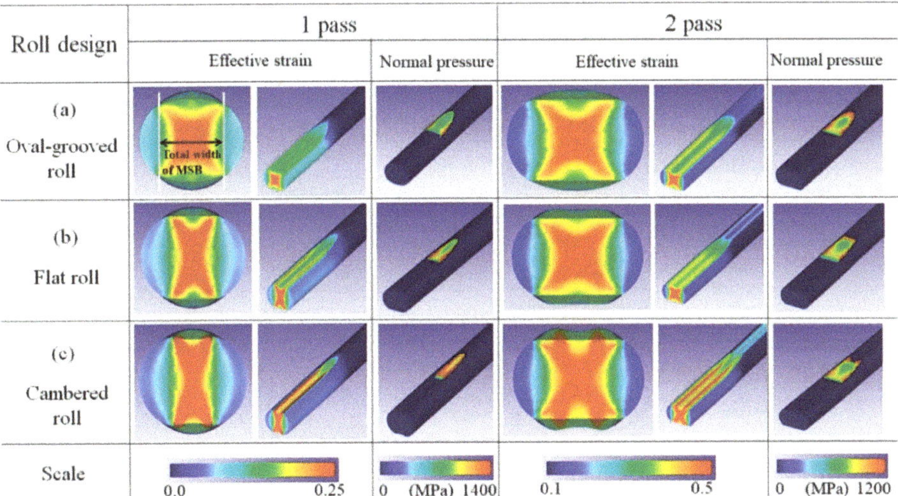

Figure 4. Contour maps of the effective strain and normal pressure of flat-rolled wires with a roll design and pass.

Interestingly, the effective strain on the flat surface of the wire increased when using the cambered roll, as shown in Figure 5c, whereas the oval-grooved roll decreased the effective strain on the flat surface of the wire. To deduce a general conclusion, a new non-dimensional indicator for the roll design (I_{RD}) was defined as follows:

$$I_{RD} = \frac{D_{wire}}{R_R} \tag{6}$$

where R_R and D_{wire} are the surface radius of the roll (Figure 1) and the diameter of the initial round wire (13 mm), respectively. To calculate the R_R value, the surface radius of the cambered roll was taken to be negative due to the reverse roll shape compared to the oval-grooved roll, as listed in Table 1. Figure 6a shows the maximum and minimum effective strains in the cross-section of the flat-rolled wire at the second pass. The maximum strain at the center area of the wire slightly increased with I_{RD}, meaning that the maximum strain increased with the decrease in the radius of the oval-grooved roll. In other words, the oval-grooved roll increased the stress concentration at the center area of the specimen due to the restriction of the plastic deformation according to roll shape, as compared to the flat and cambered rolls, as shown in Figure 2. In addition, it is well known that the strain is highly concentrated in the center area during bar and rod rolling with the oval-round roll pass sequence [26–28]. The minimum effective strain had a constant value with I_{RD}, indicating that the oval-grooved roll and camber roll did not affect the level of strain in the free surface area of a wire. In summary, the cambered roll slightly reduced the strain inhomogeneity of the specimen during the flat rolling process.

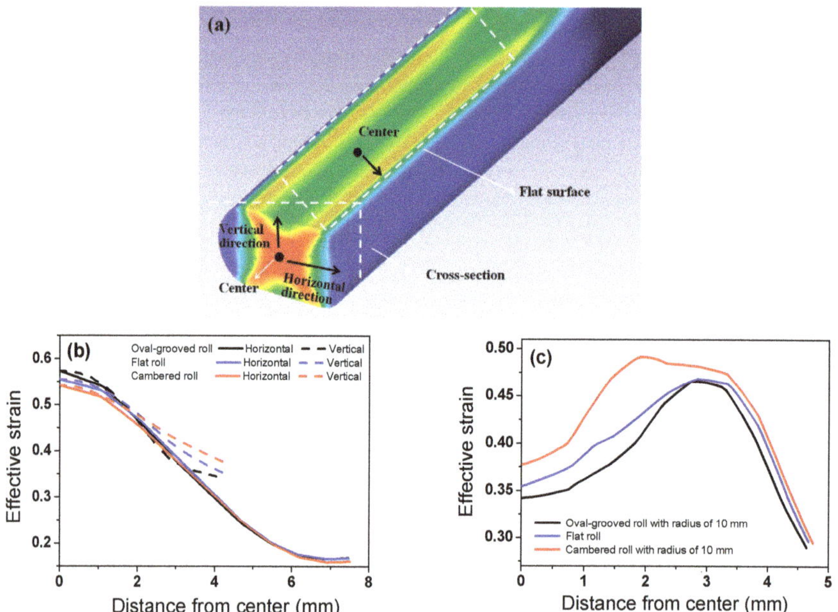

Figure 5. (**a**) Schematic showing the terminologies used in this study. Comparison of the effective strain profiles in (**b**) cross-sections of flat-rolled wires along the horizontal and vertical directions, and (**c**) the flat surface of a flat-rolled wire with the representative roll design.

Table 1. Comparison of the seven roll designs and related values during the flat rolling process used in this study.

Roll Design	Values		
	Surface Radius (mm)	R_R	I_{RD}
Oval-grooved roll	10	10	1.3
	20	20	0.65
	30	30	0.43
Flat roll	∞	∞	0
Cambered roll	30	−30	−0.43
	20	−20	−0.65
	10	−10	−1.3

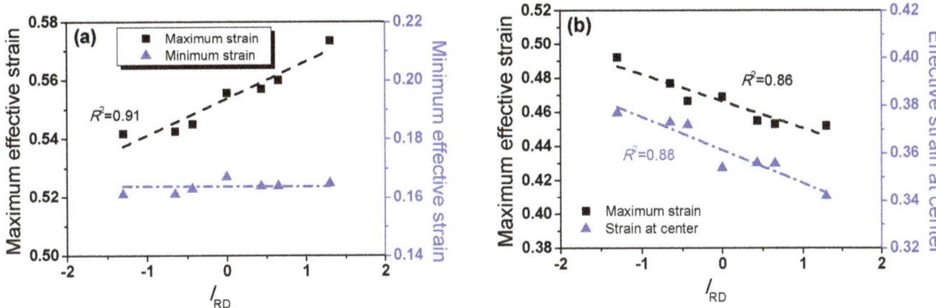

Figure 6. Variations in (**a**) the maximum and minimum effective strains in the cross-sections of the flat-rolled wires and (**b**) the maximum and centered effective strains on the flat surface of a wire with I_{RD}.

The maximum effective strain on the flat surface of the wire decreased with I_{RD}, as shown in Figure 6b, meaning that the camber roll with the small radius increased the strength on the flat surface of the wire. This was due to the strong concentration of the normal pressure on the wire surface, as shown in Figure 4c. Namely, the strain distribution on the flat surface of the wire was highly dependent on the normal contact pressure. Based on a similar mechanism, the strain of the center area on the flat surface of the wire was improved when using the cambered roll with a small radius. This result is attractive for industrial plants because many process designers want to improve the hardness of the flat surfaces in a wire owing to the strict customer demands.

4.2. Shape of Flat-Rolled Wire with Roll Design

To apply the shaped rolls to the flat rolling process, the effect of the roll design on the shape of the final product needs to be considered. Figure 7 shows the variation in b and W values with I_{RD} after the second pass. Both values decreased linearly with I_{RD}, indicating that the lateral spreading of the wire decreased when using the oval-grooved roll with a small radius, because the spreading in the free surface area of the wire was highly restricted by the roll shape and curvature, as shown in Figure 1b. In contrast, the spreading in the surface area increased during the rolling process with the cambered roll due to the reduced restriction of the metal flow in the free surface area, as shown in Figure 1c. These results practically imply that the size of the initial wire needs to be changed when using the shaped roll during the flat rolling of wire due to the different wire spreading results with roll designs.

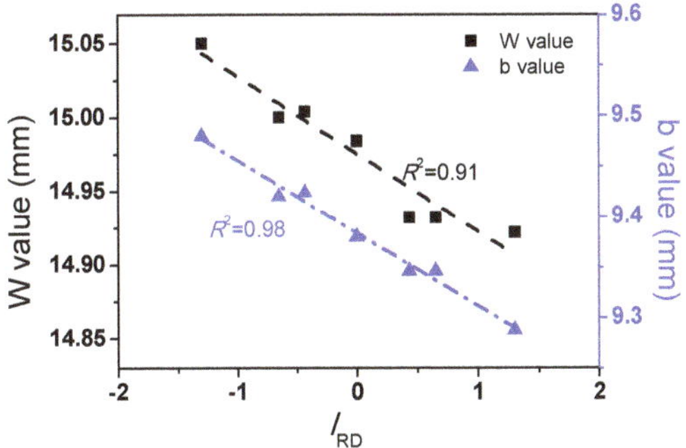

Figure 7. Variations in the b and W values with I_{RD}.

4.3. Design Concept for a High-Quality Flat-Rolled Wire

To make high-quality flat-rolled wires, a practical strategy was designed, as shown in Figure 8, based on the results achieved in the above comparative study. The strain on the flat surface of the wire increased with decreasing I_{RD}, while the strain inhomogeneity in the cross-section of the wire was slightly reduced with decreasing I_{RD}, meaning that we can produce higher quality flat-rolled wire products using a cambered roll with a small radius. In contrast, the use of an oval-grooved roll is not a good idea when fabricating a flat-rolled wire. The proposed design concepts can provide process designers with seeding ideas to choose the optimal process conditions for high-quality flat-rolled wires. This is beneficial for industrial plants because a cambered roll can be easily applied in flat rolling plants. It is worth noting that the proposed practical strategy was valid for most materials, since it was derived from the general plastic forming conditions. However, the effects of the kinematic hardening, the strain rate sensitivity in the material, and the friction coefficient should be considered to obtain more reliable results. In addition, it should be noted that the effects of the roll design on the b and W values should be considered to obtain a wire with accurate dimensions and tolerance (Figure 7). For mass production, the wear issue needs to be considered when applying the cambered roll in the flat rolling of wire. Additionally, the present results need to be confirmed by experimental data, such as a hardness analysis and microstructure evolution data. In this respect, additional research is necessary.

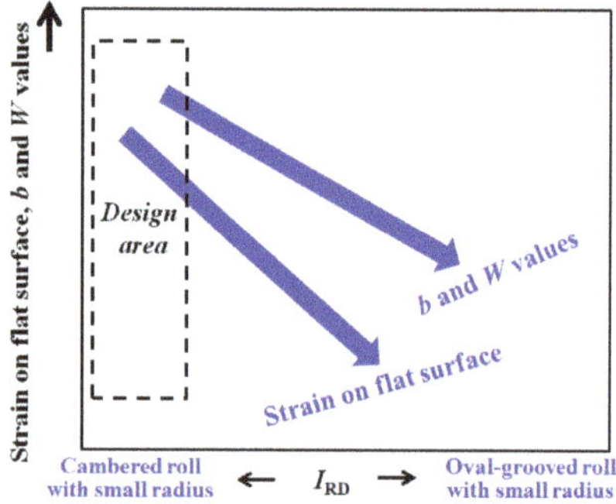

Figure 8. Schematic description of the effective strain on the flat surface and shape of a flat-rolled wire with I_{RD}.

5. Conclusions

The effect of the roll design on the strain distribution of the flat surface and on b, and W values; and the strain inhomogeneity of flat-rolled wires were systematically investigated during the flat rolling process. The major conclusions are as follows:

1. The effective strain on the flat surface of the wire increased when using a cambered roll due to the highly intensified contact pressure, while the effective strain on the flat surface of the wire decreased when using an oval-grooved roll;

2. The b and W values decreased when using an oval-grooved roll with a small radius because the spread in the free surface area of the wire was highly restricted by the roll shape. In contrast, the spread in the surface area increased when using a cambered roll due to the less-restricted metal flow at the free surface;

3. A new practical strategy was proposed for fabricating high-quality flat-rolled wires. A cambered roll with a small radius can improve the surface quality of flat-rolled wires. This is beneficial for industrial plants because cambered rolls can be applied easily in flat rolling plants.

Funding: This work was supported by the National Research Foundation of Korea (NRF) grant funded by the Korea government (MSIT) (NRF-2018R1D1A1B07050103).

Conflicts of Interest: The author declares no conflicts of interest.

References

1. Kazeminezhad, M.; Karimi Taheri, A. A theoretical and experimental investigation on wire flat rolling process using deformation pattern. *Mater. Des.* **2005**, *26*, 99–103. [CrossRef]
2. Kobayashi, K.; Asakawa, M.; Kobayashi, M. Deformation behavior of round wire in compression using a cylindrical tool and the analysis of width spreading in flat rolling. *Wire J. Int.* **2005**, *38*, 74–79.
3. Kazeminezhad, M.; Taheri, A.K. An experimental investigation on the deformation behavior during wire flat rolling process. *J. Mater. Process. Technol.* **2005**, *160*, 313–320. [CrossRef]
4. Kazeminezhad, M.; Taheri, A.K.; Tieu, A.K. A study on the cross-sectional profile of flat rolled wire. *J. Mater. Process. Technol.* **2008**, *200*, 325–330. [CrossRef]
5. Massé, T.; Chastel, Y.; Montmitonnet, P.; Bobadilla, C.; Persem, N.; Foissey, S. Impact of mechanical anisotropy on the geometry of flat-rolled fully pearlitic steel wires. *J. Mater. Process. Technol.* **2011**, *211*, 103–112. [CrossRef]
6. Utsunomiya, H.; Hartley, P.; Pillinger, I. Three-Dimensional Elastic-Plastic Finite-Element Analysis of the Flattening of Wire Between Plain Rolls*. *J. Manuf. Sci. Eng.* **2000**, *123*, 397–404. [CrossRef]
7. Parvizi, A.; Pasoodeh, B.; Abrinia, K.; Akbari, H. Analysis of curvature and width of the contact area in asymmetrical rolling of wire. *J. Manuf. Process.* **2015**, *20*, 245–249. [CrossRef]
8. Kazeminezhad, M.; Taheri, A.K. The prediction of macroscopic shear bands in flat rolled wire using the finite and slab element method. *Mater. Lett.* **2006**, *60*, 3265–3268. [CrossRef]
9. Kazeminezhad, M.; Taheri, A.K. Deformation inhomogeneity in flattened copper wire. *Mater. Des.* **2007**, *28*, 2047–2053. [CrossRef]
10. Kazeminezhad, M.; Taheri, A.K. The effect of 3D and 2D deformations on flattened wires. *J. Mater. Process. Technol.* **2008**, *202*, 553–558. [CrossRef]
11. Massé, T.; Chastel, Y.; Montmitonnet, P.; Bobadilla, C.; Persem, N.; Foissey, S. Mechanical and damage analysis along a flat-rolled wire cold forming schedule. *Int. J. Mater. Form.* **2011**, *5*, 129–146. [CrossRef]
12. Vallellano, C.; Cabanillas, P.; García-Lomas, F. Analysis of deformations and stresses in flat rolling of wire. *J. Mater. Process. Technol.* **2008**, *195*, 63–71. [CrossRef]
13. Iankov, R. Finite element simulation of profile rolling of wire. *J. Mater. Process. Technol.* **2003**, *142*, 355–361. [CrossRef]
14. Hwang, J.-K. Deformation Behaviors of Flat Rolled Wire in Twinning-Induced Plasticity Steel. *Met. Mater. Int.* **2019**, *26*, 603–616. [CrossRef]
15. Paul, H.; Driver, J.; Tarasek, A.; Wajda, W.; Miszczyk, M. Mechanism of macroscopic shear band formation in plane strain compressed fine-grained aluminium. *Mater. Sci. Eng. A* **2015**, *642*, 167–180. [CrossRef]
16. Eom, J.; Son, Y.; Jeong, S.; Ahn, S.; Jang, S.; Yoon, D.; Joun, M. Effect of strain hardening capability on plastic deformation behaviors of material during metal forming. *Mater. Des.* **2014**, *54*, 1010–1018. [CrossRef]
17. Lambiase, F.; Di Ilio, A. Deformation inhomogeneity in roll drawing process. *J. Manuf. Process.* **2012**, *14*, 208–215. [CrossRef]
18. Inoue, T. Optimum pass design of bar rolling for production bulk ultrafine-grained steel by numerical simulation. *Mat. Sci. Forum* **2010**, *654–656*, 1561–1564. [CrossRef]
19. Hwang, J.-K.; Kim, S.J. Effect of reduction in area per pass on strain distribution and microstructure during caliber rolling in twinning-induced plasticity steel. *J. Iron Steel Res. Int.* **2019**, *27*, 62–74. [CrossRef]
20. Choung, J.; Cho, S.R. Study on true stress correction from tensile tests. *J. Mech. Sci. Technol.* **2008**, *22*, 1039–1051. [CrossRef]
21. Koc, P.; Štok, B. Computer-aided identification of the yield curve of a sheet metal after onset of necking. *Comput. Mater. Sci.* **2004**, *31*, 155–168. [CrossRef]
22. Hwang, J.-K.; Yi, I.-C.; Son, I.-H.; Yoo, J.-Y.; Kim, B.; Zargaran, A.; Kim, N.J. Microstructural evolution and deformation behavior of twinning-induced plasticity (TWIP) steel during wire drawing. *Mater. Sci. Eng. A* **2015**, *644*, 41–52. [CrossRef]
23. Hwang, J.-K. The microstructure dependence of drawability in ferritic, pearlitic, and TWIP steels during wire drawing. *J. Mater. Sci.* **2019**, *54*, 8743–8759. [CrossRef]
24. Carlsson, B. The contact pressure distribution in flat rolling of wire. *J. Mater. Process. Technol.* **1998**, *73*, 1–6. [CrossRef]

25. Kazeminezhad, M.; Taheri, A.K. Calculation of the rolling pressure distribution and force in wire flat rolling process. *J. Mater. Process. Technol.* **2006**, *171*, 253–258. [CrossRef]
26. Kwon, H.-C.; Lee, Y.; Im, Y. Experimental and Numerical Prediction of Austenite Grain Size Distribution in Round-oval Shape Rolling. *ISIJ Int.* **2003**, *43*, 1967–1975. [CrossRef]
27. Jung, K.; Lee, H.W.; Im, Y.-T. A microstructure evolution model for numerical prediction of austenite grain size distribution. *Int. J. Mech. Sci.* **2010**, *52*, 1136–1144. [CrossRef]
28. Hwang, J.-K. Effects of caliber rolling on microstructure and mechanical properties in twinning-induced plasticity (TWIP) steel. *Mater. Sci. Eng. A* **2018**, *711*, 156–164. [CrossRef]

Article

Model and Algorithm for Planning Hot-Rolled Batch Processing under Time-of-Use Electricity Pricing

Zhengbiao Hu, Dongfeng He *, Wei Song and Kai Feng

Department of Ferrous Metallurgy, School of Metallurgical and Ecological Engineering, University of Science and Technology Beijing, Beijing 100083, China; liupingze2019@163.com (Z.H.); songweiustb@163.com (W.S.); WeiRunWeiUSTB@126.com (K.F.)
* Correspondence: hedongfeng@ustb.edu.cn

Received: 25 October 2019; Accepted: 18 December 2019; Published: 1 January 2020

Abstract: Batch-type hot rolling planning highly affects electricity costs in a steel plant, but previous research models seldom considered time-of-use (TOU) electricity pricing. Based on an analysis of the hot-rolling process and TOU electricity pricing, a batch-processing plan optimization model for hot rolling was established, using an objective function with the goal of minimizing the total penalty incurred by the differences in width, thickness, and hardness among adjacent slabs, as well as the electricity cost of the rolling process. A method was provided to solve the model through improved genetic algorithm. An analysis of the batch processing of the hot rolling of 240 slabs of different sizes at a steel plant proved the effectiveness of the proposed model. Compared to the man–machine interaction model and the model in which TOU electricity pricing was not considered, the batch-processing model that included TOU electricity pricing produced significantly better results with respect to both product quality and power consumption.

Keywords: hot rolling; TOU electricity pricing; hot rolling planning; genetic algorithm

---

1. Introduction

As a key link in steel production, hot rolling refers to a process in which the steel slabs sustain heating in a furnace, rough rolling, and fine rolling before becoming steel products (Figure 1). The main task of the batch-processing planning of hot rolling is to determine an appropriate sequence for the processing of multiple slabs, to achieve low power consumption, low cost of stack transfer, high efficiency, and high product quality, while the technical requirements of rolling are met [1]. Throughout the steel production process, the batch-processing plan of hot rolling directly determines the product quality and production efficiency of the steel plant. In the past, in most studies on the batch-processing planning of hot rolling, the penalty items relating to the differences (such as width, thickness, hardness, and delivery time) were only taken into consideration among adjacent slabs, while the differences in power consumption between slabs of difference sizes were ignored [2–4]. In fact, apart from guaranteeing product quality and production efficiency, a sound processing plan of rolling also served to reduce production cost through electricity cost cut-backs, thereby maximizing the economic benefits [5].

Throughout actual hot rolling, most electricity is used to drive the motor of the rolling mill and power various auxiliary electrical devices. The power consumption in hot rolling is closely related to the sizes and conditions of the devices, process parameters, as well as the types and sizes of the steel products [6]. Under the condition of time-of-use (TOU) electricity pricing, the sequence with which the slabs are processed has a major impact on power consumption. In the research area of hot rolling, only few studies were carried out regarding the impact investigation of TOU electricity pricing on the batch-processing planning of hot rolling. Most studies were focused on methods of better connection and matching during steelmaking, of continuous casting, and hot rolling processes. A charge-rolling plan coordination model based on the optimal furnace charge plan and optimal rolling plan models was presented [7,8], in which the tabu search algorithm was proposed for the corresponding solution. A steelmaking continuous-casting hot-rolling integrated planning model was proposed to enable an effective connection among the rolling unit plan [9], the furnace charge plan, and the casting plan. An integrated model on the production and logistics planning level was constructed [10], based on an analysis of the characteristics of steelmaking and hot rolling processes, as well as on the transition between the two processes, where the constraints of capacity and the conflicts between the two stages were taken into consideration. A mathematical model was presented for batch-processing optimization in the steelmaking and hot rolling processes [11]. This model was proposed for the model solution with the neighborhood search algorithm based on heuristic rules. The batch-processing planning problem of hot rolling was treated as a constraint compliant problem [12], while the vehicle routing problem with soft time windows (VRPSTW) constraint of uncertain plan number was constructed to satisfy the model. In certain studies, the effects of other factors on the rolling plan of hot-rolling planning were taken into consideration. As an example, the problem of slab stack transfer in the rolling plan was taken into consideration [13,14], whereas the solution was conducted with the improved genetic algorithm, resulting in reduced handling costs. The impact of TOU electricity pricing on the rolling planning was taken into consideration [6], but in the proposed model, the power consumptions of individual rolling units were only considered, instead of the influence of the rolling sequence on the electricity cost under TOU electricity pricing. Mao et al. [5] considered the influence of TOU electricity pricing when constructing a rolling unit as the production load unit, and a multi-objective optimization model for hot rolling was established and a multi-objective optimization algorithm was applied to solve this problem. However, all the goals were classified into one objective function in this paper.

In summary, in most previous studies, optimal rolling-plan models were proposed to be established, to produce products with the highest possible quality and production efficiency prior to the delivery deadline, indifferently to electricity cost reduction throughout production. Because the main characteristic of genetic algorithms is to directly operate the structural objects, only the objective function and the corresponding fitness functions that affect the search direction are needed. Therefore, the genetic algorithm provides a method for solving complex system problems. It does not depend on the specific fields and types of problems, and has strong robustness, so it has been widely used in many scientific fields. [15]. On the basis of previous studies, in this paper, a model for batch-processing planning of hot rolling under TOU electricity pricing was presented, resulting in a method of solving the model with an improved genetic algorithm.

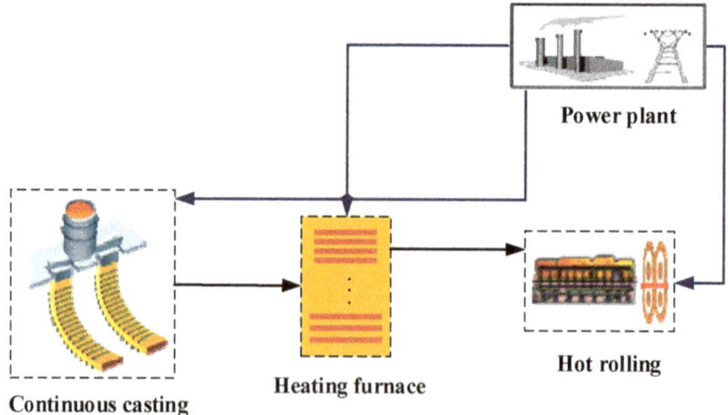

Figure 1. Flow chart of the rolling process.

2. Problem Description

2.1. Basic Principles of Hot Rolling Planning

Hot rolling constitutes the last link in the steelmaking continuous-casting hot-rolling process. The raw materials are the slabs resulting from continuous casting, while the products are strip steels (steel plates or steel coils). The strip steel could be directly sold as a finished product, or be further processed into high-quality cold-rolled strip steel. A key issue in hot rolling is the rolling plan formulation. A complete rolling unit generally consists of two parts: "roll heating pieces" and "principal pieces" (Figure 2). The roll-heating pieces are usually easy-to-process slabs, such as low-carbon steels as well as relatively thick and narrow strip steels. The number of slabs in this part is relatively low, while the main purpose of this part is to heat the rolled steel and lead it to reach heat equilibrium, creating conditions for the subsequent processing of principal pieces. Usually, the width of the slab increases gradually throughout rolling. The principal pieces are usually hard-to-process slabs, for thinner and wider products, whereas the width of the slab decreases gradually throughout hot rolling. Therefore, a complete rolling unit would have a dual trapezoidal structure, with the positive trapezoid as the roll-heated pieces and the inverted trapezoid as the principal pieces. In view of the fact that the numbers of roll-heated pieces in a rolling unit are small and the quality requirement is low, in this study, the principal pieces were focused on. The principal pieces processing met the following requirements [16,17].

(1)   The length of each principal piece had a certain limit;
(2)   When the thickness of the principal piece changed in the non-increase direction, the step must be lower than 25 cm;
(3)   When the width changed inversely, the step did not generally exceed 15 cm;
(4)   When slabs of the same width were continuously processed, the total length should not exceed a certain limit;
(5)   The width, thickness and hardness were not allowed to jump at the same time;
(6)   The thickness jump should be smooth and repeated jumps were not allowed. Changes in the non-decrease direction were preferred.
(7)   The hardness change should be smooth. Both gradual increase and gradual decrease were allowed. Repeated jumps were not allowed.

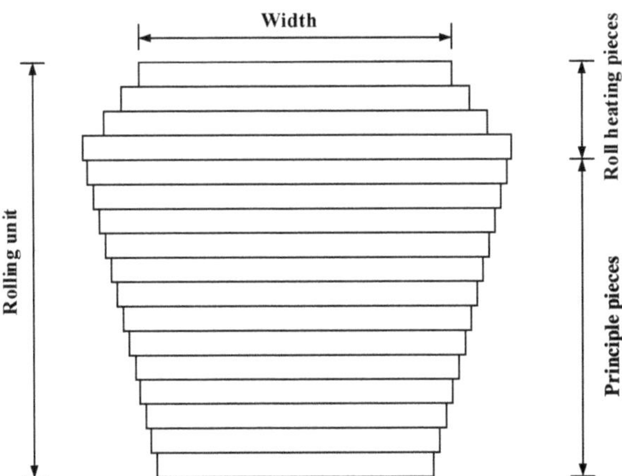

Figure 2. Schematic diagram of rolling unit.

Figure 3 presents the batch-processing plan formulation of hot rolling. Multiple slabs of different sizes were combined into a rolling unit in accordance with the process requirement, while multiple rolling units constituted a hot-rolling batch-processing plan (Figure 4). The hot-rolled batch processing plan was a rolling sequence of slabs arranged in accordance with the roll usage cycle. One rolling unit corresponded to the usage of one set of rolls. The hot-rolling batch-processing plan's successful execution directly determines the product quality, delivery date, and production efficiency, and a sound hot-rolling batch-processing plan could improve production efficiency, reduce energy consumption (such as costs), and enhance the firm's competitiveness.

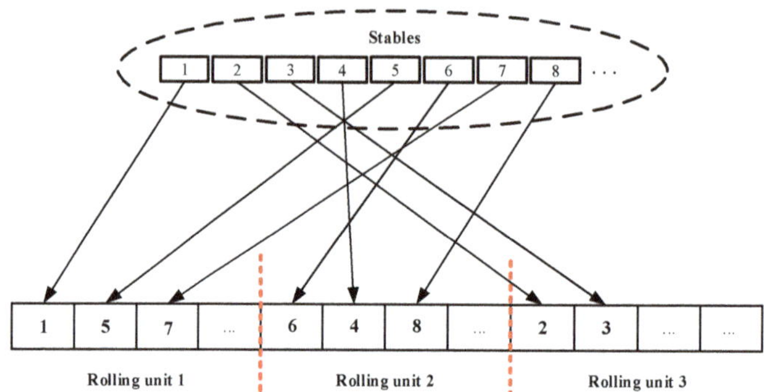

Figure 3. Batch-processing plan formulation of hot rolling.

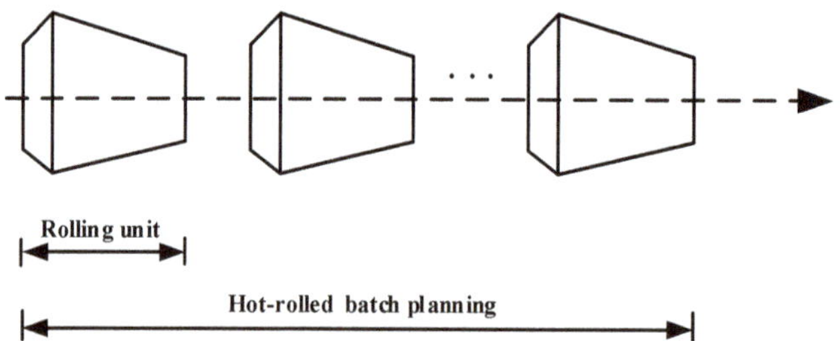

Figure 4. Schematic diagram of a batch-processing plan for hot rolling.

2.2. Analysis of Energy Consumption in Rolling

Hot rolling is a typical energy-intensive process in steel production, mainly requiring electric energy and heat energy from gas consumption. The gas is mainly consumed in the heating furnace to heat the slabs to the temperature required for hot rolling. The gas used in the heating furnace is usually a mixture of blast furnace gas, Linz–Donawitz gas, and coke oven gas, while the calorific value of the mixed gas must meet the production requirement. The electric energy is mainly used to drive the rolling mill, while the power consumption in the rolling process varies depending on the slab size, steel type, and other factors. In a hot-rolling batch-processing plan, the rolling sequence change or positions exchange of two rolling units will change the energy consumption pattern. At many locations, the electricity price varies with time. As demonstrated with the TOU electricity pricing scheme presented in Figure 5, the highest electricity price was 2.43 times that of the lowest electricity price. Consequently, the rolling sequence in the rolling plan directly affected the electricity cost of hot rolling.

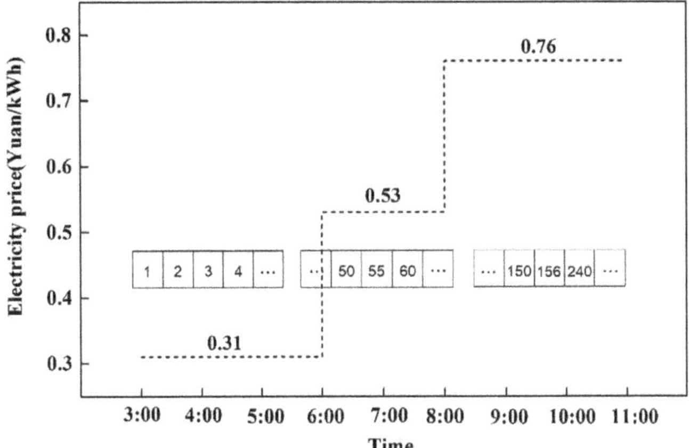

Figure 5. Schematic diagram of production scheduling under time-of-use (TOU) electricity pricing.

At present, many batch-processing plans of hot-rolling steel plants are mainly prepared by human planners in human–computer interaction interfaces, based on their experience. The impact of the batch-processing plan on the power consumption has been often ignored. Besides, this mode of planning has the drawbacks of low efficiency, poor accuracy, and high labor cost. To compensate for this deficiency, in this paper, a model for batch-processing planning of hot rolling under TOU electricity pricing was proposed, with the aim of ensuring high quality and reducing power consumption at the same time.

3. Mathematical Model and Solution Method

3.1. Mathematical Model

3.1.1. Model Assumptions

(1) The time required to process a single slab was roughly the same. In this paper, the processing time used was 2 min;
(2) Basic properties, such as width, thickness, hardness, and length of each slab were known;
(3) The change of the total attribute in the rolling unit could be determined through calculation.

3.1.2. Objective Function

In this paper, the optimizations of quality and power consumptions from different time periods were taken into consideration. On the basis of setting a certain weight coefficient, an objective function was formulated with the goal of minimizing the differences in width, thickness, and hardness among adjacent slabs of the same rolling unit, as well as the electricity cost of the rolling process. The objective function is presented as

$$\min \ F = \sum_{j=1}^{N} \sum_{i=1,i\neq j}^{N} \left\{ P^{W}\Delta W_{i,j} + P^{G}\Delta G_{i,j} + P^{H}\Delta H_{i,j} \right\} \cdot X_{ijk} + \sum_{t=1}^{T} \sum_{i=1}^{N} \left\{ C_{ele,t}(P_{ele,i}) \right\} \cdot Y_{i} \qquad (1)$$

where $\Delta W_{i,j} = \left| W_{i} - W_{j} \right|$, $\Delta G_{i,j} = \left| G_{i} - G_{j} \right|$, $\Delta H_{i,j} = \left| H_{i} - H_{j} \right|$.

3.1.3. Constraints

Based on the above description, the constraints of the hot-rolled batch processing planning model are as follows:

$$\sum_{k=1}^{N} y_{ik} = 1$$
$$i \in \{1,2,3,\ldots,N\}, \quad k \in \{1,2,3,\ldots,M\} \tag{2}$$

$$\sum_{i=1}^{N} (L_i \times y_{ik}) \leq L_{\max}$$
$$i \in \{1,2,3,\ldots,N\}, \quad k \in \{1,2,3,\ldots,M\} \tag{3}$$

$$\sum_{i=1}^{N} \left(Z_{ijk} \times L_i \times y_{ik}\right) \leq R_{\max}$$
$$i,j \in \{1,2,3,\ldots,N\}, \quad k \in \{1,2,3,\ldots,M\} \tag{4}$$

$$0 \leq X_{ijk} \times \left(W_i - W_j\right) \leq W_{\max}$$
$$i,j \in \{1,2,3,\ldots,N\}, \quad k \in \{1,2,3,\ldots,M\} \tag{5}$$

$$0 \leq X_{ijk} \times \Delta G_{i,j} \leq G_{\max}$$
$$i,j \in \{1,2,3,\ldots,N\}, \quad k \in \{1,2,3,\ldots,M\} \tag{6}$$

$$0 \leq X_{ijk} \times \Delta H_{i,j} \leq H_{\max}$$
$$i,j \in \{1,2,3,\ldots,N\}, \quad k \in \{1,2,3,\ldots,M\} \tag{7}$$

$$\sum_{i=1}^{N} X_{ijk} \cdot \sum_{j=1}^{N} X_{ijk} = 1$$
$$i,j \in \{1,2,3,\ldots,N\}, \quad k \in \{1,2,3,\ldots,M\} \tag{8}$$

$$C_{ele,t} = \begin{cases} C_0, & t \in (0, T_0] \\ C_1, & t \in (T_0, T_1] \\ C_2, & t \in (T_1, T_2] \end{cases} \tag{9}$$

$$X_{ijk} = \begin{cases} 1, & \text{Slab } j \text{ is behind slab } i, \text{ and belong to same rolling unit } k \\ 0, & \text{else} \end{cases} \tag{10}$$

$$Z_{ijk} = \begin{cases} 1, & \text{Slab } j \text{ is behind slab } i \text{ and } j \text{ have same width and belong to same rolling unit } k \\ 0, & \text{else} \end{cases} \tag{11}$$

$$y_{ik} = \begin{cases} 1, & \text{Slab } i \text{ belongs to rolling unit } k \\ 0, & \text{else} \end{cases} \tag{12}$$

where F represents the target penalty value, N represents the number of slabs, T represents the rolling period. P^W, P^G and P^H are the penalty coefficients for the differences in width, thickness, and hardness among adjacent slabs of the same rolling unit, respectively. In the equations, Y_i represents the penalty coefficient for the electricity spent on the slab i, $C_{ele,t}$ represent the prices of the electricity spent on the slab i, $P_{ele,i}$ is the electricity spend on the slab i, $L_{\max}$ is the length limit for each rolling unit, $R_{\max}$ is the length limit for the continuously processed slabs of the same width in the same rolling unit, $W_{\max}$, $G_{\max}$, and $H_{\max}$ are the upper limits for the differences in width, thickness and hardness among adjacent slabs from the same rolling unit. Constraint (2) ensured that each slab was assigned to one rolling unit; constraints (3) and (4) limited the total length of a single rolling unit and the length of continuously processed slabs of the same width; constraint (5) ensured that slabs of the same rolling unit were arranged in the descending order of width and that the width difference between two adjacent slabs did not exceed the upper limit; constraints (6) and (7) ensured that the differences in thickness and hardness between two adjacent slabs did not exceed the upper limits, respectively;

constraint (8) signified that each slab could only be processed once; constraint (9) represented the prices of electricity from different time periods; and constraints (10)–(12) were the decision variables of values of 0, 1, respectively.

3.2. Method for Model Solution

3.2.1. Brief Description of Algorithm

The genetic algorithm (GA) is an algorithm that mimics the processes of inheritance, mutation, and natural selection in the evolution of biological organisms. Figure 6 presents the solution of the genetic algorithm. First, a set of initial feasible solutions were randomly selected, according to the characteristics of the problem. Following this, a new chromosome was obtained through crossover and mutation operations involving genetic operators. Next, the fitness of the new generation chromosome was calculated and evaluated. The chromosome with good adaptability would be passed on to the next generation. This solution process would be repeated until the preset number of iterations was reached or the convergence condition was met. The GA has the characteristics of randomness, implicit parallelism, and global optimization in the operation process. In the actual operation of steel plant, a requirement to prepare a large number of batch-processing plans exists. Consequently, it was difficult to achieve the expected result through the common search methods. To tackle this problem, in this paper, it was proposed that the batch-processing model was to be solved with the GA with improved genetic operators.

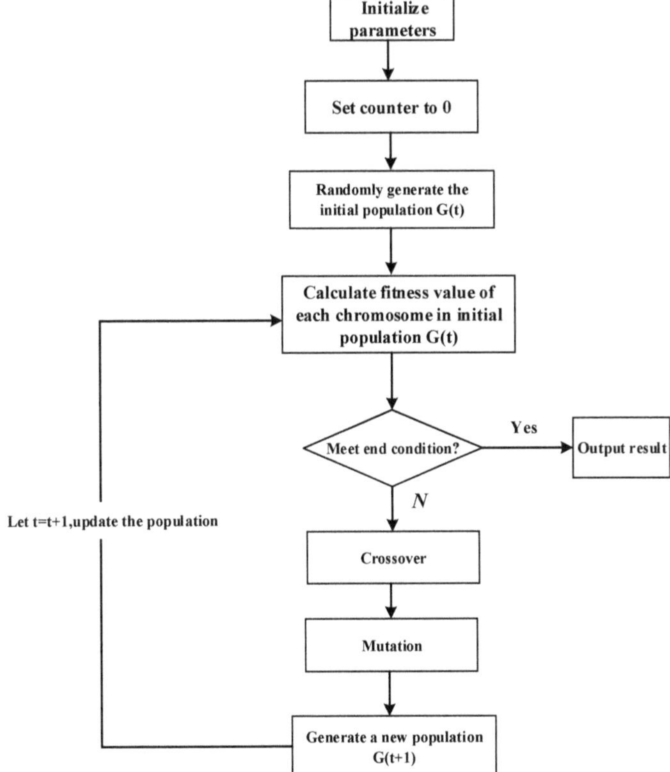

Figure 6. Process of model solution through genetic algorithm.

3.2.2. Chromosome Coding

Throughout the hot-rolling planning model solution with GA, the chromosome was represented by a coding sequence consisting of decimal natural numbers. For a rolling plan, the serial numbers of the slabs represented the genes in the chromosome coding. The chromosome coding could be expressed as the sequence of $\{a_1, a_2 \ldots a_i\}$, in which the chromosome length was the total number N of pre-processed slabs. Figure 7 presents a schematic diagram demonstrating the chromosomal coding of the rolling plan, in which, a_1, a_{50} and a_{100} were the serial numbers of the slabs; A_0, A_1 and A_2 were the slab sequences constituting the rolling units.

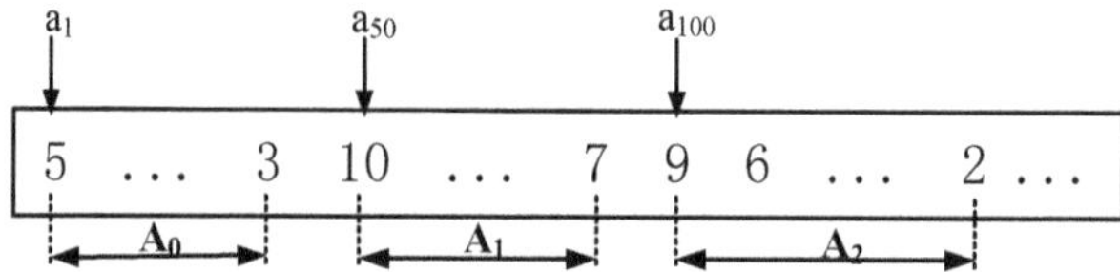

Figure 7. Schematic diagram of rolling-plan chromosome coding.

3.2.3. Chromosome Decoding

In different chromosomes, the numbers represented the serial numbers of the slabs, while the order of appearance was the order of processing. Figure 8 presents the flowchart of chromosome decoding. The procedure of rolling plan chromosome decoding was as follows:

Step 0: $M = 0$, $A_j = a_i$, $l = 0$;

Step 1: Sequentially traversed the slab numbers a_i from the *i*th position of the chromosome, recording the length l_i of the slab a_i and recording a_i in the arranged slab sequence A_j, doing $l = l + l_i$, and going to Step 2;

Step 2: If l was larger than the length L of the rolling unit, the last element in the sequence A_j was removed, turning the sequence A_{j-1} into a rolling unit, and letting $M = M + 1$; if l was equal to the length L of the rolling unit, the sequence A_j was turned into a rolling unit, and letting $M = M + 1$; otherwise, Step 1 was selected;

Step 3: When i was equal to the length of the chromosome, the program exited.

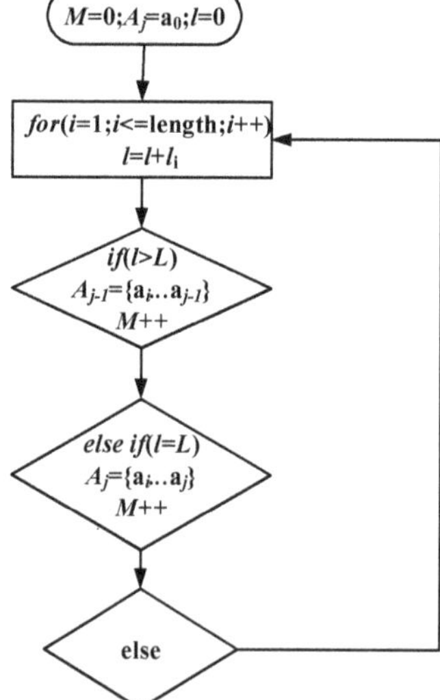

Figure 8. Flow chart of chromosome decoding.

3.2.4. Selection of Fitness Function

In this paper, a sorting-based allocation function was used as the fitness function. The equation was as follows:

$$Fit(P) = 2 - Y + 2(Y-1)(P-1)/(N-1)$$

where N is the number of chromosomes in the population, P is the sorting number of the chromosome in the population, Y is the selection pressure with a value in the range of [1.0, 2.0]. When the fitness function was applied, it was necessary to calculate at first the penalty value of each chromosome in the population, incurred by the differences in the width, thickness, and hardness among adjacent slabs, as well as the TOU electricity pricing. Following this, sorting was performed according to the penalty values, and the P value of the corresponding chromosome was obtained. Finally, the fitness values of different chromosomes were calculated. The smaller the objective function value was, the greater the fitness of the corresponding chromosome was. In this way, the high-quality chromosomes were picked from the population.

3.2.5. Genetic Operators

Genetic operators include selection operators, crossover operators, and mutation operators. In a rolling plan, each slab could only occupy one position in the sequence. In other words, no duplicate genes would appear in the same chromosome, which is different from the conventional GA. Therefore, it was necessary to improve the crossover and mutation operators. The selection operator used in the model solution method proposed in this paper was used to realize a selection operation similar to roulette according to the above-mentioned fitness function. The crossover operator was a secondary crossover operator. As illustrated in Figure 9, the crossover operator worked as follows. (1) Three contiguous genes on the chromosome were randomly selected and their positions are recorded; (2) the three genes were led to exchange positions with three corresponding genes on another chromosome without changing the original order. Finally, the mutation operator exchanged the positions of two randomly selected genes on the same chromosome, as presented in Figure 10.

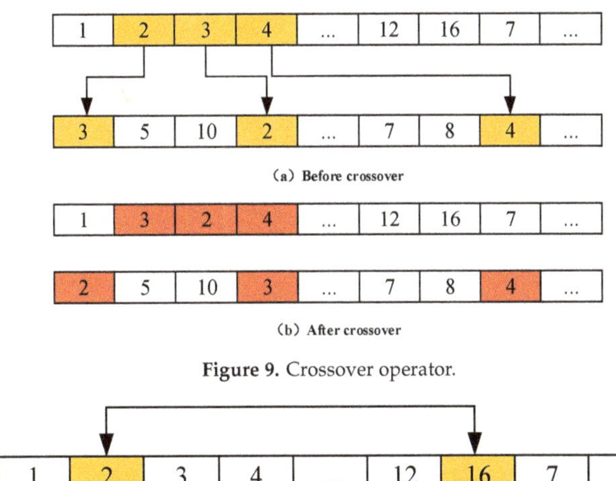

Figure 9. Crossover operator.

Figure 10. Mutation operator.

4. Case Study

4.1. Basic Parameters

The total lengths of the slabs in a rolling unit were limited to 10 km, the lengths of the slabs of the same width were limited to 1 km, while the jump penalties for width, thickness, and hardness are presented in Tables 1–3 [18]. Throughout the model solution with genetic algorithm, the initial population size was 20, the crossover probability was 0.85, and the mutation probability was 0.15. The genetic algorithm was implemented through Visual C# programming.

Table 1. Penalty of width jump.

Jump down/mm	0~25	26~55	56~90	91~150
Penalty	1.0	3.0	5.0	7.0

Table 2. Penalty of hardness jump.

Factor Change	1	2	3	4	5
Penalty	10	16	20	24	30

Table 3. Penalty of thickness jump.

Jump up/mm	0~0.06	0.0601~0.15	0.1501~0.24	0.2401~0.45	0.4501~3.00
Penalty	200.00	300.00	400.00	800.00	1000.00
Jump down/mm	0~25	26~55	56~90	91~150	0.4501~3.00
Penalty	400.00	600.00	800.00	1000.00	2000.00

Table 4 presents the different electricity prices charged to a typical steel plant during different periods within one day [19]. Each day was divided into six electricity price periods with three electricity prices (peak, flat, trough).

Table 4. TOU electricity pricing in steel plant (24-h clock).

Electricity Price Range	Period	Electricity Price, Yuan/kWh
Peak periods	08:00-11:00, 16:00-21:00	0.76
Flat periods	06:00-08:00, 11:00-16:00, 21:00-22:00	0.53
Trough period	22:00-06:00 (next day)	0.31

4.2. Analysis of Optimization Results

The subject of this case analysis was 240 slabs actually processed by a certain steel plant. The rolling operation started from 03:00. The genetic algorithm was used to solve the rolling batch-processing model under two conditions: considering the TOU electricity pricing, and not considering the TOU electricity pricing. Figure 11 presents the iteration curves under the two conditions. As it could be observed from Figure 11a, when TOU electricity pricing was not taken into consideration, the penalty value remained basically unchanged after 10 iterations. The initial penalty value was 23,537, and the value became 22,029 after optimization, amounting to an improvement of 6.4%. As presented in Figure 11b, when the TOU electricity pricing is taken into consideration, the penalty value remained basically unchanged after nine iterations. The initial penalty value was 27,880, while the value became 26,238 after optimization, amounting to an improvement of 5.9%. This indicated that a feasible solution with a low penalty value could be obtained through model optimization.

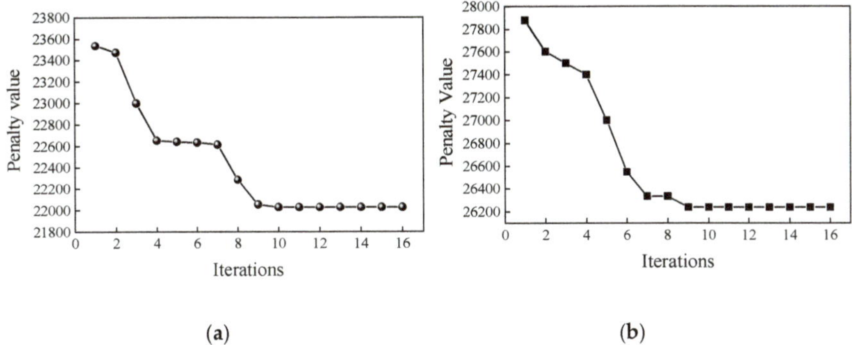

Figure 11. (**a**) TOU electricity pricing not considered (**b**) TOU electricity pricing considered Convergence curve resulted from genetic algorithm iterations.

Table 5 presents the penalty values of three hot-rolling batch processing plans prepared with different methods. It could be observed that the two rolling plans formulated through the hot-rolling batch-processing model under two electricity pricing conditions (TOU plan, non-TOU plan) had apparently lower penalty values, compared to the rolling plan formulated by the man–machine interaction method (MMI plan), even though all three rolling plans had the same number of rolling units. This indicated that better slab arrangement (width, thickness, and hardness) could be achieved after the model-based optimization, which would further lead to a smoother hot-rolling operation. The optimization model, in which the TOU electricity pricing was considered, had higher penalty value compared to the optimization model without the TOU electricity-pricing consideration. This occurred because the penalty incurred by the electricity cost had been added to the objective function. The introduction of the electricity cost penalty created the condition for simultaneous quality and power consumption optimization for the hot-rolling process.

Table 5. Penalty values of three hot-rolling batch-processing plans.

Slab Number /Piece	Preparation Method	Number of Rolling Units /Piece	Penalty Value
	Human interaction	5	31,266
240	Model-based optimization (not considering TOU electricity pricing)	5	22,029
	Model-based optimization (considering TOU electricity pricing)	5	26,238

Figure 12 presets the electricity costs during rolling. The different colors represented different rolling units. The MMI plan induced an electricity cost of 11,230 yuan, the non-TOU plan induced 10,770 yuan and the TOU plan induced 10,230 yuan. In other words, the TOU plan induced 8.9% less electricity cost than the MMI plan and 5.0% less electricity cost than the non-TOU plan. Because the model considering the TOU electricity pricing is good for reducing peaks and filling valleys, the cost of electricity consumption is effectively reduced.

Figure 13 presents the power consumption data during the period from 04:00 to 10:00. The electricity price was 0.31 yuan/kWh in the period of 04:00–06:00, 0.53 yuan/kWh in the period of 06:00–8:00 and 0.76 yuan/ kWh in the period of 08:00–10:00. It could be observed that when the electricity price was 0.31 yuan/kWh, the MMI plan, non-TOU plan, and the TOU plan consumed 4293 kWh, 5134 kWh and 4808 kWh of electricity, respectively. When the electricity price was 0.76 yuan/kWh, the TOU plan consumed 4132 kWh, while the MMI plan consumed 5574 kWh. The former was 25.9% lower than the latter. This demonstrated the benefit of model-based optimization. After the model-based optimization, the high-load rolling units were arranged in the trough-price and flat-price periods, while the low-load rolling units were arranged in the peak-price period.

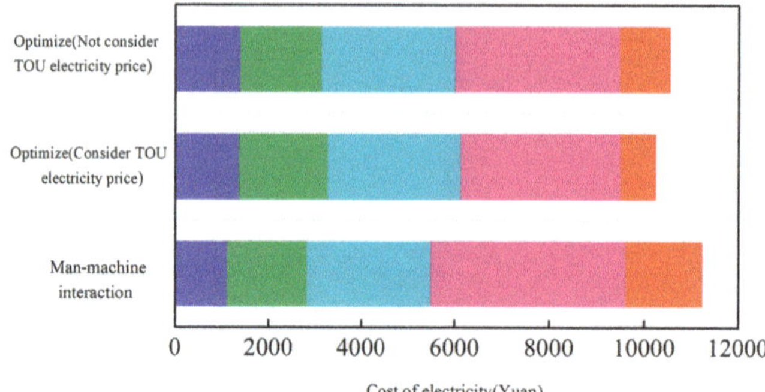

Figure 12. Electricity costs of three rolling plans.

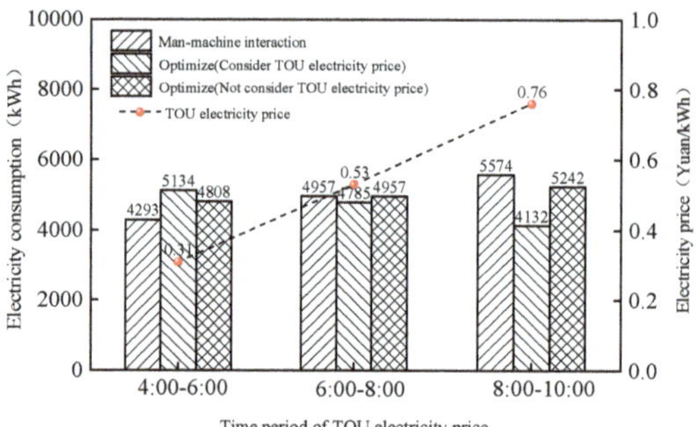

Figure 13. Electricity consumptions at different time periods.

Figure 14 presents the electricity costs of three kinds of rolling plans at different time periods. It could be observed that when the electricity price was low (0.31 yuan/kWh), the MMI plan, TOU plan, and non-TOU plan induced electricity costs of 1331 yuan, 1592 yuan and 1490 yuan, respectively. When the electricity price was medium (0.53 yuan/kWh), the electricity costs induced by the three rolling plans were very close. When the electricity price was high (0.76 yuan/kWh), the MMI plan, non-TOU, and TOU plans induced costs of 4236 yuan, 3984 yuan and 3140 yuan, respectively. In other words, the TOU plan induced a 21.2% lower electricity cost than the non-TOU plan, and 25.9% lower than the MMI plan. Therefore, the TOU plan could effectively reduce the electricity cost and contribute to the power grid fluctuation reduction of electricity consumption.

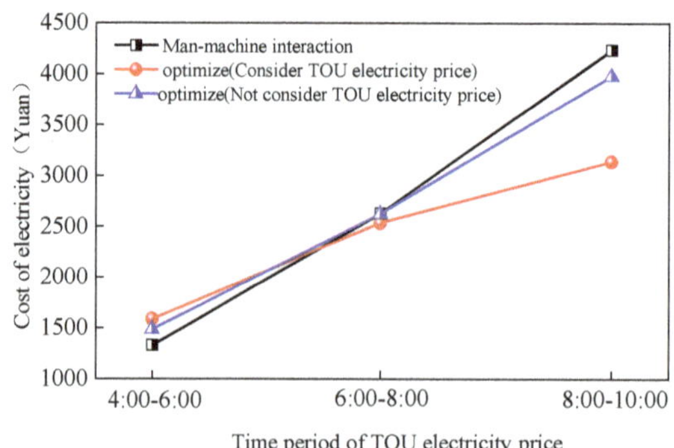

Figure 14. Electricity costs at different TOU electricity pricing periods.

5. Conclusions

(1) A hot-rolling batch-processing plan optimization model, in which TOU electricity pricing was taken into consideration was established. An objective function was used in the model that was aimed at the jump penalties minimizations of the jumps among adjacent slabs in width, hardness, and thickness, as well as in the electricity costs. The optimization method reduced the electricity cost of hot rolling, while ensuring the product quality and production efficiency, in addition to the extra benefit of power consumption fluctuation reduction.

(2) In the proposed method, the crossover and mutation operators of the improved genetic algorithm were used to solve the model for batch-processing plan production for hot rolling. The algorithm was characterized by strong search ability and good convergence. The penalty value after optimization was significantly lower than before optimization, proving the actual value of the proposed method.

(3) An experimental verification was carried out to check the electricity costs of processing of 240 slabs through three hot rolling batch processing plans, formulated under "TOU electricity-pricing consideration", "TOU electricity-pricing absence" and "man—machine interaction" models. The results demonstrated that the TOU plan induced 8.9% lower electricity cost than the MMI plan and 5.0% lower electricity cost than the non-TOU plan.

Author Contributions: Data curation, Z.H.; Formal analysis, Z.H.; Funding acquisition, D.H.; Methodology, K.F.; Project administration, D.H.; Resources, W.S. Software, W.S. and Z.H.; Writing—original draft, Z.H.; Writing—review & editing, D.H. and K.F. All authors have read and agreed to the published version of the manuscript.

Funding: The authors are grateful for the financial support provided by the General Programme of the National Natural Science Foundation of China (Grant No. 51574032, 51674030).

References

1. Dong, G.J.; Li, K.T.; Wang, B.L. Adjustment model and algorithm of rolling plan based on real-time warehouse. *Syst. Eng. Theory Pract.* **2015**, *35*, 1246–1255.
2. Chakraborti, N.; Kumar, B.S.; Babu, V.S.; Moitra, S.; Mukhopadhyay, A. A new multi-objective genetic algorithm applied to hot-rolling process. *Appl. Math. Model.* **2008**, *32*, 781–1789. [CrossRef]
3. Tang, L.X.; Liu, J.Y.; Rong, A.Y.; Yang, Z.H. A review of planning and scheduling systems and methods for integrated steel production. *Eur. J. Oper. Res.* **2001**, *133*, 1–20. [CrossRef]
4. Yadollahpour, M.R.; Bijari, M.; Kavosh, S.; Mahnam, M. Guided local search algorithm for hot strip mill scheduling problem with considering hot charge rolling. *Int. J. Adv. Manuf. Technol.* **2009**, *45*, 1215–1231. [CrossRef]
5. Tan, M.; Duan, B.; Su, Y.X.; He, F. Optimizing production scheduling of steel plate hot rolling for economic load dispatch under time-of-use electricity pricing. *Math. Probl. Eng.* **2017**, *145*, 913–923. [CrossRef]
6. He, D.F.; Liu, P.Z.; Feng, K.; Xu, A.J. Collaborative optimization of rolling plan and enemy dispatching in steel plants. *China Metall.* **2019**, *29*, 75–80.
7. Cowling, P.; Ouelhadj, D.; Petrovic, S. A multi-agent architecture for dynamic scheduling of steel hot rolling. *J. Intell. Manuf.* **2003**, *14*, 457–470. [CrossRef]
8. Cowling, P.; Rezig, W. Integration of continuous caster and hot strip mill planning for steel production. *J. Sched.* **2000**, *3*, 185–208. [CrossRef]
9. Zheng, Z.; Liu, Y.; Chen, K.; Gao, X.Q. Unified modeling and intelligent algorithm of production planning for the process of steelmaking, continuous casting and hot rolling. *J. Univ. Sci. Technol. Beijing* **2013**, *35*, 687–693.
10. Luo, Z.H.; Tang, L.X. Modeling and solution to integrated production and logistics planning of steel making and hot rolling. *J. Manag. Sci. China* **2011**, *14*, 16–23.
11. Zhang, C.S.; Li, K.T. Optimization Model and Algorithm for Lot Planning Coordination between Steelmaking and Hot Rolling. *Inf. Control* **2017**, *46*, 122–128.

12. Li, K.T.; Guo, D.F. Model and algorithm for hot-rolling batch plan based on constraint satisfaction. *Control Decis.* **2007**, *22*, 389–393.

13. Tang, L.X.; Liu, J.Y.; Rong, A.Y.; Yang, Z.H. Modelling and a genetic algorithm solution for the slab stack shuffling problem when implementing steel rolling schedules. *Int. J. Prod. Res.* **2002**, *40*, 1583–1595. [CrossRef]

14. Singh, K.A.; Srinivas; Tiwari, M.K. Modelling the slab stack shuffling problem in developing steel rolling schedules and its solution using improved Parallel Genetic Algorithms. *Int. J. Prod. Econ.* **2004**, *91*, 135–147. [CrossRef]

15. Han, K.H.; Zhang, Q.; Shi, H.B.; Zhang, J.Y. An Improved Compact Genetic Algorithm for Scheduling Problems in a Flexible Flow Shop with a Multi-Queue Buffer. *Processes* **2019**, *7*, 302. [CrossRef]

16. Hu, K.J.; Zhu, F.X.; Chen, J.F.; Noda, N.; Han, W.Q.; Sano, Y. Simulation of Thermal Stress and Fatigue Life Prediction of High Speed Steel Work Roll during Hot Rolling Considering the Initial Residual Stress. *Metals* **2019**, *9*, 966. [CrossRef]

17. Yang, Y.J.; Jiang, Z.Y.; Zhang, X.X. Mathematical model and solving algorithm for the lot planning of slab hot rolling. *J. Univ. Sci. Technol. Beijing* **2012**, *34*, 457–463.

18. Lu, Y.M.; Xu, A.J.; He, D.F.; Tian, N.Y. Hot-rolling batch planning method available to improve DHCR proportion. *J. Univ. Sci. Technol. Beijing* **2011**, *33*, 1301–1306.

19. Zhao, X.C.; Bai, H.; Shi, Q.; Lu, X.; Zhang, Z.H. Optimal scheduling of a byproduct gas system in a steel plant concerning the time-of-use electricity pricing. *Appl. Energy* **2017**, *195*, 100–113. [CrossRef]

Article

Flow Behavior and Hot Processing Map of GH4698 for Isothermal Compression Process

Rongchuang Chen [1,*], Haifeng Xiao [1], Min Wang [1,*] and Jianjun Li [2]

[1] School of Materials Science and Engineering, Hubei University of Automotive Technology, Shiyan 442002, China
[2] School of Materials Science and Engineering, Huazhong University of Science and Technology, and State Key Laboratory of Materials Processing and Die & Mould Technology, Wuhan 430074, China
* Correspondence: crc@hust.edu.cn (R.C.); minwang@126.com (M.W.); Tel.: +86-134-7628-4413 (R.C.); +86-134-7730-6746 (M.W.)

Received: 1 July 2019; Accepted: 29 July 2019; Published: 1 August 2019

Abstract: An in-depth understanding of the flow behaviors of materials deformed at high temperatures is of paramount significance. However, insufficient research on the nickel-based GH4698 alloy has resulted in inaccurate material flow prediction or even cracking in the practical billet opening of GH4698 large forgings. In this study, hot compressions were performed at 950–1150 °C and 0.001–3 s^{-1}. Single-peaked strain-stress curves were obtained under various conditions, owing to dislocation motions in dynamic recrystallizations. The Arrhenius model was formulated to accurately describe the flow stress evolutions and the mean prediction error of the flow stress was 5.90%. Processing maps were constructed at various hot working conditions. It was found that the hot working ability of GH4698 markedly decreased under lower temperatures (950–1080 °C) and higher strain rates (0.1–3 s^{-1}). Optimal thermal processing parameters were suggested. In sum, this study systematically investigated the flow behaviors and hot working ability of GH4698 in isothermal compressions.

Keywords: processing maps; nickel-based alloy; flow behavior; arrhenius equation

1. Introduction

As a type of precipitation-reinforced nickel-based high-temperature alloy, the GH4698 alloy has excellent strength, toughness, fatigue resistance, and corrosion resistance at up to 750 °C. Thus, this alloy has been widely used to manufacture machine parts working at high temperatures, such as airplane engine compressor disks, guide vanes, and gas turbine disks. However, this material is extremely sensitive to thermal processing parameters, and cracking could easily occur in the billet opening of GH4698 large forgings, owing to the addition of aluminum and titanium [1]. One solution could be to place the billet in a sleeve and forge as a whole, so that the material is under a three-dimensional compressive stress state, but this would increase the cost. A more economical method is to deform under optimized parameters, but thus far, hot processing maps of GH4698 have not been established, hindering the hot working parameters optimization in practical production. Therefore, there is an urgent need for systematic research on the flow behaviors and hot working maps of GH4698.

Various flow stress models have been proposed to describe the flow behaviors of alloys at high temperatures. The phenomenological Johnson-Cook model was successfully used to describe the exponential stress-strain relationships of GH4133B by Wang et al. [2]. However, the Johnson-Cook model was inadequate for materials with nonexponential type stress-strain curves, and the Zerilli-Armstrong model was established for an NiTi alloy by Shamsolhodaei et al. [3]. To achieve a higher prediction accuracy, the Arrhenius model was modified to incorporate the influence of strain, and based on the modified Arrhenius model, flow stress models were established for various nickel-based alloys, e.g., the GH4169 alloy by Chen et al. [4], N08028 alloy by Wang et al. [5], and 80A alloy by Gu et al. [6].

The results by Lin et al. [7] and Wang et al. [8] indicated that the accuracy of the flow stress models could be further improved by a neural network, but the applications were limited, owing to the difficulty in finite elemental integration. Moreover, physical-based models were proposed to investigate the underlying mechanisms of the influences of creep, dislocation motion, and grain size on flow stresses by Lin et al. [9], Haan et al. [10], and Zhou et al. [11]. By comparing the above-mentioned models, the Arrhenius model showed an advantage in applicability and accuracy, and thus it has been widely used in the flow stress modeling of nickel-based alloys [4–6].

The processing maps of nickel-based alloys have also been investigated intensively in recent years. Hot working maps of a nickel-based alloy for power plant applications were established by Wu et al., and it was revealed that the different recrystallization mechanisms could be reflected by the hot working maps [12]. The microstructures on the different domains of the processing maps of the IN028 alloy were inspected by Wang et al. [13], and the study showed that the deformation mechanism maps agreed well with the processing maps. The processing maps of the N08028 alloy [14], the 617B alloy [15], and the GH4169 alloy [16] showed that the efficiency peaks of the processing maps were associated with dynamic recrystallization nucleation and dramatic grain growth of the N08028 alloy, whereas incomplete recrystallization, twinning, and adiabatic shear bands occurred in the instability domain. By comparing the different instability criterions from Gegel et al. [17–20], the different shapes of the deformation instability domains of GH79 alloy were compared by Ge et al. [21], and it was noted that the deformation instability domains from Prasad's criterion could effectively predict the deformation instability of GH79. It was shown by the result of Chen et al. [4] that the optimal hot working parameters of GH4169 were located at areas whose dissipative efficiencies were 30–35%. Specifically, for the GH4698 alloy, the flow behaviors were investigated by Zhang et al. [22], and a flow stress model was established. Nevertheless, hot working maps of GH4698 have not been established, forming a barrier for hot working parameter optimization in large forging production. Thus, systematic research is required on the flow behaviors and hot workability of the GH4698 alloy at high temperatures.

Therefore, in this study, the hot deformation behaviors of the GH4698 alloy were studied via hot compressions. An Arrhenius model was established to calculate the flow stresses. Processing maps at various thermal processing conditions were constructed, and an optimal hot working parameter range for GH4698 was recommended. This study provides a reference for hot working parameter optimization of GH4698 large forgings during the forging process.

2. Materials and Experiments

2.1. Materials

The as-forged GH4698 alloy ingot used in this study had a size of $\Phi300$ mm $\times$ 1000 mm. The chemical composition of the GH4698 alloy was quantified via x-ray fluorescence (XRF1800, Shimadzu Inc., Kyoto, Japan) and an infrared carbon-sulfur analyzer (CS2800, Eltra Inc., Haan, Germany), as shown in Table 1 [23,24]. It should be noted that the carbon content of the material used in this research was at the upper limit allowed for GH4698. The specimens for elemental analysis were prepared by wire electrode cutting, turning, and mechanical polishing to $\Phi33$ mm $\times$ 12 mm cylinders. The hot compression specimens were prepared by wire electrode cutting and turning to $\Phi8$ mm $\times$ 12 mm cylinders.

Table 1. Composition of the GH4698 alloy.

Composition	C	Al	Fe	Cr	Nb	Zr	Mo	Ti	Ni
Mass Percentage (%)	0.08	1.55	1.51	14.43	1.95	0.05	2.90	2.62	Balance

2.2. Experiments

Hot compressions were conducted on a compression machine (Gleeble3500, Dynamic Systems Inc., Austin, TX, USA). The compression temperatures were determined to be 950 °C, 1000 °C, 1050 °C, 1100 °C, and 1150 °C, respectively. These temperatures covered the usual hot working temperature range of GH4698 [1]. To cover the strain rate range of the billet forging process, preforging process, and final-forging process of large forgings on hydraulics, the strain rates were selected to be 0.001 s^{-1}, 0.01 s^{-1}, 0.1 s^{-1}, 1 s^{-1}, and 3 s^{-1}, respectively. The experimental procedure is shown in Figure 1. The specimens were heated to 1150 °C at 3.3 °C/s, held for 180 s, cooled to deformation temperatures, held for 180 s, compressed to the strain of 0.95, and quenched. The true stress and logarithmic strain were calculated accordingly. The compressive strains, which had negative values, were written as positive for simplicity in this research. Smoothing was applied on the stress-strain curves.

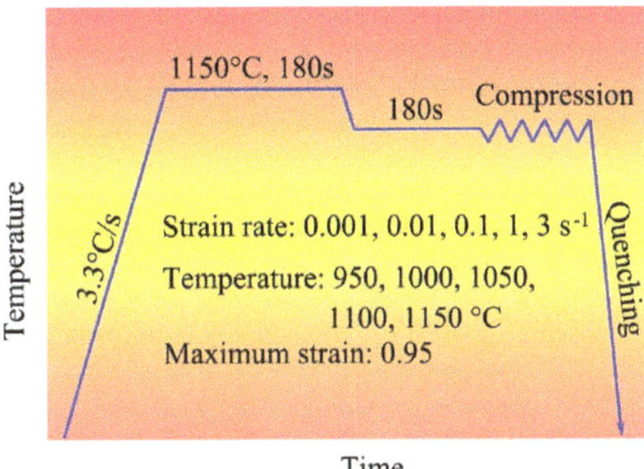

Figure 1. Illustration of the experiment procedure.

3. Results and Discussion

3.1. Flow Behaviors

The stress-strain curves of GH4698 at various strain rates are shown in Figure 2. Basically, single-peak flow stress curves were obtained. In the starting stage of deformation, the flow stress increased because of the contradicting effect of work hardening by dislocation pile ups and cross slips, as well as dynamic softening, by creep and recovery. As the deformation proceeded, the dislocation densities at the grain boundaries reached the critical value for recrystallization, and the dynamic recrystallized grains nucleated, resulting in the softening and a gradual drop in the flow stress [25]. When the dynamic recrystallization was completed, the flow stress remained nearly constant, as the hardening effect and the softening effect were balanced. It can also be seen from Figure 2 that the flow stress decreased with increasing temperature. This could be explained by the observation that the thermal movement of atoms was more intense at higher temperatures. The creep and dynamic recrystallization were more likely to occur owing to fewer obstacles for the dislocation motion to overcome, resulting in the overall softening of the GH4698 alloy. With the increasing strain rate, the flow stress increased. This is because at a faster strain rate, it was more difficult for the dynamic recovery to fully occur, leading to a higher average dislocation density and a greater deformation resistance. It is worth noting that the peak flow stress shown in Figure 2 agreed well with the results of Zhang et al. [22], but the curve shapes were different, which may be attributed to differences in the composition of as-received material. As mentioned in Section 2.1, the carbon content of the material used in this research, 0.08%, was obviously higher than that in the literature [22], 0.048%. For nickel-based alloys, Chen et al. [4] and Lin et al. [7] proved that dynamic recrystallization occurred during hot compressions. Therefore, a refined and uniform microstructure could be obtained by selecting optimal hot working parameters. In this way, the hot working ability could be improved, and cracking defects of large forgings in the billet opening process could be avoided.

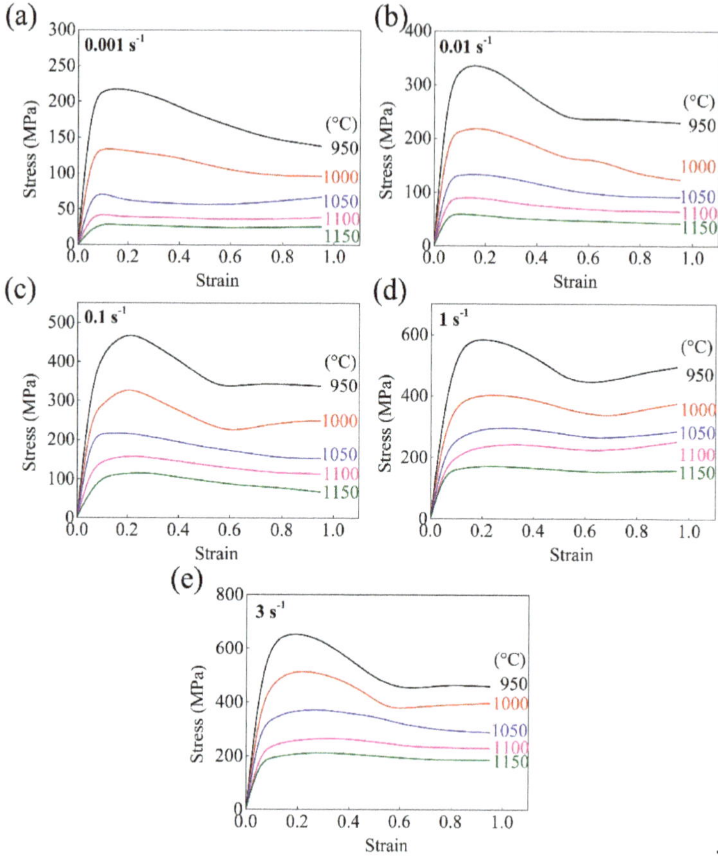

Figure 2. Stress strain curves of GH4698 compressed at the strain rates of (**a**) 0.001 s^{-1}, (**b**) 0.01 s^{-1}, (**c**) 0.1 s^{-1}, (**d**) 1 s^{-1}, and (**e**) 3 s^{-1}.

3.2. Flow Stress Modeling

Although a flow stress model of the GH4698 alloy was established by Zhang et al. [22], different shapes of stress-strain curves were obtained under various compression conditions, as shown in Figure 2. Therefore, a flow stress model describing the stress-strain relationships of GH4698 should be established. The Arrhenius equation was expressed as in [26]:

$$\dot{\varepsilon}\exp\left(\frac{Q}{RT}\right) = A(\sinh(\alpha\sigma))^{n}. \tag{1}$$

Here, A, n, and α were material constants, was the strain rate in s^{-1}, Q was the thermal activation energy in kJ/(mol·K), R was the gas constant, which equaled 8.314 J/(mol·K), and T was the deformation temperature in K. As a constraint, the following could be applied for different stress levels [27,28]:

$$\dot{\varepsilon} = A_1\sigma^{n}\exp\left(-\frac{Q}{RT}\right) \qquad (\alpha\sigma < 0.8) \tag{2}$$

$$\dot{\varepsilon} = A_2 \exp(\beta\sigma) \exp\left(-\frac{Q}{RT}\right) \qquad (\alpha\sigma > 1.2). \tag{3}$$

A_1, A_2, n, and β are constants, which follows:

$$\alpha = \frac{\beta}{n} \tag{4}$$

Taking logarithms of both sides of Equations (2) and (3) gives:

$$\ln\dot{\varepsilon} = \ln A_1 + n\ln\sigma - \frac{Q}{RT} \tag{5}$$

$$\ln\dot{\varepsilon} = \ln A_2 + \beta\sigma - \frac{Q}{RT}. \tag{6}$$

Thus, n and β could be calculated by the slopes of the fitted line of $\ln\sigma$ versus $\ln\dot{\varepsilon}$, and σ versus $\ln\dot{\varepsilon}$. The value of α is calculated according to Equation (4). Taking a logarithm of Equation (1) results in:

$$\ln\dot{\varepsilon} = \ln A + n\ln(\sinh(\alpha\sigma)) - \frac{Q}{RT}. \tag{7}$$

Therefore, the value of Q could be obtained via the slope of the fitted line of $\frac{1}{T}$ versus $\ln(\sinh(\alpha\sigma))$, and the value of $\ln A$ could be obtained via the intercept of the fitted line of $\ln(\sinh(\alpha\sigma))$ versus $\ln\dot{\varepsilon}+Q/RT$. Taking $\varepsilon = 0.05$ as example, the calculation process of the model parameters is shown in Figure 3. For effectiveness, the calculations were made using MATLAB software at various strains. The values of the model parameters with the increasing strain are shown in Figure 4.

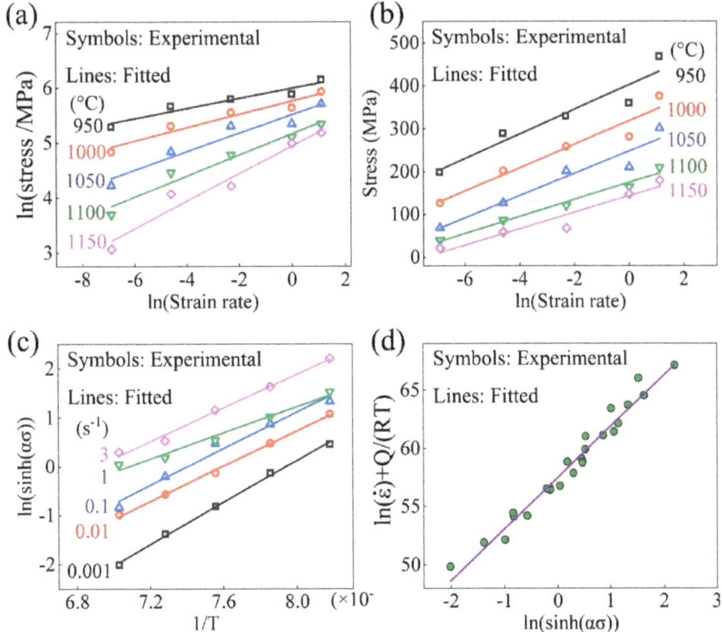

Figure 3. The calculation process of n, β, Q, and $\ln A$ from the slope of the fitted line of (**a**) $\ln\sigma$ versus $\ln\dot{\varepsilon}$, (**b**) σ versus $\ln\dot{\varepsilon}$, (**c**) $\frac{1}{T}$ versus $\ln(\sinh(\alpha\sigma))$, and from the intercept of the fitted line of (**d**) $\ln(\sinh(\alpha\sigma))$ versus $\ln\dot{\varepsilon} + \frac{Q}{RT}$.

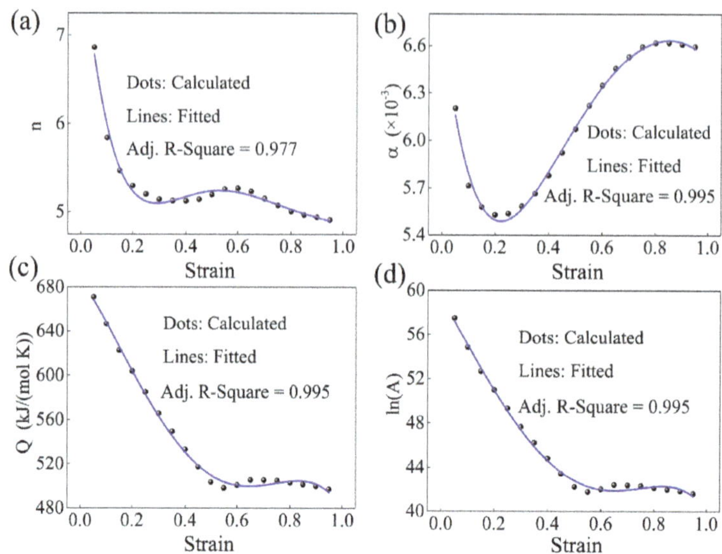

Figure 4. The Arrhenius model parameter values of GH4698 alloy at various strains. (**a**) n, (**b**) β, (**c**) Q, and (**d**) $\ln A$.

Fifth-order polynomial fittings were then applied to describe the relationships between the model parameters and the strain, resulting in the following:

$$G = k_0 + k_1\varepsilon + k\varepsilon^2 + k_3\varepsilon^3 + k_4\varepsilon^4 + k_5\varepsilon^5. \tag{8}$$

Here, G denotes n, β, Q, and $\ln A$, respectively. $k_0 - k_5$ denote the coefficients, whose values are determined by polynomial fitting in the *Origin* software, shown in Table 2.

Table 2. Coefficients of the polynomials.

Coefficients	k_0	k_1	k_2	k_3	k_4	k_5
n	7.993×10^0	-2.922×10^1	1.083×10^2	-1.825×10^2	1.426×10^2	-4.233×10^1
α	6.731×10^{-3}	-1.372×10^{-2}	5.024×10^{-2}	-7.115×10^{-2}	4.724×10^{-2}	-1.283×10^{-2}
Q	6.870×10^5	-3.307×10^5	-7.459×10^5	1.830×10^6	-7.715×10^5	-1.887×10^5
$\ln A$	5.492×10^1	-4.353×10^1	-1.359×10^0	3.554×10^1	4.416×10^1	-5.413×10^1

The comparisons between the calculated and experimental flow stress values are shown in Figure 5. Basically, the flow stress could be predicted with good accuracy, showing that the Arrhenius model was capable of accurately calculating the flow behaviors of GH4698 under various strain rates. The scatter plots of the calculated versus experimental flow stress are shown in Figure 6. The calculated stresses fit well with the experimental stresses. The adjusted R square value to describe the goodness of fit was 0.990. The average calculation error was 10.96 MPa (5.90%). The errors may be attributed to the following two reasons. First, there were still computational deviations in the fitting of model parameters by the fifth-order polynomial, which could be improved by trying better fitting expressions. Additionally, an abnormal softening occurred in the final part of the stress–strain curve under 1000 °C and 0.01 s^{-1}, which may be caused by experimental deviations. The number of repetitions under the same hot compression conditions could be increased to reduce such deviations.

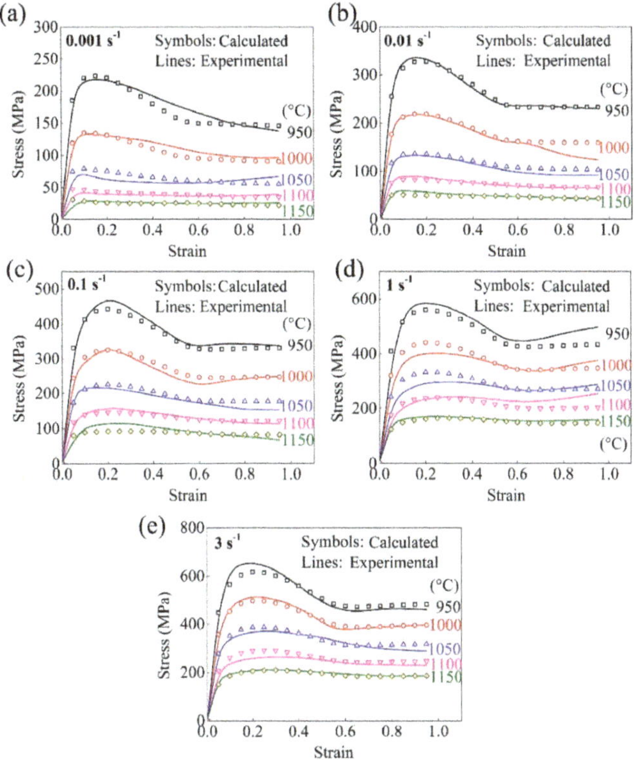

Figure 5. Comparisons of the calculated and experimental flow stress at the strain rates of (**a**) 0.001 s^{-1}, (**b**) 0.01 s^{-1}, (**c**) 0.1 s^{-1}, (**d**) 1 s^{-1}, and (**e**) 3 s$^{-1.}$

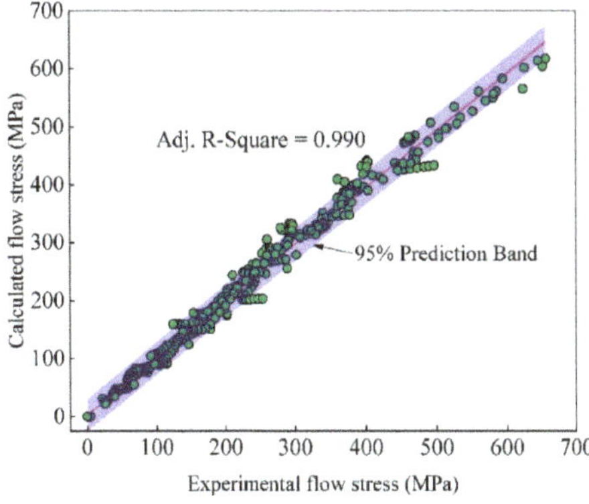

Figure 6. Scatter plots of the calculated versus experimental flow stress.

3.3. Processing Maps

The processing maps were significant references for process parameter optimizations of GH4698 large forgings. It was suggested by Prasad et al. [26] that the deformation work (P) during forging is consumed by temperature rising (G) and microstructure evolution (J), as represented by:

$$P = G + J = \int_0^{\dot{\varepsilon}} \sigma \, d\dot{\varepsilon} + \int_0^{\sigma} \dot{\varepsilon} \, d\sigma. \tag{9}$$

The power dissipation coefficient (η) quantifying the fraction of energy by microstructure evolution was calculated by [4]:

$$\eta = \frac{2\frac{\partial \ln \sigma}{\partial \ln \dot{\varepsilon}}}{\frac{\partial \ln \sigma}{\partial \ln \dot{\varepsilon}} + 1}. \tag{10}$$

An instability coefficient ζ for predicting deformation instability was expressed as in [4]:

$$\zeta = \frac{\partial \log(\eta/2)}{\partial \log(\dot{\varepsilon})} + \frac{\partial \ln \sigma}{\partial \ln \dot{\varepsilon}}. \tag{11}$$

The value of ζ was negative when deformation instability occurred according to previous research [4,12–15]. The value of σ could be obtained from the flow stress model. Thus, the thermal processing maps of GH4698 were established according to Equations (10) and (11), as shown in Figure 7. The deformation instability was shown as shaded areas, and the dissipation coefficients were shown as contours. At a strain of 0.2 (Figure 7a), three deformation instability domains were found. One was located at 950–1150 °C and 0.25–3 s^{-1}, another at 985–1015 °C and 0.001–0.002 s^{-1}, and the third at 1060–1140 °C and 0.001–0.004 s^{-1}. As the strain increased to 0.4, the deformation instability domain at high strain rates split into two, one at 950–1025 °C and 0.5–3 s^{-1}, and the other at 1060–1150 °C and 0.5–3 s^{-1}. Another deformation instability domain was located at 960–1010 °C and 0.002–0.025 s^{-1}. A comparison of data presented in Figure 7a,b showed that the majority of the deformation instability domain had dissipation coefficients lower than 35%, and the dissipation coefficients of the other parts of the map were generally greater than 35%. One reason for this could be that at higher strain rates and lower temperatures, dynamic recrystallization was insufficient, and deformation instability occurred more easily. Thus, a much larger proportion of the deformation work was converted into heat. At a strain of 0.6 (Figure 7c), two deformation instability domains were found. One was located at 950–975 °C and 0.3–3 s^{-1}, and the other at 1070–1140 °C and 0.5–3 s^{-1}. It is worth noting that the dissipation coefficients were high in the lower right part of the map in Figure 7c, which corresponds to high temperature and low strain rate conditions, and this may be because of the dramatic grain growth after the dynamic recrystallization was completed. At a strain of 0.8 (Figure 7d), two deformation instability domains occurred. One existed at 950–1070 °C and 0.05–0.63 s^{-1}, and the other at 1080–1150 °C and 0.4–3 s^{-1}. The overlapping of the processing maps in Figure 7a–d reveals that to avoid deformation instability, the hot working parameters should drop in the area of 1080–1150 °C and 0.004–0.05 s^{-1}.

The processing maps at various temperatures or strain rates are extremely important for the hot working parameter optimization when the deformation temperature or the punch speed are restrained. The processing maps of the GH4698 alloy at various temperatures could be obtained by slicing and interpolating the processing maps at specific temperatures, as shown in Figure 8. A deformation instability domain at a strain of 0.2–0.8 and at a strain rate of 0.1–3 s^{-1} was found under 950 °C (Figure 8a). The dissipation coefficients were below 25%. A comparison with data provided in Figure 8a indicated that much smaller deformation instability domains existed under 1050 °C, as shown in Figure 8b. At 1150 °C (Figure 8c), the deformation instability domain disappeared. Figure 8 shows that the workability of the GH4698 alloy was improved by dynamic recrystallization under higher hot working temperatures. It is worth noting that the dissipation coefficients generally increased with increasing temperature, because the dynamic recrystallization was more complete at higher temperatures, thereby consuming a much larger proportion of energy.

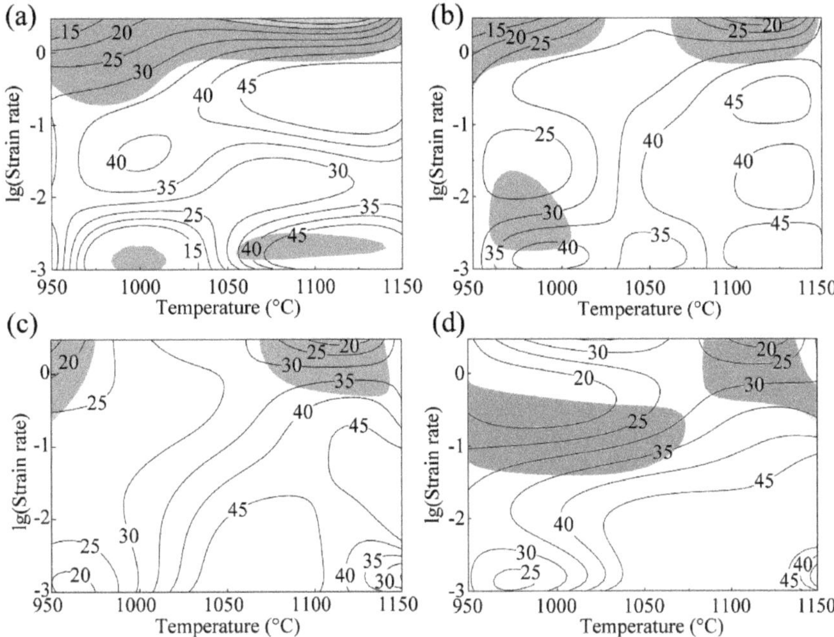

Figure 7. Processing maps of GH4698 at the strains of (**a**) 0.2, (**b**) 0.4, (**c**) 0.6, and (**d**) 0.8.

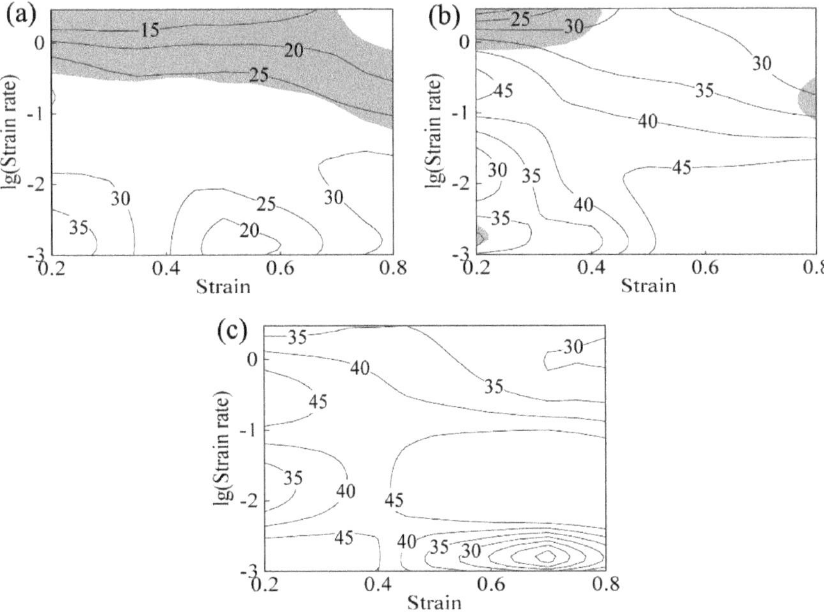

Figure 8. Processing maps of GH4698 at the temperatures of (**a**) 950 °C, (**b**) 1050 °C, and (**c**) 1150 °C.

Hot working maps of a GH4698 alloy at various strain rates are shown in Figure 9. Good hot working abilities were obtained at 0.001 s^{-1}, as no deformation instability domain was found, as shown in Figure 9a. A dissipation coefficient of ~45% at 1050–1150 °C and at the strain of 0.4–0.8 showed that a full dynamic recrystallization followed by dramatic grain growth might have occurred. A deformation instability domain at 950–1050 °C and at the strain of 0.75–0.8 could be observed at 0.1 s^{-1}, as shown in Figure 9b, and a relatively high dissipation coefficient of ~40% could still be obtained, meaning that the thermal processing parameters were still acceptable. However, the deformation instability domain occupied the majority of the processing map, as shown in Figure 9c, indicating that the GH4698 alloy should not be hot-formed at such a fast strain rate, as deformation instability and local material flow would occur.

Based on the results presented in Figures 7–9, the optimal hot working parameters were suggested as 1080–1150 °C and 0.004–0.1 s^{-1}. The forging temperature should be no lower than 1050 °C to ensure that dynamic recrystallization occurs completely, and the strain rate should be neither too high to avoid deformation instability, nor too low to prevent grain coarsening. The visible differences in the shape of the stress-strain curves in this research and in the literature [22] might lead to the difference of the shape of the processing maps. Therefore, further research should be carried out to quantify the influence of chemical compositions. In the practical production of GH4698 large forgings, the forging temperature can be determined by the pressure that the forging device can provide. The results also showed that the GH4698 alloy was suitable for low-speed isothermal forgings at high temperatures (1080–1150 °C).

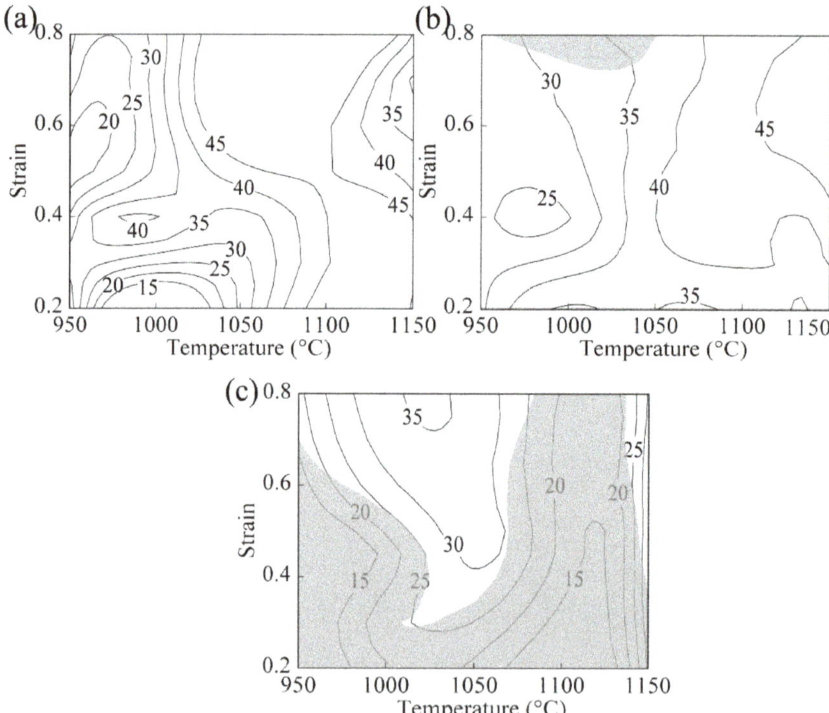

Figure 9. Processing maps of GH4698 at the strain rates of (**a**) 0.001 s^{-1}, (**b**) 0.1 s^{-1}, and (**c**) 3 s^{-1}.

4. Conclusions

Hot compression tests were performed at 950–1150 °C and 0.001–3 s^{-1}, and the following conclusions could be drawn:

(1) The stress-strain curves of GH4698 were single-peaked at various compression conditions, owing to dislocation motions during dynamic recrystallizations. The peak flow stress agreed well with the results in the literature, but the stress-strain curve shapes were different, which may be attributed to differences in the as-received material states.

(2) An Arrhenius model was established to calculate the stress evolutions under various strain rates, temperatures, and strains. The average calculation error was 10.96 MPa (5.90%), showing that the model could accurately describe the flow behaviors of GH4698 at high temperatures.

(3) The processing maps of GH4698 revealed that the hot working ability of GH4698 markedly decreased under lower temperatures (950–1080 °C) and high strain rates (0.1–3 s^{-1}). The optimal hot working parameters were suggested as 1080–1150 °C and 0.004–0.1 s^{-1}. Moreover, according to this study, the GH4698 alloy was suitable for low-speed isothermal forgings at high temperatures (1080–1150 °C).

Author Contributions: Writing—original draft preparation, R.C.; writing—review and editing, H.X.; funding acquisition, M.W. and J.L.

Funding: This study was funded by the National Natural Science Foundation of China (51435007), the Natural Science Foundation of Hubei province (2017CFB587), the State Key Laboratory of Materials Processing and Die & Mould Technology in Huazhong University of Science and Technology (P2020-015), the Outstanding Young Scientific &Technological Innovation Team Plan of Colleges and Universities in Hubei Province (T201518), the Hubei Key Laboratory of Automotive Power Train and Electronic Control in Hubei University of Automotive Technology (ZDK1201601), and the Doctoral Scientific Research Fund of Hubei University of Automotive Technology (BK201901).

Conflicts of Interest: The authors declare that they have no conflicts of interest.

References

1. Huang, B.; Li, C.; Shi, L. Nonferrous Metal materials and engineering. In *China Materials Engineering Canon*; Chemical Industry Press: Beijing, China, 2005; Volume 4, pp. 378–391.

2. Wang, J.; Guo, W.; Li, P. Modified Johnson-Cook description of wide temperature and strain rate measurements made on a nickel-base superalloy. *Mater. High Temp.* **2017**, *34*, 157–165. [CrossRef]

3. Shamsolhodaei, A.; Zarei-Hanzaki, A.; Ghambari, M. The high temperature flow behavior modeling of NiTi shape memory alloy employing phenomenological and physical based constitutive models: A comparative study. *Intermetallics* **2014**, *53*, 140–149. [CrossRef]

4. Chen, R.; Zheng, Z.; Li, J. Constitutive Modelling and Hot Workability Analysis by Microstructure Examination of GH4169 Alloy. *Crystals* **2018**, *8*, 282. [CrossRef]

5. Wang, L.; Liu, F.; Cheng, J. Arrhenius-Type Constitutive Model for High Temperature Flow Stress in a Nickel-Based Corrosion-Resistant Alloy. *J. Mater. Eng. Perform.* **2016**, *25*, 1394–1406. [CrossRef]

6. Gu, S.; Zhang, L.; Zhang, C. Constitutive Modeling for Flow Stress Behavior of Nimonic 80A Superalloy During Hot Deformation Process. *High Temp. Mater. Process.* **2016**, *35*, 327–336. [CrossRef]

7. Linl, Y.C.; Nong, F.Q.; Chen, X.M. Microstructural evolution and constitutive models to predict hot deformation behaviors of a nickel-based superalloy. *Vacuum* **2017**, *137*, 104–114. [CrossRef]

8. Wang, T.; Wan, Z.; Sun, Y. Dynamic Softening Behavior and Microstructure Evolution of Nickel Base Superalloy. *Acta Metall. Sin.* **2018**, *54*, 83–92. [CrossRef]

9. Lin, Y.C.; Wen, D.X.; Li, X.H. Hot deformation characteristics and dislocation substructure evolution of a nickel-base alloy considering effects of δ phase. *J. Alloys Compd.* **2018**, *764*, 1008–1020. [CrossRef]

10. Haan, J.; Bezold, A.; Broeckmann, C. Interaction between particle precipitation and creep behavior in the NI-base Alloy 617B: Microstructural observations and constitutive material model. *Mater. Sci. Eng. A* **2015**, *640*, 305–313. [CrossRef]

11. Zhou, H.; Zhang, H.; Liu, J. Prediction of Flow Stresses for a Typical Nickel-Based Superalloy During Hot Deformation Based on Dynamic Recrystallization Kinetic Equation. *Rare Met. Mater. Eng.* **2018**, *47*, 3329–3337. [CrossRef]

12. Wu, Y.; Zhang, M.; Xie, X. Hot deformation characteristics and processing map analysis of a new designed nickel-based alloy for 700 °C A-USC power plant. *J. Alloys Compd.* **2016**, *656*, 119–131. [CrossRef]

13. Wang, L.; Liu, F.; Zuo, Q. Processing Map and Mechanism of Hot Deformation of a Corrosion-Resistant Nickel-Based Alloy. *J. Mater. Eng. Perform.* **2017**, *26*, 392–406. [CrossRef]

14. Wang, L.; Liu, F.; Cheng, J. Hot deformation characteristics and processing map analysis for Nickel-based corrosion resistant alloy. *J. Alloys Compd.* **2015**, *623*, 69–78. [CrossRef]

15. Jiang, H.; Dong, J.; Zhang, M. Hot deformation characteristics of Alloy 617B nickel-based superalloy: A study using processing map. *J. Alloys Compd.* **2015**, *647*, 338–350. [CrossRef]

16. Kang, F.; Zhang, X.; Sun, J. Hot Deformation Behavior and Processing Map of a Nickel-Base Superalloy GH4169. *Adv. Mater. Res.* **2013**, *834*, 432–436. [CrossRef]

17. Gegel, H.L. *Synthesis of Atomistics and Continuum Modeling to Describe Microstructure: Computer Simulation in Material Science*; ASM Park: Geauga, OH, USA, 1984; pp. 291–344.

18. Malas, J.C.; Seetharaman, V. Using material behavior models to develop process control strategies. *JOM* **1992**, *44*, 8–13. [CrossRef]

19. Prasad, Y.V.R.K.; Seshachryuly, T. Modelling of hot deformation for microstructural control. *Int. Mater. Rev.* **1988**, *43*, 243–252. [CrossRef]

20. Murty, S.V.S.N.; Rao, B.N. On the development of instability criteria during hot working with reference to IN 718. *Mater. Sci. Eng. A* **1998**, *254*, 76–82. [CrossRef]

21. Zhou, G.; Ding, H.; Cao, F.R.; Han, Y.B.; Zhang, B.J. Flow instability criteria in processing map of superalloy GH79. *Trans. Nonferrous Met. Soc. Chin.* **2012**, *22*, 1575–1581. [CrossRef]

22. Zhang, P.; Hu, C.; Zhu, Q. Hot compression deformation and constitutive modeling of GH4698 alloy. *Mater. Des.* **2015**, *65*, 1153–1160. [CrossRef]

23. Chen, R.; Zheng, Z.; Li, N. In-situ investigation of phase transformation behaviors of 300M steel in continuous cooling process. *Mater. Charact.* **2018**, *144*, 400–410. [CrossRef]

24. Chen, R.; Zheng, Z.; Li, J. In Situ Investigation of Grain Evolution of 300M Steel in Isothermal Holding Process. *Materials* **2018**, *11*, 1862. [CrossRef] [PubMed]

25. Chen, R.; Guo, P.; Zheng, Z. Dislocation Based Flow Stress Model of 300M Steel in Isothermal Compression Process. *Materials* **2018**, *11*, 972. [CrossRef] [PubMed]

26. Prasad, Y.V.R.K.; Gegel, H.L.; Doraivelu, S.M. Modeling of dynamic material behavior in hot deformation: Forging of Ti6242. *Metall. Trans. A* **1984**, *15*, 1883–1892. [CrossRef]

27. Yin, F.; Hua, L.; Mao, H. Constitutive modeling for flow behavior of GCr15 steel under hot compression experiments. *Mater. Des.* **2013**, *43*, 393–401. [CrossRef]

28. Zhang, X.; Huang, L.; Li, J. Flow behaviors and constitutive model of 300M high strength steel at elevated temperature. *J. Cent. South Univ.* **2017**, *48*, 1439–1447. [CrossRef]

MDPI

St. Alban-Anlage 66

4052 Basel

Switzerland

Tel. +41 61 683 77 34

Fax +41 61 302 89 18

www.mdpi.com

Processes Editorial Office

E-mail: processes@mdpi.com

www.mdpi.com/journal/processes